J.L. Burch • V. Angelopoulos
Editors

The THEMIS Mission

Previously published in *Space Science Reviews* Volume 141, Issues 1–4, 2008

Springer

J.L. Burch
Space Science and Engineering Division
Southwest Research Institute (SwRI)
San Antonio, TX, USA

V. Angelopoulos
Department of Earth and Space Sciences, and
Institute of Geophysics and Planetary Physics
University of California
Los Angeles, CA, USA

Cover illustration: Courtesy of NASA GSFC/CI Lab

Library of Congress Control Number: 2009920771

DOI: 10.1007/978-0-387-89820-9

ISBN-978-0-387-89819-3 e-ISBN-978-0-387-89820-9

Printed on acid-free paper.

1

springer.com

Contents

Foreword

J.L. Burch · V. Angelopoulos

Originally published in the journal Space Science Reviews, Volume 141, Nos 1–4, 1–3.
DOI: 10.1007/s11214-008-9474-5

The Earth, like all the other planets, is continuously bombarded by the solar wind, which is variable on many time scales owing to its connection to the activity of the Sun. But the Earth is unique among planets because its atmosphere, magnetic field, and rotation rates are each significant, though not dominant, players in the formation of its magnetosphere and its reaction to solar-wind inputs. An intriguing fact is that no matter what the time scale of solar-wind variations, the Earth's response has a definite pattern lasting a few hours. Known as a magnetospheric substorm, the response involves a build-up, a crash, and a recovery. The build-up (known as the growth phase) occurs because of an interlinking of the geomagnetic field and the solar-wind magnetic field known as magnetic reconnection, which leads to storage of increasing amounts of magnetic energy and stress in the tail of the magnetosphere and lasts about a half hour. The crash (known as the expansion phase) occurs when the increased magnetic energy and stresses are impulsively relieved, the current system that supports the stretched out magnetic tail is diverted into the ionosphere, and bright, dynamic displays of the aurora appear in the upper atmosphere. The expansion and subsequent recovery phases result from a second magnetic reconnection event that decouples the solar-wind and geomagnetic fields.

While often appearing only as isolated events, multiple intense substorms occur during global magnetic storms, which can last several days. At times of increased power input, magnetic storms wreak havoc on astronauts, satellites, space-based telecommunication systems and ground power distribution systems. The exact means of solar wind energy transformation and release within the magnetosphere are not yet known adequately enough to result in high-fidelity predictive space weather models that can become operational. During the

J.L. Burch (✉)
Southwest Research Institute (SwRI), San Antonio, TX, USA
e-mail: JBurch@swri.edu

V. Angelopoulos
Space Physics Research Group, University of California, Berkeley, CA, USA
e-mail: vassilis@ucla.edu

past four decades the Sun–Earth system has been studied from various vantage points, using sophisticated *in situ* or imaging missions. Those missions have resulted in an understanding of magnetospheric dynamics akin to climatology. Simultaneous multi-point observations are very difficult to design and execute within the vast expanse of the magnetosphere and can only take place with focused investigations aimed at unraveling the physics of well defined processes.

Spacecraft on the night side of the Earth have found that magnetic reconnection occurs in the distance range from about 20–30 Earth radii, and once initiated the site moves rapidly down the tail. On the other hand, the rapid reduction of the cross-tail current occurs closer to the Earth at about 8–10 Earth radii. Finally, the bright auroral displays occur as a result of one or the other of these phenomena, but because of the dynamic nature of the magnetic field during these events it is not possible to determine where the magnetic field lines threading the aurora cross the equatorial plane in the tail. The field of substorm research, after over 40 years of intense effort, had reached an impasse. What occurs first—the current disruption or reconnection? Which of these two phenomena is responsible for the aurora? What causes the substorm? The answer to this final question depends on the answers to the first two. The difficulty has been the lack of spacecraft alignments along the substorm meridian near substorm onset. Because a substorm starts from a single point in space and within a few minutes evolves over the entire magnetotail past the moon's orbit, a single satellite alone cannot identify the precise substorm onset time and point of origin. Multiple satellites, in tightly choreographed orbits to ensure frequent Sun–Earth alignments are needed to answer this question.

The grand experiment set up by THEMIS involves a vast array of ground-based magnetometers and auroral imagers that can locate the initial brightening and development of the substorm aurora in both space and time. Simultaneously the five THEMIS spacecraft (identically instrumented for measurements of charged particles, electric fields and magnetic fields) are lined up in the magnetic tail over the critical distances from 8 to 30 Earth radii. In this way when and where the aurora, the current disruption, and magnetic reconnection first appear can be observed. It is a simple, elegant, and unambiguous experiment, and first indications are that it has been very successful. In addition, the THEMIS unique orbits and instrumentation on the dayside enable studies of the solar wind evolution prior to its interaction with the magnetosphere while in the inner magnetosphere THEMIS is able to study for the first time the energization of radiation belt electrons during storms by sampling the radial phase space density of the radiation belts at a recurrence rate commensurate with storm main phase and recovery.

From the THEMIS selection until its launch in February 2007, a large team of scientists and engineers dedicated themselves to designing the mission and developing the five spacecraft, the 25 scientific instruments, and the ground-based observatory array. The resounding success of the mission is testament to the hard work, dedication and expertise of all involved in the effort. Since launch, the instruments were commissioned over a period of three months, and following a six month coast-phase in a string-of-pearls configuration, the satellites were placed in their first magnetotail configuration for studying the substorm onset question. The data collected from the first several months in orbit were invaluable for inter-spacecraft calibration and for validating the in-flight characteristics of the instrumentation.

This special issue of Space Science Reviews documents the design, development and capabilities of all aspects of the THEMIS mission. First science results, instrument nuances and in-flight calibration using data from the first few months in orbit are also described. Specifically, papers on mission overview (Angelopoulos), science overview (Sibeck and Angelopoulos), mission design (Frey et al.), mission operations (Bester et al.), and spacecraft development and processing (Harvey et al.) describe the top-level THEMIS science drivers and mission implementation. Papers on the instrument data processing unit (Taylor et al.), the magnetic cleanliness program (Ludlam et al.), the instrument boom mechanisms (Auslander et al.), and the ground based observatory system design (Harris et al.) describe the space-borne and ground-based instrument suite design philosophy and development. Papers on the fluxgate magnetometer (Auster et al.), the search-coil magnetometer (Roux et al.), the electrostatic analyzer (McFadden et al.), the electric field instrument (Bonnell et al.), the digital fields board (Cully et al.), the all sky imagers (Mende et al.), and the new and pre-existing ground magnetometers (Russell et al.; Mann et al.) describe the THEMIS instrumentation design, implementation and data quality. First results from the instrument suite on the day side and night side towards the primary and secondary mission objectives (Angelopoulos et al.), on the electrostatic analyzer (McFadden et al.), on the search-coil magnetometer (LeContel et al.), on the hand-in-hand use of simulations with data to arrive at comprehensive understanding of Sun-Earth interactions during substorms (Raeder et al.), and from use of mid-latitude stations for substorm analysis and education and public outreach purposes (Peticolas et al.) provide examples of scientific use of the mission capabilities with adequate discussion of the salient features of the instrumentation. We hope that the volume will be useful for researchers to understand fully the published scientific results from the mission and to advance their own investigations of the Sun-Earth system using the openly available THEMIS data.

Great appreciation is due the editorial staff of Space Science Reviews, particularly Mr. Randy Cruz, who rapidly and efficiently managed the efforts of the large number of authors, reviewers and editorial staff in producing a very comprehensive and high-quality review of this fascinating new mission.

The THEMIS Mission

V. Angelopoulos

Originally published in the journal Space Science Reviews, Volume 141, Nos 1–4, 5–34.
DOI: 10.1007/s11214-008-9336-1

Abstract The Time History of Events and Macroscale Interactions during Substorms (THEMIS) mission is the fifth NASA Medium-class Explorer (MIDEX), launched on February 17, 2007 to determine the trigger and large-scale evolution of substorms. The mission employs five identical micro-satellites (hereafter termed "probes") which line up along the Earth's magnetotail to track the motion of particles, plasma and waves from one point to another and for the first time resolve space–time ambiguities in key regions of the magnetosphere on a global scale. The probes are equipped with comprehensive in-situ particles and fields instruments that measure the thermal and super-thermal ions and electrons, and electromagnetic fields from DC to beyond the electron cyclotron frequency in the regions of interest. The primary goal of THEMIS, which drove the mission design, is to elucidate which magnetotail process is responsible for substorm onset at the region where substorm auroras map ($\sim$10 R_E): (i) a local disruption of the plasma sheet current (current disruption) or (ii) the interaction of the current sheet with the rapid influx of plasma emanating from reconnection at $\sim$25 R_E. However, the probes also traverse the radiation belts and the dayside magnetosphere, allowing THEMIS to address additional baseline objectives, namely: how the radiation belts are energized on time scales of 2–4 hours during the recovery phase of storms, and how the pristine solar wind's interaction with upstream beams, waves and the bow shock affects Sun–Earth coupling. THEMIS's open data policy, platform-independent dataset, open-source analysis software, automated plotting and dissemination of data within hours of receipt, dedicated ground-based observatory network and strong links to ancillary space-based and ground-based programs. promote a grass-roots integration of relevant NASA, NSF and international assets in the context of an international Heliophysics Observatory over the next decade. The mission has demonstrated spacecraft and mission design strategies ideal for Constellation-class missions and its science is complementary to Cluster and MMS. THEMIS, the first NASA micro-satellite constellation, is a technological pathfinder for future Sun-Earth Connections missions and a stepping stone towards understanding Space Weather.

V. Angelopoulos (✉)
IGPP/ESS UCLA, Los Angeles, CA 90095-1567, USA
e-mail: vassilis@ucla.edu

Keywords THEMIS · Magnetosphere · Substorms · Radiation belts · Magnetopause

PACS 94.30.-d · 94.30.cl · 94.30.cb · 94.30.ch · 94.30.cj · 94.30.C- · 94.30.cp · 94.30.Lr · 94.30.Va · 94.30.Xy · 96.50.Fm

1 Introduction

A substorm is an avalanche of small-scale magnetotail energy surges (Lui et al. 2001) feeding from solar wind energy previously stored in the magnetotail lobes. During its course, auroral arcs intensify, move poleward and break up into smaller formations (Akasofu 1976). Substorms are ubiquitous at all solar phases and appear within all types of magnetospheric responses to solar wind input: Embedded within large storms they influence storm development (Daglis et al. 2000) and geo-effectiveness (Siscoe and Petschek 1997). They bind the beginning and end phases of magnetospheric convection bays (Sergeev et al. 1996a) and are closely related to pseudo-breakups (Aikio et al. 1999). Understanding the substorm process is a prerequisite to understanding the geo-magnetospheric response to all levels of solar wind energy throughput. However, the objective of deciphering the mechanism of substorm instability transcends its geophysical interest. It relates intimately to broader scientific questions, because it addresses basic plasma physics processes, such as cross-scale coupling between MHD and kinetic plasma instabilities (Shinohara et al. 2001; Voronkov et al. 1999). Beyond purely scientific applications are matters of more practical value to society, related to space weather processes (such as storms), which affect satellite communications and ground electrical distribution, and are inextricably linked to substorms. In summary, *substorms represent a fundamental mode of global magnetospheric circulation, a macroscopic instability whose phenomenological and theoretical understanding is crucial for space science, basic plasma physics and space weather.*

A substorm has well-demarcated global evolutionary phases corresponding to unique stages of the instability of the coupled solar wind-magnetospheric circulation of energy and magnetic flux. These unique stages include energy storage (growth phase), explosive release (onset) and eventual ionospheric dissipation (late expansion and recovery phases). Thus a substorm represents a fundamental mode of global circulation of energy and magnetic flux transport throughout Geospace. This global, macroscopic instability is as central to space physics and space weather as the extratropical cyclone is to meteorology and weather. Despite the elemental nature of the substorm process, the lack of appropriate spacecraft conjunctions from previous missions resulted in a contentious set of theories for its description. The question is not simply which is the operant plasma micro-instability at onset. Rather, even the location, onset time, extent and motion of the magnetotail energization process leading to the macroscopic substorm phenomenon are still unknown (Spence 1996).

Resolving the substorm problem requires accurate timing of three disparate but well-defined processes: ground auroral onset, current disruption onset at 8–10 R_E and reconnection onset at 20–30 R_E. Since these processes expand rapidly with time, knowledge of the onset location is as important as timing. THEMIS is the first mission specifically designed to determine the onset and evolution of the substorm instability. Towards this primary objective, THEMIS utilizes tail-alignments (conjunctions) between 5 identical probes on near-equatorial orbits, with periods that are multiples of each other.

THEMIS has a two-year design life, mainly driven by its radiation environment (Harvey et al. 2008), and 100% total ionization dose margin; but due to a launch vehicle delay it was launched just past the center-tail encounter of 2007. To ensure that baseline objectives remain intact and avoid a radical mission redesign late into the program, a post-launch coast-phase was prefixed to the baseline mission, giving it a total duration of 29 months. The tail mission phases last from early January to late-March in 2008 and in 2009. Radiation belt science objectives are addressed by the frequent probe traversals of the radiation belts on orbits whose periods and mean anomalies are dictated by the desire to align them in the tail. Natural evolution of the orbits in a Sun–Earth aligned system brings the probes to the dayside six months later, where THEMIS is able to address its dayside science objectives with a similar orbit strategy as in the tail. Thus, both dayside and radiation belt objectives are achieved with a mission design that is driven mainly by the tail science.

Table 1 describes the THEMIS science objectives. The THEMIS science and science closure is described in Sect. 2, and expanded upon in Sibeck et al. (2008). The THEMIS mission design is described in Sect. 3 and expanded upon in Frey et al. (2008). Figure 1 shows the orbital configurations during the first year of the baseline THEMIS mission. In the tail, reconnection is monitored by probes P1 (4 day period, $\sim$30 R_E apogee) and P2 (2 day period, $\sim$19 R_E apogee); while current disruption is monitored by probes P3 and P4 (1 day period, $\sim$12 R_E apogee). Four probes are required to accomplish the minimum mission (goals G1, G2 in Table 1); the fifth probe, P5, also on an approximately near-day period orbit, is an on-orbit spare that enhances mission reliability, but under nominal operations is used to perform timing and measure spatial gradients in yet one more dimension than would otherwise be possible with only two probes at the inner edge of the plasma sheet. Thus the fifth probe is required to satisfy other baseline requirements listed in Table 1, but is not required for the minimum mission. A network of ground observatories over the North American continent (from Eastern Canada to Western Alaska) monitors the aurora and space currents with white-light all-sky imagers and magnetometers, to provide accurate substorm onset timing, and is described in Mende et al. (2008). Probe alignments are designed to occur

Table 1 THEMIS science objectives

Mission driver	Science objective	Science goal
Primary	At the magnetotail: Onset and evolution of substorm instability	**G1** Time history of auroral breakup, current disruption, and lobe flux dissipation at the substorm meridian by timing: • Onset time of auroral breakup, current disruption and reconnection within $<$10 s. • Ground onset location within 0.5° in longitude and in space within 1 R_E. **G2** Macroscale interaction between current disruption and near-Earth reconnection. **G3** Coupling between the substorm current and the auroral ionosphere. **G4** Cross-scale energy coupling between the macroscale substorm instability and local processes at the current disruption site.
Secondary	At radiation belts: Production of storm-time MeV electrons	Source and acceleration mechanism of storm-time MeV electrons.
Tertiary	At dayside: Control of solar wind-magnetosphere coupling by upstream processes	The nature, extent and cause of magnetopause transient events.

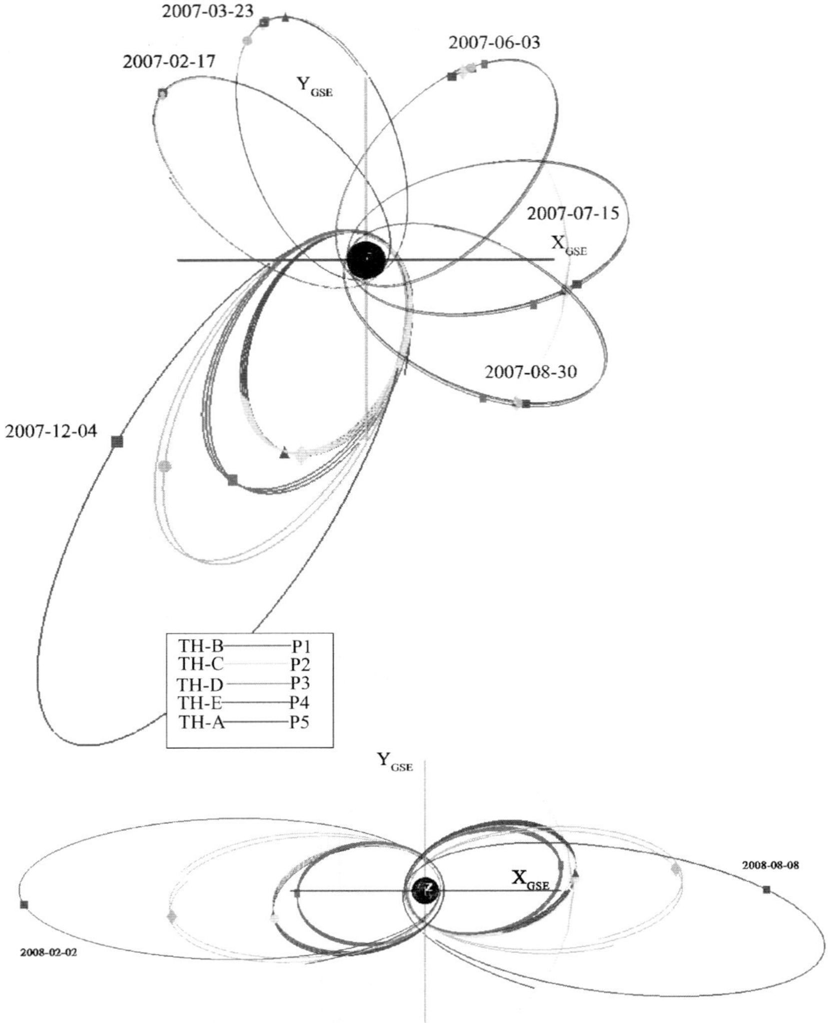

Fig. 1 *Top*: THEMIS coast phase actual orbits; *Bottom*: THEMIS 1st year baseline orbit predicts. (Axis size is 10 R_E)

year-round over the same point on the ground, i.e., over the North American sector. For the tail phase this corresponds to conjunctions between 00:30UT and 12:30UT and for the dayside it corresponds to conjunctions approximately 12 hrs later each day. This approach also optimizes utilization of GOES satellite magnetometer data sets and ground-based assets in North America.

THEMIS's five identical probes are equipped with five instruments each, measuring ions and electrons from ~5 eV to ~1 MeV and electromagnetic waves from DC to > 4 kHz. The instruments are summarized in Sect. 4 and are individually expanded upon in separate papers within this issue. Mission operations are performed by the Mission Operations Center (MOC) at SSL/UCB, as summarized in Sect. 5 and expanded upon in Bester et al. (2008). Science operations, data processing and analysis software is also summarized in Sect. 5 and expanded upon in Phan et al. (2008). All instruments and spacecraft are operating nominally and are expected to last for many years past their nominal lifetime. Plans for an extended mission are currently under way.

2 Science Objectives

Previous missions and fortuitous spacecraft conjunctions have provided a wealth of information regarding the substorm process but have been unable to determine where and how the substorm instability starts because of their un-optimized vantage points. Previous missions and fortuitous spacecraft conjunctions have been unable to determine where and how the substorm instability starts because of their unoptimized vantage points. For example, POLAR-Cluster radial conjunctions in the near-Earth plasma sheet result in less than 30 hrs of plasma sheet observations in rough Sun–Earth alignment, during which only a couple of times the near-neutral sheet location is sampled by both spacecraft. Geotail plasma sheet observations from 30 R_E in conjunction with inner magnetospheric probes (e.g., LANL satellites) results in significant observation time of substorms but without a satellite at the inner edge of the plasma sheet, where current disruption is expected to initiate. What is needed therefore is a dedicated multisatellite mission to measure with common instrumentation and with prolonged residence in the plasma sheet at 10, 20 and 30 R_E the process of current disruption and reconnection and their relative timing, as well as relationship to ground onset. THEMIS answers that need. This section is an outline of the key scientific objectives of the THEMIS mission that affect mission requirements.

2.1 Primary Objective: Substorm Causality

The components of the substorm instability i.e., Auroral Break-up, Current Disruption and Reconnection, evolve on a meso-scale range but interact over macroscales. High-sensitivity all sky imagers (ASIs) show (Friedrich et al. 2001) that the pre-onset equatorward arcs undergo large-scale undulations with wavelengths of hundreds of kilometers (Fig. 2). This is ~6° in longitude, which maps to a region of $\delta Y \sim 1$ R_E at the inner edge of the plasma sheet. Onset erupts in 10 s at a folding of one such undulation.

An intense cross-tail current (Lui 1996) (tens of nA/m^2), mainly supported by a duskward anisotropy in thermal ions (2–10 keV), provides substantial free energy at growth phase at ~10 R_E. At substorm onset the *current wedge* forms there (McPherron et al. 1973). This is an abrupt increase in the Z_{GSM} component of the magnetic field, accompanied by plasma heating. This morphological change of the field (Fig. 3) is consistent with a current-carrying particle distribution change (Mitchell et al. 1990). A current wedge is modeled as a partial disruption of the cross-tail current and diversion *along* the field lines, *into* the auroral ionosphere (Atkinson 1967; McPherron et al. 1973) where it feeds into the break-up arc. It is often termed the current disruption (CD) process (Lui et al. 1988). The hot, dipolar plasma originates in a small (Ohtani et al. 1991) equatorial area (~1 R_E^2) and expands azimuthally (Nagai 1982) up to ~10° of magnetic local time (MLT) per min and radially (Jacquey et al. 1991; Ohtani et al. 1992a, 1992b) at ~200 km/s.

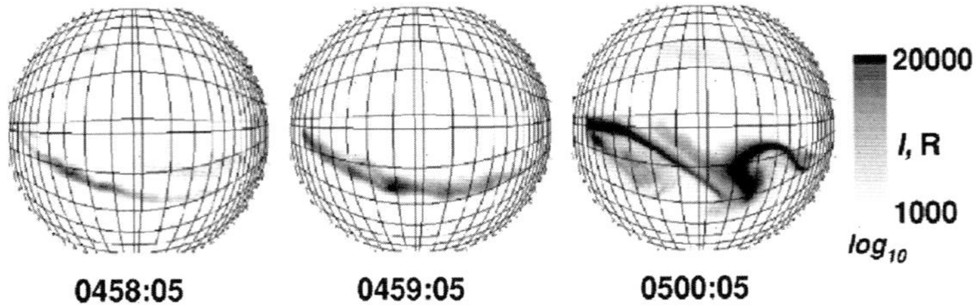

Fig. 2 Substorm onset as seen from a ground all sky camera station (Friedrich et al. 2001). Each line is 0.5 degrees in latitude (or 56 km) and in longitude (or 31 km)

Fig. 3 Development of the substorm current wedge through a reduction of the cross-tail current at 8–10 R_E in the equatorial plasma sheet

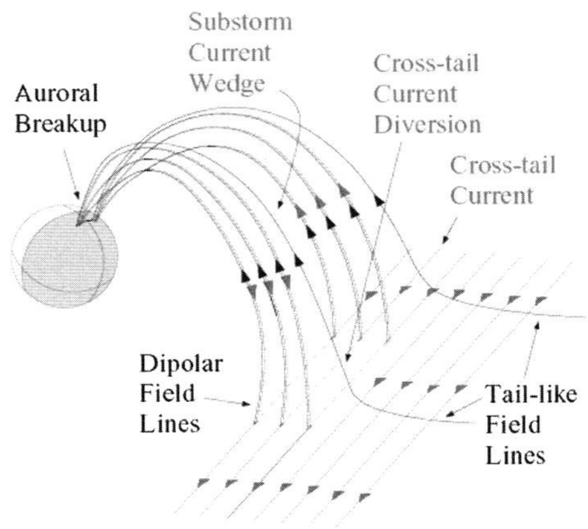

Further downtail, at ~25 R_E, there is evidence that *magnetic reconnection* takes place (Nagai et al. 1998). Fast, bursty, bulk ion flows presumably emanating from the reconnection site at Earthward speeds comparable to the Alfven velocity (1000 km/s), are also interpreted (Hones 1976; Nagai et al. 1997) as evidence of that process. Seen as close to Earth as 10 R_E (Fairfield et al. 1998, Angelopoulos et al. 1999) such flows are often localized to within 1–3 R_E (Sergeev et al. 1996b; Angelopoulos et al. 1997a) but are very efficient in energy and flux transport (Angelopoulos et al. 1994).

Presently, all possible causal sequences involving auroral break-up, Rx onset, CD onset and external triggers are viable hypotheses (Kennel 1992). In particular, CD and Rx might be causally linked, or may proceed independently of each other. As an impartial and experienced researcher summarizes:

> "Observations are gradually leading to a coherent picture of the interrelations among these various onset phenomena, but their cause remains a controversial question. The abrupt nature of substorm onsets suggests a magnetospheric instability, but doubt remains as to its nature and place of origin. Measurements increasingly suggest the region of 7–10 R_E near midnight as the likely point of origin" (Fairfield et al. 1992).

A number of substorm onset paradigms exist, but two of them can help epitomize the main ideas and reveal the primary observational requirements. These are the *"current disruption"* and the *"Near-Earth Neutral Line" (NENL)* paradigms.

According to the current disruption paradigm, an instability *local* to the current disruption region (8–10 R_E) is responsible for substorm onset (Lui 1996). The paradigm stems from two

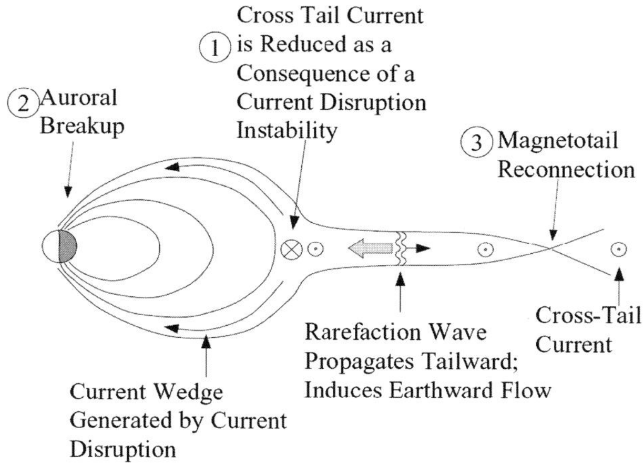

Fig. 4 Time-history of events at the substorm meridian according to the Current Disruption model for substorms (adapted from Lui 1991). *Numbers* indicate proposed chronological and causal sequence

Table 2 CD model event chronology

Order	Time (s)	Event
1	$t = 0$	Current disruption
2	$t = 30$	Auroral breakup
3	$t = 60$	Reconnection

basic observations: First, the break-up arc maps near Earth (Lui and Burrows 1978). This has been reinforced through advanced mapping of auroral images from Viking (Elphinstone et al. 1995), POLAR (Frank et al. 1998; Frank and Sigwarth 2000) and by ground-based observations (Samson 1992; Voronkov et al. 1999). Second, the cross-tail current density reaches tens of nA/m^2 and peaks near 8–10 R_E prior to substorm onset (Kaufmann 1987). This happens explosively (Ohtani et al. 1992a, 1992b) suggesting that it is in *that* region that the free-energy source and trigger for the substorm auroral surges reside. This paradigm suggests (Fig. 4) that Rx and fast Earthward flows are triggered by a CD-initiated fast mode rarefaction wave ($V_x = -1600$ km/s) once it reaches ~25 R_E. Flows cause neither the CD nor the auroral break-up itself. This rarefaction wave has not been conclusively reported before, as natural plasma sheet oscillations and the resultant diamagnetic effect, cause large amplitude, background magnetosonic waves. The relevant substorm component chronology appears in Table 2.

Recent experimental evidence in support of this paradigm comes from the observation that the particles energized first at the CCE spacecraft (located at 8–9 R_E) at onset are those with gyrocenters Earthward of CCE (Lui et al. 1988; Ohtani 1998). Finite gyroradius remote sensing applied on equatorial pitch angles produces the CD expansion's speed and direction ($\mathbf{V}_{xy}$). However, performing accurate CD onset timing requires knowledge of the CD expansion velocity at two probes which bracket the onset location. The probes should be at the neutral sheet (± 2 R_E) and near the CD location itself (± 2 R_E) so that the expansion speed will not vary significantly during its motion. Such timing has not been performed to date.

According to the Near Earth Neutral Line (NENL) paradigm (Hones 1976; Baker et al. 1996), bursty flows generated by near-Earth reconnection (Baumjohann et al. 1989) (~ 25 R_E) are responsible for substorm onset (Fig. 5). Observations pivotal for this model's development at the substorm meridian include fast tailward/Earthward flows (Hones 1976;

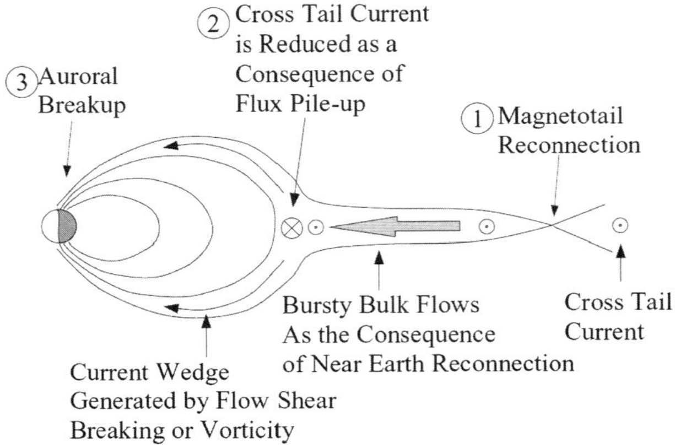

Fig. 5 Similar to Fig. 4, but from the viewpoint of the NENL model for substorms (adapted from Shiokawa et al. 1998b). Note the difference in the sequence of events

Table 3 NENL model event chronology

Order	Time (s)	Event
1	$t = 0$	Reconnection
2	$t = 90$	Current disruption
3	$t = 120$	Auroral breakup

Nagai et al. 1997) and plasmoid ejection (Hones et al. 1984; Slavin et al. 1992) both timed to start within 1–2 minutes from ground onset. This paradigm suggests that the flow kinetic energy is converted to particle thermal energy at the CD region. While heating generates a steep pressure gradient, the flow decelerates and deflects around Earth. The field-aligned current created locally by these processes (Hesse and Birn 1991; Shiokawa et al. 1998a, 1998b; Birn et al. 1999) leads to current disruption and auroral breakup. The recent observation that fast Earthward flows at 12–18 R_E occur within 1 min from substorm onset (Angelopoulos et al. 1997b; Shiokawa et al. 1998a, 1998b; Sergeev et al. 1995; Petrukovich et al. 1998; Yamade et al. 2000) has spurred renewed interest in field-aligned current generation in the NENL context. The NENL substorm component chronology differs from the current disruption model's (Table 3).

The NENL-predicted fast flow protrusion at 8–10 R_E has been rarely reported at substorm onset, but has been seen during pseudo-breakups, auroral streamer events (Henderson et al. 1998; Sergeev et al. 2000) and at substorm recovery (Nakamura et al. 2001a, 2001b). This has led to the suggestion (Ohtani 2001) that pseudo-breakup flows are CD onset triggers/substorm precursors. Alternatively: (i) The incoming flow may decelerate to compensate for the increasing magnetic field (Schodel et al. 2001) or (ii) The flow may dissipate through field-aligned Poynting flux (Wygant et al. 2000) along high latitude field lines (Zesta et al. 2000; Angelopoulos et al. 2001). The flow evolution and causal relationship (if any) to substorm onset is unclear, largely due to a lack of tail-aligned spacecraft conjunctions.

As in the case of CD onset detection, accurate Rx onset timing requires *two* probes at the plasma sheet, or its boundary, measuring velocity dispersed, field aligned, 30–300 keV particles. A strictly temporal interpretation of the dispersion provides L, the distance to the source (Sarris et al. 1976, 1996). A spatial interpretation (Richardson et al. 1987; Richardson and Cowley 1985) provides $L \cdot V_E / V_B$. Here, V_E is the convection velocity along the flight path of the particles (inferred by the dawn-dusk electric field component or measured by the plasma detector). V_B is the Z_{GSM} component of the boundary velocity measured by finite gyroradius remote sensing on East–West particle fluxes. The latter is the more general interpretation (when $V_E = V_B$ we retrieve the temporal one), but can only be used if the Rx site is nearby (within ~5–10 R_E), because the locally measured V_E/V_B is not necessarily constant along distant flight paths. Thus two probes at distances of 5–10 R_E from each other should *bracket* the nominal Rx site. Oppositely-directed fluxes at the probes establish that the reconnection site is between them (nearby), justifying the assumption of a constant V_E/V_B. The two probes should observe the particles as the boundary expands over. Thus the two Rx monitors need not be at the neutral sheet but within $\delta Z_{GSM} \sim 5$ R_E of it. Plasma sheet Z-fluctuations affect little the timing capability because the active plasma sheet expansions are large relative to those fluctuations. Such accurate Rx timing has not been performed to date.

Distinguishing between the CD and NENL models imposes similar observational requirements on timing and location as distinguishing between all substorm models. For example the Magnetosphere-Ionosphere (MI) coupling model (Kan 1998) suggests that the substorm starts due to the breaking of the Earthward flows at a rate >3 mV/m/R_E, and the ensuing Alfven wave bouncing. Contrary to the current disruption model, the flows come first, as a result of mid-tail or distant tail processes and the remaining sequence of events is similar to the current disruption scenario. As in the current disruption model, Rx is not a necessary condition for onset triggering.

Spontaneous (Henderson et al. 1996) onsets and externally triggered (McPherron et al. 1986; Lyons 1996) onsets (stimulated by sudden impulses, northward turnings or rotational discontinuities (Sergeev et al. 1990)) may exhibit different destabilization scenarios (Lyons 1995). It is possible, e.g., that external triggers result in an NENL-like path to substorm onset, whereas spontaneous onset substorms follow the CD paradigm prescription. It is thus important to classify substorms according to the external conditions in order to distinguish between different scenarios.

The science goals and objectives of Table 1, and the previous discussion on substorm phenomenology lead to a set of Mission Requirements (MR). These requirements are tabulated in Table 4. For example, ground onset timing should be performed along the substorm onset meridian (δMLT ~ 6°, which corresponds to 1 R_E at the CD site) and must be better than the time scale of interaction of those processes (30 s). Since CD onset is limited in $\delta XY \sim 1$ R_E^2 the CD monitors should be no more than δY~ δX~ ±2 R_E apart. Rx monitors should be around 19 R_E and 30 R_E, i.e., within ±5 R_E of the nominal Rx site to ensure constancy of the measured V_E/V_B ratio. The neutral sheet location (maximum Z_{GSM} distance in winter solstice) determines the orbit inclination of both the CD and the Rx monitors. Diurnal fluctuations at 10 R_E ($\delta Z \pm 2$ R_E) have little effect on the capability of the CD monitors to determine CD expansion speeds. Plasma sheet diurnal fluctuations at 20 and 30 R_E ($\delta Z \pm 3$ R_E) are small compared to the ±5 R_E tolerance. Additionally, the two inner probes in combination should permit cross-tail ($\delta Y \sim 0.5$–5 R_E) or cross-sheet ($\delta Z \sim 1$ R_E) conjunctions (not necessarily simultaneously).

Mission Closure of these requirements is shown at the right-hand column of Table 4. The objective to time auroral onset using < 30 s time resolution ASIs in Alaska/Canada necessitates that the probe apogees are in the US winter season, at central US midnight, i.e., ~6:30 UT (best performance of ASIs is in winter). This in turn calls for orbit periods that are multiples of a day. Remote sensing requirements for both CD and Rx monitors necessitate that they reside in near-equatorial orbits. Benign attitude Control System (ACS) requirements (better than 11.25°) are derived from the SST technical specifications (to control the

Table 4 THEMIS mission requirements and capability

Goal	Baseline [minimum] requirement	Mission capability
G1. Time history of breakup, CD, Rx at the substorm onset meridian	**MR1i** An ASI, two (a high- and a mid-latitude) GMAG stations per MLT hr, pin-point onset at δMLT$< 1°$, $t_{res} < 30$ s [one GMAG, δMLT $< 6°$].	**MC1i** 2 ground ASIs, 2 (high lat./mid lat.) magnetometers provide onset detection within δMLT $< 0.5°$, $\delta t = 1$ s. Even when cloudy, PiBs provide $\delta t = 1$ s and mid-lat. gmags determine onset meridian within δMLT $< 5°$.
	MR1ii 2 equatorial probes at 10 R_E, separated by $\delta XY \sim 2$ R_E monitor CD onset at $t_{res} < 30$ s [same].	**MC1ii** P3 & P4 ($\delta XY \sim 2$ R_E) time CD onset at $t_{res} < 10$ s.
	MR1iii Two orbits bracketing Rx region, separated by $\delta Y \sim 2$ R_E and at apogee within 5 R_E of neutral sheet (at 19 R_E, inc ~9° and at 30 R_E, inc ~7°) measure reconnection onset at $t_{res} < 30$ s [same].	**MC1iii** P1 & P2 at required orbits time once per 4 days at $\delta Y \sim 2$ R_E ($\delta X \sim$ 6–10 R_E) fast flow onset at $t_{res} < 10$ s.
	MR1iv CD and Rx monitors align (within ±2 R_E) during >10 [>5] substorms near winter (±2 mo.).	**MC1iv** P1, P2, P3 & P4 align once per 4 days. P5 also part of alignment strategy (average ~12 hours/alignment). 80 substorms/yr; 16 substorm-alignments/yr).
	MR1v SST to measure on ecliptic plane (axis control $\sim \pm 30°$) i^+/e^- fluxes (40–100 keV) at $t_{res} = 10$ s [same].	**MC1v** Spin-plane-mounted SST (20 keV to >1 MeV) on all probes at $t_{res} =$ 3 s, covers required FOV at all seasons. Spin axis normal to ecliptic. ACS control~0.5°.
	MR1vi $\delta B/B \sim 10\%$, or $\delta B \sim 1$ nT absolute [same].	**MC1vi** $\delta B \sim 0.6$ nT absolute, routinely at 4 vectors/s.
G2. CD–Rx coupling.	**MR2i** Track rarefaction wave (1600 km/s) in B.	**MC2i, 2 ii** P3 & P2 measure time delays at $\delta X/\delta t = 6$ R_E/3 s = 12000 km/s during 160 substorms (32-alignments)/yr.
	MR2ii Track earthward flows (400 km/s) in V.	**MC2iii** $\delta B \sim 0.6$ nT absolute and $\delta V/V \sim 10\%$.
	MR2iii $\delta B \sim 1$ nT absolute, $\delta V/V \sim 10\%$.	

Table 4 *(Continued)*

Goal	Baseline [minimum] requirement	Mission capability
G3. Substorm coupling to auroral ionosphere	**MR3i** Measure radial/cross-sheet pressure gradients ($\delta P/\delta XY \sim 0.1$ nPa/R_E); flow vorticity/deceleration ($\delta V/\delta XY \sim 100$ km/s R_E). Requires 10% accuracy in δV, δP over 1 R_E scales ($\delta P/P \sim \delta V/V \sim 1$).	**MC3i** δXY conjunctions between P3, P4, P5 over ranges of 0.3–10 R_E provide δP, δV with 10% absolute accuracy. Modeling provides curlV, gradP.
	MR3ii Measure $J_{current\_sheet}$ (planar approximation, $\delta J/J \sim 10\%$, $\delta B/B \sim 10\%$ or $\delta B \sim 1$ nT absolute, 0.1 nT relative, over $\delta Z \sim 0.5$ R_E) and incoming flows.	**MC3ii** P4 & P5 δZ-conjunctions provide $\delta B \sim 0.6$ nT absolute, 0.03 nT relative while P2 measures flows.
	MR3iii E field ($t_{res} = 10$ s) for non-MHD part of flow.	**MC3iii** E field measured at 4 vectors/s routinely.
	MR3iv Study > 10 events in each δX, δY, and δZ.	**MC3iv** Cross-tail, cross-sheet or tail-aligned separations: 320 substorms/yr. P2 (incoming flows) available during 160 of those. Simultaneity in δX–δY or δX–δZ observations (not required) is possible.
G4. Substorm coupling to local modes at 10 R_E	**MR4i** Cross-tail pairs to measure FLRs, KH and ballooning waves in B, P, V and E at $\delta Y \sim 0.5$–10 R_E, $t_{res} = 10$ s.	**MC4i** P3, P4, P5 measure B, P, V and E at separations $\delta Y \sim 0.3$–10 R_E, at $t_{res} = 3$ s or better.
	MR4ii Cross-sheet pairs to measure $J_{current\_sheet}$ (as before) as free energy for cross-field current instabilities at 6 Hz, on E field @ spin-plane (3D), B-field in 3D.	**MC4ii** P3 & P5 δZ-conjunctions measure sheet density ($B \sim 0.6$ nT absolute).
	MR4iii Study 10 substorms or more.	**MC4iii** 160 substorms/yr (P3, P5). P2 aligns and times flows for 64 substorms/yr.

SST detectors that will be affected by Sun pulse). $\delta B/B \sim 10\%$ requirements arise from the need to monitor the rarefaction wave (also the cross-tail current within $\delta J/J \sim 10\%$, given a ΔB ~B between probes at separation $\delta Z \sim 1$ R_E). In a minimum field of 10 nT this renders the absolute stability requirement of 1 nT.

THEMIS desires to measure at least a few solar wind-triggered and a few spontaneous onset substorms (assuming equal chances to observe each). At least 5 substorms should be observed in each probe conjunction configuration. This defines the baseline mission requirement. Given a 3–6 hr recurrence time for substorms (Borovsky et al. 1993), this necessitates 30 hrs of useful data in each year of conjunctions. THEMIS's orbit strategy accounts for >260 hrs of conjunctions in each year (Frey et al. 2008). Clear evidence that tail-aligned spacecraft equipped with THEMIS-like instrumentation can indeed monitor the progression of the incoming flows despite their $\delta Y \sim 1$–3 R_E localization comes from fortuitous ISTP conjunctions during north–south arcs at late substorm recovery (Henderson et al. 1998; Sergeev et al. 2000).

While significant losses of useful events may occur due to plasma sheet fluctuations, lack of solar wind data, possible extreme event localization, and early evening/late morning substorms, the mission can easily satisfy the requirement to capture at least a handful of substorms from a tail-aligned vantage point and resolve the pressing question of substorm causality. Of those, a few high quality, clear and effective conjunctions will receive attention by a large number of people (like CDAW events).

The mission design is stable to the J2 terms of the geo-potential and sufficient fuel exists to counteract lunar perturbations. THEMIS is immune to the differential precession of the line of apsides between the high and low altitude orbits, because it relies on mean anomaly phasing to obtain tail alignments. Further information on how the THEMIS orbits, probe design, attitude, instruments, data rates and cadence satisfy mission objectives is presented in Sibeck et al. (2008).

2.2 Secondary Objective: Radiation Belt Energization

At storm main phase, MeV energy electrons are abruptly (1–4 hrs) lost; they reappear also abruptly at storm recovery with fluxes higher than prior to the storm (Fig. 6). This MeV electron flux increase represents the main electron flux increase of electrons during a storm. The observed *rapid* increase of MeV electron flux inside of geosynchronous altitude cannot be accounted for by the *relatively slow* diffusion of solar wind plasma. The "Dst effect" *alone* cannot account for this process either, since the electrons reappear at much higher fluxes than before the storm. Electron fluxes are therefore likely enhanced at $L = 11$ before being transported inwards. Daily variations of MeV electrons are modeled successfully under that assumption (Li et al. 2001), but it is unclear whether such an electron source is indeed present beyond geosynchronous altitude at storm recovery or whether local acceleration of inner magnetosphere cold/warm electrons by ULF or VLF waves may play a role (e.g., Friedel et al. 2002; Millan and Thorne 2007).

The instantaneous radial profile of the electron flux at constant μ and the transport process fully determine the evolution of the outer belt. But no single satellite traversing the equatorial radiation belt and its sources (i.e., L-values from 3.5 to 11) can measure the radial profile of the electron fluxes faster than once per ten hours, due to its orbital period. Low altitude (polar) satellites measure near-loss-cone fluxes and underestimate the true equatorial flux value which peaks at 90° at active times. Multiple satellites on eccentric, equatorial orbits are needed in order to provide repetitive cuts through the radiation belt. The satellites should be displaced sufficiently along their orbit. THEMIS's P3 and P4 probes are separated

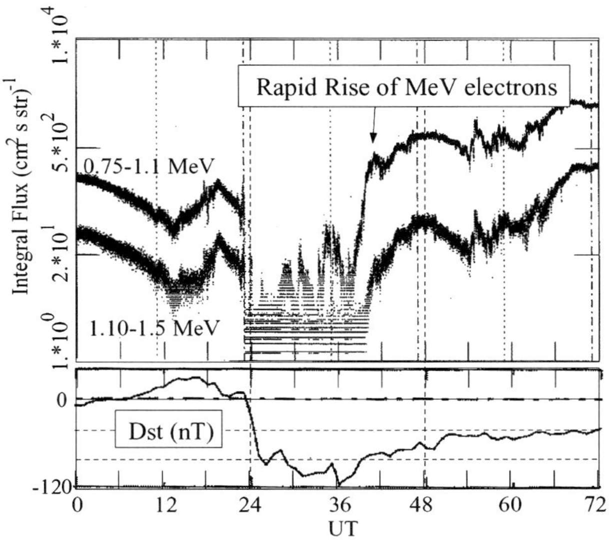

Fig. 6 LANL satellite data from a storm on November 3, 1993 exemplify the rapid loss and reappearance of storm time electron flux at geostationary orbit at storm onset and rapid (1–4 hr) reappearance at recovery (Li et al. 1997)

by several Earth radii along track when they traverse the radiation belts, due to their desired separation in mean anomaly for achieving tail science objectives. THEMIS's P5 probe is in a slightly different orbital period than sidereal. Probes P1 and P2 also traverse the radiation belts during the inbound and outbound passes. Together the five THEMIS probes traverse the inner magnetosphere from $L = 3.5$ to $L = 11$ with a median rate of recurrence of 3.8 hours. They can provide the needed radial profiles of the radiation belt electrons without any modification to the mission design. If the slope of the electron phase space density is inconsistent with a radially inward diffusion of killer electrons, THEMIS has the fields instrumentation to detect the presence and distribution of electromagnetic wave power to determine if such waves play a role in rapid electron acceleration at storm recovery.

2.3 Tertiary Objective: Upstream Processes

Observations near the equatorial magnetopause provide strong evidence for the predicted signatures of transient solar wind-magnetosphere coupling, namely fast flows (Paschmann et al. 1979) and flux transfer events (FTEs) (Russell and Elphic 1978). These may be either triggered by solar wind features (Lockwood and Wild 1993) or occur in response to intrinsic instabilities (Le et al. 1993). A number of other externally driven transient phenomena also contribute to the variations observed on single spacecraft. *Efforts to discriminate between the causes of magnetopause transients and determine the significance of each phenomenon to the solar wind-magnetosphere interaction have been hampered by several obstacles:*

First, observations near the L1 point or several 10 s of R_E off the Sun–Earth line are of limited use because solar wind features transverse to the Sun–Earth line are on the order of $\sim$20 R_E (Crooker et al. 1982; Paularena et al. 1998) and lag time uncertainties increase with distance (Collier et al. 1998). *Second*, foreshock and magnetosheath processes affect the magnetopause. These cannot be observed within the pristine solar wind (Fairfield et al. 1990; Thomas and Brecht 1988) and must be observed in place. Examples are: Hot flow anomalies transmitted across the bow shock (Völk and Auer 1974; Lin et al. 1996) and sheath (Paschmann et al. 1988; Sibeck et al. 1997); externally-driven, propagating slow shocks (Song et al. 1992) or standing slow shocks (Southwood and Kivelson 1995) in the magnetosheath. *Third*, the significance of individual events depends upon their azimuthal dimensions. FTEs range from 0.5 to 5 R_E (Phan and Paschmann 1995). Events with similar features include solar wind/foreshock pressure-driven waves (Sibeck et al. 1989) or Kelvin-Helmholtz (Farrugia et al. 2001) waves.

Thanks to its unique Sun–Earth aligned probe conjunctions, THEMIS will overcome the aforementioned obstacles and determine the response of the coupled dayside solar wind-magnetosphere system to varying incident conditions. With particle and magnetic field instrumentation similar to that flown on AMPTE/IRM, THEMIS probes P1 and P2 will not only characterize the solar wind but also will determine its modification within the foreshock (Paschmann et al. 1988; Sibeck et al. 1989). Hundreds of hours of conjunctions will enable us to conduct statistics of event occurrence patterns and characteristics as a function of the solar wind conditions.

Further information of how THEMIS's instruments and orbits meet the mission requirements and address the science objectives, including preliminary results showing the efficacy of the THEMIS mission to perform its task is provided in Sibeck et al. (2008).

3 Mission Design

Minimum science closure can be achieved with four probes in one year of tail crossings. Inclusion of the fifth probe reduces risk and increases science return towards a baseline mission. Any of the inner probes have sufficient fuel reserves to replace either of the outer probes during the mission, but P5 was designated the "replacement probe". Science increase from the presence of the replacement probe allows probe pair measurements simultaneously in both X and Y GSM dimensions the first year (Y and Z dimensions the second year). Azimuthal separations are very desirable in both years in order to maintain an adequate baseline to determine the location and timing of the current disruption. Simultaneous radial and cross-sheet (X and Z GSM) separations await an extended mission.

The spinning probes ($T_{\text{spin}} = 3$ s) are designed to be dynamically stable even under worst-case scenarios (as demonstrated by fault tolerance analyses). A single-string probe design was further simplified by a minimal hardware complement, by inherent functional redundancy, strong instrument heritage and with the instruments and probe bus designed for graceful degradation. A probabilistic risk assessment and contingency analysis demonstrates that, with P5 ready to replace any other probe during the mission, even a single string design results in $> 93\%$ reliability for achieving the minimum mission. Redundancy is a key feature of any constellation of satellites and THEMIS is the first NASA mission to take advantage of it.

3.1 Probe Conjunctions

THEMIS was launched on February 17, 2007 and the probes were released on a highly elliptical, 14.716 R_E geocentric apogee, 437 km altitude perigee, 15.9 deg inclination, 31.4 hr period orbit by a Delta-II 7925 rocket from Cape Canaveral, with their line of apsides pointing at apogee towards the pre-midnight sector (Right Ascension of Perigee = 288.8 deg). The probes were checked out and placed in a stable, coast-phase orbit, traversing the dayside magnetosphere in a string-of-pearls configuration near their launch orbit. After instrument commissioning, the probes, initially named by their letters A–E, were assigned their target orbits and were designated a probe number based on their on-orbit performance (mainly antenna performance) as follows: B, C, D, E, and A were assigned constellation positions P1, P2, P3, P4 and P5 respectively. In the initial part of the coast phase, concurrent with spacecraft and instrument commissioning (February–May 2007) the probe relative dispersion from the launch vehicle resulted in a C-DBA-E series configuration, with DBA clustered near each other at 100 s of km separation and C and E leading and training respectively by several 1000 s of km (Fig. 1). Magnetometer booms were deployed only days after launch but electric field antennas on TH-C (P2), TH-D (P3) and TH-E (P4) were deployed by mid-June 2007. Electric field booms on TH-B (P1) and TH-A (P5) were deployed in November and January respectively, i.e., after the placement maneuvers on those probes, in order to facilitate operations. After probe number assignment, it was decided that the probes with electric field antennas deployed during the coast phase would be maneuvered in the middle of the constellation, at small-scale separations, while the other two probes would become leading (TH-B) and trailing (TH-A). Inter-spacecraft separation was on the order of 100 km between inner probes and several 1000 s of km between the outer probes. Coast phase science in this optimal string-of-pearls configuration continued until the end of August 2007. Placement into the baseline orbits occurred between September and December 2007. Following on-orbit calibration and a period of bonus (non-baseline) coast-phase science, the probes were placed by December 4th 2007 into their final orbits in anticipation of the baseline mission. The baseline orbit strategy follows naturally from the requirements in Tables 1 and 4. Probe elements as function of year are tabulated in Table 5.

In the 1st tail season, P1 has apogee ~30 R_E and a ~4 day period, while probe P2 has apogee ~19 R_E and a ~2 day period. Once per 4 days these probes align near apogee and bracket the reconnection site. Probes P4 and P3 have apogees at 12 R_E and differ in their mean anomaly such that at apogee they are separated by ~1 R_E. At or near apogee these probes routinely monitor the CD using the finite gyroradius technique. The third innermost probe (P5) has initially an apogee of ~10 R_E: During the first tail season it has a faster-than-synchronous period, gaining 6 hrs/day along is orbit relative to P3, P4. Once every four days the inner probes cluster near apogee. Cross-tail separations between P3/4 and P5 range between 0.3 and 10 R_E and permit long wavelength studies of low frequency MHD waves. P5 is given an inclination change of 5° relative to P3,4. This affects the apogee conjunctions very little during the 1st tail season, when the argument of perigee (APER) is small; but creates a Z-separation of the inner probes in the 2nd tail season, when APER is large (note that the inner probes drift in APER by ~90°/yr due to J2 terms).

The comprehensive THEMIS approach to solving the substorm problem calls for monitoring the nightside auroral oval with fast-exposure (1 s), low cost and robust white-light all sky imagers (ASIs) and high-time resolution (0.5 s) ground magnetometers to achieve faster than 3 s cadence measurements of the auroral break-up. A map of the sites is shown in Mende et al. (2008). The THEMIS Ground-Based Observatories (GBOs) cover a 12 hr local time sector, over the North American continent, from Eastern Canada to Western Alaska. ASIs provide a global view of the Northern Hemisphere aurora at unprecedented temporal and spatial resolution.

The THEMIS mission design optimizes substorm capture at probe conjunctions for the tail phases of the mission. A conjunction is defined as a four (or five) probe alignment within $\delta Y_{GSM} = \pm 2$ R_E, at the plasma sheet. Plasma sheet encounters are near-neutral sheet occurrences for the inner probes, P3,4,5 ($Z_{NS} = \pm 2$ R_E) but more relaxed for the outer probes, P1,2 since those probes' function is to determine Rx location using timing of boundary layer beams ($Z_{NS} = \pm 5$ R_E). Those major conjunctions occur once per four days, due to the orbit period of P1. Minor conjunctions are alignments between P2 and two inner probes. They recur once per two days (due to the orbit period of P2). When P1 is also available the minor conjunction becomes a major one. Daily conjunctions are those between P3, P4 (on a

Table 5 Probe elements as function of season (last 4 columns are predicts)

		Date:					
		Release: 2007-02-18	Coast: 2007-07-15	Tail #1: 2008-02-02	Dayside #1: 2008-08-08	Tail #2: 2007-02-17	Dayside #2: 2007-08-13
P1	R_A [R_E, geocentric]	14.72	14.72	31.0	30.7	31.1	30.4
(TH-B)	R_P [km, alt]	437.	699.9	1275.	3824.	1402.	5418.
	INC [deg]	16.0	13.5	0.7	15.5	7.1	1.2
	RAP (=APER+RAAN) [deg]	288.8	302.9	312.	318.	322.	327.
	APER [deg]	319.6	344.2	270.	189.	354.	300.
P2	R_A [R_E, geocentric]	14.72	14.71	19.5	19.3	19.3	19.5
(TH-C)	R_P [km, alt]	437.	761.3	1976.	2613.	2039.	1594.
	Inc (deg)	16.0	13.5	5.6	0.9	13.6	11.
	RAP (=APER+RAAN) [deg]	288.8	302.9	314.	322.	331.	348.
	APER	319.6	344.3	3.0	94.	9.0	39.
sP3	R_A [R_E, geocentric]	14.72	14.72	11.8	11.7	11.6	11.6
(TH-D)	R_P [km, alt]	437.	720.5	2677.	3059.	4143.	4206.
	Inc (deg)	16.0	13.5	6.7	5.2	4.8	4.8
	RAP (=APER+RAAN) [deg]	288.8	302.6	322.	336.	341.	4.0
	APER	319.6	343.8	20.	60.	99.	125.
P4	R_A [R_E, geocentric]	14.72	14.73	11.8	11.7	11.6	11.6
(TH-E)	R_P [km, alt]	437.	626.6	2677.	3059.	4079.	4270.
	Inc (deg)	16.0	13.5	6.1	4.7	4.1	4.6
	RAP (=APER+RAAN) [deg]	288.8	303.2	322.	336.	341.	3.0
	APER	319.6	344.9	19.	59.	98.	122.
P5	R_A [R_E, geocentric]	14.72	14.70	10.0	10.8	11.7	12.9
(TH-A)	R_P [km, alt]	437.	881.9	2868.	3088.	3187.	3760.
	Inc (deg)	16.0	13.4	11.2	10.0	9.4	10.6
	RAP (=APER+RAAN) [deg]	288.8	302.4	318.	336.	343.	2.0
	APER	319.6	342.8	13.	55.	94.	118.

sidereal day period) and the ground-based observatories. By definition those occur at the nightside between 00:30 and 12:30 UT. Once per two days they include P2 and constitute minor conjunctions. Once per four days they include P1, P2 and P5 and constitute major conjunctions.

In preparation for 1st *year dayside observations* P5's apogee is raised to $\sim$11 R_E with the same perigee, thereby increasing its period T_{P5} to 7/8 of $T_{P3,4}$. This has two advantages: First, it reduces the rate of differential precession of P5 relative to P3 and P4, in anticipation of the upcoming 2nd tail season. Second, it optimizes azimuthal separations between P5 and P4/5 at several R_E scales near the subsolar magnetopause to address the extent and lateral motion of boundary layer phenomena: Once per 8 days the subsolar magnetopause is encountered by all three inner probes with separations as close as 1–2 R_E, with P3 and P4 typically at the *magnetosheath* when P5 is at the *magnetopause*. During other periods of major, minor or daily probe conjunctions, separations between P5 and P3,4 at the magnetopause can be as large as 6 R_E or more.

For the 2nd *year tail season*, P5's apogee and mean anomaly are made identical to P3 and P4's. P5's inclination difference (5°) relative to P3/4 (achieved by a cost-effective maneuver in the 1st year), and orbit design considerations for a common inner probe APER ($\sim$90°) ensures a $\sim$1 R_E difference in the Z-direction at apogee between P3,4 and P5. This permits studies of the thin cross-tail current during substorms (assuming a planar approximation) during the second tail season.

For the 2nd *year dayside season*, P5's apogee is increased to 13 R_E (same perigee as other probes), thereby increasing its period, T_{P5} to 9/8 of $T_{P3,4}$. During the 2nd year inner probe conjunctions at the magnetopause occur also once per 8 days, like the 1st year. There are, however, two differences: First, During the 2nd year P5 is a *magnetosheath* monitor at the subsolar region, whereas P3,4 are *magnetopause* monitors; Second, the average distance of magnetopause encounters is about 1 R_E further away from Earth than during the 1st year, which permits magnetopause observations in the pre- and post-noon sectors, further away from the subsolar point.

3.2 Inertial Location and Attitude of the Constellation

As the choice of a target date for center-tail observations moves from winter solstice into vernal equinox, the Sun–Earth line moves closer to the apsidal line near the equatorial plane and the maximum shadow duration of P1's and P2's near-Equatorial orbits increase. A given target date is characterized by the inertial location of the mission orbits' semi-major axis, whose longitude in inertial space is the least affected by lunar and J2 terms. The orbit inertial longitude is measured by the Right Ascension of Perigee (RAP), the sum of the argument of perigee (APER) and the right ascension of the ascending node (RAAN). RAP is fixed for each choice of a target center-tail date (e.g., it is 330 degrees for Feb-21 and moves a degree per day). Mission design seeks center-tail target dates that minimize shadows and maximize conjunctions. The choice of RAP = 312 deg for the 1st tail season was deemed optimal considering the entire end-to-end mission design, as it ensures $>$ 188 hrs of tail-aligned conjunctions and $<$ 3 hrs of shadows. This choice of inertial pointing of the constellation line of apsides in the sky determined the choice of our launch elements, such that the orbits drift into the desired inertial location in the sky by 1st year mid-tail (properly accounting for precession through the coast phase season). The RAP drifts by about 11°/year for P1, 22°/yr for P2 and 33°/yr for the inner probes. Thus, differential precession naturally limits the duration of useful conjunctions (and the mission lifetime) to approximately two years. Lunar perturbations affect mostly the outer probe inclinations and need to be balanced by inclination change maneuvers. These, in addition to the orbit placement maneuvers, were the main drivers on probe fuel.

Additional considerations for the selection of the center-tail observation season were: (1) the substorm recurrence rate, which maximizes around equinoxes; (2) the dipole tilt, which affects neutral sheet hinging and reduces simultaneous residence in the plasma sheet at apogees of 10, 20, 30 R_E, by our near-equatorial probes; (3) the angle of the tail magnetic field to the nominal, near-ecliptic probe spin plane. This angle is desired to be $> 10°$ to enable computation of the spin axis electric field from the spin plane components, under the $\mathbf{E} \cdot \mathbf{B} = 0$ approximation, and increases away from solstice; (4) the ASI viewing conditions (cloud cover in Alaska and Western Canada reduces in late winter months with optimal viewing in mid-February); (5) the dark-sky duration at polar latitudes, which optimizes in mid-winter. Those considerations were satisfied by our choice of a tail season centered in mid-February.

The probes' nominal attitudes are nearly normal to the ecliptic plane. Driven by the desire for an efficient probe design (Harvey et al. 2008), only four thrusters are used in the probe design: Two co-aligned side thrusters, used for spin-up and spin-down, with thrust vectors located on a plane containing the probe center of gravity; two co-aligned axial thrusters, used for pulsed reorientations and for long duration, efficient orbit-change thrusts opposite to the spin axis. Axial thrusters are mounted at two bottom deck corners of the probes, pointing opposite to the spin vector. Therefore, the probes can thrust sideways and opposite to the spin axis but not along the spin axis. Large thrusts are nominally planned using the axial thrusters, after reorienting the probe spin axis along the desired deltaV direction. Once electric field booms are deployed reorientations are very costly in deltaV due to the large moment of inertia. Thus most large thrusts were executed during the orbit placement maneuvers prior to the electric field boom deployments. During the course of the baseline observation period, required thrusts by the inner probes are either on the ecliptic or near ecliptic-south. Thus the inner probes have to continue to point towards near-ecliptic north for the remainder of their lifetime, as they have since the early part of the mission and all through the coast phase. The outer probes, P1, P2, on the other hand, each have to perform one large inclination change maneuver prior to the second tail season, aimed at orbit corrections to lunar perturbations. Those maneuvers require thrusting near the ecliptic-north direction. This necessitates placing the outer probes with their spin axes pointing near ecliptic-south. Probes P1 and P2 were placed into an ecliptic-south attitude near the end of the placement maneuvers, in the Fall of 2007, and are expected to remain in that approximate attitude throughout the rest of their lifetime.

Finally, to ensure a large angle of the nominal magnetic field and the spin plane, the inner (outer) probe attitudes were tipped an additional 8° towards (away from) Sun on reference day of Feb 7, 2008. These attitudes also ensure that the electric field instrument spin plane spheres will not be shadowed by the probe body during most of the year, except for a period of several days at dawn and dusk when accurate electric field observations are less critical.

4 Instrumentation

The five spin-stabilized ($T_{\text{spin}} = 3$ s) THEMIS probes carry identical instruments that meet or exceed the requirements of the baseline science objectives. Table 6 summarizes the instruments and their specifications. A centralized parts-procurement, instrument development, quality assurance, safety and verification program resulted in an efficient production and integration of the THEMIS instruments. A production-line approach was utilized to increase

Table 6 THEMIS instruments

	Provider	Specifications	
FGM	TUBS& IWF	Stability	<1 nT/yr (0.02 nT/12 hrs)
	Auster et al. (2008)	Resolution	3 pT (digitization: 12 pT)
		Noise	10 pT/sqrt(Hz) @ 1 Hz
		Frequency	DC – 64 Hz§
ESA	SSL/UCB	Energy	i: 5 eV – 25 keV
	Carlson et al. (2008)		e: 5 eV – 30 keV
		$\delta E/E$	Inherent: i: ∼19%; e: ∼ 15%
			Transmitted: 35% (32 steps)
		g-factor, per anode	i: 0.875×10^{-3} cm^2 str keV/keV
			e: 0.313×10^{-3} cm^2 str keV/keV
		eflux, per anode	i: 10^3–10^9 eV/ (cm^2 s str eV)
			e: 10^4–10^{10} eV/ (cm^2 s str eV)
		Angular Res.	Inherent: e: (22.5° × 11.25°)
		Elev. × Azim.,	i: (SW: 5.625° × 5.625°max)
		FOV [deg]	Typical: 22.5° × 22.5°, 4π str
SST	SSL/UCB	Energy	i: 25 keV – 6 MeV
	Larson et al. (2008)		e: 25 eV – 1 MeV
		Energy steps	16 (transmitted)
		g-factor (closed/open)	i: $8 \times 10^{-4}/0.1$ cm^2 str
			e: $8 \times 10^{-4}/0.1$ cm^2 str
		eflux, per detector	i, e: 0.5–5 × 10^8 keV/ (cm^2 s str keV)
		Angular Res.	Inherent: 30° × 11.25°
		Elev. × Azim.,	Transmitted: 30° × 22.5°
		FOV	4 elevations ×16 azimuths
SCM	CETP	Frequency range	1 Hz – 4 kHz§
	Roux et al. (2008)	Sensitivity	0.8 pT/sqrt(Hz) @ 10 Hz
		Dynamic range	10^{-5}–1 nT/sqrt(Hz) [spectra]
EFI	SSL/UCB	Frequency range	DC–8 kHz§; AKR band: 100–300 kHz
	Bonnell et al.	Sensitivity	10^{-4} mV/m/sqrt(Hz) @ 10 Hz
	(2008)	Dynamic range	10^{-4}–10^2 mV/m/sqrt(Hz) [spectra]
		Time series:	±300 mV/m; 0.009 mV/m [DC coupled]
		range; resolution	±100 mV/m; 0.003 mV/m [AC coupled]
		Noise	3×10^{-6} mV/m (SpB); 3×10^{-5} mV/m (AxB)
		DC offset error	0.1 mV/m (SpB); 1 mV/m (AxB)
		Antenna lengths	50 m (12), 40 m (34), 7 m (56) tip-to-tip
GBO:	SSL/UCB	Sensitivity	< 1 kRayleigh
ASIs	Mende et al. (2008)	Resolution	> 250 pixels ASI dia.; 0.5° thumbnail
	Harris et al. (2008)	FOV	170 deg
		Spectral band	400–700 nm (white light)
		Cadence	3 s image rate / 1 s exposures
GBO/EPO:	UCLA	Noise	10 pT/sqrt(Hz) at 1 Hz
gmags	Russell et al. (2008)	Range/Resolution	±72,000 nT/0.01 nT
		Rate	2 samples/sec

§Nyquist

efficiency. A common Instrument Data Processing Unit (IDPU) housed all instrument electronics and interfaced with the Bus Avionics Unit (BAU) for data relay and instrument commanding. IDPU and BAU simulators developed early in the program allowed efficient parallel development of the probe and the instruments. The IDPU is described in Taylor et al. (2008). The BAU is described in Harvey et al. (2008). The FluxGate Magnetometer (FGM), measures the DC magnetic field up to 128 S/s and is described in Auster et al. (2008). The sensors were built by TUBS, and the electronics design and testing were performed by IWF. The flight electronics were implemented at UCB under a common-parts buy-and-make program. In-flight calibration was be performed by TUBS and UCLA. A magnetic cleanliness program was implemented jointly by UCLA and UCB. It encompassed parts selection and testing for DC and AC fields, modeling and compensation of solar panel and power system currents, propulsion system and SST magnets, i.e., the principal known offenders. Testing and verification took place at UCLA, UCB and JPL. Details of the magnetic cleanliness program are provided in Ludlam et al. (2008).

The ElectroStatic Analyzer (ESA), built at UCB to the recent heritage of the FAST ESA and the Cluster HIA instruments, measures ions and electrons between 5 eV and 25 keV. It is described further in Carlson et al. (2008). On-board moment computations on an FPGA permit subtraction of photoelectron fluxes and routine data collection and transmission of moments at spin period resolution. Careful ESA ground and in-flight inter-calibration, intra-calibration and absolute calibration are described in McFadden et al. (2008) and result typically in better than 10% accuracy moments.

The Solid State Telescope (SST), built at UCB to the recent heritage of similar instruments flown on the WIND and STEREO spacecraft, measures ions and electrons between 25 keV and >1 MeV. It consists of two units (heads) per probe, each unit measuring ions and electrons in two directions. Quadrupole fields resulting from matched and paired electron broom magnets reduce magnetic contamination. A mechanical attenuator results in a factor-of-100 increase in instrument dynamic range, which enables it to avoid saturation at the high fluxes near the inner edge of the plasma sheet, and have superior sensitivity at ~30 R_E in the mid-tail and in the solar wind. The SST is described further in Larson et al. (2008).

The triaxial Search Coil Magnetometer (SCM) was built by CETP to the recent heritage of similar units flown on Cluster and Interball, and measures AC magnetic fields from ~1 Hz to 4 kHz. The instrument is described in Roux et al. (2008). The signals from the three SCM sensor axes are pre-amplified in a highly integrated electronics module and then processed together with signals from the electric field instrument by a Digital Fields Board (DFB). The DFB processes the analog signal, digitizes the signal at high time resolution and produces a number of waveform and spectral data products available for triggers and memory storage and transmission. The board was designed by the University of Colorado and is described in Cully et al. (2008).

The three-dimensional Electric Field Instrument, EFI, consists of four spherical sensors, mounted on two pairs of 20 m and 25 m long Spin-plane Booms (SpB), built to the recent heritage of Cluster, and two axial tubular sensors, each ~1 m long and mounted on an Axial Boom (AxB), a 3.5 m long stacer element, built to the recent heritage of similar units on FAST and POLAR. The boom electronics board in the IDPU controls the voltages on the sensors, preamp (ushers and guards), and braids while the aforementioned DFB conditions, digitizes and processes the signals and data products.

Figure 7 shows the THEMIS instruments in stowed and deployed configuration. Figure 8 shows the probe coordinate systems and the relative spin phase of the various instruments and in a deployed configuration. The FGM is on a deployable, articulated ~2 m hinged boom and the SCM on a deployable ~1 m boom (Pankow et al. 2008). The deployed booms

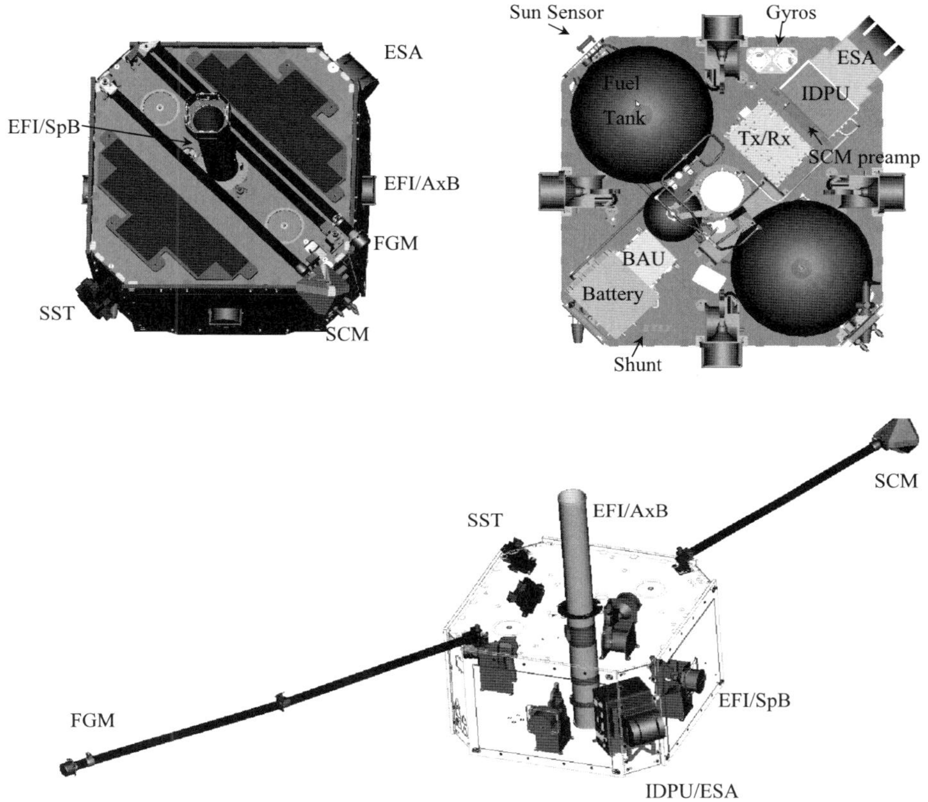

Fig. 7 (**a**) *top left*: Top view of THEMIS probe indicating locations of the instruments in a stowed configuration; AxB is Axial Boom, SpB is Spin plane Boom; (**b**) *top right*: Internal instrument accommodation (**c**) *bottom*: Side view of THEMIS probe with magnetometer booms in deployed configuration, revealing instrument locations

are at an angle to the spin plane, such that even after the aforementioned 8° spin axis tip off the ecliptic normal, there is no boom shadowing of the solar arrays and thus stray currents and magnetic interference from the booms is minimal.

The Spinning Probe Geometric (SPG) coordinate system is also depicted in Fig. 8. The probe geometric axis, Z_{SPG}, was measured during environmental testing to be within 0.25° from the principal axis of inertia (which is the direction of the momentum vector, **L**). Therefore, the Spinning Sun-sensor L-vector (SSL) coordinate system, which is defined with the X_{SSL} axis along the field of view, the Sun sensor, Z_{SSL} axis pointing along the spin axis, or momentum vector **L**, and the Y axis completing the orthogonal system, is approximately (within much less than 0.25°) derived by a 135°rotation of the SPG system about the Z axis. The DSL system (not shown) is defined as a De-spun, Sun-pointing, L-vector system, such that Z_{DSL} is identical to Z_{SSL}, the X_{DSL}–Z_{DSL} plane contains the Sun direction (X_{DSL} positive towards the Sun), and the Y_{DSL} axis completes the orthogonal system. This is obtained from SSL to first order by a rotation opposite to the spin phase that elapsed since the last Sun crossing by the Sun sensor. Under nominal attitude, the inner probes (having a spin axis near ecliptic north) have a DSL system roughly along the GSE system (within 8°), whereas

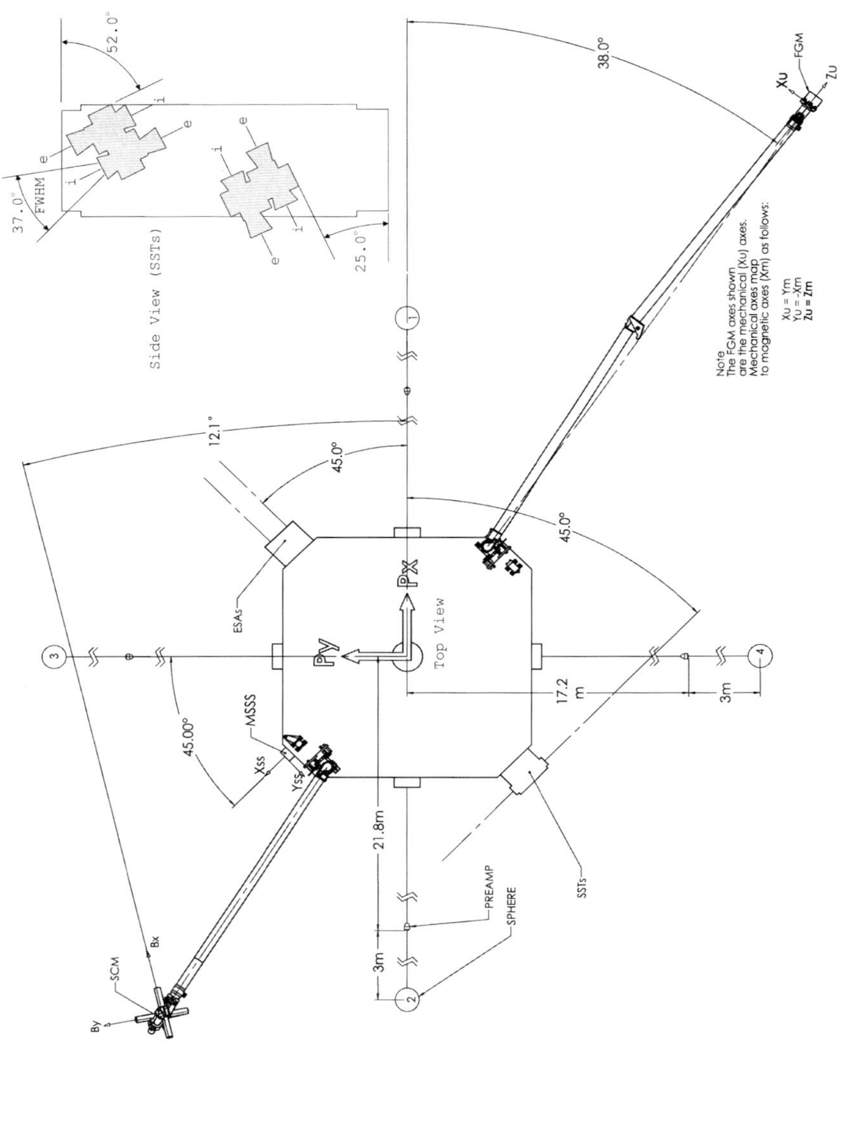

Fig. 8 Top view of THEMIS probe showing locations relative to each other and to the Sun sensor

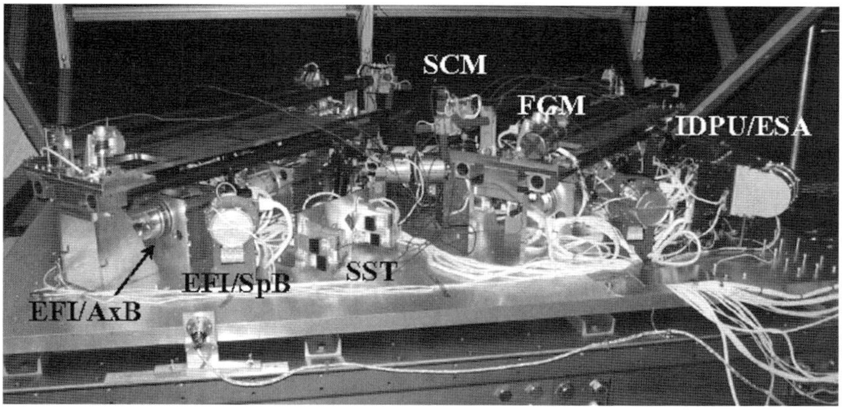

Fig. 9 THEMIS instrument suites FM2 (*left pellet*) and FM3 (*right pellet*), prior to undergoing thermal vacuum testing at the Space Sciences Laboratory, UCB

Fig. 10 THEMIS probe FM2 undergoing vibration testing at JPL. SSTs are covered by a thermal box, as in flight

the outer probes (having a baseline mission spin axis near ecliptic south) have a DSL system that is rotated 180°about the X-axis from GSE.

Figure 9 shows a picture of the THEMIS instruments undergoing environmental tests as a suite prior to integration with the spacecraft. Figure 10 shows an integrated probe during environmental testing.

The THEMIS ground-based observatory (GBO) instruments were designed to meet the mission requirements under minimum maintenance. The GBO systems were built at UCB based on recent experience with the Automated Geophysical Observatories (AGOs) in Antarctica. Each station includes an auroral all sky camera imager (ASI), designed and developed at UCB based on commercially available components. A UCLA-provided GPS receiver card and magnetometer are also part of the integrated GBO design. UCLA magnetometers use a new electronics design, based on heritage from the UC-LANL, MEASURE, and SMALL ground magnetometer networks. Site installation at Canadian sites and data retrieval is done in collaboration with the University of Calgary. Existing magnetometer sites in Canada were reconfigured by the University of Alberta to produce data at 0.5 s resolution, and feed into, and be commensurate with, the standard THEMIS data flow. The THEMIS GBO program is described in Mende et al. (2008). The ASI imager is described in Harris et al. (2008). The GBO magnetometer stations are described in Russell et al. (2008), while Canadian-built THEMIS magnetometers and ancillary ground magnetometer datasets are described in Mann et al. (2008). The THEMIS team at UCLA produced and installed a network of mid-latitude stations to promote science education in rural schools. The way those tie into THEMIS's Education and Public Outreach program is described in Peticolas et al. (2008).

5 Mission Operations and Data Analysis

The THEMIS mission is operated by the Mission Operations Center (MOC) at the Space Sciences Laboratory, UCB (Bester et al. 2008). The MOC performs mission planning functions in accordance to science, flight dynamics, orbit and attitude determination, maneuver planning, commanding and state-of-health monitoring of the five probes, recovery of science and engineering data, data trending and anomaly resolution. Science operations comprise the generation of instrument schedules, data processing and archiving functions. The THEMIS ground systems are the Ground Station Network required for communications, the Mission Operations Center (MOC), the Science Operations Center (SOC) and the Flight Dynamics Center (FDC). The primary ground station for the THEMIS ground station network is the 11 m Berkeley Ground Station (BGS). Additional ground stations utilized are: Wallops Island (WFF), Merritt Island (MILA), Santiago Chile (AGO), and Hartebeesthoek (HBK) in South Africa. Recently, Universal Space Network stations in Hawaii (South Point) and Australia (Dongara) are coming on board to relieve the heavy load of routine contacts and provide backup capability. During the early part of the mission, the Deep Space Network and the TDRSS satellites have been also used for communications with the probes. Data are routinely transferred between various stations over secure network segments of NASA's IONet. The MOC, SOC, FDC and BGS are co-located at Space Sciences Laboratory, enabling efficient operations.

The probes are operated in store-and-forward mode. Transmissions are initiated by time sequence commands stored on-board. These commands are part of an Absolute Time Sequence (ATS) load generated individually for each probe with the Mission Planning System (MPS). ATS loads are uploaded up to a few times per week. Real Time Sequence (RTS) commands are also utilized when necessary. The command and control system for THEMIS is ITOS, the Integrated Test and Operations System, which allowed seamless transitioning of operations personnel from the Integration and Test phase pre-launch into flight operations post-launch.

Probe orbits were designed to meet science requirements using tools and training provided by NASA/GSFC. Specifically, the Goddard Trajectory Determination System (GTDS), a high-fidelity orbit integrator was ported into UNIX systems and was made callable from an Interactive Data Language (IDL) higher-level data analysis and visualization language. This facilitated end-to-end mission design trades in order to optimize science (conjunctions) and reduce shadows and fuel. The General Maneuver program (GMAN), built also by NASA/GSFC and ported into the UCB mission design and mission operations tool-chest, permits finite-maneuver targeting functions with built-in propagation capability using GTDS. Calls to GMAN were also enabled from within IDL, such that a unified Maneuver Design Tool (MDT) encompassed the entire mission design effort calling interchangeably GTDS or GMAN depending on the fidelity and speed required. An IDPU engineering unit and two copies of the Bus Avionics Unit (BAU) unit equipped with an IDPU flight spare or an IDPU simulator are used to test commands before they are uploaded for execution in light, and are able to predict the vehicle behavior through special software. These "Flatsat" systems are used to test commands to instruments or spacecraft prior to execution in flight. This suite of programs and hardware provides an efficient and robust command generation and verification to meet science requirements, mission design requirements, operational requirements, and probe requirements in the pre- and post-launch phases.

Science operations are designed to accommodate an average of ~750 Mbits per orbit, which is required from the probes. Instrument-specific loss-less compression is applied to reduce data volume by approximately a factor of two. Baseline primary science can be accomplished with routine data accumulation, which transitions from Slow Survey (SS) into a "Fast Survey" mode (FS) during conjunctions. Higher time resolution particles and fields accumulation is possible by burst mode collection, which is enabled by evaluating on-board trigger quantities.

Burst mode can be of two types: *particle* or *wave*. *Particle bursts* collect high-resolution distributions and low frequency waveforms. They are aimed at capturing the components of the global magnetospheric substorm instability (from -3 min to $+6$ min since burst trigger). They are triggered in the tail by dipolarizations, and in the dayside by density changes. Other trigger quantities are also possible. Since substorms occur ~10% of the time (10 min collection/3 hr substorm recurrence time) which is similar to the occurrence rate of bursty flows (Angelopoulos et al. 1994, 1999) and current disruption in the region of primary interest ($X > -13\ R_E$) our memory allocation of 10% of the conjunction time to particle bursts leads to full coverage of all surge intervals by this mode. *Wave* bursts are intended to capture the **E&B** field waveforms of the waves anticipated within the disruption region. Broadbanded low frequency waves occur nominally 10–20% of the bursty flow time (and proportionately less at higher frequencies). Memory allocation to wave bursts (10% of the particle burst time) results in waveform accumulation during most onset-related waves. Table 7 shows the data allocation per instrument assuming realistic compression levels, and Table 8 shows the resultant allocation per data collection mode. These apply for the inner probes. Outer probes have the same duration of particle burst per orbit as the inner probes and rely on additional contacts and compression to relieve the memory.

Upon receipt, and after quality checks and file statistics, automated file processing of the raw ("Virtual Channel" or VC) files takes place. The processing performs decompression, extracts housekeeping information, performs time-ordering and overlap-deletion, sorts by "Application Identifier" or "APID" code, containing data of an individual instrument or of an individual type and produces level zero (L0) files. These L0 APID files are generated as 24 hr files and can already be used directly for data analysis and visualization. However, further automated processing produces "Level 1" (L1) un-calibrated data files which are in machine-independent "Common Data Format" (CDF format), typically within an hour of downlink. L1 CDF files contain raw data from all instruments at the highest temporal resolution. Science team validated data are updated daily on the web along with standardized-format plots (.gif and .ps).

Table 7 THEMIS memory distribution, per instrument and mode, for the inner probes. Assumes realistic on-board compression (average compression of factor of 2)

P3/4/5	Mbits/	24	hrs			
Instrument	Slow Survey	Fast Survey	Particle Burst	Wave Burst	ACS	Total (Mbits)
FGM	1	6	14	0	2	24
SCM	1	17	19	9	0	46
EFI	2	31	25	18	0	75
SST	10	66	13	0	0	88
ESA	14	87	42	0	0	143
Volume	27	207	113	27	2	376

Table 8 Duration of various operational modes for the inner probes. Assumes realistic on-board compression (average compression of factor of 2)

T_orbit =	24 hrs	**P3/4/5**						Data Vol
S/C Mode	Rel. Allocation		Abs. Allocation		Hours	Seconds	Rate (bps)	(Mbits)
SS	50%	T_orb	12.5%	T_orb	12.0	43200	623	27
FS	50%	T_orb	11.3%	T_orb	10.8	38923	5324	207
PB	10%	T_fs	1.2%	T_orb	1.188	4277	26479	113
WB	1%	T_pb	0.0%	T_orb	0.012	43	629170	27
CAL	8.3%	T_orb	8.3%	T_orb	2.000	7200	261	2
								376

Data visualization and calibration is performed "on-the-fly" by routinely available IDL code, which uses the aforementioned L1 APID CDF files and instrument calibration files. The analysis code distributed also provides a Graphical User Interface (GUI) that allows users unfamiliar with command line IDL coding quick access to the data. The GUI is also accessible by IDL's Virtual Machine, which is free of charge. The IDL calibration and analysis code is disseminated to the community via the THEMIS web site; tutorials are routinely conducted at coI sites and during GEM or AGU meetings.

Further automated data processing performs standard calibration within hours of receipt, producing "Level 2" (L2) calibrated daily data files, also in CDF format. Those files do not require further calibration and can be read by any software that is able to access CDF files, such as Fortran, C, Matlab and IDL. Standard overview plots are also produced to facilitate data quality evaluation, and quick event selection, especially in conjunction with other missions. Plots, data, documentation and on-line tutorials are also available on the web (http://themis.ssl.berkeley.edu). Further description of the THEMIS science operations and data handling can be found in Phan et al. (2008).

6 Summary

THEMIS, the first NASA micro-satellite constellation, is a focused investigation to determine the onset and evolution of the macroscale substorm instability, a fundamental mode of mass and energy transport throughout Geospace. This primary objective drives mission design: resonant orbits of a minimum of four satellites bracket the reconnection and current disruption regions in the magnetotail to determine where the first energization during a substorm occurs. A fifth probe is an on-orbit spare, which increases mission reliability and enables baseline science that far exceeds the minimum objective of substorm timing. The five-probe, 2 year baseline mission utilizes conjunctions with dedicated Ground-Based Observatories to monitor the auroral break-up and place the spacecraft observations in the context of global and local geomagnetic activity. Additionally, THEMIS's radial traversals of the radiation belts and their surroundings will address what causes the production of storm-time MeV electrons, and through its alignments in the dayside magnetosphere, will determine how the bow-shock and upstream processes affect the pristine solar wind and thus the solar wind-magnetosphere energy coupling. Understanding the ubiquitous substorm process, storm time electron energization and solar wind–magnetosphere coupling are essential for understanding and predicting space weather. While the currently operating Cluster mission and the upcoming MMS mission study in tetrahedral configuration local plasma boundaries from 100 s to 1000 s of km scales, THEMIS provides the necessary macro-scale vantage point, in the range of 1000 km–10 R_E, to study the global evolution of the magnetosphere during substorms. THEMIS is operating at a time when unprecedented coverage of the solar wind input is possible by WIND, ACE and STEREO. The THEMIS orbits are ideal for conjunctions with Cluster, other NOAA, DOE, NASA and international missions (such as GOES, LANL geosynchronous satellites, FAST and Geotail), and ancillary ground-based observatories (such as AMISR, SuperDARN, Sonderstrom and ULTIMA). THEMIS has an open data policy and readily provides data, documentation, plots, analysis software and training to the community at large, in order to maximize the benefit for the Heliophysics Great Observatory over the next decade.

Acknowledgements The mission became a reality because of the diligent efforts and dedication of a large number of individuals: P.R. Harvey led the mission implementation from selection through launch, providing inspiring leadership, technical excellence, project team- (including PI-) training and a fantastic team spirit that permeated the entire project during development. M. Bester led the THEMIS ground systems and mission operations center development and has been operating the mission impeccably and efficiently since launch. F.S. Mozer, R.P. Lin, C.W. Carlson and S. Mende were invaluable to the start of the mission and to sustaining a rational, well-advised leadership through development, launch and operations. M. Cully led the THEMIS probe bus, probe carrier and probe release system development and test at ATK (formerly Swales Aerospace Inc.) with tenacity, commitment to excellence, and determination. P. Turin and E.R. Taylor led the mechanical and systems design and implementation of what in retrospect would have seemed an unfathomable proposition. Instrument technical leads U. Auster, J. Bonnell, C. Carlson, J. McFadden, D. Larson, O. LeContel, M. Ludlam, W. Magnes, A. Roux designed, produced and tested the THEMIS instruments with attention to detail, resulting in the highest quality data in orbit. Many thanks to the THEMIS teams at: the Space Sciences Laboratory, for taking this "bull by the horns"; ATK Space (formerly Swales Aerospace Inc.) for their excellence in design and implementation of a highly integrated science-craft and probe carrier; University of Colorado's Laboratory for Atmospheric and Space Physics for the smart design and timely delivery of an optimal fields processing solution; JPL's Environmental Test Laboratory for their informal but highly professional support during our verification through their facilities; NASA/GSFC for project management and representation, oversight, quality assurance and safety engineering—they were an integral part of the team however hard it was for them to admit; ASO and NASA/KSC for a great launch processing experience; ULA for a safe ride to space; UCLA for making magnetic cleanliness seem easy, and for support in flight and ground-based systems development and EPO program; CETP, TUBS, IWF for high quality flight hardware development and testing; University of Calgary for GBO installation, data retrieval and continued GBO site maintenance in Canada. THEMIS was made possible by NASA, under contract NAS5-02099.

References

A.T. Aikio et al., Characteristics of pseudobreakups and substorms observed in the ionosphere, at the geosynchronous orbit, and in the midtail. J. Geophys. Res. **104**, 12263 (1999)

S.-I. Akasofu, *Physics of Magnetospheric Substorms* (Reidel, Dordrecht, 1976)

V. Angelopoulos et al., Statistical characteristics of bursty bulk flow events. J. Geophys. Res. **99**, 21257 (1994)

V. Angelopoulos et al., Magnetotail flow bursts: association to global magnetospheric circulation, relationship to ionospheric activity and direct evidence for localization. Geophys. Res. Lett. **24**, 2271 (1997a)

V. Angelopoulos et al., Multipoint analysis of a bursty bulk flow event on April 11, 1985. J. Geophys. Res. **101**, 4967 (1997b); also see correction: J. Geophys. Res., **102**, 211 (1997b)

V. Angelopoulos et al., On the relationship between bursty flows, current disruption and substorms. Geophys. Res. Lett. **26**, 2841 (1999)

V. Angelopoulos et al., Plasma sheet electromagnetic power generation and its dissipation along auroral field lines, J. Geophys. Res. (2001, in press)

G. Atkinson, The current system of geomagnetic bays. J. Geophys. Res. **23**, 6063 (1967)

U. Auster et al., Space Sci. Rev. (2008, this issue)

D.N. Baker et al., Neural line model of substorms: Past results and present view. J. Geophys. Res. **101**, 12975 (1996)

M. Bester et al., Space Sci. Rev. (2008, this issue)

W. Baumjohann et al., Average plasma properties in the central plasma sheet. J. Geophys. Res. **94**, 6597 (1989)

J. Birn et al., Flow braking and the substorm current wedge. J. Geophys. Res. **104**, 19895 (1999)

Bonnell et al., Space Sci. Rev. (2008, this issue)

J.E. Borovsky et al., The occurrence rate of magnetospheric-substorm onsets: random and periodic substorms. J. Geophys. Res. **98**, 3807 (1993)

C.W. Carlson et al., Space Sci. Rev. (2008, this issue)

M.R. Collier et al., Timing accuracy for the simple planar propagation of magnetic field structures in the solar wind. Geophys. Res. Lett. **25**, 2509 (1998)

N.U. Crooker et al., Factors controlling degree of correlation between ISEE 1 and ISEE 3 interplanetary magnetic field measurements. J. Geophys. Res. **87**, 2224 (1982)

C.M. Cully et al., Space Sci. Rev. (2008, this issue)

I.A. Daglis et al., "Fine structure" of the storm-substorm relationship: ion injections during Dst decrease. Adv. Space Res. **25**, 2369 (2000)

R.D. Elphinstone et al., Observations in the vicinity of substorm onset: implications for the substorm process. J. Geophys. Res. **100**, 7937 (1995)

D.H. Fairfield et al., Upstream pressure variations associated with the bow shock and their effects on the magnetosphere. J. Geophys. Res. **95**, 3773–3786 (1990)

D.H. Fairfield et al., Advances in magnetospheric storm and substorm research, 1989–1991. J. Geophys. Res. **97**(A7), 10865–10874 (1992)

D.H. Fairfield et al., Geotail abservations of substorm onset in the inner magnetotail. J. Geophys. Res. **103** (1998)

C. Farrugia et al., Viscous-type processes in the solar wind–magnetosphere interaction. Space. Sci. Rev. **95**(1/2), 443–456 (2001)

L.A. Frank, J.B. Sigwarth, Findings concerning the positions of substorm onsets with auroral images from the Polar spacecraft. J. Geophys. Res. **105**, 12747 (2000)

L.A. Frank et al., in *Proceedings of the International Conference on Substorms - 4 (ICS-4)* (Terra Scientific, Tokyo, 1998), p. 3

S. Frey et al., Space Sci. Rev. (2008, this issue)

Friedel et al., J. Atmospheric Sol. Terr. Phys. **64**, 265–282 (2002)

E. Friedrich et al., Ground-based observations and plasma instabilities in auroral substorms. Phys. Plasmas **8**, 1104 (2001)

S. Harris et al., Space Sci. Rev. (2008, this issue). doi:10.1007/s11214-007-9294-2

P.R. Harvey et al., Space Sci. Rev. (2008, this issue)

M.G. Henderson et al., Observations of magnetospheric substorms occurring with no apparent solar wind/IMF trigger. J. Geophys. Res. **101**, 10773 (1996)

M.G. Henderson et al., Are north-south aligned auroral structures an ionospheric manifestation of bursty bulk flows? Geophys. Res. Lett. **25**, 3737 (1998)

M. Hesse, J. Birn, On dipolarization and its relation to the substorm current wedge. J. Geophys. Res. **96**, 19417 (1991)

E.W. Hones Jr., The magnetotail: its generation and dissipation, in *Physics of Solar Planetary Environments*, ed. by D.J. Williams, AGU, vol. 558, 1976

E.W. Hones Jr. et al., Detailed examination of a plasmoid in the distant magnetotail with ISEE 3. Geophys. Res. Lett. **11**, 1046 (1984)

C. Jacquey et al., Location and propagation of the magnetotail current disruption during substorm expansion: analysis and simulation of an ISEE multi-onset event. Geophys. Res. Lett. **3**, 389 (1991)

J.R. Kan, A globally integrated substorm model: tail reconnection and magnetosphere-ionosphere coupling. J. Geophys. Res. **103**, 11787 (1998)
R.L. Kaufmann, Substorm currents: growth phase and onset. J. Geophys. Res. **92**, 7471 (1987)
Kennel, 1992, The Kiruna conjecture: The strong version, in *ICS-1 Proceedings*. ESA SP-335, 1992, p. 599
Larson et al., Space Sci. Rev. (2008, this issue)
G. Le, C.T. Russell, H. Kuo, Flux transfer events–Spontaneous or driven? Geophys. Res. Lett. **20**, 791 (1993)
X. Li et al., Multisatellite observations of the outer zone electron variation during the November 3–4, 1993, magnetic storm. J. Geophys. Res. **102**, 14123 (1997)
Li et al., Quantitative prediction of radiation belt electrons at geostationary orbit based on solar wind measurements. Geophys. Res. Lett. **28**, 1887 (2001)
Y. Lin, D.W. Swift, L.C. Lee, Simulation of pressure pulses in the bow shock and magnetosheath driven by variations in interplanetary magnetic field direction. J. Geophys. Res. **101**, 27251 (1996)
M. Lockwood, M.N. Wild, On the quasi-periodic nature of magnetopause flux transfer events. J. Geophys. Res. **98**, 5935 (1993)
M. Ludlam et al., The THEMIS magnetic cleanliness program. Space Sci. Rev. (2008, this issue)
A.T.Y. Lui, Extended consideration of a synthesis model for magnetospheric substorms. AGU Mon. Ser., vol. 64, 1991, p. 43
A.T.Y. Lui, Current disruption in the Earth's magnetosphere: Observations and models. J. Geophys. Res. **101**, 13067 (1996)
A.T.Y. Lui et al., A multiscale model for substorms. Space Sci. Rev. **95**, 325 (2001)
A.T.Y. Lui, J.R. Burrows, On the location of auroral arcs near substorm onsets. J. Geophys. Res. **83**, 3342 (1978)
A.T.Y. Lui et al., A case study of magnetotail current sheet disruption and diversion. Geophys. Res. Lett. **7**, 721 (1988)
L.R. Lyons, A new theory for magnetospheric substorms. J. Geophys. Res. **100**, 19069 (1995)
L.R. Lyons, Substorms: Fundamental observational features, distiction from other disturbances, and external triggering. J. Geophys. Res. **101**, 13011 (1996)
J.P. McFadden et al., Space Sci. Rev. (2008, this issue)
R. McPherron et al., Satellite studies of magnetospheric substorms on Aug 15th, 1968. J. Geophys. Res. **78**, 3131 (1973)
R.L. McPherron et al., Solar wind triggering of substorm onset. J. Geomagn. Geoelectr. **38**, 1089 (1986)
S. Mende et al., Space Sci. Rev. (2008, this issue)
Millan, Thorne, J. Atmos. Solar Terr. Phys. **69**, 362–377 (2007)
D.G. Mitchell et al., Current carriers in the near-Earth cross-tail current sheet during substorm growth phase. Geophys. Res. Lett. **17**, 583 (1990)
T. Nagai, Observed magnetic substorm signatures at synchronous altitudes. J. Geophys. Res. **87**, 4405 (1982)
T. Nagai et al., Substorm, tail flows, and plasmoids. Adv. Space Res. **20**, 961 (1997)
T. Nagai et al., Structure and dynamics of magnetic reconnection for substorm onsets with Geotail observations. J. Geophys. Res. **103**, 4419 (1998)
R. Nakamura et al., Flow bursts and auroral activations: Onset timing and foot point location. J. Geophys. Res. **106**, 10777 (2001a)
R. Nakamura et al., Earthward flow bursts, auroral streamers, and small expansions. J. Geophys. Res. **106**, 10791 (2001b)
S.-I. Ohtani, Earthward expansion of tail current disruption: dual-satellite study. J. Geophys. Res. **103**, 6815 (1998)
S.-I. Ohtani, Substorm trigger processes in the magnetotail: recent observations and outstanding issues. Space Sci. Rev. **95**, 347 (2001)
S.-I. Ohtani et al., Tail current disruption in the geosynchronous region, in *Magnetospheric Substorms*. AGU Mongr. Ser., vol. 64, 1991, p. 131
S. Ohtani et al., Radial expansion of the tail current disruption during substorms: A new approach to the substorm onset region. J. Geophys. Res. **97**, 3129 (1992a)
S.-I. Ohtani et al., Initial signatures of magnetic field and energetic particle fluxes at tail reconfiguration: explosive growth phase. J. Geophys. Res. **97**, 19311 (1992b)
Pankow et al., Space Sci. Rev. (2008, this issue)
G. Paschmann, G. Haerendel, N. Sckopke, E. Möbius, H. Lühr, C.W. Carlson, Three-dimensional plasma structures with anomalous flow direction near the Earth's bow shock. J. Geophys. Res. **93**, 11279 (1988)
Paschmann et al., Plasma acceleration at the magnetopause: evidence for reconnection. Nature **282**, 243 (1979)
K.I. Paularena et al., Solar wind plasma correlations between IMP 8, INTERBALL-1, and WIND. J. Geophys. Res. **103**, 14601 (1998)
Peticolas et al., Space Sci. Rev. (2008, this issue)

A.A. Petrukovich et al., Two spacecraft observations of a reconnection pulse during an auroral breakup. J. Geophys. Res. **103**, 47 (1998)
T. Phan et al., Space Sci. Rev. (2008, this issue)
T.D. Phan, G. Paschmann, The magnetosheath region adjacent to the dayside magnetopause, in *Physics of the Magnetopause*. AGU Monograph, vol. 90 (1995)
I.G. Richardson, S.W.H. Cowley, Plasmoid-associated energetic ion bursts in the deep geomagnetic tail: properties of the boundary layer. J. Geophys. Res. **90**, 12133 (1985)
I.G. Richardson et al., Plasmoid-associated energetic ion bursts in the deep geomagnetic tail: properties of plasmoids and the postplasmoid plasma sheet. J. Geophys. Res. **92**, 9997 (1987)
Roux et al., Space Sci. Rev. (2008, this issue)
Russell et al., Space Sci. Rev. (2008, this issue). doi:10.1007/s11214-008-9337-0
C.T. Russell, R.C. Elphic, Initial ISEE magnetometer results: magnetopause observations. Space Sci. Rev. **22**, 681 (1978)
J.C. Samson, Proton aurora and substorm intensifications. Geophys. Res. Lett. **19**, 2171 (1992)
E.T. Sarris et al., Location of the source of magnetospheric energetic particle bursts by multispacecraft observations. Geophys. Res. Lett. **3**, 437 (1976)
E.T. Sarris et al., Detailed observations of a burst of energetic particles in the deep magnetotail by Geotail. J. Geomagn. Geoelectr. **48**, 649 (1996)
R. Schodel et al., Rapid flux transport in the central plasma sheet. J. Geophys. Res. **106**, 301 (2001)
V.A. Sergeev et al., Triggering of substorm expansion by the IMF directional discontinuities: Time delay analysis. Planet. Space Sci. **38**, 231 (1990)
V.A. Sergeev et al., In situ observations of magnetic reconnection prior to the onset of a small substorm. J. Geophys. Res. **100**, 19121 (1995)
V.A. Sergeev et al., Steady magnetospheric convection: a review of recent results. Space Science Reviews **75**, 551 (1996a)
V.A. Sergeev et al., Detection of localized, plasma-depleted flux tubes or bubbles in the midtail plasma sheet. J. Geophys. Res. **101**, 10817 (1996b)
V.A. Sergeev et al., Multiple-spacecraft observation of a narrow transient plasma jet in the Earth's plasma sheet. Geophys. Res. Lett. **27**, 851 (2000)
I. Shinohara et al., Rapid large-scale magnetic field dissipation in a collisionless current sheet via coupling between Kelvin-Helmholtz and lower-hybrid instabilities. Phys. Rev. Lett. **87**, 095001 (2001)
K. Shiokawa et al., Azimuthal pressure gradient as driving force of substorm currents. Geophys. Res. Lett. **25**, 959 (1998a)
K. Shiokawa et al., High-speed ion flow, substorm current wedge, and multiple Pi2 pulsations. J. Geophys. Res. **103**, 4491 (1998b)
D.G. Sibeck et al., The magnetospheric response to 8-minute period strong-amplitude upstream pressure variations. J. Geophys. Res. **94**, 2505–2519 (1989)
D.G. Sibeck, K. Takahashi, S. Kokubun, T. Mukai, K.W. Ogilvie, A. Szabo, A case study of oppositely propagating Alfvén fluctuations in the solar wind and magnetosheath. Geophys. Res. Lett. **24**, 3133 (1997)
D.G. Sibeck et al., Space Sci. Rev. (2008, this issue)
G.L. Siscoe, H.E. Petschek, On storm weakening during substorm expansion phase. Ann. Geophys. **15**, 211 (1997)
J.A. Slavin et al., CDAW 8 observations of plasmoid signatures in the geomagnetic tail: An assessment. J. Geophys. Res. **97**, 8495 (1992)
P. Song, C.T. Russell, M.F. Thomsen, Slow mode transition in the frontside magnetosheath. J. Geophys. Res. **97**, 8295 (1992)
D.J. Southwood, M.G. Kivelson, Magnetosheath flow near the subsolar magnetopause: Zwan-Wolf and Southwood-Kivelson theories reconciled. Geophys. Res. Lett. **22**, 3275 (1995)
H.E. Spence, The what, where, when and why of magnetospheric substorm triggers. EOS **77**, 81 (1996)
Taylor et al., Space Sci. Rev. (2008, this issue)
V.A. Thomas, S.H. Brecht, Evolution of diamagnetic cavities in the solar wind. J. Geophys. Res. **93**, 11341–11353 (1988)
H.J. Völk, R.-D. Auer, Motions of the bow shock induced by interplanetary disturbances. J. Geophys. Res., 40-48, 1974
I. Voronkov et al., Shear flow instability in the dipolar magnetosphere. J. Geophys. Res. **104**, 17323 (1999)
J.R. Wygant et al., Polar spacecraft based comparisons of intense electric fields and Poynting flux near and within the plasma sheet-tail lobe boundary to UVI images: an energy source for the aurora. J. Geophys. Res. **105**, 18675 (2000)
Y. Yamade et al., Field-aligned currents generated in magnetotail reconnection: 3D Hall-MHD simulations. J. Geophys. Res. **27**, 1091 (2000)
E. Zesta et al., The auroral signature of earthward flow bursts observed in the magnetotail. Geophys. Res. Lett. **27**, 3241 (2000)

THEMIS Science Objectives and Mission Phases

D.G. Sibeck · V. Angelopoulos

Originally published in the journal Space Science Reviews, Volume 141, Nos 1–4, 35–59.
DOI: 10.1007/s11214-008-9393-5

Abstract The five THEMIS spacecraft and a dedicated ground-based observatory array will pinpoint when and where substorms occur, thereby providing the observations needed to identify the processes that cause substorms to suddenly release solar wind energy stored within the Earth's magnetotail. The primary science which drove the mission design enables unprecedented observations relevant to magnetospheric research areas ranging from the foreshock to the Earth's radiation belts. This paper describes how THEMIS will reach closure on its baseline scientific objectives as a function of mission phase.

Keywords THEMIS · Magnetosphere · Substorms · Radiation belts · Magnetopause

1 Introduction

THEMIS (Time History of Events and Macroscale Interactions during Substorms) is NASA's fifth MIDEX mission, following in the successful footsteps of FUSE, IMAGE, WMAP, and Swift. THEMIS will provide the multipoint and multi-instrument observations needed to determine why the transfer of solar wind energy to the Earth's inner magnetosphere and ionosphere generally occurs via geomagnetic substorms. As the building blocks of solar wind-magnetosphere interaction, substorms encompass a wide array of magnetospheric phenomena including sudden reconfigurations of the nightside magnetospheric magnetic field, jetting plasmas, injections of energetic ions and electrons into the Earth's radiation belts, field-aligned beams of energetic particles and currents directed from the magnetotail to the ionosphere, auroral displays, and associated disturbances in surface magnetic fields.

Of the many proposed substorm models, two have become widely accepted on the basis of their ability to explain the full panoply of observed phenomena. The current disruption

D.G. Sibeck (✉)
Code 674, GSFC/NASA, Greenbelt, MD 20771, USA
e-mail: david.g.sibeck@nasa.gov

V. Angelopoulos
IGPP, UCLA, Los Angeles, CA 90095, USA
e-mail: vassilis@ucla.edu

model predicts that substorms begin with a disruption of the cross-tail current about 8 to 10 Earth radii (R_E) from Earth, while the reconnection model predicts that substorms begin with the onset of reconnection in the current sheet some 20 to 30 R_E from Earth. Despite their strikingly different predictions concerning when and where substorms begin, the lack of coordinated high time resolution multipoint and multi-instrument observations has precluded efforts to discriminate between these two (and other) models.

NASA's THEMIS mission, managed by the University of California at Berkeley, will enable researchers to pinpoint when and where substorm onsets occur. The identical instruments on each of the five THEMIS spacecraft were selected to identify both substorm signatures and the physical processes that trigger them, while the orbits of the THEMIS spacecraft were chosen to bound proposed locations where substorms begin. A dedicated array of ground observatories located throughout Canada, Alaska, and the northern regions of the contiguous United States supplies the global observations needed to place the multipoint, but isolated, spacecraft observations in context.

While in the magnetotail, the THEMIS spacecraft will also provide important observations concerning the consequences of magnetospheric substorms, including the generation of field-aligned currents by vortical plasma flows and pressure gradients, and the coupling of substorm disturbances to local instabilities responsible for geomagnetic pulsations and ballooning modes. THEMIS observations of the Earth's radiation belts will be used to determine the means by which ions and electrons are energized, transported, and lost. Finally, the THEMIS spacecraft will spend many months on the dayside, where their observations will enable researchers to discriminate between various modes of steady and transient solar wind-magnetosphere interaction.

Angelopoulos (2008) provides a mission overview, summarizing the overall scientific objectives, orbits, instrumentation, mission and science operations that are expanded upon in individual papers in this compendium. In this paper we address how the above mission elements will enable the THEMIS team to reach closure on its science objectives as function of mission phase. In terms of organization, Sect. 2 summarizes the THEMIS science objectives and Sect. 3 describes the THEMIS mission elements. Sections 4 to 6 outline THEMIS science objectives as a function of mission phase: the nightside magnetotail phases, the dayside phases and ongoing observations throughout the mission life within the Earth's radiation belts. Section 7 presents conclusions.

2 THEMIS Science Objectives

2.1 Substorms

Substorms represent a fundamental mode of solar wind-magnetosphere interaction, an interaction that transfers energy from the solar wind and deposits it in the Earth's ionosphere, atmosphere, and radiation belts (Akasofu 1977). Substorms follow a clear and repeatable cycle. During the growth phase, southward turnings of the interplanetary magnetic field (IMF) initiate reconnection on the dayside magnetopause. The removal of newly reconnected magnetic flux reduces dayside magnetospheric magnetic field strengths and pressures, while the addition of the same flux to the nightside magnetosphere enhances magnetotail magnetic field strengths and pressures. As a result, the dayside magnetopause moves inward, the magnetotail magnetopause flares outward, the plasma sheet thins, and cross-tail currents increase to accommodate a stretched, tail-like, magnetic field configuration. The region of open magnetic field lines over the polar caps expands and quiet time aurorae move equatorward.

Just prior to substorm onset, strong (10's nA m^{-2}) cross-tail currents, generated primarily by duskward anisotropies of 2 to 10 keV ions, appear in the near-Earth ($\sim$10 R_E from Earth) magnetotail (Lui 1996). Just after substorm onset, these cross-tail currents are diverted to flow through the ionosphere, and stretched magnetotail magnetic field lines suddenly snap back towards more dipolar configurations. Plasma flows rapidly both sunward and anti-sunward away from reconnection sites in the plasma sheet typically some 20 or 30 R_E from Earth. Beams of energized particles flow sunward along the boundaries of the plasma sheet, reflect upon reaching near-Earth mirror points and flow back into the plasma sheet at slightly lower latitudes. Particles that do not reflect precipitate into the ionosphere, brightening pre-existing auroral arcs. The auroral brightening expands rapidly poleward into the polar cap as reconnection converts open-lobe magnetic field lines into closed plasma sheet magnetic field lines. The collapse of nightside magnetic field lines into more dipolar configurations injects energetic particles into near-Earth geospace.

Two models seek to explain the sequence of events that occur during geomagnetic substorms. The current-disruption or near-Earth initiation model, shown in the top panel of Fig. 1, predicts that the strong currents that appear within the near-Earth magnetotail at substorm onset trigger instabilities that result in the collapse of the stretched magnetotail magnetic field to a more dipolar orientation, current diversion to and through the ionosphere, the injection of a heated plasma into the inner magnetosphere, and the launching of a fast rarefaction wave that propagates down the magnetotail and initiates reconnection at greater distances (Lui 1996). Advocates interpret (1) the explosive growth and then collapse of cross-tail currents in the near-Earth magnetotail at substorm onset, (2) the initial brightening of the most equatorward pre-existing auroral arc, which maps to locations deep within the magnetosphere, (3) the initial appearance of enhanced particle fluxes earthward of spacecraft located in the near-Earth magnetotail, and (4) the initial appearance of enhanced cosmic noise absorption caused by precipitating energetic electrons at the equatorward edge of the auroral oval as evidence for the current disruption model paradigm.

By contrast, the reconnection or mid-tail initiation model shown in the bottom panel of Fig. 1 invokes a current-driven instability that triggers reconnection some $\sim$25 R_E down the magnetotail (Hones 1976; Baker et al. 1996; Shiokawa et al. 1997). Reconnection launches fast, often bursty, flows that transport plasma and magnetic flux sunward towards the inner magnetosphere. The inner magnetosphere poses an obstacle that brakes and deflects these flows and converts their kinetic energy to thermal energies and enhanced magnetic field strengths in a dipolar configuration. Precipitating particles and field-aligned currents associated with the flow braking generate aurorae and magnetic field disturbances. Advocates interpret (1) the tendency of sunward flows to be associated with northward magnetic fields and anti-sunward flows to be associated with southward magnetic fields, (2) the appearance of magnetic field perturbations expected for Hall current effects, and (3) the anti-sunward motion of plasmoid bubbles released at substorm onset as evidence for the reconnection model.

Accurate, multipoint measurements that pinpoint the location(s) where substorms begin and the timing of the phenomena that follow can distinguish between these two models (e.g. Baker et al. 2002). In the current disruption model, the sequence should be: (1) current disruption some 8 to 10 R_E from Earth, (2) auroral breakup, and (3) magnetic reconnection at greater distances down the magnetotail. By contrast, in the reconnection model, the sequence of events should be: (1) reconnection some 25 R_E from Earth, (2) current disruption in the near-Earth tail, and (3) auroral breakup. It is important to characterize substorms as a function of solar wind conditions, not least because abrupt northward IMF turnings or solar wind dynamic pressure increases may trigger current disruption or magnetic reconnection (e.g. McPherron et al. 1986).

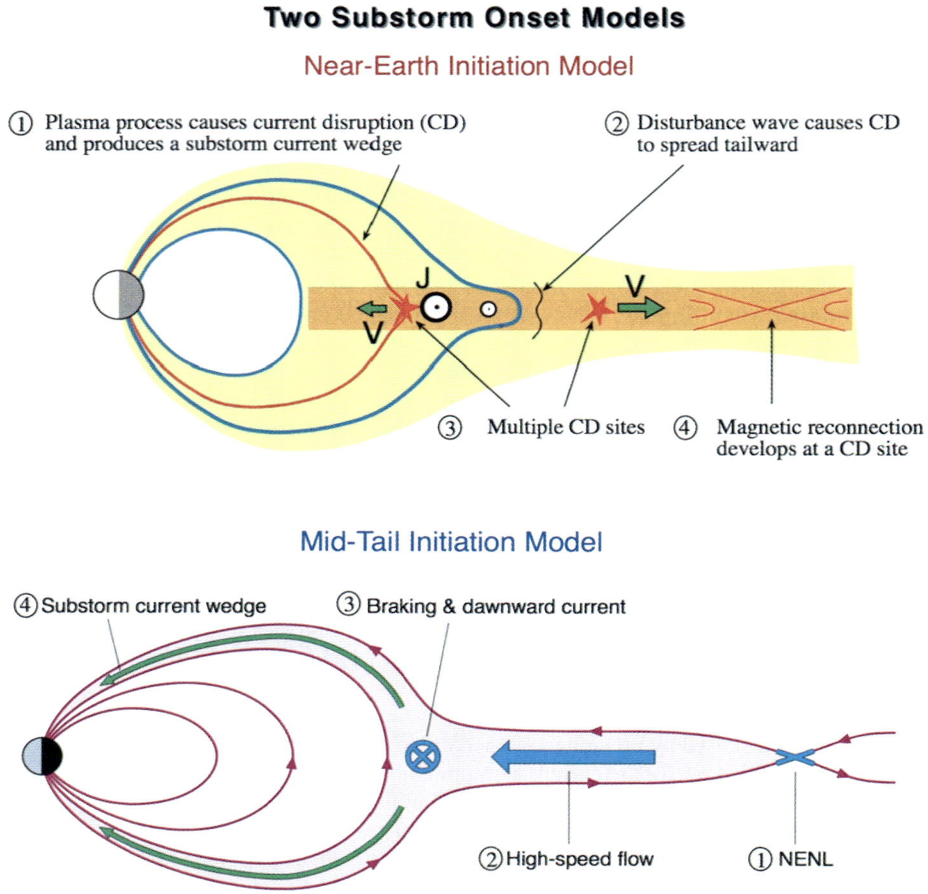

Fig. 1 Current disruption (near-Earth initiation) and reconnection (mid-tail initiation) models for geomagnetic substorms. In the current disruption model, a sudden disruption of the cross-tail current in the near-Earth magnetotail launches a tailward propagating rarefaction wave that initiates reconnection further down the magnetotail. The disrupted current flows into the ionosphere along magnetic field lines and the aurora brightens. In the reconnection model, reconnection in the mid-magnetotail launches sunward propagating flows and a fast mode compressional wave that cause magnetic flux to pile up in the near-Earth magnetotail. Flow shears or flow breaking launch currents that flow along magnetic field lines into the ionosphere and the aurora brightens

In the process of determining the mechanism(s) that drive geomagnetic substorms, THEMIS will also address both fundamental plasma physics questions and space weather forecasting needs. The cross-scale coupling and particle energization processes that occur during substorms may be ubiquitous throughout the plasma universe. Microphysical instabilities at the cross-tail current sheet trigger mesoscale flows within the Earth's magnetotail, which in turn result in macroscale reconfigurations of the entire magnetosphere. Stressed current sheets like those that occur within the Earth's magnetotail can also be found in fusion devices, throughout the heliosphere, at the Sun and other stars, and in other planetary magnetospheres. Consequently, the multipoint THEMIS measurements offer an opportunity to test conflicting model (and simulation) predictions against in situ observations and apply the knowledge gained to other systems where this is not possible.

The practical impact of THEMIS on space weather forecasting is equally important. For a variety of operational purposes, it is important to predict the occurrence of substorm-generated geomagnetic disturbances and corresponding auroral displays. THEMIS will provide the scientific understanding needed to advance our forecasting capabilities towards the ability to predict substorm onset time, extent, and amplitude.

2.2 Radiation Belts

Despite more than 40 years of study, the processes that generate, transport, and remove energetic particles in the Earth's radiation belts remain unclear. Proposed acceleration and injection mechanisms include direct energization by electric fields associated with large- and small-scale storm-time convective flows into the inner magnetosphere (Khazanov et al. 2004), impulsive energization via substorm-launched inward-propagating injection fronts (Mithaiwala and Horton 2005), radial diffusion and energization by storm-time ULF waves (Loto'aniu et al. 2006) or solar wind dynamic pressure variations (Ukhorskiy et al. 2006); energization by VLF waves followed by pitch angle anisotropization (Horne et al. 2005), impulsive trapping of solar energetic ions during geomagnetic storms (Kress et al. 2005), and prompt acceleration by compressional wave fronts associated with strong interplanetary shocks (Hudson et al. 1995). Proposed loss mechanisms are no less numerous than energization mechanisms. They include pitch angle scattering into the ionosphere via cyclotron and Landau resonant interactions with plasmaspheric hiss, whistler-mode chorus, and EMIC waves, scattering by interaction with the magnetotail current sheet, and magnetopause shadowing (Green et al. 2004; Millan and Thorne 2007).

As illustrated in Fig. 2 (Li et al. 2001), recurrent geomagnetic storms generate dramatic variations in energetic particle fluxes within the inner magnetosphere. THEMIS will supply the multipoint and multi-instrument observations needed to discriminate between the various mechanisms proposed to account for the appearance, transport, and loss of ions and electrons with energies less than 1 MeV. The results obtained by THEMIS will be used to help plan NASA's forthcoming LWS radiation Belt Storm Probe mission, which is dedicated to understanding the processes governing the more energetic (and more hazardous) particles found within the Earth's radiation belts.

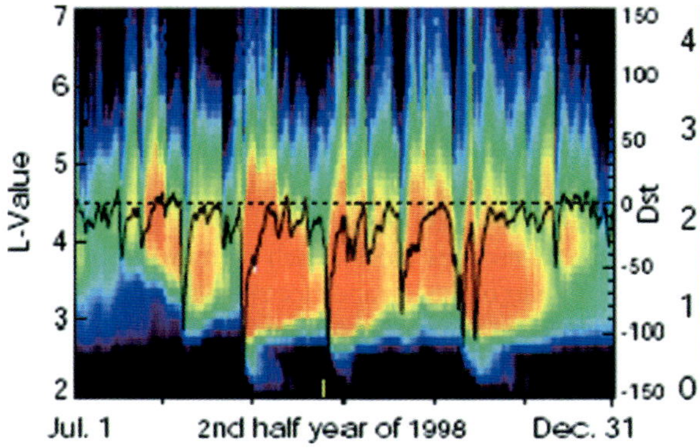

Fig. 2 Daily averaged SAMPEX electron measurements of 2–6 MeV electrons (#/cm^2-s-sr) and Dst index with 1-day window-average for the second half year of 1998 (Li et al. 2001). The *black line* shows the Dst index, a measure of geomagnetic storm activity

2.3 Dayside Interactions

Many aspects of the dayside interaction between the solar wind and magnetosphere remain poorly understood. For example, we do not know when our use of solar wind monitors far upstream or far off the Sun-Earth line to measure the solar wind input into the magnetosphere is valid, because we do not know the scale sizes for solar wind features and understand poorly the processes that modify these features within the Earth's foreshock and magnetosheath. Nevertheless, in recent years it has become apparent that kinetic processes within both these regions can result in drastic, albeit transient, perturbations to the incoming solar wind density, velocity, and magnetic field strength. Figure 3 shows hybrid code model predictions for the distribution of plasma ion densities and temperatures in the vicinity of the dayside bow shock attending the passage of a solar wind tangential discontinuity (Omidi and Sibeck 2007). Kinetic processes result in the development of a hot flow anomaly within the solar wind at the intersection of the discontinuity with the bow shock (Thomsen et al. 1988). The density variations associated with such structures launch the full spectrum of waves when they strike the bow shock and magnetopause. The propagation paths of these waves remain unclear but may provide important information concerning the distribution of plasmas in and around the magnetosphere.

The magnetopause is constantly in motion. The motion may be directly driven by transmitted solar wind and foreshock-generated pressure variations (Farrugia et al. 1989), result from the Kelvin-Helmholtz instability (Otto and Fairfield 2000), or be triggered by bursts of reconnection (Song et al. 1988). As illustrated in Fig. 4, bursts of reconnection generate twisted ropes of interconnected magnetospheric and magnetosheath magnetic field lines (Russell and Elphic 1978). Discriminating between these possibilities for any individual event requires simultaneous solar wind, magnetosheath, and magnetopause observations, while statistical studies designed to determine the significance of each proposed mechanism require surveys of event occurrence patterns, dimensions, growth, and decay as a function of solar wind conditions. THEMIS will provide the observations needed to address these dayside science topics.

3 Mission Elements

The five spin-stabilized THEMIS spacecraft have 3 s spin periods and carry identical high-heritage instruments. The triaxial fluxgate magnetometer (FGM) measures DC and low-frequency perturbations of the magnetic field (Auster et al. 2008), times the propagation of waves and plasma structures between spacecraft, and provides information concerning currents flowing between two or more probes. Accuracies, stabilities, and interference from spacecraft systems are 0.3 nT or better. A pair of back-to-back top hat hemispherical electrostatic analyzers (ESA) measures the distribution functions of thermal ions (0.005 to 25 keV) and electrons (0.005 to 30 keV) over 4π-str to determine accurate 3 s time resolution plasma moments and instantaneous gradients in these parameters between probes (Carlson et al. 2008; McFadden et al. 2008).

Solid-state telescopes (SST), each comprising two sensors, measure the superthermal (0.02–1 MeV) part of the ion and electron distributions over 3π str (Larson et al. 2008). Mechanical attenuators diminish the geometric factor within the radiation belts (radial distances from Earth below 8 R_E) by a factor of $\sim$100, thereby limiting damage to the silicon detectors from intense fluxes of low energy ions. The telescopes will be used to remotely sense the current disruption region and time the arrival of particles energized by reconnection.

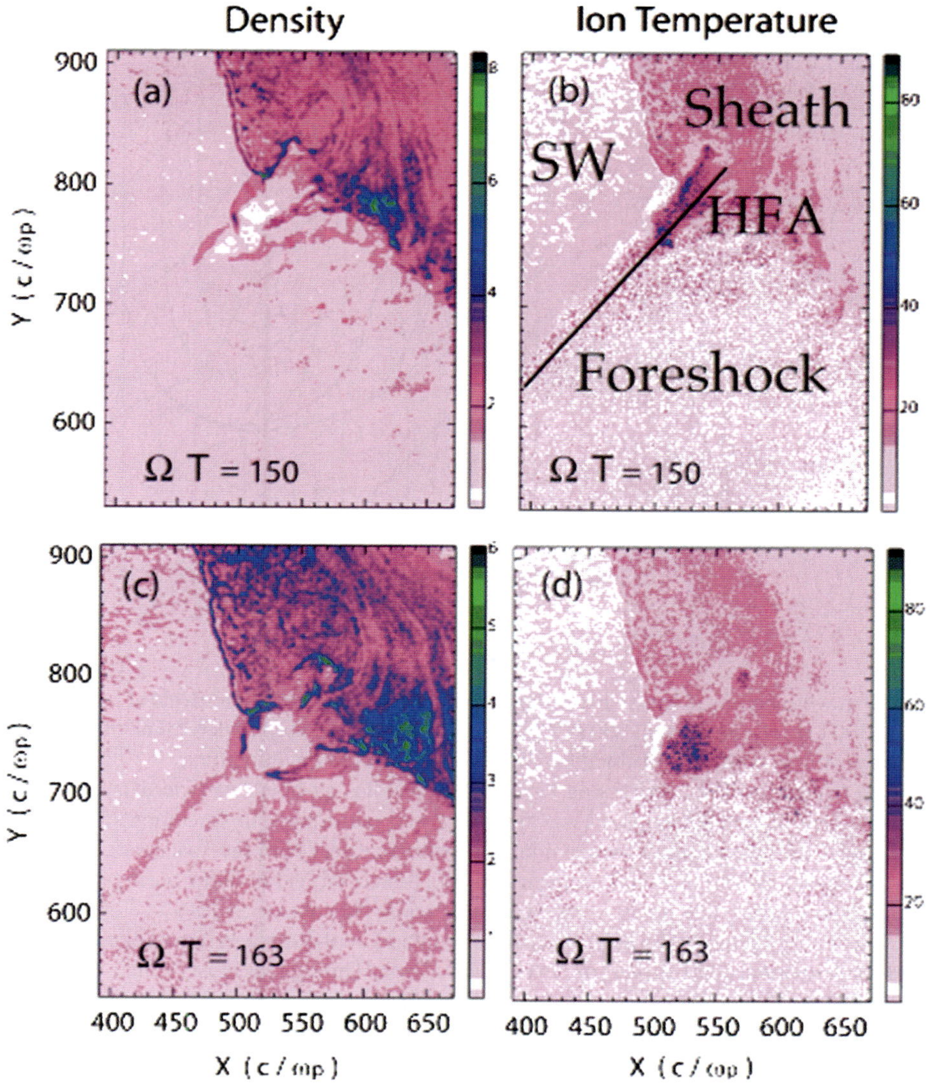

Fig. 3 Hybrid code simulation results for the formation of a HFA at the intersection of an interplanetary discontinuity with the Earth's bow shock (Omidi and Sibeck 2007). The panels on the *left* show densities, while those on the *right* show ion temperatures. Two stages in the development of the HFA are shown, at $\Omega T = 150$ and 163, where Ω is the ion gyrofrequency. The *solid black line* in the *upper right* panel indicates the location of the IMF discontinuity

Search coil magnetometers (SCM) extend the measurements of the FGM from 0.1 Hz to frequencies of 4 kHz (Roux et al. 2008). Sensitivities of $0.8 pT/\sqrt{\mathrm{Hz}}$ at 10 Hz and $0.02 pT/\sqrt{\mathrm{Hz}}$ at 1 kHz suffice to measure the waves that accompany cross-tail current disruptions some 8 R_{E} from Earth. Periodic calibration using the discrete signals generated by coils ensures accurate measurements. Each electric field instrument (EFI) employs two pairs of spherical sensors on 20 and 25 m deployable cables and a pair of axial tubular sensors

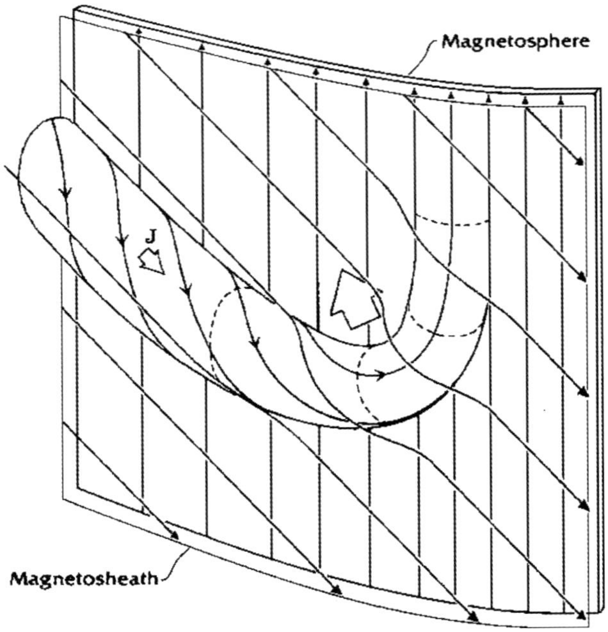

Fig. 4 Flux transfer events (FTEs) are twisted ropes of interconnected magnetospheric and magnetosheath magnetic field lines (Russell and Elphic 1978)

on 3.0 m whip booms to make three-dimensional electric field measurements at frequencies up to 300 kHz (Bonnell et al. 2008). EFI observations can be used to determine the ambient density, plasma convection velocities, wave modes, and the Poynting flux.

The Instrument Data Processing Unit (IDPU) handles on-board particle data collection and moment computations, processes field data and performs spin fit computations, executes memory storage and compression, and communicates with the spacecraft computer for data transmission. Both independent and inter-spacecraft calibrations during quiet time conjunctions ensure that errors in moment calculations do not exceed 10%.

Stored or triggered commands cause the instruments on the spacecraft to operate in one of four modes: slow or fast survey, particle or wave burst. Throughout most of their orbits, the spacecraft operate in slow survey mode, returning magnetic field vectors, plasma moments, and other parameters with 3 s time resolution. Near apogee in the magnetotail and in regions of interest like the dayside magnetopause, stored commands trigger the instruments to operate in fast survey mode. In fast survey mode, FGM samples the magnetic field 16 times per spin, SCM and EFI sample 32 times per spin, and SST and ESA provide observations with greater spatial resolution. Encounters with the bow shock, magnetopause, bursty bulks flows within the magnetotail, magnetic field reconfigurations, and other phenomena trigger burst mode operations. In burst mode, FGM can sample the magnetic field at up to 128 Hz, while SCM and EFI can sample at up to 4096 Hz.

A dedicated dense network of 20 all-sky white light imagers and ground magnetometers (when no pre-existing magnetometer is located nearby) covering the Arctic and mid-latitude regions of North America ensures accurate determination of substorm onset locations to within 0.5 hours of magnetic local time (Mende et al. 2008). The magnetometers have 0.5 s time resolution, while the imagers take snapshots of duration $\sim$1 s each 3 s.

The five THEMIS spacecraft operate in highly elliptical, near-equatorial, orbits that precess about the Earth. At 23:01 UT on February 17, 2007, the spacecraft were launched into common initial orbits with geocentric apogee at 14.7 R_E and 2100 LT, geocentric perigee at 1.07 R_E, inclination of 16°, and orbital periods of 31 hours. During the coast phase of the mission, from launch until September 2007, apogees precessed through the dusk and dayside magnetosphere. Interspacecraft separations ranged from a few 100 km to 2 R_E along track. This orbit phase was prefixed on the baseline mission design to avoid a mission redesign late in the program after the launch vehicle provider announced a launch delay that caused the mission to slip past its ability to perform tail studies in the winter of 2007. From September 15 to December 15, 2007, the spacecraft were on the dawnside of the Earth's magnetosphere, where their apogees were moved to 31.0, 19.6, 11.8, 11.8 and 9.9 R_E in anticipation of the tail season. The corresponding orbital periods during the first magnetotail phase of the mission from December 15, 2007 to April 15, 2008 were $\sim$1, 2, and 4 days, enabling the radial alignments of the spacecraft apogees needed to address the baseline substorm science once every 4 days. During this period probe P5 was separated in apogee from P3 and P4 by 2 R_E, resulting in an orbital period that was 4/5 that for P3 and P4. Probe P5 thus participated in a major conjunction once every 5 orbits, or 4 days, but was separated from the other probes by several R_E the rest of the time. This enables a wide range of azimuthal separations, but also provides for 1–3 R_E clustering during major conjunctions along the orbit plane.

Six months after the first tail season, the first dayside observations commence. The orbital configuration remains the same, but the apogee of P5 is reduced to lie only 1 R_E away from those of P3 and 4 to reduce differential precession. Major conjunctions between 4 spacecraft occur once every 4 days, but P5 is tightly clustered with P3/4 only once per 8 days, and therefore scans a large range of azimuthal separations with P3 and 4 the rest of the time. The spacecraft return to the Earth's magnetotail for a second season of substorm observations from December 15, 2008 to April 15, 2009. During this second season, the orbits of the three inner spacecraft will have common apogees of 11.6 R_E, but due to a 5° difference in orbital inclinations, P5 will be separated from P3 and 4 by $\sim$1 R_E in Z_{GSM} at apogee. The separation is designed to enable researchers to determine the properties of the near-Earth current and plasma sheet. Azimuthal separations of $\sim$ 1 R_E between P3 and P4 are designed to identify simultaneous current disruptions. Figure 5 illustrates representative four-day-long orbital segments throughout the nominal mission.

4 Science Closure in the Magnetotail

THEMIS will provide the observations needed to discriminate between conflicting models for geomagnetic substorms by determining when, where, and why substorm onset occurs. Every four days the THEMIS spacecraft will line up within the Earth's magnetotail, affording opportunities to conduct timing studies of substorm features as a function of distance down the magnetotail. As illustrated in Fig. 6, spacecraft P1 and P2 will bound the expected location of the reconnection line. Observations of high-speed sunward flows and northward magnetic field orientations at P2, but anti-sunward flows and southward field orientations at P1, will demonstrate the appearance of a reconnection line between the two spacecraft.

THEMIS probes P3–5 lie near the expected location of current disruption within the Earth's magnetotail. These spacecraft will be used to time the changes in plasma and magnetic field configuration that occur in the near-earth magnetotail for comparison with the times determined for reconnection from probe P1 and P2 observations. Timing and remote sensing techniques employing the energetic particles observations will be used to discriminate between sunward-propagating compressional events that begin in the distant

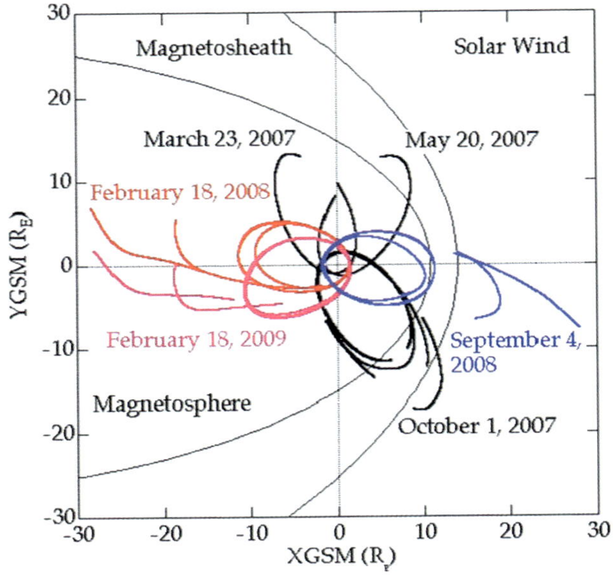

Fig. 5 Day-long intervals of ephemeris at representative stages of the mission. In March and May of 2007, the spacecraft were in nearly-identical orbits with apogee on the duskside of the Earth. By October 2007, apogee separation had begun and the spacecraft were on the dayside of the Earth. In February 2008, the spacecraft apogees will be fully separated and in the Earth's magnetotail. By September 2008, the spacecraft will be on the dayside again. In February 2009, the orbits of the three innermost spacecraft will have similar apogees, but different inclinations

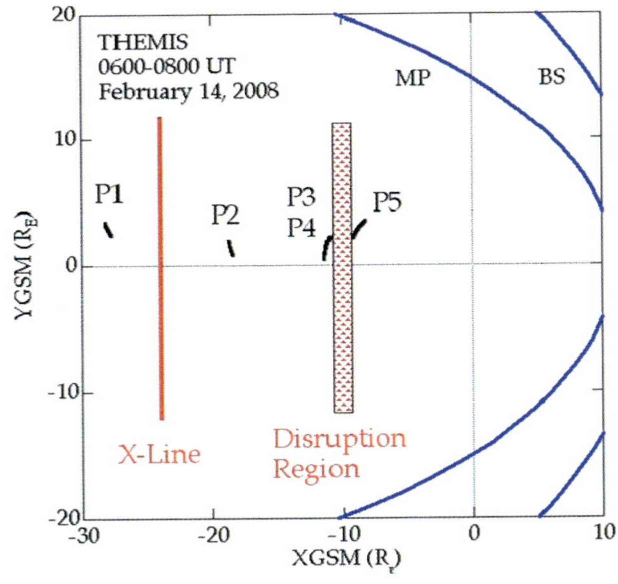

Fig. 6 The trajectories of the THEMIS spacecraft from 06:00 to 08:00 UT on February 14, 2008. Spacecraft P1 and P2 bound the reconnection line. Spacecraft P5 and closely-spaced spacecraft P3 and P4 bound the current disruption region

magnetotail and anti-sunward-propagating rarefaction events that begin in the near-Earth magnetotail.

4.1 Time History of Events

To distinguish between proposed models for substorms, THEMIS must time their onset. Based on current orbit predictions (Frey et al. 2008), the THEMIS probes will accumulate more than 250 h/year of tail-aligned conjunctions (no spacecraft separated by more than 2 R_E in the cross-tail or Y-direction from P1). With alignments lasting ~12 hours, and substorms recurring every ~3 hours, each alignment should result in 3–4 substorms for a mission total of ~80 substorms. At least a dozen of these should be observed from the pre-midnight substorm onset meridian. With these data THEMIS will be able to delineate the time history of the events that comprise the substorm process. THEMIS will rely upon WIND, ACE and, on occasion, Cluster and Stereo to define the external solar wind conditions and distinguish between the different paths to substorm onset. Specific timing techniques include:

Current disruption (CD) onset determination At speeds of 200 km s^{-1}, a current disruption onset 1 R_E away expands over the THEMIS probes within 30 s. THEMIS probes P3 and 4 will employ remote sensing (finite gyroradius) techniques to obtain timing information from energetic ion observations (Lui et al. 1988; Ohtani 1998) applied to energetic ions. This method provides boundary expansion speeds to within 10 km s^{-1} and directions measured to a fraction of the angular resolution of the ion detector (Daly et al. 1984; Kettmann and Daly 1988). Onset times will be determined from the expansion velocities on the two nearby probes to within the 3 s probe time resolution, ensuring current disruption timing to within 10 s or better.

Reconnection (Rx) onset determination THEMIS will time the onset of reconnection by monitoring the arrival times of field-aligned energetic particles from the reconnection site at its two outer probes, located within 5 R_E from the nominal site of reconnection some 25 R_E from Earth. Ancillary timing information will be obtained from measured flow speeds and local observations of electrons, waves, and the MHD pulse from the reconnection process (Sarris et al. 1976; Petrukovich et al. 1998). THEMIS will provide reconnection onset timing to within 10 s or better.

Auroral breakup onset determination Although mid-latitude ground magnetometers have long been used to identify global Pi2 pulsations and thereby determine the time of substorm onset, much more accurate timing can be obtained from high-latitude imagers and ground magnetometers (Olson 1999; Liou et al. 2000). The network of white light all sky imager and ground magnetometer stations in Alaska, Canada and the US will ensure accurate determination of onset to within 0.5° in magnetic local time (Mende et al. 2008). Cloudy skies or moonlight sometimes obscure the images. At these times, PiB (1–40 s period, 3 s nominal) pulsations (Heacock 1967), which are good substorm indicators (Troitskaya 1961; Bösinger et al. 1981), will determine onset time to within a few seconds. Substorm current wedge modeling from a dense North American network of auroral and mid-latitude magnetometer stations provides determination of the substorm meridian to within 5° or better (still fulfilling the science goal of 6°). Such modeling is routinely performed using data from the existing network of mid-latitude stations (Clauer and McPherron 1974; Kubyshkina 1999) and has been validated using global imaging (Sergeev et al. 1996). In short, THEMIS's ground network of all sky imager and ground magnetometer stations has the density and time resolution to detect auroral breakup onset meridian and onset time nominally within $\delta\text{MLT} < 0.5°$, $\delta t < 10$ s. Figure 7 combines information from the THEMIS arrays of all-sky cameras and ground magnetometers to depict an instant (06:19:36 UT on December 23, 2006) and a location (marked by the white box) when the aurora brightened and there was a strong shear in the magnetic field perturbations.

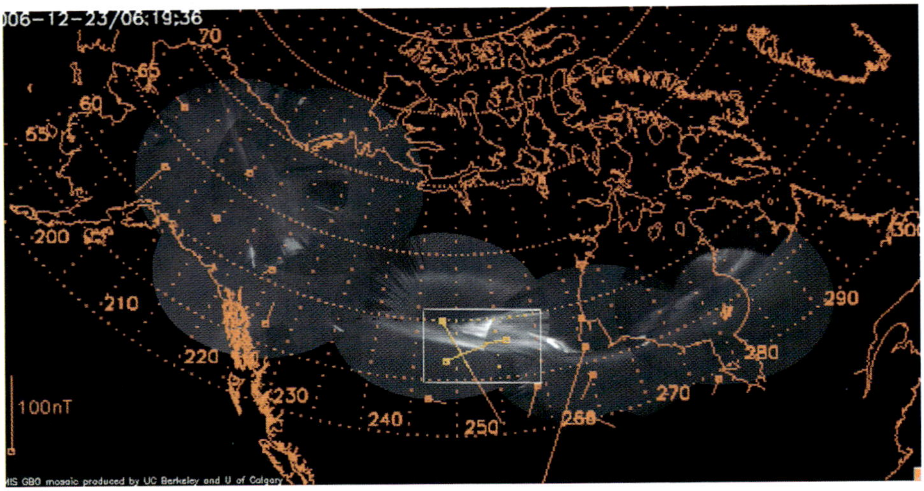

Fig. 7 Observations from the THEMIS ground-based observatories at 06:19:36 UT on December 23, 2006. The *white box* encompasses an auroral brightening accompanied by a strong shear in the directions of the magnetic field perturbations (red pointers extending from individual ground stations)

4.2 Macroscale Interactions

THEMIS will determine how the various localized, mesoscale substorm components interact over macroscale ranges for both the CD and Rx paradigms.

In the context of the CD paradigm, THEMIS will measure the velocity, magnetic, and plasma pressure perturbations associated with rarefaction waves propagating rapidly ($\sim$1600 km s^{-1}) antisunward at speeds comparable to the local fast mode speed (Chao et al. 1977). At onset, probes P3 and P4 should observe fast Earthward flows, followed by P2 some 20 s later. P1 will observe no Rx signature for at least another 20–25 s. The same THEMIS probes should track the outward motion of the rarefaction wave that links lobe flux dissipation to current disruption.

In the context of the Rx paradigm, THEMIS will track the sunward motion of the fast ($\sim$400 km s^{-1}) flows ejected from the reconnection region (Baumjohann et al. 1990; Angelopoulos et al. 1994). The anticipated delay time from P2 to P3 or P4 is greater than 90 s. In the second year, probe P5, separated in Z_{GSM} from the other inner probes, will determine if flow-driven boundary layer waves carry substantial Poynting flux. THEMIS probes P2, P3 and P4 will monitor the Earthward flow and establish the link between current disruption onset and reconnection.

Global MHD and particle codes will be used to model both specific events driven by measured solar wind parameters and idealized scenarios (Raeder et al. 2001, 2008). Particle modeling in prescribed E and B fields will validate the outgoing rarefaction wave or the incoming flow hypothesis (Li et al. 2000). Using MHD and particle simulations will strengthen closure on the macroscale interaction of components of the substorm instability.

By using the array of THEMIS spacecraft to determine the time and location of substorm onset, we have implicitly assumed that disturbances associated with substorms extend over a wide range of azimuths and propagate nearly radially towards or away from Earth. In fact, multipoint measurements in the magnetotail and on the ground indicate that the transient high-speed plasma flows that play such an important role in mass, energy, and magnetic flux transport have spatial extents in the dawn-dusk direction limited to 2–5 R_E (Angelopoulos et al. 1997; Kauristie et al. 2000; Nakamura et al. 2004). By providing observations that map to a wide range of locations in the magnetotail, the dedicated array of THEMIS ground observatories will play a crucial role in selecting events with the wide azimuthal extents needed to ensure that all the THEMIS spacecraft observe the same phenomenon. In turn, with the help of appropriate magnetic field models, the ground observations can be used to determine the extent, and consequently the significance, of individual events.

4.3 Ionospheric Coupling

THEMIS will remotely infer (i) cross-tail current evolution and (ii) field-aligned current generation. By studying 100–200 substorms from various perspectives and local times relative to the onset meridian, THEMIS will establish the macroscale coupling between the global substorm instability and auroral arc formation.

During their second season in the magnetotail, THEMIS probes P4 and P5 will routinely straddle the current sheet at interspacecraft separations ranging from 0.2 to 1 R_E to measure the cross tail current strength (one to tens nA/m^2) and its evolution using a planar approximation. Tail flapping due to solar wind buffeting (Sergeev et al. 1998; Runov et al. 2005a, 2005b) and diurnal effects (Lopez 1990) ensure multiple neutral sheet crossings. Probe P2 observations will be used to determine the relationship between the current sheet motion and incoming flows.

When the inner THEMIS probes are away from the neutral sheet, they will obtain magnetic field measurements across and along the tail. The current disruption process can then be remotely sensed using methods established on ISEE-1/2 and Interball-1 observations (e.g., Jacquey 2000).

According to MHD theory, flow vorticity, flow braking, radial and cross-sheet pressure (p) gradients can generate field-aligned currents (Haerendel 1992; Shiokawa et al. 1998; Hesse and Birn 1991). Pairs of conjunctions between probes P4, P3, and P5 across the path of laterally expanding flow channels or tailward expanding pressure gradients will determine the vorticity and pressure gradients over scale sizes commensurate with the flow shear and expected pressure gradients. Observations by P2 further down the magnetotail will place the field-aligned currents within their global context.

The incoming flow interacts with the Earth's dipole in a region where there are strong gradients in the magnetic field strength and high ion temperatures. As a result, ion diamagnetic drifts are pronounced and non-MHD effects are readily apparent. Whereas ions gradient-curvature drift duskward, electrons follow $\mathbf{E}\times\mathbf{B}/B^2$ drifts. The distinction requires Hall MHD or hybrid simulations, and profoundly affects the generation of field-aligned currents (Yamade et al. 2000). In addition to plasma observations, electric field measurements become essential (Angelopoulos et al. 1999). THEMIS will measure both the plasma and $\mathbf{E}\times\mathbf{B}$ flows independently and will therefore determine the (non-MHD) component of the ion drifts.

Theory predicts current sheet structures with single-centered, single-off-centered, triple, and bifurcated current density peaks, limits on the minimum current sheet thickness, and the existence of flapping waves in the presence of anisotropic and non-gyrotropic plasmas (Sitnov et al. 2006). THEMIS observations of multiple current sheet crossings at different distances from the current sheet can be used to reconstruct the current sheet structure, including the identification of embedded flux rope structures, as a function of time or distance from the reconnection site (Nakamura et al. 2002, 2006; Runov et al. 2003, 2005b, 2006).

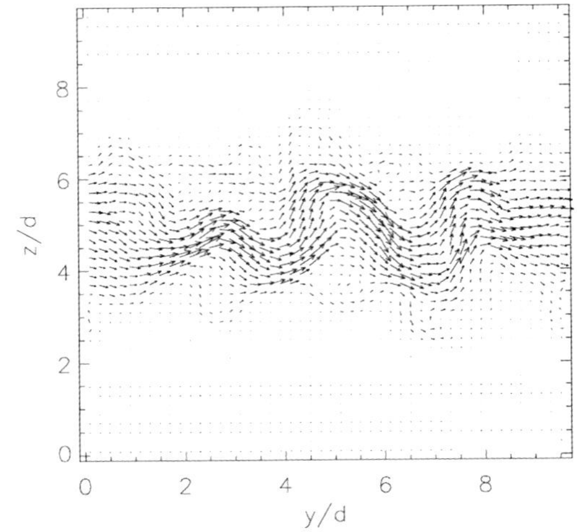

Fig. 8 A plot of the current density vectors in the plane perpendicular to the magnetotail axis from a full particle code simulation (Sitnov et al., 2006). Here d is the ion inertial scale length based on the density

During the second magnetotail season, the north-south displacement of spacecraft P5 relative to P3 and P4 will permit studies that track the thinning of the current sheet prior to substorm onset and the thickening that occurs thereafter.

Figure 8 presents model predictions for quasi-rectangular flapping waves superimposed upon the cross-tail current layer structure (Sitnov et al. 2006). When the spacecraft are azimuthally-separated, their observations can be used to determine the amplitude, wavelength, and propagation velocity and direction of these waves (Runov et al. 2005a). Similarly, azimuthal separations can be used to determine the characteristics of waves generated by ballooning-mode instabilities (Roux et al. 1991), thereby distinguishing between models invoking strong pressure gradients (Hurricane et al. 1999) and those based upon the velocity shears and the Kelvin-Helmholtz instability (Voronkov et al. 2000).

Finally, event-specific hybrid simulations will be used in conjunction with observations from probes P3, P4 & P5 to determine if observed CDs are due to electron acceleration accompanied by flux transport or a reduction in the ion drift rate.

4.4 Cross-scale coupling to local modes at 10 R_E

Substorms operate over a wide range of coupled scale-lengths (see Table 1). Identifying these coupling processes and determining when and where they occur is an essential aspect of substorm studies.

Ballooning modes Both geosynchronous (Roux et al. 1991) and ionospheric (Elphinstone et al. 1995) observations provide evidence for ballooning modes. Their free energy source is the pressure gradient in the near-Earth magnetotail (1 nPa R_E^{-1}). The modes have wavelengths ~2000–12000 km, move azimuthally at the 50–100 s km s^{-1} ion drift speed, and have Doppler-shifted periods of $T \sim 0.3$–2 min. Spacecraft traversing the near-Earth region at the ~1 R_E^2 large onset location should observe coherent waves (Ohtani et al. 1993). Classical ballooning occurs near marginal stability for typical tail parameters (Lee and Wolf 1992; Hurricane et al. 1999). This has led to non-linear ballooning mode theories (Samson et al. 1996) and predictions for linear but absolute instabilities (Hurricane et al. 1999).

Shear-flow ballooning, an alternative approach, suggests that ballooning is part of a larger cross-scale coupling process (Voronkov et al. 1999, 2000). Field line resonances ($\lambda \sim 2$–10 R_E, $T \sim 5$ min, e.g. Fenrich and Samson 1997) drive Kelvin-Helmholtz waves

Table 1 Scales of processes at substorm onset

Scale	Size (R_E)	Process
Macro	10	Rx/CD coupling. Current Wedge formation. Field line resonances
Meso	1	CD onset size. Ballooning modes. Kelvin-Helmholtz waves
Micro	0.1	Cross-field current instabilities. Alfvén waves

($\lambda \sim 0.2$–$1\ R_E$), which in turn become non-linearly unstable within ~1 min. The Kelvin-Helmholtz waves then drive smaller ($\delta Y \sim 0.1\ R_E$) Alfvénic currents that dissipate energy in the ionosphere. The azimuthal cross-field flow shear is on the order of $\delta V \sim 200\ \mathrm{km\,s^{-1}}$, while the waves have phase speeds $V_\phi \sim 50\ \mathrm{km\,s^{-1}}$. Independent Poynting vector calculations provide evidence for the bouncing Alfven waves (Maynard et al. 1996; Erickson et al. 2000), but their association with ballooning has not been confirmed.

During the second magnetotail season, azimuthal separations will enable probes P4 and P5 to study the coherent waves and resonances associated with ballooning modes using cross-spectral, wave-telescope (Motschmann et al. 1998) and Poynting vector techniques. Phase speeds calculated from observations by both probes at separations ranging from 0.3 to 10 R_E will be compared with flow speeds measured by each probe, and the properties of the waves will be compared with results from MHD simulations (e.g. Voronkov et al. 2000). Once again, simultaneous observations by probe P2 will provide the observations needed to determine the nature of coupling to the global substorm instability.

Cross-field current instabilities Cross-field current instabilities occur when the cross-tail current exceeds an instability threshold on the order of $10\ \mathrm{nA\,m^{-2}}$ or $100\ \mathrm{mA\,m^{-1}}$ (Lui 1996). They have frequencies of 0.01–$0.1 f_{LH}$ (where the lower hybrid frequency $f_{LH} \sim 60$ Hz at 8 R_E), wavelengths on the order of 300–2000 km, and exhibit no cross-tail spectral coherence. Observations of the phase relationships between the electric and magnetic field observations will identify the mode and propagation direction of the unstable wave. Cross-tail probe pairs (P3, P4, P5) will determine the degree (if any) of spectral coherence. Particle-in-cell simulations (e.g., Büchner et al. 1998) will establish if the observed wave amplitudes and particle streaming compare favorably with non-linear saturation amplitudes of the unstable modes. P2 monitors coupling to the global substorm process.

4.5 Additional Tail Science

THEMIS can contribute towards understanding other important phenomena indirectly related to substorms. As these are not primary mission goals, they do not drive the mission design.

Ionospheric mapping Despite many years of research, establishing the relationship(s) between phenomena in the Earth's magnetotail and those in the high-latitude nightside auroral oval remains an important objective. Phenomena of interest in the ionosphere include flow channels, arcs, omega bands, and transpolar arcs, while those of interest in the magnetotail include bubbles, waves, and flux ropes. Azimuthal and latitudinal separations of the THEMIS spacecraft provide the gradients in velocity, pressure, and magnetic field strength and direction needed to calculate vorticity and field-aligned currents. These in turn can be mapped to the ionosphere and compared with auroral structures, and field-aligned currents calculated from ground-based magnetometer and radar observations. Success in this endeavor would lead to the ability to routinely detect and understand substorm-related phenomena from ground-based observations alone.

Flux-tube evolution along streamlines Adiabatic convection cannot produce the observed average radial gradient in the lobe magnetic pressure, resulting in the so-called "pressure balance crisis" (Goertz and Baumjohann 1991; Erickson and Wolf 1980). Bubbles generated by uneven density loading in the magnetotail and propagating rapidly earthward have been proposed as the solution to this crisis (Pontius and Wolf 1990; Chen and Wolf 1993), but have only been observed during the late substorm phase (Sergeev et al. 1996). Working in conjunction with each other, THEMIS probes P4/P3, P2 and P1 will define the evolution of flux tubes as they move along streamlines, determine whether our ideas concerning bubble evolution are applicable to all flux tubes moving rapidly earthward, and provide an estimate of the importance of such bubbles in resolving the pressure balance crisis.

High frequency modes Waves in the Pi1 pulsation range (Perraut et al. 1998) or beyond (Shinohara et al. 1998) have been observed during substorms. They may be driven unstable by 0.5–2 keV electrons (Sugiyama et al. 1997) or free energy sources resulting from the kinetic structure of a thin plasma sheet (Le Contel et al. 1998). Bursty and broad-banded, they extend to $f \sim 4 f_{\mathrm{LH}}$ about 10–20% of the time. They are occasionally accompanied by whistlers at $1\text{–}10 f_{\mathrm{LH}}$. Burst waveform collection of **E** and **B** data at frequencies up to $10 \times f_{\mathrm{LH}}$ (or 600 Hz) will help identify their modes and place these waves in the context of substorm evolution.

5 Science Closure in the Outer Radiation Belt

Discriminating between the various source mechanisms for radiation belt particles requires equatorial measurements of radial phase space density profiles, particle spectra, and particle pitch angle distributions as a function of solar wind and geomagnetic conditions, including the occurrence patterns for both ULF and VLF waves. As illustrated in Fig. 9, in situ acceleration mechanisms generate local maxima in radial profiles for the phase space density, convective processes generate flat profiles, and diffusive processes generate positive radial gradients. Radial diffusion results in pitch angle distributions that peak perpendicular to the magnetic field, while wave-particle interactions keep particle distributions outside the loss cone nearly isotropic. Finally, flux enhancements propagate inward towards Earth from the point of injection, but both inward and outward from the point of energization.

Energetic particles drift in response to both transient and steady-state electric fields as well as in response to gradients in the magnetic field. Consequently, particle transport studies require comprehensive observations of the magnetospheric electric and magnetic field configuration, and the distribution of wave activity, as a function of geomagnetic and solar wind conditions. Proposed loss mechanisms are no less numerous than energization mechanisms. They include loss via scattering at the magnetopause or cross-tail current sheet and wave-particle scattering by resonant interaction with plasmaspheric hiss, whistler-mode chorus, and EMIC waves (Millan and Thorne 2007). Efforts to identify the principle loss mechanisms necessarily require accurate determination of particle drift paths to determine whether or not they intersect the current sheet or magnetopause, as well as information concerning the locations and intensity of plasma wave activity that can scatter particles into the loss cone and cause them to precipitate.

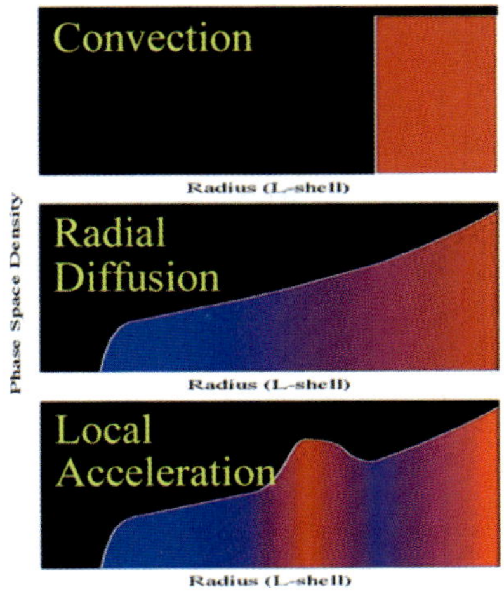

Fig. 9 Radial profiles of energetic particle phase space densities at constant first adiabatic invariant can be used as a tool to help distinguish between the various processes that energize particles within the Earth's radiation belts. In this figure *red shading* highlights potential source regions

The observed rapid increase of MeV electron flux inside of geosynchronous altitude cannot be accounted for by the relatively slow diffusion of solar wind plasma. The "Dst effect" alone cannot account for this process either, since the electrons first disappear and then reappear at much higher fluxes than before the storm. Electron fluxes are therefore likely enhanced at $L = 11$ before being transported inwards, but it is unclear if sufficient flux of electrons exists at such distances just prior to storm recovery.

The THEMIS probes traverse the inner magnetosphere from $L = 3.5$ to 11 with a median recurrence rate of 3.8 hours during their nominal phase (ranging from several up to 8 radial cuts per day). Thus, THEMIS will determine the radial profile of the electron phase space density at constant magnetic moment μ on time scales commensurate with the storm-time radiation belt MeV electron loss and re-appearance. Within the THEMIS mission, particle observations of the radiation belts are primarily the responsibility of the SST instrument. Although ESA can measure source populations at distances beyond geosynchronous orbit, penetrating radiation results in high background counts that dominate ESA measurements within the radiation belts.

Based on the slope of the obtained phase space density profiles versus L-shell, THEMIS will determine whether there is a sufficient source of electrons at the outer boundary. If the answer to this question is affirmative, THEMIS will identify the primary transport mechanism. The Dst-effect will be readily evaluated from individual radial flux profiles. The radial diffusion coefficient will be obtained from first order differencing of consecutive profiles, while the plasma convection will be directly measured on each probe. If radial transport alone cannot account for the MeV electron enhancement (Brautigam and Albert 2000), THEMIS, will determine whether other proposed mechanisms (e.g. in situ waves) are responsible for local electron heating.

As illustrated in Fig. 10, during the coast phase of the mission inter-spacecraft separation distances were on the order of 1–2 R_E within the radiation belts (radial distances from 2 to 6 R_E from Earth). Rapid consecutive visits by the different spacecraft to the same radial distance provide numerous opportunities to intercalibrate the particle instruments. Once the instruments were intercalibrated, the short separation distances enabled the calculation of radial gradients in the phase space density, a key discriminator between proposed particle energization mechanisms.

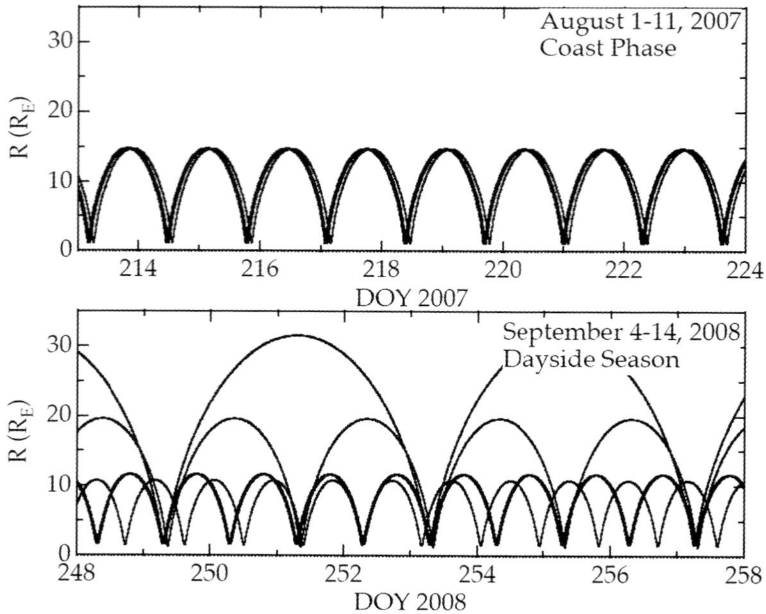

Fig. 10 The radial distances of the THEMIS spacecraft from Earth as a function of time during the coast phase and second dayside season

During later phases of the mission, illustrated in the lower panels of Fig. 10, separation distances of 1–2 R_E will remain common, but there will also be opportunities to compare energetic particle observations over the full range of radial distances from perigee to the 30 R_E apogee of the outermost spacecraft, P1. Large azimuthal and radial separation distances will enable researchers to track the motion of magnetic field dipolarizations and associated particle injection fronts, and determine whether processes other than adiabatic inward convection are needed to explain the particle populations seen near Earth (Fox et al. 2006).

Over the course of the mission, the THEMIS spacecraft will survey magnetospheric particle populations, electric and magnetic field configurations, and wave activity over the full range of local times for a wide range of solar wind and geomagnetic conditions. This information will be of critical importance in the development of new theoretical and empirical models for the Earth's radiation belts. Finally, THEMIS's ground observatories and its tail flow monitor P2 along with the radiation belt monitors P3, P4 and P5 promise to advance our knowledge of storm-substorm relationships.

6 Science Closure at the Dayside

6.1 Science Closure during the Baseline Dayside Mission

THEMIS's four probe conjunctions at the dayside recur once every 4 days and allow simultaneous measurements at the magnetopause, the foreshock and the pristine solar wind. Five-probe conjunctions recur once per 8 days, enabling simultaneous P3 and P4 measurements of the magnetopause with corresponding P5 measurements of the magnetosphere during the first dayside season, and simultaneous P3 and P4 measurements of the magnetopause with corresponding P5 measurements of the magnetosheath in the second dayside season. As the P5's apogee moves outward by $\sim 1\ R_E$ from the first to the second dayside season, the focus of attention for the three probes will move from the subsolar to the flank magnetopause. P1 serves as an upstream monitor in the pristine solar wind, P2 observes conditions within the foreshock, while P3 through P5 will determine the response of the magnetopause and magnetosphere.

These orbits offer an opportunity to identify solar wind or foreshock triggers for the full range of transient events observed at the bow shock, magnetopause, and in the outer dayside magnetosphere. Possible studies include determining the conditions under which hot flow anomalies or foreshock cavities attain large amplitudes (Sibeck et al. 2001), the impact of these upstream phenomena upon the magnetosphere and ionosphere, evidence for the initiation of reconnection at solar wind discontinuities transmitted into the magnetosheath (Phan et al. 2007), the search for southward IMF turnings (Lockwood and Wild 1993; Le et al. 1993) or solar wind/foreshock-generated pressure pulses (Potemra et al. 1992) as triggers of flux transfer events, as well as a determination of the predominant cause of magnetopause boundary motion (e.g. Borodkova et al. 1995).

Figure 11 presents global MHD model predictions for the interaction of an interplanetary shock with the Earth's magnetosphere (Samsonov et al. 2007). The figure shows profiles for the magnetic field strength, density, temperature, and velocity along the Earth-Sun line as a function of time. Following the arrival of the interplanetary shock, the bow shock (BS), magnetopause (MP), and a transmitted fast mode wave (FS) move Earthward. After the interval shown in the figure, the transmitted fast wave reflects from the plasmasphere/ionosphere, moves outward, and reverses the inward motion of the bow shock and magnetopause. At apogee, the innermost THEMIS spacecraft, indicated by asterisks in the bottom panel, are well situated to observe the motion of the bow shock and magnetopause. When deeper within the magnetosphere, the spacecraft will observe the passage of the transmitted and subsequent reflected fast mode waves as a set of well-defined correlated plasma and magnetic field perturbations.

6.2 Additional Science during the Coast Phase

The orbits of the 5 THEMIS spacecraft during the coast phase of the mission were ideal to study the structure of the magnetopause and bow shock, and the transients superimposed upon this structure. With a common apogee of 14.7 R_E, spacecraft apogees grazed the dusk magnetopause and the sub-solar bow shock. Separation distances ranging from 0.1 to 3 R_E permitted timing studies of flux transfer event, hot flow anomaly, and boundary wave motion along these boundaries, while similar separation distances perpendicular to the dayside magnetopause enabled researchers to measure the amplitude of boundary waves, test jump conditions, and determine how the layered structure of this boundary varies as a function of time and external conditions.

By definition, the magnetopause is a sharp ($\sim$600 km thick) boundary across which the magnetic field rotates from magnetosheath (shocked interplanetary) to magnetospheric magnetic field orientations. Past work suggests that the transition in plasma parameters is often far more gradual. A layer of magnetosheath plasma, known as the depletion layer, is often found immediately outside the magnetopause in the magnetosheath (Crooker et al. 1979). Similarly, there is often a low-latitude boundary layer of magnetosheath-like plasma on magnetospheric magnetic field lines just inside the magnetopause (Sckopke et al. 1981). The properties of these boundary layers are not well-known, but there are reasons to believe that the spatial extent and density variations associated with them become more pronounced during periods of strongly northward IMF orientation.

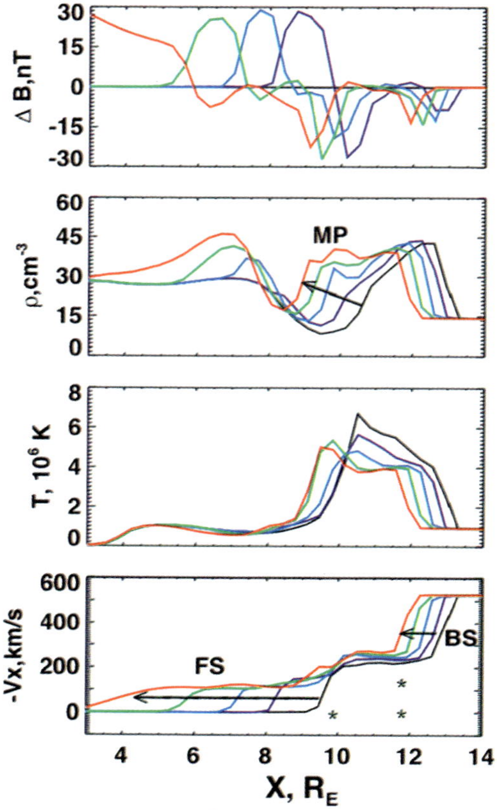

Fig. 11 Results from a global MHD simulation for the interaction of an interplanetary shock with the magnetosphere. The shock launches a transmitted fast mode wave (FS) that propagates through the magnetosheath and magnetosphere along the Earth-Sun line as a function of time (*black*, *blue*, *red* indicating successively later times during the simulation) (Samsonov et al. 2007). MP indicates the magnetopause, separating the low density magnetosphere from the high density magnetosheath. BS indicates the bow shock, separating superalfvénic flows in the solar wind from subAlfvénic flows in the magnetosheath. From *top* to *bottom*, the panels show the perturbation of the magnetic field strength, the density, the temperature, and the component of the velocity along the Earth-Sun line. Following their interaction with the transmitted waves, both the bow shock and magnetopause move inward towards the Earth. Stars in the *lower panel* indicate the apogees of THEMIS spacecraft P3–5 during the second dayside season

Several models have been proposed to account for the low-latitude boundary layer. During periods of southward IMF orientation, reconnection on the equatorial magnetopause should generate thin boundary layers on open magnetic field lines, marked by accelerated plasma flows and the loss of magnetospheric ions and electrons. During periods of northward IMF orientation, simultaneous reconnection at both cusps may append magnetosheath flux tubes with relatively uniform densities to the magnetosphere, resulting in sluggishly-moving boundary layers whose density varies little with radial distance. Diffusion should produce a boundary layer moving at speeds less than those in the nearby magnetosheath with a strong outward radial density gradient. The nonlinear Kelvin-Helmholtz instability should produce a complex structured boundary layer with rolled-up vortices of intermixed magnetosheath and magnetospheric plasma.

Figure 12 illustrates how the THEMIS spacecraft can be used to distinguish between these possibilities. When aligned nearly perpendicular to the nominal magnetopause, there will frequently be occasions when the spacecraft straddle the magnetopause, with two or more located within individual boundary layers. In this configuration, the spacecraft can be used to discriminate between boundary layers with near-uniform parameters, boundary layers exhibiting strong radial gradients, and boundary layers marked by complex local structures. Simultaneous magnetosheath, magnetospheric, and boundary layer observations can be used to compare boundary layer densities and velocities with predictions based on simultaneous magnetosheath and magnetospheric conditions.

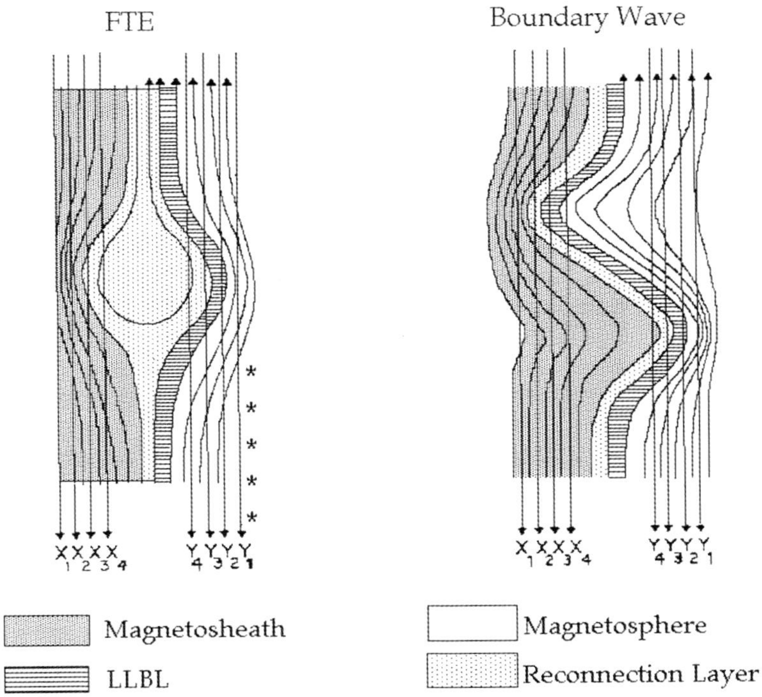

Fig. 12 A comparison of the FTE and pressure pulse models for transient events at the magnetopause (Lockwood 1991). Spacecraft at radially-separated locations in the magnetosheath and magnetosphere observe the events along the cuts labeled X1 to X4 and Y1 to Y4, respectively. Spacecraft arrayed parallel to the magnetopause (*asterisks*) can be used to time the motion and determine the velocity of the transient events

Transient structures, including both intrinsic Kelvin-Helmholtz and pressure-pulse driven boundary waves and flux transfer events (FTEs), are often superimposed upon the boundary layer structure. Whereas FTEs cause the magnetopause current layer to broaden and bulge outward simultaneously into both the magnetosheath and magnetosphere, boundary waves simply displace the magnetopause. Boundary waves driven by pressure pulses should be associated with pressure pulses in the magnetosheath, whereas those driven by the Kelvin-Helmholtz instability should not. Finally, FTEs should move with a velocity determined by the balance of pressure gradient and magnetic curvature forces, boundary waves driven by the Kelvin-Helmholtz instability should move antisunward in the direction of the magnetosheath flow, while those driven by pressure pulses should move in a direction determined by the orientation of the discontinuity associated with the pressure pulse sweeping across the magnetosphere.

As illustrated in Fig. 12, multipoint THEMIS observations will prove ideal in distinguishing between the predictions of these models for transient events. When arrayed perpendicular to the nominal magnetopause, the spacecraft can be used to distinguish between boundary waves and two-regime FTEs (Farrugia et al. 1987). Simultaneous observations of the magnetosheath and magnetosphere can be used to test whether conditions favor the occurrence of the Kelvin-Helmholtz instability during intervals when boundary waves are observed. When the spacecraft lie arrayed along the magnetopause (asterisks in Fig. 12), their observations can be used to determine FTE and boundary wave velocities, and whether they move with the magnetosheath flow in a direction consistent with pressure pulses sweeping across the magnetosphere, or in a direction consistent with predictions for the interconnected magnetosheath and magnetospheric magnetic field lines expected within FTEs (Fear et al. 2007).

Many other topics regarding the detailed structure of the dayside magnetopause remain to be addressed. While four spacecraft can be used to determine the orientation, velocity, and thickness of discontinuities (Dunlop et al. 2002a, 2002b; Paschmann et al. 2005), observations by the five THEMIS spacecraft afford an opportunity to determine boundary acceleration and curvature. Furthermore, five spacecraft place greater constraints than four on efforts to recover boundary plasma and magnetic field structure via the Grad-Shafranov reconstruction technique (Walthour et al. 1993; Hasegawa et al. 2005). Finally, THEMIS observations may help researchers determine why boundary waves on the magnetotail flanks exhibit non-sinusoidal shapes that are inconsistent with the Kelvin-Helmholtz instability (Chen and Kivelson 1993).

7 Concluding Remarks

Substorms represent a fundamental mode of the solar wind-magnetosphere interaction, one involving a series of well-defined and repeatable steps leading to the abrupt release of solar wind energy stored within the Earth's magnetotail. Some of the energy is released into the Earth's ionosphere, some into the Earth's radiation belts, and some flows anti-sunward down the magnetotail. Despite their fundamental importance to magnetospheric physics, the absence of coordinated observations has long prevented a determination of the reasons for the sudden release of energy that occurs during substorms. In conjunction with theory and modeling, the array of ground- and space-based observations provided by THEMIS will offer an opportunity to pinpoint when and where this release occurs, precisely the information needed to understand the mechanisms driving substorms.

However, observations from THEMIS can be used to address a host of research problems in magnetospheric physics. These include the overall solar wind-magnetosphere interaction, as well as the mesoscale phenomena that occur within the foreshock and magnetosheath, reconnection, diffusion, and instabilities of the magnetopause, magnetosphere-ionosphere interactions, the characteristics of the aurorae and geomagnetic pulsations, and the processes that energize and remove particles in the Earth's radiation belts.

With its implementation of a complex mission involving multipoint ground- and space-based observations via a compact team, THEMIS will serve as a pathfinder for future NASA missions, including Magnetospheric Multiscale (MMS), Radiation Belt Storm Probes (RBSP), and Magnetospheric Constellation. The lessons learned from THEMIS, on topics ranging from multiple spacecraft and instrument builds, spacecraft commissioning and operations, through data analysis tools and open data systems, will be applied to these future missions. THEMIS inaugarates a new era in space research, one of cooperation and inclusiveness.

Acknowledgements Work at UCB and UCLA was supported by NASA Contract NAS5-02099. Work at NASA/GSFC was supported NASA's Explorer program and THEMIS MO&DA.

References

S.I. Akasofu, *Physics of Magnetospheric Substorms* (Reidel, Dordrecht, 1977)
V. Angelopoulos et al., J. Geophys. Res. **99**, 21257 (1994). doi:10.1029/94JA01263
V. Angelopoulos et al., J. Geophys. Res. **102**, 211 (1997). doi:10.1029/96JA03217
V. Angelopoulos et al., Geophys. Res. Lett. **18**, 2841 (1999). doi:10.1029/1999GL900601
V. Angelopoulos, Space Sci. Rev. (2008, this issue)
H. Auster et al., Space Sci. Rev. (2008, this issue)
D.N. Baker, T.I. Pulkinnen, V. Angelopoulos, W. Baumjohann, R.L. McPherron, J. Geophys. Res. **101**, 12975 (1996). doi:10.1029/95JA03753
D.N. Baker et al., Geophys. Res. Lett. **29** (2002). doi:10.1029/2002GL015539
W. Baumjohann et al., J. Geophys. Res. **95**, 3801 (1990)
T. Bösinger, K. Alanko, J. Kangas, H. Opgenoorth, W. Baumjohann, J. Atmos. Terr. Phys. **43**, 933 (1981)
J. Bonnell et al., Space Sci. Rev. (2008, this issue)
N.L. Borodkova, G.N. Zastenker, D.G. Sibeck, J. Geophys. Res. **100**, 5643 (1995)
D.H. Brautigam, J.M. Albert, J. Geophys. Res. **105**, 291 (2000)
J. Büchner et al., in *Geospace Mass and Energy Flow: Results from the ISTP Program*, ed. by A. Nishida, D.N. Baker, S.W.H. Cowley (AGU, Washington, 1998), p. 313
C. Carlson et al., Space Sci. Rev. (2008, this issue)
J.K. Chao et al., Planet. Space Sci. **25**, 703 (1977)
C.-X. Chen, R.A. Wolf, J. Geophys. Res. **98**, 21409 (1993)
S.-H. Chen, M.G. Kivelson, Geophys. Res. Lett. **20**, 2699 (1993)
C.R. Clauer, R.L. McPherron, J. Geophys. Res. **79**, 2811 (1974)
N.U. Crooker, T.E. Eastman, G.S. Stiles, J. Geophys. Res. **84**, 869 (1979)
P.W. Daly et al., Geophys. Res. Lett. **11**, 1070 (1984)
M.W. Dunlop, A. Balogh, K.-H. Glassmeier, J. Geophys. Res. **107** (2002a). doi:10.1029/2001JA005089
M.W. Dunlop, A. Balogh, K.-H. Glassmeier, P. Robert, J. Geophys. Res. **107** (2002b). doi:10.1029/2001JA005088
R.D. Elphinstone et al., J. Geophys. Res. **100**, 7937 (1995)
G.M. Erickson, R.A. Wolf, Geophys. Res. Lett. **7**, 897 (1980)
G.M. Erickson et al., J. Geophys. Res. **105**, 26265 (2000)
C.J. Farrugia, D.J. Southwood, S.W.H. Cowley, R.P. Rijnbeek, P.W. Daly, Planet. Space Sci. **35**, 737 (1987)
C.J. Farrugia, M.P. Freeman, S.W.H. Cowley, D.J. Southwood, M. Lockwood, A. Etamadi, Planet. Space Sci. **37**, 589 (1989)
R.C. Fear et al., Ann. Geophys. 1669 (2007)
F.R. Fenrich, J.C. Samson, J. Geophys. Res. **102**, 20031 (1997)
N.J. Fox, B.H. Mauk, J.B. Blake, Geophys. Res. Lett. **33** (2006). doi:10.1029/2006GL026598
H. Frey et al., Space Sci. Rev. (2008, this issue)
C.K. Goertz, W. Baumjohann, J. Geophys. Res. **96**, 20991 (1991)
J.C. Green, T.G. Onsager, T.P. O'Brien, D.N. Baker, J. Geophys. Res. **109** (2004). doi:10.1029/2004JA010579
G. Haerendel, in *ICS-1*, ed. by C. Mattock (European Space Agency, Noordwijk, 1992), p. 417
H. Hasegawa, B.U.Ö. Sonnerup, B. Klecker, G. Paschmann, M.W. Dunlop, H. Réme, Ann. Geophys. **23**, 973 (2005)
R.R. Heacock, J. Geophys. Res. **72**, 3905 (1967)
M. Hesse, J. Birn, J. Geophys. Res. **96**, 11513 (1991)
E.W. Hones Jr., in *Physics of Solar Planetary Environments*, ed. by D.J. Williams (AGU, Washington, 1976), p. 558
R.B. Horne, R.M. Thorne, S.A. Glauert, J.M. Albert, N.P. Meredith, R.R. Anderson, J. Geophys. Res. **100** (2005). doi:10.1029/2004JA010811
M.K. Hudson, A.D. Kotelnikov, X. Li, M. Temerin, J. Wygant, J.B. Blake, M.S. Gussenhoven, Geophys. Res. Lett. **22**, 291 (1995)
O.A. Hurricane, B.H. Fong, S.C. Cowley, F.V. Coroniti, C.F. Kennel, R. Pellat, J. Geophys. Res. **104**, 10221 (1999)
C. Jacquey, in *Magnetospheric Currents*, ed. by S.-I. Ohtani, R. Fujii, M. Hesse, R.L. Lysak (AGU, Washington, 2000), p. 275
K. Kauristie, V.A. Sergeev, M. Kubyshkina et al., J. Geophys. Res. **105**, 10677 (2000)
G. Kettmann, P.W. Daly, J. Geophys. Res. **93**, 7376 (1988)
G. Khazanov, M. Liemohn, T. Newman, M. Fok, A. Ridley, Ann. Geophys. **22**, 497 (2004)
B.T. Kress, M.K. Hudson, P.L. Slocum, Geophys. Res. Lett. **32** (2005). doi:10.1029/2005GL022373
M.V. Kubyshkina, J. Geophys. Res. **104**, 24977 (1999)

D. Larson et al., Space Sci. Rev. (2008, this issue)
G. Le, C.T. Russell, H. Kuo, Geophys. Res. Lett. **20**, 791 (1993)
O. Le Contel et al., in *ICS-4 Proceedings* (Kluwer Academic, Dordrech, 1998), p. 425
D.-Y. Lee, R.A. Wolf, J. Geophys. Res. **97**, 19251 (1992)
X. Li et al., Geophys. Res. Lett. **27**, 1447 (2000)
X. Li, D.N. Baker, S.G. Kanekal, M. Looper, M. Temerin, Geophys. Res. Lett. **28**, 3827 (2001)
K. Liou et al., J. Geophys. Res. **105**, 2495 (2000)
M. Lockwood, J. Geophys. Res. **96**, 5497 (1991)
M. Lockwood, M.N. Wild, J. Geophys. Res. **98**, 5935 (1993)
R.E. Lopez, Geophys. Res. Lett. **17**, 1671 (1990)
T.M. Loto'aniu, I.R. Mann, L.G. Ozeke, A.A. Chan, Z.C. Dent, D.K. Milling, J. Geophys. Res. **111** (2006). doi:10.1029/2005JA011355
A.T.Y. Lui, J. Geophys. Res. **101**, 13067 (1996)
A.T.Y. Lui et al., Geophys. Res. Lett. **7**, 721 (1988)
J. McFadden et al., Space Sci. Rev. (2008, this issue)
R.L. McPherron, T. Terasawa, A. Nishida, J. Geomagn. Geoelectr. **38**, 1089 (1986)
N.C. Maynard et al., J. Geophys. Res. **101**, 7705 (1996)
S. Mende et al., Space Sci. Rev. (2008, this issue)
R.M. Millan, R.M. Thorne, J. Atmos. Sol. Terr. Phys. (2007). doi:10.1016/j.jastp.2006.06.019
M.J. Mithaiwala, W. Horton, J. Geophys. Res. **110** (2005). doi:10.1029/2004JA010511
U. Motschmann et al., Multi-spacecraft filtering: Plasma mode recognition, in *Analysis Methods for Multi-Spacecraft Data* (Kluwer Academic, Dordrecht, 1998), p. 79
R. Nakamura et al., Geophys. Res. Lett. **29** (2002). doi:10.1029/2002GL016200
R. Nakamura et al., Geophys. Res. Lett. **31** (2004). doi:10.1029/2004GL019558
R. Nakamura et al., J. Geophys. Res. **111** (2006). doi:10.1029/2006JA011706
S.-I. Ohtani et al., J. Geophys. Res. **98**, 19355 (1993)
S.-I. Ohtani, J. Geophys. Res. **103**, 6815 (1998)
J.V. Olson, J. Geophys. Rev. **104**, 17499 (1999)
N. Omidi, D.G. Sibeck, J. Geophys. Res. **112** (2007). doi:10.1029/2006JA011663
A. Otto, D.H. Fairfield, J. Geophys. Res. **105**, 21175 (2000)
G.S. Paschmann et al., Ann. Geophys. **23**, 1481 (2005)
S.A. Perraut et al., in *ICS-4 Proceedings* (Kluwer Academic, Dordrecht, 1998), p. 349
A.A. Petrukovich et al., J. Geophys. Res. **103**, 47 (1998)
T.-D. Phan et al., Geophys. Res. Lett. **34** (2007). doi:10.1029/2007GL030343
D.H. Pontius, R.A. Wolf, Geophys. Res. Lett. **17**, 49 (1990)
T.A. Potemra, L.J. Zanetti, R. Elphinstone, J.S. Murphree, D.M. Klumpar, Geophys. Res. Lett. **19**, 1615 (1992)
J. Raeder et al., J. Geophys. Res. **106**, 381 (2001)
J. Raeder et al., Space Sci. Rev. (2008, this issue)
A. Roux et al., J. Geophys. Res. **96**, 17697 (1991)
A. Roux et al., Space Sci. Rev. (2008, this issue)
A. Runov et al., Ann. Geophys. **24**, 247 (2006)
A. Runov et al., Geophys. Res. Lett. **30**, 1579 (2003). doi:10.1029/2002GL016730
A. Runov et al., Ann. Geophys. **23**, 1391 (2005a)
A. Runov et al., Planet. Space Sci. **53**, 237 (2005b)
C.T. Russell, R.C. Elphic, Space Sci. Rev. **22**, 681 (1978)
J.C. Samson et al., in *ICS-3*, ed. by E.J. Rolfe, B. Kaldreich (European Space Agency, Noordwijk, 1996), p. 399
A.A. Samsonov, D.G. Sibeck, J. Imber, J. Geophys. Res. **112** (2007). doi:10.1029/2007JA01262
E.T. Sarris et al., Geophys. Res. Lett. **3**, 437 (1976)
N. Sckopke et al., J. Geophys. Res. **86**, 2099 (1981)
V.A. Sergeev, J. Geophys. Res. **101**, 2615 (1996)
V.A. Sergeev et al., J. Geophys. Res. **101**, 10817 (1996)
V. Sergeev et al., J. Geophys. Res. **103**, 9177 (1998)
I. Shinohara et al., J. Geophys. Res. **103**, 20365 (1998)
K. Shiokawa, W. Baumjohann, G. Haerendel, Geophys. Res. Lett. **24**, 1179 (1997)
K. Shiokawa, W. Baumjohann, G. Haerendel, Geophys. Res. Lett. **25**, 959 (1998)
D.G. Sibeck, R.B. Decker, D.G. Mitchell, A.J. Lazarus, R.P. Lepping, A. Szabo, J. Geophys. Res. **106**, 21675 (2001)
M.I. Sitnov, M. Swisdak, P.N. Guzdar, A. Runov, J. Geophys. Res. **111** (2006). doi:10.1029/2005JA011517
P. Song, R.C. Elphic, C.T. Russell, Geophys. Res. Lett. **15**, 744 (1988)

T. Sugiyama et al., Geophys. Res. Lett. **24**, 1651 (1997)
M.F. Thomsen, J.T. Gosling, S.J. Bame, K.B. Quest, C.T. Russell, J. Geophys. Res. **93**, 11311 (1988)
V.A. Troitskaya, J. Geophys. Res. **66**, 5 (1961)
A.Y. Ukhorskiy, B.J. Anderson, K. Takahashi, N.A. Tsyganenko, Geophys. Res. Lett. **33** (2006). doi:10.1029/2005GL024380
I. Voronkov et al., J. Geophys. Res. **104**, 17323 (1999)
I. Voronkov et al., J. Geophys. Res. **105**, 18505 (2000)
Y. Yamade et al., Geophys. Res. Lett. **27**, 1091 (2000)
D.W. Walthour et al., J. Geophys. Res. **98**, 1489 (1993)

Orbit Design for the THEMIS Mission

S. Frey · V. Angelopoulos · M. Bester · J. Bonnell · T. Phan · D. Rummel

Originally published in the journal Space Science Reviews, Volume 141, Nos 1–4, 61–89.
DOI: 10.1007/s11214-008-9441-1

Abstract THEMIS, NASA's fifth Medium Class Explorer (MIDEX) mission will monitor the onset and macro-scale evolution of magnetospheric substorms. It is a fleet of 5 small satellites (probes) measuring in situ the magnetospheric particles and fields while a network of 20 ground based observatories (GBOs) monitor auroral brightening over Northern America. Three inner probes ($\sim$1 day period, 10 R_E apogee) monitor current disruption and two outer probes ($\sim$2 day and $\sim$4 day period, 20 R_E and 30 R_E apogees respectively) monitor lobe flux dissipation. In order to time and localize substorm onsets, THEMIS utilizes Sun–Earth aligned conjunctions between the probes when the ground-based observatories are on the nightside. To maintain high recurrence of conjunctions the outer orbits have to be actively adjusted during each observation season. Orbit maintenance is required to rearrange the inner probes for dayside observations and also inject the probes into their science orbits after near-simultaneous release from a common launch vehicle.

We present an overview of the orbit strategy, which is primarily driven by the scientific goals of the mission but also represents a compromise between the probe thermal constraints and fuel capabilities. We outline the process of orbit design, describe the mission profile and explain how mission requirements are targeted and evaluated. Mission-specific tools, based on high-fidelity orbit prediction and common magnetospheric models, are also presented. The planning results have been verified by in-flight data from launch through the end of the first primary science seasons and have been used for mission adjustments subject to the early scientific results from the coast phase and first tail season.

Keywords THEMIS · Mission design · Orbit · Magnetosphere · Substorm · Maneuver · Neutral sheet · Aurora · Launch · Re-entry · Fuel budget · Orbital perturbation · Lunar force · Geopotential · Orbit maintenance · Magnetotail · Reconnection · Solar wind · Propulsion

S. Frey (✉) · M. Bester · J. Bonnell · T. Phan · D. Rummel
Space Sciences Laboratory, University of California, Berkeley, CA 94720, USA
e-mail: sfrey@ssl.berkeley.edu

V. Angelopoulos
IGPP/ESS UCLA, Los Angeles, CA 90095, USA

1 Introduction

THEMIS is the first mission to study the onset and evolution of the magnetospheric substorm instability in a macro-scale constellation. Substorms are considered the fundamental mechanism within the Earth's magnetic field environment that shields Earth from the impact of magnetized plasma clouds from outer space such as the solar wind. The main components of substorm instabilities are the giant auroral eruption at the ionosphere near the Earth poles, and near the equator, the disruption of plasma sheet currents where the magnetosphere undergoes the transition from the stretched tail to a dipole fieldlike shape, and even further out in the tail, the reconnection of magnetic fields in the neutral sheet, triggering plasma flows. For the first time, all three main substorm components that can expand over more than 30 R_E are monitored simultaneously with time resolutions that match the dynamics of the substorm-related processes. Twenty all-sky cameras and ground magnetometers completely covering North America determine the timing of auroral breakups within an accuracy of 3 s, while the five identical THEMIS probes equipped with field and particle instruments determine onsets of current disruption and plasma flow originating from the tail reconnection to an accuracy of 10 s. The synchronized measurements of the ground and space segments, taken where the equatorial tail region maps along the magnetic field onto the substorm auroras in the polar regions, allow one to correlate the onset of substorms with the macroscopic interaction of the substorm components. This is achieved by aligning all five probes along the Sun-Earth line in the magnetotail near the neutral sheet once per four days over North America during local night times, and maintaining these conjunctions while the orbits intersect with the magnetotail. A comprehensive outline of the THEMIS mission and its science objectives is given in Angelopoulos (2008) in this issue. Further outline on how the various segments address the science goals is provided by Sibeck et al. (2008), also in this issue. The THEMIS mission is operated by the Mission Operations Center located at the UC Berkeley Space Sciences Laboratory (SSL) and a description of mission operations is given by Bester et al. (2008) in this issue.

In this paper, we explain in more detail how the selected orbit strategy addresses mission requirements and how it is has been developed. We also describe the mission phases, how the constellation was configured, and how meeting the science requirements is evaluated. Locations and instrumentation of the ground-based observatories are provided by Mende et al. (2008), Harris et al. (2008), and Russell et al. (2008) in this issue. For specific information about flight instruments, we refer to the various articles in this issue.

2 THEMIS Orbit Parameters

2.1 Orbit Requirements and Constraints

In order to determine the sequence of events within, substorms the THEMIS orbit design has to combine the space segment of five probes with the ground segment of 20 observatories based along the auroral oval in northern America, the Ground Based Observatories or GBOs. The five probes are to line up in 1-, 2-, and 4 day orbits in the magnetotail in times of optimal night sky conditions for the GBOs. Prior to the first primary science season (tail season), all probes have to be placed in their science orbits, and orbital periods have to be adjusted periodically to optimize the amount and quality of the constellation's alignment. The inclinations of the outer probes need to be restored prior to the second year tail season to counteract drifts due to orbital perturbations.

In addition to the science-driven targets, the orbital design has to comply with NASA orbital debris guidelines of re-entry by selecting a launch trajectory from which LV stages and probe carrier will re-enter and by including end-of mission re-entry maneuvers in the nominal mission plan. Furthermore, the orbits should avoid long eclipses in order to achieve the energy balance for which the probes are designed. Also, mission redundancy is achieved by in-orbit replacement ability. Any replacement option needs to be considered in the orbit design. Finally, the orbit design must allow a feasible maneuver plan and is constrained by the finite fuel supply. Given the number of maneuvers it takes to set up and maintain the THEMIS mission, it is essential to consider fuel efficiency and fault tolerance in the implementation. Once the long antennas of the electric field instruments are deployed, large maneuvers to reorient a probe are to be avoided. Upon meeting all engineering requirements and constraints, the validity of the orbit design is measured as the accumulated times at which the probes are aligned within small stripes along the Sun-Earth line and in the vicinity of the neutral sheet (conjunctions) as defined in Table IV in Angelopoulos (2008).

2.2 Selection of Apogee Distances and Orientation in Inertial Space

The space and time targets are solely driven by the primary science goal to study the sequence of events during substorms. At the time target all apogee passes line up along the Sun–Earth line in the magnetotail, which at that time crosses the center meridian of the ground based observatories. The time target defines the center epoch of the first primary THEMIS season and was set to early February, balancing between shorter eclipse duration, visibility of the night sky, and substorm occurrence. The geocentric apogee distances and orbital periods are selected to cover current disruptions near $-10\ R_E$ as well as tail reconnection between -20 to $-30\ R_E$. To ensure mapping of substorm-related events in the tail, observed by the space segment, into the field of view of the ground-based observatories in the night sky (for exact locations see Mende et al. 2008, Fig. 4), the apogee passes of the inner probes are locked over the center of the ground segment at local midnight by sidereal-day period. In Fig. 1, the top left panel depicts the THEMIS constellation at the center epoch (WD) for the first tail season in sun-referenced GSM coordinates with the z-axis following the magnetic dipole axis and overlaid with a neutral sheet model, the predominant plasma feature on the night side. At that instance all apogees are aligned near the x-axis in the tail (see also Fig. 6). With respect to the sun the orbits drift around so that half a year later apogees align near the x-axis pointing to the Sun at the center epoch (at local noon). For that instance orbits are shown in the top right panel of Fig. 1 with sun-referenced GSE coordinates with the z-axis pointing to ecliptic north and overlaid with magnetopause and bow shock models, the predominating interfaces on the dayside. This natural evolution of the orbits in the sun-referenced system allows to address the secondary and tertiary science goals by simply following the same alignment strategy. Hence, the orbit design is solely driven by the primary science goals.

The strategic position each probe has within the constellation is reflected by the constellation IDs (CIDs). CIDs start with the outermost probe and count Earthward, whereas the flight models are referenced by letter IDs A through E. As shown in Fig. 1, P1 always points to the four day orbit, P2 always points to the two day orbit, P3 and P4 point to the two orbits with sidereal-day periods, and P5 points to the remaining fifth orbit, complementing the baseline mission. The CIDs have been assigned to the flight models after launch to best match the probe in-flight properties with the different orbit demands, such as propulsion systems and instrument ranges. Mainly based on communication systems, the assignments are P1PB, P2PC, P3PD, P4PE and P5PA.

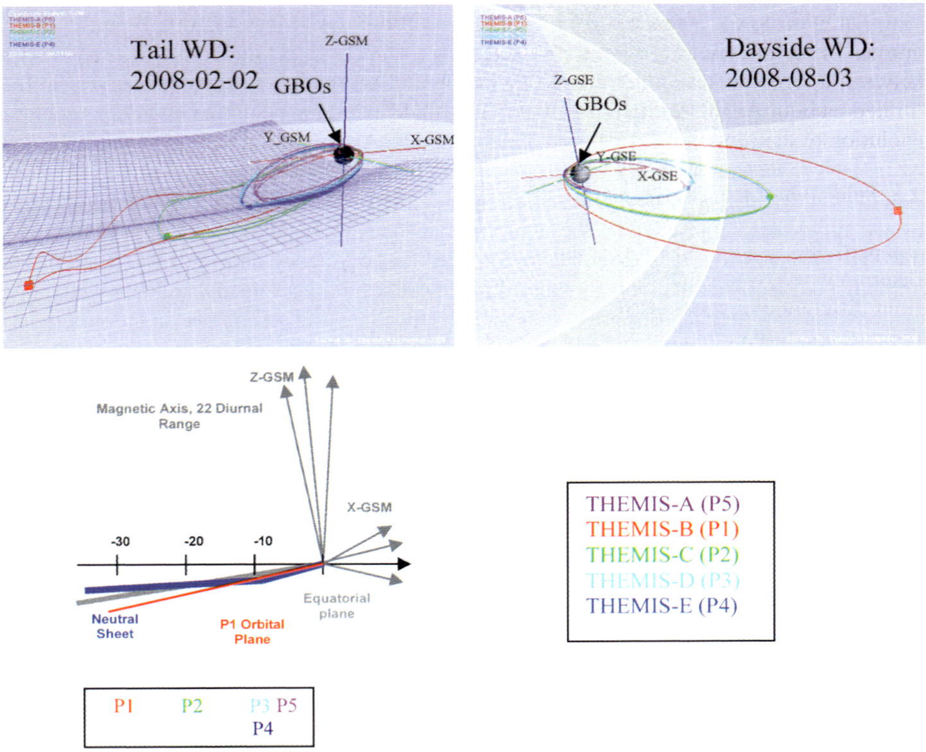

Fig. 1 *Upper two panels* show THEMIS orbits at center epochs (WD) for the first year with an axis scale of 10 Re (using SSCweb 3D orbit viewer). *Top left*: Tail season with the neutral sheet in GSM coordinates, GBO positions are indicated on the northern night side; *Top right*: Dayside with magnetopause and bow shock in GSE coordinates and GBOs indicated on the northern dayside; *Bottom left*: Sun referenced observational geometry at the center epoch of the tail season, superimposed are equatorial plane, P1 orbital plane, and a simple neutral sheet model. The CIDs (P1,P2,P3,P4,P5) are positioned to indicate the apogee distance at that moment of crossing the Sun–Earth line during the tail season

Inclination and argument of perigee are the main drivers to balance conjunctions and eclipse durations with the neutral sheet as described in Angelopoulos (2008). For the inner probes, the inclination is mainly determined to limit eclipses in the second year and to establish a z-separation between P3, P4 and P5 also for the second year. Increasing apogee distances make cutting through the daily and seasonal variations of the magnetospheric processes more difficult with the relatively fixed orbits. Figure 1, bottom left panel is a snap shot of the observational geometry in the Sun-referenced meridional planes (GSE, GSM coordinates) at the time the orbits are at the tail season center epoch. The neutral sheet starts forming along the magnetic equator but then follows the bending of the magnetotail towards the ecliptic induced by the solar wind at a geocentric distance of about 10 R_E downtail. The relative location of the neutral sheet to the ecliptic dramatically varies with the season along the Earth's orbit around the Sun, and the offset between the spin axis and the magnetic dipole axis leads to significant diurnal fluctuations relative to the orbital planes. As Fig. 1 indicates, a low inclination is key to bringing P1 close to the neutral sheet, especially around its scientifically strategic positions near apogee. However, placing the probes in the vicinity of the ecliptic in the anti-Sunward hemisphere in winter sets up the condition to encounter long Earth shadows, which are only to be avoided by larger inclinations. The choice of argument of perigee (APER) can mitigate the effect the inclination has on the relative position between apogee passes and the neutral sheet. As will be shown later, this option is rather limited as both APER and inclination drift significantly on the larger orbits. For the entire evolution of the orbital elements through the first tail season see also Fig. 13.

2.3 Selection of Perigee Distances and Orbital Perturbations

Once orbital periods and approximate apogee distances have been established, the drivers to determine perigee altitudes become orbit stability, Orbital Debris Requirements, the need to limit differential precession, and fuel consumptions. All of these aspects are heavily related to orbital perturbations caused by the additional forces beyond the Keplerian two-body problem that lead to various periodic and secular variations of orbital elements with time. Within the scope of this paper, we will focus on the specific drivers of the THEMIS orbit design and illustrate the most significant effects and refer to Vallado (1997) for a more comprehensive discussion of orbital perturbations. For the high altitude THEMIS orbits we have to consider third-body perturbations by the Sun and Moon, the non-spherical mass distribution of the Earth, and solar radiation, while atmospheric drag needs to be considered for launch trajectory and re-entry targets. Across the THEMIS constellation, these forces act with different magnitude and impact on the THEMIS orbit design in three ways (1) a drift between inner (P3, P4, P5) and outer orbits (P1, P2) over time (differential precession), (2) significant increases and fluctuations of the outer probes' perigee altitudes, and (3) dramatic change of inclination and argument of perigee of the outer probes.

Orbital precession depends strongly on semi-major axis and perigee, lesser on inclination, and to an even smaller extent on lunar perturbations. Hence, differential precession between inner and outer probes evolves over time due to the declining effect of the non-spherical mass distribution of the Earth with increasing orbital altitude. Figure 2 shows the drop in rotation rate of the line of apsides with increasing apogee distances as encountered by the THEMIS orbits. The different curves vary perigee altitudes representative for THEMIS orbits. Each such curve is repeated at two inclinations that are representative for the first tail season. In order to yield sufficient THEMIS conjunctions (see Table IV in Angelopoulos 2008) during the two-year nominal life time differential precession should ideally stay below 25 deg and not exceed 45 deg. As shown in Fig. 2, the rotation rates between inner and outer probes can be equalized by higher perigees for the inner probes and lower ones for the outer probes. Ideally, one would like to set perigee targets that substantially suppress differential precession, but other restrictions on perigee altitudes prevent complete optimization.

While perturbations by the geopotential decrease with increasing apogee heights far beyond a geosynchronous orbit, the Moon becomes an increasingly significant perturbing force. Figure 3 compares the lunar effect on APER (upper plot) and perigee altitude (lower plot) for science orbits of P1, P2 and P3 with the respective apogee distances of 30 R_E, 20 R_E, and 10 R_E. The thin lines are high-fidelity orbit propagations with full force models including Sun and Moon, whereas for the thick lines lunar perturbations were turned off. After one year, lunar perturbations are still negligible for the inner probes but become visible for the outer probes already within the first and severe after 3 (P1) and 4 (P2) lunar months. In addition to the rise of perigee, the orbital planes are swung around until equilibrium is established with an APER at roughly 180 degrees. The corresponding dramatic change in inclination can be seen in Fig. 13.

When left in their final science orbits, no probe would re-enter within 25 years as required, and re-entry must be forced by reducing perigee altitudes at the end of mission.

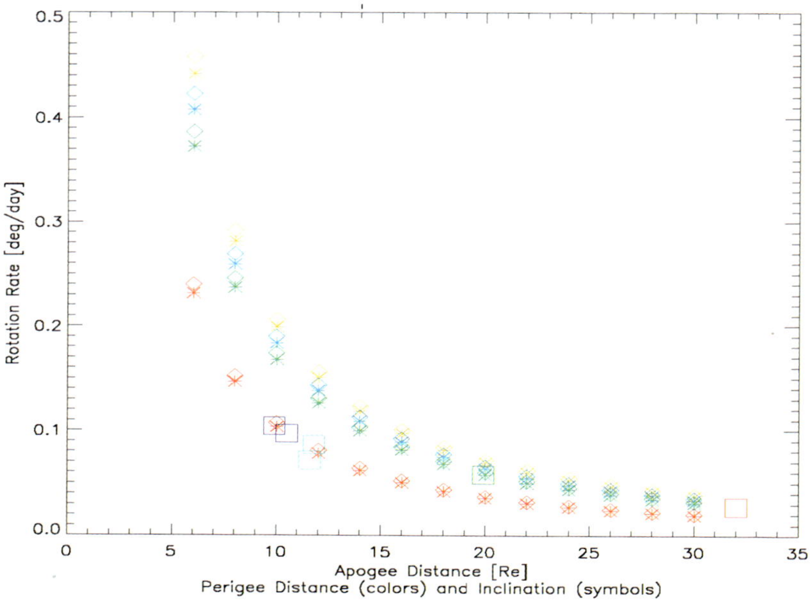

Fig. 2 Rotation of line of apsides shown as sum of argument of perigee and right ascension of node due to the oblateness of Earth as a function of apogee distances for various perigee altitudes and inclinations. Perigee altitudes are 3189 km (*red*), 1072 km (*green*), 733 km (*blue*), 446 km (*yellow*); inclinations are 7 (*) and 1 degree (◊). *Squares* mark average values for each THEMIS probe for the first year based on numbers from Table 2

Naturally, the re-entry requirement contradicts high perigees for orbital stability and the prevention of premature re-entry. Re-entry maneuvers can be significantly large and hence end-of-life maneuvers must be accounted for in the fuel budget. In particular, the perigee altitude for the outer most probe P1 must be high enough to tolerate lunar perturbations of perigee altitudes by up to 3200 km. Perigee altitudes of around 1900 and 3800 km correspond to the breakpoint between short (<5 years) and long (>20 years) P1 lifetimes (Berry 2005). A perigee altitude of 3200 km results in lifetimes of >10 years under all lunar phases. Considering differential precession and re-entry requirements, we chose 3200 km as the nominal perigee altitude for P1, about 1070 km for P2, and 730 km for P3, P4, and P5, respectively. To our advantage due to lunar perturbations, the toll in fuel spending for re-entry is much less for the outer probes than for the inner ones, thus equalizing fuel consumption between the probes. Targeting a re-entry perigee altitude of 320 km for the inner probes almost doubles their total fuel allocation. Whereas maneuvers to reach the target for the outer probes of 3000 km (P1) and 640 km (P2) take a rather small fraction of the fuel amount allocated for their ascent and conjunction maintenance.

The squares in Fig. 2 mark the targeted average rotation rate for each THEMIS probe through the first year based on the nominal perigee altitudes, which are listed in Table 1 together with apogee distance. For P5, and (P3, P4) the two sets reflect the changes in their orbits after the first tail season.

The change in apogee distance for P5 and the large difference in its inclination relative to (P3, P4) are both driven by science goals (Angelopoulos 2008). In order to keep the inner probe orbits close, the large difference in inclination could only be accommodated by a smaller perigee for (P3, P4) during the first tail season.

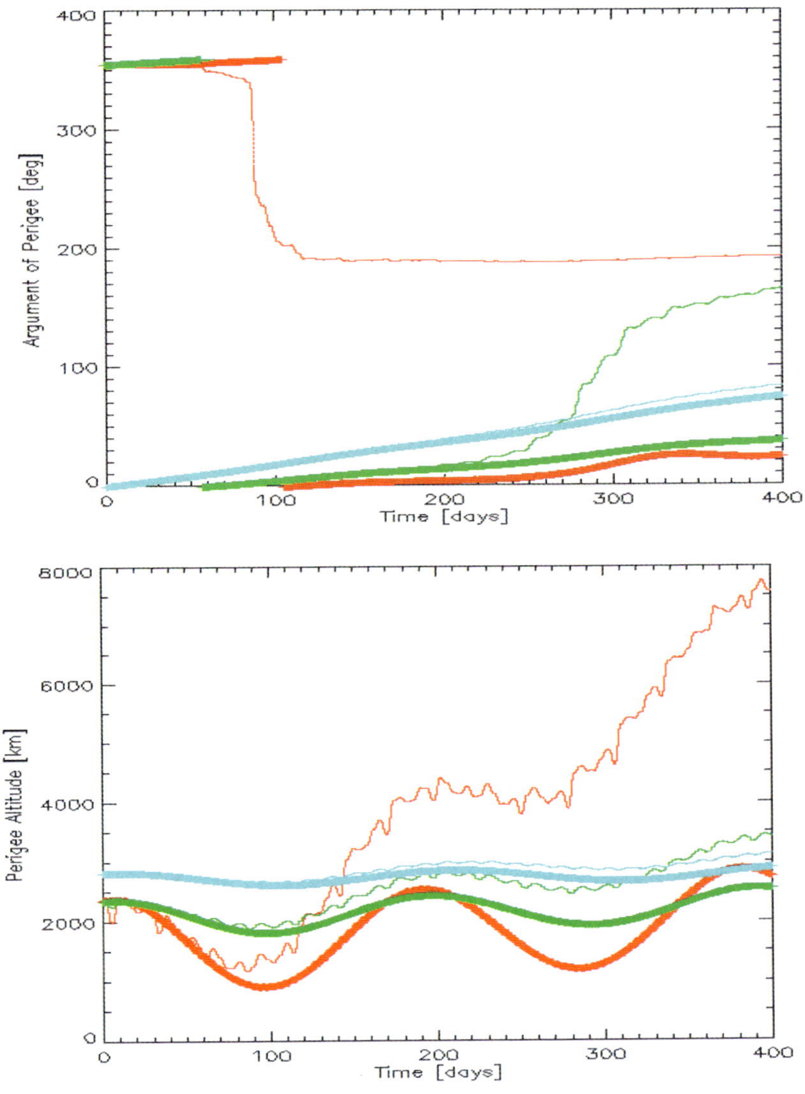

Fig. 3 Build-up of lunar perturbations in APER (*upper plot*) and perigee altitude (*lower plot*) over one year for inner and outer science orbits. Orbit propagations from high fidelity force models including (*thin lines*) and excluding (*thick lines*) the Moon. Inner orbits of P3 in light *blue*, P2 orbits in *green*, and P1 orbits in *red*

Since the perigees start low at launch, driven by an orbital debris requirement to re-enter the second and third stage, and have to be reduced at end-of-mission for re-entry, the final perigee target to minimize differential precession is ultimately dictated by the amount of fuel margin. Also, there is an additional fuel efficiency incentive to keeping the perigee low, as apogee changes and inclination changes are more efficient the lower the perigee. Shown in Fig. 3, the Moon raises the perigee for P1 and P2 significantly. Thus perigee "station keeping" is recommended to maintain a low perigee both for an efficient re-entry maneuver and for optimizing the delta V (the amount of energy needed to change the orbit is expressed

Table 1 Apogee distances, perigee altitudes, and inclination used to determine probe specific rotation rates shown by the squares in Fig. 2. The last two rows show the relative rotation rates referenced to P3, P4 during Tail 1 and afterwards. All values are average values over the first year. For P3, P4, and P5 the two sets reflect the changes after the first tail season

	P5		P3, P4		P2	P1
	Tail 1	After tail 1	Tail 1	After tail 1		
Ra [R_E]	9.9	10.5	11.8	11.6	19.9	32.0
Rp [R_E]	1.46	1.46	1.44	1.64	1.17	1.5
Rp [km]	2934	2934	2806	4082	1071	3827
In [deg]	12.0	10.5	7.0	5.0	7.5	5.0
Rel. rot. [deg/day]	0.018		0		–0.028	–0.071
Rel. rot. [deg/day]	(0.034)	0.027	(0.016)	0	–0.012	–0.054

Table 2 Perigee target constraints imposed by competing science and engineering requirements

Orbit requirements	Inner probes		Outer probes	
	P5	P3, P4	P2	P1
Re-entry of 2nd, 3rd stage	Low	Low	Low	Low
Stability	High	High	High	High
End-of mission re-entry	Low	Low	Low	Low
Differential precession	High	High	Low	Low
Fuel efficiency	Low	Low	Low	Low
Fuel consumption	Low	Low	Low	High

in total change of velocity or delta V) required for various apogee changes. Since it is done most efficiently if coupled with an inclination change, the perigee reduction for P1 and P2, needed to ensure re-entry, is scheduled immediately after the first year and combined with the reversion of the lunar pull on the inclinations that is necessary to keep eclipse durations in the second year below 3 h.

Table 2 shows how differently science requirements and fuel consumption compete in determining perigee altitudes for each probe. While the driver for (P3, P4) perigee altitudes is differential precession, the perigee altitudes of P1 and P2 are driven by lunar perturbations. For P5 to have the same average precession rate as (P3, P4), perigee selection is driven by its inclination. The difference of 7 degrees relative to the (P3, P4) inclination will naturally produce the z-separation between P5 and (P3, P4) in the second year required to achieve mission science goals.

3 Process of Orbit Design

3.1 Parameter Study

In Phase-B a large parameter study was undertaken to determine optimal orbital parameters and period tweaks necessary to meet science requirements. Conjunctions were initially simulated and selectively tabulated as a function of right ascension of perigee (RAP) which is defined as the sum of the right ascension of the ascending node (RAAN) and the argument of perigee (APER), referenced to the 1st year tail season center epoch using a high fidelity orbit propagator Goddard Trajectory Determination System (GTDS), provided by the Goddard Space Flight Center (GSFC), with an IDL-based wrapper for self-sufficient run setups and orbit post-processing. A small maneuver adjusting the period was at 36 days before and after the center epoch, and various values for APER at center epoch were used to account for any launch day of the year. From those runs, an initial determination of the RAP and launch APER was made, and the perigee and inclination drifts were tabulated for all orbits as a function of time. In the next step of runs with the selected orbit parameters, the optimal phasing schedule was derived (see Table 4). These runs were not "forward" runs from launch thereafter, but rather started at center-tail and propagated backwards and forwards from that reference time.

3.2 Design Flow

In conjunctions with that effort, ManCalc, a Microsoft Excel®-based tool, was developed to allow fast, efficient analysis of the five fuel budgets and margins as a function of launch day, launch vehicle errors, thruster inefficiencies as well as replacement strategy and trade-offs for perigee targets. It simulates the maneuvers needed to transform from insertion to the science orbits as well as the tweak maneuvers. The burns are modeled as impulsive maneuvers, using standard formulae for the J2 perturbations by the non-spherical mass distribution of the Earth (Wertz 2001), and linear drift rates for perigee and inclination as determined by the high-fidelity orbit propagations with GTDS. The main deterministic fuel inefficiency for the mission is the finite arc loss that is associated with long burns near perigee. Those were modeled separately using a GSFC-provided tool, the General Maneuver Program (GMAN), which was also made callable through IDL. After modeling maneuvers as impulsive burns, as finite fixed attitude burns, and adjustable attitude burns, we were able to include a "linearized" loss term (1–4% loss/degree in mean anomaly of the burn) to properly account for finite arc inefficiency in ManCalc.

ManCalc was thus able to simulate differential precession and RAP evolution. With reasonable assumptions on tweak maneuvers, inclination changes and re-orientation fuel requirements could be modeled for each burn. Additionally, ManCalc was able to model all realistic inefficiencies and related losses towards a computation of a true margin available for operator error at launch, and any possible extended mission science. Finally, it outlined the entire baseline mission profile as an initial guess for the higher-fidelity GTDS-based "forward" runs that were necessary in order to determine conjunction hours and shadow durations.

The mission-long end-to-end computation of the conjunctions, shadows and deltaV can only be done with "forward" runs that account accurately for the cumulative lunar perturbations and differential precession since launch. This can be done using GTDS and GMAN. The IDL-based wrapper, integrating GTDS and GMAN, that determine those quantities in a self-consistent manner, has been termed Mission Design Tool (MDT). This formed the basis of the accurate mission design, and resulted in a first order self-consistent validation of the launch elements that meet mission requirements.

By the time the mission design was almost complete and frozen, in February of 2006, the two complementary tools were both in routine use: ManCalc and MDT. ManCalc, as a Microsoft Excel®-based model of impulsive maneuvers and inclination changes, has a quick turnaround and is able to compute all deterministic inefficiencies albeit with pre-evaluated assumptions about the effects of lunar perturbations on orbit elements. ManCalc is used for quick mission redesign scenarios (e.g., if the launch month were to change, delaying first

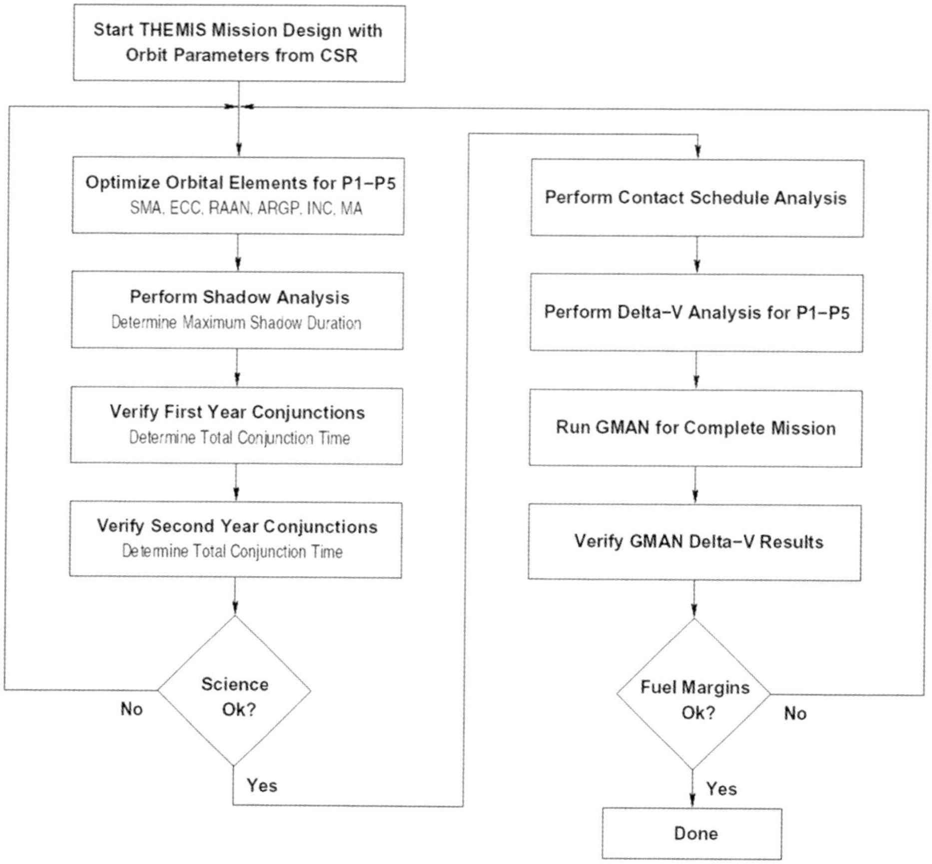

Fig. 4 Orbit design flow

prime season by one year, and extended mission concepts), for replacement scenario fuel tracking, and for accounting for deterministic inefficiencies. The input is either the mission deltaV directly from ManCalc or the input from MDT.

MDT, the GTDS and GMAN-based tool with IDL-wrappers is used for mission setup and long and short term maneuver planning. It is used for self-consistent and high-fidelity computation of conjunction hours subjected to shadow and maneuver duration limitations.

The results from the two tools are validated by frequently running identical mission profiles. Pressure, thrust, maneuver duration and deltaV curves are cross-checked and inconsistencies corrected or inconsistencies analyzed and approved as necessary. Additionally, following the Mission Critical Design Review, a mission design validation was performed over a period of 6 months by an independent team from JPL, and using JPL propagators an agreement was established. Independent validation of the P1 ascent, plane changes, and use of GTDS and GMAN, were accomplished in addition by GSFC. Finally, SatTrack (Bester et al. 2008) is used for processing mission operations-related output from MDT, i.e., for communications analysis, pass scheduling, orbit events analysis and 3D orbit visualization.

The orbit design flow is summarized in Fig. 4. Once orbital parameters set by impulsive maneuvers satisfy mission requirements, the design is reiterated with simulating finite tra-

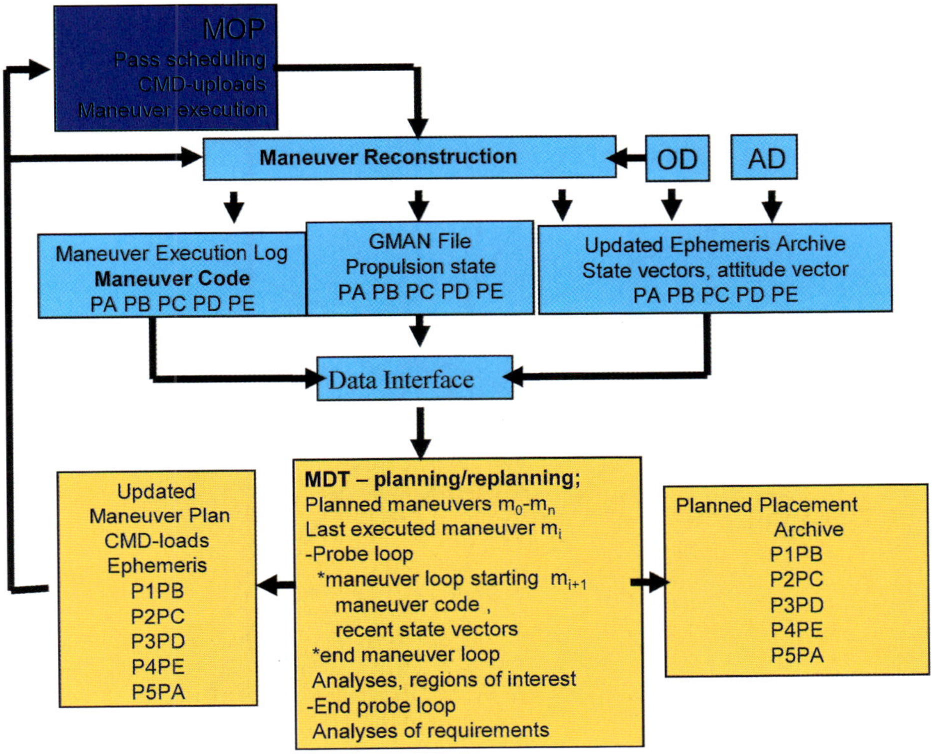

Fig. 5 Data flow between MDT, mission operations (MOP), and status updates by orbit determination (OD), as well as attitude determination (AD)

jectories, and analyzing step-by-step all operational aspects of the actual maneuvers (ground contacts, burn times, etc.). Since the MDT was developed to provide the flight-ready trajectories and maneuver command loads with finite burns end-to-end for any season, in a final step it was modified to allow maneuver re-planning as the mission is progressing. Figure 5 illustrates how the highly automated data flow is organized. Using one single interface, launch trajectories used in pre-launch mission planning are replaced by state vectors from the archive which is constantly updated after launch. The very last or any previous update is fetched by time reference. A maneuver code is the reference to properly parse the actual orbit data into the maneuver sequence (maneuver loop) for each probe. Human interference is focused on verification rather than data entry.

4 Mission Profile

4.1 Nominal Science Phases: Tail Seasons and Dayside Seasons

In a Sun reference frame (GSM or GSE coordinates) the THEMIS mission falls essentially into two observational seasons per year, the tail season with the apogee passes in the anti-Sun hemisphere, and the dayside season with apogee passes in the Sunward hemisphere (see Fig. 6). Center epochs, ideally defined as crossing the Sun–Earth line in the tail or on the

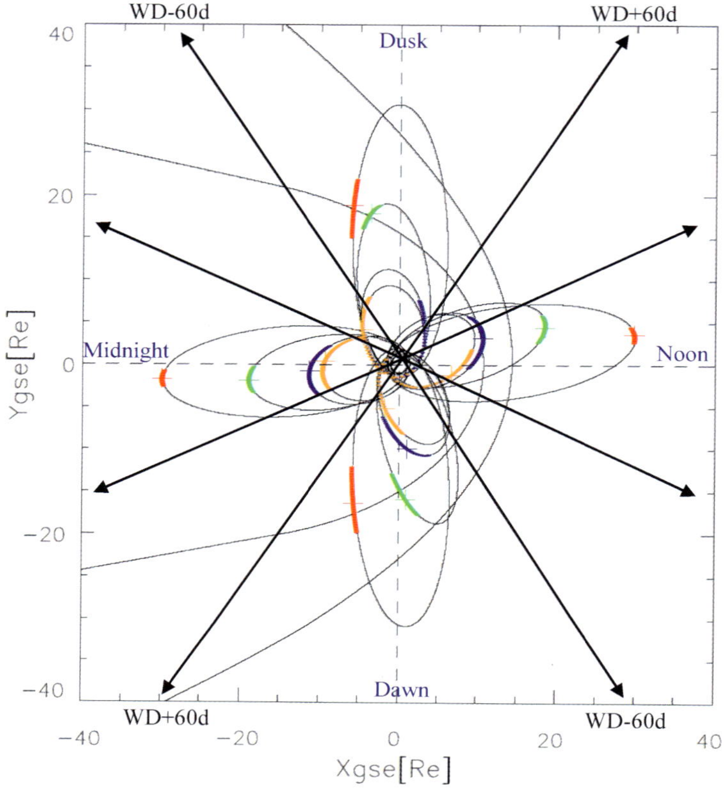

Fig. 6 The THEMIS orbits in Sun-referenced coordinates for the first year with magnetopause and bow shock. *Arrows* cut out the observational intervals according to the tweak schedule. *Colored orbit tracks* are 3-hour intervals centered at the WDs (midnight for tail season and noon for dayside) and at magnetopause crossings of P2

dayside, also referred to as Wedding Days (WD), are separated by about 6 months. Each tail season is confined to the time the orbits maintain conjunctions along the X-axis in the magnetotail. Between WD – 60 days and WD + 60 days, the outer probes will sufficiently monitor the reconnection zone while the inner probes pass through the current disruption zone. For the dayside, the 120 days centered at WD translate into a separation of the outer probes along the upstream solar wind direction from the magnetosheath, through foreshock into the pristine solar wind. In the first year, each season has about 10 to 14 WDs (about three P1 orbits) with more than 200 h of conjunctions. The actual WD is selected as the one with the highest return in the midnight interval and low deltaV. As Fig. 7 shows, once in their science orbits, the high return of conjunctions and the delta V fluctuate with the 4-day orbital period of P1.

Preparation for science operations was done during the checkout phase (early orbit operations, health and safety checks, probe constellation assignment), and the placement phase (a series of probe constellation specific maneuvers and the stepwise EFI boom deployments; for more details, see Bonnell et al. 2008 and Pankow et al. 2008). Instrument commissioning was spread over those phases at times when probes did not perform maneuvers.

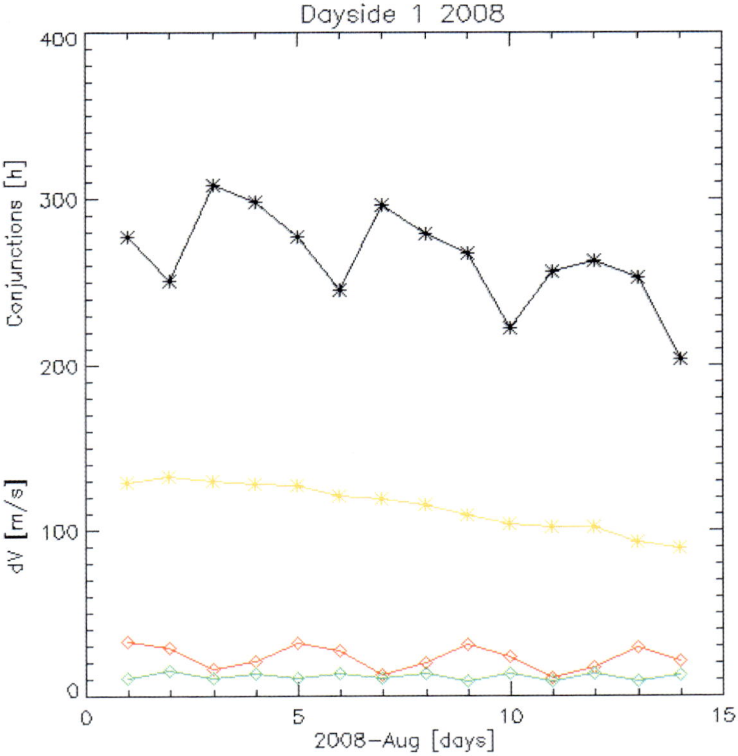

Fig. 7 Dayside 1 conjunctions (*black*) are shown for WDs August 1 to 14, in *yellow* conjunctions from the noon interval are shown. At the *bottom* the deltaV of all tweak maneuvers of this season is shown in *red* for P1 and *green* for P2

The small in-season maneuvers (tweak maneuvers) between WD – 60 days and WD + 60 days that realign the probes by small apogee adjustments to account for differential precession among inner and outer probes split each science season into three intervals (tail: dawn, midnight, dusk; dayside: dusk, noon, dawn) according to the precession in the Sun reference frame (see Fig. 6).

The schedule of the tweak maneuvers fixed relative to the WD was carefully designed to optimize conjunctions through the entire science season and minimize interruption of science data collection. This recurring timeline is:

WD – 88d, WD – 60d, WD – 24d, WD + 24d, WD + 60d. While the first one of each series is to re-align the probes after 28 days, the other three cut the entire science season into the three intervals centered around the WD, and WD + 60d marks the end of a season.

Once probes are successfully aligned for the first tail season, the tweak maneuver schedule, inner probe orbit maintenance, and P5 apogee variations are repeated for each following season. In order to comply with NASA requirements, after the nominal mission, we will be able to ensure re-entry of all probes by a series of maneuvers. Table 3 gives an overview of all phases of the entire mission.

The tweak maneuver schedule and the observational intervals for all four seasons are listed in Table 4.

Table 3 THEMIS nominal mission profile

Mission phase	Approx. time range	Periods P5->P1
Launch	Feb. 17, 2007	
Checkout	Feb., 2007–May, 2007	
Coast phase	May, 2007–Sep., 2007	String of Pearls, 31 h
Placement phase	Sep., 2007–Nov., 2007	
Tail Season1 observation	Dec., 2007–Apr., 2008	4./5d, 1d, 1d, 2d, 4d
Dayside 1 set up	Apr., 2008–May, 2008	
Dayside 1 observation	Jun., 2008–Oct.,2008	8./9.d, 1d, 1d, 2d, 4d
Tail Season 2 set up	Oct., 2008–Nov., 2008	
Tail Season 2 Observation	Dec., 2008–Apr., 2009	1d, 1d, 1d, 2d, 4d
Dayside 2 set up	Apr., 2009–May, 2009	
Dayside 2 observation	Jun., 2009–Oct., 2009	8./7.d, 1d, 1d, 2d, 4d
End of nominal mission	October, 2009	

Table 4 THEMIS tweak maneuver schedule split up into pre-season alignment and tweak intervals. First year data are based on in-flight data through the end of the inner probe setup for the dayside

Schedule	Season	Tail 1	Dayside 1	Tail 2	Dayside 2
WD –88d	Alignment	Nov. 06, 2007	May 08, 2008	Nov. 11, 2008	May 2009
WD –60d	1st Tweak	Dec. 04, 2007	Jun. 04, 2008	Dec. 09, 2008	Jun. 2009
	Observations	dawn	dusk	dawn	dusk
WD –24d	2nd Tweak	Jan. 09, 2008	Jul. 10, 2008	Jan. 14, 2009	Jul. 2009
	Observations	midnight	noon	midnight	noon
WD		Feb. 02, 2008	Aug. 03, 2008	Feb. 07, 2009	Aug. 2009
	Observations	midnight	noon	midnight	noon
WD +24d	3rd Tweak	Feb. 26, 2008	Aug. 27, 2008	Mar. 03,2009	Sep. 2009
	Observations	dusk	dawn	dusk	dawn
WD +60d	End of season	Apr. 02, 2008	Oct. 02, 2008	Apr. 08, 2009	Oct. 2009

4.2 Launch Days

Although the THEMIS mission is hinged to a fixed schedule once the tail center epoch (WD) is chosen, the design allows for launch on any day of the year. Any shift in launch day will be accommodated by an increase of RAAN to ensure proper orbit alignment in the magnetotail. However, as soon as the apogee distances of the outer orbits exceed about 20 R_E, their inclination rapidly drifts due to lunar forces. The effect is most severe for the outermost probe, with an inclination drift of about 0.1 deg/day. In order to keep the orbital plane in the required vicinity of the neutral sheet, the placement of the outer probes can only start within three month of the tail season. With the WD in early February, launches between August and October 2006 would have allowed all probes to assume their place for the first tail season in the upcoming winter of 2006/2007. For that short placement concept, the schedule for early orbit checkouts, placement maneuvers, and EFI deployment was kept flexible to ensure probe readiness for the first observational season. Launches after mid-October would have required cutting the tail season short by one tweak interval (running from WD – 24 d to WD + 60 d), but still would have provided >200 hrs of conjunctions. For any other launch day the placement of P1 and P2 needed to be delayed until early September. This in turn triggers the delay of the placement of the prospective inner probes to avoid the build-up of differential precession before the first primary season. In order to efficiently set inclinations, the launch trajectory also needed to be adjusted for APER drifts. Its inclination had to be chosen to minimize fuel consumptions for P1 and P5. Compliance with the re-entry guaranty for probe carrier also required adjustments of perigee and apogee altitudes as well as APER as a function of launch day.

Over the last 4 years milestone launch days have been the 19th of August 2006, the 19th of October 2006, the 27th of November 2006, the 20th of January 2007, and the 15th of February 2007, always meeting mission requirements within deltaV constraints. For the case of February 15, 2007, Fig. 8 shows the launch window analysis over 14 days. For each day, over the entire launch window of 20 minutes, conjunctions, and shadow durations at the end of the first tail season are well within the requirements, and the impulsive delta Vs accumulated by each probe (shown is P1 with the largest delta V account) through the first tail season remain well below the constraints by the fuel budget. Eventually, on February 17, 2007 the THEMIS probes were launched on a Delta-II 7925 from KSC and released quasi-simultaneously, setting the first primary science season for winter 2007/2008.

4.3 Coast Phase

The coast phase covers the period after early orbit check-out including constellation assignment until the start of the placement phase, during which all probes were on essentially the same orbit, with periods slightly dispersed by the release mechanism and thus drifting apart. The simultaneous launch of all probes in February, 2007 provided the opportunity to enhance the dayside science by maintaining a close formation of all five probes. While orbits drifted into the dayside, the focus was set on the consecutive magnetopause crossings by a string-of-pearls formation with three probes at small-scale inner separations and enclosed by a leading and trailing probe to provide large-scale context (Fig. 1 in Angelopoulos 2008, in this issue). The coast formation set up was primarily driven by the constraint to avoid additional fuel consumption that would compromise the primary mission. The release mechanism almost accomplished the string-of-pearls formation, except that not all inner probes would have had deployed spin plane booms of the electric field instrument. Before periods could be locked at the appropriate separations, the probes had to be rearranged from the original order of probes C <–> D, B, A <–> E to B <–> C, E, D <–> A, making small separations below 100 km over 10 to 20 orbits almost impossible under the given constraints. The reshuffle was choreographed as a sequence of maneuvers, one or more per probe, to have the coast phase formation fully in place in June, 2007 with an almost constant separation between outermost probes B and A of about 6000 km. In order to maintain the fuel budget, these maneuvers had to work towards the eventual perigee and apogee targets (Fig. 9). A common perigee target for the coast phase was needed to limit differential precession and was capped at 1.16 R_E, the final target of P2. Only for probe C, assigned to P2, which has the shortest ascent into its 2-day orbit, the fuel reserves were considered sufficient to allow for one small reversal of an apogee raise for the final phasing. Therefore, the phasing periods had to be accommodated by small apogee dispersions. Since time over which to set it up was also limited, the final coast phase periods for the inner probes were selected to allow the probes to pass each other slowly (Fig. 10). That way separations at magnetopause crossings below 500 km could be maintained for 60 to 90 days with shorter durations below 300 km and on some orbits with less than 100 km in June, 2007.

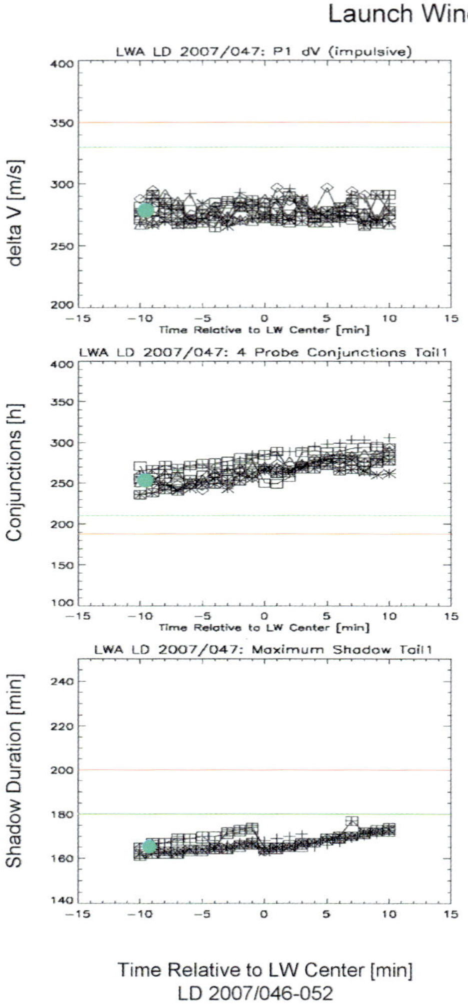

Fig. 8 THEMIS launch window analysis February 15 until 21, 2007 (*left panels*) and February 22 until 28, 2007 (*right panels*). *Upper panels* show the predicted accumulated impulsive delta V for P1 which has the largest delta V throughout the mission. Panels in the middle row show predicted accumulated conjunctions of four probes as required for the minimum mission after the first tail season. Panels at the *bottom* show predicted maximum shadow durations encountered by P1 (all other probes will experience less) during the period of peak shadows towards the end of the first tail season. Highlights are data from first season

4.4 Placement Phase

From September till November in 2007, the outer probes ascended and the inner probes descended independently into their science orbits in time for the first primary science season (tail season 1), targeting WD – 88 days for the alignment of the outer probes, with the inner probes properly locked with their apogee passes over the center of the GBOs in Central Canada (CCA). The challenge was to compromise between science targets, fuel efficiency, time constraints set by unavoidable orbital element drifts, time needed for post-maneuver

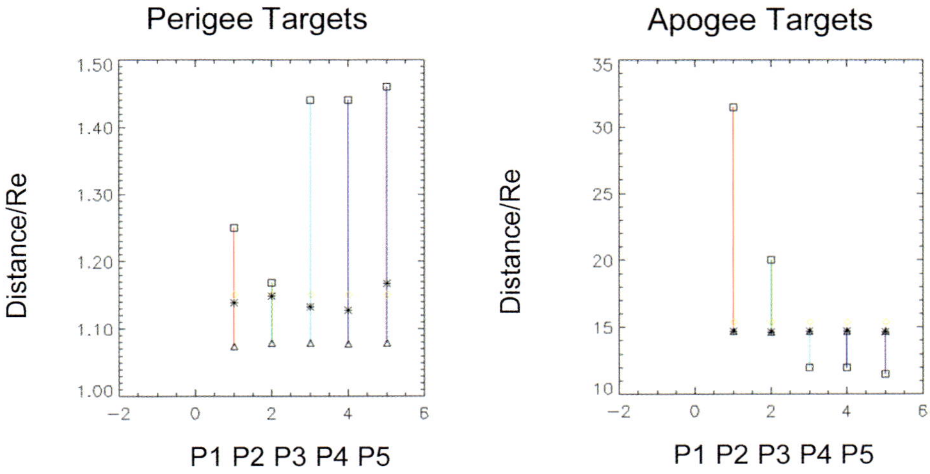

Fig. 9 Shown are perigee (*left*) and apogee (*right*) geocentric distances at probe release (Δ), during the coast phase (*), and at the start of science orbits (□) for all five probes

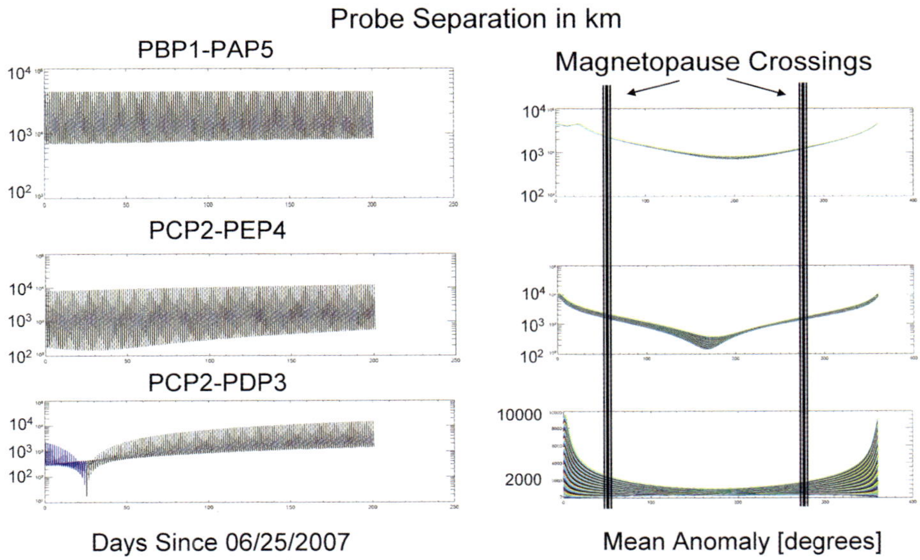

Fig. 10 Probe separation during coast phase; *left panels* show separation over time for leading to trailing probes (*top*) and pairs of inner probes (*middle*, *bottom*), *right panels* show separation over mean anomaly for same probe pairs. The change in color in the *bottom left* panel indicates passing

updates, time required for maneuver preparation, and the capability of the propulsion system with a yet flexible and robust maneuvering scenario.

The key to mission success is the center epoch (WD) of the first tail season. As it is fixed in time and space, any launch day is supported by a flexible ascent phase that leads into the fixed tweak maneuver schedule (Table 4). Based on the estimates of the orbital parameters from the ManCalc, the MDT determines the actual time and size of the ascent maneuvers for any given time span between launch day and center epoch, taking into account time slots for prescience phases and operational constraints such as orbit determination and maneuver planning. The ascent goal for the outer probes is to have their perigee passes for the 1st tweak maneuver around WD – 60 days, and at periods close to phasing period.

To reach those targets within a few weeks with a series of maneuvers per probe, the periods of the intermediate orbits determine the time of maneuvers according to (1),

$$t_{\text{final}} = t_{\text{begin}} + \sum_{i=l}^{k} n_i \cdot T_i \quad \text{with } T_k \approx T_{\text{align}} \tag{1}$$

where t_{final} is time target after k numbers of maneuvers, t_{begin} the beginning of the placement phase, and T_i periods of intermediate orbits as a result of either a perigee or apogee change. T_{align} is the period needed to maintain phasing.

The descent target for the inner probes is to place their apogee passes near the center longitude of the ground-based observatories (GBO's) at midnight. Similarly to (1), (2) relates the number of orbits with the drift of the apogee passes in geographic longitude (apogee drift),

$$lon_{\text{final}} = lon_{\text{begin}} + \sum_{j=p}^{q} m_j \cdot dlon_j \quad \text{with } dlon_q < 10 \text{ deg} \tag{2}$$

where lon_{begin} is the geographic longitude of the apogee pass at the start of placement phase and lon_{final} is the center of the GBO range. The drift rate of the last intermediate orbits determines the precision of meeting the target longitude as well as the separation between P3 and P4. The drift rate of apogee passes, $dlon_j$ is a function of period, with a value of zero for sidereal-day period. While apogee drifts of the intermediate orbits drive the time of maneuvers, the intermediate periods are constrained by the time limit for the inner probe placement,

$$t_{\text{final}}^{\text{inner}} = t_{\text{begin}} + \sum_{j=p}^{q} m_j \cdot T_j \quad \text{and } t_{\text{final}}^{\text{inner}} < t_{\text{final}}^{\text{outer}} \tag{3}$$

Figure 11 shows this for P3PD, where no apogee drift corresponds to the sidereal-day period. At about 50 days after launch it is set for the coast phase, drifting roughly 30 degrees per orbit. At the end of the placement phase, around 240 days after launch, the drift rates consecutively decrease. After tail season 1 the rise to the final perigee altitude and the reset to sidereal-day period can be seen starting around 410 days after launch. The order and times of the apogee and perigee changes have been set up to cause opposite drift rates in order to maintain alignment with CCA.

The number of maneuvers depends on the number of orbital parameters to be changed. For each such parameter, it is dictated by the size of the change of each orbital parameter between the starting point and the final target and the satellite system, as those steps usually have to be cut into pieces with feasible burn times. If a few maneuvers are necessary to reach a target, one would like to start out with big steps and continuously decrease of maneuver duration to reduce the effects of thrust variability. The feasibility of burn times depends on a variety of factors of which many are system-specific. From an operational point of view and in order to keep the placement phase as short as necessary, the fewer maneuvers, the better. For the THEMIS probes, the major burn time restrictions have been transmitter-on times since real time contacts are required during maneuvers, fuel efficiency and targeting over extended finite maneuver arcs. After launch, with in-flight evaluation, the original

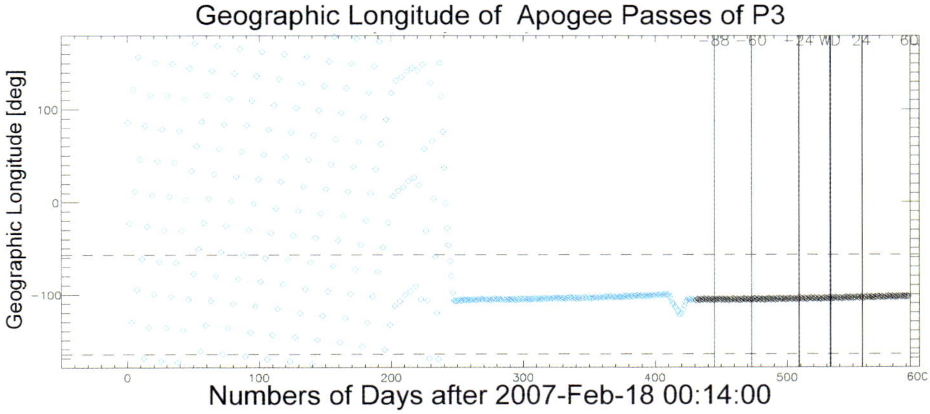

Fig. 11 Geographic longitudes of each apogee pass of P3PD is shown from launch through first year tail and dayside seasons. *Vertical lines* indicate dayside tweak schedule. Changes in apogee drift rates are due to intentional changes in period by maneuvers. *Horizontal dashed lines* frame the longitude range of the GBOs

constraint to 30-min transmitter-on time could be extended to more than 1 hour and fuel efficiency, targeting and operational aspects became the limiting factors. Once burn times have been established, the characteristics of the reaction control system (RCS) compete with constraints from (1) and (2) about maneuver size.

The THEMIS RCS (Sholl et al. 2007) is a complex design of a blow-down hydrazine system of two spherical tanks, a pyro-activated helium gas repressurization system, and four 4.5 N thrusters. Allocating 4 kg of fuel for attitude control, the fuel load at launch of 49 kg provides the equivalent of approximately 930 m/s of the total delta V capacity for each probe.

Two thrusters are located at the bottom of the probes for axial thrust, parallel to the spin axis, and the other two are located at the sides for thrusting tangentially to the spin plane (side thrust). The tangential thrusters are shown in Fig. 10 in Angelopoulos (2008). For orbit change maneuvers the axial thrusters are fired continuously, while the tangential thrusters are fired in a pulsed mode and the maneuver duration (referred to as burn time) becomes much longer than the actual thruster-on time depending on spin rate and pulse duration (for more on maneuver modes see Bester et al. 2008). Consequently, the two different sets of thrusters have a significant effect on maneuver size and thus number of maneuvers. While axial thrusts have much shorter burn times compared to same-deltaV side thrusts, they require large reorientations of the probes before and after the maneuvers. On THEMIS, those attitude changes become very costly as soon as the 40 and 50 m EFI wire booms are deployed and side thrust mode is the only feasible way to accommodate all orbit changes within the orbital plane increasing the number of maneuvers by a factor of two or even three.

In addition to thrust mode, the characteristic decrease in thrust with increasing fuel consumption of the blow-down system means significantly smaller and smaller maneuvers. Figure 12 shows how the blow-down system limits maneuver size for the side thrust mode after repressurization with a pulse width of 60 degrees. As we progress through the mission, the delta V of a 25-minute side thrust burn drops quickly from 20 m/s to 10 m/s, which is the range of the tweak maneuvers. The delta V of a continuous axial thrust of 25 minutes has dropped from 115 m/s to about 60 m/s after a total delta V of about 300 m/s (not shown). On THEMIS, the combined effect on burn time by the type of thrust and the fuel status were much larger than could be compensated by a small increase of burn time by a few

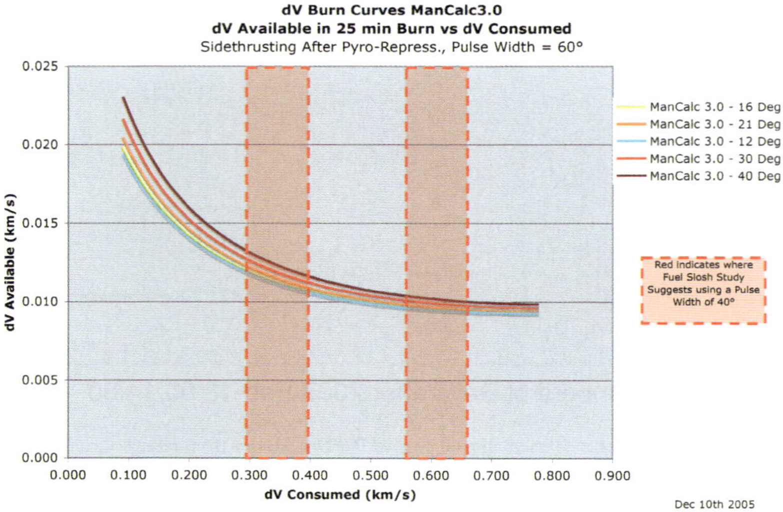

Fig. 12 Blow down curve for 25-minute side-thrust burns with a pulse width of 60 degrees for various temperatures over the THEMIS temperature range as a function of cumulative delta V. In *red* are zones where a pulse width of 40 degrees is recommended to avoid fuel sloshing

minutes. For a feasible and robust placement scenario and replanning capability, maneuvers are planned based on burn time rather than delta V using charts like Fig. 12 and a flexible number of maneuvers.

The RCS thrust performance is limited to pressures between 0.51 and 2.8 MPa. The time of the first recharge had to be chosen so that the system pressure is maintained within these limits. This added additional constraint of the size of maneuvers leading up to the first recharge. In fact, starting with large pressure drops and having to fit early maneuvers into the recharge window overwrote many of the guidelines based on orbital dynamics one typically follows to minimize the change of velocity for each maneuver.

Last but not least, the placement maneuver concept was kept very flexible and robust to ensure a high level of fault tolerance. Maneuvers have been designed smaller towards the final placement of each parameter in order to be able to account for underperformance of a previous maneuver without the penalties of very large finite arcs with the exception of P3 and P4. For them, the fully deployed EFI booms increased the number of maneuvers so much that maintaining the time target for their placement pushed burn times to the upper limit set by finite arc losses. Whenever possible, the time between maneuvers was increasing towards the final targeting, and placeholders for each final target were also part of the nominal schedule for each probe. This way, short-term rescheduling of individual maneuvers was possible without impact on the final placements.

During the placement phase, 42 orbit change maneuvers, not counting attitude and spin rate changes, have been performed to bring each of the five THEMIS probes into its science orbits well in time. For more details about THEMIS mission operations see Bester et al. (2008). After each maneuver, orbit and attitude updates as well as results from maneuver reconstruction were fed into the MDT to evaluate mission criteria and to adjust the remaining maneuver targets and maneuver times following (1) and (2). For each probe, the actual status was compared with predictions. For the outer probes, deviations in maneuver time for the next maneuver were converted in to period adjustments for intermediate orbits following all

Table 5 Numbers of executed placement maneuvers from launch through first tail season

Numbers of maneuvers	P1PB	P2PC	P3PD	P4PE	P5PA
Coast phase	3	6	2	4	4
Placement	8	10	9	9	6
Tail 1 tweaks	3	3	–	–	–
Total after tail 1	14	19	11	12	10

remaining apogee changes. For the inner probes, the number of orbits with small apogee drift rates was adjusted. Each such run always processed all five probes and quite often, while maneuvers on more than one probe were up for maneuver preparation, all previous maneuvers have been evaluated as well.

The mission-significant orbital elements since launch and through the first tail season are shown in Fig. 13. The left column characterizes the orientation of the orbital plane and the right column monitors orbit size. Changes in period can be traced as either apogee or perigee changes. Differential precession, shown as differences in RAP, starts as soon as the outer probes have started their ascent. Maneuvers are indicated by an almost instantaneous jump, though these data are based on finite burns. All inclination changes are probe-specific and are done at near-zero APER for fuel efficiency, verifying that the offset of –10 deg at launch is sufficient to account for the checkout phase.

4.5 Orbit Maintenance

The focus of orbital maintenance is to keep the constellation within science requirements. Per-probe amount and purpose differs according to the different orbits and strategic roles. Once brought into their science orbit with a sidereal-day period, the two inner probes P3, P4 need to maintain conjunction over CCA and might have to readjust for small apogee drifts and reset to sidereal-day period. Prior to each season this can be achieved by two or three small maneuvers.

The outer probes P1, P2 need to maintain their conjunctions with the inner probes P3 and P4 as well as the neutral sheet in the magnetotail. During each season, the set of four tweak maneuvers (see Table 4) will account for differential precession by small changes of the apogees. The offset in the GSM-Y-component between the center of the P3 and P4 apogees and the position of P1, P2, respectively at the end of each such interval is converted into a time offset, which is then spread over all orbits in that interval as a change to orbital period as expressed in (4)

$$T_{\text{new}}^{\text{outer}} = T_{\text{old}}^{\text{outer}} + dT \quad \text{with } dT = (t_{P3,4} - t_{P1(2)})/n \text{ and } T_{\text{old}}^{\text{outer}} \approx T_{\text{phase}} \tag{4}$$

where $t_{P3,4}$ is the time of center of apogee passes of P3 and P4 and $t_{P1(2)}$ is the time where $Y_{P1(2)} = (Y_{P3} + Y_{P4})/2$ and n is the number of orbits of P1, (P2) during that interval. Independently for P1 and P2, the correction dT is determined in an iterative process, invoking high fidelity orbit propagation until the GSM-y components of inner and outer probes are within the science criteria. Figure 14 shows the variations in period for all probes for the remaining three seasons, with the tweak schedule overlaid. The variation for the P1 period is in the order of 2 to 4 hours per tweak, and for P2 it is in the order of 20 to 30 minutes.

The distance to the neutral sheet in the tail is set by APER and inclination. Both are heavily perturbed by lunar forces as seen in Fig. 13. The flip in APER as soon as the inclination is very low (for P2 happening during the following dayside) must be reversed for the

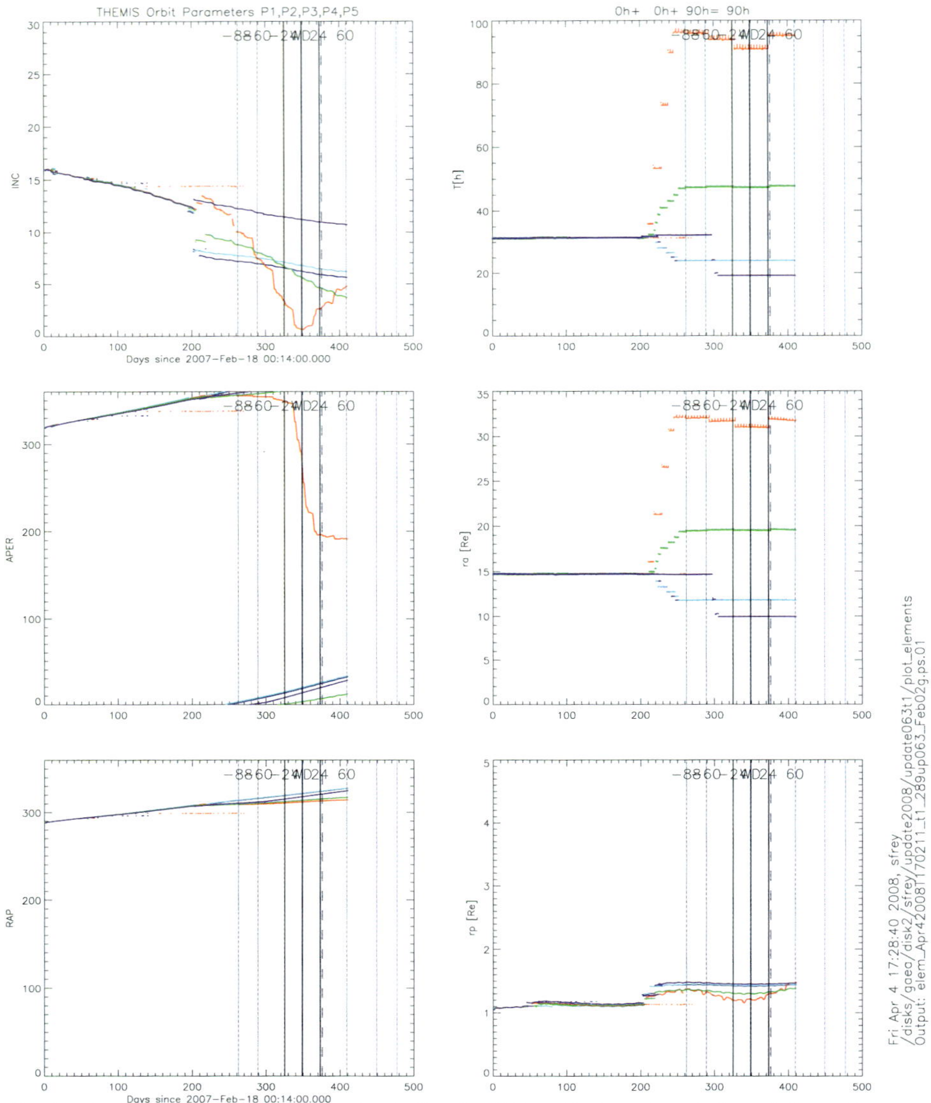

Fig. 13 Evolution of orbital elements for all five THEMIS orbits from launch through first tail season. *Left column* from top shows inclination, APER, RAP, all in degrees, *right column* from top shows period, geocentric apogee distance, geocentric perigee distance, not shown are eccentricity and RAAN. *Black vertical lines* mark the tail season schedule of WD + [−88, −60, −24, 0, 24, 60] days, the *dashed line* marks mission elapsed days and data left of it are definitive

outer probes prior to the second year tail season in order to keep probe conjunctions near the neutral sheet and also shadows below 3 hours. As discussed earlier, for fuel efficiency, this inclination change will be combined with a perigee reduction to support end-of-mission maneuvers.

The fifth probe, once relieved from its replacement role, undergoes larger changes in orbital period, aiming to greatly enhance science data in the current disruption zone as well

Table 6 Projected numbers of all maneuvers through two years of the mission

Numbers of maneuvers	P1PB	P2PC	P3PD	P4PE	P5PA
Total after tail season 1	14	19	11	12	10
Dayside 1 set up	–	–	5	5	6
Dayside 1 tweaks	4	4	–	–	–
Tail season 2 set up	2	2	3	3	6
Tail season 2 tweaks	4	4	–	–	–
Dayside 2 set up	–	–	3	3	6
Dayside 2 tweaks	4	4	–	–	–
Total after 2 years	28	23	22	23	28

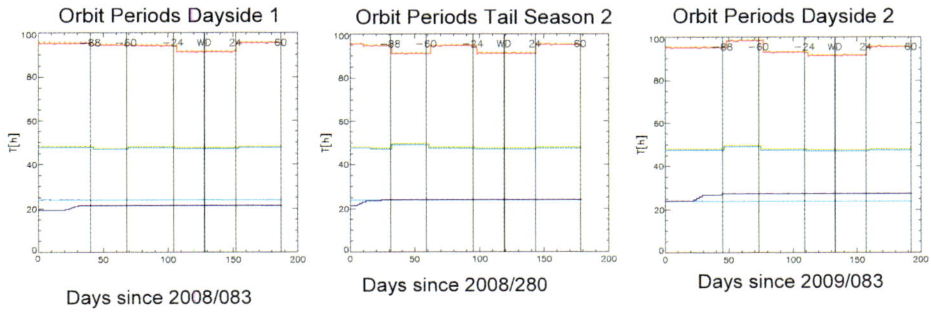

Fig. 14 Orbital periods for all five probes through seasons after first tail season. *Vertical lines* mark tweak maneuver schedule

as around magnetopause crossings in the dayside as outlined in Angelopoulos, (2008) in this issue. Figure 14 shows how P5 period goes from 4/5 of sidereal-day period in the first tail season, to 8/9 of sidereal-day period in the first dayside, to sidereal-day period in the second tail season, and is finishing off with 8/7 in the second dayside, always changing apogee.

5 Assessment of Mission Requirements

5.1 Conjunctions

The challenge of the THEMIS mission is to catch the various components of substorm events simultaneously at high time resolution with only 5 probes and a partial coverage of the auroral oval. The complexity of the tasks is broken down into a multitude of baseline science requirements in Angelopoulos (2008) in this issue and is listed in Table IV therein. For orbit design and mission planning these requirements are summarized into much simpler conjunction criteria and only probe conjunctions that fulfill those criteria are counted towards the accumulated conjunctions of four probes of at least 188 hours per season. This allows efficient, predictive evaluation of orbit solutions as an integrated part of the planning process at any stage. Likewise integrated are assessments of eclipse durations as well as the total of all velocity changes and the equivalent in fuel consumption for each probe throughout the mission.

The simplified conjunction criteria are that 1) inter probe separations of the GSM-y coordinates be within two R_E, 2) and in tail seasons only, the distance between the neutral sheet be within two R_E for the inner probes and within five R_E for the outer probes, and 3) event times during the 12 hours are centered at 6:30 UT for tail seasons and 18:30 UT for day sides. For the midnight interval of the first season, Fig. 15 shows where along the orbits conjunctions occur. The thick black markings in the left plot, showing all orbits in the GSM-XY plane, are those instances where all three criteria are met. Showing the distance to the neutral sheet along the GSM-X component for the corresponding black tracks from above, the right plot assesses where conjunctions are cut off by the neutral sheet distance criteria. Predicting the neutral sheet is rather difficult, as the transition from the magnetic equator into the bended neutral sheet varies in position and angle with dipole tilt, magnetic activity, and solar wind pressure. Current research relies on models that yet need more data-based verification and improvements. For our integrated routine analysis, we composed a neutral sheet model (THEMIS) that works for inner as well as outer probes and is focused on the magnetospheric dynamics during substorm onset in the winter season. For the outer probes where the separation becomes crucial further out in the near-tail, we apply the Hammond modification of the Fairfield model (Hammond model), Hammond et al. (1994) which is optimized for near-tail distances outside the transition region between -15 and -35 R_E. On the near-Earth side the Hammond model is not applicable and we replace it with the magnetic equator for the inner probes. A comparison with an alternative global magnetospheric data-based model optimized for the near magnetosphere inside -15 R_E (T96), Tsyganenko (1995), is shown in Fig. 16 where the upper plot shows the neutral sheet models corresponding to an inner (P3) and an outer (P1) orbit at WD. Both models agree reasonably at the tail distances, while the T96 models the inner neutral sheet fairly well by adjusting corresponding parameters to measurements of the actual magnetospheric condition. However, for our predictive analysis the lower end of the range rather than the exact position of the bend are of concern regarding the inner probes. As the lower plot indicates, for that purpose the magnetic equator is a good estimate and justifies the simpler model to avoid time-consuming computations. As conjunction instances for the outer probes are outside -10 R_E, the limits of the Hammond model through the transition region can be neglected. Furthermore, the lower plot, comparing the probe distance to either model, shows that differences between the models do not affect the evaluation of the z-criteria. Whether the chosen criteria are sufficient will be assessed by data analysis during tail season 1. Figure 17 gives an overview of the 4-probe conjunctions through the entire first tail season with time running from right to left. At WD – 60 days, orbits start to enter the magnetotail and conjunctions are happening at the outbound flank. Approaching WD conjunctions move to center around apogees. Most

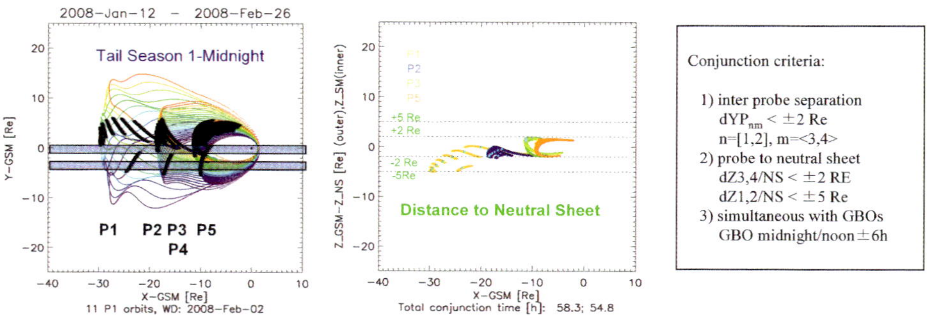

Fig. 15 Illustration of conjunction criteria for conjunctions of all five probes during the midnight interval. *Left plot* shows in *blue* the 2 R_E of inter probe separation for a given orbit parallel to the Sun–Earth line. Instances that meet all three criteria are marked in *black*

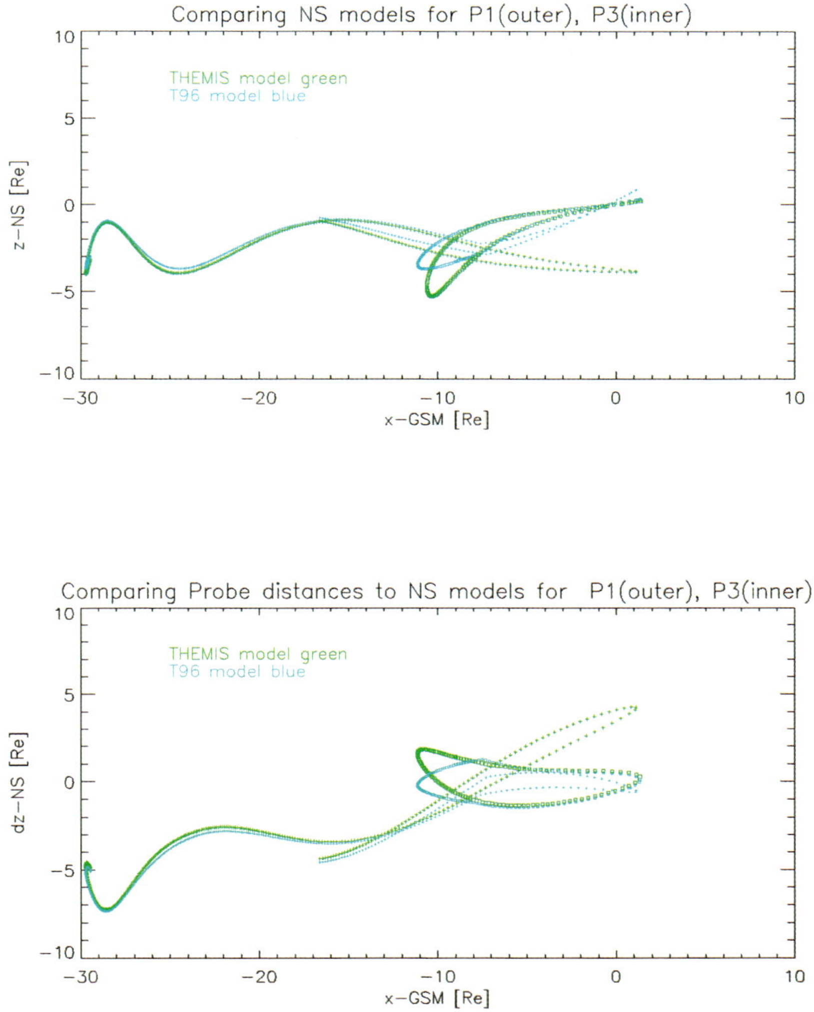

Fig. 16 *Upper plot*, neutral sheet models are compared for P1 (*crosses*) and P3 (*squares*) orbit on WD; *lower plot*, probe distances to neutral sheet models around WD

likely reconnection zones are best bracketed during the midnight interval (WD ± 24 days), where on the other hand, the actual waving of the neutral sheet becomes the limiting factor. Past WD conjunctions move towards the inbound flank and as the plots at the bottom show, all orbits are almost embedded in the neutral sheet near vernal equinox.

5.2 Shadow Duration

Driven by power and thermal properties, the upper limit for total shadows is three hours. Post launch experience have found this to be a rather conservative limit, thus easing maneuver target constraints. Typical eclipses encountered on the THEMIS orbits are either short umbrae near perigee or long shadows on the inbound flank in spring. Partial lunar shadows

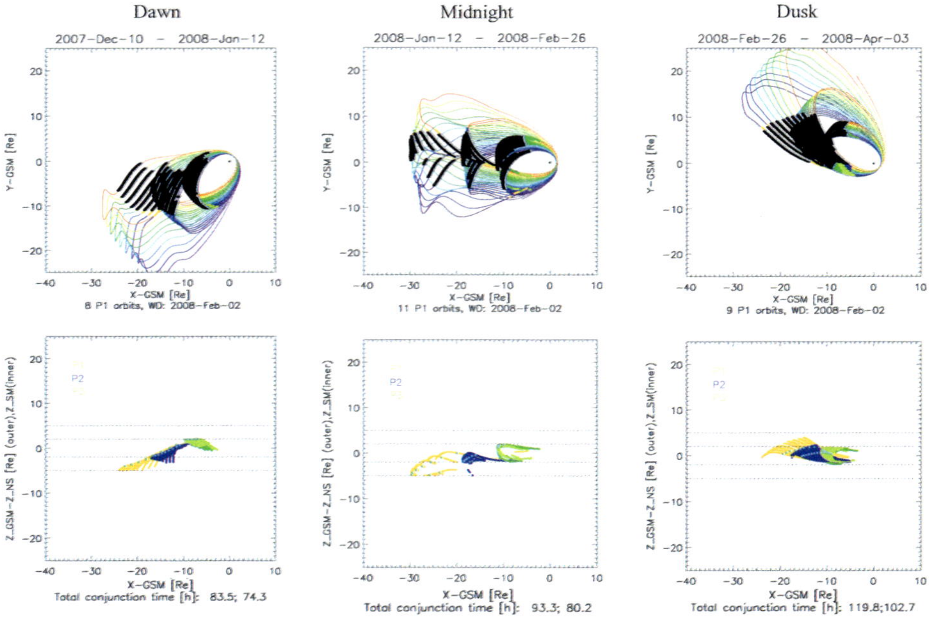

Fig. 17 Conjunctions of four probes for each interval. *Upper panels* show orbits projected into the GSM-XY plane; conjunction instances are marked in *black*, *lower panel* shows distance to the neutral sheet only for conjunction instances; *dotted lines* frame the dZ-limits

are scattered sporadically around the orbit, lasting from only a few minutes to 2 hours with varying depth. Hence, shadow analysis includes lunar and Earth shadows. For the predictive routine analysis with the MDT, which checks for upper limits of eclipse duration not to exceed 3 hours and possible overlaps with maneuver times, shadows are modeled as cones, including entry and exit of the penumbra using extended spherical objects. For applications on modeling thermal properties and for final product data generation in SatTrack, a more sophisticated shadow analysis is run that includes atmospheric effects, geoid approximation and defines entry and exit at 99% of full sunlight. This method allows determination of the depth for individual shadows and is employed for case studies such as long lunar partial shadows. The inclination is the key parameter to keep shadows below the 3-hour limit in both tail seasons and is set prior to the first tail season. In addition, the outer probes need to reverse the lunar effect on inclination and argument of perigee at the start of the second tail season. Since the inclination target of the inner probes is driven by the 2nd year, their placement maneuvers are based on verification by the shadow analysis of the 2nd year. The outer probes, P1 more than P2, derive the inclinations for each season separately to compromise between conjunctions and shadow length. Without shadow avoidance maneuvers in spring of the 3rd year, shadows would dramatically increase for the outer probes and can become of critical duration for the inner probes.

5.3 Total delta V Budget

THEMIS is a very active mission, changing orbits significantly and often during mission lifetime and constantly monitoring the delta V budget becomes essential. The largest contributors are of course the placement phase and the launch trajectory as a starting point. By

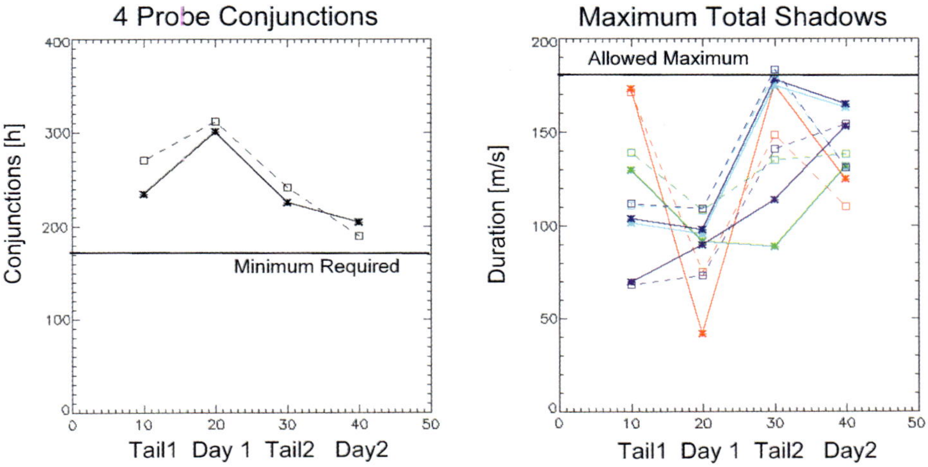

Fig. 18 Comparing evolution of mission requirements from launch through first tail season and predictions for the remaining seasons, the plot on the *left side* shows 4-probe conjunctions, the one on the *right side* shadow durations, pre-launch data (*dashed*) post-launch data with executed maneuvers until 2007-May-18 (*solid line*)

raising the launch trajectory from 12 to 14 R_E, the delta V needed for the placement is rather well equally split over inner and outer probes as inner probes lower their apogees and outer probes raise their apogees. While inner probes need more delta V for possible re-entry maneuvers, the outer probes will have to use that reserve to counter lunar perturbations. The delta V is checked in two ways as velocity change of impulsive maneuvers only and as the sum of velocity change through the finite arc maneuver and the equivalent imparted by probe reorientation and/or spin rate changes. The contributions from orbit changes, probe reorientation, and spin rate changes are also recorded separately in order to assess predictions. The total fuel allocated was based on the sum of all velocity changes from finite arc maneuver modeling, additional inefficiencies due to thruster alignments and all reorientations and spin rate changes, including those during boom deployments, plus the required 15% margin at launch.

6 Summary

Prior to launch the orbit design provided solutions for a wide range of conditions such as launch days or launch vehicle dispersion, but was centered at nominal targets, whereas ever since launch, the orbit design could be optimized for the actual launch trajectory with finalizing the maneuvers leading up to the first repressurization, based on the best WD for the first tail season. The seasons following the first tail season have frequently been updated based on in-flight data. The evaluation of the three main mission requirements, conjunctions, shadow duration, and delta V budget since launch and maneuvers through the set-up of the inner probes for the first dayside, shown in Figs. 18 and 19 and listed in Table 7, confirms the orbit design strategy.

The THEMIS orbit design and its realization is very complex and challenging in many ways and has been successfully put to test. Since launch in February 2007, we retrieved excellent science data during the coast phase and the first tail season.

Table 7 Overview of mission requirements, conjunctions are split into the three intervals. Data for the first year tail season after all maneuvers and for the first dayside after inner probe setup. Data for second year based on current estimate of WDs; delta V is adjusted to account for ACS fuel usage

Season	WD	Conjunctions [h] (4probes)	Max. shadow [min]					dV [m/s]				
			P1	P2	P3	P4	P5	P1	P2	P3	P4	P5
Tail 1	02-02-2008	$72 + 77 + 102 = 251$	166	113	109	114	79	373	287	307	300	345
Day 1	08-03-2008	$80 + 120 + 95 = 304$	63	97	108	106	80	392	299	389	378	396
Tail 2	02-07-2009	$57 + 96 + 85 = 238$	159	130	180	180	168	705	552	390	377	448
Day 2	08-09-2009	$33 + 99 + 93 = 225$	129	137	164	165	158	731	568	394	381	498

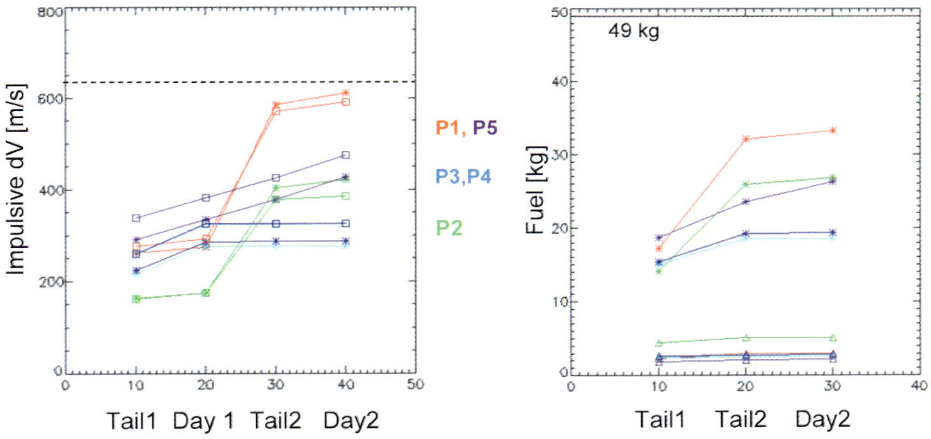

Fig. 19 Comparing the evolution of mission requirements from launch through first tail season and predictions for the remaining seasons, using post-launch data with executed maneuvers until 2007-May-18 (*). Accumulated delta V per probe is shown on the *left side*, *squares mark* pre-launch data. On the *right side* the accumulated fuel usage is shown, *triangles* are ACS fuel usage

The ambitious series of maneuvers during the placement phase went entirely according to plan, and on time. We are entering the first dayside season with all inner probes well in place.

Acknowledgement The THEMIS mission is funded by NASA contract NAS5-02099. Microsoft Excel® is a registered trademark of the Microsoft group of companies.

References

V. Angelopoulos, The THEMIS mission. Space Sci. Rev. (2008, this issue). doi:10.1007/s11214-008-9336-1

K. Berry, Orbital decay analysis for THEMIS at GSFC (2005)

M. Bester et al., Space Sci. Rev. (2008, this issue)

J. Bonnell et al., Space Sci. Rev. (2008, this issue)

C.M. Hammond et al., Imaging the effect of dipole tilt on magnetotail boundaries. J. Geophys. Res. **99**, 6079 (1994)

S.E. Harris et al., THEMIS ground based observatory system design. Space Sci. Rev. (2008, this issue)

S.B. Mende et al., The THEMIS array of ground-based observatories for the study of auroral substorms. Space Sci. Rev. (2008, this issue). doi:10.1007/s11214-008-9380-x

D. Pankow et al., Space Sci. Rev. (2008, this issue). doi:10.1007/s11214-008-9386-4

C.T. Russell et al., Space Sci. Rev. (2008, this issue). doi:10.1007/s11214-008-9337-0

D.G. Sibeck et al., Space Sci. Rev. (2008, this issue). doi:10.1007/s11214-008-9393-5

M. Sholl, M. Leeds, J. Holbrook, THEMIS reaction control System—From I&T through early orbit operations, in 43rd AIAA/ASME/SAE/ASEE Joint Propulsion Conference & Exhibit, Cincinnati, OH, July 8–11, 2007

N.A. Tsyganenko, Modeling the Earth's magnetospheric magnetic field confined within a realistic magnetopause. J. Geophys. Res. **100**, 5599 (1995)

D.A. Vallado, *Fundamentals of Astronomics and Applications*. Space Technology Series (McGraw-Hill, New York, 1997)

J.R. Wertz, *Mission Geometry; Orbit and Constellation Design and Management, Space Technology Library*. (Microcosm Press and Kluwer Academic, Dordrecht, 2001)

THEMIS Operations

M. Bester · M. Lewis · B. Roberts · J. McDonald · D. Pease · J. Thorsness · S. Frey · D. Cosgrove · D. Rummel

Originally published in the journal Space Science Reviews, Volume 141, Nos 1–4, 91–115.
DOI: 10.1007/s11214-008-9456-7

Abstract THEMIS—a five-spacecraft constellation to study magnetospheric events leading to auroral outbursts—launched on February 17, 2007. All aspects of operations are conducted at the Mission Operations Center at the University of California at Berkeley. Activities of the multi-mission operations team include mission and science operations, flight dynamics and ground station operations. Communications with the constellation are primarily established via the Berkeley Ground Station, while NASA's Ground Network provides secondary pass coverage. In addition, NASA's Space Network supports maneuver operations near perigee. Following a successful launch campaign, the operations team performed on-orbit probe bus and instrument check-out and commissioning tasks, and placed the constellation initially into a coast phase orbit configuration to control orbit dispersion and conduct initial science operations during the summer of 2007. Mission orbit placement was completed in the fall of 2007, in time for the first winter observing season in the Earth's magnetospheric tail. Over the course of the first 18 months of on-orbit constellation operations, procedures for instrument configuration, science data acquisition and navigation were refined, and software systems were enhanced. Overall, the implemented ground systems at the Mission Operations Center proved to be very successful and completely adequate to support reliable and efficient constellation operations. A high degree of systems automation is employed to support lights-out operations during off-hours.

Keywords THEMIS · Satellite constellation · Satellite ground systems · Satellite tracking · Satellite navigation · Mission operations · Flight operations

1 Introduction

The *Time History of Events and Macroscale Interactions during Substorms* (THEMIS) mission is a National Aeronautics and Space Administration (NASA) Medium-class Explorer

M. Bester (✉) · M. Lewis · B. Roberts · J. McDonald · D. Pease · J. Thorsness · S. Frey · D. Cosgrove · D. Rummel
Space Sciences Laboratory, University of California, Berkeley, CA 94720-7450, USA
e-mail: mbester@ssl.berkeley.edu

(MIDEX) mission to study magnetospheric events leading to auroral outbursts (Angelopoulos 2008). The space segment consists of five small, identical spacecraft called *probes*, each carrying a suite of five science instruments. The probes were launched on February 17, 2007 on a single Delta II launch vehicle from Cape Canaveral Air Force Station (CCAFS) into a highly elliptical insertion orbit with an orbital period of 31.4 hours at an inclination of 16 deg. Significant magnetospheric science observations were already made in March 2007, only little more than one month after launch. Following an initial 30-day on-orbit check-out and science instrument commissioning period, the probes were placed into nearly identical, temporary coast phase orbits to control orbital dispersions. Mission orbit placement in preparation for the first winter observing season commenced in early September 2007, and the constellation was fully deployed by mid January 2008 (Bester et al. 2008; Frey et al. 2008).

THEMIS is NASA's first scientific constellation mission. This paper describes aspects of mission operations conducted by the University of California at Berkeley's Space Sciences Laboratory (UCB/SSL), and covers ground systems, operational software tools, navigation, planning of science observations and data recovery.

1.1 Concept of Operations

The concept of operations for the THEMIS constellation involves launching five small, identical spacecraft into low-inclination, highly elliptical Earth orbits with harmonic orbital periods, aligned in such a way that orbital conjunctions occur periodically within the magnetospheric tail of the Earth. During these conjunctions, measurements of electric and magnetic fields as well as distributions of plasma particle fluxes are made to study events leading to auroral outbursts (Angelopoulos 2008).

The THEMIS probes are robust, spin-stabilized instrument platforms with nominal operational spin rates of 20 rpm. During normal science operations, the probes are oriented such that their spin axes point towards either the ecliptic north or south pole, providing a stable and safe power and thermal environment. Monopropellant hydrazine propulsion systems are used for orbit and attitude control. Each probe carries an identical suite of five science instruments comprising two magnetometers, two particle detectors to measure the energy distribution of electrons and ions, and an electric field instrument (Angelopoulos 2008).

For communications at S-band, ten telemetry data rates allow for high-rate data recovery near perigee and closing of the telemetry and command links with 11-m class ground antennas out to the farthest apogee at a range of 200,000 km. All probes share the same radio frequencies, but have unique spacecraft identifiers that are hard-coded in each probe bus. Telemetry and command frame formats are compatible with the Consultative Committee for Space Data Systems (CCSDS) Version 1 standard. Telemetry links employ concatenated Reed-Solomon and rate-1/2 convolutional coding with Viterbi decoding for error correction. In addition to seven ground stations, special operations near perigee are also supported by NASA's Tracking and Data Relay Satellite System (TDRSS), a.k.a. the Space Network (SN). Orbit determination is based on two-way Doppler tracking, and attitude determination on sun sensor and three-axis magnetometer data.

The THEMIS constellation operates in store-and-forward mode. Science and engineering data are recorded in on-board solid-state memory and are recovered primarily near perigee at the highest data rate compatible with the predicted link margin for any given pass. Instrument configuration for science data acquisition is based on modeled crossing times of magnetospheric regions of interest in combination with various on-board trigger algorithms (Frey et al. 2008; Taylor et al. 2008).

All aspects of THEMIS constellation management and operations are performed at the Mission Operations Center (MOC) at UCB/SSL and include mission and science operations, flight dynamics and ground station operations. A high degree of automation and autonomy is achieved using a number of novel software tools that are integrated into a coherent ground system to perform all required operations functions. These tools are discussed in more detail further below and support routine task execution, flight dynamics products generation, pass support and ground station operations, networking, telemetry processing and archiving, and spacecraft limit monitoring with error detection and operator notification. Virtually all state-of-health monitoring, tracking and data recovery passes are conducted in lights-out mode.

1.2 Mission Timeline

An overview of the THEMIS mission timeline, beginning with launch on February 17, 2007, is shown in Fig. 1. Since all five probes were released into nearly identical insertion orbits and with identical configurations and fuel loads, any probe could in principle be placed into any of the five constellation orbits. Following on-orbit check-out and detailed characterization of all probe buses and science instruments, the probe placement decision was made to determine which probe was to be maneuvered into which of the five mission orbits. This decision was based on knowledge gained from pre-launch testing and on-orbit performance during the first 38 days. Once the probe placement decision was made, it was clear as to which probes could deploy their Electric Field Instrument (EFI) and which ones had to keep their wire booms stowed to maintain a low moment of inertia required to perform a series of efficient attitude and ΔV maneuvers in order to complete mission orbit placement. Details are discussed further below in Sect. 4.

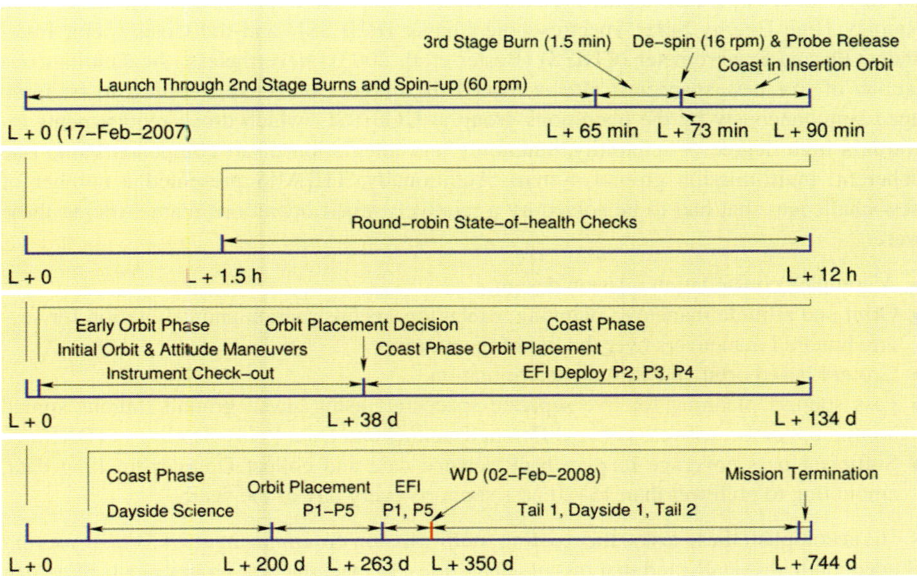

Fig. 1 THEMIS mission timeline from launch to nominal mission termination. The five probes are designated by their mission orbits as P1–P5. Deployment of the Electric Field Instrument (EFI) on each probe affects the moment of inertia and hence the maneuvering capabilities, and is therefore tied to the orbit placement sequence. Wedding Day (WD) of the first tail observing season falls on February 2, 2008

As a result of several launch slips from August 2006 to February 2007, a coast phase was inserted into the timeline in order to control orbital dispersion prior to mission orbit placement in the fall of 2007 (Frey et al. 2008). The coast phase also allowed for two months of additional science observations with the five probes arranged in a string-of-pearls configuration. Next in the timeline was the mission orbit placement campaign, followed by the first tail season (T1), the first dayside season (D1), the second tail season (T2) and an anticipated, nominal mission termination in early 2009. For the purpose of designing the mission orbits and their relative alignment in the presence of Earth and lunar gravity perturbations, the concept of using a reference date called *Wedding Day* (WD) was introduced for each observing season (Frey et al. 2008). As an example, Wedding Day of the first tail season was defined as February 2, 2008.

2 Ground Systems

Ground systems supporting the THEMIS constellation include the Mission and Science Operations Centers (MOC/SOC), the Flight Dynamics Center (FDC) and the primary Berkeley Ground Station (BGS), all co-located at UCB/SSL, plus a number of external elements, such as the secondary Ground Network (GN) stations and the Space Network (SN) with their interconnecting network links.

2.1 Ground System Elements

From the earliest stage of development, THEMIS flight operations were able to take advantage of much of the existing ground system architecture already developed to operate three other NASA missions, the Fast Auroral Snapshot Explorer (FAST), the Reuven Ramaty High Energy Solar Spectroscopic Imager (RHESSI) and the Cosmic Hot Interstellar Plasma Spectrometer (CHIPS) (Bester et al. 2003). Nevertheless, the simultaneous launch of five new spacecraft almost tripled the number of active satellites to be managed simultaneously by the operations group at UCB/SSL, which drove requirements towards a high degree of reliability, autonomy and integration of all components into one coherent, multi-mission ground system. Additionally, THEMIS presented a number of new challenges that had to be solved by a relatively small operations team. Among these were:

- A complex constellation mission design.
- Orbit and attitude maneuver planning, execution, reconstruction and calibration for several hundred maneuvers over the life of the mission.
- Ground-based orbit and attitude determination.
- Pass support planning for five separate spacecraft using seven ground stations spread around the globe, as well as five different Tracking and Data Relay Satellites (TDRS).
- Sufficient pass coverage to play back science data and collect Doppler tracking data, amounting to no fewer than 15–20 pass supports every day of the year.

To accomplish these tasks, the existing multi-mission environment at UCB/SSL was expanded, both physically and in terms of capabilities, to meet the new requirements related to operating a satellite constellation. All of the already integrated software tools were reused, while a number of new tools were added, particularly for handling the complex demands for mission design and navigation. These new tools are described in more detail in the following sections. A block diagram of the functional elements of the THEMIS ground system is

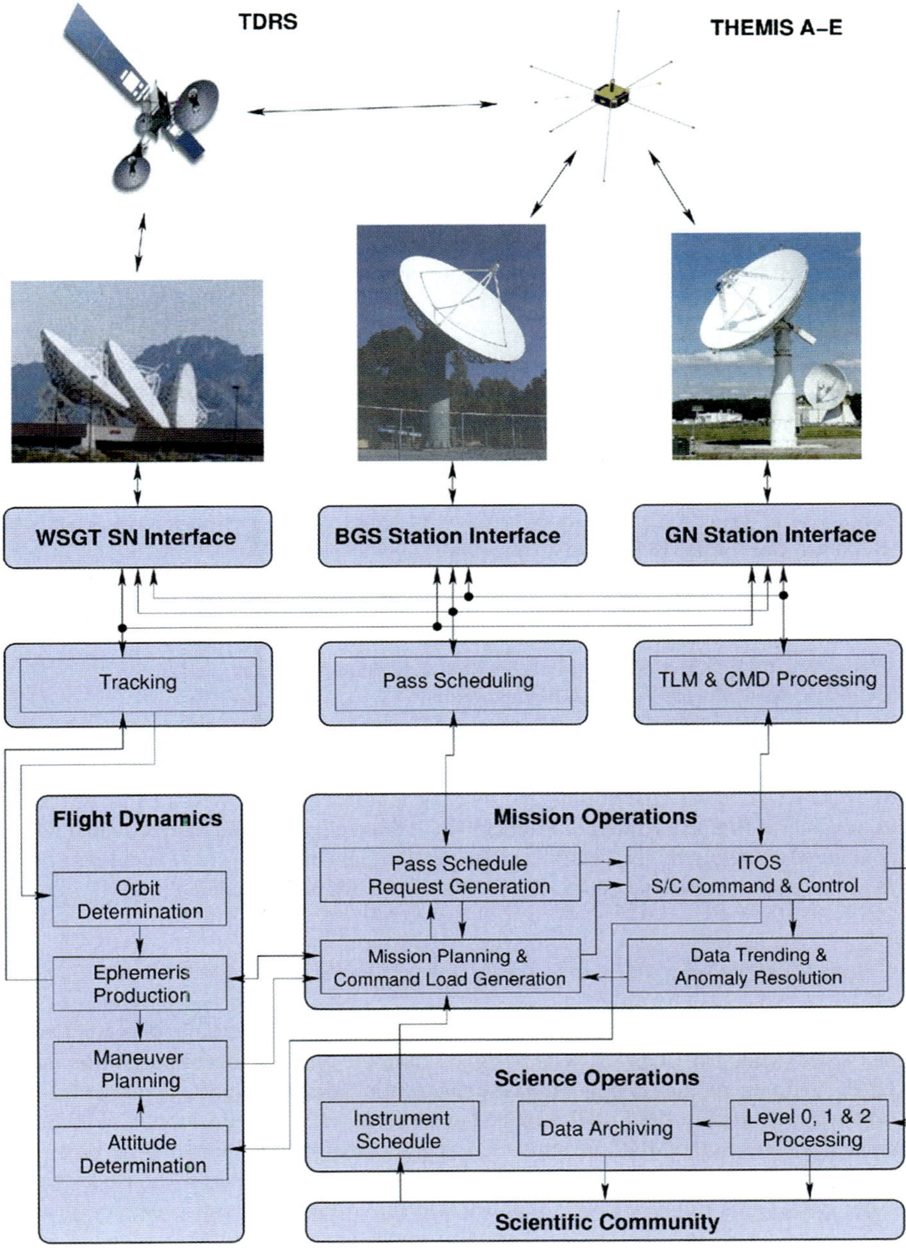

Fig. 2 THEMIS ground system functions and operational interfaces

shown in Fig. 2. The spacecraft command and control system for THEMIS is the Integrated Test and Operations System (ITOS) (Pfarr et al. 2008), which is also used for FAST and RHESSI mission operations. Other tools within Mission Operations support functions such as pass scheduling, mission planning, data trending and anomaly resolution. Details of the science data processing and archiving systems are covered elsewhere.

2.2 Communications Network

Stored data playback must occur when each probe is near perigee, and since the probes' onboard solid-state memory is not large enough to store more than one orbit's worth of data, the communications network requires a sufficient selection of ground stations so at least one will be available to cover every perigee pass of every probe. The coordinated nature of the probes' orbits often causes several probes to approach perigee at nearly the same time, requiring separate ground stations to support simultaneous data playback passes. In addition to data recovery, accurate Doppler-based orbit determination requires that several measurement arcs be taken from different ground stations on every orbit. The communications links were designed to be closed with 11-meter class ground stations at the lowest telemetry rates out to the farthest apogee at a 200,000 km range, allowing execution of maneuvers as well as recording two-way Doppler data and monitoring state-of-health telemetry in real-time anywhere along mission orbits. The seven ground stations currently used by THEMIS are:

1. Berkeley, California (BGS)—11-m antenna
2. Wallops Island, Virginia (WGS)—11-m antenna
3. Merritt Island, Florida (MILA)—two 9-m antennas
4. Santiago, Chile (AGO)—9 and 12-m antennas
5. Hartebeesthoek, South Africa (HBK)—10 and 12-m antennas
6. Dongara, Australia (USNAU)—13-m antenna
7. South Point, Hawaii (USNHI)—13-m antenna.

An overview of the THEMIS mission control network is shown in Fig. 3. Network connectivity is achieved via a frame routing and relay system that can be envisioned as the Transmission Control Protocol / Internet Protocol (TCP/IP) equivalent of a matrix switch which is configured remotely by the centralized, automated pass and network scheduling system to facilitate secure command and telemetry data flows for any scheduled pass (Bester and Stroozas 2007).

2.3 Multi-Mission Control Center

Established in 1998 to support the RHESSI and FAST missions, the Multi-Mission Operations Center (MOC) at UCB/SSL was designed from the onset to function as a true multi-mission environment, and is now the nerve center for THEMIS flight operations, as well as ongoing operation of the FAST, RHESSI and CHIPS satellites (Bester et al. 2003). All computer systems in the MOC are supported by a secure, isolated operations network with centralized, redundant file servers. Electrical power for critical computers and electronics is backed up by uninterruptible power supplies (UPS) as well as a diesel generator, guaranteeing that the MOC and the Berkeley Ground Station can operate through extended power outages.

The bulk of the 850 square foot facility, shown in the floor plan in Fig. 4, is taken up by equipment racks needed to operate the Berkeley Ground Station (BGS), and by THEMIS ITOS workstations. Each THEMIS probe has a dedicated ITOS workstation for command and control operations (OPS 1-5) to allow for simultaneous communication pass supports with all five probes. A second row of five telemetry-only ITOS workstations (OPS 6-10) accommodates instrument and spacecraft engineering and flight dynamics staff during critical

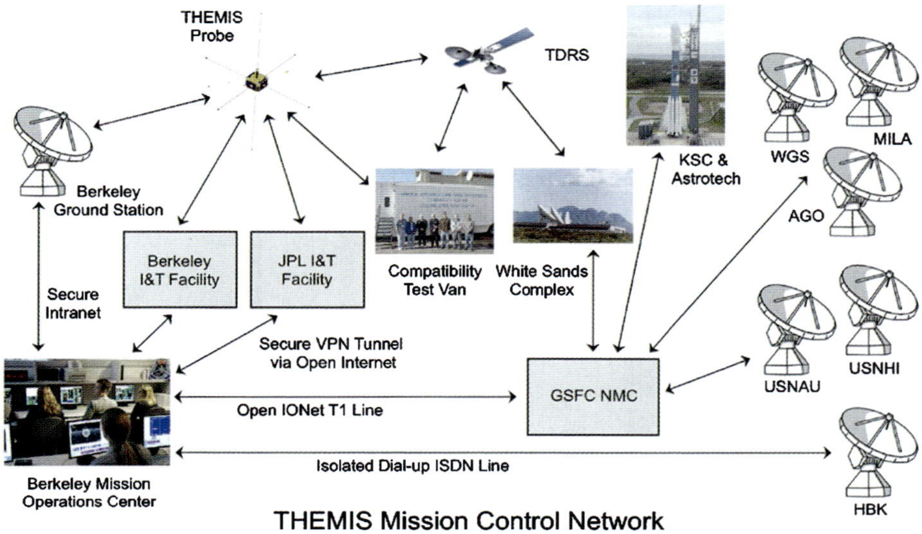

Fig. 3 The THEMIS communications network includes seven ground stations and the Tracking and Data Relay Satellite System (TDRSS). Connections to NASA's Internet Operational Network (IONet) are routed through the Network Management Center (NMC) at Goddard Space Flight Center (GSFC). Secure communications via the Hartebeesthoek Ground Station (HBK) utilize an Integrated Services Digital Network (ISDN) line that is brought up separately for each pass support

operations and maneuvers. Though not used for spacecraft commanding, these workstations can nevertheless be reconfigured as command consoles in case one of the primary ITOS systems fails.

The Flight Dynamics Center (FDC) is co-located with the MOC and is responsible for timely mission design, maneuver planning and reconstruction, as well as orbit and attitude determination. Servers for science data processing and storage are located in the adjacent building.

2.4 Software Tools

The software needed to operate the THEMIS mission is a blend of government off-the-shelf (GOTS) and commercial-off-the-shelf (COTS) products, heritage software that had already been developed in-house for other missions and underwent further upgrades, plus additional software written specifically for the THEMIS mission. The major components are briefly described below, and their interaction is outlined in Fig. 5.

1. ITOS—the Integrated Test and Operations System, a NASA/GSFC-developed system already in use for the FAST and RHESSI missions, is used for THEMIS real-time telemetry monitoring and command and control as well as limit-level-based health and safety monitoring (Pfarr et al. 2008). In-house development at UCB/SSL extends ITOS and integrates it with other components of the ground system allowing pass supports, including recovery of stored science and engineering telemetry data to be conducted in a hands-off and lights-out manner with a high degree of reliability.
2. SatTrack—a comprehensive COTS software suite that controls, monitors, coordinates and automates most aspects of the ground system, such as the completely automated

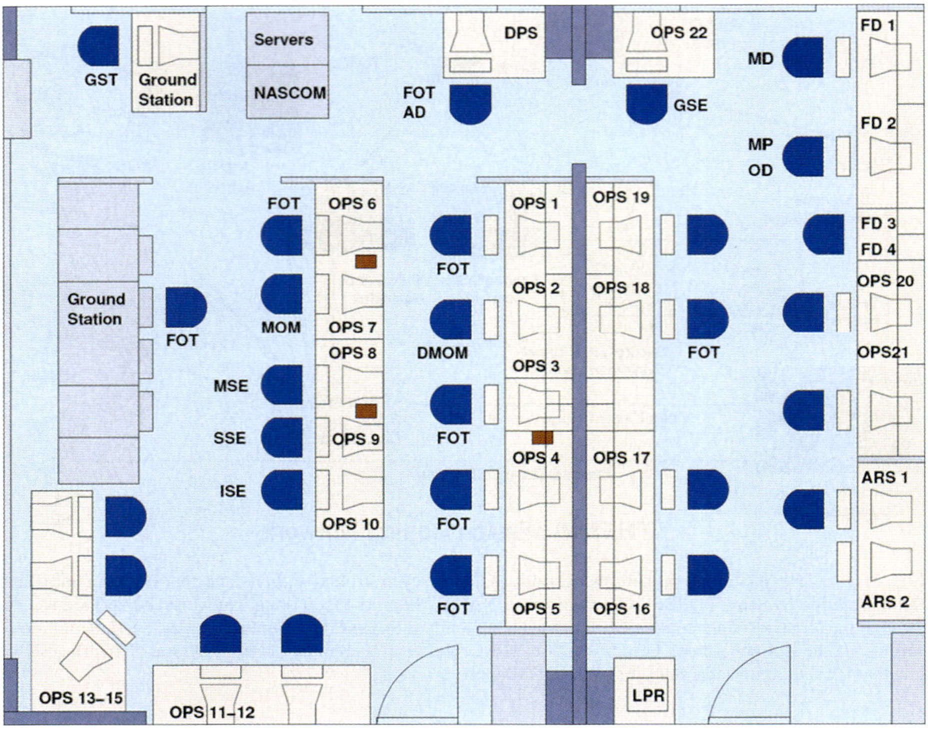

Fig. 4 Floor plan of the multi-mission operations center at UCB/SSL, showing workstations as well as equipment racks for the Berkeley Ground Station and the NASA Communications Network (NASCOM). Workstations are labeled according to their technical functions (OPS: Operations; FD: Flight Dynamics; ARS: Anomaly Response System; DPS: Data Processing System; LPR: Color Laser Printer). Also indicated are team roles and responsibilities (FOT: Flight Operations Team; MOM: Mission Operations Manager; DMOM: Deputy Mission Operations Manager; MSE: Mission Systems Engineer; SSE: Spacecraft Systems Engineer; ISE: Instrument Systems Engineer; GSE: Ground Systems Engineer; AD: Attitude Determination Lead; MD: Mission Design Lead; MP: Mission Planning Lead; OD: Orbit Determination Lead; GST: Ground Station Support)

operation of the Berkeley Ground Station, scheduling, maintaining and disseminating the operational pass schedule for all active satellites, routing command and telemetry streams between ground stations and command and telemetry workstations, and autonomously configuring ITOS and other software clients prior to each pass support (Bester et al. 2008). The SatTrack Suite also generates and distributes daily-updated orbital data products necessary for science planning, pass support, and maneuver execution.

3. MDT—the Mission Design Tool is a suite of Interactive Data Language (IDL) programs that was developed in-house at UCB/SSL and was purposely written for overall THEMIS mission orbit design as well as fast, iterative maneuver re-planning (Frey et al. 2008). MDT calls GTDS for orbit propagation and GMAN for finite maneuver targeting (see below).
4. GTDS—the Goddard Trajectory Determination System is a NASA/GSFC GOTS software package used for THEMIS orbit determination and orbit propagation with a high-fidelity force model.

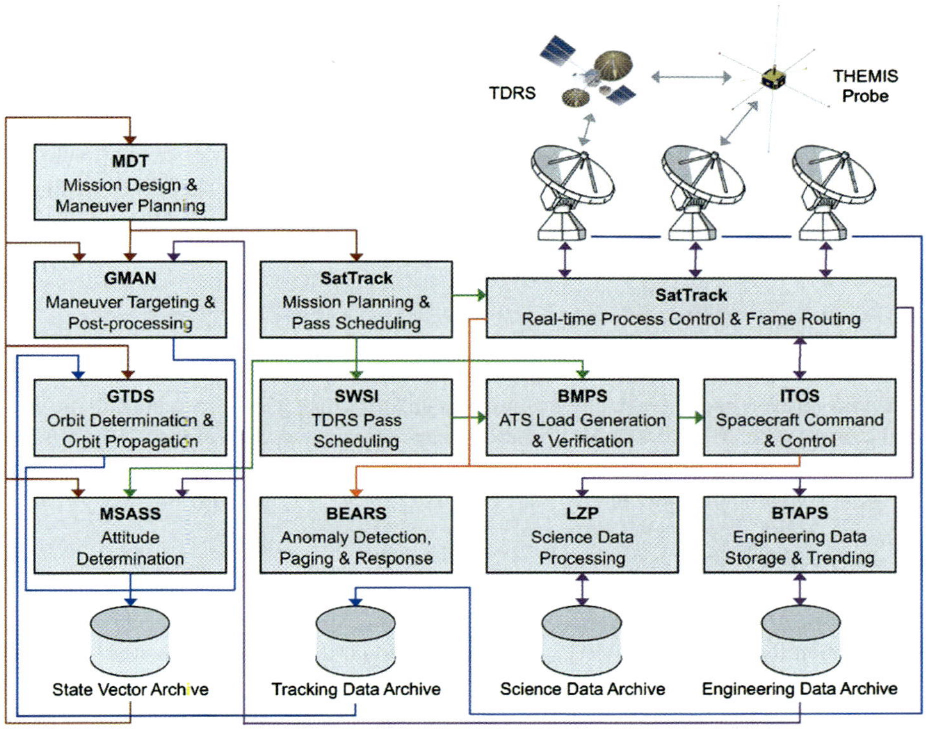

Fig. 5 Interaction and data flows between various operational software tools. Acronyms are explained in the preceding and following paragraphs

5. GMAN—the General Maneuver Program, another NASA/GSFC GOTS software package, is invoked by the MDT to perform high-accuracy, finite maneuver targeting.
6. MSASS—the Multi-mission Spin-axis Stabilized Spacecraft attitude determination system is a suite of MATLAB-based tools used to perform batch and real-time ground-based attitude determination. MSASS was inherited from NASA/GSFC and extended in-house at UCB/SSL.
7. BTAPS—the Berkeley Trending and Plotting System is another in-house developed software suite that decommutates and converts all real-time and post-pass engineering telemetry data and stores these in a MySQL database, allowing both real-time strip charting as well as archival plotting, trending and anomaly detection (Cruce et al. 2007). BTAPS also provides maneuver and attitude related engineering telemetry data for attitude determination and maneuver reconstruction.
8. BMPS—the Berkeley Mission Planning System, developed in-house at UCB/SSL, incorporates orbital data products generated by the MDT and SatTrack, and builds Absolute Time Sequence (ATS) command tables that autonomously control science data collection and spacecraft operation outside of real-time passes.
9. BEARS—the Berkeley Emergency and Anomaly Response System, developed in-house at UCB/SSL, detects spacecraft and ground system anomalies and broadcasts messages to the Flight Operations Team (FOT) members by way of electronic mail and two-way paging messages, until the problem is resolved.

10. LZP—the Level Zero Processing software for science telemetry data, developed in-house at UCB/SSL, verifies post-pass delivery of expected telemetry files, performs quality checking and extracts CCSDS packets from received telemetry transfer frames. Packets are ordered by acquisition time and archived in individual files for each Application Process Identifier (APID), spanning 24 hours of observations.
11. SWSI—the Space Network Web Services Interface is a remote access tool provided by NASA and allows scheduling and remote monitoring of end-to-end link performance between the White Sands Ground Terminal (WSGT) and a THEMIS probe during pass supports via a Tracking and Data Relay Satellite (TDRS).

3 Mission Operations

THEMIS mission operations include all aspects and activities related to managing the constellation on orbit. Members of the operations team participated in all phases of the mission life cycle, beginning with the earliest stages of proposal writing throughout the mission development and integration phases to prepare and plan for on-orbit operations.

3.1 Pre-Launch Testing

In preparation for on-orbit operations, the THEMIS team developed and executed an extensive Mission Readiness Test (MRT) program. Individual test catalog items were designed to exercise all aspects of on-orbit operations as close as possible—where practical—to a *Test-like-you-fly* configuration, and were categorized in the following scheme:

0xx—Ground Systems
1xx—Flight Systems
2xx—Interfaces and Data Flows
3xx—Ground Operations
4xx—Launch Operations
5xx—Special Operations
6xx—Maneuver Operations
7xx—Normal Science Operations
8xx—Contingency Operations
9xx—Operational Readiness Tests.

All of the nearly 300 individual tests with an increased level of complexity and involvement of different systems elements were executed successfully at least once to obtain a *pass* mark. Many tests, such as end-to-end data flows between the ground stations and the MOC, were repeated multiple times to shake out any remaining issues and to allow the operations team to gain a high level of proficiency and confidence. This comprehensive and meticulous test approach paid off many times over after launch, as it allowed the operations team to focus on operating the constellation rather than being forced to spend precious time with debugging ground systems issues.

Since final integration of the probes occurred in the same building at UCB/SSL where the MOC and BGS are located, the operations team had a unique opportunity to perform radio frequency (RF) and data compatibility tests with full end-to-end data flows. Communications were established via a low-power RF path by pointing the 11-m antenna at the integration facility and commanding the five probes from the MOC as if they were on orbit already. A full cycle of round-robin state-of-health checks with all five probes could be simulated in the same way it would occur during the first acquisition after orbit insertion.

Leading up to launch, formal mission simulations and dress rehearsals were conducted (Harvey et al. 2008). These involved the MOC, the THEMIS probes and personnel at CCAFS, United Launch Alliance (ULA), the NASA Ground and Space Networks (GN/SN), NASA's Integrated Services Network (NISN), and the Flight Dynamics Facility (FDF) at GSFC. Simulations were conducted six times for Launch Day (LD) and twice for Launch Day + 1 (LD+1).

3.2 Launch and Early Operations Phase

The THEMIS launch was originally scheduled for Thursday, February 15th, but was moved to Friday, February 16th because lightning storms near the launch pad delayed the fueling operations at L − 2 days. On February 16th, the countdown was in the 4-minute built-in hold just prior to the opening of the launch window at 23:05:00 UTC when excessive high-altitude wind speeds forced the launch to be scrubbed with a 24-hour turn-around. THEMIS finally launched aboard a Delta II 7925-10 from Space Launch Complex (SLC) 17B at CCAFS on Saturday, February 17, 2007 at 23:01:00.384 UTC, right at the opening of the 19-minute launch window.

Following burn-out of the STAR48 third stage solid rocket motor, the five probes separated from the Probe Carrier, beginning with THEMIS A, mounted at the top of the stack, and followed 3 seconds later by the simultaneous release of THEMIS B–E into a $435 \times 91{,}958$ km predictive, post-launch insertion orbit at an inclination of 16.0 deg. Definitive orbital elements are summarized in Table 1. The average achieved apogee height was 4,632 km lower than the predicted (nominal) value, resulting in an orbital period of 1884 min, or 115 min shorter than the expected 1999 min, but within the projected 3-σ spread of ± 180 min. The Delta II launch sequence is depicted in Fig. 6.

To monitor the separation event, a communications link with THEMIS A was established via TDRS West three minutes prior to the scheduled time of separation. As launch

Table 1 Insertion orbital elements and attitudes

Parameter	THEMIS A	THEMIS B	THEMIS C	THEMIS D	THEMIS E
Perigee height [km]	466.9	466.8	465.4	466.3	469.5
Apogee height [km]	87349.9	87329.7	87089.3	87310.6	87548.9
Inclination [deg]	15.9	15.9	15.9	15.9	16.0
RAAN [deg]	329.1	329.1	329.1	329.1	328.9
Arg. of perigee [deg]	319.8	319.8	319.8	319.8	320.0
Anomalistic period [h]	31.174	31.164	31.052	31.155	31.267
Predicted spin rate [rpm]	16.0 ± 2.0	16.0 ± 2.0	16.0 ± 2.0	16.0 ± 2.0	16.0 ± 2.0
Observed spin rate [rpm]	17.1	16.1	16.1	16.1	16.1
Predicted Sun aspect angle [deg]	47.0 ± 5.0	47.0 ± 5.0	47.0 ± 5.0	47.0 ± 5.0	47.0 ± 5.0
Observed Sun aspect angle [deg]	45.6	47.3	46.2	41.8	43.4

Note: Orbital elements are given in Earth Centered Inertial (ECI) True-of-date (TOD) coordinates and correspond to the first orbit solution for each probe with epochs on February 20, 2007. Uncertainties in the predicted sun aspect angle include launch vehicle dispersions, characteristics of the probe release mechanisms and post-separation nutation

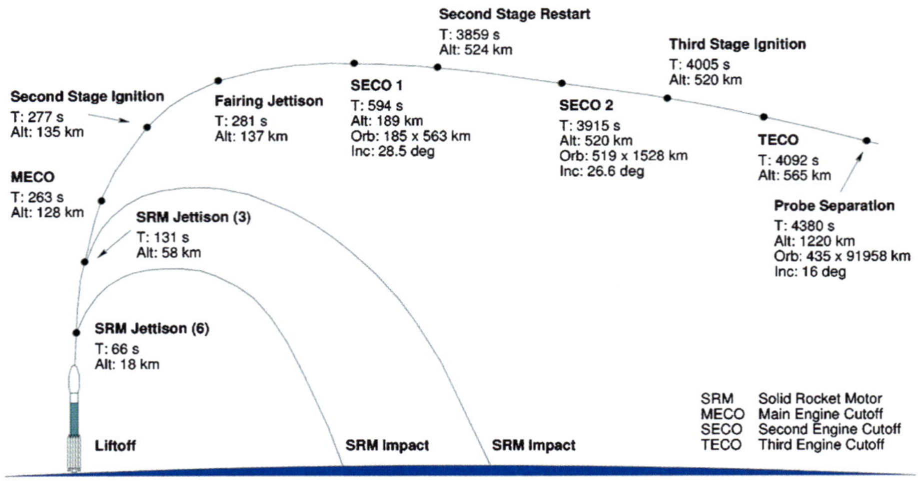

Fig. 6 THEMIS Delta II flight profile on February 17, 2007

occurred on time at the opening of the launch window, vector rotation was not required and the nominal pre-launch state vector was used for acquisition. The spacecraft transmitter was successfully commanded on via blind acquisition 70 min after lift-off at 00:11:00 UTC, and the return link came up nominally at a telemetry rate of 1.024 kbps. Separation occurred right on time at 00:14:00 UTC. Since all probes share the same radio frequency, separation of only one probe could be confirmed in real-time telemetry. However, there was a high level of confidence that separation of the other four probes had occurred as well.

Shortly after the initial acquisition of THEMIS A and verification of its release, all five probes were contacted via BGS to verify the release of THEMIS B–E, to check their state of health and to record two-way Doppler data in order to obtain an early orbit solution. Insertion attitude parameters are summarized in Table 1. Despite an insertion attitude that was very challenging from a communications perspective, a sufficient number of telemetry frames were received to verify probe separation, attitude and spin rate, and to establish good state of health across the constellation. Figure 7 shows photographs of the post-launch activities at the MOC.

3.3 Instrument Commissioning

Flight operations procedures and Spacecraft Test and Operations Language (STOL) scripts for on-orbit instrument commissioning, control and configuration were developed during the mission integration and test phase. Changes were validated on a flight simulator called *FlatSat*, connected to a complete and fully functional instrument suite that was built as a flight spare.

Instrument commissioning started on LD+5 with powering the Instrument Data Processing Units (IDPUs) and Fluxgate Magnetometers (FGMs) on (Taylor et al. 2008; Auster et al. 2008). The strategy was to keep all five probes in a similar state and to perform corresponding operations on back-to-back passes, where practical. In this case, all five IDPUs and FGMs were powered on and checked out during five consecutive passes, spanning 6 hours total. This approach worked very well as on-console staffing could be optimized and the operations and engineering support teams concentrated on one set of procedures at a

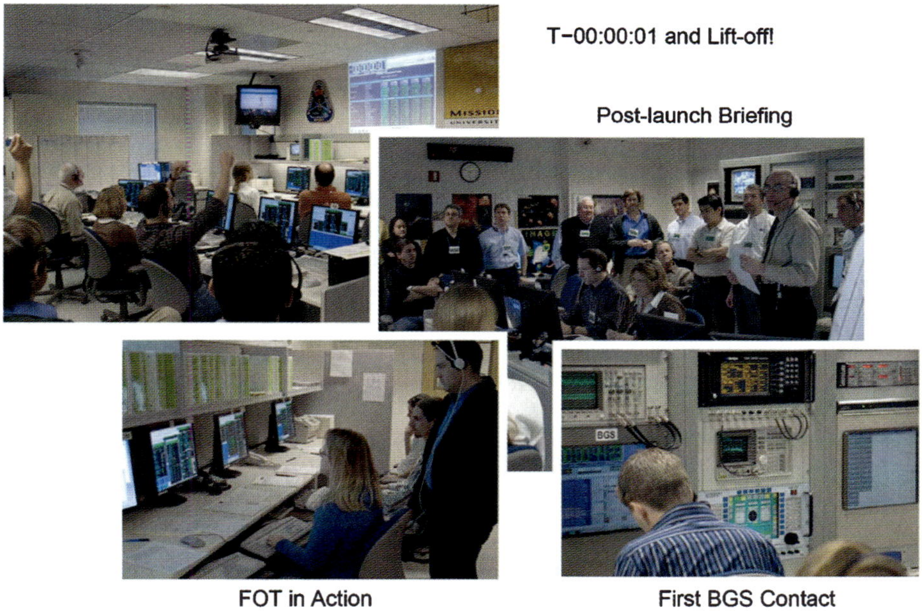

Fig. 7 THEMIS launch team at the Mission Operations Center at UCB/SSL

time. The remaining instruments, namely the Search Coil Magnetometers (SCMs) (Roux et al. 2008), Electric Field Instruments (EFIs) (Bonnell et al. 2008), Electrostatic Analyzers (ESAs) (McFadden et al. 2008), and Solid State Telescopes (SSTs) (Angelopoulos 2008), were powered on and checked out in a similar assembly-line fashion. The magnetometer booms were deployed on all probes between LD+7 and LD+9 (Auslander et al. 2008).

The first probe to deploy its EFI spin-plane and axial booms was THEMIS C, beginning on LD+81 (Bonnell et al. 2008). Detailed analyses showed that reeling out a section of the wire booms followed by a pulsed spin-up maneuver with two short pulses per spin revolution would not compromise dynamic stability (Auslander et al. 2008). Nevertheless, a great deal of care was used to gradually deploy the booms. Once initiated by ground command, the X and Y wire boom pairs were deployed autonomously by on-board software in the IDPU, controlling the deploy motors in such a way that a symmetrical deploy state was maintained at any time (Taylor et al. 2008). Once the wire booms were fully deployed to their end-to-end lengths of 50 m in $\pm X$ and 40 m in $\pm Y$, the $\pm Z$ axial stacer booms were released. No issues with excitation of wire boom bending modes and/or fuel slosh oscillations were encountered—the amplitudes were small, as predicted. The entire EFI deployment sequence of the first probe, involving 13 steps of alternating deploy, spin-up and sensor diagnostic tests, was completed by LD+88, within 7 working days.

Once the EFI deployment procedures were successfully executed on the first probe, the next two probes, THEMIS D and E, were deployed in back-to-back operations that started on LD+103 and completed on LD+110, involving only 5 working days. Again, it turned out to be very efficient to group deployment activities on the two spacecraft in such a way that the number of shift changes for on-console support of instrument and maneuver operations was minimized, allowing the overall operations schedule to be accelerated. The EFI booms on THEMIS A and B were deployed after their mission orbit placement had been completed, so that the low moments of inertia allowed attitude precession maneuvers and ΔV maneuvers in axial firing mode to be used efficiently.

3.4 Routine and Special Operations

Operations that are repeated at least weekly are considered routine operations. Examples include monitoring state-of-health and progression of automated data recovery from the constellation during normal working hours, and uploading Absolute Time Sequence (ATS) tables. These activities are conducted by flight controllers at the command and control consoles. Clock correlation is also performed manually, using a special software tool that measures the probe clock offset by comparing time tags inserted into transfer frames on the spacecraft and on the ground station side, also taking into account a range dependent propagation delay. The on-board clock offset is then adjusted accordingly. The requirement for each probe's clock is to always match Coordinated Universal Time (UTC) within 0.5 s.

One of the supporting software tools developed in-house during the first year of on-orbit operations is an *autopilot* system that allows for autonomous end-to-end interaction between the flight and ground segment during real-time passes. As a result, virtually all state-of-health and tracking passes, as well as most data recovery passes near perigee, are conducted reliably without operations personnel at the console.

Special operations include preparing and loading of flight parameter tables, probe bus and instrument configuration changes, all operations of the propulsion system, and recovery from probe bus or instrument anomalies. Passes involving special operations are always supported by flight controllers at the console.

4 Navigation

Navigation tasks for the THEMIS constellation include executing ΔV and attitude maneuver plans, performing post-maneuver processing, calibrating thruster performance, determining and archiving the probe states, and maintaining accurate knowledge of the on-board fuel loads.

4.1 Maneuver Planning and Operations

Spacecraft trajectories, maneuver plans and thruster command sheets for finite thrust maneuvers are generated by the Mission Design Tool (MDT) with calls to GTDS and GMAN (Frey et al. 2008). Command sheets are included in ATS loads and are verified on a flight simulator prior to on-orbit execution. There are five maneuver types that can be executed on the spacecraft, using a combination of thrusters in different firing modes, as summarized in Table 2.

Spin rate adjustments are performed by firing one of the two tangential thrusters (T1 or T2) in a pulsed firing mode with pulses phased 180 degrees from each other to minimize the torque on the EFI wire booms. The direction of spin rate change, up or down, is controlled by whichever tangential thruster is activated. Targeted attitude precession maneuvers are performed by phased firing of one of the two axial thrusters, with pulse widths selected to avoid oscillatory resonances of fuel motions and wire boom bending modes.

ΔV maneuvers use either an axial thrust or a side-thrust mode. Axial thrusts are used when the probe's spin-axis is aligned in the direction of the desired velocity change. This mode fires both axial thrusters (A1 and A2) in a continuous burn to achieve large velocity

Table 2 Maneuver modes

Maneuver type	Typical maneuver goal	Thruster firing mode
Axial thrust	ΔV maneuver with stowed EFI spin-plane booms or with deployed EFI booms when no large attitude precession is required	A1 and A2 continuous firing
Side thrust	ΔV maneuver with deployed EFI spin-plane booms or with small ΔV goals	T1 and T2 sun synchronous pulsed firing
Beta thrust	ΔV maneuver with deployed EFI spin-plane booms	A1 and A2 continuous firing alternating with T1 and T2 sun synchronous pulsed firing
Attitude precession	Attitude change	A1 or A2 sun synchronous pulsed firing
Spin-up/spin-down	Spin rate adjustment	T1 or T2 continuous or pulsed firing

Note: Beta thrust maneuvers are executed as a segmented sequence of alternating axial and side-thrust maneuvers

changes in a short time. Axial thrusts were utilized for large ΔV maneuvers on all probes while their EFI booms were still undeployed, as these maneuvers usually required attitude maneuvers into and out of the axial firing attitude that would become expensive in fuel once the EFI booms were deployed. ΔV maneuvers in side-thrust mode utilize both tangential thrusters (T1 and T2) in a pulsed firing mode. Pulse widths are selected as either 40 or 60 deg for different regimes of remaining fuel mass to avoid excitation of fuel slosh resonances. The direction of velocity change in the spin plane of the spacecraft can be controlled by adjusting the phase of the thruster firing relative a sun pulse, allowing orbit changes to be executed without expensive attitude precessions.

Beta thrust maneuvers are executed as a segmented sequence of axial and side-thrust burns. While more complex and less efficient, this firing mode may still be advantageous in certain cases to achieve ΔV maneuver goals without precessing the spacecraft to a firing attitude that would be preferable for either axial or side-thrust maneuvers.

The first maneuvers executed during the THEMIS mission, and also the first maneuver on each spacecraft, were attitude maneuvers to precess the inertial attitude in such a way that the sun aspect angle changed from 49.0 to 15.0 deg, providing a more stable power and thermal environment, and better communications. These maneuvers were executed as so-called Attitude Recovery Maneuver to Sun Normal, using an ITOS STOL script that specifies all thrust parameters via ground command. This type of maneuver procedure does not require extensive planning and FlatSat simulations, as it fires one axial thruster, A1 or A2, at a fixed angle of 90 or 270 deg from the sun pulse with a pre-determined number of pulses.

Maneuver operations were typically supported by the Mission Systems Engineer and at least one of the propulsion subsystems engineers from UCB or Alliant Techsystems (ATK) on console. Eventually, the operations and navigation team gained a high level of proficiency so that system engineering support was no longer required. Progression of maneuvers was closely followed using real-time trend plots of critical subsystems parameters, such as the plot shown in Fig. 8.

Maneuver reconstruction involves analysis of recorded engineering and thrust history telemetry data that are downloaded from the probes once a maneuver is completed. The primary quantities taken into account are tank temperatures and pressures, and the exact thruster firing times.

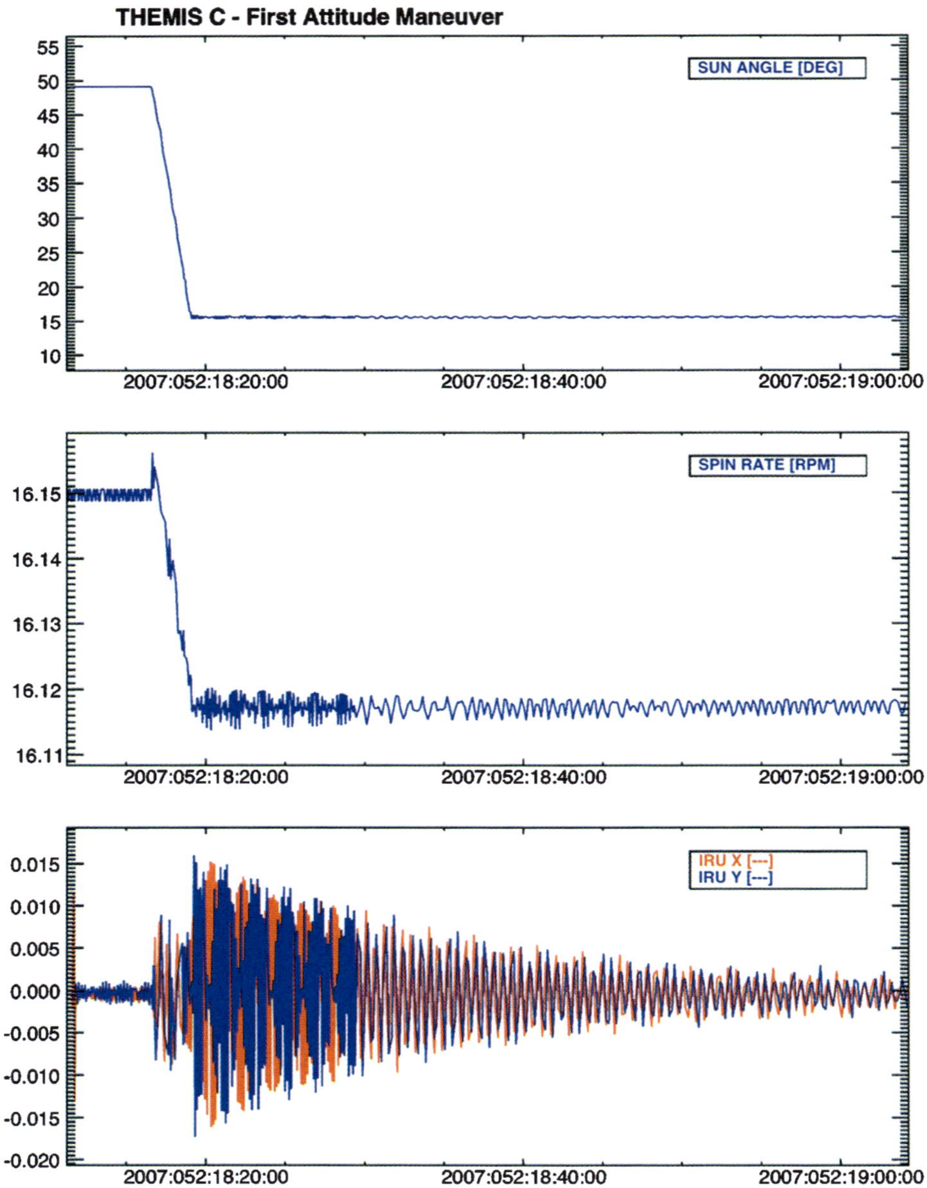

Fig. 8 The first maneuver of the mission was a THEMIS C attitude precession maneuver towards sun normal. The *top panel* of this plot, generated with BTAPS, shows the targeted change in sun aspect angle from 49.0 to 15.0 deg, the *center panel* the undesired, but unavoidable small change in spin rate from 16.15 to 16.118 rpm, and the *bottom panel* the accelerations in X and Y probe body coordinates, as measured by the Inertial Reference Units (IRUs), indicating an onset of nutation caused by fuel slosh at the beginning and more so at the end of the thrust maneuver, and decaying significantly after 40 min

Undesired but unavoidable changes in spin rate and orbital elements are experienced with attitude precession maneuvers. These changes are caused by small differences in thrust efficiencies up to 4% and vary from one maneuver to another. The mass models of the probes were continually refined based on dynamic data obtained during the maneuvers. Once the spin-plane booms were deployed, the moments of inertia were much larger, and the probes became less sensitive to these effects. The maneuver calibration procedure includes models for the tank stretch as a function of pressure and temperature, and appears to work very well.

4.2 Coast Phase

After completing the first two months of on-orbit operations, the THEMIS constellation continued to function in a very good state of health. All five spacecraft were in stable orbits and attitudes with solid power and thermal conditions. All science instruments were operational and collected data, although the EFI spin-plane and axial booms were not yet deployed.

The probe placement decision that relates the probe bus names to the constellation orbit identifiers was made on March 27, 2007 in the following way:

$$\begin{array}{lcl} \text{THEMIS A} & \rightarrow & \text{P5 Orbit} \\ \text{THEMIS B} & \rightarrow & \text{P1 Orbit} \\ \text{THEMIS C} & \rightarrow & \text{P2 Orbit} \\ \text{THEMIS D} & \rightarrow & \text{P3 Orbit} \\ \text{THEMIS E} & \rightarrow & \text{P4 Orbit.} \end{array}$$

This decision was primarily based on the performance characteristics of the telecommunications subsystems, since the five probes were otherwise essentially equivalent. Mission orbit placement in preparation of the first tail observing season was planned to commence in late August 2007 and be completed in early November 2007 when the probe orbits would align with their lines of apsides with the Earth's magnetospheric tail. Meanwhile all five probes were maintained in temporary *coast-phase* orbits, designed to prevent differential drifts of their orbits (Frey et al. 2008). The coast phase also provided additional opportunities to collect interesting science data.

The separation from the launch vehicle had placed the five probes into nearly identical orbits in a string-of-pearls configuration with C leading and E trailing the group D–B–A with differential orbital periods of ±5 min, respectively:

$$<\!\!—\, \text{C} \,———\, \text{D} - \text{B} - \text{A} \,———\, \text{E} \,—.$$

A snapshot of the probe orbits in April 2007 is shown in Fig. 9. As probes C, D and E had their EFI booms deployed by then, the desired orbit configuration for the coast phase was as follows:

$$<\!\!—\, \text{B} \,———\, \text{C} - \text{E} - \text{D} \,———\, \text{A} \,—.$$

The THEMIS E orbit served as the reference for the coast phase orbit placement. To achieve this coast phase configuration, several small orbit maneuvers were performed to initiate a drift into the desired orbit positions. These drifts were stopped between late May and early June 2007 to maintain the coast phase configuration during the 2007 summer observing season. The relatively small orbit and attitude maneuvers required to arrange the orbits for the coast phase counted towards the mission orbit placement and were included in the fuel budget.

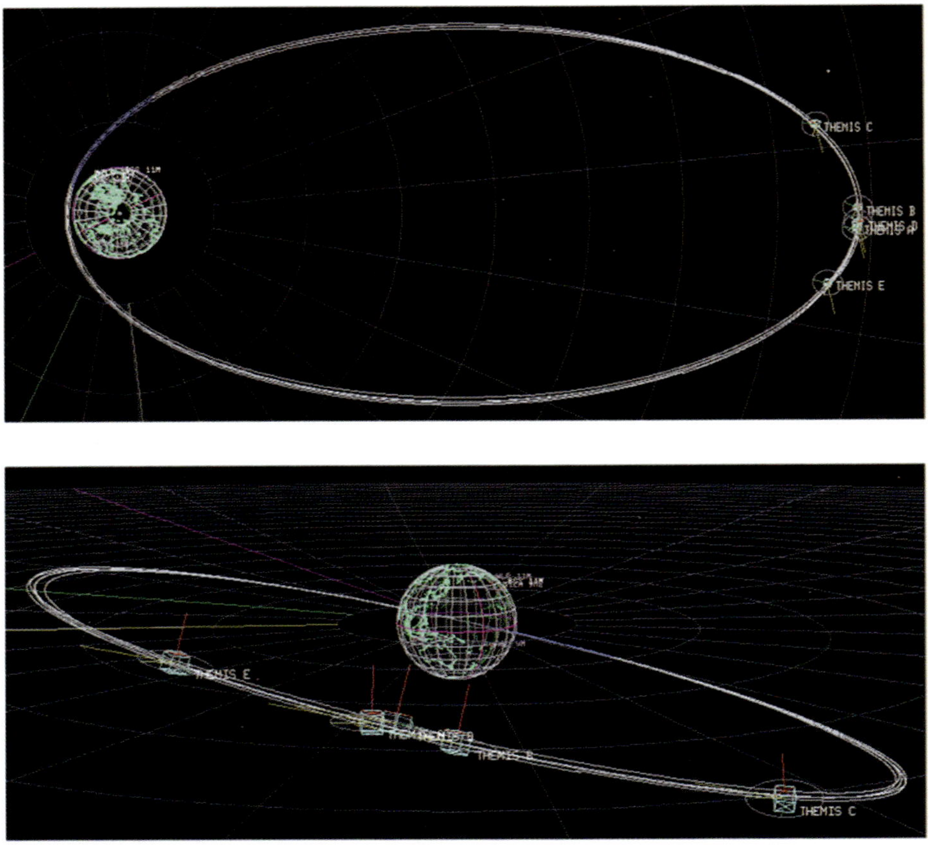

Fig. 9 THEMIS orbit views from north (*top panel*) and apogee (*bottom panel*) on April 22, 2007, shortly after beginning to rearranging the constellation for the coast phase. Equatorial grid circle spacing is 2 Earth radii

4.3 Mission Orbit Placement

The mission orbit placement phase involved maneuvering the five probes from their nearly identical coast phase orbits with a period of approximately 32 hours into their final mission orbits, and was easily as complex as the launch campaign from an operations perspective, if not more demanding in many ways. Up to the end of the coast phase, 68 individual thrust maneuvers had been executed. For the mission orbit placement and remaining EFI deployments, another 108 thrust operations had to be performed. Some of the required orbit maneuvers applied a ΔV of more than 10% of a probe's total fuel budget. The achieved THEMIS orbit configuration at the center of the first observing season in the magnetospheric tail in early February 2008 is shown in Fig. 10, and corresponding orbital elements of the constellation are summarized in Table 3.

Early maneuvers in the orbit placement sequence were very difficult to plan in terms of magnitude and timing, as the mission design team had to work with a rather narrow window of opportunity in the blow-down pressure profile for fuel tank repressurization (Frey et al. 2008; Sholl et al. 2007). Tank repressurization had to be coordinated with the fuel consumption on each probe in such a way that neither firing of the thrusters at too low an

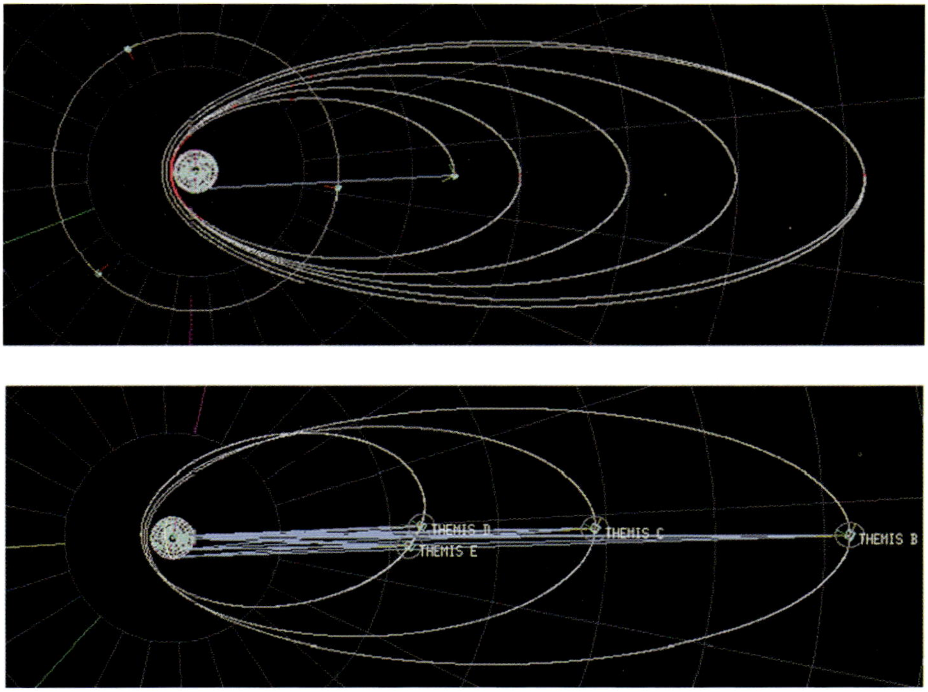

Fig. 10 Depiction of the maneuver sequence for placement of THEMIS B (P1) into its mission orbit (*top panel*), and the achieved orbit constellation of THEMIS B–E (P1-P4) on February 2, 2008, the Wedding Day of the first tail observing season (*bottom panel*)

Table 3 Orbital elements and attitudes on February 2, 2008

Parameter	THEMIS A P5	THEMIS B P1	THEMIS C P2	THEMIS D P3	THEMIS E P4
Perigee height [km]	2873.0	1281.8	1935.6	2678.0	2713.9
Apogee height [km]	57063.5	191226.7	117971.1	68897.8	68862.5
Inclination [deg]	11.2	0.7	5.6	6.2	6.8
RAAN [deg]	304.4	54.9	310.4	303.2	302.1
Arg. of perigee [deg]	13.7	257.1	3.5	18.6	19.8
Anomalistic period [h]	19.2	90.9	47.2	23.9	23.9
Spin axis RA [deg]	281.0	103.6	103.7	276.8	277.5
Spin axis Dec [deg]	60.2	−60.0	−60.8	60.1	60.1
Spin rate [rpm]	20.0	20.0	20.0	20.0	20.0

Note: Orbital elements and attitudes are given in Earth Centered Inertial (ECI) True-of-date (TOD) coordinates

Table 4 Maneuver and propellant summary February 2007–July 2008

Parameter	THEMIS A P5	THEMIS B P1	THEMIS C P2	THEMIS D P3	THEMIS E P4
Initial fuel load [kg]	48.800	48.780	48.810	48.810	48.820
Expended fuel [kg]	20.717	19.641	15.871	20.134	19.900
Remaining fuel [kg]	28.083	29.139	32.939	28.676	28.920
Total ΔV [m/s]	396.331	383.401	296.669	389.318	376.724
Attitude precession maneuvers	18	18	8	10	9
Spin rate change maneuvers	14	16	13	10	9
ΔV maneuvers	16	17	22	14	18
Total number of maneuvers	48	51	43	34	36

Note: Total ΔV includes targeted ΔV maneuvers plus contributions imparted by attitude precession and spin rate change maneuvers

inlet pressure, nor over-pressurization of the fuel tanks could occur. A pyrotechnic valve had to be fired at a time when the ullage volume determining the fuel tank pressure met all low and high-pressure constraints at all expected temperatures. Working around these critical constraints, fuel tank repressurization was successfully accomplished on all probes by LD+227.

Additional maneuvers were required to maintain orbital conjunctions during the first tail season, to prepare the constellation for the first dayside season, and then to maintain conjunctions during the first dayside season. Overall maneuver statistics for 212 individual thrust operations and the fuel budget for the first 18 months of on-orbit operations are summarized in Table 4.

4.4 Orbit Determination

Orbit determination (OD) for the constellation is based on two-way Doppler tracking data, obtained from all ground stations supporting THEMIS. These data are processed by the Goddard Trajectory Determination System (GTDS). Arc lengths are typically 7 days long, but shorter arcs are usually selected to obtain a quick orbit solution following a ΔV maneuver. For a single-station solution, based on BGS tracking data only, the required number of passes is typically three times higher than for a multi-station solution to achieve convergence and comparable accuracy. For operational purposes, the quality of THEMIS orbit solutions is characterized by comparing the differences in orbit periods from one orbit solution to the next. Orbit solutions are routinely generated three times per week, and more frequently during maneuver campaigns. An OD summary is provided in Table 5.

4.5 Attitude Determination

Attitude determination for THEMIS is based on data from the Miniature Spinning Sun Sensor (MSSS) and the three-axis Fluxgate Magnetometer (FGM)—one of the science instruments. The FGM provides data suitable for attitude determination when the magnetic field strength is greater than 5 mG. Since the vector components of the magnetic field vary rapidly with spacecraft position in the near-Earth region, accurate orbit knowledge is essential. Therefore, attitude determination requires orbit determination as a prerequisite.

Table 5 Orbit determination summary February 2007–July 2008

Parameter	THEMIS A P5	THEMIS B P1	THEMIS C P2	THEMIS D P3	THEMIS E P4
Typical arc length [d]	7	7	7	7	7
Typical number of passes per arc	25	25	25	25	25
Typical pass duration [min]	20–30	20–30	20–30	20–30	20–30
Typical achieved accuracy in orbit period [s]	0.05	0.8	0.2	0.05	0.05
Total number of orbit solutions	256	270	269	256	260

Table 6 Attitude determination summary February 2007–July 2008

Parameter	THEMIS A P5	THEMIS B P1	THEMIS C P2	THEMIS D P3	THEMIS E P4
Typical arc length [min]	20	20	20	20	20
Typical achieved accuracy [deg]	0.5	0.5	0.5	0.5	0.5
Total number of attitude solutions	76	78	76	74	72

Attitude sensor data are processed by the Multi-mission Spin-axis Stabilized Spacecraft (MSASS) software developed at NASA/GSFC, and utilizes its Kalman filter attitude determination estimator. Attitude solutions are typically generated once per week using a single 20-min data arc centered on the probe's perigee transit where the magnetic field strength is at its maximum. Attitude determination results are summarized in Table 6.

5 Science Operations

Science operations include all operational activities related to the planning of observations, science instrument configuration, data acquisition, data recovery and subsequent ground processing. Most of these tasks are carried out by operations team members in collaboration with and under the guidance of the science team.

5.1 Science Planning

Due to the complex nature of the scientific observations made with five instruments aboard five identical probes, each in a different orbit, and thus sampling different regions of the magnetosphere, the instrument configuration differs from probe to probe, and has also changed with the different mission phases. The Instrument Systems Engineers (ISEs) have coordinated with the science and operations teams to support the necessary instrument configurations as the THEMIS mission has progressed. Instrument configuration has evolved since the first deployments after launch, through the final EFI deployments after the last major orbital placement maneuvers, and has continued to be refined as science data are processed and interpreted. As the probes pass through different regions of the magnetosphere on each orbit, on-board timed commands configure the instruments for data acquisition, including Fast Survey and Burst data collection, Slow Survey and data compression, as well as autonomous actuations of the SST attenuators. Each week, the operations team loads a new

ATS table aboard each probe. The ATS table also supports maneuver and calibration configurations, such as the pre-maneuver ramp down of instrument high-voltage power supplies, or monthly gain toggle tests for the ESA instrument that are executed during certain parts of the orbit for calibrations and confirmation of configuration changes.

Members of the science team coordinate with the ISEs by passing on desired configuration changes that are documented in Instrument Configuration Change Requests (ICCRs). The ISEs devise the necessary command sequences and test these on the FlatSat and VirtualSat instrument and flight dynamics simulators. Once the commands are tested, following approval by the science team and the operations manager, the real-time commanding is executed during subsequent ground station contacts with the probe or probes in question. Temporary or permanent changes are executed by the operations team via ground commands and are sometimes integrated into regular ATS loads. Changes are also reflected in updates of mission operations procedures and on-board patches to the instrument flight software and flight parameter tables.

5.2 Telemetry Requirements

The THEMIS constellation captures data by a store-and-forward operation. Science and engineering data are recorded in on-board solid-state memory and are played back to the ground segment primarily near perigee, where the highest downlink data rates are achieved. The required science data volume is 750 Mbits per orbit for each probe, and periods of data acquisition at different cadencies are selected such that one complete orbit of data can be downlinked during each perigee passage. The IDPU flight software provides three different compression algorithms that were optimized to compress different types of science data (Taylor et al. 2008). Whenever possible, science data are compressed prior to transfer to the ground, reducing the required transmission time by as much as a factor of two.

State-of-health telemetry from the instruments is part of the normal telemetry stream acquired during each ground station contact. Instrument configuration and status changes are typically monitored via specific IDPU mnemonics in real-time, or by plotting a corresponding data history via BTAPS. Analysis of the configuration changes, however, is best achieved in the science data that are normally evaluated by the science team after downlink and subsequent processing of the instrument science telemetry.

5.3 Science Data Acquisition

The baseline for on-orbit science data acquisition is Slow Survey mode. Special data acquisition schemes such as Fast Survey or Burst data collection are selected based on predicted passages through magnetospheric regions of interest, and also by configurable on-board trigger logic that takes input from different science instruments to autonomously detect and record interesting events. Burst data are sampled at a higher cadence and stored in dedicated memory segments.

Regular changes in instrument configuration are part of the normal daily operation of the probes as they acquire data in various scientifically distinct regions of their orbits. Also, on a larger time scale, there are instrument configurations that correspond to the long-term scientific periods of the THEMIS mission, namely post-launch coast phase, magnetospheric tail, dawn/dusk and dayside season.

Daily or per-orbit changes are made via ATS commands stored aboard the probes. The ATS table also contains commands for probe bus operations, such as transmitter cycling for ground station contacts, and thruster commands for maneuvers. The instrument commands select data acquisition modes and periods of data compression. The transition from Slow Survey into and out of Fast Survey and Particle Burst collection, for example, is achieved by ATS activation of on-board relative time sequences (RTS), which include commands for the transition and execution of IDPU programs called *scripts* that properly configure instruments for that particular period of data acquisition (Taylor et al. 2008). Once configured, internal triggering mechanisms that are programmed per science team requirements via memory settings in the IDPU autonomously trigger specific data acquisition routines.

ATS tables are built with BMPS, incorporating data products from the mission planners and flight dynamics group. These products are used to determine orbital periods of interest for differing instrument configurations, such as passage through the radiation belts, magnetopause crossings, and apogee and perigee passages. These *regions of interest* are different for each of the probes in their different orbits, but also are coordinated such that conjunctions between multiple probes and orbital geometries allow for a wide variety of magnetosphere data sampling. The system is versatile and responds quickly to most configuration changes requested by the science team, usually within days and sometimes within a single day.

5.4 Pass Scheduling and Data Recovery

Ground station pass scheduling is based on the predictions of dynamic link margins, taking into account probe range, attitude, antenna gain pattern, and ground station figure of merit (G/T) to optimally select the highest available data rate—up to 1048.576 kbps—for each pass. The pass scheduling software applies a number of rules and constraints to generate a strawman schedule. Confirmed passes are ingested by the SatTrack Gateway Server that drives the entire operations center (Bester et al. 2003). This includes the configuration of frame routers to enable telemetry and command data flows via secure network connections between the MOC and any supporting ground station, and to initiate pass supports from the Berkeley Ground Station. Much of the pass scheduling and data downlink procedure was automated by the first anniversary of the THEMIS launch, with redundancies built in, and has proven to be very reliable overall.

As mentioned earlier, the THEMIS instrument telemetry stream containing Survey and Burst data comprises approximately 750 Mbits per orbit (differing on duration of each probe's orbit, depending upon season between 0.8 and 4 days). The two probes in larger orbits, and thus with less frequent perigee passages, have smaller margin for data storage and are usually scheduled with backup data recovery contacts on each orbit. On very infrequent occasions, when primary data recovery opportunities were missed and backup passes were not available, a decision had to be made to either remove the data from the on-board solid-state recorder (SSR) or to allow the SSR to fill up during the subsequent orbit. Causes for missed passes could usually be traced to ground station equipment failure or misconfiguration, software issues or network outages. Statistics for on-orbit science data acquisition and recovery to the ground for the first nine months of 2008 are summarized in Table 7. Recovered science data volumes exceeded mission requirements on all probes.

5.5 Ground Data Processing and Archiving

Once downloaded to the ground, a science data processing pipeline first matches the arriving files against the pass schedule to detect missed passes and then performs quality checking for data gaps. If an error condition occurs, the operations team is automatically notified and data replay is requested for a subsequent pass. Science data are then passed into an automated processing pipeline to generate Level 0, 1, and 2 data sets.

Table 7 THEMIS science data recovery statistics January–September 2008

Parameter	THEMIS A P5	THEMIS B P1	THEMIS C P2	THEMIS D P3	THEMIS E P4
Recovered/acquired data volume (average) (required: 95%)	99%	98%	98%	99%	99%
Recovered/required data volume	117%	147%	139%	112%	111%

Note: Recovered and required data volumes assume telemetry data free of bit errors

The science data processing system utilizes several MySQL databases to track the data recovery and processing status. These database systems also provide critical feedback to the operations team regarding completeness and quality of telemetry data sets downloaded from the constellation.

6 Summary

Generally, all aspects of THEMIS on-orbit operations have been very successful. By mid-January 2008, all five probes were completely commissioned and the constellation was fully deployed in its mission orbits for science data acquisition in the first tail observing season. There were no on-orbit failures, and all 5 probe buses and 25 science instruments functioned very well. All critical operations such as deployment of 10 magnetometer booms, 20 spin-plane and 10 axial booms, 5 releases of the ESA instrument covers and the firing of 5 pyrotechnic valves for fuel tank repressurization were performed flawlessly.

The flight dynamics and flight operations teams planned and executed 212 thrust operations across the constellation within the first 18 months of the mission. As many as 4 ΔV maneuvers were performed on different probes within a single 24-hour period. Mission orbit placement and fuel consumption are close to projections, and no planning or operational mistakes were made that could have led to reducing fuel reserves or delaying the commissioning schedule. Towards the end of the mission orbit placement phase, most of the flight dynamics operations had transitioned into routine activities.

All of the ground systems and operational software worked as expected, and all functional elements of the multi-mission operations facility at UCB/SSL worked essentially flawlessly. During the first 18 months of on-orbit operations, the THEMIS ground systems supported more than 7,500 passes. By the end of the first tail season, all of the routine operations had transitioned to a fully automated lights-out mode.

Acknowledgements The authors wish to thank Dr. Vassilis Angelopoulos for giving us the opportunity to contribute to this exciting project. We also like to thank UCB team members Peter Harvey, David King, Dr. Ellen Taylor, Stewart Harris, Richard Sterling, Michael Ludlam, Hillary Richard, Dr. Michael Sholl, Chris Smith, Dr. David Pankow, Paul Turin, Martha Eckert, Linda Croton, Renee Dumlao, James Wheelwright, Timothy Quinn, James Lewis, Thomas Clemons, Jonathan Loran, Robert Boyd, Clarina Quan, Bruce Satow, Kevin Edgecomb, and ATK and Hammers team members Kevin Brenneman, Robert Kraeuter, Rommel Zara, Michael Leeds, Craig Woodruff, Chris Xenophontos and Greg Greer for their support of on-orbit operations. The authors also wish to thank the Instrument Lead Scientists, Drs. John Bonnell (EFI), Charles Carlson (ESA), Davin Larson (SST), Karl-Heinz Glassmeier (FGM) and Alain Roux (SCM), for their support of on-orbit instrument commissioning operations.

THEMIS was made possible by NASA, under contract NAS5-02099.

References

V. Angelopoulos, The THEMIS mission. Space Sci. Rev. (2008, this issue). doi:10.1007/s11214-008-9336-1

D. Auslander et al., Instrument boom mechanisms on the THEMIS satellites; magnetometer, radial wire, and axial booms. Space Sci. Rev. (2008, this issue)

H.U. Auster et al., The THEMIS fluxgate magnetometer. Space Sci. Rev. (2008, this issue). doi:10.1007/s11214-008-9365-9

M. Bester et al., Automation of operations and ground systems at U.C. Berkeley, in *Proc. 5th International Symposium on Reducing the Cost of Spacecraft Ground Systems and Operations (RCSGSO)*, Pasadena, CA, USA, July 8–11, 2003

M. Bester et al., Ground systems and flight operations of the THEMIS constellation mission, in *Proc. IEEE Aerospace Conference*, Big Sky, MT, USA, March 1–8, 2008, Paper 12.0502

M. Bester, B. Stroozas, Telemetry and command frame routing in a multi-mission environment, in *Proc. 42nd International Telemetring Conference (ITC)*, Las Vegas, NV, USA, October 22–25, 2007, Paper 07-23-04

J.W. Bonnell et al., The Electric Field Instrument (EFI) for THEMIS, Space Sci. Rev. (2008, this issue)

P. Cruce et al., A database centered approach to satellite engineering data storage, access, and display, in *Proc. 21st Annual AIAA/USU Conference on Small Satellites*, Logan, UT, USA, August 12–17, 2007, Paper SSC07-XII-5

S. Frey et al., Orbit design for the THEMIS mission. Space Sci. Rev. (2008, this issue). doi:10.1007/s11214-008-9441-1

P.R. Harvey et al., The THEMIS constellation. Space Sci. Rev. (2008, this issue). doi:10.1007/s11214-008-9415-2

J.P. McFadden et al., The THEMIS ESA plasma instrument and in-flight calibration. Space Sci. Rev. (2008, this issue). doi:10.1007/s11214-008-9440-2

B. Pfarr et al., Proven and robust ground support systems—GSFC success and lessons learned, in *Proc. IEEE Aerospace Conference*, Big Sky, MT, USA, March 1–8, 2008, Paper 12.0504

A. Roux et al., The search coil magnetometer for THEMIS. Space Sci. Rev. (2008, this issue). doi:10.1007/s11214-008-9455-8

M. Sholl, M. Leeds, J. Holbrook, THEMIS reaction control system—from I&T through early orbit operations, in *Proc. 43rd AIAA/ASME/SAE/ASEE Joint Propulsion Conference & Exhibit*, Cincinnati, OH, USA, July 8–11, 2007

E. Taylor et al., THEMIS instrument data processing unit. Space Sci. Rev. (2008, this issue)

The THEMIS Constellation

P. Harvey · E. Taylor · R. Sterling · M. Cully

Originally published in the journal Space Science Reviews, Volume 141, Nos 1–4, 117–152.
DOI: 10.1007/s11214-008-9416-2

Abstract The Time History of Events and Macroscale Interactions during Substorms (THEMIS) mission is the fifth NASA Medium-class Explorer (MIDEX), launched on February 17, 2007 to determine the trigger and large-scale evolution of substorms. The mission employs five identical micro-probes (termed "probes"), which have orbit periods of one, two and four days. Each of the Probes carries five instruments to measure electric and magnetic fields as well as ions and electrons. Each probe weighs 134 kg including 49 kg of hydrazine fuel and measures approximately 0.8 × 0.8 × 1.0 meters ($L \times W \times H$) and operates on an average power budget of 40 watts. For launch, the Probes were integrated to a Probe Carrier and separated via a launch vehicle provided pyrotechnic signal. Attitude data are obtained from a sun sensor, inertial reference unit and the instrument Fluxgate Magnetometer. Orbit and attitude control use a RCS system having two radial and two axial thrusters for roll and thrust maneuvers. Its two fuel tanks and pressurant system yield 960 meters/sec of delta-V, sufficient to allow Probe replacement strategies. Command and telemetry communications use an S-band 5 watt transponder through a cylindrical omni antenna with a toroidal gain pattern. This paper provides the key requirements of the probe, an overview of the probe design and how they were integrated and tested. It includes considerations and lessons learned from the experience of building NASA's largest constellation.

Keywords THEMIS · Microsatellite · Probe · Constellation

PACS 94.30.-d · 94.30.Cl · 94.30.Cb · 94.30.Ch · 94.30.Cj · 94.30.C- · 94.30.Cp · 94.30.Lr · 94.30.Va · 94.30.Xy · 96.50.Fm

P. Harvey (✉) · E. Taylor · R. Sterling
Space Sciences Laboratory, University of California at Berkeley, Berkeley, CA, USA
e-mail: prh@ssl.berkeley.edu

M. Cully
ATK Space Division, Beltsville, MD, USA

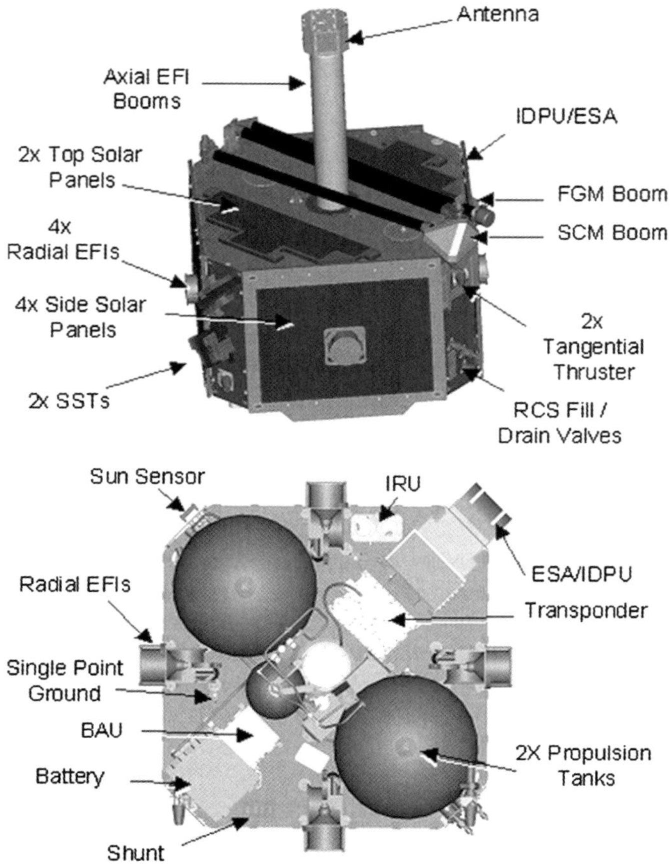

Fig. 1 Probe bus and instrument subsystems

1 Introduction

The THEMIS mission employs five simple, identical probes, shown in Fig. 1 that fly independent and synchronized orbits around earth. The orbit periods are designed to produce a combined measurement set resulting from the apogee region conjunctions due to the natural evolution of the orbits. The probes communicate independently with the operations center that operates each probe in a round-robin serial fashion during normal operations. While the probes are highly autonomous, attitude and orbit determination is maintained by the ground operations center with all orbit and attitude maneuvers nominally taking place during ground contacts. Thorough discussions of the mission design and instruments are presented in Angelopoulos et al. (2008) and the operations are described in Bester et al. (2008). The ground-based observatories measuring the aurora are described in Harris et al. (2008).

All five THEMIS probes were launched together on a Delta II 2925-10 ELV from the Eastern Range into a stable 1.07×15.4 R_e orbit. This orbit was near the planned orbits of the inner probes, called P3, P4 and P5. From this orbit, probes P1 and P2 would need to accelerate to their final orbits, while P3 through P5 would decelerate into their final orbits. The Probe Carrier (PC), a simple mechanical fixture bolted to the 3rd stage, dispensed the probes, directly into this common initial orbit spin-stabilized at 16 ± RPM. An on-board hydrazine propulsion Reaction Control System (RCS) performed the final placement of each probe into its final orbit with minor trimming prior to the prime science tail season.

The spinning probes are passively stable, even under worst-case scenarios. The single-string design is simplified by a minimal hardware complement, inherent functional redundancy, with the instruments and the bus designed for graceful degradation. Analyses showed that probes in the P3 or P4 orbits would have sufficient fuel to accelerate to the P1 or P2 orbits, or decelerate to P5, allowing a replacement of a failed probe. Since four probes define the minimum mission, THEMIS benefits from "constellation redundancy" with a reliability of 80% for the nominal 2-year life and 93% for the minimum 1-year life (Frey et al. 2008).

The five flight instruments include a Electro-Static Analyzer (ESA), Solid State Telescope (SST), Fluxgate Magnetometer (FGM), Search Coil Magnetometer and Electric Field Instrument (EFI) (Angelopoulos et al. 2008). All instruments are identical on all five probes and were built using production methods. The instruments have adjustable data rates to suit different orbit profiles and utilize heritage burst-data collection strategies incorporated in the Instrument Data Processing Unit (IDPU), which has the single electrical interface to the bus (Taylor et al. 2008).

2 Systems Engineering

The engineering of a constellation of probe, rather than a single probe, influenced nearly every discipline involved in the mission, from the number of probes, their level of redundancy, their power and mass, their magnetic and surface charging cleanliness, to how to integrate and test them, and even how to implement the NASA review process. From the project start, engineers and managers understood they needed a Probe design which was simple to build, test and operate. To implement "standard" spacecraft features on multiple probes would likely exceed the schedule and cost caps of the program.

Redundancy. Figure 2 illustrates the electrical block diagram for the Probe. Probes use a single-string design, taking advantage of inherent redundancy and having added redundancy only when mission critical and practical. Probes have only one Bus Avionics Unit (BAU) and one IDPU since it would be impractical to make these redundant. On the other hand, probes have redundant heater circuits since this was practical to implement. Spin plane booms are inherently redundant with one another every quarter rotation, and this is sufficient to meet the E-Field timing requirements. The axial boom and magnetometer boom have redundant firing circuits because of mission criticality and minimal increase in complexity.

Robust Design. Probes are both simple and robust as possible. Solar Arrays cover all six sides and the BAU simply shed power loads automatically if the battery gets too low, assuring a positive power configuration in any attitude. Probes, therefore, have no on-board maneuver capability and their RCS "cat bed" heaters are nominally off. All maneuvers are calculated in detail by Mission Operations, simulated on the ground and then uploaded to each Probe as needed (Bester et al. 2008). In addition, Probe attitude stability during on-orbit deployments was extensively analyzed and was one fault tolerant in most cases (i.e. Probe deployment and EFI deployment). This architecture simplified the design, implementation and cost of the Probes while keeping flight operators in complete control.

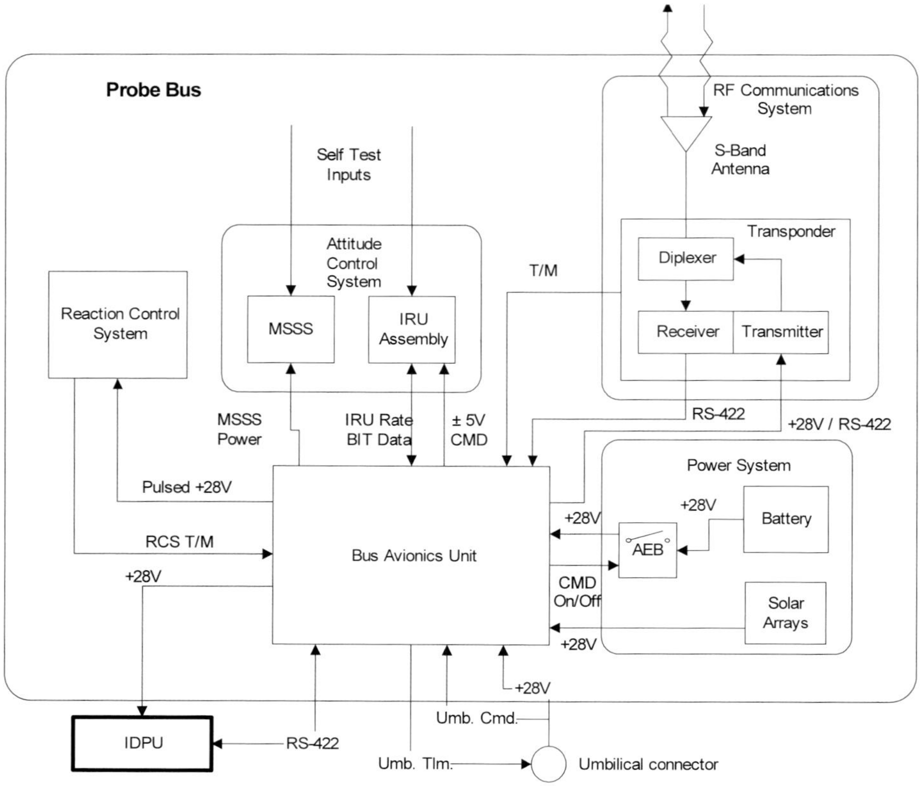

Fig. 2 Probe electrical block diagram

Fault Detection. Probes incorporate on-board fault detection and correction (FDC) sequences to keep the Probes both thermally safe and power positive at all times. Real Time Sequences (RTSs) react to table driven Limit Monitors (LMs) to turn on heaters when components get too cold and turn off loads in response to under-voltage or over-temperature conditions. Solar Arrays cover all six sides and the BAU steps down power loads as the battery drops lower in voltage, assuring a positive power configuration in any attitude. Probes, therefore, have no need for on-board maneuver capability to stay power positive and their RCS "cat bed" heaters are nominally off. Table 1 shows the load shedding sequence, the voltage and the depth of discharge the load shedding begins at, and the loads that are shed. All non-essential loads (e.g. instrument payload, pressure transducer and inertial reference unit (IRU)) are shed first, in Loadshed 1. Primary heaters for the Instrument and then the Bus are turned off next if the voltage continues to drop. Secondary heaters are set to come on ~ 5 degrees lower than the primary ones, saving some power by letting components get a little colder. The primary circuits of critical heaters such as the RCS heaters are never turned off.

All maneuvers are calculated in detail by Mission Operations, simulated on the ground and then uploaded to each Probe as needed (Bester et al. 2008). In addition Probe attitude stability during on-orbit deployments was extensively analyzed and was one fault tolerant in most cases (i.e. Probe deployment and EFI deployment). This architecture simplified the

Table 1 THEMIS load shed sequence

LM	Description	Limit exceeded	RTS	RTS description
LM 02	Loadshed 1	Battery voltage < 29.0 V (30% state of charge)	RTS04	IDPU, Catbed Htrs, ISO valve, IRU, Press Transducer are powered off
LM 03	Loadshed 2	Battery voltage < 28.0 V (25% state of charge)	RTS05	Instrument primary heaters and transmitter are powered off
LM 04	Loadshed 3	Battery voltage < 27.0 V (20% state of charge)	RTS06	Bus primary heater is powered off

design, implementation and cost of the Probes while keeping flight operators in complete control.

Low-Power Design. Fitting multiple probes within the constraints of the launch vehicle fairing limited the overall size of the probe solar arrays. This restricted available power to almost all subsystems, except ones with relatively low duty cycle such as the transmitter. Chief power users such as the flight computers have their clock frequencies as low as possible to keep power to a minimum. The Catalyst Beds and Inertial Reference Units are powered off until maneuvering. In order to survive earth eclipse of up to 3 hours, the surface materials passively bias all temperatures up several degrees so that heater power is minimized in eclipse.

Power Loads by Mode. The IDPU has a number of independent configurations, which mainly affect Instrument data storage rate. Three basic modes, as described below, effect power consumption and dissipation. A fourth mode, engineering mode, affects the IDPU data rate only:

- SAFE POWER MODE—IDPU Power-On State. Core Systems (LVPS, PCB, and DCB) are powered on, all instruments off. Mode is entered on reset (power-on), by ground command, or in response to flag in Probe status field (power-down imminent) in preparation for IDPU load shed. Saves power and the contents of SRR.
- LOW POWER MODE—IDPU Core Systems (LVPS, PCB, and DCB) and FGM are powered on, all other instruments off. Mode is entered by ground command in preparation for maneuvers (FGM data for attitude determination) or in response to flag in Probe status field (low-power) in case of low power condition.
- SCIENCE MODE (Nominal)—Normal operating state, full science data collection. IDPU Core Systems, instrument sensors and associated electronics are powered on. Mode is entered by ground command (instruments are powered on one at a time during early operations).
- ENGINEERING MODE—Higher engineering rate and additional telemetry points telemetered. Operational only, typically during ground contact. Mode is typically entered by ground command in preparation for early operations (instrument health and safety diagnostics) and special case instrument operations (boom deploy, high voltage turn-on).

Table 2 provides the measured power consumption by instrument modes.

Separation Design. The design of the separation sequence and carrier changed several times during the project, driven by both engineering and safety concerns. Initially, Probes would be commanded autonomously by the individual Probe Bus Avionics Unit computer

Table 2 Instrument power consumption by mode

Instrument mode	IDPU	EFI	FGM	SCM	ESA	SST	Power
SAFE MODE	ON	OFF	OFF	OFF	OFF	OFF	6.8 W
LOW POWER MODE	ON	OFF	ON	OFF	OFF	OFF	7.7 W
SCIENCE MODE	ON	ON	ON	ON	ON	ON	15.5 W

to release from the carrier. A second concept incorporated a separation system timer with a dedicated battery hosted on the PC to release the Probes. Both concepts were identified as high-risk developments due to the nature of controlling explosives at the range and the critical aspect of requiring the near simultaneous firing of the four lower Probes from the Probe Carrier to avoid collisions. NASA management at GSFC and KSC recommended, and then implemented, the extension of the launch vehicle third stage separation event onto the PC. The extension lines were provided by the launch vehicle provider and qualified as if part of the vehicle itself. Finally, at GSFC recommendations (based on previous mission lessons learned), a Probe separation status system was added by ATK Space to diagnose if any probe had not separated. This data was relayed to the ground through the launch vehicle telemetry system and this, in fact, verified that all systems separated within 1 millisecond of the commanded time.

Reviews. THEMIS conducted a thorough review program with a NASA-provided review team composed of GSFC and HQ-selected members called the Integrated Independent Review Team (IIRT). A total of 39 reviews were conducted at the system and subsystem levels, 27 of which formally run by the IIRT. These reviews resulted in 269 Requests for Action (RFA) and all actions were formally documented and closed by the IIRT prior to launch.

THEMIS benefited greatly by the experience and insights provided by the IIRT. Given that the constellation presented new and unique challenges for the THEMIS team, the bilateral discussions proved effective in improving the design and implementation of the constellation.

Spares. Based upon past experience in the Cluster I&II programs, in which most of the instruments relied upon their spares at one time or another, THEMIS chose to build at least one spare of each instrument. This decision proved wise as three of the five instruments swapped out sensors for the spare unit. Spare bus components included a battery, side solar panel and top/bottom solar array although none of these were actually used.

Radiation. THEMIS has a two-year design life, mainly driven by its radiation environment and 100% total ionization dose margin. Originally planned for launch in Fall 2006 and having two winter campaigns in 2007 and 2008, the launch vehicle was delayed until just past the winter campaign of 2007. To ensure that baseline objectives remain intact and avoid a radical mission redesign late into the program, a 5-month coast-phase was inserted into the mission giving it a total duration of 29 months, and cutting the radiation margin to about 65%.

Figure 3 shows the approximate annual radiation dose encountered by the P3/P4 electronics inside increasing amounts of Aluminum shielding. Calculations were made by Innovative Concepts. The P1 and P2 orbits have less radiation exposure since they spend less time in the electron belts, while P5 has a slightly higher dose rate based upon more time in these belts. Based upon these simulations, THEMIS enveloped these radiation requirements and

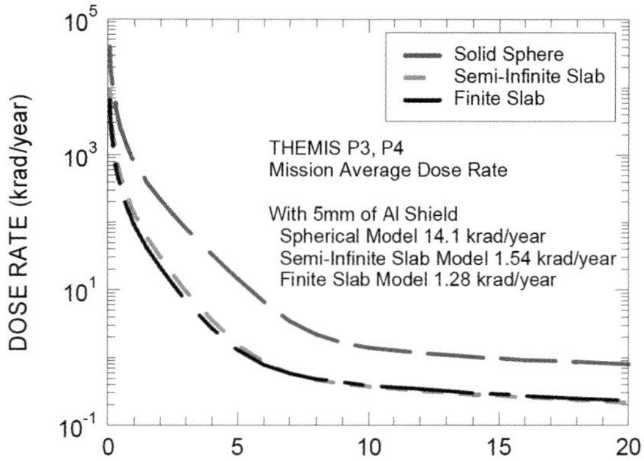

Fig. 3 The dose depth curve for probe 3 and 4 orbits in millimeters of Al

baselined using 5 mm of aluminum (or equivalent) and 66 krad tolerant electronic parts. Together these requirements were found to provide an achievable balance between parts costs and Probe mass.

Single Event Upsets. Parts were selected to be SEL-immune to a LET of >37 MeV cm^2/mg, or else protected against damage by protection circuitry. Both the bus and instrument systems implemented SEU recovery schemes. In the bus, flight software kept three copies of all data in memory and continuously scrubbed out bit errors. The instrument data processor implemented a three-second watchdog timer, which resets the processor if it does not "pet" the Watchdog in that amount of time. The instrument processor also includes a hardware Error Detection and Correction (EDAC) circuit and scrubber subsystem, which uses the upper sections of the instrument Solid State Recorder to store the error correction codes. The scrubber operates on data 4 bytes at a time, generating a byte of check-bits in the ECC segment for every 32 bits in the data section. All FPGAs in the instrument used a Triple-Modular-Redundancy scheme to avoid any single SEU causing an error.

3 Payload

The THEMIS science payload combined the science instruments into a single package with shared data processing and storage capabilities. The instrument suite was designed, built, tested and delivered as a single item for integration with each probe. While providing greater scientific capabilities in on-board power and logic sharing, the approach also provided a single electrical interface to the probe, allowed completely parallel instrument and bus development schedules, while greatly simplifying probe I&T. The only exception to this rule was the axial EFIs, which were contained in cylindrical composite tube that was integrated with the Bus structure at ATK following test verification at UCB.

3.1 The Instrument Complement

The instrument complement consists of five instruments: the Fluxgate Magnetometer (FGM), the Search Coil Magnetometer (SCM), the Electrostatic Analyzer (ESA), the Solid State Telescope (SST) and the Electric Field Instrument (EFI). These sensors are controlled by, and data is returned through the Instrument Data Processing Unit (IDPU), which has the

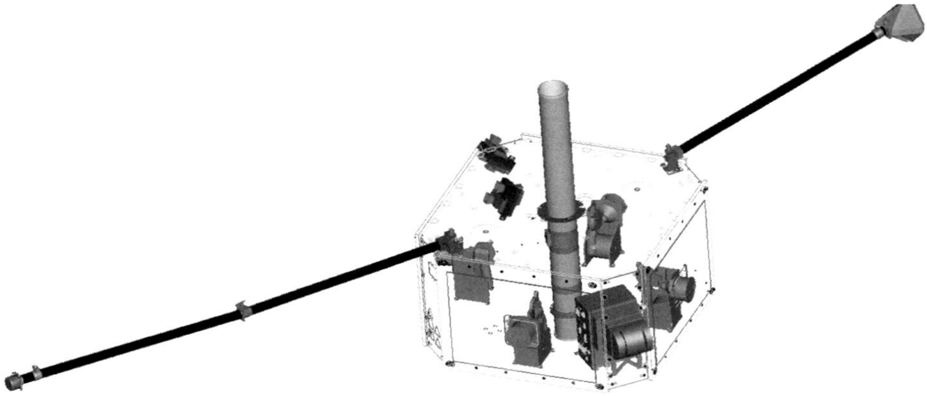

Fig. 4 Payload accommodation showing FGM and SCM deployed

electrical interface to the Bus Avionics Unit. All harnessing from the sensors to the IDPU was built and tested prior to Probe delivery, further simplifying subsequent I&T.

An overview of the instrument performance characteristics and mission requirements is given in (Angelopoulos et al. 2008). Details of each instrument and IDPU are provided in a series of companion papers by Auster et al. (2008), Bonnell et al. (2008), McFadden et al. (2008), Larson et al. (2008), Roux et al. (2008), and Taylor et al. (2008).

Mechanical. Though none of the instruments have critical pointing requirements, integration of the boom-mounted magnetometers nevertheless required precision measurements both for balance calculations and for attitude determination Pankow et al. (2008). The FGM and SCM booms are one-shot deployment mechanisms responsible for holding the sensors still with respect to the probe chassis. Thus, after the booms were mounted to the top deck, precise measurements were made using a portable coordinate measuring machine (CMM). Instrument accommodation is shown in Fig. 4.

The EFI provides 3D coverage once its Spin-Plane and Axial booms are deployed. These boom systems are located on the Probe Center of Gravity (CG) so that the deployed wires are orthogonal. The fuel tanks composite CG was aligned with EFI so that the wires will stay orthogonal even as fuel is depleted. The boom deployment sequence had Spin Plane Booms deploy first, followed by the Axial Booms in order to maintain spin stabilization. Details of this sequence are given in Pankow et al. (2008).

The ESA and SST sensors poke through the corner panels in mid span. While the SST heads were light enough to mount to the panel, the ESA was internally mounted to the IDPU chassis for support at that elevation. Table 3 lists properties of the instruments.

Electrical. The IDPU-to-Probe electrical interfaces consist of a low speed bi-directional serial interface for commands, housekeeping, and status information exchanged between the Probe and instrument, as well as a high-speed serial Clock and Data lines for science telemetry. An 8 MHz clock and 1 Hz tick line combined with a Probe UTC message provides synchronization of the two systems. The 8 MHz clock was used to synchronously sample science quantities in the IDPU.

The BAU provides instrument commands, time and probe status to the IDPU every second using the serial interface. A buffered sun-sensor pulse is used by the IDPU for spin-sectoring the SST and ESA data.

Table 3 Instrument mass, power and on-orbit temperature predictions

Instrument	Mass (kg)	Avg power (W)	Min temp (C)	Max temp (C)
IDPU	4.25	8.00	−37.3	49.2
EFI w/6 Booms	12.22	0.24	−32.5	40.7
FGM w/Boom	1.33	0.85	−54.9	15.1
SCM w/Boom	1.76	0.09	−57.1	14.6
ESA	2.87	1.70	−37.3	45.8
SST	1.35	1.20	−20.0	11.6
TOTAL	23.78	12.08		

Table 4 THEMIS contamination requirements

Sensor	Key requirement on probe
FGM	Mag < 1 nT at 2 m
SCM	Mag low AC fields
EFI	ESC < 10e–5 ohms/cm^2
ESA	Molecular < 0.01 μg/cm^2
SST	Molecular < 0.1 μg/cm^2

The IDPU provides instrument housekeeping packets to the BAU, which is combined with its data into CCSDS frames for downlink. Stored science data is transmitted over the high-speed link when commanded to do so.

Power. The Probe provides the IDPU a Main power service and an Actuator service. The Actuator service is used for deployments.

Thermal. Since each probe is very small, body mounted instruments were expected to experience larger thermal extremes than in previous missions. Temperatures for the instrument components were set at very wide ranges of −50°C to +65°C survival and −50°C to +50°C operational. Predicted on-orbit extremes are shown in Table 3 and are based upon probe-level thermal balance testing.

Contamination. While several THEMIS sensors are sensitive to contamination, they were designed for easy handling and simple integration to the probe. The key contamination requirements imposed on the Probe are provided in Table 4.

The ESA and SST sensors are sensitive to molecular and particulate contamination at the sub-micron level. Both have covers and an external purge provided by the instrument for integration and test.

The EFI sensors are sensitive to handling issues; i.e. asymmetries in the reflective properties of the sphere, which would generate a spin-period photo-emission. Deployment testing at I&T required a clean room environment and handling with gloves. The EFI also required that all Probe surfaces be electro-statically clean to 10^{-8} ohms/cm^2, which equated to a requirement of limiting the total Probe exposed surface to have less than 1 cm square of insulating surface. All exterior surfaces and apertures, which are eclipsed had to meet this requirement.

Typical sources of contamination on the Probe were mitigated to a satisfactory level for THEMIS instruments. The solar panels required particular attention to ensure they were both electrostatically and magnetically clean. To be electrostatically clean, the panels were designed such that: all cover glasses were conductively coated; areas between cells and all perimeter areas were covered by conductive layers; cell interconnects were covered by conductive shielding; no exposed insulators were present; and each panel substrate was directly connected to the spacecraft single point. To be magnetically clean, cell laydown and wiring of each panel met the derived requirement that the resultant magnetic field be less than 24 pT as measured at SCM, with the predicted field strength having a peak of 9 pT. This was accomplished by orienting the cells, strings and back wiring so their magnetic fields cancel and by configuring the power and return wires in twisted pairs to minimize the stray magnetic fields generated by the current flow. For molecular contamination, wire harnesses, solar array panels, thermal blankets and heaters were baked out prior to instrument I&T. The Thermal Vacuum chamber was baked prior to probe insertion, its contamination level monitored during the test, and the chamber backfilled with dry nitrogen at the end of the test.

3.2 Fabrication and Test

Parts. Parts selection required GSFC-311-INST-001 and GSFC PPL-22, Appendix B derating practices. In general, we used grade 3 parts with some up-screening to grade 2 for key items in critical sub-systems such as the Bus Avionics Unit. We organized the "Common Buy" program, purchasing parts for all instruments both for the raw cost efficiency as well as to limit the number of different part types and purchasing lots needed. By limiting part types and lots, we lowered the probability that we would have to open the Probes and replace parts due to a NASA Alert.

Manufacturing. The sheer number of subsystems drove manufacturing decisions towards automated circuit board fabrication as well as internal test functions for each subsystem. For example, the field instruments can stimulate all sensors and the particle instruments can simulate counts and energies. We arranged for manual work to be performed by the same technician for all units of the same design. This included hand soldering, harnessing, thermal blankets and thermal taping. These actions yielded remarkably uniform performance and substantially accelerated the flight test program.

Flight Software. UCB developed the IDPU software using PC-based assembler and linker products developed in-house. The 208 software requirements, specification and test verification were actively reviewed by NASA Independent Verification and Validation (IV&V), and their recommendations were very helpful. For the most part, FSW performance analyses and data products used Excel spreadsheets, and the modular software was tested on the IDPU engineering model. See Taylor et al. (2008).

3.3 Instrument Suite Integration

Instrument Integration and Test (II&T) was a two step process, in which sensors were tested at the box level for unique functions, then integrated to the IDPU and flight harness forming the Instrument Suite. This maximized the instrument level test time while minimizing personnel and facility resources.

For II&T, we used the UCB/SSL E125 clean room. Instrument harnessing was built using mockups of the Probe deck, and Multi-Layer Insulation (MLI) blankets were made at GSFC using instrument mockups. All harnessing and MLI were baked out in UCB/SSL vacuum chambers.

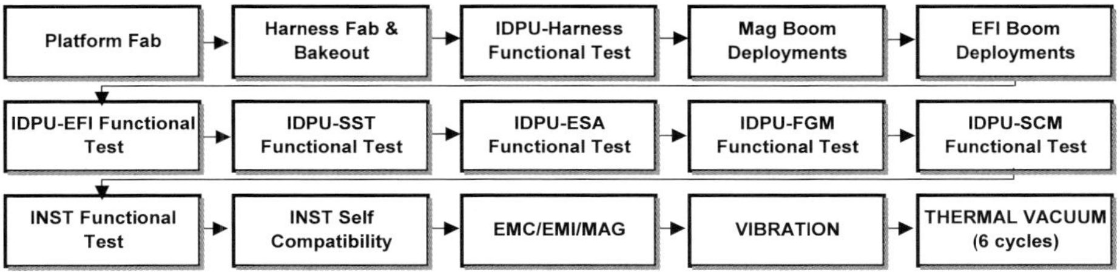

Fig. 5 Fabrication/flow for engineering and flight suites

Fig. 6 Instrument suite in TV preparation

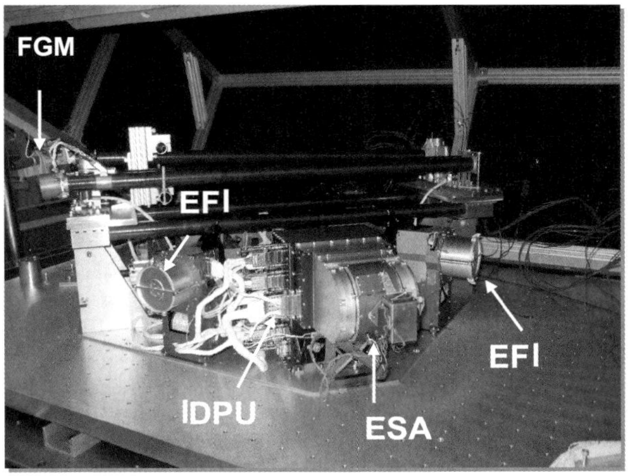

As expected, testing the first flight model (FM1) was pivotal in the maturation of procedures for subsequent models. We tested subsequent models in pairs of instrument suites subsequently; i.e. FM2, FM3 and FM4, FM5. The Instrument Ground Support Equipment (IGSE) used a language nearly identical to the ITOS used by the probe bus. Thus procedures which were developed at II&T flowed with only minor modifications into Probe I&T. This flow is shown in Fig. 5 and a photo of a suite in Thermal Vacuum is shown in Fig. 6.

4 Probe Bus and Carrier

The THEMIS Probe is a highly optimized system that met the extreme challenges posed by the mission. As summarized in Table 5, each probe consists of the bus subsystems and the instrument suite, consisting of four EFI radial instruments, two EFI axial instruments, one ESA, one pair of SSTs, one SCM, one FGM, and an IDPU. The bus subsystems include Structural/Mechanical, Thermal, Reaction Control Subsystem, Attitude Control Subsystem, Power, Communications, and Avionics.

In order to implement the concept of "constellation redundancy," each of the five probes is identical in design and capable of being placed in any of the THEMIS orbits. The probe design was driven by a number of requirements including

- All five had to be small enough and light enough to be launched on a single launch vehicle;
- Assuming small solar arrays, each probe had to be power efficient;

Table 5 Observatory facts

Number of probes	Five
Mass	Probe bus dry mass: 51 kg Instrument mass: 26 kg Probe dry mass: 77 kg Propellant: 49 kg Probe wet mass: 126 kg Allowable mass: 134 kg
Power	Probe bus power: 11 W Instrument power: 15 W Heater power (EOL/24 hr orbit/3 hr eclipse): 11 W Probe power: 37 W Available power: 40.5 W Battery capacity (BOL): 12 AHr
Communications	S band EIRP: 2.4 dB W Two-way Doppler tracking Uplink command rate: 1 kbps Downlink telemetry rates: 1 kbps to 1.024 Mbps
C&DH	Command and telemetry format: CCSDS Version 1 Engineering data storage: 64 MB, 5 days worth Timing: 8 MHz, 1 Hz and UTC distribution
ACS	Spin rate (Science): 20 rpm Spin axis orientation: $< 1°$ (knowledge), $< 3°$ (control) Spin phase knowledge: $< 0.1°$ Ground based attitude determination
Propulsion	Monopropellant hydrazine system Number of thrusters: 4 (4.4N ea.) Total ΔV: 940 m/s Propellant: 49 kg
Probe carrier	Probe carrier mass: 147 kg Total payload mass: 777 kg Mass to orbit capability: 829 kg
Science instruments	Instruments Flux gate magnetometer Search coil magnetometer Electrostatic analyzer Solid state telescope ($\times 2$) Electric field instrument radial ($\times 4$) and axial ($\times 2$) Booms 5-m axial booms ($\times 2$) 20-m radial booms ($\times 4$) 1-m SCM boom 2-m FGM boom Instrument data processing unit
Science Data Volume	Data volume: ~ 400 Mbits per day 5 days worth of storage
Radiation Environment	Total dose: 66 krads (2 years, 5 mm Al shielding, RDM of 2)
Reliability	Observatory Ps = 0.91 (2 years) Mission Ps = 0.94 (4 of 5 s/c required for mission success)

- Given the low mass, each probe had to use radiation-hardened electronics;
- To implement the orbits, the design had to maximize its fuel carrying capacity;
- To avoid contaminating the magnetic measurements, the design and components had to be non-magnetic and non-permeable;
- To reduce surface charges that would impact EFI, the exterior materials were conductive and grounded;
- Given the 3-hour shadows while operating the instruments, the design included considerable thermal blanketing, thermostatically controlled heaters, and careful selection of surface materials;
- In order to operate in any attitude/orientation, the structure design had to tolerate extreme temperature swings from $-115°C$ to $+105°C$.

The major subsystem designs and how these subsystem designs achieve the mission objectives are described in the following sections.

4.1 Structure and Mechanical Subsystem

The THEMIS Probe Bus structure provides mechanical support for all other subsystems and consists of ultra lightweight panels constructed of composite graphite epoxy face sheets and an aluminum honeycomb core. All panels have embedded fittings of either titanium and/or aluminum that have been machined to minimize mass. The sandwich panels have M55J/RS-3 facesheets and Aluminum 5056 honeycomb core. The core/facesheet bond is unsupported FM-73. Aluminum and titanium inserts are bonded into the sandwich to provide component interfaces and mating patterns for edge joints. The panels are joined at their edges using threaded fasteners and shear pins. High strength A286 fasteners in locking Phosphor Bronze helicoils are used. The probe is rectangular in shape with overall dimensions of approximately $82 \times 82 \times 45$ cm in order to provide simplicity and minimizing costs in the solar arrays. The structure is divided into a lower deck, an upper deck, four corner and side panels. The lower deck is the primary mounting surface for the instruments, propulsion system and Probe components. It also interfaces to the probe separation system. The side panels double as substrates for the solar cells. The exterior surface of the upper deck provides inserts for mounting the two magnetometer booms. The probe structure is designed so that all four side panels could be removed during I&T to allow access to the internal components.

The instrument and Probe components are mounted to the lower deck simplifying the load-bearing structure design and facilitating integration. The lower deck and separation adapter fitting is the probe primary structure, carrying the load from all the internal components, side panels and upper deck into the probe separation fitting and ultimately through to the Probe Carrier and launch vehicle. Strength margins were assessed for all structural and thermal design load cases using safety factors of 1.25 on yield and 1.4 on ultimate. Design limit loads of 10.2 G lateral and 6.03 G lateral, applied simultaneously, were used. A detailed structural FEA model was created and verified. The model was used to assess normal modes and strength.

The primary structure must also withstand the extreme temperature swings during early orbit operations and eclipses. The design employed low conductance composite structure and isolated solar panels in order to minimize internal thermal swings between full-sun and shadow operations. Extensive analysis and development testing was performed on the new composite elements of the structure. These environments are simulated via vibration testing and panel level thermal cycling at the subsystem level prior to delivery of the probe structure to Integration and Test (I&T). The mass of the entire structure and mechanical subsystem

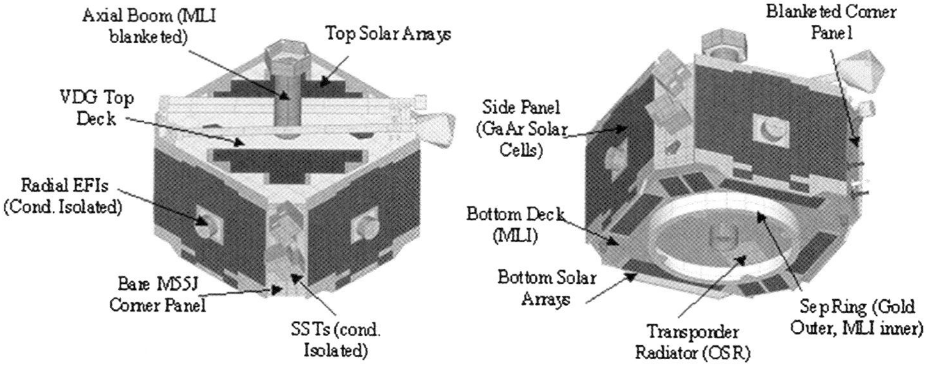

Fig. 7 Probe thermal features

including mounting hardware is 15 kg and represents approximately 19.5% of the Probe dry mass.

A proper grounding scheme was essential to minimize generation of conducted and radiated noise and to ensure predictable system level performance. The probes used a "modified" Single Point Grounding (SPG) system. The SPG is located within the BAU. It ties primary power returns and signal grounds to its chassis ground at one point. This SPG is then connected with a thick braid wire to the separation ring, which is used to terminate chassis grounds from all components. The separation ring was used instead of the probe structure because it provided a lower impedance connection than the composite structure.

4.2 Thermal Control Subsystem

The Probe thermal design was a challenge given the 3-hour eclipses, the need for maneuverability and the probe's low mass. Its thermal subsystem employs a hot-biased design using solar heat to bias component temperatures upward so the probe can survive long eclipses with minimal heater usage (less than 12 watts orbit average). Additionally, the design allows the probe to be thermally safe in nearly all sun aspect angles. With the exception of the transponder, all components either radiated directly to space or were coupled by a standard bolted interface to the spacecraft structure. Electronics boxes with significant power dissipation were painted black to radiatively coupled to the spacecraft interior as well. The transponder was similarly mounted and coated but its base was also covered with Optical Solar Radiators (OSRs) that had a direct view to space through an opening in the spacecraft deck. Components are blanketed with Multi-Layer Insulation (MLI) and have simple thermo-statically controlled film heaters. Thermistors are used for temperature monitoring. High-efficiency MLI blankets minimize heat loss from the hydrazine Reaction Control System, which must always remain above 5°C to keep the fuel from freezing in the lines. The probe includes external coatings with high solar absorptance-to-emittance ratios, such as Vapor Deposited Gold (VDG). In order to reject the transponder heat, Optical Solar Reflectors (OSR's) are used on the bottom of the probe. See Fig. 7.

4.3 Reaction Control Subsystem (Propulsion)

Each THEMIS Probe includes a Reaction Control Subsystem (RCS) to correct launch vehicle dispersion errors, inject each probe into its respective mission orbit, maintain the orbits,

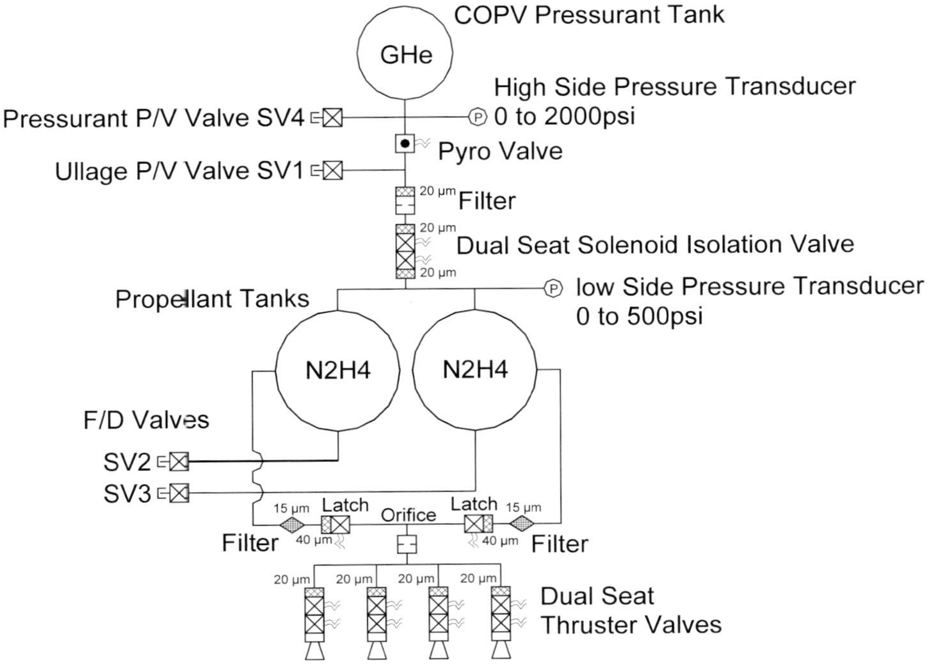

Fig. 8 RCS block diagram

adjust spin-axis pointing and maintain a nominal spin rate. The fundamental robustness of the mission design is due to the capability of probes 3 or 4 (P3, P4) to fully replace any probe, should it fail. Thus, the RCS has been sized for a nominal mission profile plus the worst-case contingency of replacing the P1 probe.

The probe is capable of both axial and side thrusting for orbit maneuvers with minimal efficiency loss allowing for operational flexibility. The tangential thrusters also act individually for spin rate adjustment. As shown in Fig. 8, the system consists of two fuel tanks, four 4.4 newton thrusters, a pressurant tank, latch valves, pyro valves, and miscellaneous hardware.

The two lightweight fuel tanks hold up to 49 kg of hydrazine (in total) and were specially designed and qualified for the THEMIS program. The tanks are made of high strength alloy (inconel) and are supported by the bottom and top panels via integral polar fittings. Tanks were verified non-magnetic by testing at UCLA. A high-pressure Carbon Over-wrapped Pressure Vessel (COPV) tank and pyrotechnic actuated valve dramatically enhance the systems capability. Once the fuel in the tanks has been depleted by approximately 25%, ground personnel command the pyrotechnic valve to open, which connects the high pressure tank to the fuel tanks. The resulting increase in pressure provides significantly more delta-V, totaling 960 meters per second.

The two axial engines provide 4.4 Newtons of thrust allowing for major orbit changes of the probe. In addition, two tangential engines of the same size provide either spin control or lateral thrust to the probe.

In order to maintain mass balance throughout the mission life, the two tanks were mechanically arranged to allow for symmetric fuel depletion. The pressurant and propellant sides of the RCS are interconnected to provide both symmetry and added probe reliabil-

Fig. 9 Miniature sun sensor

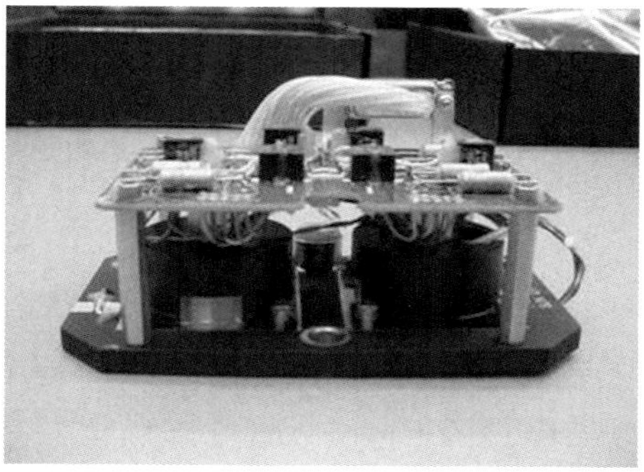

Fig. 10 IRU assembly

ity. Latch valves are located strategically to prevent adverse propellant migration during the launch phase of the mission. Following launch, both latch valves were opened to take advantage of *stabilizing* propellant migration inherent in this configuration.

Tank, line and thruster heaters are thermostatically controlled to maintain the hydrazine propellant comfortably above its freezing point. Thruster catalyst bed heaters are controlled by the BAU. The Flight Operations Team preheats the catalyst bed 30 minutes prior to firing in order to prevent cold-start degradation. The entire RCS weighs only 12 kg without fuel and is approximately 15% of the Probe dry mass.

4.4 Attitude Control Subsystem

The Attitude Control Subsystem (ACS) measures sun pulses and vehicle motions needed to support maneuvers, spin rate control and science data analysis. The ACS components are the Miniature Spinning Sun Sensor (MSSS) and the Inertial Reference Unit (IRU). The MSSS provides the sun elevation once per spin and assists in the calculation of spin rate. Using multiple spin pulses, flight software is able to determine the spin rate. The IRU is a solid-state assembly which measures angular rate of motion of the probe in X and Y axes. While these devices provide probe-relative data, the near Earth FGM data are used to verify probe absolute attitude once per orbit.

ACS telemetry is transmitted to the ground where it is processed into physical coordinates. If maneuvers are required, ground systems calculate the necessary commands to be sent to the probe, and these commands are verified on a ground-based avionics simulator prior to application in space. The ACS Bus components together weigh only 0.6 kg. Figures 9 and 10 show the Miniature Sun Sensor and IRU assembly.

Probes are stable spinners by design, as long as the radial booms are deployed before the axial booms. Even if severely perturbed, probes naturally return to their spin stable position without intervention.

The ACS design depends completely upon both the spin stability of the probes throughout all mission phases and the knowledge of the FGM sensor with respect to the probe body. To achieve spin stabilization, the probes are configured to have their center of mass closely aligned to the geometric axis. This alignment is accomplished through painstaking placement of components and by adjusting balance masses prior to launch. Careful design and measurements of the FGM boom, its repeatability and stiffness in thermal extremes were essential in providing accurate attitude knowledge to mission operations.

4.5 Power Subsystem

The Power subsystem is designed to provide all of the necessary power for the bus and instrument subsystems for the life of the mission in both sunlight and during eclipse. The power system is a Direct Energy Transfer (DET) system with the battery and solar array connected directly to the power bus. The solar array power control and battery charging are performed using linear and sequential switching shunts.

Each probe has eight solar arrays that provide power generation in any orientation. There are two arrays mounted on the bottom and top decks and there are four side panels. The arrays use high efficiency cells that are bonded to the composite substrates. The side panels are also primary structure that adds to their design complexity since they have to transfer loads between the top and bottom decks. In order to reduce surface charging, all the cover glass is electrically grounded to a common ground on each panel. This is accomplished by bonding a highly conductive grid onto the panels following cell placement.

The probe is highly efficient in power usage with approximately 36.85 watts required in full science mode for a 24-hour orbit, which includes a 3-hour eclipse and a 30-minute transmitter turn-on. The capability for that orbit at the mission End of Life (EOL) is 40.35 W. The top and bottom panels are intended to provide approximately 20 W at EOL, which is sufficient power to enable the probe bus components to survive anomalous attitudes in a low power condition.

High efficiency triple-junction Gallium Arsenide solar cells are used. The cells are approximately 4×6 cm, and they have an average BOL efficiency of 27% at room temperature. Each side panel has four strings and each top and bottom panel has two strings. Each string consists of 20 series cells with integral bypass diodes. Strings are carefully arranged on the panels to cancel the magnetic field generated by cells. Cover glasses are 8 mm thick with UV reflective coating. The cover glasses also have ITO coating to provide electrical conductivity and electrostatic cleanliness. The cover glasses are inter-connected by conductive epoxy to provide a bleed-path for surface charges to chassis.

Power is stored onboard by a lithium-ion battery that maintains full probe power during eclipse. The battery is lightweight and has a 12.0 A h capacity. The major power subsystem components weigh approximately 10.3 kg and represent approximately 13% of the total Probe bus dry mass.

4.6 Communication Subsystem

The Communication subsystem consists of an S-Band transponder, diplexer and circularly polarized antenna mounted to the center boom structure.

The antenna consists of six receiver/transmit stack patch antennas and a power divider. These antennas are extremely lightweight and must have a conductive surface in order not to

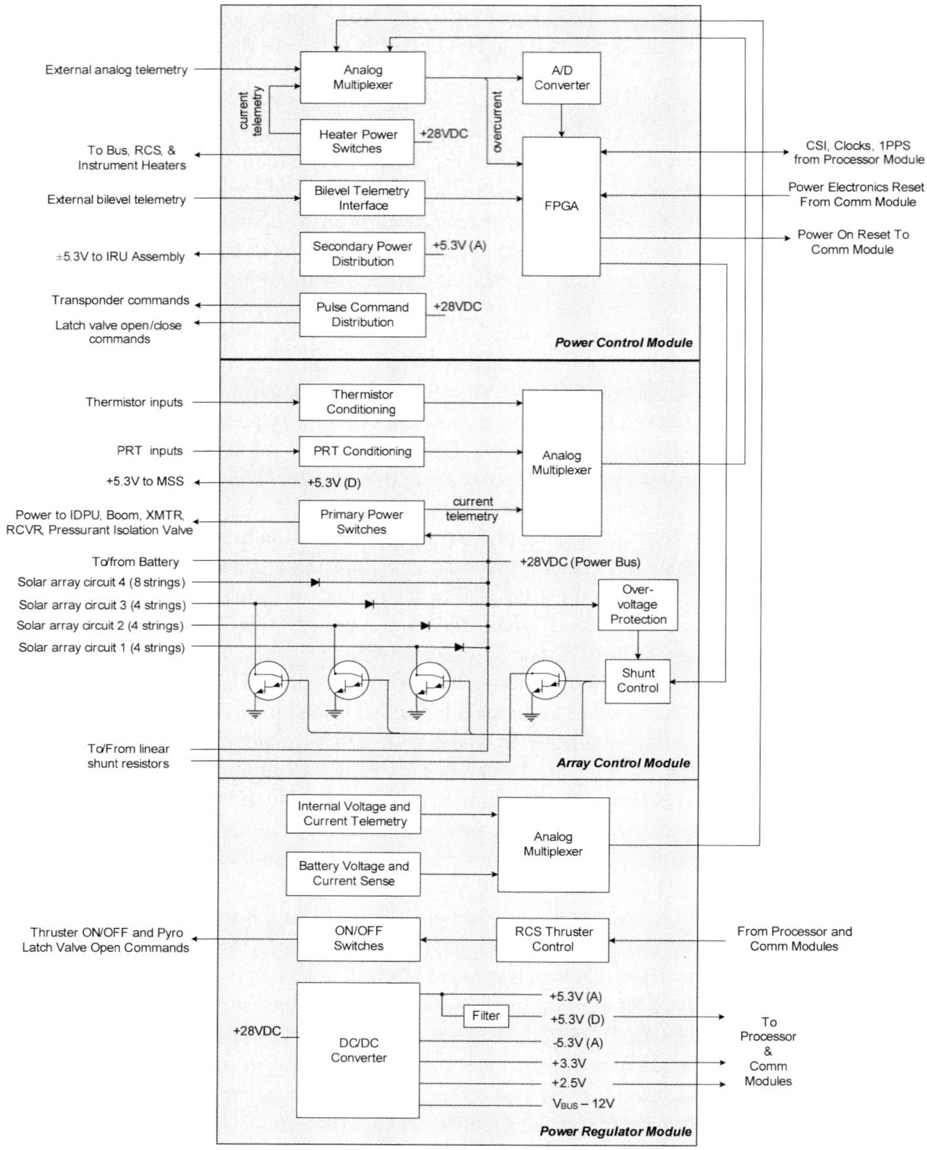

Fig. 11 Electrical power system block diagram

build up surface charge. For high data rate communication, they provide increased gain in a ±45° band about the plane perpendicular to the spin axis. Although reliable communication is possible outside this region, there is a small null region about the antenna boresight, and if the probe orientation is such that the line-of-sight to ground falls within this region, communication is not possible. However, since the probe is inertially pointed, communication outages of this kind would last for only small fraction of an orbit.

Fig. 12 S-band transponder

Fig. 13 S-band antenna

The antenna is connected to the transponder via the diplexer. The CXS-610 transponder contains a receiver and a 5 W transmitter. The antenna is always coupled to the receiver with no switches in the receive path, and the receiver is always powered. It demodulates command signals and outputs both data and timing to the BAU. For telemetry, the BAU provides baseband signals to the transmitter which phase modulates them onto the carrier. The transponder can be operated in a coherent mode that provides turn-around ranging capability.

Robust link margins exist for the uplink, even for the case of Probe 1 at apogee. For the downlink, multiple telemetry rates ranging from 1 kbps to 1024 kbps are provided. The total mass of the communication subsystem is 3.2 kg and represents 4% of the Probe dry mass. Figure 12 shows the transponder. The S-Band flight antenna is shown in Fig. 13.

4.7 Avionics

The Bus Avionics Unit (BAU) provides numerous functions for the probe bus and contains the flight computer. The BAU provides the communication interface, instrument electrical interface, data processing and power control for the probe bus. It contains a set of stacked modules with a total weight of 3.0 kg and average power of 7 watts. The electrical block diagram is shown in Fig. 14 and a photo is provided in Fig. 15.

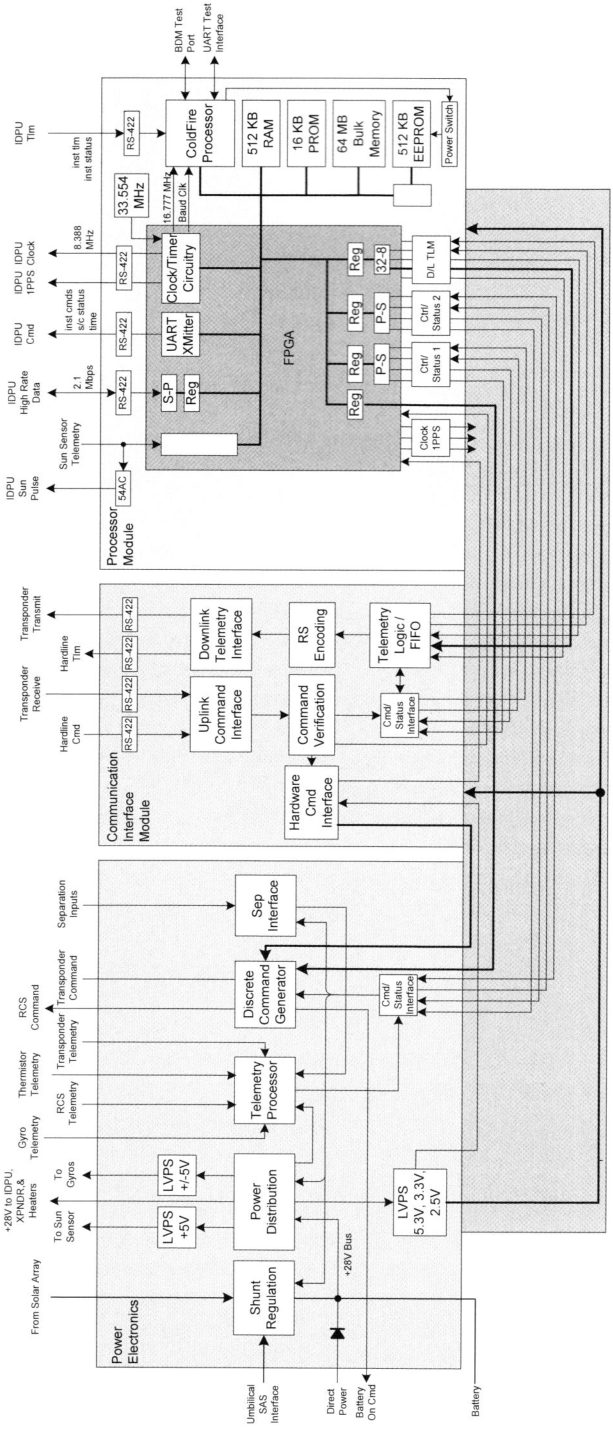

Fig. 14 BAU electrical block diagram

Fig. 15 Flight BAU

The Data Processor Module (DPM) contains a radiation-hardened computer featuring a Cold Fire processor operating at 16 MHz. This module performs all the onboard processing and data handling using 64 MB of bulk memory and a 2.1 Mbps data interface with the instrument electronics. The BAU hosts the RTEMS real-time operating system and the application control and data handling software for the probe bus. Instrument and bus housekeeping data is stored in the local bus memory with science data stored in the IDPU. During a ground pass the housekeeping data is transmitted directly by the processor, while science data is copied out of the IDPU to the transmitter.

The Communications Interface Module (CIM) receives the command bit stream from the receiver and provides CCSDS blocks to the processor. The card also processes a limited number of hardware commands that may be received from the ground and executed without the intervention of the processor. The card provides downlink telemetry data to the transmitter for transmission to the ground. The data may be real-time engineering, playback engineering data, and playback science data from the IDPU. Multiple telemetry rates are provided, ranging from 1 kbps to 1024 kbps. All data are encoded with rate 1/2 convolutional and Reed-Solomon encoding. The communications card also provides a hardline telemetry data stream for ground testing.

The Power Control Electronics (PCE) has 3 modules needed to control the solar array shunts, regulate battery charge, generate and distribute secondary voltages, generate and distribute discrete commands, and monitor separation signals from the launch vehicle and initiate probe separation from the probe carrier. The PCE also contains circuitry needed to condition and digitize most of the analog signals on the probe including IRU rate signals and temperature sensors.

The battery is charged at a fixed rate until the battery voltage reaches a commandable limit, at which point the charge current switches to trickle charge. The upper voltage limit can be selected conservatively so that no cell balancing is required. The BAU has the ability to autonomously remove power from the IDPU in case of over-current or battery undervoltage.

The BAU receives uplink commands at a fixed rate of 1000 bps. Commands are received using CCSDS protocols that guarantee correct, in-sequence delivery of variable-length command packets. All command transfer frames undergo several authentication checks. Invalid frames are rejected and the rejection is reported in telemetry. Commands sent to the probe will either be executed in real time or stored for later execution. Two kinds of stored com-

Fig. 16 Probe carrier ready for probes

mands are provided: Absolute Time Sequence (ATS) commands and Relative Time Sequence (RTS) commands. ATS commands have time tags expressed in UTC times, with a resolution of 1 second, specifying an absolute time of day. RTSs are command sequences that include relative delays between commands.

Since THEMIS orbits have long periods between contacts as well as radiation belt exposures, the BAU provides 64 MB of engineering data storage with error detection and correction (EDAC). Playback data stored in bulk memory is formatted into multiple segments, called virtual recorders, which allows for easy segregation into different types of data such as bus engineering data, instrument engineering data, event files, etc. The size of the virtual recorders is adjustable, allowing the memory to be remapped in the event of failed locations. The integrity of the data stored in bulk memory is maintained by memory scrubbing software, which uses the EDAC to correct single bit errors. Operating at a low priority, the memory scrub task cycles through all the data stored in bulk memory once per orbit.

The BAU maintains a precision onboard clock and distributes time to the IDPU in Universal Time Code (UTC). Time synchronization between the bus and the payload is achieved by the use of synchronous 8 MHz and 1 Hz clock signals sent to the IDPU. The IDPU uses the 8 MHz to collect science data. Once per second, the BAU sends the IDPU the UTC that will be valid at the next 1-Hz pulse. Together these actions allow the bus and instrument to properly time-tag all science and engineering data.

The BAU provides several autonomous functions that insure the health and safety of the probe while out of ground contact. A watchdog timer is provided which continuously monitors processor operations, and should a processor malfunction be detected, restarts the processor automatically. A checksum routine runs at low priority, checking static areas of memory. A telemetry and statistics monitoring function is provided which performs "limit check" operations on the data and which maintains telemetry statistics. If pre-specified conditions occur, it can initiate the execution of a stored command sequence.

The BAU utilizes system tables to implement operational controls and to ease ground system operations. System table operations constitute the primary ground interface for probe control functions such as stored command operations and modifications of on-board parameters. The BAU also has the ability to build "memory dwell" packets to monitor any memory location for diagnostic support.

Fig. 17 Probe F2 in EMC

4.8 Probe Carrier and Separation System

The THEMIS Probe Carrier and Separation Systems met the challenge to launch all five probes on a single Delta II, modified to accommodate access to five payloads and provide five redundant separation signals. The payload challenges included imparting an initial stabilizing spin rate to each probe, maintaining a positive separation between probes and doing so even if one probe failed to separate.

The launch vehicle design had the Probe Carrier bolted to the third stage of the Delta II. At the end of the launch sequence, the third stage de-spun to 16 RPM and initiated separation pyrotechnics. As designed, the top probe deployed first and the lower four probes deployed simultaneously three seconds later. Launch vehicle analog telemetry confirmed the correct release profile and probe telemetry confirmed expected probe motions. Initial spin rates for all five were between 16 and 17 RPM while sun angles were within 6 degrees of each other.

There were a number of significant challenges in the separation system design. First, since the Probe Carrier was spinning at release, the four lower separation systems had to operate reliably with a side load. Second, in order to avoid collisions between probes and/or the carrier, the separation system had to move the probe quickly away from the separation plane while imparting a low tip off rate.

The Probe Carrier is predominately aluminum alloy, is weight optimized, and includes a patch panel that manifolds all of the umbilical electrical and control circuit cabling from the probes to the launch vehicle. The separation system was extensively analyzed and tested to properly characterize its performance and to verify all of the mechanical parameters

Fig. 18 Probe F2 in magnetics

that drive the overall Probe and Probe Carrier system clearance verification analysis. Figure 16 shows the Probe Carrier at Astrotech Space Organization (ASO) launch site payload processing facility.

5 Integration and Test

5.1 Probe Integration at UCB

Integration of the Probe buses and Instrument Suites was performed at UCB/SSL in a class 10000 clean room. Each instrument suite, complete with flight harnessing, was rolled up to a Probe and electrically connected via extension cable. Following interface verification, all instrument components were mechanically integrated and alignments verified. Due to the proximity of the Berkeley Ground Station (BGS) and Mission Operations Center (MOC), end to end verification of RF communications with BGS and MOC were verified at this point.

5.2 Environmental Verification Testing

Environmental testing of the THEMIS probes was conducted at the Jet Propulsion Laboratory (JPL) in Pasadena, California, in two stages: first with a single "pathfinder" Probe F2 and later with all five Probes. While other probe buses and instrument suites were still in subsystem test, F2 proceeded to JPL and through environmental testing. The purpose was to uncover problems before fully integrating the other Probes. See Figs. 17–19.

Fig. 19 Probe F2 in vibration

Pathfinder Environmental Tests. The Pathfinder schedule and test sequence is shown in Table 6. The test team included staff from UCB, ATK Space and JPL. Daily teleconferences were held to coordinate staff and plans, and to discuss and resolves questions and issues as they arose. During the pathfinder test, we verified the mechanical ground support equipment used for lifting, rotating and manipulating the probe. We also tested and verified the electrical ground support equipment used to monitor and command the probe.

The magnetics survey took place late at night to avoid interference from vehicles passing outside the building. Three axis magnetic field measurements were taken: (1) all instruments off; (2) all instruments turned on; and (3) after 15 gauss de-perm. The Probe had measurements taken in both vertical and horizontal positions, with several rotations in each position. Measurements showed the Probe to be well within the specified requirement of 5 nT at 2.5 meters from the center.

Full Payload Environmental Tests. While the pathfinder was in environmental test at JPL the other four probe buses and instrument suites were in assembly and test at ATK Space and UCB. The final buses were delivered in May and June, enabling completion of Probes 3, 4, 1 and 5, a combined Pre-Environmental Review and return to JPL by the end of June.

Table 7 summarizes the activities of the five Probes and carrier in environmental testing. After arriving at JPL, the Probe Carrier was tested with the probe mass dummies, proving

Table 6 Probe F2 environmental test sequence

DATE	ACTIVITY
Mar 20–21	Arrival, offload and set up
Mar 22	Magnetics testing
Mar 23–24	Move to vibration facility and set up
Mar 27–30	Vibration tests X, Y & Z axes
Mar 31	Separation shock test
Apr 3	Move to EMC facility and set up
Apr 4–6	EMC tests
Apr 7	Move to thermal vacuum facility and set up
Apr 10–11	Thermal vacuum closeouts and blanketing
Apr 12	Install probe in TV chamber
Apr 13–16	Thermal balance
April 17–21	Thermal vacuum cycles (4×)
Apr 24	Move to magnetics facility, final mag test
Apr 25	Transport to UCB/SSL

Table 7 Full payload environmental test sequence

Week	Major Tests and Activities	
July 10	Post-ship probe functional Tests	
	Probe tank pressurization	
	PCA integration	
	PCA X and Y axis Vibration	
July 17	PCA Z axis vibration	
	PCA acoustics test	
	Probe separation-shock (5×)	
	Probe tank depressurization	
July 24	F3, F4 thermal vacuum set up	F1, F5 magnetics survey
	F3, F4 thermal balance	F1, F5 spin balance
	F3, F4 thermal vacuum cycles (4×)	F1, F5 alignment measurements
July 31	F2, F3, F4 magnetics survey	F1, F5 thermal vacuum set up
		F1, F5 thermal balance
Aug 7	F2, F3, F4 spin balance	F1, F5 thermal vacuum cycles (4×)
	F2, F3, F4 alignment measurements	F1, F5 move to bldg 179
Aug 14	F2, F3, F4 functional tests	F1, F5 spin balance (repeat)
		F1, F5 functional tests

that the carrier was capable of sustaining the worst-case masses of the fueled probes. After probe post-ship functional tests, the Probes were integrated to the Carrier while still on the vibration table (see Fig. 20). Environmental testing of the five probes was completed on schedule with no major anomalies. See Figs. 21 and 22.

Fig. 20 Integrating the probe carrier assembly on vibration table

5.3 Launch Site Activities

Following a four-month delay due to launch vehicle issues, the project received the green light to proceed to launch in December. Initial pre-launch activities took place at Astrotech Space Organization (ASO) located very near KSC. The flow of activities is shown in the chart below (Fig. 23).

As illustrated in Figs. 24 to 28, important activities were as follows:

- The five probes and carrier arrived at Astrotech on December 11, 2006.
- Functional tests, pressurization tests, and bolt cutter installation were carried out in the Payload Processing Facility (ASO1);
- The Probe Carrier was electrically checked in a "side-by-side" test with the 3rd stage, and integrated with its Separation System pyrotechnic lines in the Hazardous Processing Facility (ASO2);
- Following the holiday break, Probes were moved to ASO2. Probes were weighed individually, fueled all at once, then re-weighed individually;
- The probes were integrated to the PCA, electrically checked and spin-balanced;
- On January 29, the PCA was bagged, lifted and mated with the Boeing 3rd stage;
- On February 3, the containerized 3rd stage and payload were transported to the pad;
- At the pad, the THEMIS team tested and charged probe batteries and practiced launch sequences while the Boeing Delta II rocket was being prepared for launch;
- After delays due to lightning and high altitude winds, THEMIS was launched on February 17, 2007.

Fig. 21 Probe carrier assembly ready for Z-axis vibration

6 Lessons Learned

The development of the Constellation proved demanding on several levels. The following are conclusions from the experience:

- THEMIS validated the effectiveness of the "pathfinder" approach, and showed dramatic improvement in performance, schedule, and cost of subsequent units;
- Keeping the Probes identical, despite the fact that the mission required the Probes to be in different orbits, greatly benefited the test sequence;
- Using the same technician for similar tasks across all Probes proved effective in maintaining similar Probe performance;
- Requiring the instruments be designed with internal test features limited the need for drag-on test equipment through I&T;
- Vibrating the full PCA rather than individual probes (or pairs of probes) was a significant improvement. It made the test more flight like, saved schedule and provided useful experience with the full PCA before heading to the launch site;
- Following the Pathfinder Probe#2, performing the thermal vacuum test of the subsequent probes in sets of two worked well and saved considerable schedule time and costs;

Fig. 22 Probes 3 and 4 (enclosed) at thermal vacuum

- Enclosing the probes within individually controlled thermal enclosures added substantial work and complexity. Despite this, the conclusion of the thermal engineers was that it resulted in a significantly better test and resultant thermal modifications.
- Environmentally testing the pathfinder was very useful and resulted in numerous improvements, especially in thermal design. Thermal blanket modifications to the transmitter radiator were suggested by the pathfinder tests. These modifications significantly improved transmit durations and were verified in subsequent thermal vacuum testing on other probes.
- Probe spin balance resulted in all five being balanced within specification with approximately 1.6 kg total balance mass. Substantial benefits were realized with the number of probes. The balancing of the first two probes was time consuming and initially indicated need for substantially more balance mass than expected (over 3 kg). Corrections made with the later probes showed that balance could be improved and the balance mass reduced. The two probes that were the first to be balanced were then re-balanced. The five probes were shown to have very high degree of uniformity and almost identical balance mass.
- The four sets of electrical GSE proved to be uniform and we could operate any Probe from any GSE. This fact greatly facilitated the schedule as the GSE often stayed in one spot while the Probes moved through the facilities and were attached to any GSE test set.

Problems in Integration and Test were concentrated on the first unit and fell off with each pair of subsystems tested. As the instrument suites and Probes were tested, Problem/Failure Reports (PFR) specified all units needing modification. For design errors or common fabrication errors, all units were modified, including units that had not yet been tested. Of the 171

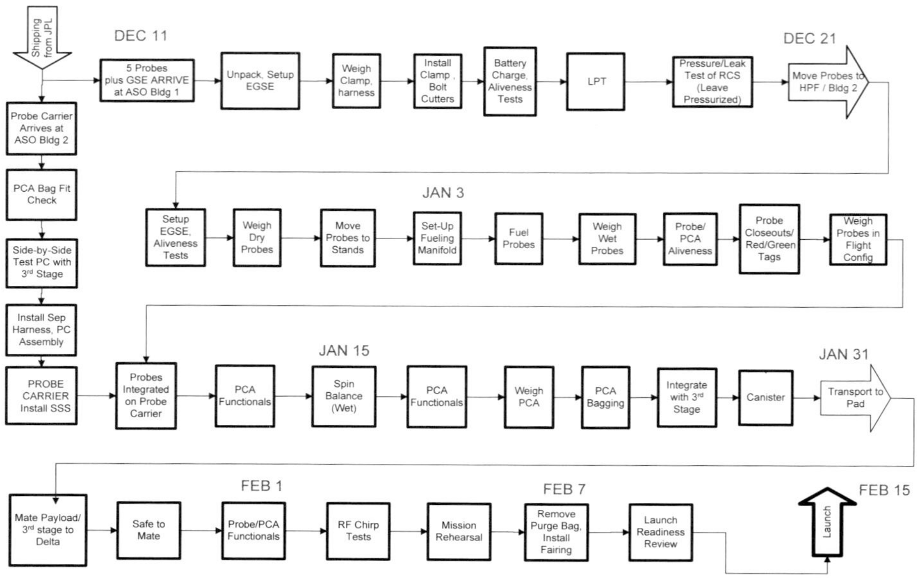

Fig. 23 Payload processing at ASO/KSC

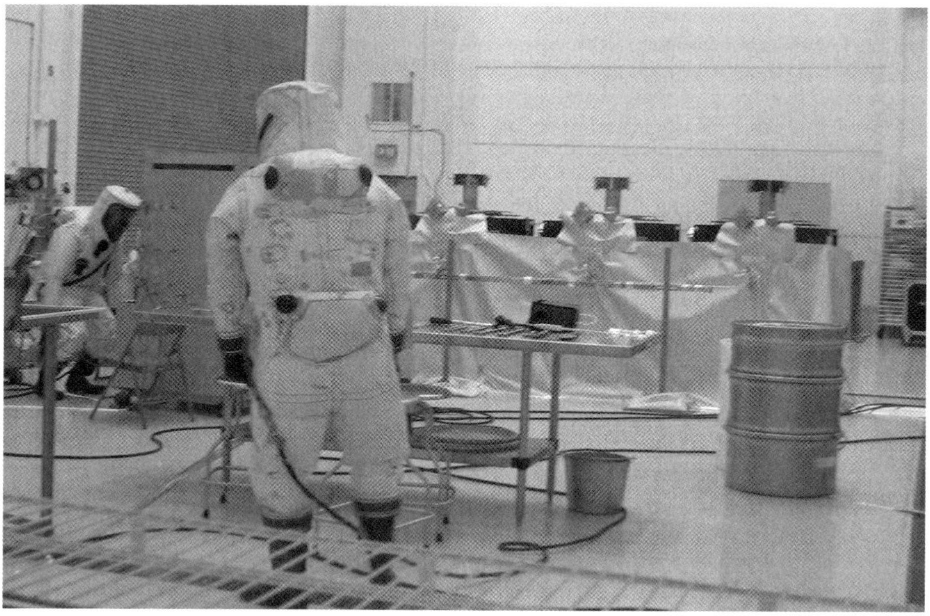

Fig. 24 Fueling probe in Hazardous Operations Facility

Bus and Instrument I&T PFR's, 63 corrected multiple units providing down stream benefits. While modifications proceeded on the current unit, changes of future units occurred in parallel, effectively advancing the schedule of future units. As shown in Table 8, the first unit

Fig. 25 Probe carrier assembly on spin balance machine

Table 8 Problems encountered at I&T (by unit order)

I&T new PFRs	1st	2nd	3rd	4th	5th	Total
Probe Bus in I&T	35	3	2	1	6	47
Inst Suite in I&T	9	3	3	3	3	21
Inst Suite	48	19	20	9	7	103
Total	92	25	25	13	16	171

found more than half the problems, followed by a rapid drop in problems in later units. Since the last unit was on the critical path, this had a real effect on the mission level schedule.

7 Project Performance

7.1 Technical Performance

The Probes and Carrier performed very well through integration and test, components accumulating an average of 810 hours overall and 250 hours in thermal vacuum conditions. Probes had an average of 350 failure free hours at launch. Of the 476 mission requirements, only 2 requirements were waived. These were due to minor variances on the EFI noise floor and clearances between Probes on the Carrier.

Final Probe dry masses were 4.7% below their 81.8 kg Not-to-Exceed limits and matched to less than 1%. Probes A–E measured 78.0, 77.6, 76.7, 76.7, and 78.1 kg, respectively. The

Fig. 26 PCA mate to the 3rd stage

Probe Carrier Assembly weighed in at 777 kg with a 6.3% margin to the launch vehicle NTE of 829 kg. The final Probe power budget at 38.7 W has a 4.9% margin in surviving a 3-hour eclipse at the end of mission.

7.2 Schedule Performance

The successful scheduling of the THEMIS project required a mixture of optimism, dogged determination and endless coordination between UCB, ATK Space and GSFC management. From project funding to launch took 46 months, including a 4-month launch vehicle delay. The first Probe began integrating instruments at 31 months and all Probes completed all testing 9 months later.

The THEMIS top-level schedule is illustrated in Fig. 29. The master schedule linked together 23 instrument and 10 probe and carrier schedules, together totaling more than 5000 tasks. These were managed by schedulers at UCB and ATK Space and milestone-tracked by GSFC.

The development of instrument suites and probe buses used a 1, 2 and 2 approach to build the five Probes. The Environmental Verification Tests (EVT) used a different approach, verifying the first Probe then all five Probes and Carrier. The latter sequence provided "test-as-you-fly" configurations and shortened the overall project schedule.

Fig. 27 Delta II launch vehicle

All levels of integration, whether instrument or bus or probe, witnessed dramatic schedule improvements with each unit. These were principally due to (1) improved test procedures and (2) reduction in newly discovered problems.

7.3 Cost Performance

The THEMIS budgetary estimate at Confirmation Review was 158.3M (FY02) assuming a payload development and operations of \$89.3M and a launch vehicle at \$69M. Actual payload costs ended 4% high at \$92.9M. Actual costs of the launch vehicle and the launch delay brought the total cost to \$172.8M, yet still beneath the MIDEX cap of \$180M. See Table 9.

The project experienced dramatic cost reductions as the units were fabricated. The design phase through CDR cost \$20M, roughly one quarter the total, while costs of the first

Fig. 28 Launch, February 17, 2007

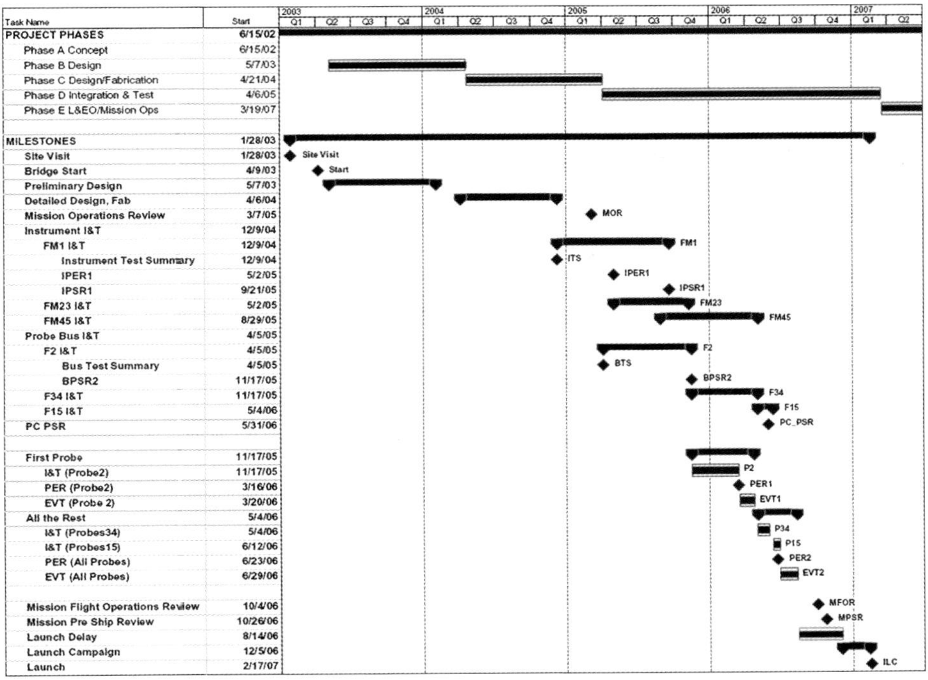

Fig. 29 THEMIS constellation schedule

Table 9 THEMIS cost in FY02 M$

THEMIS budget performance FY02 $M	Phase A–D	Phase E	Total
UCB/SAI Probe & carrier development	80.3	10.3	90.6
JPL environments and simulations	1.1	0.0	1.1
GSFC thermal and data analysis	0.4	0.9	1.2
Payload (w/o launch delay)	81.7	11.1	92.9
Launch vehicle (AO)	69.0		69.0
Launch vehicle additional costs	4.9		4.9
UCB/SAI impact of launch delay	3.1	3.0	6.0
Total	158.7	14.1	172.8

flight- Probe were another $20M. Remarkably, the four remaining Probes were completed at roughly half the cost of the first.

Acknowledgements The THEMIS mission is that rare combination of inspiration and imagination that challenges scientists, engineers, and managers alike. Without question, the success of the project is due to the indefatigable efforts and contagious optimism of the PI Vassilis Angelopoulos, who not only convinced all of us that it could be done, but that *we could do it*.

Dr. D. Pankow provided vehicle dynamics and flawlessly led the vibroacoustics and balance efforts. Dr. Auslander and UCB graduate students simulated vehicle dynamic behavior due to fuel slosh and wire booms. D. Curtis and S. Harris carefully verified each bus as C. Chen, H. Richard and M. Ludlam did the same for the instrument suites. Dr. M. Sholl verified the propulsion system throughout integration and led probe fueling, and M. Leeds provided RCS training and support. Dr. M. Bester verified communications with BGS and the Mission Ops Center.

Deputy Project Manager D. King and Project Scheduler D. Meilhan tracked an unbelievable number of tasks and kept it all under control as Financial Manager K. Harps kept costs in line. Mission Assurance Manager R. Jackson and quality personnel J. Fisher and C. Scholz managed to get all the parts, inspect all the components and track all the problems to closure.

ATK Space system engineers K. Brenneman and W. Chen, propulsion designer M. McCullough, separation system designer D. Jarosz and thermal engineer R. Zara were instrumental in the success of the probes and carrier. The Hammers company for excellent support of Bus flight software. ATK Space Vice President F. Hornbuckle committed the company to the project at a time when it was probably unpopular to do so.

Mission Manager F. Snow demonstrated great patience and support particularly while the project went through difficult times. The Explorers team of D. Lee, R. Miller, J. Thurber and D. Gates provided support where needed in government services, scheduling, RF expertise and communications. Finally, IIRT co-chairman Mark Goans and Brian Keegan provided valuable reviews, personal counsel and much-appreciated confidence in the THEMIS design and implementation.

References

V. Angelopoulos et al., The THEMIS mission. Space Sci. Rev. (2008, this issue)

U. Auster et al., The THEMIS fluxgate magnetometer. Space Sci. Rev. (2008, this issue)

M. Bester et al., Mission operations for THEMIS. Space Sci. Rev. (2008, this issue)

J. Bonnell et al., The electric field instrument for THEMIS. Space Sci. Rev. (2008, this issue)

S. Frey et al., THEMIS orbit design. Space Sci. Rev. (2008, this issue)

S. Harris et al., THEMIS ground based observatory system design. Space Sci. Rev. (2008, this issue)

D. Larson et al., The solid state telescope for THEMIS. Space Sci. Rev. (2008, this issue)

J. McFadden et al., The THEMIS ESA plasma instrument and in-flight calibration. Space Sci. Rev. (2008, this issue)

D. Pankow et al., THEMIS booms: Design, deployment and stability. Space Sci. Rev. (2008, this issue)

A. Roux et al., The search coil magnetometer for THEMIS. Space Sci. Rev. (2008, this issue)

E. Taylor et al., Instrument data processing unit for THEMIS. Space Sci. Rev. (2008, this issue)

Instrument Data Processing Unit for THEMIS

E. Taylor · P. Harvey · M. Ludlam · P. Berg · R. Abiad · D. Gordon

Originally published in the journal Space Science Reviews, Volume 141, Nos 1–4, 153–169.
DOI: 10.1007/s11214-008-9459-4 © Springer Science+Business Media B.V. 2008

Abstract The Time History of Events and Macroscale Interactions during Substorms (THEMIS) mission is a NASA Medium-class Explorer (MIDEX) mission, launched on February 17, 2007. The mission employs five identical micro-satellites, or "probes," which line-up along the Earth's magnetotail every four days in conjunctions to determine the trigger and large-scale evolution of magnetic substorms. The probes are equipped with a comprehensive suite of instruments that measure and track the motion of thermal and super-thermal ions and electrons, and electric and magnetic fields, at key regions in the magnetosphere. Primary science objectives require high data rates at periods of scientific interest, large data volumes, and control of science data collection on suborbital time scales. A central Instrument Data Processing Unit (IDPU) is necessary to organize and prioritize the data from the large number of instruments into a 200 MB solid state memory. The large data volume produced by the instruments requires a flexible memory capable of both high resolution snapshots during conjunctions and coarser survey data collection throughout the orbit. Onboard triggering algorithms select and prioritize the snapshots based on data quality to optimize the science data that is returned to the ground. This paper presents a detailed discussion of the hardware and software design of the THEMIS IDPU, describing the heritage design that has been fundamental to the THEMIS mission success so far.

Keywords THEMIS · Space instrumentation · Data processing · Power distribution and control · Instrument processing · Flight software

1 Introduction

The Time History of Events and Macroscale Interactions during Substorms (THEMIS) mission, the fifth NASA Medium-class Explorer (MIDEX), launched on February 17, 2007 to determine the onset and large-scale evolution of substorms. Flying in synchronous orbits

E. Taylor (✉) · P. Harvey · M. Ludlam · P. Berg · R. Abiad · D. Gordon
Space Sciences Laboratory, University of California at Berkeley, Berkeley, CA, USA
e-mail: ertaylor@ssl.berkeley.edu

within the earth's magnetosphere, the five THEMIS satellites, or Probes, track the particle and field processes responsible for eruptions of the aurora. A simple Instrument Data Processing Unit (IDPU) manages the entire instrument suite, collecting and organizing the data before transmission to the ground. The IDPU controls a large number of instruments: 2 Solid State Telescopes (SSTs), 1 Electrostatic Analyzer (ESA), 6 Electric Field Instruments (EFIs), 1 Search Coil Magnetometer (SCM) and 1 Flux Gate Magnetometer (FGM). It also performs critical actuator and deployment tasks such as wire boom deployments (4 per Probe), rigid boom deployments (4 per Probe), one-shot cover openings (1 per Probe) and attenuator control (2 per Probe) (see Pankow et al. 2008). In addition, support electronics include instrument signal conditioning, power control and distribution, command routing, data optimization and analysis, and housekeeping monitoring.

The high resolution science data required during scientifically interesting events drove a design philosophy to autonomously change data rates based on parameters of interest, or "trigger" functions. To accommodate the different data rates, the THEMIS team developed four levels of data collection: Slow Survey, Fast Survey, Particle Burst and Wave Burst. The IDPU controls the data collection rates based upon trigger data, and prioritizes the high resolution burst data for selective transmission to the ground. To reduce the work load on the low-power 8085 IDPU processor, numerous tasks normally performed by a processor were delegated to custom designed circuits using Field Programmable Gate Arrays (FPGAs). The IDPU software is then able to focus on data prioritization and compression schemes to maximize the quality and quantity of the science data. This data optimization scheme has led to high value, scientifically interesting data being collected and downlinked every orbit.

The IDPU circuitry is packaged in a single box to reduce mass, to simplify harnessing and interfaces, and to reduce duplication in power converters and other common services. Electronic parts for the IDPU were selected according to criteria for low-power, functionality, reliability, and radiation hardness. Moderate part reliability levels (to MIDEX standards) were coupled with extensive board and instrument-level testing to assure reliability. Parts were specified to and/or lot tested and shielded to achieve a minimum total dose radiation tolerance requirement of 66 Krads.

This paper is organized into two main sections providing a detailed discussion of the hardware and software design of the THEMIS IDPU.

2 Hardware Description

The IDPU consists of a 7-slot 6U VME chassis. The IDPU hardware boards can be grouped into two main functions: core electronics boards including a Data Controller Board (DCB), Low Voltage Power Supply (LVPS), and Power Controller Board (PCB); and instrument specific support boards including a Digital Fields Board (DFB), Boom Electronics Board (BEB), Fluxgate Magnetometer Electronics (FGE), ESA/SST Interface Circuit (ETC), and SST Instrument Digital to Analog Processing Board (DAP). The DCB collects, formats, frames, and stores data from the instrument specific boards and then sends it to the Probe's C&DH processor in the Bus Avionics Unit (BAU) for transmission to ground. It also accepts commands from the BAU and controls instrument operations. The solid state recorder (SSR), resident on the DCB, is a 200 MB error-corrected SDRAM that can be configured to store multiple data types (survey, burst) and data rates up to 630 Kbps. The power system (LVPS and PCB) provides stable, regulated voltages through DC/DC conversion to the instrument support boards and sensors. The LVPS, connected to the VME Chassis through a Hypertronics connector, conditions and converts probe power for the instrument electronics and mechanisms. The PCB monitors current draw and voltages, switches instrument

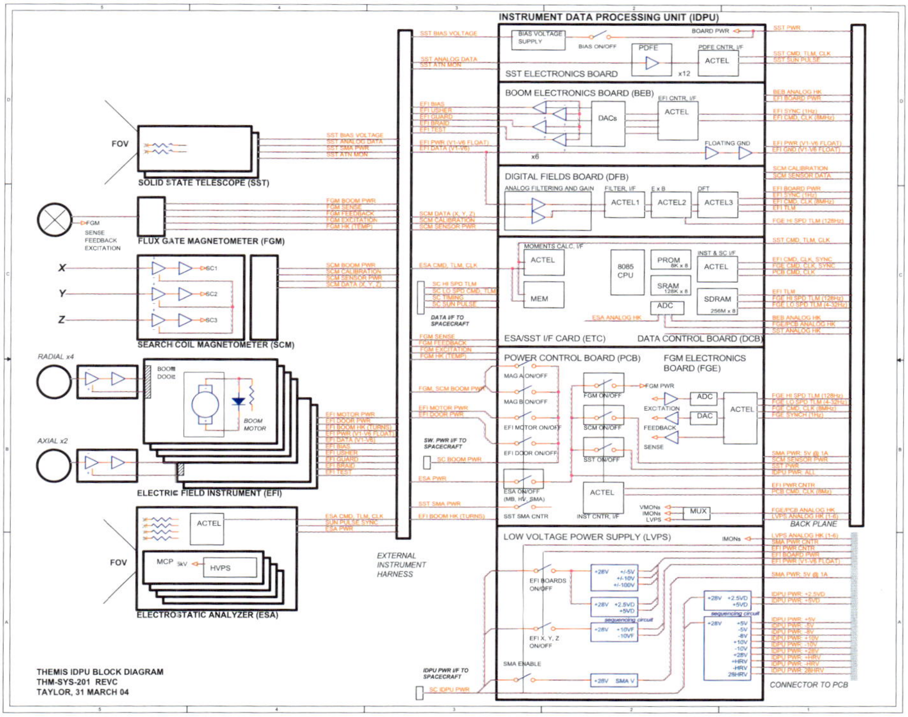

Fig. 1 THEMIS IDPU block diagram

power services, and provides current limiting. The instrument support boards (DFB, BEB, FGE, ETC, and DAP) collect and condition instrument specific data. The instrument support boards are briefly described here and more extensively in separate papers (Auster et al. 2008; Bonnell et al. 2008; Cully et al. 2008; Larson et al. 2008; McFadden et al. 2008). Figure 1 shows a detailed block diagram of the IDPU.

The IDPU has a number of independent operating configurations, which mainly affect instrument science data accumulation rates. Only three basic modes (Safe, Low Power, and Science Mode) affect power consumption and dissipation. A fourth mode, Engineering Mode, affects the IDPU housekeeping data rate only. In Safe Mode, only the core systems are powered. This mode is entered on reset (power-on), by a ground command, or in response to a flag in the spacecraft computer signaling that power is imminent. Safe Mode preserves power to and saves the contents of the SSR. In Low Power Mode, the core system and the FGM are powered on, but all other instruments are off. This mode is entered by a ground command in preparation for maneuvers, as FGM data is used for attitude determination, or in response to a flag in the spacecraft computer signaling a low power condition. Science Mode is the normal operating state. In this mode, the instrument payload is ready for full science data collection and is autonomously controlled by flight software using onboard triggers, as described in the Software Section of this paper. Finally, Engineering Mode enables safety critical operation. This mode can be entered by command only, typically during ground contact and in preparation for early operations (instrument health and safety diagnostics) and special case instrument operations (boom deploy and boom releases).

2.1 IDPU Core Electronics

2.1.1 Data Control Board (DCB)

The THEMIS DCB provides intelligence for the IDPU, controlling the flow of raw data through the memory system. It receives commands from the Probe's BAU processor and manages the instrument payload. A solid state recorder (SSR) is included on the DCB to eliminate a high bandwidth interface between the BAU and the IDPU, and to allow more control of the data recording by the IDPU. Science data volume drove the data storage requirement of 200 MB. The DCB formats data directly into packets and frames so it is ready for transmission without further processing. The processor is an 8085 microcontroller, running at 2.0 MHz. It is supported by an 8 K × 8 Boot ROM, SEU-Immune Static RAM, SSR SDRAM, and EEPROM. The 128 K × 8 SRAM is segmented and shared by the FPGA subsystems. The processing resources required for the IDPU are modest because the bulk of the high-speed data handling is done by the FPGA. Support circuitry in the FPGA includes address/data demultiplexing, memory decoding, and spacecraft interface logic. The DCB communicates with the instruments or their interface subsystems via dedicated serial command data interfaces (CDIs), receiving/storing data and forwarding instrument commands and register loads. An Analog to Digital Converter (ADC) on the DCB is used by the processor to collect analog housekeeping. Figure 2 shows an overview of the main subsystems comprising the DCB. The DCB resides on a 6U VME board, shared with the ETC (SST & ESA control subsystem). It connects to a backplane via a standard VME 96 pin connector. Power is received through the backplane connection, which is also used to communicate with the instrument support boards. A harness is used to connect the DCB to the Probe's BAU.

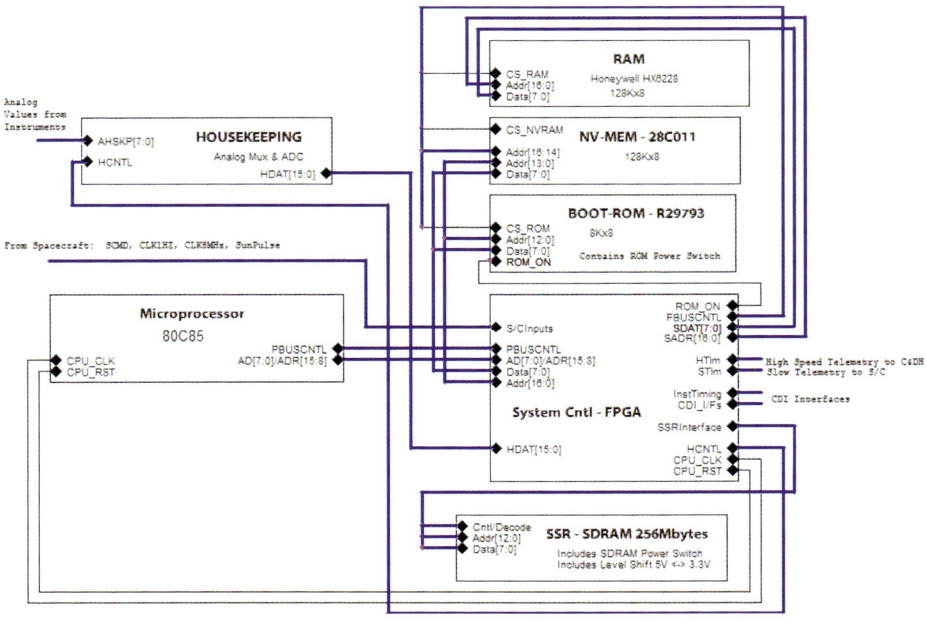

Fig. 2 THEMIS Data Control Board (DCB) block diagram

2.1.1.1 Processor The DCB central processing unit (CPU) is an 8-bit radiation hardened processor 80C85RH with very low power requirements (< 100 mW). It provides control and monitoring of the instruments and data flow, collection of instrument housekeeping data, and is the sole interface between the instruments and the BAU. The processing function is shared by the DCB FPGA which handles the processor bus control and provides registers for accessing the various sections of memory. Included in the FPGA are the Instrument Interface Logic, the SDRAM controller, Error Detection and Correction (EDAC) for the SDRAM, the S/C Interface Logic, direct memory access (DMA) and data management control, analog housekeeping control, and timing/time-tagging support.

The IDPU has an overall reset capability, generated by the DCB FPGA. The reset is an OR of two sources: a power-on hardware reset (generated by an RC delay network and AC14 gate) and a watchdog reset. The watchdog reset pulse is generated if the CPU does not write to the Watchdog Reset Clear register for a period of 3 seconds. An external jumper allows the watchdog to be disabled during testing or debug. During test, a debug interface allows for the connection of external diagnostic peripherals and monitoring of the CPU Bus.

2.1.1.2 Memory The CPU memory bus directly connects to the DCB FPGA and boot ROM (8 K $\times$ 8 Raytheon R29793). An external SRAM (Honeywell HX6228) resides on a private memory bus controlled by the FPGA and available to the CPU. An EEPROM (128 K $\times$ 8) provides non-volatile storage. A write-protect bit in the Control Register prevents spurious accesses. The EEPROM also contains internal write-protection mechanisms. The EEPROM resides on the CPU bus, and is not accessible to anything but the processor. At reset, EEPROM starts booting from address 0, where boot ROM is mapped. After boot-up, the CPU turns off power to the boot ROM via an on-board switch controlled by the FPGA, and executes directly out of RAM.

The address space of the processor (64 Kbytes) is much smaller than the total memory available. The processor accesses all the various memory locations by a combination of mapping and switching. When the ROM is off, the lower 32 K of address space is mapped to SRAM. The upper 32 K is switched between the remaining SRAM locations, EEPROM, and the SSR. Two page registers in the FPGA define the paging for the upper half (32 K) of CPU address space.

2.1.1.3 Spacecraft Interface The IDPU-spacecraft (BAU) interface contains a 2 MHz High-Speed Telemetry Interface, a bidirectional UART operating at 38.4 Kbaud for Low Speed Telemetry and Commands, timing signals or Clocks, and a “Sun Pulse” signal from the BAU to the DCB. The UART signals are conditioned with differential RS-422 drivers/receivers. The transmission lines are shielded twisted pairs with the individual shield grounded at the driver end only.

Command Interface The Command interface is a serial interface used to send data from the BAU to the IDPU. Data is transmitted at 38.4 Kbaud. Data is exchanged once per second and includes instrument specific commands, probe status information, and time. The probe status information includes instrument current draw, transponder status, maneuver status, eclipse flags, and thermal information.

Low Speed Telemetry Interface The Low Speed Telemetry interface is a serial interface used to send housekeeping data from the IDPU to the BAU. The data is transmitted using UART encoding, and a differential interface. The UART is a standard 8-bit bidirectional UART operating at 38.4 Kbaud. Low-speed data is exchanged once per second and includes a housekeeping packet, containing instrument state of health data, and FGM data, used for attitude determination on the ground.

High Speed Telemetry Interface The High Speed Telemetry (HST) interface is a bit serial interface used to transfer CCSDS-formatted instrument data telemetry packets from the SSR to the BAU for transmission to the ground. The HST interface transfers data in a single direction, from the IDPU to the BAU. Both the BAU and the IDPU must be enabled, by command, to participate in a transfer session. Packetization is in conformance with CCSDS Packet Telemetry recommendations, and includes primary and secondary headers followed by a stream of data. Data is transmitted to the high-speed link on a frame by frame basis. When the BAU is ready for a frame, it asserts a ready signal, at which point the DCB clocks out a bit-stream along with a clock synch corresponding to each data bit. The CPU is responsible for setting up the DMA transfer on a packet by packet basis. The FPGA inserts the variable sized CCSDS packets into fixed sized frames and generates the Transfer Frame Headers (both primary and secondary). The flow of data is not continuous; rather, it's broken into discrete transfers, or "Transfer Frames." The Transfer Frame has no Sync Word or Reed Solomon Block. These blocks are added to each frame in the BAU to produce the Master Frame, which is then ready for transmission to ground.

Timing The IDPU receives Probe time in the form of two clock signals at 1 Hz and approximately 8.4 MHz plus a periodic synchronizing command. The Probe BAU has a stable, oven-controlled oscillator (OCXO) that provides timing for all the subsystems. The DCB internal time-base is set by a local oscillator at 20 MHz which is used by the FPGA as the overall system clock. The DCB generates a 1 Hz Clock directly from the Probe 1 Hz Sync when in "external clock mode" or via its own internal counter when in "internal clock mode." The time code is in UTC spacecraft time. The time value sent is 32 bits of integer seconds; the fractional seconds are zero at the time of the 1 Hz Clock pulse. This time interface is used to synchronize and time tag all instrument data, providing relative timing accuracy to 1 μs, and absolute time accuracy to about 1 ms.

Sun Pulse A Sun Pulse is provided to the DCB once per spin, indicating the sun crossing. The Spin-Synch timing consists of a 14-bit counter, clocked by a programmable pulse generator. The upper 5 bits of the counter are used to generate the "SpinSector Pulse," while the full 14-bits are used to generate a "SpinSynch Pulse." Timing is supervised by the CPU, via a programmable pulse generator.

2.1.1.4 Instrument Interfaces The processor communicates with the instrument support boards using a custom bi-directional serial Command and Data Interface (CDI) over the IDPU backplane. The backplane CDI is a serial protocol, synchronized to the 8.4 MHz clock provided by the spacecraft. Each subsystem receives its own set of CDI signals (Clock and Command) and returns message data via Telemetry signals. Each Clock is a continuous 2^{23} Hz signal (approximately 8.4 MHz) provided by the DCB in order to synchronize data transfers and to provide the basis of the common sampling clock. The CDI transfers data at 1 Mbps in 24-bit words, which include an 8-bit destination address and a 16-bit data value. No handshaking is required. The DCB is designed to ingest data as fast as the instrument boards can provide it. The instruments buffer commands as needed to keep up with several back-to-back commands.

2.1.1.5 Housekeeping For Instrument housekeeping data, there is an analog-to-digital converter on the DCB board, along with an 8-input analog mux. The mux channel is selected via an ADC Control Register. Typically the CPU sets the mux channel to the next channel to be sampled, waits until the switch and low-pass filter have settled, and then starts a conversion. In order to optimize for low-power, the ADC is kept in shutdown (nap mode, the default at reset) until the CPU is ready to perform a conversion. The inputs to the ADC come via the backplane. Each board has an analog housekeeping multiplexer attached to a common analog housekeeping signal on the backplane, which is routed to the ADC. The IDPU controls this distributed multiplexer tree via registers on the boards controlled over the backplane.

2.1.2 Power System

The IDPU power system was designed to be highly efficient (>75%) while meeting the typical requirements of isolation, regulation, control and monitoring. The IDPU receives electrical power from the Probe on four separately switched services. Each service provides unregulated 28 V (28 ± 6 V). The primary BAU controlled 28 V service supplies twelve separate DC-DC converters containing both isolated and regulated outputs. A second BAU controlled 28 V service is used for actuator power and boom deployments. The third and fourth services are used for primary and secondary heater power respectively.

The primary IDPU converters are always powered and supply power to the IDPU and the instruments through individual switches. The Low Voltage Power Supply (LVPS) takes unregulated spacecraft 28 V power and generates the secondary voltages used by the IDPU and instruments. The normal power bus is regulated and converted into a number of secondary voltages, including +5 V and +2.5 V digital, ±5 V analog, ±8 V analog, ±10 V analog, ±12 V analog, +28 V regulated, +4 V analog, ±10 V floating, and ±100 V. Most of the secondaries are provided with approximately 0.5 V over-voltage so that they can run low-drop-out regulator/current limiters downstream on the boards. This provides clean power on the boards to reduce cross-talk, and also provides power isolation for fail-safe requirements.

Actuator power goes to the Power Control Board (PCB) where FET switches control its distribution to the boom mechanisms and motors. This service is switched on for boom deployments only and has the appropriate lock-out capability (through a separate enabling plug) required by safety during ground operations and launch vehicle processing.

Heater power is simply routed through the PCB to the instrument heater/thermostat circuits without conditioning. The heater services are switched on for all modes and controlled by Instrument thermostats.

Figure 3 shows an overview of the main components comprising the IDPU power system (LVPS and PCB). The Instrument payload enable switches are current limited and provide necessary isolation between the different instruments and the IDPU core subsystems. All other switches are simple FETs. The PCB incorporates a single FPGA that handles all power control functions and the CDI interface to the DCB. Analog housekeeping is multiplexed, addressed by commands to the PCB logic, and fed up the backplane to the DCB for A/D conversion.

2.1.2.1 Low Voltage Power Supply (LVPS) The IDPU LVPS generates the various voltages required by the THEMIS instrument suite from the unregulated 28 V Probe power. Regulated voltages are Pulse Width Modulated (PWM) and regulated to ±1%. Voltages regulated by similarity are regulated to ±5%. The output ripples were designed and tested to be less than ±10 mV rms. The current monitor ranges are 2.5 V full scale for each monitored voltage. Each power supply is current limited on its primary side and is galvanically isolated primary to secondary. The input from the 28 V Probe power is soft started and filtered to meet EMI requirements.

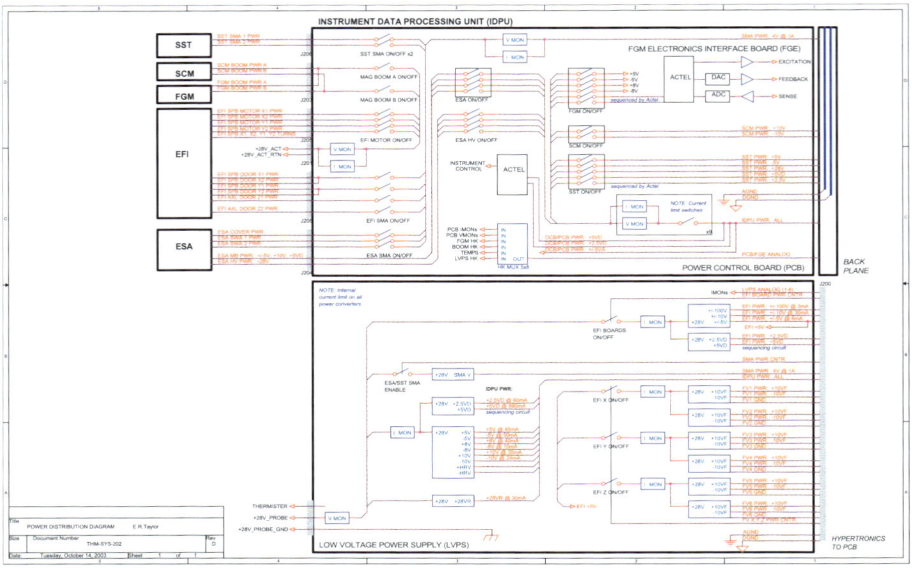

Fig. 3 LVPS/PCB block diagram

Distributed IDPU Voltages The LVPS was required to provide analog voltages (±5 VA, ±8 VA, ±10 VA) and digital voltages (+2.5 VD, +5 VD) for distribution on the backplane to the other IDPU boards. The input current for each service is monitored and reported to the DCB. All voltages appear with the presence of 28 V Probe power and are then separately switched on the PCB for distribution to the instruments. Positive and negative headroom voltages (±12 VA) are also provided to operate current monitors and limiters on the PCB. The +2.5 VD, +5 VD, and +5 V are regulated; the other voltages regulated by similarity. The presence of the +2.5 VD supersedes the presence of the +5 VD for FPGA power. The sequencing is in compliance with the FPGA requirements as specified by the manufacturer.

Regulated 28 V *ESA Voltage* The LVPS provides a separate converter for regulated 28 V that is used to power the ESA electronics and detector high voltage supplies. The high voltage supplies provide a programmable output up to 5000 V.

SMA Voltage The LVPS provides a regulated 4 V used to power actuators resident in many of the THEMIS mechanisms. The actuators are Shape Memory Alloy (SMA) systems, which provide a force when heated by passing current through them. The SMA voltage is switched on at command from the DCB.

EFI Floating Voltages The LVPS provides each axis (X, Y, Z) of the EFI instrument with pairs of ±10 V, and each set a separate "floating" return. A separate line from the DCB commands each pair. The voltages are derived from a regulated +5 V and their regulation are proportional to it.

EFI Voltages The LVPS provides ±5 VA, ±10 VA, +2.5 VD, and +5 VD to be used by the fields experiment BEB and DFB. The input current is monitored and reported to the DCB. The voltages are enabled by a command from the DCB. The +5 V, +5 VD and the 2.5 VD are regulated; the other voltages regulated by similarity. The presence of the +2.5 VD supersedes the presence of the +5 VD for the DFB FPGA.

The LVPS resides on a 6U shielded and pocketed board. It connects to the rest of the IDPU via a Hypertronics connector. Power is received from the Probe BAU through a standard, redundant 9-pin connector.

2.1.2.2 Power Control Board (PCB) Following the LVPS DC-DC converters is a power controller switch bank on the PCB which consists of 45 switches, controlled by 28 logic signals (one for each instrument or actuator). Switched power services resident on the PCB include: instrument control (9); operational heater power switching (5); and actuator power switches for the instrument actuators (SST attenuators (2), ESA cover (1), EFI doors (4), SCM and FGM booms and back-up (4), and EFI Axial Boom (AXB) and back-up (3)). The current limiters latch (exceeding the current limit will interrupt the service until it is reset by the DCB) to limit power dissipation in the transistors. Both current and voltage are monitored on all voltage services and these values are read by the processor via the DCB housekeeping system, and stored for instrument health and safety.

The PCB resides on a 6U VME board, shared with the FGE. It connects to the backplane via a standard VME 96 pin connector. Power is received and distributed through the backplane connection. Actuator and heater power is received from the Probe BAU through separate standard connectors.

2.2 IDPU Instrument Support Electronics

In addition to the core instrument electronics, the IDPU houses the instrument specific electronics which consists of the boards as described below.

The BEB and DFB support the fields experiment and share one switched power service. Their CDI Interface is also shared. The BEB does not output messages, but uses the CDI for command reception. The DFB connects directly to the analog outputs of the EFI and SCM sensors, performs processing and programmable filtering (see Cully et al. 2008). The Boom Electronics Board (BEB) contains power supplies and DACs to control the EFI sensors bias settings (see Bonnell et al. 2008).

The FluxGate Magnetometer Electronics (FGE) controller for the FGM connects to the FGM sensor via an external harness, processes data and generates messages. (FGE shares a board with the PCB, but each subsystem has its own CDI since the FGE power service is switched) (see Auster et al. 2008).

The ETC Subsystem shares a board with the DCB, but communicates with the DCB using an interface similar to the other IDPU subsystems. The ETC receives commands and timing signals, and generates messages from the particles experiment. The ETC subsystem controls the ESAs (Electrostatic Analyzers) and SST-DAP board. It acts as a router during data collection and generates trigger inputs such as moments (see Larson et al. 2008).

The Solid State Telescope Analog-to-Digital Processing Board (DAP) houses the SST LookUp tables, accumulation RAM and ADCs. The DAP receives commands and timing signals from the DCB and returns telemetry, which is processed by the ETC (see Larson et al. 2008).

3 Software Description

The THEMIS IDPU Flight Software (FSW) is responsible for instrument power control, time and attitude determination, mass memory control, science instrument control, command distribution, telemetry formatting and boom deployments. The flight code follows a

Fig. 4 FSW module connections

CSC#	Function	Name	Segment
1	Executive	EXEC	PROM
2	Background	BKG	PROM
3	Commands	CMD	PROM
4	Telemetry	TM	PROM
5	Housekeeping	HSK	PROM
6	Loader	LD	PROM
7	Utilities	UTIL	PROM
8	Input/Output	IO	PROM
9	Power Manager	PWR	PROM
10	SSR Manager	SSR	PROM
11	ACS	ACS	PROM
12	EFI Manager	EFI	PROM
13	ETC Manager	ETC	PROM
14	FGM Manager	FGM	PROM
15	SCM Manager	SCM	PROM
16	EFI Deployment	DEP	PROM
17	EFI/FGM Fit Manager	FIT	EEPROM
18	Spin Fit Calculator	SPIN	EEPROM
19	Matrix Solver	MATRIX	EEPROM
20	Trigonometrics	TRIG	EEPROM
21	Fast Floating Point	FFP	EEPROM
22	Compression Algs	CMP	EEPROM
23	Science Optimization	SCI	EEPROM
24	Software Changes	EEP	EEPROM

long line of software products now flying on a number of spacecraft, most closely resembling the Fast Auroral Snapshot (FAST) IDPU (see Harvey et al. 2001).

The THEMIS software is comprised of the 24 modules, totaling just over 19250 lines of assembly code. It requires 16.8 Kbytes of code space and 14 Kbytes of RAM. The FSW was developed in four phases, basically paralleling the instrument electronics development.

The 24 software modules can be functionally grouped into four main elements, as seen in Fig. 4 and described by module in more detail below. Modules 1–8 provide the core processor functions. Modules 9–15 interface to other instrument cards in the IDPU and control instrument sensors. Modules 16–21 are data analyzers and one-time use, and Modules 22–24 are optimizers. Figure 5 shows how the modules are connected and the information that passes between them.

The IDPU software splits the workload into Foreground (EXEC) and Background (BKG). The Executive runs the long-term tasks, anything requiring more than 2 milliseconds to perform. The BKG module splits a 256 Hz interrupt into a number of low-frequency interrupts for modules depending upon their requirements. The Command (CMD) module decodes, checks and distributes commands to other modules within the program. The Loader (LD) module provides loading and dumping capabilities. The Housekeeping (HSK) module samples all the A/D values for internal use and for use by the Telemetry (TM) module in generating telemetry. The TM module formats low speed and high speed telemetry. The Utility

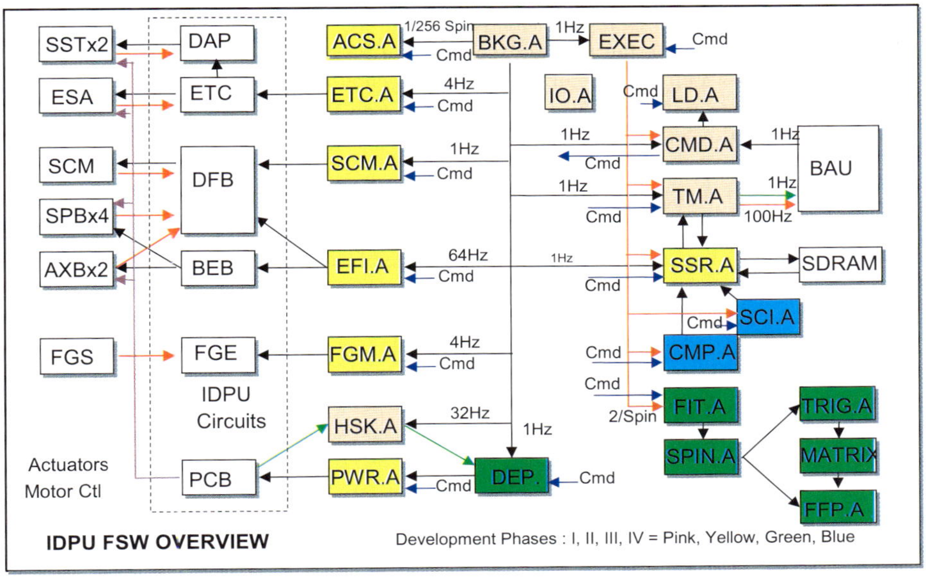

Fig. 5 FSW module connections

(UTIL) module provides common utilities for all the modules and the IO module provides device independence.

The Power Manager (PWR) controls to the instruments. The SSR module stores and retrieves data from the 200 MB memory. The ACS module provides a phase-locked-loop control of the Spin Sectoring for the sake of the ESA and SST instruments. Device drivers include the EFI, ETC, FGM, and SCM modules which communicate with their respective instrument electronics. And, the Deployment (DEP) module deploys the EFI Spin Plane Booms in a balanced fashion.

The next 5 modules are data analyzers, performing the necessary mathematical computations on the data.

The last three modules; CMP, the SCI, and the EEP modules provide data compression, burst data collection optimization and a host of small change requests, respectively.

3.1 Core Processor Functions

Executive (EXEC) The EXEC module is responsible for system initialization, mode implementation and foreground coordination. Specifically, it handles ROM Execution, EEPROM Selection and Execution. For radiation tolerance, a bootstrap version of the flight code is stored in ROM and later versions are kept in an EEPROM. Whenever the IDPU is powered on, the ROM is mapped to the start of the memory address space, and the flight software operates briefly from the ROM. Upon initialization, the ROM immediately copies itself to a specific segment of RAM and then, through a hardware select circuit, swaps the RAM into low memory. The ROM is then powered off, leaving the bootstrap version of the flight program running in RAM. This process takes only a few milliseconds. For the first ten seconds, the IDPU runs the bootstrap program. This program initializes all the internal program modules, sets default values, and begins communicating with the Probe BAU immediately. During this time, the IDPU adopts a minimum power level, with only the core systems (DCB, LVPS and PCB) powered on, and all the instruments powered off. Unless commanded to stop within the first ten seconds, the FSW checks over the four available EEPROM programs and automatically loads and executes the latest flight software version. Thus, the flight code runs entirely in RAM, requires no ROM or EEPROM power, and is directly patchable by ground command.

Background (BKG) The BKG module is the timing coordinator for the IDPU software. Its job is to service and distribute the interrupts of the processor so that the system is responsive to physical events. Thus, the other modules are isolated from the details of the CPU interrupt hardware, and the background manager is able to level the load. The background module uses the clock interrupt to receive and maintain Universal Time (UT) and to ensure that all time stamps have the correct time. During each second, the IDPU software receives 256 interrupts per second based upon the spacecraft-provided 2^{23} Hz clock. Using these interrupts, the Time register is maintained to 1/256ths of a second. For packets requiring the most precise time possible, the input clock register may be read by FSW and stamped on each packet header.

Command (CMD) The CMD module is the process by which all commands enter the IDPU. It sets up the DMA transfers, receives the packets, decodes them and distributes commands to appropriate modules. For command and control functions, the IDPU communicates with the BAU using a low rate serial line. Once per second the spacecraft and instrument exchange fixed-length blocks of data over this serial interface. This instrument side of the serial interface is connected to the processor via Direct Memory Access (DMA). The DMA transfers the data directly from/to processor memory. There are several types of command blocks executed by the FSW, each identified by their Application ID and Function Code.

Telemetry (TM) The TM module is the process by which all telemetry is generated by the IDPU. It sets up the DMA transfers. The TM module is the central coordinator of telemetry generation and playback. Once per second, the TM module collects and formats IDPU engineering data, both digital and analog status, into an SOH packet. Double-buffering is used to put a new SOH packet into one buffer while the previous data is being transmitted.

Housekeeping (HSK) The HSK module is responsible for collecting A/D samples for the flight software. Using a 32 Hz interrupt time, the HSK cycles through a list of multiplexer addresses and collects the data. Both 8-bit and 16-bit data are collected in separate lists for convenience. The ordering of the samples is defined by a PROM table which calls out the subsystem and the multiplexer within that subsystem.

Loader (LD) The LD module is responsible for loading (patching) SRAM and EEPROM from the ground, as well as dumping blocks of data to housekeeping.

Utility (UTIL) The UTIL module is a collection of general purpose routines which extend the capabilities of the 8085 processor. This module provides support functions for the flight software including memory clearing and copying, bit manipulation, 16-bit math and array functions.

Input/Output (IO) The IO module is the logical-to-physical separation layer of the software.

3.2 Instrument Control Functions

Power (PWR) Manager The PWR module provides power supply and actuator control under both direct ground command and internal calls from software modules. The module controls the LVPS and PCB, the latter through the use of the CDI. This module controls a number of one-time actuators, plus the SST attenuator multi-use actuators. The PWR module controls one actuator at a time and rejects other requests for activation while another activation sequence is in progress. It verifies that the selected actuator is enabled before allowing it to be fired. If not enabled, the FIRE command will result in an error message and no actuation takes place.

Solid State Recorder (SSR) Manager The SSR module is responsible for the maintenance of the SSR system, including error scrubbing, memory segmentation, and memory pointer management. The SSR module turns on SDRAM power and defines an initial memory configuration, as well as the minimum number of packets needed in a segment to allow transmission. The SSR module uses the SDRAM exclusively for variable length CCSDS packets ranging from 1 to 4 KB in length. In order to make the transmission of these data easier, the header and data sections of these packets are contiguous in memory, and all packets begin on a 4 KB boundary. The SDRAM is divided among a number of storage areas, each managed by separate logic in the SSR module. The SSR module can be commanded to reconfigure memory with different allocations of Engineering, Quick Look, Survey, Particle Burst and Wave Burst data.

The ECC scrubber in the SSR module is hardwired to operate on the lower 200 MBytes of SDRAM. (The upper quarter of SDRAM is reserved for the check bits). Single Bit errors are automatically corrected and counted. Multiple-bit errors are counted. The two counters (Single Bit and Multiple Bit Errors) are read and reset via the FPGA Register Interface. Each counter is allocated 8-bits, telemetered in housekeeping, and can be cleared by the CPU. The upper bits of scrubber current addresses are also available as status so that CPU can monitor the error counts periodically, and determine if SDRAM failures are address dependent.

Attitude Control System (ACS) The ACS module is responsible for the spin period and spin phase control of the instrumentation. It is responsible for determining precision spin information using the sun pulse signal from the spacecraft and an accurate clock. ACS software controls the DCB spin sectoring circuit which provides 2^{16}, 2^5 and 2^0 pulses per spin. Each timing register is 16-bits of subsecond time information. The FSW must properly apply UT to generate the correct time of these events. The ACS module is also able to read the current spin phase to 8-bit resolution for use in fine-tuning the spin synchronization.

Instrument Managers (EFI, ETC, FGM, SCM) After reset, the Instrument manager modules set up the initial configuration of the DMA channels, set the I/O configuration for DMA swaps, and generate default telemetry headers for each instrument (EFI, ETC, FGM and SCM). The modules do not send commands at this time, since all instrument circuits are turned off by reset. The modules process commands in 2 ms or less, per the system requirement. For CDI lists, the modules start a command list processor going which uses subsequent interrupts to execute commands in the list until all are exhausted.

The EFI module manages the DFB and BEB interfaces for commands and telemetry. The ETC Module is responsible for controlling the ETC circuit, the DAP board, the SST sensor and the ESA sensor. In addition to housekeeping functions, this module controls the ESA HV registers, stepping the HV up to a target position at a programmed rate. The FGM module is responsible for the FGE circuit and the FGS sensor. The SCM module manages the SCM instrument interface for calibration and engineering status (filter banks). To ensure the instrument does not stay in calibration mode, the SCM module uses an 8-bit maximum value for the calibration mode timer and counts at 1 Hz or greater to guarantee that the signal is turned off in less than 256 seconds. The SCM module does not need to direct SCM science measurements to the SSR since this function is handled by the DFB board under the control of the EFI module.

EFI Deployment The DEP module is responsible for deploying the EFI Spin Plane Boom (SPB) units. Deployment of the spin plane booms systems is normally performed in pairs by the DEP module. The operator selects which pair of booms to deploy, and then gives the deployment length. The rest is automatic. If need be, deployment of one boom at a time can be performed by either using the DEP commands or direct CDI Motor control commands. Each boom unit is equipped with a turns-counter microswitch which is sampled by the IDPU software to track the length deployed. Since the booms deploy at slightly different rates, software monitors the lengths and if one boom gets more than 2 'clicks' ahead of the opposite boom, the longer boom is paused until the shorter boom catches up.

3.3 Data Analyzers

EFI/FGM Fit Manager (FIT) The FIT module is responsible for collecting samples from the EFI spin plane boom and FGM instruments, and performing Sine-Wave Least Squares Fits of these data. Each fit provides the Electric Field and Magnetic Field vectors in the spin plane along with the averaged Z-axis component and standard deviation of the fit. The result is in 4 floating point scalars, A, B, C and Sigma where the vector is the waveform is $A + B\cos(\omega t_i) + C\sin(\omega t_i)$. The terms of the fit are shown in Fig. 6 and the matrix shown in Fig. 7. Each fit requires approximately 0.4 seconds.

Spin Fit Calculator (SPIN) The SPIN module is the calculator of the Sine-Wave-Least-Squares fit function. The Spin Fit calculator uses this function to determine the Electric Field and Magnetic Field strength and direction.

$$F = \sum_{i=1}^{N} [E(t_i) - (A + B\cos(\omega t_i) + C\sin(\omega t_i))]^2$$

$$\frac{\Delta F}{\Delta A} = \sum_{i=1}^{N} -2[E(t_i) - (A + B\cos(\omega t_i) + C\sin(\omega t_i))]$$

$$\frac{\Delta F}{\Delta B} = \sum_{i=1}^{N} -2[E(t_i) - (A + B\cos(\omega t_i) + C\sin(\omega t_i))] - \sin(\omega t_i)$$

$$\frac{\Delta F}{\Delta C} = \sum_{i=1}^{N} -2[E(t_i) - (A + B\cos(\omega t_i) + C\sin(\omega t_i))]\cos(\omega t_i)$$

$$\Gamma = \sqrt{F/(N-1)}$$

Fig. 6 Spin fit formulae

Fig. 7 Spin fit matrix

A	B	C	
N	$\sum_{i=1}^{N} \cos$	$\sum_{i=1}^{N} \sin$	$\sum_{i=1}^{N} E(t_i)$
$\sum_{i=1}^{N} \cos$	$\sum_{i=1}^{N} \cos^2$	$\sum_{i=1}^{N} \sin\cos$	$\sum_{i=1}^{N} E(t_i)\cos$
$\sum_{i=1}^{N} \sin$	$\sum_{i=1}^{N} \sin\cos$	$\sum_{i=1}^{N} \sin^2$	$\sum_{i=1}^{N} E(t_i)\sin$

Matrix Solver (MATRIX) The MATRIX module is a general 2×3 or 3×4 matrix solver using a Fast Floating Point format. It uses a standard process of diagonalization, and uses a practical zero of 10E–40.

Fast Floating Point (FFP) The FFP module is a collection of Fast Floating Point routines developed by Dave Curtis and Peter Harvey in 1980 and flown on numerous UCB spaceflight instruments for the last 24 years. The source code and description was written by UCB for the AMPTE and CRRES projects. The FFP module is used for on-orbit data analysis (sine wave least squares fit subroutine with sufficient range and precision of floating point) of the DC electric and magnetic fields.

3.4 Optimizers

Compression Algorithms (CMP) The CMP module is responsible for the compression of science and engineering data in the SSR. The CMP software requests packets from the SSR, compresses each packet based on the APID, and marks it as compressed. For the vast majority of the time, the survey packets are compressed right after they are stored in the SSR. When all the survey is compressed, completed Burst segments are compressed (highest value first).

The CMP module runs at a variable rate through the memory since the compression rate is dependent upon the specific data set. Typically, compression runs around 100 Kbps. Compression is enabled to operate on a given segment of memory which is not simultaneously enabled for telemeter to ground. The CMP module comes up disabled and will not disturb memory unless enabled to do so. The CMP module operates in the Executive level of the processor, but does not have to meet interrupt timing requirements.

For a given packet, the CMP module decodes the APID and references an APID-to-Algorithm list to determine the proper compression algorithm to use for that data. Generally speaking, Huffman works best on counter data (ESA/SST) and DeltaMod works best on Field data (EFI, FGM, SCM).

Science Optimization (SCI) The SCI module is responsible for the science level operation of the Instrument. If information is shared between two instruments, or the operation of one instrument is virtually controlled on the outputs of another instrument, the science module is responsible for making this inter-instrument connectivity. It provides the optimum configuration of the electronics and sensors to return the best science data.

Most importantly, the SCI module samples science and engineering data in order to trigger on significant events and saves that data in the SSR. The IDPU electronics provides a number of data sets which are considered useful for triggering, including: ESA and SST ion and electron full distribution and reduced distribution data sets; EFI DFB Filter-Bank outputs; and magnetic field spin fit data. The trigger data is used to change instrument modes from survey (Slow or Fast) to burst (particle or wave). Particle Bursts are slow processes and the data is gathered in a matter of spin periods. Wave Burst phenomena are quick and the data must be collected and evaluated quickly, e.g. several times per second.

In addition to trigger functions, the SCI module averages the voltage inputs from the EFI sensors and produces the spacecraft potential each spin. This value is made available for the ETC module to send to the ETC chip in order to adjust its accumulations.

Software Changes (EEP) The EEPROM module provides a collection of software changes to the boot software. As the first module of the EEPROM memory, this software is executed after the EEPROM code is loaded and is therefore responsible for installing patches for the EEPROM module and calling the initialization routine for SCI module. Patching the ROM area is possible since the ROM area of memory is copied to RAM and electronically swapped into the memory map at address 0. Thus, by the time that the EEPROM is executed (at reset plus 10 seconds), the EEPROM can simply modify the ROM area at will.

4 Conclusion

The THEMIS probes required a sophisticated, central Instrument Data Processing Unit (IDPU) to operate the large instrument suite and to collect the high resolution data necessary for the scientific objectives. The IDPU routes commands to the various instrument support boards, controls the power system, collects instrument housekeeping, controls boom deployments, directs science data to the mass memory, and optimizes the data downlinked to the ground by prioritizing data selection and incorporating triggering algorithms. A design philosophy was employed that provides autonomous instrument data accumulation rate control with minimal commanding and a data recording system with minimal processor interaction. Custom designed FPGAs perform numerous tasks normally delegated to a processor. With a reduced work load for the processor, the software focuses on data optimization and compression schemes that maximize the science return. The simple, flexible design of the THEMIS IDPU has been essential to the success of the mission. The hardware and software design description discussed here can be used to help integrate multiple instruments into a single experiment on future constellation missions that are typically mass and power constrained.

Acknowledgements The successful design, fabrication, development, integration and test of the THEMIS IDPU required significant time and effort from a large group of individuals, not all listed as authors on this paper, but crucial to its flawless operation to date on-orbit. Specifically, we would like to thank H. Richard and C. Chen for their meticulous integration, test and performance verification of not only the IDPU, but the entire instrument suite; S. Heavner for her long hours and week-ends of testing the LVPS; J. Fischer and C. Scholtz for their tireless job of obtaining, testing, and tracking every electrical part (especially for their willingness to do extra leg-work to flight qualify some plastic parts, allowing us to get much better performance at a lower power); J. Lewis for his work on a user-friendly IDPU GSE; J. Potts for her careful and timely layout work; H. Bersch and P. Turin for their work on the mechanical box design; and B. Dalen, H. Yuan, M. Colby and Y. Irwin for their diligent work on cabling and populating more than 36 flight boards.

This work was made possible by NASA, under contract NAS5-02099, and we would like to specifically thank NASA Mission Manager F. Snow, the Explorers Team, and the IIRT Review Teams for their shared expertise and knowledge. Finally, none of this work would have been possible, of course, without the unrelenting effort and dedication of the THEMIS PI, V. Angelopoulos, to whom we owe the on-going success of the THEMIS project from its inception to now.

References

U. Auster et al., The THEMIS fluxgate magnetometer. Space Sci. Rev. (2008, this issue). doi:10.1007/s11214-008-9365-9

J. Bonnell et al., The electric field instrument for THEMIS. Space Sci. Rev. (2008, this issue). doi:10.1007/s11214-008-9469-2

C.M. Cully et al., The THEMIS digital fields board. Space Sci. Rev. (2008, this issue). doi:10.1007/s11214-008-9417-1

P.R. Harvey et al., The FAST instrument data processing unit (2001)

D. Larson et al., The solid state telescope for THEMIS. Space Sci. Rev. (2008, this issue)

J. McFadden et al., The THEMIS ESA plasma instrument and in-flight calibration. Space Sci. Rev. (2008, this issue). doi:10.1007/s11214-008-9440-2

D. Pankow et al., THEMIS booms: design, deployment and stability. Space Sci. Rev. (2008, this issue). doi:10.1007/s11214-008-9386-4

The THEMIS Magnetic Cleanliness Program

M. Ludlam · V. Angelopoulos · E. Taylor · R.C. Snare · J.D. Means · Y.S. Ge · P. Narvaez · H.U. Auster · O. Le Contel · D. Larson · T. Moreau

Originally published in the journal Space Science Reviews, Volume 141, Nos 1–4, 171–184.
DOI: 10.1007/s11214-008-9423-3

Abstract The five identical THEMIS Spacecraft, launched in February 2007, carry two magnetometers on each probe, one DC fluxgate (FGM) and one AC search coil (SCM). Due to the small size of the THEMIS probes, and the short length of the magnetometer booms, magnetic cleanliness was a particularly complex task for this medium sized mission. The requirements leveled on the spacecraft and instrument design required a detailed approach, but one that did not hamper the development of the probes during their short design, production and testing phase. In this paper we describe the magnetic cleanliness program's requirements, design guidelines, program implementation, mission integration and test philosophy and present test results, and mission on-orbit performance.

Keywords THEMIS · Magnetic cleanliness · Spacecraft cleanliness

PACS 94.05.-a · 94.80.+g · 95.40.+s · 07.87.+v

1 Introduction

As with other space missions where a good measurement of the magnetic field is a primary mission requirement (Anderson et al. 2008; Kugler 2001; Narvaez 2004), the need to limit

M. Ludlam (✉) · V. Angelopoulos · E. Taylor · D. Larson · T. Moreau
Space Sciences Laboratory, University of California, Berkeley, CA 94720-7450, USA
e-mail: mludlam@ssl.berkeley.edu

V. Angelopoulos · R.C. Snare · J.D. Means · Y.S. Ge
IGPP/ESS, University of California, Los Angeles, CA 90095-1567, USA

P. Narvaez
NASA/JPL, 4800 Oak Grove Dr., MS 179-220, Pasadena, CA 91109, USA

H.U. Auster
TUBS, Braunschweig, 38106, Germany

O. Le Contel
CETP/IPSL, 10-12 Avenue de l' Europe, 78140 Velizy, France

and calculate the spacecraft induced magnetic field is critical. Due to the small size of the THEMIS bus and the short length of the magnetometer booms, approximately 2 m for the Fluxgate Magnetometer (FGM), 1 m for the Search Coil Magnetometer (SCM) (Angelopoulos 2008), it was necessary to review subsystems and components early on in the program. To do this a Magnetics Review Board (MRB) was established that set out a Magnetics Control Plan (MCP).

The objectives of the plan were to; establish overall responsibility for magnetic cleanliness; state the system-level magnetic requirements; establish a magnetic moment budget; list special considerations and requirements for worst offender subsystems and assemblies; provide generic subsystem and assembly design requirements and guidelines; describe magnetic test methods and procedures for performing tests on subsystems and assemblies; and provide methods for preventing subsystems/assemblies from becoming magnetically contaminated.

Although the requirements did allow a small level of remnant spacecraft induced field, this was small enough to require that even unlikely items needed to be checked and recorded. It was realized early on, that even if each subsystem's magnetic moment was a small fraction of the magnetics budget, all together could easily add up and be greater than the requirement. Therefore care was taken to measure components, alleviate problems and compare all subsystems performance in order to achieve a low cost, scientifically optimal solution that impacted the project development the least.

2 Requirements

The requirements for the spacecraft to meet were levied on the instruments and the spacecraft contractor early in the program. These requirements were based on a trade-off between science objectives and engineering possibilities. The requirement for the DC magnetics was that the magnetic field generated by the Probe should not exceed 5 nT at the location of FGM sensor. The 5 nT requirement is derived from the stability requirement that the magnetic field measurement to be stable and known at the sensor to within 0.2 nT over 12 hours with a reasonable thermal fluctuation of the tentative error sources on the spacecraft. The stability of the probe induced field was set to be 0.1 nT over 12 hours.

The requirement for the AC magnetic noise generated by the Probe was driven by the location of the SCM sensor, the expected instrument sensitivity, and the amplitude of relevant geophysical phenomena in the regions of interest. Accounting for the locations of the sensor, the AC noise requirement referenced on a common 1 m distance from the spacecraft is shown in Fig. 1. Further information about the sensitivity and performance of the magnetometers is contained in the respective instrument papers (Auster et al. 2008; Roux et al. 2008).

3 Parts Selection, Design, Modeling and Early Testing

Key to the success of the magnetic cleanliness effort was early identification of potential sources of magnetic contamination. Starting this work early with the spacecraft contractor, Swales Aerospace (now ATK Space), enabled the magnetics requirements to be inserted in the procurement process. Working with vendors who understood the magnetics issue or by educating them in the importance of correct material choice was key. Having knowledgeable personnel who had experience from past programs helped enormously by providing advice

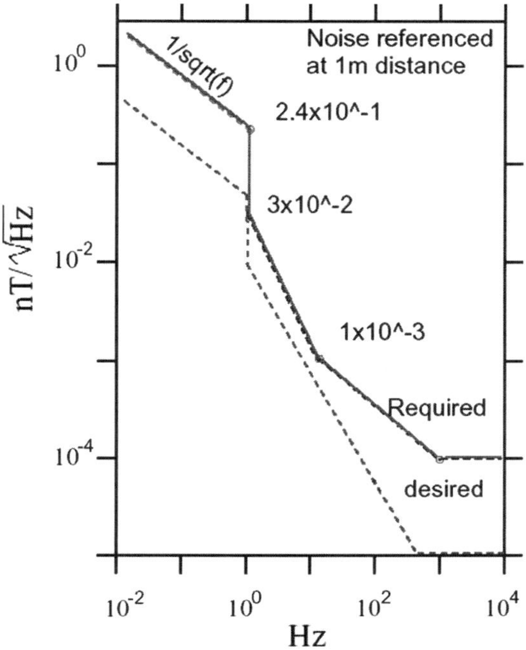

Fig. 1 AC magnetic noise level requirement (*solid curves*) and goal (*dashed curves*) at 1 m from the spacecraft. *Abscissa* is frequency in Hz. *Ordinate* is amplitude spectral density in $nT/\sqrt{Hz}$

and reassurance that the goals set out were obtainable. The MCP was distributed amongst the team and the magnetic items were identified and tracked. A survey of THEMIS components identified the main offenders. These were grouped into three categories; Hard Perm Fields including SST magnets, EFI Motors, Latch valves, Thruster valves, Tanks; Soft Perm Fields including Mu metal shielding, Welding, kovar cell interconnects; and AC Fields including Solar panels, Current loops, Battery, RF components and Power converters. Special considerations and plans were then outlined for each item. Examples of magnetic subsystem items are detailed below.

3.1 Solid State Telescope Magnets

As part of the Solid State Telescope instrument, Sm-Co permanent magnets were used to deflect electrons from the ion sensor. By matching the magnets closely, it was possible to attain the necessary field inside the sensor and have the field outside mostly cancel in the dipole regime. The non-canceling field was made up of the small unbalanced dipole field and quadrupole field which falls off as $1/r^4$. This resulted in a field of approximately 1 nT at 2 m. Pairs of SST sensors were also matched to ensure that the remnant field at the sensor for each spacecraft fell below the requirement. The Sm–Co magnets are extremely stable over time and temperature and so this field will not drift significantly over the course of the two-year mission lifetime. The testing also showed that orientation of the sensors would also help to reduce the DC field at the location of the FGM sensor.

3.2 Electric Field Instrument Motors

The second potentially large magnetic source in the instrument suite was from the Electric Field Instrument (EFI) Spin Plane Booms (SPB) that house brushed motors used to deploy wire booms from the probe. The magnets inside the booms were sufficiently strong to require shielding. Several shielding schemes were tried and tested before selecting a combination of Co-Netic AA and Netic S3-6 materials. This resulted in a field of less than 1 nT at 2 m from the sum of all SPB contributions.

3.3 Spacecraft Reaction Control System

The probe thrusters were selected to operate with a soft internal core. Redundant solenoids within them were wired anti-parallel, such that their operations would produce a remnant field in the soft cores that had a net quadrupole field, and thus a sharp drop off at distances comparable to the thruster dimension. Latch valves also were selected and wired anti-parallel to the internal solenoid wiring. However, due to their design the remnant field depended on the latch valve state, and thus changing the latch valve position changes the spacecraft magnetic field. Additionally a permanent small magnet (for position sensing) was a contributor to the total field from the propulsion system. Anti-parallel mounting of the latch valve sensing magnet, in its open (nominal) position was designed into the propulsion system tubing, in order to eliminate the total field at a distance, again by imparting a quadrupole field to the combined, two-latch valve system of the THEMIS propulsion system design. Inconel 718 propulsion tanks were approved for use on the THEMIS mission before the preliminary design review, based on analytical calculations of Inconel properties, and testing of scrap tanks at UCLA. Finally, structural welding on components such as the propulsion system pipes and tanks were done in accordance with mil standards, and fill material selection was based on fracture toughness and other mechanical properties, not driven by magnetic requirements but welds came out magnetically clean when tested. The propulsion system pipes were built from non-magnetic 304L stainless steel.

3.4 Spacecraft Power System

Current loops were minimized, by using the standard method of twisting power and return lines together. This was also extended to the design of the heaters and thermostat wiring. In the case of the solar arrays this required considerable effort during the design of the panels. By backwiring two return wires (return trace laid under the solar array cells on the forward lines) the field was reduced to first order quite significantly. Nearby strings (four per panel) were designed to conduct current in a way that when a panel was illuminated the four adjacent strings had such polarity so as to reduce the total field at the SCM and FGM sensors. Modeling of the stray field caused by the panels was performed by the vendor (COI-ATK). Analysis and modeling was repeated at UCLA. This analysis showed the field to be approximately 12 pT at the SCM sensor, which is commensurate with requirements. The panels were then tested by running current opposite to the cell at 2 kHz, at matching phases in all four cells, and measuring the response of the panel circuit at a mockup of the sensor location. When this was tested against measurements taken using a qualification panel, the results showed that the actual noise level was higher than expected by a factor of three. The discrepancy between model and test measurement was never understood to the team's satisfaction. Nonetheless the tests verified that the back-wiring and adjacent-string nulling was performing well because individual strings were tested separately and the noise was shown to decrease per model, when the strings were conducting in tandem.

Modeling of the magnetometer booms shadows and the EFI open door shadow on the solar arrays was also conducted. This was to determine the effect on string current and associated magnetic noise from one or multiple arrays turning off in the course of a single

Fig. 2 Proper cell layout (above) and backwiring of solar panels reduced the total field generated by the spinning probe

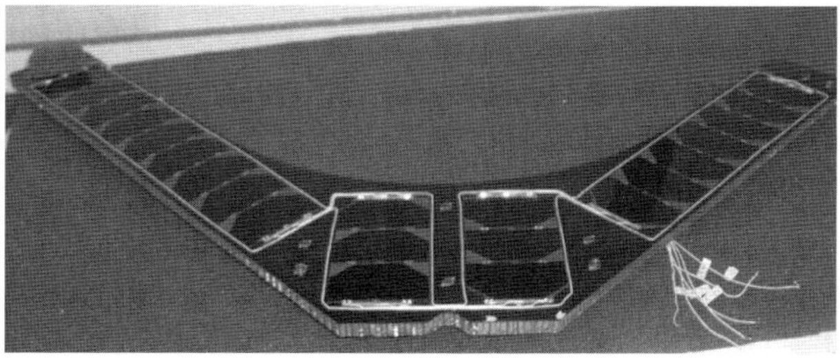

Fig. 3 The top (above) and bottom panels are single string and thus front-wiring was used there to run the return current and null the total field

spin. From this analysis it was found that the side panel design required two additional cells in the center strings, in order to withstand partial boom shadowing by the EFI snout and the magnetometer. It was found that at an angle of spin axis $<10^\circ$ to the sun these effects were minimized, and in that case both magnetic noise was reduced and power input to the probes became optimal. This placed a desire to operate the probes at an angle to the sun that was around 8°. The reduction of the spin ripple as function of the spin-angle due to the minimization of boom shadowing was later validated in orbit.

Similar to the solar panels, the layout of the battery cells was designed to cancel each other when operating. The battery strings were connected in a "horseshoe" arrangement, in order to ensure cancellation of magnetic contributions from nearby neighbors (Fig. 4) and wiring paths were designed to minimize induced field. The battery supplier, ABSL Space Products, identified the soft materials in the battery and was especially careful in the battery design. After successful degaussing of a test battery, all flight models were duly depermed before integration to the probe, and were handled very carefully thereafter.

3.5 Vigilance and Testing

The level of vigilance for magnetic items was kept high throughout all phases of the project. Personnel knew to pass on material data to the magnetics lead engineer every time a new component or part was identified. This caught several items, including proposed balance masses that were made from magnetic materials. Parts used in the magnetometer boom construction including the blankets were checked with a magnetometer in a shielding can to verify a non-magnetic boom. During transport to the launch site from JPL, 'witness plates'

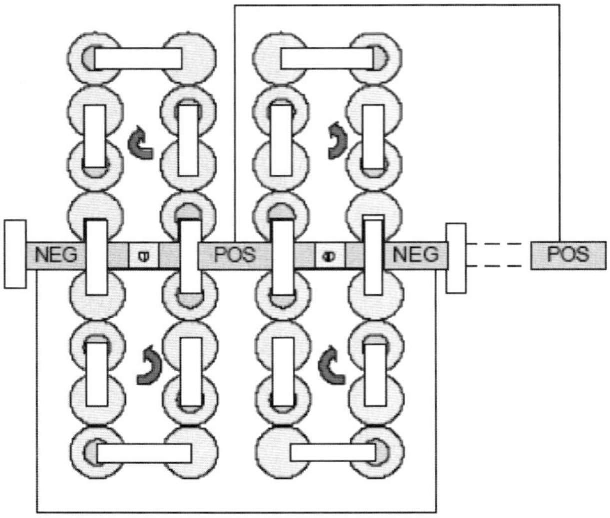

Fig. 4 Battery cells arranged in low magnetic, horseshoe configuration

made from soft perm able material traveled alongside the probes to detect if they were subjected to high magnetic fields. As the project progressed towards launch it became more and more important to protect the hardware from high magnetic fields. Personnel were regularly reminded of the requirements to survey the depermed (magnetically clean) tools and equipment that would come in close proximity of the probes and probe carrier. This continued right through to the white room and the fairing closure.

4 Unit Testing

All instrument units, except for the axial booms, for all five probes were tested using the Magnetic Coil Facility (MCF) loaned to UC Berkeley by Imperial College London (see Fig. 5). The coil arrangement allowed the Earth's field to be nulled and then a magnetic mapping of each unit was conducted by rotating it on a turntable. Most units were verified to be non-magnetic, other than as described above. One surprise was that the Instrument Data Processing Unit (IDPU) had a strong field associated with it when first tested. Troubleshooting of this lead to identification of transistors used extensively on one board. The cans of these transistors had sufficient amounts of Nickel in them that had been permed up in the manufacturing process. Deperming the boards individually lead to a significant decrease in the magnetic moment of the whole unit, this was then repeated on all units. The results of the measurements taken are tabulated in Table 1.

During development and flight model construction the SST sensors also were measured using an in-house test jig. Magnets were matched and yokes were trimmed to ensure lowest weight while there was sufficient material to avoid flux leakage from the assembly. The modeling was done in two dimensions (Fig. 6(a)), with a commercial fluxgate magnetometer. The magnetic moment was computed and numerical superposition of the results, taking into account the locations/orientations of the SSTs resulted in sensor matching to reduce the total SST-related field at the FGM sensor location.

As the spacecraft bus was assembled at Swales Aerospace the individual units did not under go unit mapping. In some instances when subsystem units were removed at UC Berkeley

Fig. 5 Magnetic Coil Facility used to map units individually

Table 1 Results from magnetic mapping of the instrument units

Instrument unit	FM1	FM2	FM3	FM4	FM5
Dipole magnetic moment (mA m^2)					
IDPU/ESA/SCM-PA	6.8	27.8*	3.9	9.4	4.3
SPB1	3.7	4.5	7.7	4.6	19.6
SPB2	8.9	4.6	2.8	4.1	7.2
SPB3	2.9	11.5	9.3	3.1	5.7
SPB4	5.0	9.7	8.2	8.8	1.5
SST1	31.2	10.1	32.5	28.7	23.8
SST2	36.2	34.6	35.0	32.6	30.8
Quadrupole magnetic moment (mA m^3)					
SST1	5561	4634	5424	5547	5455
SST2	5476	4681	4302	5343	5582

*Note the FM2 IDPU shows a higher moment than the others, this is the pre-deperm of the magnetic transistors

Table 2 FM2 measurements taken for some of the spacecraft bus subsystems

Probe Bus Units	FM2
Battery	8.4 mA m^2
Spacecraft computer	23.3 mA m^2
Transponder	28.2 mA m^2
Auxiliary Electronics Box	43.6 mA m^2

during integration, the opportunity was used to verify that the units were within the magnetic allocation. In the one instance when a flight battery was shipped independently it was checked at UC Berkeley prior to integration and found to be permed-up. It was successfully depermed and then installed in the spacecraft.

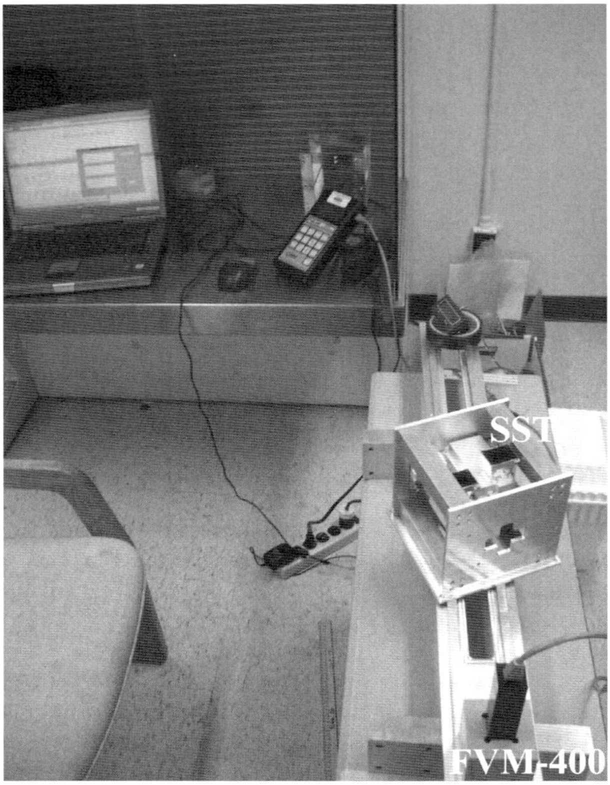

(a)

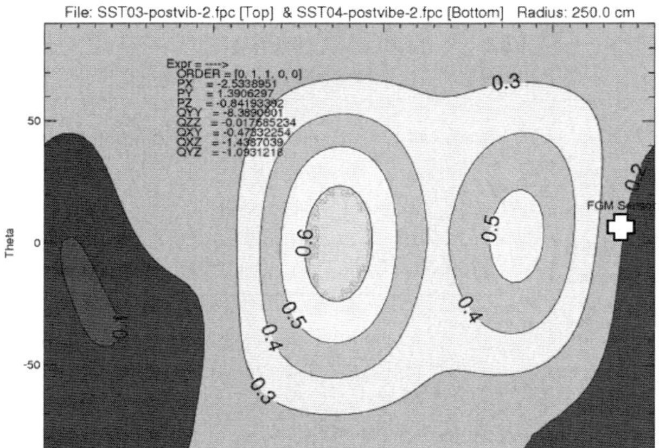

(b)

Fig. 6 (**a**) SST test configuration, with SST inside the manipulator, rotating on the horizontal plane. (**b**) Modeling the superposition of two SST sensors in nT on a phi-theta sphere, 2.5 meters away from the probe center, and recording the result at the location of the FGM instrument (*white cross*)

5 Spacecraft Testing

The spacecraft tests were carried out at JPL (Huang and Narvaez, May 2006a; Huang and Narvaez, Sept 2006b; Narvaez 2006; Ruff et al. 2006). Due to the small size of the THEMIS probes it was possible to fit them inside the JPL Helmholtz coil facility. The coil facility, built for the Mariner family of spacecraft in the 1960s, but recently refurbished, uses 3 pairs of coils to cancel out the Earth's magnetic in the center of the enclosed volume and has a single axis of deperming coils. See Fig. 7. The spacecraft magnetics tests were an important verification that the probe met the requirements. The 5 nT requirement was the most important and easiest requirement to test for, the 0.1 nT stability requirement was more difficult but tests were conducted that attempted to verify this requirement.

To verify the 5 nT requirement the spacecraft was rotated in the X–Y plane and then in the X–Z plane using a fixture that kept the center of the spacecraft in the center of the coil system. At each 5 degree interval a sample was acquired from the facility magnetometers located at 1.5 m and 2.5 m from the center of the probe. This data was then fed into in-house software that calculated the dipole and quadrupole magnetic moment of the spacecraft. The spacecraft was then depermed with a 1.5 mT (15 gauss) field with a linearly increasing then decreasing field while the probe was rotated around the X–Y plane followed by the X–Z plane. This was done with the SST sensor removed so the field from the sensor was not frozen into the soft materials in the solar arrays. Another survey with the SST installed gave the final spacecraft magnetic map which magnetic fields at the FGM location could be calculated. The details of this are presented in Table 3. The coil system was not actively compensated during the surveys and so it was important to verify no large shifts had taken

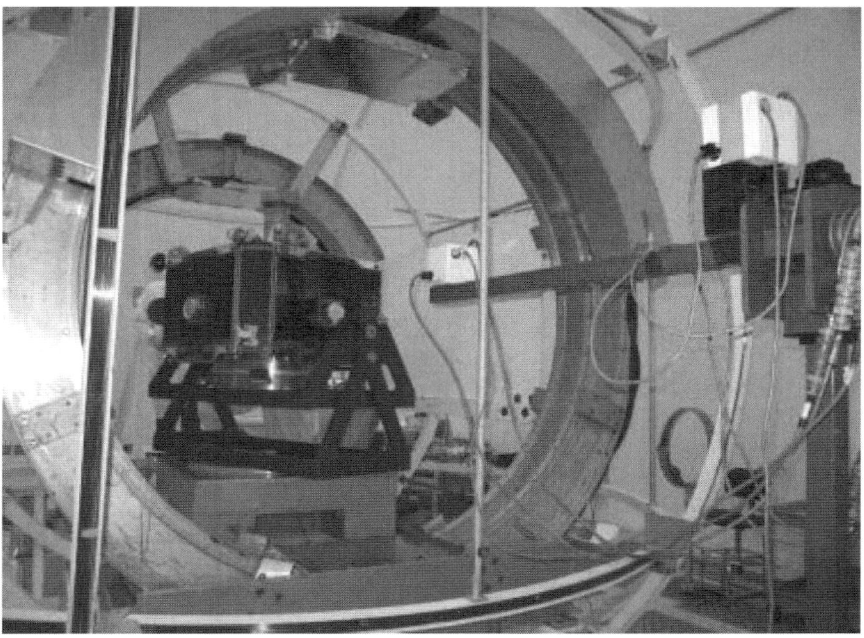

Fig. 7 Showing the coil facility with a THEMIS probe under test. The stand to the right supports the two test magnetometers

Table 3 Results of DC magnetics tests in dipole magnetic moment and calculated field at FGM sensor. The FGM sensor was located at [188 cm, 132 cm, 26 cm] from the moment location (top center of the spacecraft)

Probe	Dipole & quadrupole moments in S/C coordinates							Calculated field at FGM			
	M_x/Q'_{xx}	M_y/Q'_{yy}	M_z/Q'_{zz}	Q'_{xy}	Q'_{xz}	Q'_{yz}	M_{total}	B_x	B_y	B_z	B_{tot}
F1 Dipole	−27.64 mA m^2	111.97 mA m^2	15.98 mA m^2				116.43 mA m^2	1.18 nT	0.21 nT	0.40 nT	1.26 nT
F1 Quad	13.88 mA m^3	−11.41 mA m^3	−2.47 mA m^3	4.89 mA m^3	−5.62 mA m^3	−9.19 mA m^3	21.65 mA m^3				
F2 Dipole	−38.46 mA m^2	52.62 mA m^2	10.37 mA m^2				66.00 mA m^2	0.50 nT	0.32 nT	0.14 nT	0.61 nT
F2 Quad	17.91 mA m^3	−16.98 mA m^3	−0.94 mA m^3	6.32 mA m^3	−1.88 mA m^3	−3.42 mA m^3	25.79 mA m^3				
F3 Dipole	−74.79 mA m^2	71.15 mA m^2	41.63 mA m^2				111.31 mA m^2	0.34 nT	0.02 nT	0.10 nT	0.36 nT
F3 Quad	24.68 mA m^3	−13.71 mA m^3	−10.97 mA m^3	0.9 mA m^3	−5.3 mA m^3	−9.96 mA m^3	32.33 mA m^3				
F4 Dipole	51.27 mA m^2	91.71 mA m^2	−44.54 mA m^2				114.12 mA m^2	1.39 nT	1.31 nT	0.78 nT	2.07 nT
F4 Quad	19.62 mA m^3	−22.3 mA m^3	2.68 mA m^3	4.98 mA m^3	0.57 mA m^3	−7.17 mA m^3	31.08 mA m^3				
F5 Dipole	−66.87 mA m^2	52.65 mA m^2	12.08 mA m^2				85.96 mA m^2	0.05 nT	0.31 nT	0.19 nT	0.37 nT
F5 Quad	28.84 mA m^3	−26.24 mA m^3	−2.6 mA m^3	2.6 mA m^3	−3.76 mA m^3	−6.54 mA m^3	39.88 mA m^3				

place in the DC field during the test. If jumps were seen in the data, the coils had to be re-zeroed and the survey re-run.

To verify the probe 0.1 nT stability requirement a magnetic survey was conducted in Earth's field (with the nulling coils off) on the F2 probe only. By comparing the results from this test with the low field survey it was possible to estimate the change in the spacecraft generated field from low field at Apogee to higher field at Perigee. The results showed that in this extreme case of zero field to Earth's field the change would be a maximum of 0.2 nT. As the field at perigee is approximately half that at the Earth's surface it was agreed that this requirement was met.

A further opportunity to look at the permeability of the probes came when the deperm of the F1 spacecraft had to be halted when the field was 1.2 mT (12 gauss). A survey of the probe at this time found the field to be around 80 nT at 1 m, this translated to a field of approximately 10 nT at the FGM sensor. Even with this extreme case with a field many times more than normal Earth's field the probe did not perm up excessively, giving encouragement that the probes had been designed without too much soft material. The probe was depermed and the field was reduced to approximately 6 nT at 1 m.

At the first set of magnetics testing, on F2, a powered test of the probe inside the coils was included. The change in field at the location of the magnetometer was less than 1 nT from the tests involving the nonpowered probe. This showed that there would be no large DC offset when the probe was powered and therefore no large current loops, however it was not an AC magnetics test and so did not show up time varying fields from the power system of the probe nor was able to show how different spacecraft modes would change the field. Although the powered test was not repeated on the other four probes, a quick check using a gradiometer was conducted during a probe functional test at JPL. Using the gradiometer positioned at the same angle of the deployed FGM sensor but closer to the probe, the change in magnetic field was measured between the probe powered off and all the instruments on. For all probes the field change was negligible at the distance of the FGM sensor. The limitation of this test was that it did not test the full power system from the solar arrays to the battery and then through the distribution of power to the subsystems. However it is extremely difficult to reproduce the conditions of the spinning spacecraft and a fully illuminated and accurately measured induced field.

6 On Orbit Results

Post launch spacecraft induced fields at the FGM sensor are difficult to determine with certainty as they are tied together with the FGM sensor offsets, which drift over time. However the absolute offset (sensor offset plus spacecraft generated field) in all axes is less than the requirement of 5 nT in all but one axis of one probe. Also more importantly, this offset is seen to be stable and meets the 0.2 nT over 12 hours stability requirement. More details can be found in the FGM Instrument Paper (Auster et al. 2008). The offsets determined in-orbit at the FGM sensors are shown in Table 4.

The on orbit raw data from the SCM instrument shows noise levels above the requirement at a number of frequencies. However, these tones are fairly constant and so it is possible to remove them with standard data analysis techniques (see Fig. 8). There are two types of noise picked up, electronic noise (not discussed here, see THEMIS SCM instrument paper (Roux et al. 2008)) and magnetic noise. The magnetic noise is assumed to be from the power system and solar arrays as it is spin synchronous or at the switching rate of the charging circuitry. Despite a careful design of the solar arrays and power system there is still sufficient

Table 4 Sample FGM offsets for all probes on March 2007

Probe	X	Y	Z
F1	−0.77 nT	−2.05 nT	1.14 nT
F2	−4.2 nT	0.04 nT	0.38 nT
F3	1.09 nT	−1.73 nT	−2.23 nT
F4	−6.15 nT	1.59 nT	2.04 nT
F5	−5.05 nT	0.85 nT	1.47 nT

Note that the sensor pre-flight offsets are in the range ±0–3 nT. The stability of these offsets is ∼0.2 nT / 6months (Auster et al. 2008)

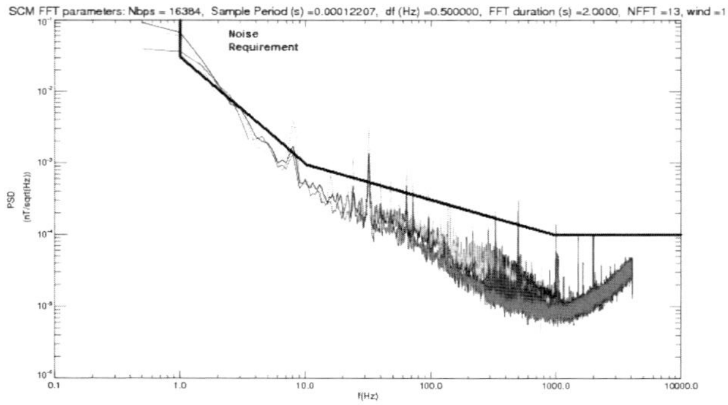

(a)

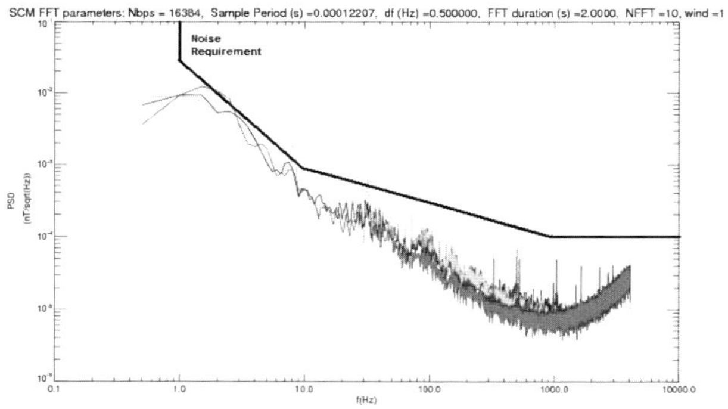

(b)

Fig. 8 (**a**) Sample SCM FFT showing noise levels relative to the requirement and (**b**) after data clean up

noise to affect the SCM measurement. A longer boom, perhaps of another half meter, would have lowered the noise levels below the requirements. However, as engineering is often a trade in competing needs, a longer boom would have resulted in much more complex spacecraft balancing and dynamics which could not be met by a mission of THEMIS's size. However, considering the short length of the boom, the complexity of the spacecraft bus and the stability of the noise levels, the result represents a considerable success.

It was determined after launch that spin tones with peak amplitudes around the fourth harmonic were present in the FGM and SCM data. By comparison the 1st harmonic was quite low. Since the power system is such that one string from each panel is in series with its counterpart on all panels (in order to minimize power losses from rectification) this immediately suggested that the spurious signal is a result of some uncompensated loop from the internal power distribution arising from all panels, rather than radiation from the panel adjacent to the FGM or SCM sensors. The power was found to be sun-pulse synchronous and was removed to the extent possible from the SCM instrument (Le Contel et al. 2008). The same technique can be applied on the FGM instrument by special processing, but because the spin tone can be both related to the ripple in the power system and the orthogonality of the sensor, the spin removal is coupled to the calibration procedure. A routine process for removal of the spin tone from the FGM data signal is currently under construction.

7 Conclusions

The magnetic cleanliness program benefited from good early work to identify risks. With an experienced knowledgeable team that was able to balance the needs of the magnetics budget with the rest of the mission, the main magnetics requirements were met. Getting all personnel on the project aware of the requirements from subcontractors through to launch site personnel helped managing the magnetic budget. Continued vigilance through the project picked up small items that could have had a big effect on the FGM and SCM measurement. Whilst proper bookkeeping of the budget was kept, a detailed multiple dipole magnetic model of the spacecraft was not made. This worked for THEMIS due to its size and relative simplicity—there were few units that produced significant magnetic fields, by and large due to early work on the subsystems to remove magnetic components. For a larger mission a more robust approach with model and full verification might be considered as was done, for example, for Cluster. However, the THEMIS approach proves that this may not be necessary. Full spacecraft DC magnetics testing helped confirm a well built and clean spacecraft, however did not verify any AC requirements. This was done at a subsystem level, although the power system only really functions when fully assembled, and to a certain extent only after launch. The difficulties in testing the AC magnetics requirements presents a difficulty to any mission that needs low AC induced fields and therefore a thorough approach. On reflection more attention could have been paid to the AC budget to attempt to reduce the noise levels below the requirements. Finally it is noted that is possible to build identical spacecraft all with a high level of magnetic cleanliness without the use of compensation magnets on a medium sized mission budget.

Acknowledgements We wish to thank Richard Schnurr, James Slavin, and Todd Bonalsky for their help during RCS and battery system testing through the GSFC magnetic cleanliness facility; the Space Magnetometer Laboratory at Imperial College London and Guenter Musmann for their assistance with the Magnetic Coil Facility used for unit testing at UCB; Betty Ruff, Nelson Huang and Al Whittlesey for their assistance during the magnetics testing through the JPL facilities; Kevin Brenneman, Warren Chen, Ginger Robinson and Michael McCullough at Swales Aerospace for their assistance in the system design and implementation of the magnetic cleanliness program; Ted Stern of COI/ATK for his diligence in the solar array design; Jamie Holbrook and the Aerojet team for excellent RCS component choices and system design; Robert Bond and Andrea Bennetti at AEAT for being sensitive to the stringent magnetic requirements of the program and Karl-Heinz Fornacon and David Fischer for error analysis of FGM Data.

References

B.J. Anderson, M.H. Acuna, D.A. Lohr et al., *The Magnetometer Instrument on MESSENGER, The MESSENGER Mission to Mercury* (Springer, New York, 2008), pp. 417–450

V. Angelopoulos, The THEMIS mission. Space Sci. Rev. (2008, this issue). doi:10.1007/s11214-008-9336-1

U. Auster et al., The FluxGate Magnetometer for THEMIS. Space Sci. Rev. (2008, this issue). doi:10.1007/s11214-008-9365-9

N. Huang, P. Narvaez, THEMIS post magnetics test report. May 16, 2006a, JPL IOM Number 5132-06-036

N. Huang, P. Narvaez, THEMIS F1-F5 magnetics test Report. September 14, 2006b, JPL IOM 5132-06-073

H. Kugler, Lessons learned during the magnetic cleanliness programs of the cluster projects, in *Proceedings 4th International Symposium on Environmental Testing for Space Programmes*. June 2001

O. Le Contel, A. Roux, P. Robert et al., First results of the THEMIS search coil magnetometers. Space Sci. Rev. (2008)

P. Narvaez, The magnetostatic cleanliness program for the Cassini spacecraft. Space Sci. Rev. **114**, 385 (2004)

P. Narvaez, THEMIS Probe DC magnetics procedure. March 16, 2006, JPL D-33980

Roux et al., The search coil magnetometer for THEMIS, Space Sci. Rev. (2008, this issue)

B. Ruff, N. Huang, P. Narvaez, THEMIS flight model P2 system level electromagnetic compatibility test report. May 12, 2006, JPL IOM Number 5132-06-035

Instrument Boom Mechanisms on the THEMIS Satellites; Magnetometer, Radial Wire, and Axial Booms

David Auslander · Joshua Cermenska · Gregory Dalton · Mauricio de la Pena · C.K.H. Dharan · William Donokowski · Robert Duck · Jonghak Kim · David Pankow · Alec Plauche · Mustapha Rahmani · Stephen Sulack · Tien Fak Tan · Paul Turin · Tyler Williams

Originally published in the journal Space Science Reviews, Volume 141, Nos 1–4, 185–211.
DOI: 10.1007/s11214-008-9386-4

Abstract The five "Time History of Events and Macroscale Interactions during Substorms" (THEMIS) micro-satellites launched on a common carrier by a Delta II, 7925 heavy, on February 17, 2007. This is the fifth launch in the NASA MeDIum class EXplorer (MIDEX) program. In the mission proposal the decision was made to have the University of California Berkeley Space Sciences Laboratory (UCB-SSL) mechanical engineering staff provide all of the spacecraft appendages, in order to meet the short development schedule, and to insure compatibility. This paper describes the systems engineering, design, development, testing, and on-orbit deployment of these boom systems that include: the 1 and 2 meter carbon fiber composite magnetometer booms, the 40 and 50 m tip to tip orthogonal spin-plane wire boom pairs, and the 6.3 m dipole stiff axial booms.

Keywords THEMIS · Magnetosphere · Radiation belts · Magnetopause · Constellation · Mechanisms

PACS 94.30.-d · 94.30.cl · 94.30.cb · 94.30.ch · 94.30.cj · 94.30.C- · 94.30.cp · 94.30.Lr · 94.30.Va · 94.30.Xy · 96.50.Fm

1 Introduction

1.1 Mission Background

The scientific objectives of this magnetospheric physics mission are to investigate many of the fundamental questions on the nature of magnetic sub-storm instabilities. The spatial nature of these activities dictates the need for multiple synchronized probes (Angelopoulos

---

D. Auslander · J. Cermenska · M. de la Pena · C.K.H. Dharan · J. Kim · M. Rahmani · S. Sulack · T.F. Tan · T. Williams
Mechanical Engineering Department, University of California, Berkeley, USA

G. Dalton · W. Donokowski · R. Duck · D. Pankow (✉) · A. Plauche · P. Turin
Samuel Silver Space Sciences Laboratory, University of California, Berkeley, USA
e-mail: dpankow@ssl.berkeley.edu

2008). Each of the five identical probes has a complete science package. Probes are coordinated by our ground based missions operations center.

The Electric Fields Instrument (EFI), Electrostatic Analyzer (ESA), and Solid State Telescopes (SST) were all provided by the University of California Berkeley (UCB). The Fluxgate Magnetometer (FGM) was provided by the Technical University of Braunschweig, (TU-BS), Germany. The Search Coil Magnetometer (SCM) was provided by Centre des Environements Terrestre et Planétaires (CETP), France. The two Magnetometer Booms were provided by UCB. The Instrument Data Processing Unit was provided by UCB. The Probe Carrier and five Probe Buses with avionics were provided by Swales Aerospace under contract to UCB. Delta II launch services were provided by United Launch Alliance under contract to NASA Kennedy Space Flight Center Dr. Vassilis Angelopoulos of UCB was the mission Principal Investigator. Peter Harvey was the UCB Project Manager. Frank Snow was the GSFC Explorers Office Mission Manager.

1.2 Spacecraft Boom Sensors Configuration

Three orthogonal dipoles with six tip mounted sensors are needed for a vector measurement of the DC and AC electric fields in the plasma. Sounding rocket and early satellite experiments used stiff, deployable booms for the dipoles. Solar-thermal bending and vehicle dynamics severely limited these stiff booms to lengths of several meters, far short of the lengths desired for more precise physical measurements. In the evolution of these instruments, the preferred practical configuration has been found to be a spinning vehicle with four limp wires in the "spin plane" and two stiff axial booms along the spin axis. The limp wires can be precisely positioned by centripetal acceleration, and are immune to the bending and buckling concerns in stiff booms. These lightweight wires allowed about a tenfold increase in the practical radial boom dipole lengths. The SCM and FGM are mounted on stiff CFRP booms for immunity to bus induced and stray fields, as well as one another. The particle instruments, ESA and SST's are probe mounted with outward looking, clear fields of view.

1.3 Spacecraft Stability Constraints

In practice, boom lengths are determined by the need for a spin stable vehicle. Briefly stated, a spinning body will be passively stable about the principal axis having the largest principal moment of inertia, based on conservation of angular momentum and body-flexing dissipation of energy to a rotational energy minimum (Meirovitch and Calico 1972). The spin stability ratio (which must be >1) is defined as the ratio of the moment of inertia about the spin axis to the larger of the two transverse axes ($I_s/I_{T\max}$), while the stability margin is defined as this ratio minus one. This means the radial wire booms improve stability and can be quite long, while the axial booms are length limited because they reduce the stability margin by increasing the transverse moment of inertia. The wire boom cables are essentially limp to any transient motions or oscillations induced by spacecraft maneuvers. The resulting pendulum behavior is mostly dependent on the wire root or hinge attachment radius, the distance from the spin axis to the wire attachment, or exit point. The deployed wire boom plane was located close to the spacecraft Z axis center of mass to avoid spin axis tilt caused by wire boom mass moment asymmetries. The axial booms must be sufficiently rigid to avoid elastic instability and subsequent collapse. As previously stated, the vehicle stability margin severely limits the axial boom length. In the mission planning stages, it was decided to include the stabilizing effect of wire booms in the overall moment of inertia calculations, to maximize the allowed axial boom length. In practice this increased the

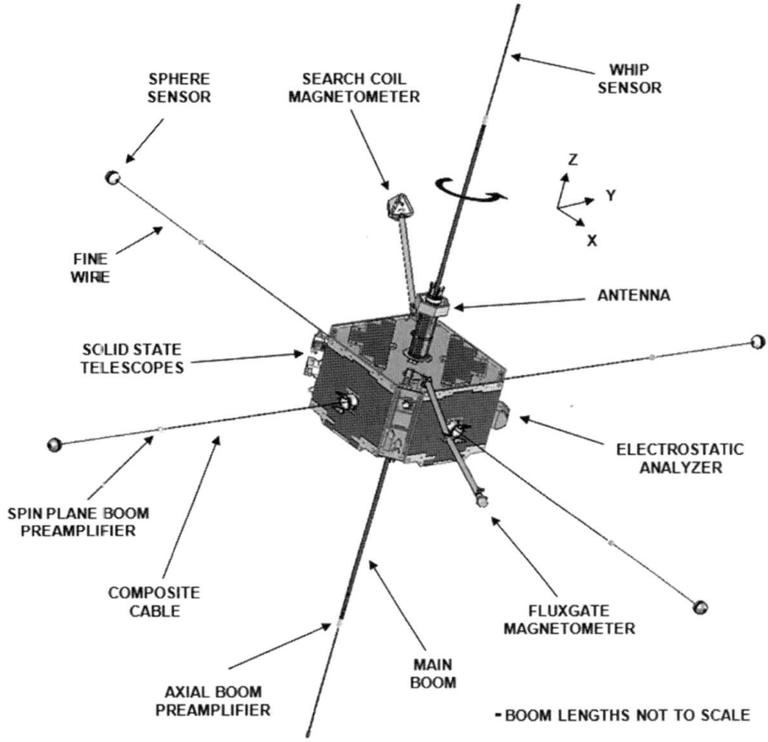

Fig. 1 On orbit deployed booms configuration

boom length from 2.6 m to 3.2 m each, which is a very significant improvement for minimizing the effects of vehicle photo-electron emission. Conventional wisdom suggests that boom length might be increased by decreasing the boom mass, which will also decrease the stiffness. However, spin induced boom flexing amplifies the 'effective' boom second mass moment (Meirovitch and Calico 1972). A boom cantilever resonance of four times the spin (as compared to a customary requirement of two) was selected to maximize boom length. A systems level concern was evaluation of the spin axis alignment budget. A list of many uncertainties, ranging from deployed boom straightness to alignment of the vehicle balance fixtures, will affect the alignment of the spin axis with the vehicle geometric axis. Simple addition of this list is far too conservative, and not warranted. If each of the uncertainties is assumed to have random clocking with respect to the spin axis, the resulting imbalance is half the root square summation (RSS) of these residual inertia products. The traditional NASA minimum requirement for the vehicle stability margin is 4%, based mostly in the uncertainties of mass moment measurements. Sensitivity of the spin axis alignment indicated that a more practical stability minimum was 8–10% for most satellites. The probe stability margins range from 16 to 25%, a function of the remaining fuel.

The on orbit deployment sequence serially released the magnetometer booms, the radial wire booms, then the axial booms. For both enhanced reliability and simplicity, these boom mechanisms are purposely designed without a retraction capability. The boom systems were manually rewound and reset after ground testing, and on orbit retraction is neither possible nor necessary.

1.4 Spacecraft Dynamic Simulations

A central feature of the Themis mission is the synchronized one, two and four day probe orbits, which were predicted to include very significant station keeping maneuvers. Each probe has only four thrusters; tangential spin and despin thrusters plus two axial thrusters pointed in the $-Z$ direction. Each of these pairs is diametrically opposed so that they may also be used in pairs. The two axial thrusters are needed for the timely and very large velocity change maneuvers needed to initially place the probes in the desired orbits. The flexible booms are not yet deployed during these early maneuvers. Many of the later maneuvers were known to be most effective at perigee, which would call for both timely and aggressive action. The probe science attitude has its spin vector close to orbit normal. The fuel needed to tip the spin axis into the orbit plane with all booms deployed is prohibitive, which meant that most of the later orbit delta velocity maneuvers would need to be performed by synchronous pulsing of the tangential and opposite spin and despin thrusters. The second vital maneuver was spin axis pointing to maintain the desired probe science attitude, which would be achieved by pulsing of one of the radial offset spin axis thrusters. In the context of probe dynamic time constants, it was expected that these two pulsed maneuvers would reach steady state, the equivalent of pulsing forever in the simulations. At launch, the probe mass was 40% fuel in two non-restrictive spherical tanks, which meant that pulse excited fuel slosh would be a major maneuvering constraint. Short pulses could reduce slosh, but are also known to reduce thruster specific impulse. One goal of these studies was to maximize pulse widths, consistent with attitude stability.

Given the critical nature of these maneuvers, two teams of graduate students developed independent parallel simulations, guided by David Auslander. One team developed simulations in Matlab-Simmechanics while the other worked in a "home brew" C++ environment. The initial ground rules were that observed modal frequencies needed to agree to a few percent and amplitudes to perhaps 25%. These simulations were both developed using techniques pioneered by Auslander (2000), where the desired multi-body dynamics were developed using only a small manifold of point masses connected by springs. Distributed mass rigid bodies were represented by six, or more, point masses inter-connected with very stiff springs. A third independent confirmation of the simulation results was also developed by David Pankow, using the published analytic results of Lai and Bhavnani (1975). Figure 2 provides the various oscillation modes. The slosh modes are similar, but with only two tanks.

In the early stages of simulations development, sub-models confirmed that the limp radial booms could be adequately represented by a simple point mass 3D pendulum with all of the actual hardware mass positioned at the computed Center of Percussion about the wire exit, or hinge point. Similar sub-models confirmed that each stiff axial boom could be represented by an ensemble of 24 properly chosen springs and 12 point masses. Published slosh damping characteristics by Franklin Dodge (2000), which is an update of NASA-SP106, were used to model the fuel mass behavior as a slug mass 3D pendulum. The spherical tank geometry dictates the pendulum length as a function of fuel fill. Both the radial wire booms and the fuel pendulums are inherently limp, which meant the apparent pendulum stiffness is provided by the probe spin forces. One dynamic simulation rule of thumb is that appendages with a first resonant mode greater than four times the spin may be considered to be rigid, with modest loss of fidelity. With this, the >3 Hz magnetometer booms were assumed rigid, lumping their mass moments into the probe hub.

The analytic and numeric results identified semi-resonant slosh conditions where the fuel pendulum slosh period is some integer multiple of the spin period. With continuous pulsed thrusting this causes the familiar, and troublesome, resonant amplification. The smooth,

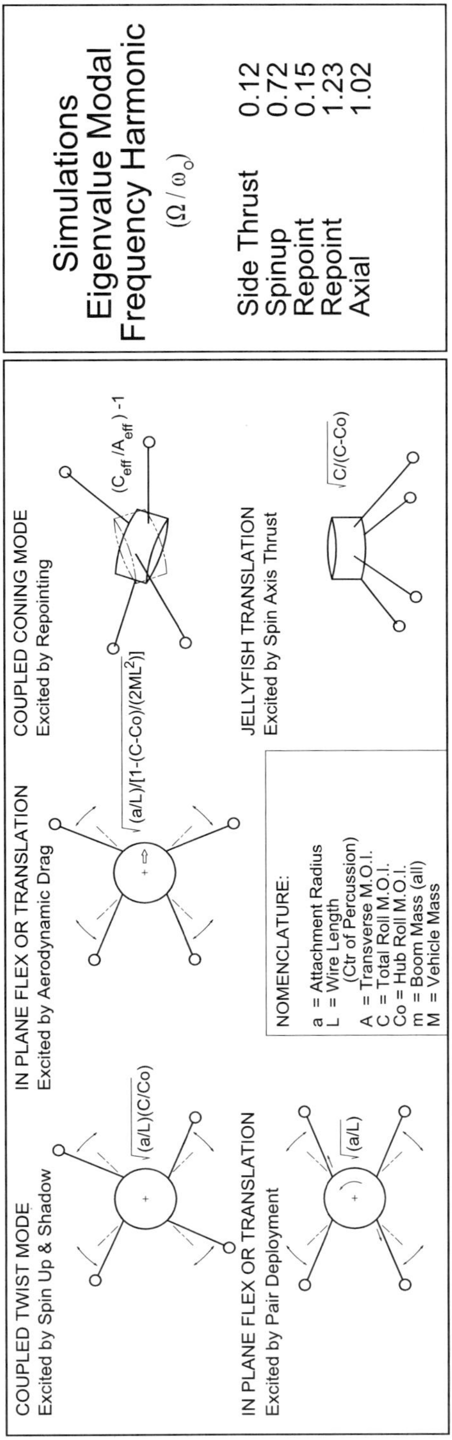

Fig. 2 Rotation and translation modes of a central hub with four wire booms

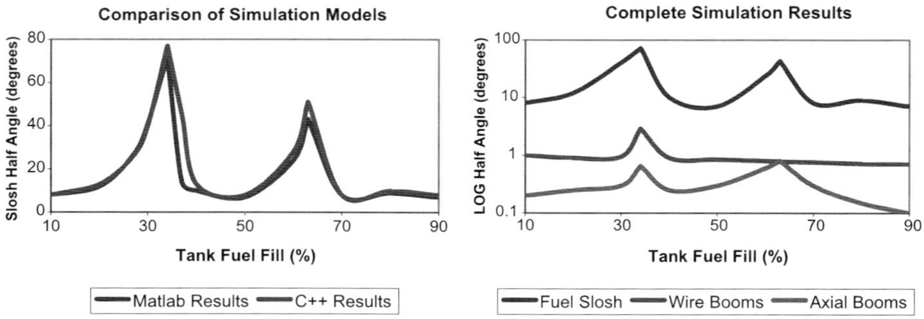

Fig. 3 Synchronous pulsed (±30°/360°) side thrusting results

spherical tanks provide very little laminar flow, viscid damping. Stated in familiar engineering terms, for a single degree of freedom system, the maximum native Q (resonant amplification) is 350 at 50% fill. For side thrusting these harmonics were at 34 and 63% fuel fill. For pointing maneuvers these harmonics are at 20 and 60% fuel fill. These are different because of the physical nature of the spin induced centripetal acceleration. In side thrusting the fuel moves in the spin plane and the centripetal acceleration vector must point radially inward to the spin center. In pointing maneuvers, the fuel moves out of the spin plane and centripetal acceleration vector must be orthogonal to the spin vector. The radial wire booms do not experience these near resonant conditions. Given the limp pendulum representations of both fuel and wires, the ratio of pendulum periods to spin period is dictated by physical geometry only. This means that probe spin rate changes cannot be used to avoid the near resonant conditions.

Figure 3 presents side thrusting results from the complete simulation models as a function of fuel fill. The left plot provides a comparison of the two independent models. The amplitudes are peak values from the models response, where the multi-mode oscillations typically showed beating patterns, as later illustrated in Fig. 4. The right plot compares the response of all probe flexible elements. The modest response peaks reflect only weak coupling of flexible elements, given adequate separation of resonant conditions. The simultaneous peaks reflect a larger hub motion, which may be viewed as a larger base input for all elements, rather than element to element coupling. The ±30 pulse width was felt to be an upper bound for maintaining high specific impulse, given the directional variations during each pulse. Hence, added investigation of pulse widths was not pursued.

Figure 4 presents the pointing maneuver model results, where the limiting factor was slosh amplitude, which is presented as a function of fuel fill. The right plot illustrates the beating behavior which was typical in all results. This plot also illustrates the modest settling time constant, which was needed for mission maneuver planning.

Figure 5 presents the pointing maneuver probe nutation and slosh model results at 20 and 40% fuel fill. These typical preliminary results were used to select a baseline 12° half pulse width for the later detailed simulations.

The full simulation results were reviewed with the Themis Mission Operations Team and used as the basis for the Maneuvering Flight Rules. Nominal pulse widths of ±30° for side thrusting and ±12° for pointing were selected. The side thrust pulse would be reduced to ±20° when the fuel is 29–39% or 58–68% to moderate the response peaks. The ±12° pointing pulse behavior was judged to be adequate at all tank levels. One additional maneuver that is sometimes used is the so called "beta thrusting" where radial and axial thrusters are

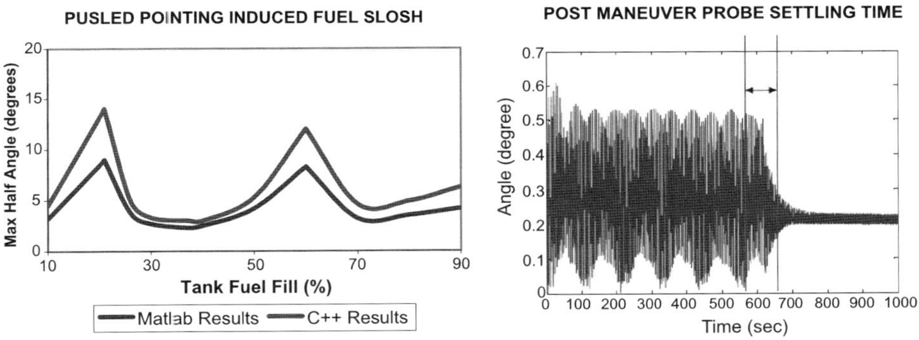

Fig. 4 Pointing induced slosh and post maneuver nutation settling results

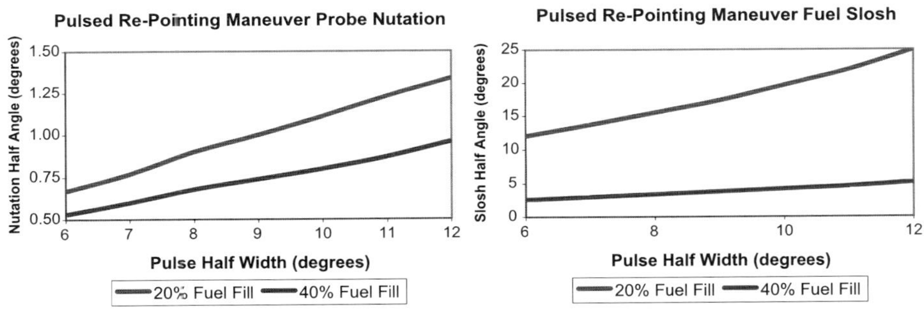

Fig. 5 Pointing induced nutation and fuel slosh as a function of pulse width

simultaneously pulsed to provide a velocity change at the angle beta to the spin plane. The simulations indicated smaller pulse widths would be needed, and small time saving as compared to sequential axial and radial thrusting. Beta thrusting was not included in the flight rules.

2 The Magnetometer Boom Mechanisms

2.1 Magnetometer Boom Science Configuration

The magnetometer booms are stowed during launch and deployed to provide rigid support for accurate pointing of the magnetometers while keeping the magnetometers far enough away from the main body of the satellite to avoid the magnetic interference from small current loops in the onboard circuitry. Each probe has two magnetometer booms. One supports the Flux Gate Magnetometer (FGM) approximately 2-m away from the probe. The other supports the Search Coil Magnetometer (SCM) approximately 1-m away from the probe. A picture of the deployed magnetometer booms is shown in Fig. 6.

The design of the magnetometer booms takes into account a variety of mission requirements. The magnetometer booms must fit on the top deck of the probe. Additionally, the probe and carrier must fit in the Delta II launch fairing. The masses of the FGM magnetometer boom (FGB) and SCM magnetometer boom (SCB) and instrument are also limited.

Fig. 6 Deployed magnetometer booms; SCB on *left* and FGB on *right*

In the deployed configuration, the largest of the three principal moments of inertia must line up with probe spin axis within 1 degree. The magnetometer boom deployment shall be repeatable to 1 degree with stability better than 0.1 degrees.

The magnetometer booms must survive the vibration loading from launch and the stresses from deployment between 2 to 18 RPM. Additionally, since the booms are mounted outside the probe, they must survive thermal cycling between 75°C to −115°C. The magnetometer booms must also meet magnetic cleanliness of less than 0.1 nT and carry the harnessing from the magnetometers to the probe.

2.2 Magnetometer Boom Design

To meet the accuracy and repeatability requirements, a rigid unfolding link design is used with one link for the SCB and two links for the FGB. The spinning dynamics of the deployment and packaging constraints dictated that the booms would be located on the upper deck of the spacecraft, unfolding along axes parallel to the spacecraft spin axis. This configuration is shown in Fig. 7.

The SCB consists of a composite boom segment with the base hinge assembly and magnetometer on opposite ends. When the SCB is stowed, it is clamped via a frangible Ti bolt (Frangibolt) to a (Deployment Assist Device) DAD tower that contains the shape memory alloy (SMA) deployment device (TiNi Aerospace, San Leandro, CA), and the Search Coil Magnetometer instrument and interface.

The FGB consists of two composite boom segments. The inner segment is attached to the base hinge and the outer segment is attached to the magnetometer. The two segments are attached together with an elbow hinge. When the FGB is stowed, the two boom segments are folded parallel and held near the base spring with a SMA deployment device. The elbow hinge is held in a bracket with disc springs to help keep the stowed boom latched and aid deployment.

Generally, the magnetometer boom's fittings and deployment mechanisms are machined out of non-magnetic metals like aluminum, bronze and beryllium copper. However, the use of composite booms instead of metallic booms is advantageous in the design of the magnetometer booms. The booms are stowed as 1 m lengths during launch to minimize the mass due to clamping and the necessary release mechanisms. The high specific stiffness of composites easily met the frequency requirements. Additionally, the composite boom is designed to have minimal thermal expansion to match that of the composite deck when exposed to on orbit temperature extremes. The composite tubes are also nonmagnetic and their low density minimizes the mass budget of the magnetometer booms.

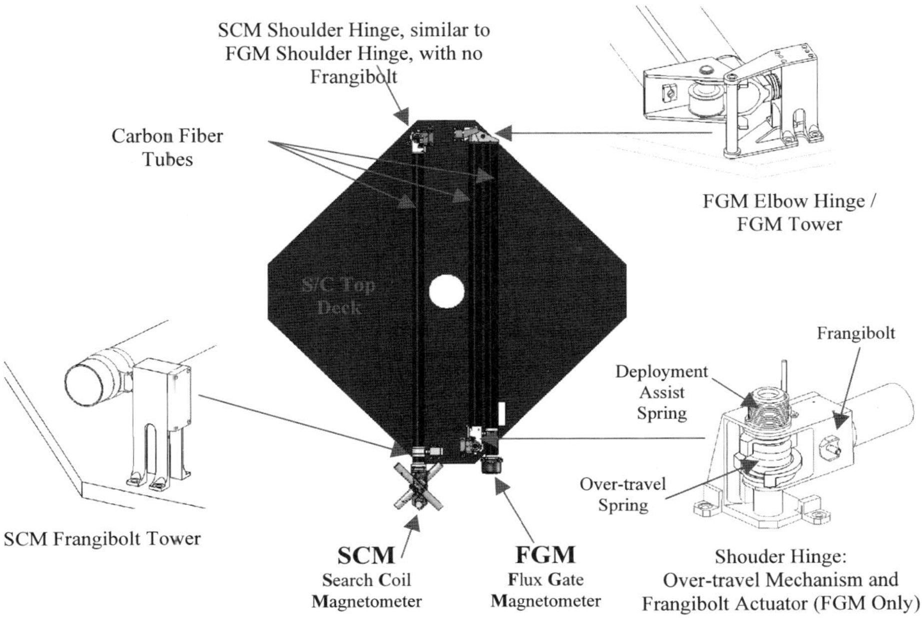

Fig. 7 Magnetometer boom configuration

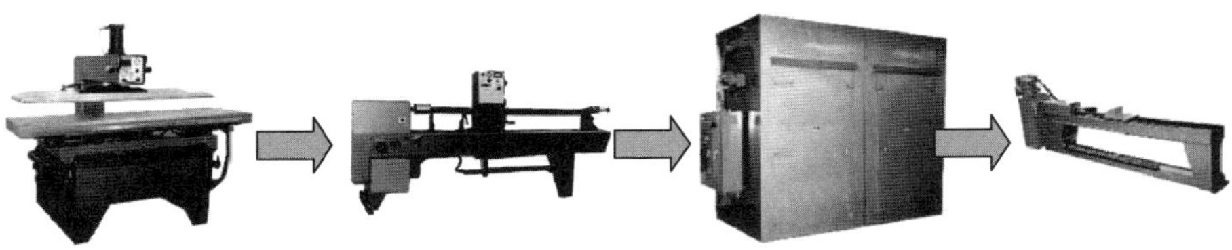

Fig. 8 Magnetometer boom CFRP tube fabrication

Carbon fiber/RS-3 prepreg were used to fabricate the magnetometer boom tubes. The booms are designed using quasi-isotropic high strength T300 carbon fiber fabric to provide shear, hoop and handling strength and ultra high modulus M55J unidirectional carbon fiber to provide longitudinal stiffness. The layup is also designed to minimize coefficient of thermal expansion. The RS-3 (YLA, Benicia, CA) cyanate ester matrix with a 170°C cure temperature was selected, to ensure low outgassing and dimensional stability of the composites.

The composite booms are fabricated by Berkeley Composites Laboratory using the tube rolling process (Century Design, San Diego, USA). The tube rolling process uses a tube roller, shrink tape wrapper, cure oven and mandrel extractor, as shown in Fig. 8. The prepreg is rolled around a hard anodized aluminum mandrel coated with a release agent using the table roller with controlled rolling speed, pressure, and lower platen heat. The mandrel with the prepreg rolled around it is then transported onto the shrink tape wrapper machine where tape is wrapped around the tube with constant tension and speed control. A convection

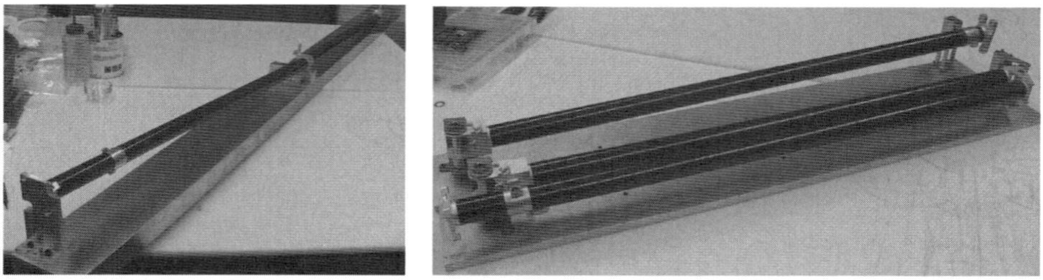

Fig. 9 Magnetometer boom in bonding fixtures

oven is used to cure the booms at 170°C for two hours. Once the booms are cooled to room temperature, a chain driven mandrel extractor is used to extract the boom from the aluminum mandrel. This process provided consistent through thickness consolidation of the composites and the minimized part to part variability.

The composite booms are then finished by high speed machining before integrating with the deployment mechanisms. Special bonding fixtures, shown in Fig. 9, are designed with tight flatness tolerances to first bond the composite booms to the end fittings in the deployed position (with a longitudinal cant and axial twist) to ensure boom accuracy. Hysol EA 9394 (Henkel, Düsseldorf, Germany) is used as the bonding adhesive to ensure high temperature stability. The composite boom tubes are bonded on the inside surfaces to aluminum fittings.

The next step is to bond the saddle rings onto the composite booms in their stowed configuration. The FGB is folded and stowed using the Frangibolt simulator, and saddle ring is bonded to the outside of the composite boom tube. Similarly, the SCB saddle ring is stowed using the clamping mechanism and bonded to the outside of the composite boom tube.

2.3 Magnetometer Boom Deployment Mechanisms

The base hinge assembly in both booms contains three custom beryllium copper springs: the deployment spring, latch pin spring and saloon door spring. During deployment, the deployment torsion spring assists centripetal forces and acts on the booms as they deploy and latch. The latch is engaged when a spring-loaded bronze pin with PTFE impregnated Acetal tip insert springs into a gap between two rotating cogs. At this point, the saloon door spring engages. After the kinetic energy is dissipated, the deployment spring holds the boom against the hard stop of these preloaded cogs. The base hinge assembly is constructed of aluminum, to save weight, and has kinematic flexure mounting points to minimize thermal stresses between the carbon fiber deck and aluminum bracket and to provide stable pointing through thermal cycles.

The magnetometer booms are deployed when the spacecraft is in orbit and spinning, in the following sequence: 1) the SMA deployment device are activated, breaking the Frangibolts which secure the booms during launch; 2) the SCB and outer segment of the FGB begin to open; 3) at approximately 20° of deployment, the elbow latch releases, freeing the inner segment of the FGB to begin deploying; 4) as the outer segment of the FGB opens along with the inner link of the FGB, it is slowed by coriolis acceleration; 5) when the booms deploy to their final position, the "saloon door" style hinges engage, and excess energy is lost in the ensuing oscillations while the booms settle to their final positions which are positively defined by the saloon door springs. The sequence is shown in Fig. 10. Detailed Matlab simulations confirmed the deployment dynamics, torque margins, and peak loads.

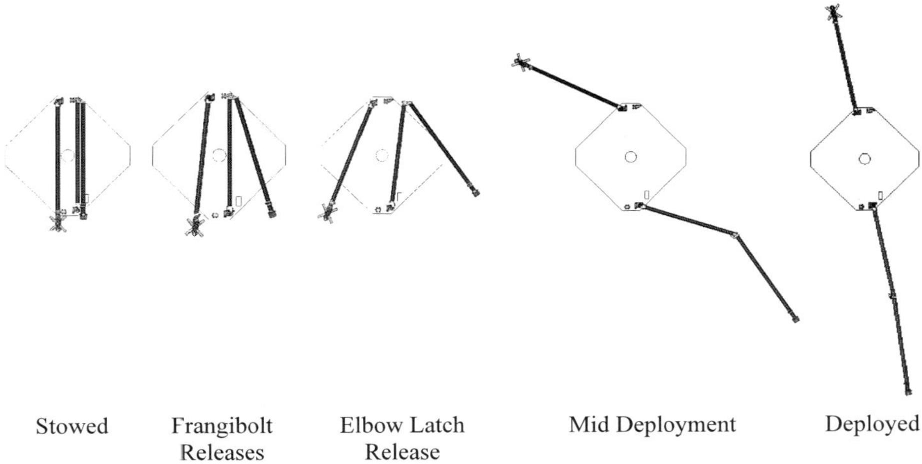

Fig. 10 Magnetometer boom deployment sequence

Fig. 11 Torque margin test fixture

2.4 Magnetometer Boom Testing

Before assembly of multiple flight units, proof testing is performed on the various components of the design. The composite booms are designed by theoretical and numerical analyses and optimized by testing on the vibration table. A key concern of the composite boom is their bonding with aluminum end and the potential large thermal mismatch. The bonding interface design and technique is proven by thermally cycling a short composite boom bonded to an aluminum end fitting.

The deployment hinge mechanisms are tested for fit by constructing rapid prototyping parts. After the flight hinge mechanism is fabricated, a torque margin test of the hinge is conducted at the deployment temperature extremes. The torque margin test fixture, shown in Fig. 11, consists of a stepper motor, which rotates the free end and a load cell which measured the torque at the fixed end. The difference in torque levels measured in stowing and deployment motions are used to determine the torque margin.

After assembling the magnetometer boom, each magnetometer boom is statically proof tested with the quasi-static equivalent load along each axis. This static test loads the mount-

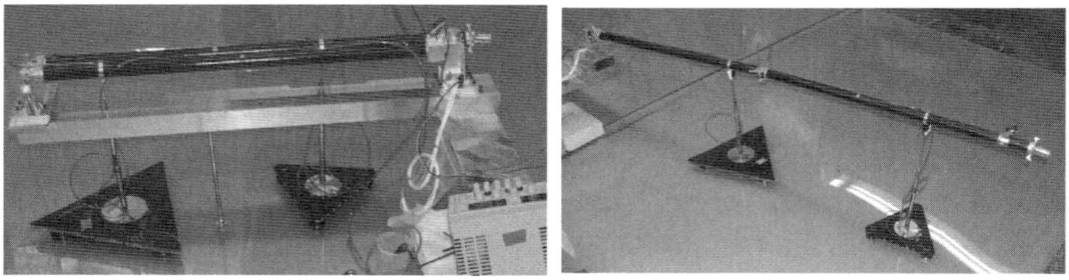

Fig. 12 Stowed (*left*) and Deployed (*right*) FGB

Fig. 13 Vibration test of magnetometer booms

ing points, clamping Frangibolts, the composite boom and the composite/aluminum interface.

Next, the deployment of the stowed magnetometer boom is functionally tested by performing a conservative 0 RPM deployment test. The friction from moment due to magnetometer boom weight and the magnetometer boom tilting up when the boom rotates complicates the deployment test. To eliminate these two problems, air pistons are use to offset the magnetometer boom's weight. The air piston's force is controlled by pressure and the constant force of air pistons as they extend allow for continuous weight offset of the magnetometer booms as they tilt up when deploying. The air pistons are mounted on a low-mass composite sandwich panel, fitted with low friction air bearings and allowed to travel along on a smooth, acrylic base as the magnetometer boom deploys. The stowed and deployed FGB with weight offset devices are shown in Fig. 12.

After deployment testing, the magnetometer booms are re-stowed and tested on a vibration table, shown in Fig. 13. A low-level ($0.5g$) sinusoidal frequency sweep (sine signature) is first performed from 5–2000 Hz to identify the resonant peaks and their quality factors. Following this, a high-level sweep, or "sine strength" test, is performed up to 50 Hz at $16g$ to verify the mechanism strengths below their natural frequency.

This is followed by a repeat of the sine signature to verify that the frequency response curve has not changed, which would indicate damage. After this, a random vibration test is performed at levels dictated by finite element vibration analysis at the probe carrier and probe level. Finally, the sine signature is performed a third time to again verify the frequency response. All three major axes are tested under sinusoidal and random vibration.

A post-vibration deployment test is performed to test functionality. The booms then stowed and then thermally cycled in vacuum.

The thermal vacuum tests, shown in Fig. 14, consisted of a hot and cold deployment verification in which the booms were first cycled from 75° to $-115°$C, with a first motion deployment at 50° and $-45°$C respectfully. The first motion test proves the functionality

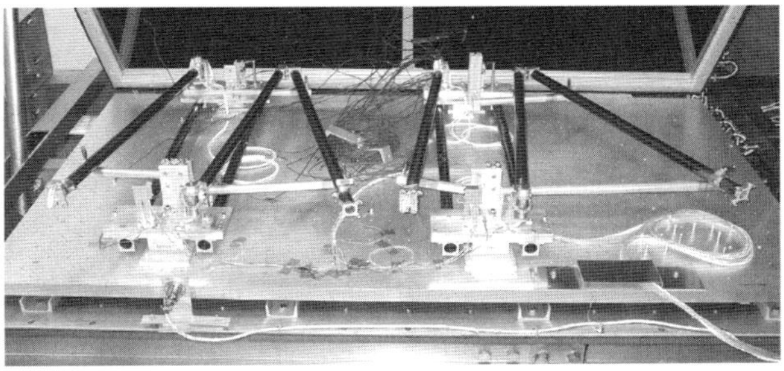

Fig. 14 Thermal vacuum test fixture for magnetometer booms

of the Frangibolt actuator and initial movement in each hinge. The mounting fixtures are designed with the same amount of thermal expansion as the magnetometer booms.

The magnetometer booms underwent three gravity drop tests, in which the boom is mounted to a wall orientated such that gravity would simulate the centripetal acceleration of extended magnetometer boom. The first test is a proof test of the hinges with the kinetic energy from deployments of the spacecraft spinning up to 18 RPM. This is done using scaled mass dummies and allowing the boom to drop from an angle translated from the predicted kinetic energy of a deployment.

The second test tests the repeatability of the final deployed position. The magnetometer boom is deployed from a more moderate angle, and the inclination of the mount is recorded from a mounted 2-axis inclinometer sensor. This test is repeated five times per boom to determine that the alignment/orientation repeatability is better than 0.01°. The third test measures alignment of the magnetometer booms using the inclinometer sensor mounted in two orthogonal positions on the magnetometer mounts and compared the measured angles to reference surfaces on the base hinge. The alignment of the booms are measured, or shimmed if necessary, to be within $\pm 0.1^\circ$ in Z axis from the predicted value. The alignment angles in the other two axes are measured to be within $\pm 0.25^\circ$ of the predicted value. Numerical simulation of the deployment environment is used to verify deployed boom resonance above 0.75 Hz.

3 The Radial Electric Fields Mechanisms

3.1 Overview

The THEMIS Spin Plane Booms (SPB's) deploy the $\pm X$ and $\pm Y$ Electric Field Instrument (EFI) sensors in a radial direction from the spinning spacecraft, as shown in Fig. 1. The deployed boom preamplifiers with attached 3 m sensors are deployed to a tip-to-tip length of 40 m in X and 50 m in Y direction. Each SPB consists of a spool, motor and meter wheel, release mechanism, chassis, Gore composite cable, SPB preamplifier enclosure, and sensor subassembly, as shown in Fig. 15. The four SPB's are mounted to the lower deck of the spacecraft with the snout protruding through a square cutout in the center of each side solar array.

3.1.1 Cable Spool

A spool mechanism safely stores the 21.5 meter sensor cable prior to deployment. The spool electrically isolates the cable from the SPB chassis, allowing the controlled electrical grounding of the outer shield of the cable to the structure to dissipate static charge. The

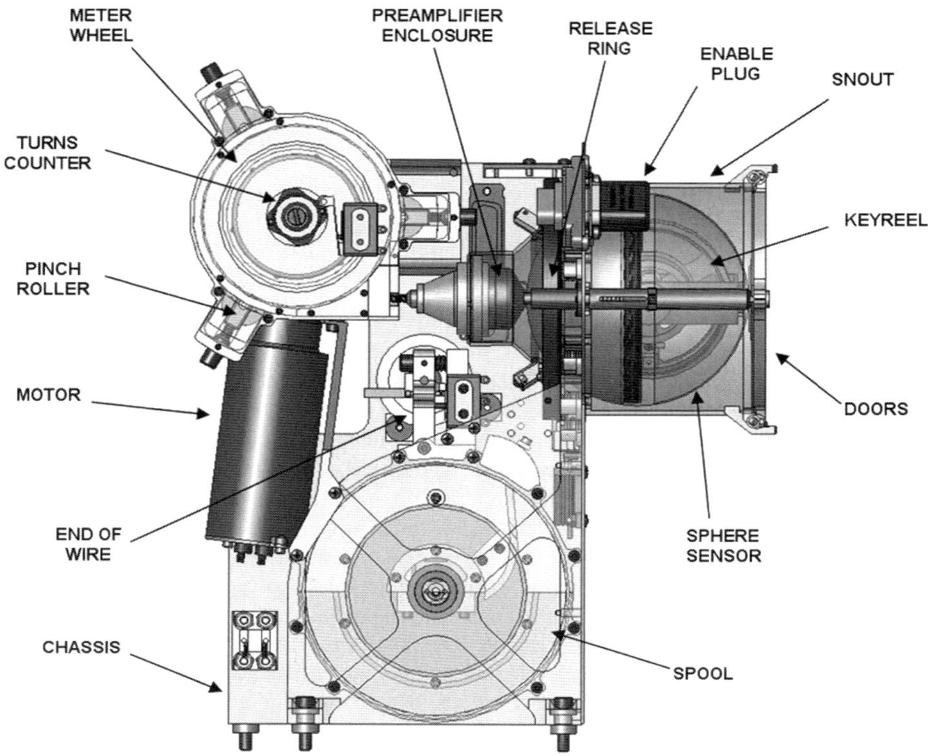

Fig. 15 SPB, transparent view

electrical signals between the rotating spool and the chassis are transferred using slip rings. Each circuit on the slip ring has redundant fingers that improve the electrical noise levels, even though the sensor is not recording accurate E-field measurements during deployment. The spool also anchors the cable and prevents the motor from breaking the cable by strain relieving the cable to a harness cable bracket. An end of wire switch actuates to terminate motor power when an over-tension condition places too much strain on the cable. This over-tension condition can occur when the cable is fully deployed, or if the cable becomes nested and tangled in itself. To ensure the cable does not become nested during launch vibration, a friction brake caliper is adjusted to apply rotational resistance to the spool.

3.1.2 Motor and Meter Wheel

A brushed, DC-powered gear motor deploys the SPB cable and sensor. The motor requires a multi-layer (magnetic attenuating), mu-metal shield to decrease the magnetic dipole created by the motor's permanent magnets. A close-out shield with EMI filters reduces the high frequency electrical noise produced when the motor is in operation, and a diode between the motor terminals helps reduce back-EMF when the motor power is turned off. Bevel gears transmit torque from the motor to the meter wheel through a 90° bend, creating a compact SPB design. The SPB cable lies in a v-shaped grove cut in the vulcanized rubber on the meter wheel outside diameter. Three pinch rollers apply inward radial pressure to the cable to ensure it remains in good contact with the meter wheel, thus creating a capstan force for the meter wheel to pull the cable from the spool. Cable metering is accomplished by a micro-switch that follows a four-lobed cam, indicating every 4.7 cm deployed. If the end of wire switch fails to prevent an over tension condition in the cable, a shear pin in the meter wheel can also shear to prevent cable damage.

3.1.3 Release Mechanism

While the SPB is stowed, the doors apply sufficient preload to cage the sphere and preamp to avoid damage during launch vibrations. The doors are held closed by two release pins under spring tension, allowing for even loading and compliance. The inboard ends of the pins rests on bearings that are mounted on a titanium release ring. The release ring rotates when voltage is applied to shape memory alloy (SMA) wire. The SMA wires thermally contract, which causes the bearings to roll off the release pins, and release springs force the release pins forward and clear of the doors' travel. An end-of-travel switch is actuated when the release ring is fully rotated, turning off power to the SMA wires to preventing overstress and subsequent loss of wire 'memory' during testing.

3.1.4 Chassis

The chassis consists of a fixed and removable side plate, front plate, support angle, and snout. The structural pieces, with exception of the snout, are made of an aluminum-magnesium alloy and are weight-relief pocketed to decrease overall mass. There are two sheet metal pieces that both support and increase stiffness of the front plate and protect the cable and release ring assembly when thermal blanketing is applied. The chassis is maintained at spacecraft potential through the SPB harness, and is thermally isolated from the spacecraft using Ultem spacers.

3.1.5 Composite Cable

The THEMIS SPB Gore composite cable consists of an inboard connector termination, cable, bead, and outboard preamplifier enclosure termination. Similar configurations are on both ends of the AXB Gore cable discussed in Sect. 4. A section view of the cable is provided in Fig. 16 illustrating how the coax and eight single conductors are held tightly together by a load carrying Kevlar braid. The Kevlar is then wrapped by an aluminized Kapton tape that acts as an electrostatic barrier. The most outer layer is a silver plated copper (SPC) braid that prevents abrasion and wear of the aluminized Kapton. The inboard connector attaches the cable to the spool electrically and serves as an anchor to stress relieve the cable through the Kevlar braid. An epoxy bead is located inboard of the preamplifier enclosure to allow the last 3 meters of cable to be biased at Distal Braid (D-braid) potential, illustrated in Fig. 19. The bead conceals an electrical discontinuity in the outer SPC braid and aluminized Kapton electrostatic barrier, but maintains the axial strength via the Kevlar braid. The outboard preamplifier enclosure termination consists of the cable holder, guard surface of the preamp, sockets and socket holder. The Kevlar strength member is extracted from the cable and anchored to the cable holder to provide strain relief. The coax and single conductors are terminated in sockets that are housed in the socket holder. This forms the inboard end of the preamplifier enclosure.

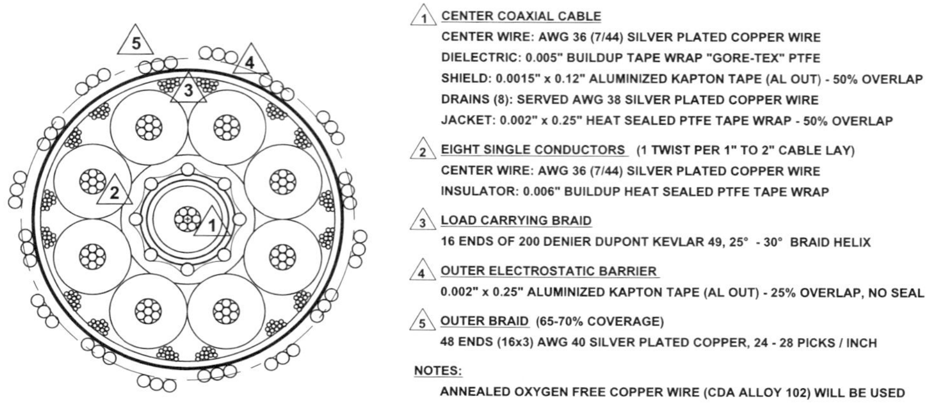

Fig. 16 Gore custom cable construction

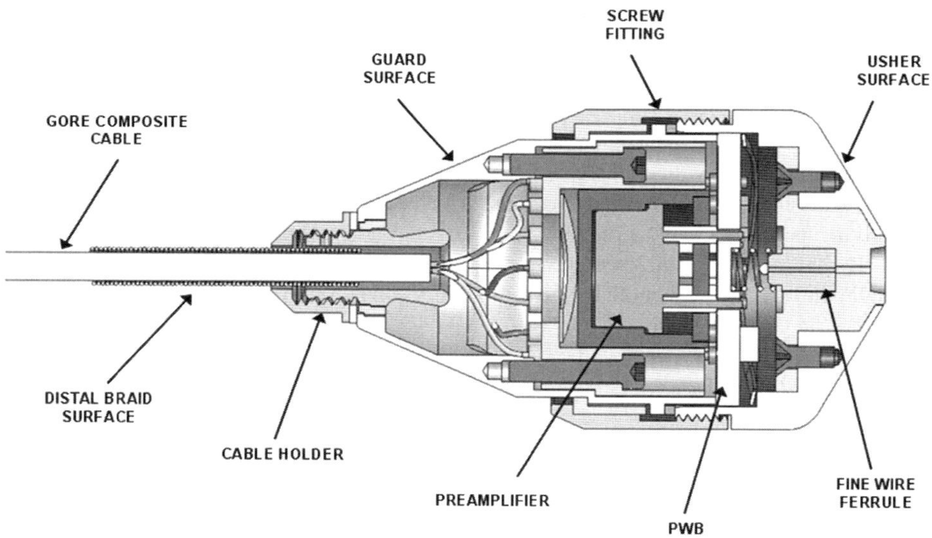

Fig. 17 Preamplifier enclosure, section view

3.1.6 SPB Preamplifier Enclosure

The preamplifier (preamp) enclosure is divided into the inboard cable connector end, the cable preamp printed wire board (PWB), and outboard end as shown in Fig. 17. The inboard preamp end was described in the composite cable section. The PWB contains a few discreet components, an op-amp, and pins that mate with sockets on the inboard end. The outboard end consists of the usher surface and fine wire ferrule. The ferrule is crimped to securely anchor the fine wire that comes from the spherical sensor, and also makes contact with the ferrule spring that is the input to the operational amplifier (op-amp). The preamp is assembled by mating the PWB with the inboard end, mating the outboard end and fine wire ferrule on the PWB, and finally tightening the screw fitting. The usher surface includes the screw

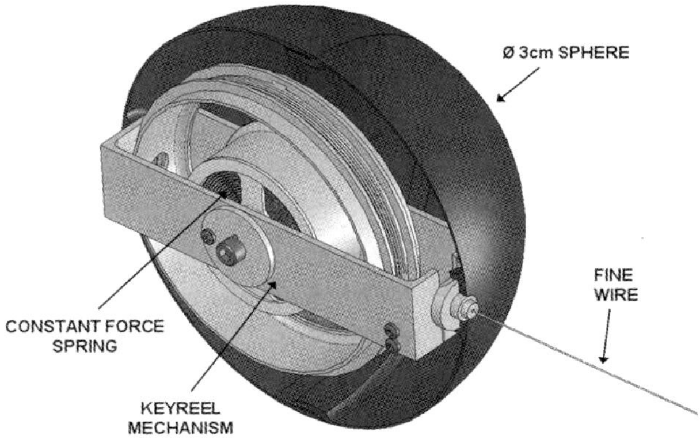

Fig. 18 Sphere sensor, sectioned view

fitting and is electrically isolated from the preamp enclosure guard surface (Guard) and D-braid. Traces on either side of the PWB make contact with the Guard and Usher surfaces and allow each to be biased at a different potential. A tantalum cover and disk enshroud the op-amp for radiation hardening, Careful analysis and design permits the extreme temperature fluctuations that are experienced while deployed well away from the thermally controlled spacecraft. Modular construction of the preamp allows for ease of preamp removal during integration and testing.

3.1.7 Sensor

An ∅8 cm aluminum sphere shell is coupled to a ∅0.25 mm stainless steel fine wire that terminates in the preamp enclosure at the preamp ferrule. Prior to deployment, a constant force spring key-reel mechanism inside the sphere stows the fine wire and prevents the fine wire from nesting and becoming tangled, as shown in Fig. 18. Owing to key-reel spring force, the sensor remains coddled next to the preamp enclosure when the SPB doors are opened. When the sensor and preamp enclosure are deployed to a pre-determined distance from spacecraft and spacecraft spin rate, the centrifugal force overcomes the constant-force key-reel mechanism spring and the sphere deploys the 3 m of fine wire smoothly.

3.1.8 Electric Field Biasing Elements

The best e-field measurements are made when the effects of photo-emissions from the spacecraft and the boom mechanisms are minimized. As discussed in the science section, the potential of various surfaces on the booms need to be controlled separately to accomplish this, illustrated in Fig. 19. The SPB chassis is at spacecraft ground potential, and the proximal Gore cable braid (P-braid) is grounded to the chassis through a 330 kΩ resistor. The D-braid, Guard, and preamp enclosure usher surface (Usher) potentials are biased by circuits through the Gore composite cable from the IDPU. These surfaces help isolate the sensor from the photoemission charge cloud surrounding the spacecraft. The sphere and preamplifier enclosure surfaces have a DAG-213 carbon based coating for a uniform work function and to moderate the on orbit temperatures.

3.1.9 Testing

THEMIS SPB mechanical testing began with a functional test of the sphere sensor key-reel spring mechanism. The sensor assembly was placed on a motorized take-up spool and load

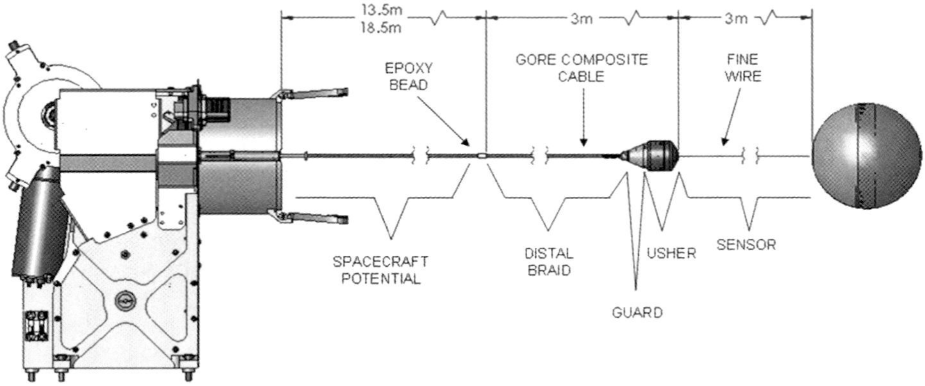

Fig. 19 SPB sensor surfaces

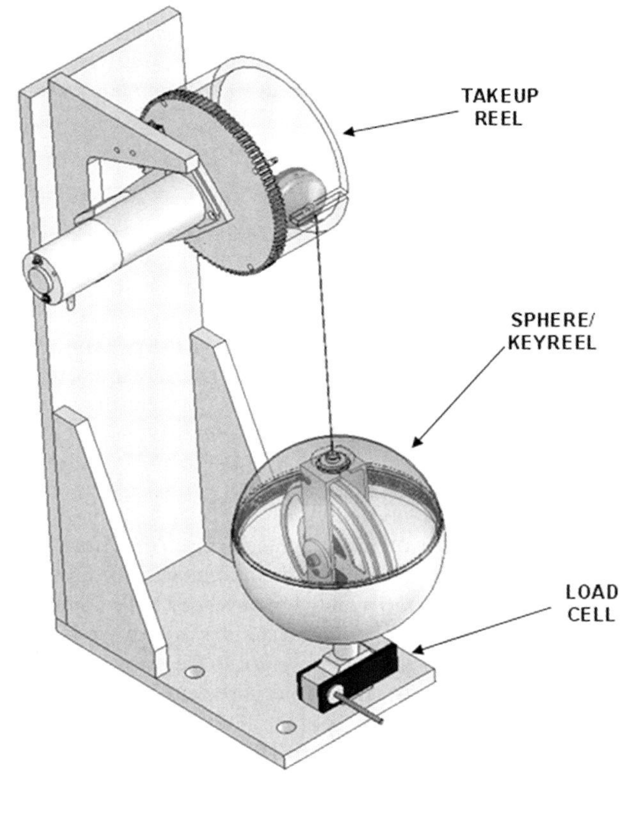

Fig. 20 SPB sphere TVAC GSE for force-deflection characterization

cell apparatus in a thermal vacuum chamber to characterize the force vs. extension of the key-reel spring over extreme operational temperature ranges, illustrated in Fig. 20. This data would indicate at what combinations of spacecraft spin rate and preamp deployed length the sphere sensor would deploy the fine wire.

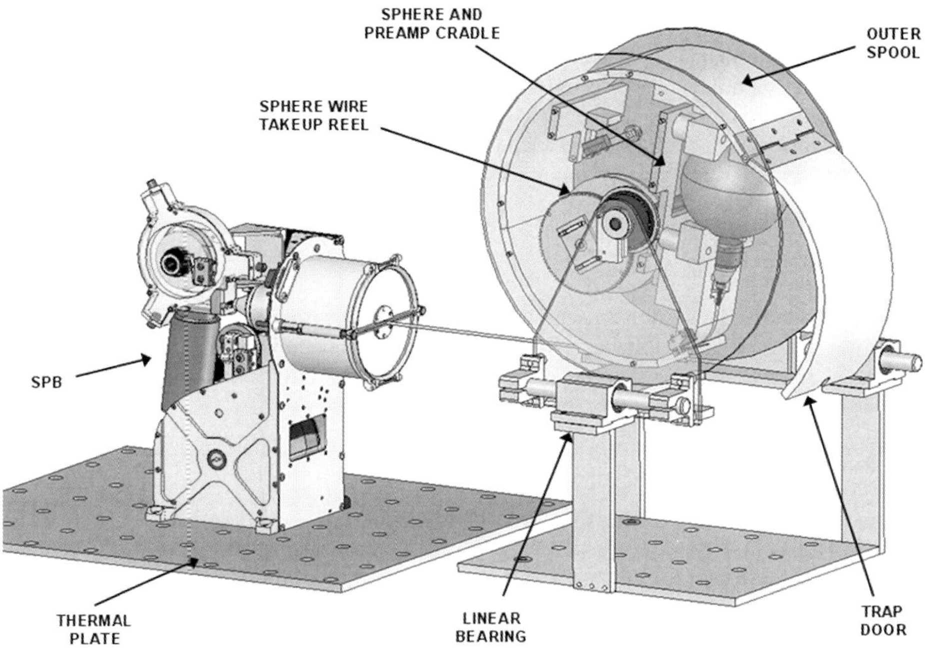

Fig. 21 SPB thermal vacuum test setup, transparent view

Once the sphere sensor was integrated with the SPB cable and preamp, the SPB was stowed and electrical continuity and isolation checks verified that all assemblies were fitted properly and electrical results were as predicted. A functional test deploy and length calibration was performed in the high bay, observing proper door firing, and cable length measurements and meter wheel turns counts were recorded. The SPB was then stowed, and then instrument level sine and random vibration tests were conducted to verify strength design and proper assembly of the SPB.

After a post-vibration electrical continuity and isolation test, the SPB was then placed in a thermal vacuum (TVAC) chamber with a sphere, cable, and preamp take-up device to conduct an end-to-end deployment of the SPB at extreme operational temperatures of -25 to $+55^{\circ}$C, as illustrated in Figs. 20 and 21. Nominal current and release time for door firing, SPB motor current, preamp quiescent current, deploy time, and meter wheel turns counts were recorded. The full length of fine wire and Gore cable was deployed to ensure the end of wire switch operated properly to prevent cable damage.

Upon completion of TVAC testing, the SPB was stowed, electrical continuity and isolation was performed, and finally delivered for integration and further instrument suite level testing.

4 The Axial Electric Fields Mechanisms

4.1 Axial Boom Science Configuration

Deployable rigid booms are provided to hold the whip sensors on the spacecraft spin axis to make the Z-axis E-field measurement. The two boom deployment mechanisms are mounted in a tube that forms the central structural element in the THEMIS bus. This tube also serves as the antenna mount. The deployed boom elements are each 2.44 m long, topped with a preamplifier module and a 0.75 m whip sensor, forming a 6.4 tip-to-tip sensor array. The main elements are grounded to the bus to hold them at spacecraft potential. A nine conductor cable runs inside each boom element to the preamp, and the whip sensor is the preamp input. The outer surface of the preamp module is divided into Guard and Usher surfaces that are each electrically driven, as in the Spin Plane Booms.

4.2 Axial Boom Design

Each axial boom consists of a Stacer boom element, a deployment assist device (DAD) with roller nozzles, a preamp, a Stacer whip sensor, and a cable and bobbin. The boom is caged for launch by a Frangibolt shape-memory alloy device. The two axial booms are mounted diametrically opposed in a carbon fiber tube assembly.

4.3 Stacers

The extendible portions of the spin axis booms and whip sensors are Stacers, shown in Fig. 22. The main boom Stacer is a tubular spring element fabricated from a 0.10 mm thick × 127 mm wide strip of Elgiloy (a non-magnetic super-alloy). It is helically formed such that when released, it forms a slightly tapered, stiff, hollow tube 1.98 m long. The main Stacer stows into a 127 mm long by 50 mm diameter canister. As it is stowed, the coils are manually

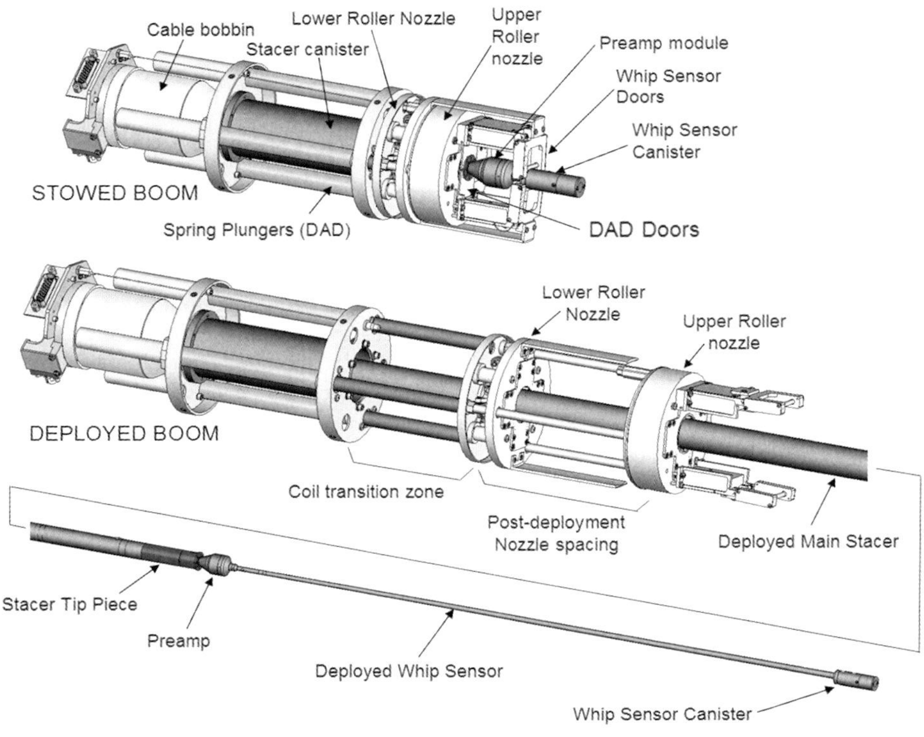

Fig. 22 Axial boom configuration

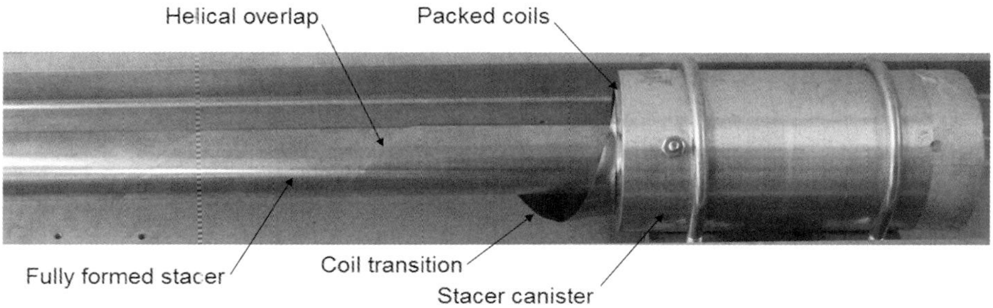

Fig. 23 Stacer deployment

packed out against the canister I.D., increasing the coil diameter. As the Stacer deploys, the coils curl inward and protrude axially, cinching around first the tip piece, and subsequently around the previously deployed coils illustrated by Fig. 23. This change in diameter during the unfurling provides the strain energy that generates the deployment push force. Thus, the Stacer deployment is self-powered. The tip piece is slightly larger than the strip free-coil diameter, so that the coils grab tightly. This generates enough friction between coils that they do not easily slip on one another, giving the Stacer has a bending stiffness similar to a tube with comparable wall thickness for small deflections. The stacking of subsequent coils causes a buildup in deployed diameter, leading to the gradual taper from the 18 mm diameter tip the 23 mm diameter base. This taper gives better bending stiffness and mass properties than constant diameter booms. The sensor Stacer is a smaller version with a strip thickness of 0.05 mm, a length of 0.75 m, and base and tip diameters of 7.9 mm and 6.4 mm. The Stacer elements mass is 270 g and 7 g for the mains and whips respectively.

A primary advantage of Stacers, as compared to other types of long rigid booms such as Stem booms, is the Stacer's thermal symmetry. The poor thermal conductivity between the two axial strips forming a Stem boom can produce significant asymmetries in its thermal bending, which has been known to cause thermal pumping on spacecraft such as Ulysses. The helical overlap of the single Stacer strip, by contrast, provides a helically symmetric path for heat flow from the sun-lit to shadowed sides of the Stacer. This leads to a small, but uniform deflection 90° from the sun line that does not excite spacecraft wobble.

4.4 Boom DAD and Roller Nozzles

Two roller nozzles are used to give the main Stacer the required cantilever stiffness. Because the Stacer coils are not well supported in the transition zone between the canister and the fully formed Stacer, it has no inherent stiffness at its root. The necessary support to make it a stiff cantilever is provided by two roller nozzle assemblies spaced approximately 130 mm apart when deployed. These are positioned beyond the Stacer canister by means of a telescoping two-stage Deployment Assist Device (DAD), included in Fig. 22. This design has a compact stowed geometry while providing the necessary separation between the nozzles to give a rigid pinned-pinned base attachment.

The expanding roller nozzle design, illustrated in Figs. 24 and 25, provides play-free positioning despite the changing boom diameter during deployment, while also minimizing friction. The four ball bearing rollers are held on pivoted rocker arms attached to a common cover that is pulled by spring cartridges towards the base plate. Idlers on each rocker arm roll on the base plate, rotating the rollers inwards in a pinching motion, providing play-free

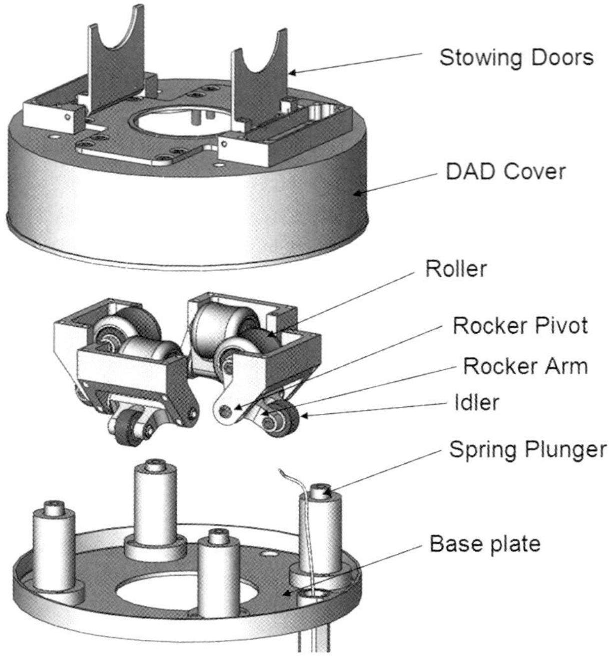

Fig. 24 Roller nozzle components

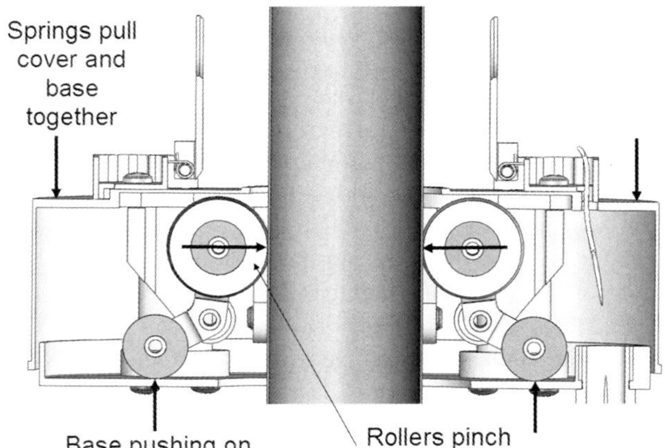

Fig. 25 Roller nozzle operation

lateral constraint and centering for the Stacer while allowing it to roll outward freely during deployment.

The roller nozzles are held together and against the canister for launch by trap doors that catch an edge on the Stacer tip piece, shown in Fig. 26. The nozzles are pushed beyond the coil transition zone and apart by telescoping spring loaded plungers when the boom is released. A second outer set of doors cages the small whip sensor Stacer. The plungers also provide an initial kick force to help the cinching of the first coil on the tip piece, ensuring correct formation of subsequent coils. After the Frangibolt initiated release, the DAD

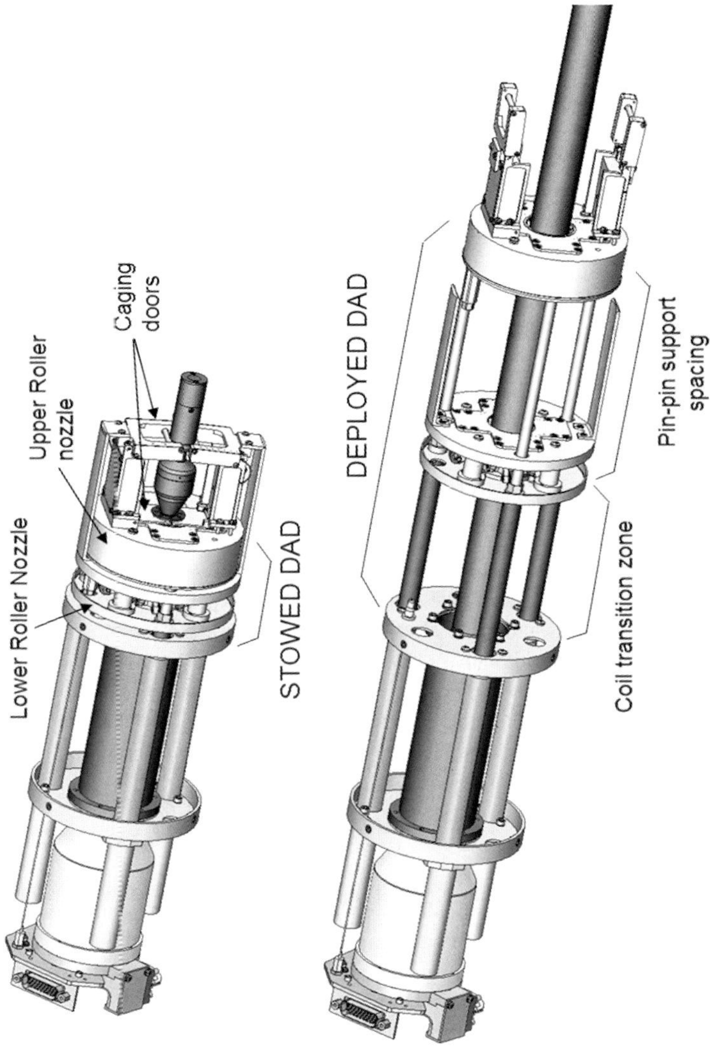

Fig. 26 DAD deployment

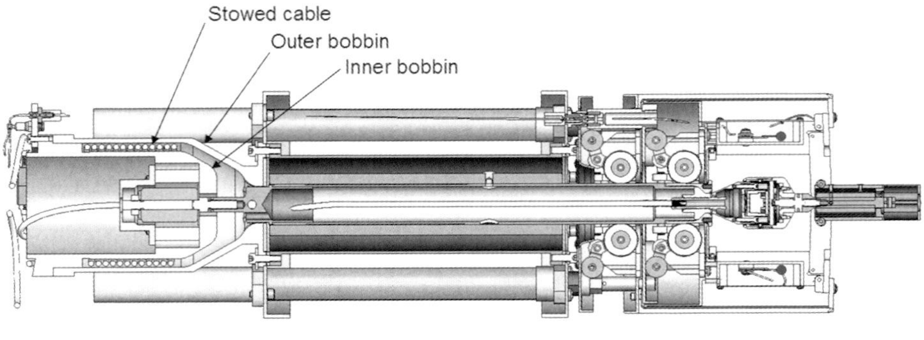

Fig. 27 Cable bobbin

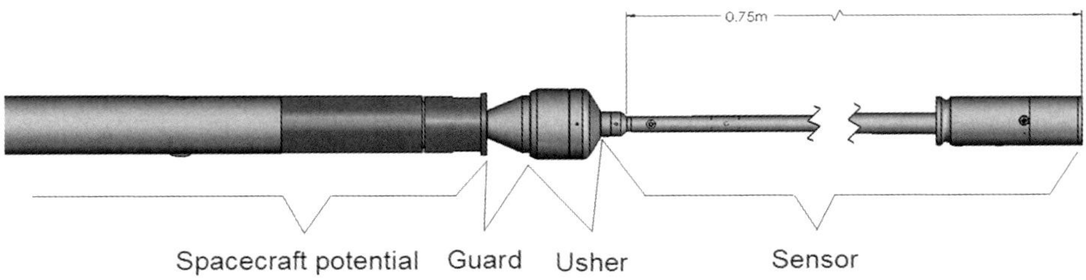

Fig. 28 Sensor surfaces

plunger pushes the tip piece to beyond the coil forming zone of the Stacer, and the trap doors are pushed open by the deploying boom. Once the deployment is initiated, the Stacer makes contact with the rollers and these allow only axial motion.

The final length of the main Stacer is controlled by a cable that is stowed in a bobbin at the aft end of the canister illustrated in Fig. 27. During deployment, this cable pulls off of the bobbin, and strops the deployment at the desired length. This cable, of the same construction as for the Spin Plane booms, provides the necessary conductors to the preamp, as well as a Kevlar layer that absorbs the end-of-stroke deployment energy.

4.5 Electric Field Biasing Elements

The best E field measurements are made when the effects of photo-emissions from the spacecraft and the boom mechanisms can be minimized. As discussed in the science section, the potential of various surfaces on the booms need to be controlled separately to accomplish this. The main Stacer and its tip piece are grounded to the spacecraft through a 1 MΩ resistor. The guard and usher potentials are controlled through lines running back to the instrument IDPU, shown in Fig. 28. These surfaces can be biased to help isolate the sensor from the charge cloud surrounding the spacecraft. The whip and its canister form the sensor surface and are the input to the preamp. All external surfaces are conductive to minimize charging, and are coated with DAG 213 carbon based paint for a uniform work function and to moderate temperatures. A UCB designed electrical connector between the preamp and Stacer tip piece allowed swapping of the sensors without disturbing the stowed boom.

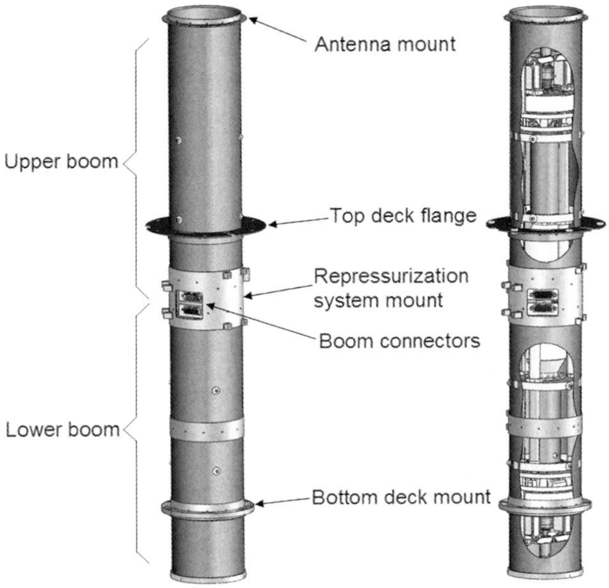

Fig. 29 Boom mounting tube

4.6 Boom Mounting Tube

The axial boom mechanisms are held in the THEMIS spacecraft by a mounting tube assembly. This tube holds the booms coaxial with the spacecraft Z axis. It also serves as a structural member in the spacecraft bus, supporting the top deck at its center and providing needed supports for the probe RCS re-pressurization system and antenna. The tube is 0.86 m long, and extends 0.33 m above the spacecraft top deck, supporting the S-band antenna well away from spacecraft.

The pair of boom assemblies is mounted in a 100 mm diameter thin wall graphite-epoxy tube that is manufactured by the UCB Composites Lab. The tube consists of 5 layers of a Fiberite 0.13 mm thick prepreg woven graphite fiber material. The inner, middle and outer layers were laid at 0°–90°–0° and the layers between were set to ±45° to the tube axis. Titanium, aluminum, and carbon fiber flanges are mounted to the outside of the tube to provide mounting points for the flanges to bolt the boom to the spacecraft top and bottom decks, the antenna, as well as to support the re-pressurization system. Because the mounts are structural components of the spacecraft, these were completed early in the program so that they could be shipped to the spacecraft contractor for incorporation onto the spacecraft structure for the bus structural testing.

4.7 Axial Boom Testing

In-air testing of the axial booms was performed in two ways. Initially the booms were deployed upward vertically to yield a non gravity-biased concentricity measurement. The unit was mounted coaxially on a turn table base, and after deployment, the unit was rotated to measure the total tip run-out. Each boom was then deployed with the tip piece attached to a low friction trolley on a horizontal track. This horizontal deployment gives a more accurate length measurement with the absence of gravity in the deployment direction. For environmental testing, the individual boom units were subjected to 3 axis sine and random vibration, and then 12 cycles of thermal vacuum with hot and cold deployments. The horizontal deployment track was also used here. After hot and cold soaks at −60°C and +75°C, each boom was deployed down the trolley in a long vacuum tight tube attached to the chamber. One boom was deployed hot (+40°C), the other cold (−35°C). A final verification deployment was performed after inspection, The final step was inspection and stowing for launch. Deployed boom straightness was found to be within 3–19 mm at its 3.2 m length, and the measured lateral resonance was 1.5–1.6 Hz.

5 On-Orbit Performance

The five THEMIS micro-satellites were launched on a common carrier by a Delta II, 7925 heavy, on February 17, 2007. The probe fuel capacity would only accommodate two years of science operations, with the prime science season in February. For these reasons, stage I (2/15/07–9/15/07) of the mission was a coast phase, where the probes were kept in their post launch orbits. The EFI booms on two of the probes were deployed for early science and for mission diagnostics. The booms on the remaining probes remained stowed, so that they could be maneuvered to higher orbits with better fuel efficiency. All probe magnetometer booms were deployed shortly after launch. Stage II (9/15/07–12/15/07) of the mission was the orbit placement period. The probes were maneuvered to their assigned 0.8, 1, 2, and 4 day orbits. Following this orbit placement, EFI booms on the remaining three probes were deployed. In all, 40 boom mechanisms were deployed on the five probes with no failures or anomalies. Careful planning by the Berkeley Mission Operations team resulted in no improper maneuvers on orbit. The probe dynamic behavior, during RCS maneuvers and boom deployments was as predicted and tested. Post maneuver settling times were near real time, when compared to next command formulation and execution. Mission stage III prime science began on 12/15/07 with all five probes fully operational.

6 Summary

The THEMIS boom mechanism designs have evolved over a period of thirty years, from a long series of successful satellite instruments flown on S3-3, ISEE, Viking, Freja, CRRES, Polar, FAST and Cluster 1 & 2. The Themis mission was a unique opportunity to advance the state of the art of the three boom systems it employed. It was predicated as a higher risk mission with a short development and fabrication cycle. The five probes are achieving their primary science objectives, and by doing so, provided an engineering development history for future missions to call upon.

Acknowledgements The authors wish to thank the Explorers Office at NASA Goddard Space Flight Center for funding this mission under contract NAS5-02099. Dr. Vassilis Angelopoulos, University of California Berkeley was the energetic and motivated THEMIS Principal Investigator. Frank Snow was the GSFC Mission Manager and Peter Harvey was the UCB Project Manager. Both managers were enthusiastically motivating and played a critical role in achieving our first proposed launch date.

References

V. Angelopoulos, The Themis mission. Space Sci. Rev. (2008, this issue). doi:10.1007/s11214-008-9336-1

D.M. Auslander, An object-oriented approach to basic mechanics, in *Proceedings of the ASME International Mechanical Engineering Conference & Exposition*, Orlando, FL, Nov. 2000, pp. 771–778

F.T. Dodge, *The New "Dynamic Behavior of Liquids in Moving Containers"* (Southwest Research Institute, San Antonio, 2000)

S.T. Lai, K.H. Bhavnani, Dynamics of satellite wire boom systems, AFCRL-TR-75-0220, 1975

L. Meirovitch, R.E. Calico, The stability of motion of satellites with flexible appendages. NASA CR-1978, 1972

THEMIS Ground Based Observatory System Design

S.E. Harris · S.B. Mende · V. Angelopoulos · W. Rachelson · E. Donovan · B. Jackel · M. Greffen · C.T. Russell · D.R. Pierce · D.J. Dearborn · K. Rowe · M. Connors

Originally published in the journal Space Science Reviews, Volume 141, Nos 1–4, 213–233.
DOI: 10.1007/s11214-007-9294-z

S.E. Harris (✉) · S.B. Mende · V. Angelopoulos · W. Rachelson
Space Sciences Laboratory, University of California, 7 Gauss Way, Berkeley, CA 94720-7450, USA
e-mail: sharris@ssl.berkeley.edu

S.B. Mende
e-mail: mende@ssl.berkeley.edu

V. Angelopoulos
e-mail: vassilis@ssl.berkeley.edu

W. Rachelson
e-mail: wrachelson@ssl.berkeley.edu

E. Donovan · B. Jackel · M. Greffen
University of Calgary, 2500 University Dr. N.W., Calgary, AB, T2N 1N4, Canada

E. Donovan
e-mail: donovan@phys.ucalgary.ca

B. Jackel
e-mail: bjackel@phys.ucalgary.ca

M. Greffen
e-mail: mgreffen@ucalgary.ca

C.T. Russell · D.R. Pierce · D.J. Dearborn · K. Rowe
Institute of Geophysics and Planetary Physics, University of California Los Angeles, Los Angeles, CA 90095, USA

C.T. Russell
e-mail: ctrussel@igpp.ucla.edu

D.R. Pierce
e-mail: dpierce@varametrix.com

D.J. Dearborn
e-mail: ddearbor@igpp.ucla.edu

K. Rowe
e-mail: krowe@igpp ucla.edu

M. Connors
Athabasca University, 1 University Dr., Athabasca, AB, T9S 3A3, Canada
e-mail: martinc@athabascau.ca

Abstract The comprehensive THEMIS approach to solving the substorm problem calls for monitoring the nightside auroral oval with low-cost, robust white-light imagers and magnetometers that can deliver high time resolution data (0.33 and 2 Hz, respectively). A network of 20 Ground-Based Observatories (GBOs) are deployed across Canada and Alaska to support the collection of data from these instruments. Here we describe the system design of the observatory, with emphasis on how the design meets the environmental and data-collection requirements. We also review the design of the All Sky Imager (ASI), discuss how it was built to survive Arctic deployments, and summarize the optical characterizations performed to qualify the design to meet THEMIS mission requirements.

Keywords All sky camera · Geophysics · Aurora · Magnetometer · Magnetosphere

PACS 94.80.+g · 94.30.Aa

1 Introduction

In the spring of 2003, the team responsible for developing the THEMIS Ground Based Observatory (GBO) network kicked off the project by deriving a set of engineering and site requirements based upon the science requirements of the THEMIS mission. These top-level requirements are:

> The GBO shall monitor the auroral light and ionospheric currents across North America in order to localize the time, location, and evolution of the auroral manifestation of the substorm.
> Determine substorm onset time and substorm meridian magnetic local time (MLT) using ground All Sky Imagers (one per MLT hr) and Ground Magnetometers (two per MLT hr) with time resolution better than 30 s and MLT resolution better than 6 degrees, across an 8 hr geographic local time sector including the US.

The team advanced GBO goals to satisfy these requirements with a network of 20 GBOs, that would provide time resolution better than 5 s and MLT resolution better than 1°. The network would span more than 10 hours of geographic local time with ASIs, and more than 14 hours of GMAG coverage, when non-THEMIS magnetometers are considered. This level of continental coverage was a primary technical driver for our efforts.

This team, consisting of researchers at the University of California Berkeley (UCB), the University of Calgary and the University of California Los Angeles (UCLA), collectively included a considerable amount of experience in fielding unattended, ground observatories. It became clear from the outset that the practical considerations of deploying GBOs in remote areas of the North American Arctic would constitute the greater challenge to the project than designing the package. For an example, see Fig. 1. Nonetheless, the combined efforts of the team resulted in a concise set of design criteria and an implementation philosophy that successfully carried the project through to completion.

UCB experience with All Sky Cameras and remote observations is derived from years of field development, deployment, and operation of such systems in remote locations, like the Automatic Geophysical Observatory (AGO) sites in Antarctica (Mende et al. 1999). Figure 2 is a photo of one of the AGO sites, from which we borrow many "lessons learned" and heritage for the GBO development. UCalgary experience is derived from developing

Fig. 1 GBO installation at Inuvik, NT, Canada. The GBO site is about 6 km outside the town of Inuvik. It is very remote, as suggested by the image on the *left*, which shows the road out to the site. The shelter is an ISO container that was outfitted by researchers from the University of Saskatchewan, who operate an ionospheric sounder at the site

Fig. 2 AGO site in Antarctica. Lacking a ready source of local power, the AGOs have been a much greater challenge to design and operate than the GBOs. Nonetheless the systems have much in common

and fielding All-Sky Cameras for the CANOPUS and Canadian GeoSpace Monitoring Programs (Donovan et al. 2003). UCLA's experience is derived from many years of development, deployment, and operation of ground magnetometers in remote sites around the world, including the US, China, Central, and South America.

In this paper we describe the details of the GBO system design, and also include some discussion of our integration and test program. Accompanying articles separately cover other aspects of the THEMIS GBO experiment design, mission requirements, and magnetometer design, so that will not be repeated here.

1.1 Implementation Philosophy

From the start, the plan was to "get in the field, early and often." In other words, the real test of our designs would be in the field, experiencing real weather and other real hazards. While normal laboratory testing is important, it is difficult to simulate the environment and hazards that the GBOs would face. We therefore planned to field a prototype GBO during the next winter season (2003–2004), and prepared to follow that up by fielding four additional units

Fig. 3 GBO Prototype at Athabasca University Geophysical Observatory (AUGO). Shown here is the environmental enclosure, designed to house the Observatory Support Electronics, which are inside. On the right side of the enclosure can be seen the shroud covering the solid state air conditioner and the cable access

during the winter of 2004–2005. We were fully aware that these early systems would have to be upgraded in the field once, or if, the design evolved and changed. And, of course, the design most certainly did change.

This approach meant a very quick development to enable us to field a prototype system by the winter of 2003–2004. The prototype testing took place in Calgary, and also at the very accessible Geophysical Observatory, operated by Athabasca University. Figure 3 shows the prototype GBO as installed in April, 2004.

With our Calgary partners spearheading the deployment effort in Canada, three sites were in operation during the winter of 2004–2005. By the following winter of 2005–2006 we had 10 sites operating, and by the fall of 2006, we had completed 19 of the planned 20 deployments. The final, 20th site was deployed in northern Quebec in fall 2007.

With more than 20 systems to build and deploy, other important aspects of our design philosophy included the following:

- Minimize custom design, and use off-the-shelf components to the greatest extent possible, thereby keeping cost low.
- Select sites that meet optical and magnetic requirements, but they must have access to power, and they must be accessible by a qualified local manager, or custodian, who can maintain the GBO.
- Some sites did not require some components, e.g., magnetometers were not required at all of the sites. Regardless of this, minimize design differences between the sites so that operating software could be maintained uniformly. Keep the design simple and modular enough that components could be easily left out.

2 System Overview

Figure 4 provides a pictorial illustration of all the components that comprise a fully deployed GBO. The primary scientific instruments are the All Sky Imager (ASI) and the Ground Magnetometer (GMAG). The remaining elements are there to support data acquisition, control and communication. GPS is used as a time reference. Iridium, a low-speed, satellite-based communication network, is used for backup communication with the site when the Internet connection is down or unavailable. The Observatory Support Equipment (OSE), mounted

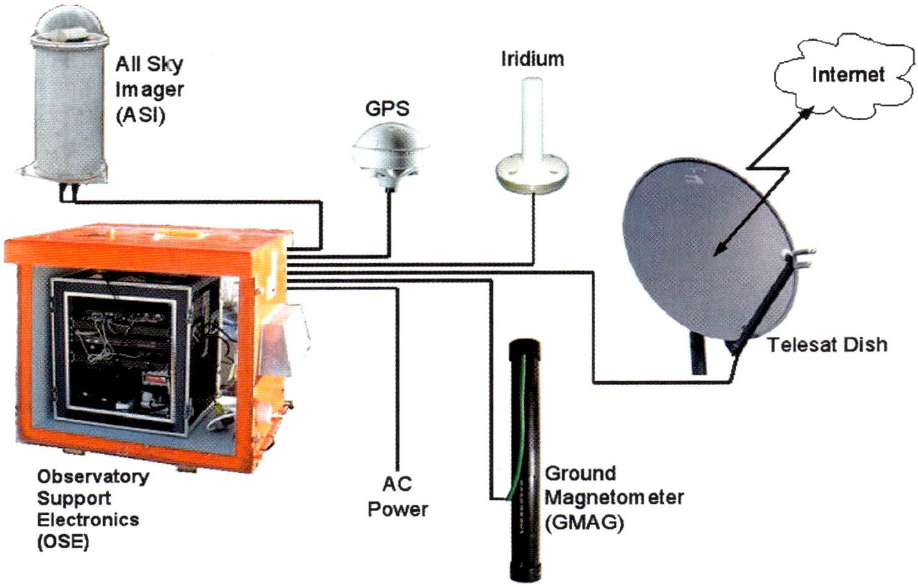

Fig. 4 A depiction of the top-level block diagram for a fully deployed GBO. The Observatory Support Electronics (OSE), assembled in rack-mount shipping case is shown situated inside its custom-built hut. Modular in design, not all components are installed at each site, but the basic design of each system is identical

in a rack, is the assortment of computers, modems, power supplies, and interface hardware needed to operate the station.

An installation is customized according to needs. Only 11 of the sites include the Ground Magnetometer (GMAG) sensor, although all sites include a UCLA-provided GMAG interface, in keeping with our modular design. In those cases, it interfaces only the GPS antenna to the system computer, thereby making the GPS interface identical for all stations.

Other optional installation items include the environmental enclosure (or "hut") for the OSE. Huts are installed at 7 sites, while the OSE is housed in an existing shelter at all the other sites. These other accommodations range from very comfortable school rooms in McGrath and Kiana, while at Gillam, the OSE is located in a helicopter hangar, and at Prince George it is located in the custodian's personal garage.

Not all sites require a satellite Internet connection. Many of the sites have wideband Internet service available, either from a local ISP or provided by the custodian's service.

Beyond these differences in components that make up each site, the logistics requirements for each site are all equally unique and sometimes formidable. By this we mean the requirements for mounting and securing the ASI, the antennas, running cables, and just coping with the logistics of shipping equipment and traveling to these remote locations, where access to hardware stores and other conveniences is not possible. This means that each installation requires a considerable amount of planning, sometimes including prior site visits, to adequately prepare for a deployment. The result is that each site has a unique, and generally different layout.

3 Environmental Requirements and Enclosure Designs

All external components needed at a site are required to operate in the ambient conditions of the far North. This means temperatures ranging from −50°C to +40°C, with all types of weather expected, including blizzards, torrential rain, dust, hot sun, and bitter cold. Fortunately some of the components, and their associated cabling, can be found off-the-shelf and ready for this environment. For example, the GPS antenna and Iridium antenna are commercial units, from Trimble and NAL Research, respectively, that meet our requirements. The same applies to the satellite dish.

The GMAG sensor is buried in the ground. This technique has been employed by our UCLA partners, with great success, for many years. It provides mechanical stability for this very sensitive sensor, and it also allows it to operate at a very constant temperature. The GMAG design is very well suited for operation in just about any terrestrial environment. The ASI and OSE, however, needed special enclosure designs, as described in the following.

3.1 All Sky Imager Enclosure

To help us design the camera enclosure, we enlisted the aid of outside experts, the Allison Park Group, who are located near Pittsburgh, PA. The resulting design is shown in Fig. 5. The ASI CCD camera and lens assembly is put together and tested as a unit, prior to mounting it in the enclosure.

The insulation for the ASI enclosure is foil-faced, polyethylene, air pillow wrap, with three layers used. This flexible insulation, selected for its low out-gassing properties, is wrapped around the internal circumference of the main cylinder. Heat is provided by silicone strip heaters, powered from 120 VAC, i.e. line voltage, which is carried out to the ASI in the multiconductor power cable. The strip heaters on the main camera bracket are rated for 180 W, while the four small heaters located on the dome heater plate are collectively rated

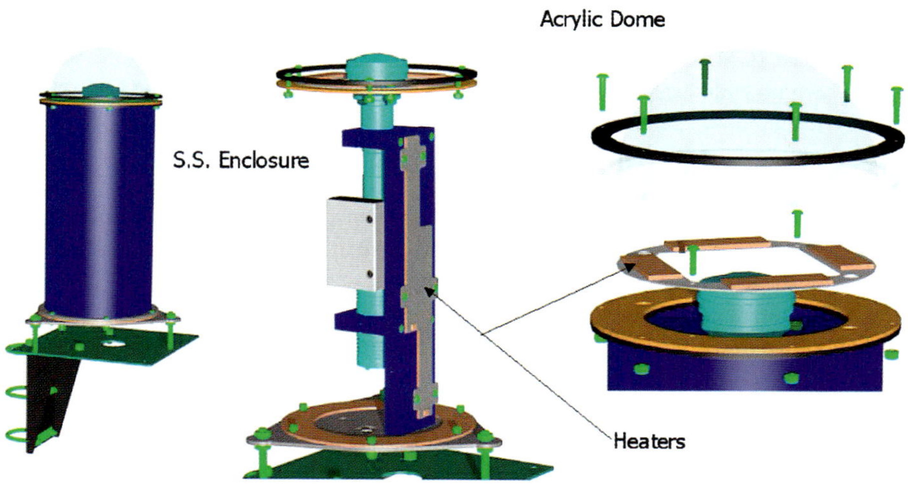

Fig. 5 Details of the ASI enclosure. Mounted inside an insulated, stainless steel case, the internal bracket assembly provides a secure mount for the ASI camera and lens, which is colored in cyan. The assembled housing can be mounted using either of two methods. It can be clamped to a vertical 2″ pipe, or the clamping feature can be removed from the base plate and the housing simply bolted to a flat surface

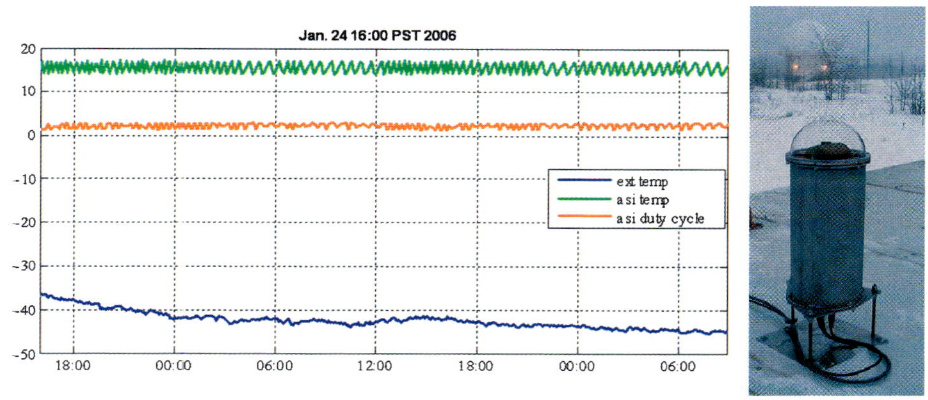

Fig. 6 ASI temperature data from Ft. Yukon. ASI duty cycle is plotted as a temperature 0–10°, corresponding to a percentage of 0–100%. The ASI at Ft. Yukon is shown on the *right*. A routine period of extreme cold at this location demonstrated the adequacy of the ASI enclosure design

for 60 W. The heater circuit is thermostatically protected to prevent a run-away situation. We have shown, referring to Fig. 6, that this level of heating is more than adequate to maintain a ΔT of 60°C, using an average power of 50 W (21% duty cycle), at ambient temperatures near -50°C.

An often-asked question is whether the dome sheds snow easily. Our experience thus far has shown that the dome heaters are effective in quickly removing snow from the dome, but we have seen some ice buildup on the flange surrounding the dome. This has been a minor issue. Generally, no intervention is required to keep the dome clear of snow or rain.

3.1.1 ASI Sun Shade

While not part of the enclosure specifically, the ASI sun shade assembly is intimately associated with it, as seen in Fig. 7. Kept closed during the day, it shields the lens and CCD from daytime sun exposure. This device was found to be necessary after the initial prototype trials in Athabasca demonstrated that:

- the fisheye lens coatings appeared to degrade after periods of sun exposure, and
- the Sony CCD used in the camera features plastic "microlenses" on each pixel, which can become discolored with constant exposure to sunlight, especially UV.

3.2 Observatory Support Electronics (OSE) Enclosure

The OSE is a collection of electronic support equipment that is assembled in a rack-mount shipping case. At most sites, the OSE is housed in an existing shelter, which provides protection from the weather and a room-temperature environment. At six of the sites, a suitable shelter was not available, and for these situations we developed an environmental enclosure referred to as the hut. One of these sites is shown in Fig. 8, and a more detailed look at the hut design is shown in Fig. 9.

The interior space of the hut measures 35″ (w) × 30″ (l) × 32″ (h), and, when empty, it weighs less than 80 lb. The construction material is fiberglass sheets bonded to 2″ thick panels of foam core insulation. This is similar to the construction of the AGO shelters used

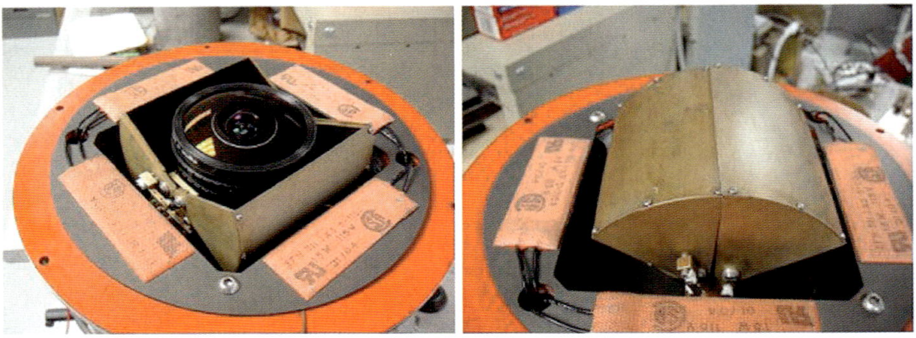

Fig. 7 ASI Sun Shade, shown both open and closed. A clamshell design, it falls open upon removal of power from the solenoid. The inside surface of the clamshell is painted flat black, to enable an ability to take dark images in the field

Fig. 8 The GBO site at Ft. Smith during installation. Our enclosure provides a dry and temperate environment for the OSE. When deployed the hut is typically mounted on cinder blocks, or some sort of platform. Tie points on the side rails provide a method to guy down the hut to prevent movement

in Antarctica. In fact, they were manufactured by the same vendor, an expert in fiberglas manufacturing, Moore Sailboats, located in Watsonville, CA. Both doors use pliable, silicone gaskets, rated for extreme cold, to seal against the main structure. Two latches on each side of the door secure it into place. Two fiberglass "awnings" cover the upper part of each door seal. Their purpose is to keep snow and water from pooling. This prevents a process, through melting and refreezing, that could force open the seal.

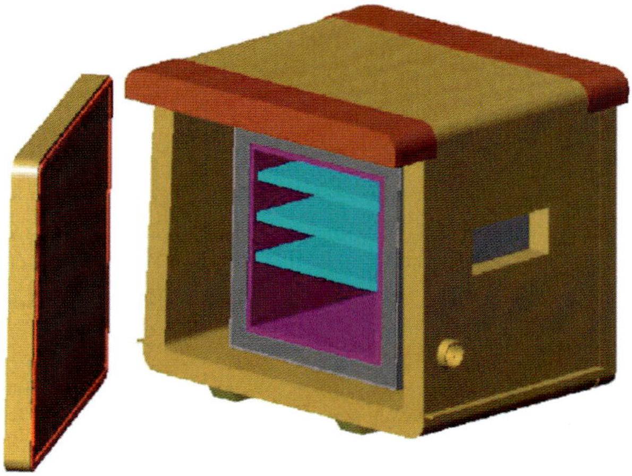

Fig. 9 A solid model of the hut. From this model, the design features are more readily apparent. A model of the rack is shown inside the hut

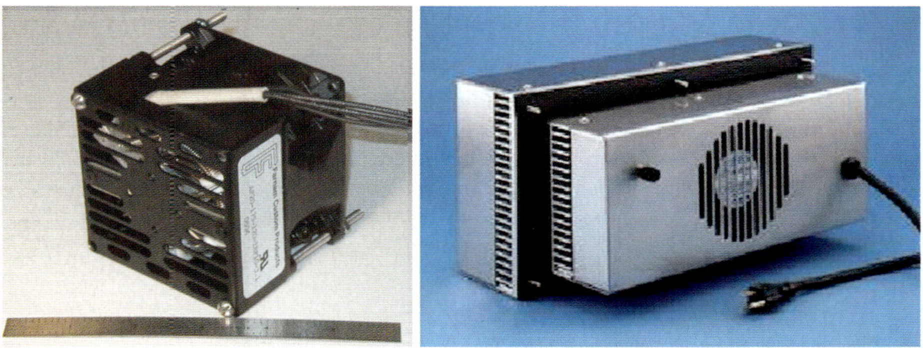

Fig. 10 An OSE heater is shown on the *left*, and the thermoelectric air conditioner is on the *right*. Two heater units are installed in each OSE, mounted to the underside of the lower rack shelf. The space heaters are thermostatically protected from overheating

The only openings in the hut, once the doors are closed, are cut outs for the air conditioner and a 3″ pipe used for cable access. Typically, a pipe elbow is mounted on the external portion of the cable pipe so that moisture is unlikely to travel inside the hut. The pipe is stuffed with insulation once the cables are installed. The air conditioner unit, when mounted on the hut, seals its opening. This makes an enclosure that is not only well insulated, but also dust free.

Active heating and cooling are both necessary, depending on ambient temperatures. Figure 10 shows the devices used for these purposes. Two small, electric space heaters, mounted inside the OSE, are available to add heat inside the hut. Each heater is rated for 175 W, and operates at 120 VAC. The air conditioner is a solid state device, which uses thermoelectric coolers to transfer heat through the wall of the hut. It has a capacity of 163 W.

The performance of the hut has generally been excellent, although the initial design turned out to be a bit too well insulated. With a typical heat load inside the hut reaching 150 W, we found that the OSE heaters were seldom, if ever called on to add heat, even at outside temperatures as low as −40°C. On the other hand, the air conditioner was turning on with external temperatures of 0°C, as seen in Fig. 11. Not only is this rather an unreasonable

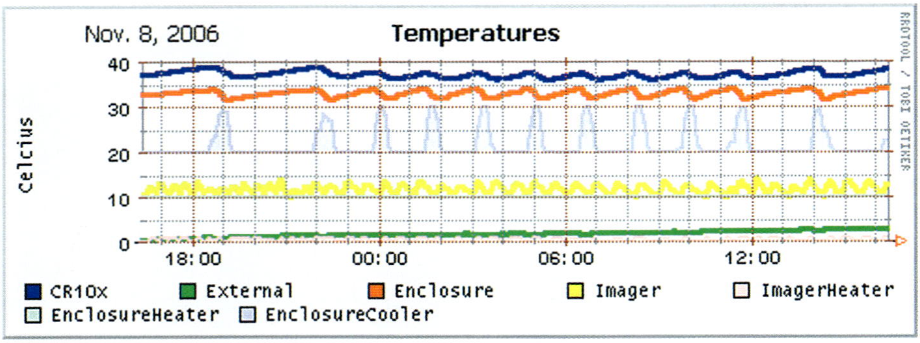

Fig. 11 GBO temperature plot from Chibougamau. This plot shows the temperatures inside the hut, colored *blue* and *red*. The external temperature, colored *green*, is increasing from 0° to 3°C. The duty cycle for the air conditioner is depicted as a temperature ranging between 20° and 30°, corresponding to 0–100%. The duty cycle value is updated every 10 minutes. In this example, the overall "on-time" of the cooler, with ambient temperature just above 0°, is about 30%. The internal temperatures are obviously modulated by the cooler action

situation, but it means that the air conditioner is operating for long periods of the year, which contributed to two failures of the devices during the summer of 2006.

At the time of this writing, we have developed a modification to the hut that involves removing some of the insulation from one of the doors. This has shown to make the hut require less cooling, with the air conditioner turning on at higher ambient temperatures.

4 Observatory Support Electronics

The Observatory Support Electronics (OSE) is based around the System Computer, a small-footprint PC comprised of a VIA Mini-ITX motherboard (model EPIA-CL). This PC is purchased from a vendor, SmallPC, located in Ontario, Canada (ref: http://www.smallpc.com). This computer was selected for two reasons. First, its small size and low power are attractive and, second, it provides a multitude of interface ports needed for the system. A block diagram of the small PC interfaces, and how they are used, is shown in Fig. 12.

Looking at Fig. 12, it is clear that interfaces and connectivity are a prime requirement for the GBO system computer. The Redhat Linux distribution, version 2.4.26, is installed on each system. This operating system is significantly pared down from the standard distribution. Support for Ethernet connections, serial ports, and generic USB is native to this distribution. It also supports the GPS interface using the Network Time Protocol (NTP) that provides accurate time stamps for our data. The "Custodian Laptop" is simply a laptop computer that is supplied with each system and left onsite. It is used as the terminal interface for the System Computer. Software provided by the GBO Team performs data acquisition from the ASI and GMAG, as well as housekeeping data that are acquired via the CR10X Datalogger. The Datalogger is used primarily as the environment monitor and power control device. More will be said about it later.

Also apparent in Fig. 12 is the two ways to communicate with the GBO station, using either the Internet or the Iridium link. The Iridium, acting as our backup mode, is connected via four-port serial switch, such that we can either login to the PC, via port ttyS1, or connect directly to the CR10X. This allows us to control and query the GBO when the Internet is

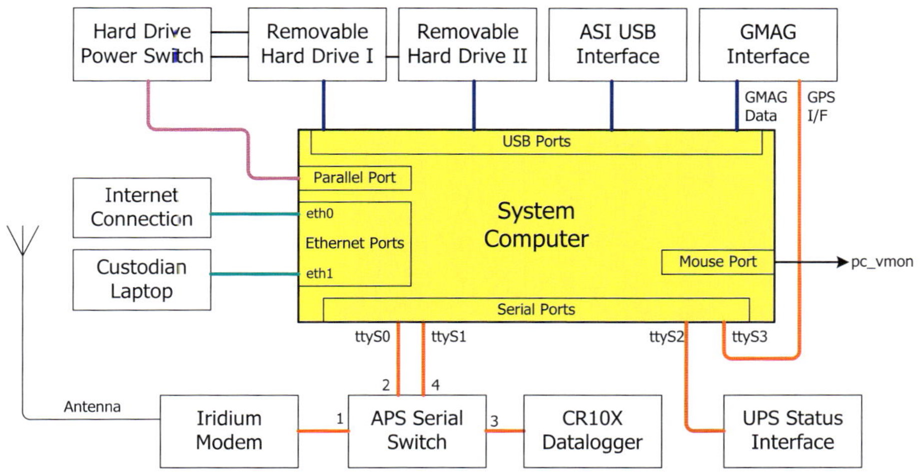

Fig. 12 Block diagram of computer port usage

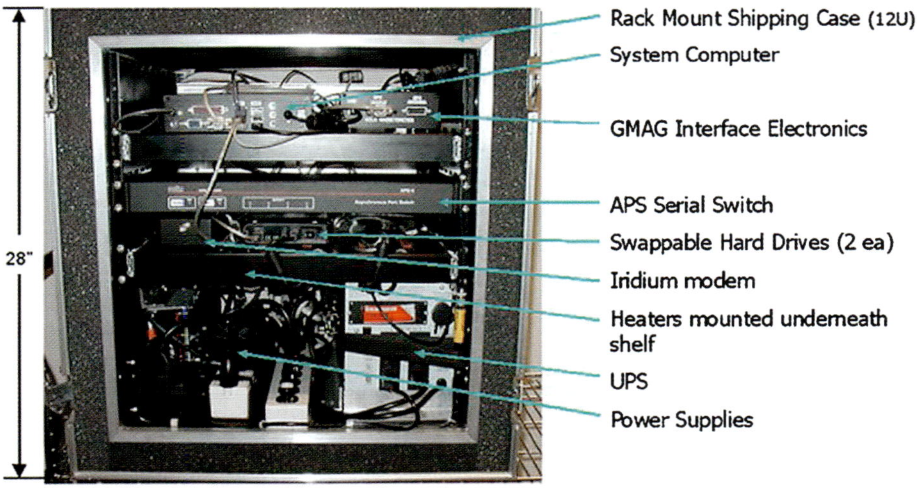

Fig. 13 The OSE, front view. Sliding shelves provide easy access for most components. All items are secured in the rack so that it survives transportation. Most items use high-strength hook-and-loop straps ("Velcro"), to hold them in position, while still allowing easy removal. Twenty systems, assembled in this manner, have been shipped to the field with no problems

down and in times of stress, e.g., station power failure, and we can query the CR10X to learn what is happening.

Views of the physical OSE are shown in Fig. 13 and Fig. 14.

Figure 13 shows the major components of the OSE as seen from the front. The Swappable Hard Drives, as the name implies, are used to both back up data acquired from the ASI and GMAG, and also to create a copy that can be physically transported to the data repository at the University of Calgary. Since the volume of data collected at the station, especially from the ASI, is so great, the only method to retrieve the high-resolution images is via shipment of

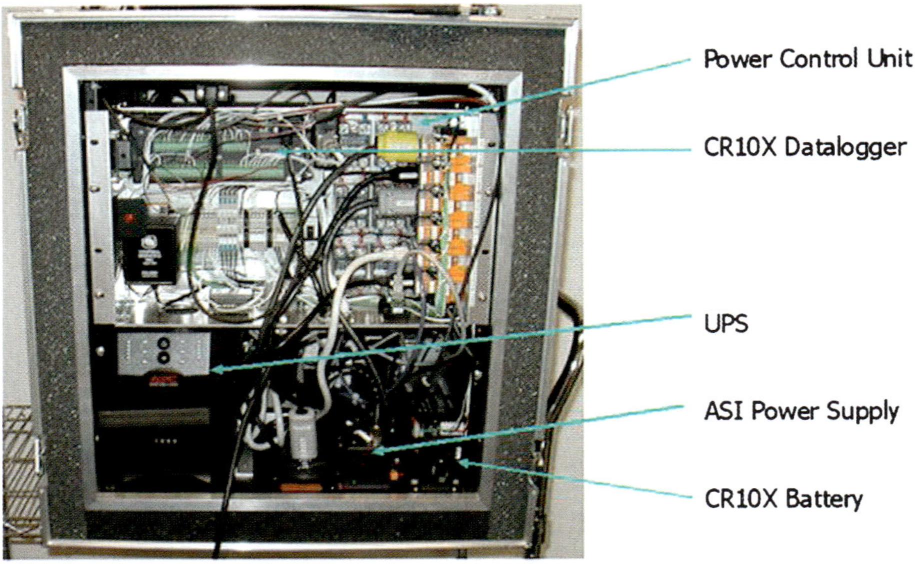

Fig. 14 The OSE rear view. From this side, the Power Control Unit (PCU) is the prominent item. All power switches are accessed from this side of the rack

the external hard drive. With two external drives available, a backup copy of the data always exists until the transfer disk is successfully received at Calgary. The heaters, instrument power, and UPS are controlled from the processor in the Power Control Unit (PCU). In the following sections, we will further describe the functions of the OSE.

4.1 Power and Temperature Control

It was recognized from the outset that power in the remote areas that would be home to the GBOs is often unreliable, and we could experience power outages that last for many hours or days. Also, maintaining a room-temperature environment for the OSE components, being primarily commercial-grade computer equipment, would need some controller that enabled the system computer to be shut down in an orderly fashion in the event of long power failures, or in situations where the system simply gets too cold or too hot. This controller is provided in the Power Control Unit (PCU), a CR10X Datalogger, manufactured by Campbell Scientific. In our application, the CR10X is used not so much as a datalogger, but as a programmable controller. The advantages of using it as the principal device to interface with the environment and control power is as follows:

- It is rated to operate over the entire operating range of the GBO, i.e. $-55°$ to $+85°$C. As such, it never needs to be turned off. It is always on, monitoring temperatures and power conditions. Operating from its own, dedicated battery pack, it can run without external power for a couple of weeks.
- It implements the graceful shutdown of the computer system, ASI and GMAG in the event of long power failures (those exceeding the capacity of the UPS), and also during loss of temperature control, either too cold or too hot.
- On boot-up, if temperatures and power are within limits, it turns on the UPS and System Power via switched outlets, which allows the System Computer to boot up. After boot-up,

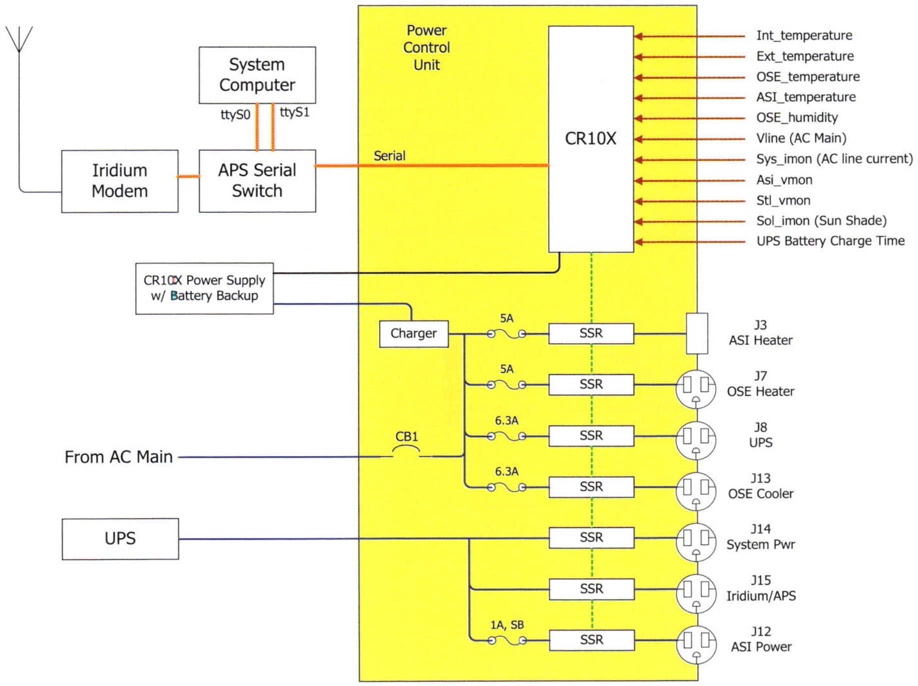

Fig. 15 Block diagram of PCU and discrete sensor interfaces on CR10X. Temperature sensors connect directly to the CR10X via screw terminals. Solid state relays (SSR) are used to control power

the CR10X acts as watchdog to the System Computer, and will reboot the system if the PC does not service the watchdog.

- Since most temperature and voltage sensors in the GBO are interfaced to the CR10X, the System Computer is able to access these data, via the serial interface connection. As such the CR10X acts as a peripheral to the System Computer, and allows us to remotely control the power.
- The CR10X is simple to program, and in the GBO it can be queried remotely via Iridium.

Figure 15 is a block diagram of the Power Control Unit, which illustrates these features.

The CR10X program monitors temperatures and key voltage and current sensors on a cycle that repeats every six seconds. Provided that the line voltage and temperatures are within acceptable limits, it maintains the power on the System Power Bar, which supplies power to the computer, ASI, GMAG, and external hard drives. If power is off for an extended period, exceeding the one-hour capacity of the UPS, then the CR10X flags the System Computer to shutdown. After an appropriate interval, the CR10X turns off power to the Power Bar and waits for the power to be restored. The same process applies to a too hot or too cold situation.

Referring to Fig. 16, ASI power is controlled from a switched outlet which is supplied from the UPS. The ASI CCD camera comes with an AC brick power supply, which after one winter of use was determined to be of marginal design. It had poor regulation with varying levels of AC line voltage and it ran too hot, which in the long term caused internal damage. To circumvent these difficulties, a separate, regulated 12 V supply has been substituted

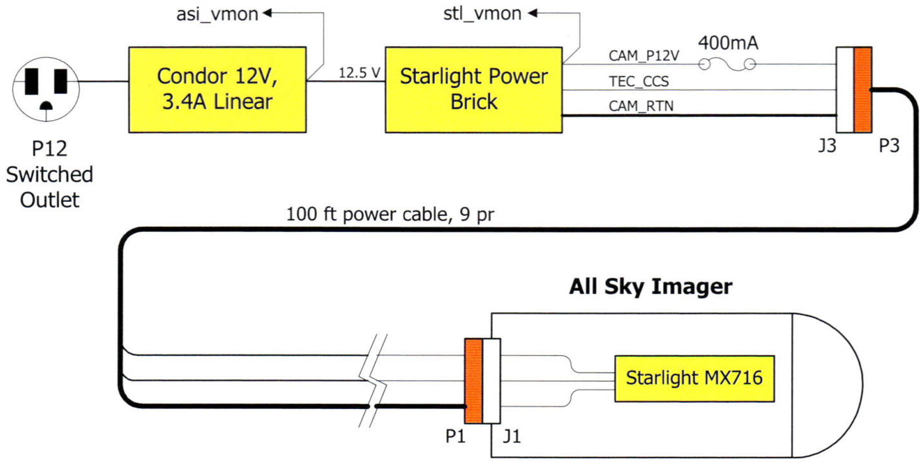

Fig. 16 ASI camera power diagram. The 100' power cable also carries conductors for heaters, sun shade control and temperature sensors in the ASI enclosure

for the camera 12-volt line, and the only function really provided by the "Starlight Power Brick" is the constant current source for the thermoelectric cooler (TEC). The power brick is modified to incorporate a small fan and large heat sink to reduce internal temperature rise to 10°C. The ASI data acquisition application, "imagerd", commands the ASI to power on by setting a flag in the CR10X. The CR10X then turns on the ASI power, provided the temperatures are acceptable.

The temperature inside the ASI is controlled by the CR10X. As the temperature drops below a programmable set point, the heater circuit is activated via SSR. The heaters are turned off when the temperature rises above the set point, plus a suitable amount of hysteresis. An internal thermostat protects against a thermal run-away. The OSE heater and cooler are operated in a similar fashion. In the ASI, the Sun Shade is also controlled by the CR10X. It is closed or opened, as requested by the System Computer, under software control.

4.2 Instrument Interfaces

Referring back to Fig. 12 again, the data interfaces for the ASI and GMAG are implemented with USB. We also collect housekeeping data, referred to as "monitor" data, from the CR10X, the UPS, and from devices and processes in the System Computer. These data are collected, stored, and transmitted back to Calgary via the Internet.

The GMAG subsystem has two interfaces with the System Computer. The USB interface is the GMAG sensor data. This is magnetic vector data, sampled at 2 Hz. Internal to the GMAG Interface Electronics is a processor that interfaces with the GPS antenna, a standard Trimble Accutime 2000. The time stamps for the GMAG data are created directly from the GPS. The GPS serial interface is also made available to the System Computer, where it is serviced by NTP, and it maintains the system clock to within 10 ms rms error. Recorded time stamps for the ASI and monitor data are derived from this system clock.

The ASI data interface is illustrated in Fig. 17. Faced with a USB interface in the Starlight Xpress camera, we were limited to a cable length of five meters, the maximum length supported by USB. This issue was solved with a USB Extender (Iogear model GUCE50), which

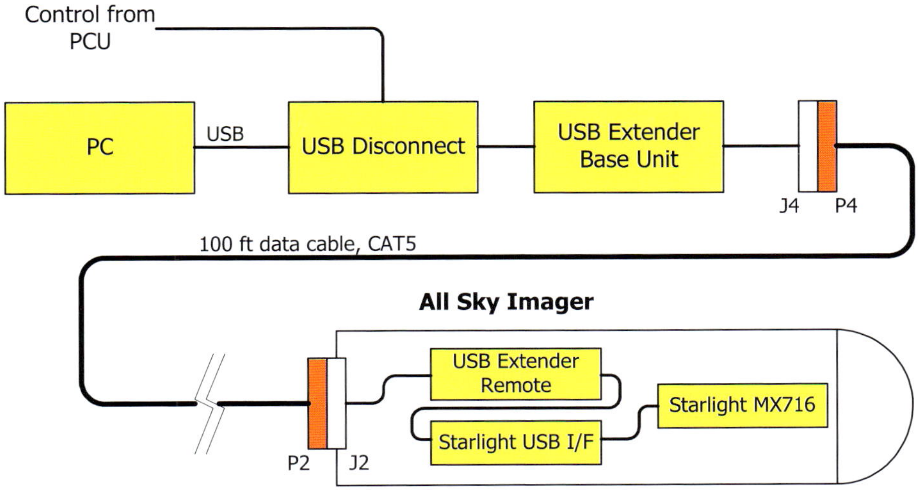

Fig. 17 Block diagram of the ASI data interface. It illustrates the cascade of devices that are used to enable operation of the ASI on long cables. The USB interface to the ASI is only powered on when the main ASI power is On, a function provided by the "USB Disconnect" module

takes the data and power pairs in the standard USB interface and transforms them into four pairs, suitable for transmission on standard CAT-5 cable. This allows USB (v1.1) operation on cable lengths up to 50 m. We also found that the Starlight Xpress USB interface circuitry itself is powered from this USB port. Since this meant we couldn't easily turn off that part of the camera, we developed a computer-controlled USB Disconnect, that allows us to power off the ASI USB interface, effectively disconnecting it from the computer.

4.3 Data Retrieval and Remote Intervention

Data acquired in the System Computer is stored and then backed up on the external hard drives. Real-time data of ASI thumbnails, GMAG, and monitor data are transmitted over the Internet connection. This connection is also used for remote intervention, as shown in Fig. 18. Typically, no intervention is required on a day-to-day basis. Data flow is described in Fig. 19.

5 Integration and Test

Besides construction, the integration and testing of all GBOs was the responsibility of UC Berkeley. An illustration of the integration and test flow is shown in Fig. 20. A brief description of the activities shown is provided in the following sections.

5.1 ASI Testing

ASI testing is done in two steps. First the Starlight Xpress camera and All Sky Lens are mated and characterized. This characterization is done at room temperature on an optical bench using a light source with known spectral radiance, and an optical path that included the acrylic dome. Narrow band filters were used to measure camera responsivity. A sample

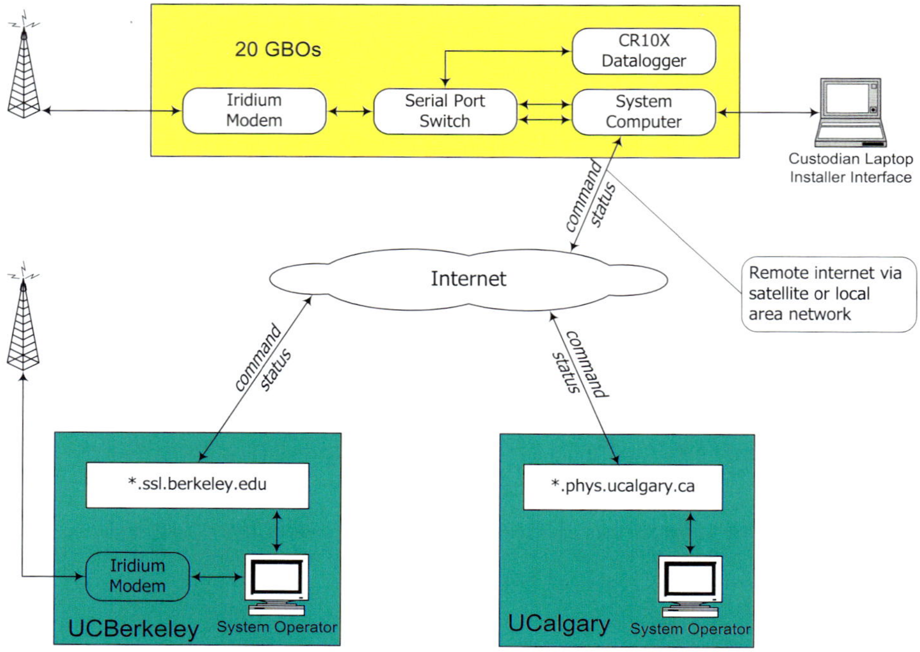

Fig. 18 Remote intervention. Operators at UC Berkeley and University of Calgary can login at any GBO to perform a variety of intervention tasks, such as software updates, troubleshooting, data download, etc. Similar access is afforded by the back-up iridium link

of these data is shown in Fig. 21. In addition, the following parameters were verified during this step in accordance with demonstrating that the ASI met THEMIS mission requirements.

- FOV > 170° (see Fig. 22 for typical data).
- Exposure time duration controllable for 1 ms to 5 s. This verifies the electronic shuttering capability of the Starlight Xpress camera.
- Measure spectral response. Verify minimum responsivity to source radiance is less than 10 kR.
- Spatial resolution >250 pixel across all sky image. Verify focus ability.
- Cadence is better than 1 image every 5 s.
- Record dark and bias images for reference (at room temperature).

Once the optical characterization is complete, the camera/lens is assembled in the ASI enclosure. This completed assembly is tested further to verify:

- Alignment of camera such that top of image is aligned to the "North" datum on housing.
- Final focus adjustment.
- Heater control functional test.
- Sun Shade functional test, which involves a one-year equivalent of open/close cycles.
- Final test and burn-in with completed OSE. This usually lasts several days.

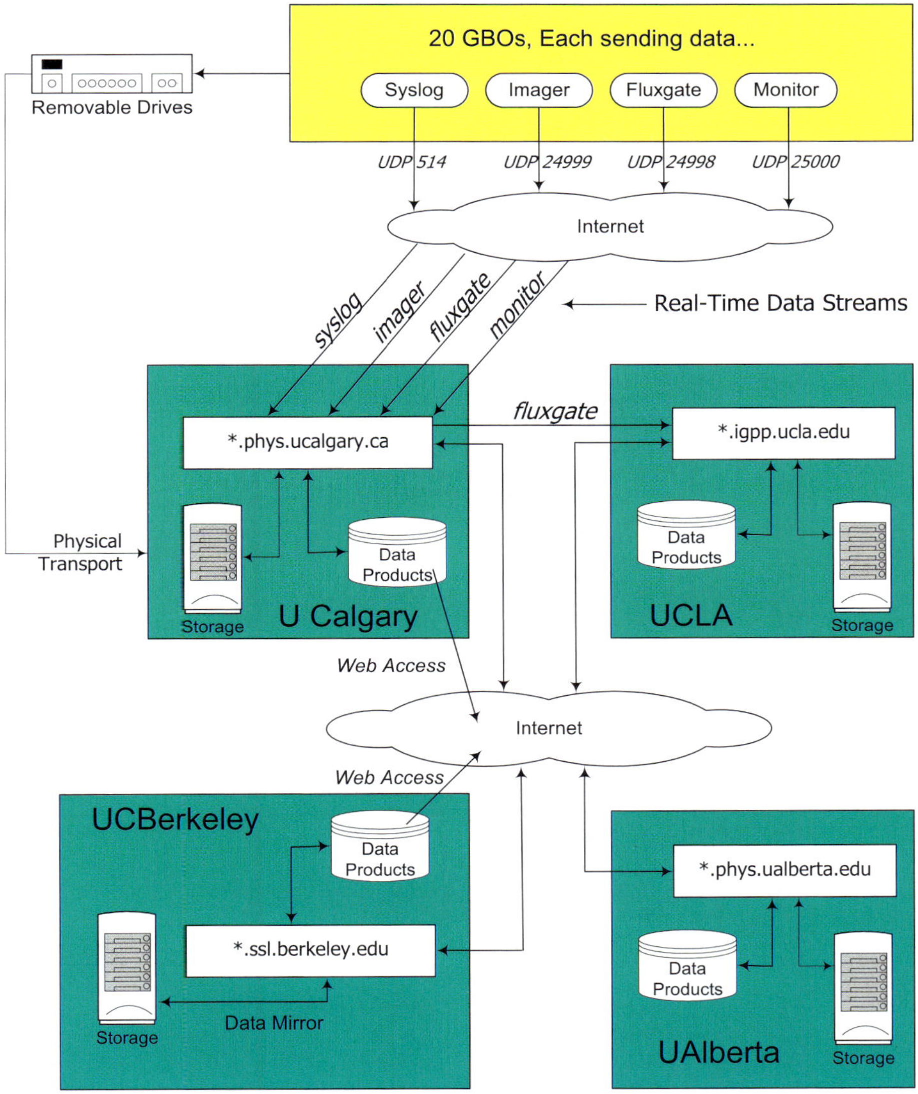

Fig. 19 Block diagram of data flow from GBOs to the Universities. The University of Calgary is the primary collection point for all data. All data are mirrored at UC Berkeley. Real-time data are transmitted as UDP packets to U. Calgary. Magnetometer data from the GBO GMAGs are validated by UCLA and then distributed to the rest of the team. For the GBO sites without magnetometers, we obtain mag data from other sources, such as the University of Alberta and the Geophysical Institute at the University of Alaska

6 Discussion of Operational Hazards

The GBOs have survived many operational hazards in the field. The CR10X embedded in the OSE has proven very reliable in handling situations of too cold (e.g., door to shack left open), too hot (e.g., stuck thermostat in the shack), and long power outages. When we notice these situations, a call to the on-site custodian usually prompts some action to deal

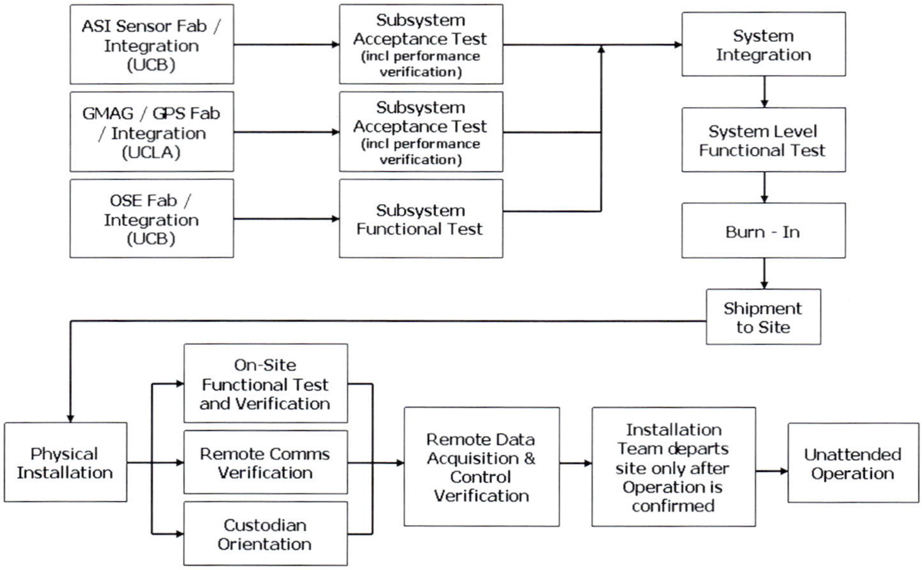

Fig. 20 GBO integration and test flow. Instruments were fabricated and tested individually prior to integration with the OSE

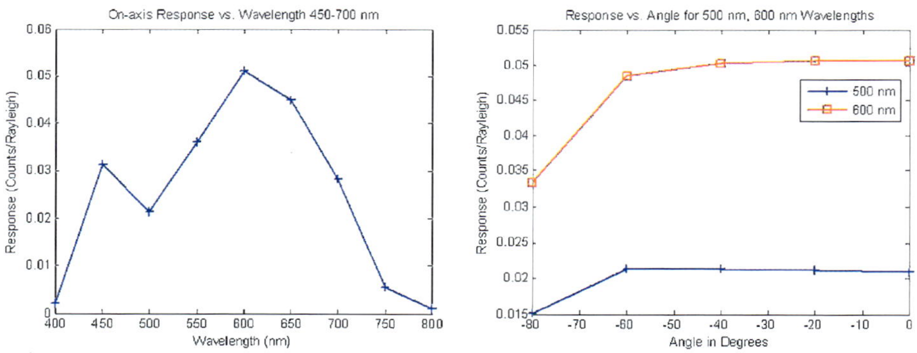

Fig. 21 Typical response curves for the All Sky Imager, one-second exposure. On-axis response is shown on the *left* for all wavelengths tested. On the *right* is response vs field angle. All cameras produced very similar results

with the problem, and the system automatically restarts when conditions are OK. The proof of this statement is backed up by certain sites (e.g., Ekati and Inuvik) that are often fairly inaccessible to the custodian. We have operated these sites, without problem, for two years, and they have needed very little custodian intervention.

During the implementation phase of this project we have had to deal with some difficult problems, some of which are ongoing and some of which have been solved. Here is a short list of such problems including what was done to mitigate the issue where that was possible.

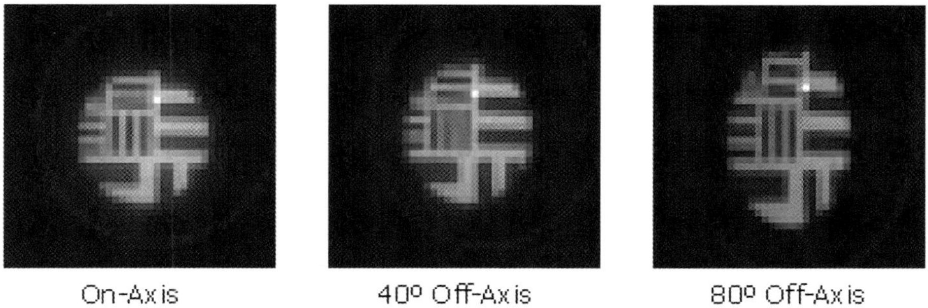

Fig. 22 Typical focus images for the All Sky Imager. A bar test pattern is imaged at various field angles, showing good focus over the entire FOV, although obvious distortion exists off-axis. These are 2X binned images, the normal mode used in the field

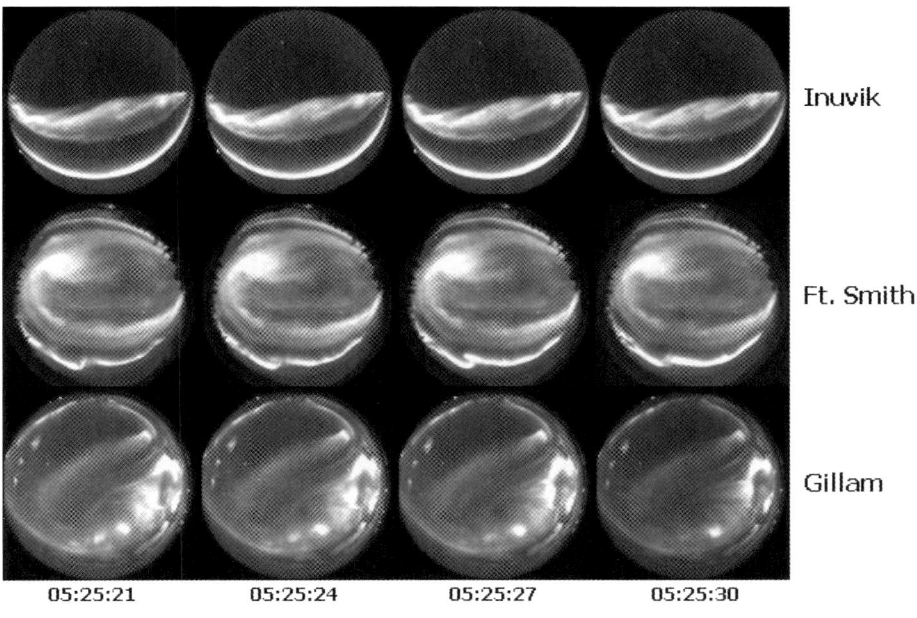

Fig. 23 Sample data from All Sky Imagers acquired on October 5, 2007. The panel shows four consecutive all-sky images taken at three different sites in the GBO network. Image cadence is one image every three seconds. This data sample shows the full-resolution images which are acquired at these sites

- RF Interference. At several sites, the GBO is colocated with existing ionospheric sounding equipment (e.g. CADI, SuperDARN). Unfortunately, RF interference at some sites has occasionally made operation impossible, primarily because it seems to interfere with the commercial implementation of the USB bus. This problem was solved with the addition of common-mode RF filters, added to the ASI data interface.
- Internet connection outages are a fact of life. Our software, however, continues to operate and store data locally, regardless of this connection loss.

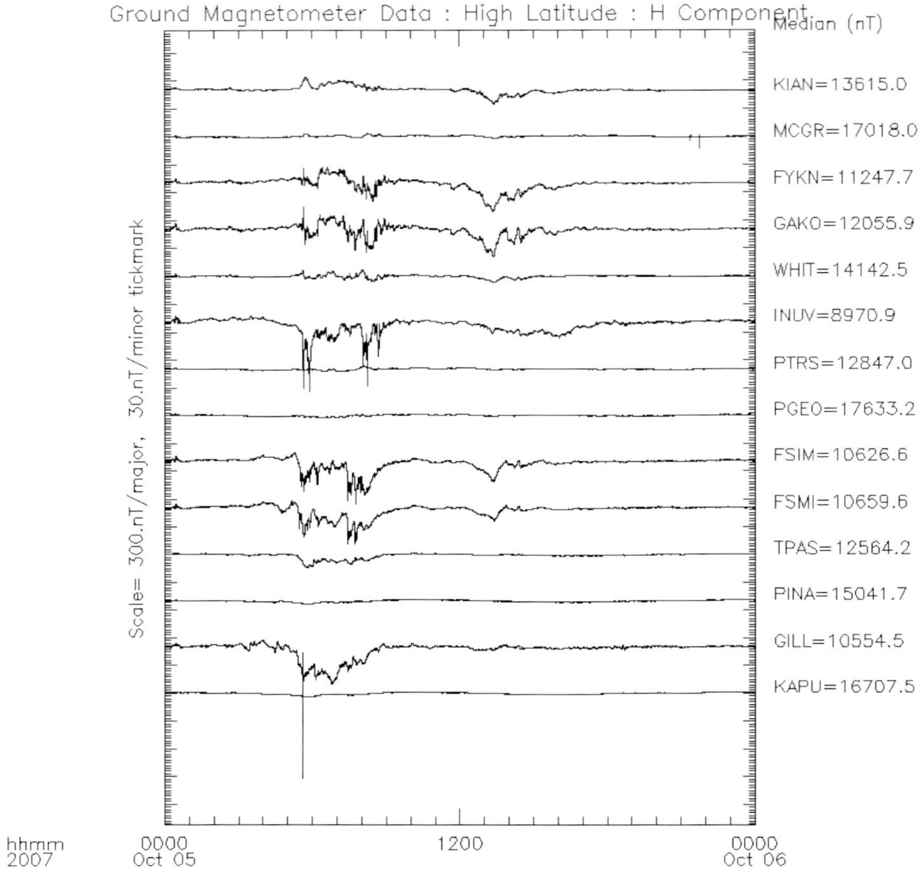

Fig. 24 Sample data from Ground Magnetometers acquired on October 5, 2007. This panel combines magnetometer data from sites outside the GBO network, as they are included in the THEMIS daily summary plot

- Some software issues seem to be not easily solved. For example, USB interfaces can hang inexplicably and some system processes can sometimes go awry (e.g., NTP daemon). These are infrequent occurrences, and they are generally handled by manual intervention.
- Lightning is an occasional occurrence. The ASI is probably the most vulnerable to this, and frankly there is not much that can be done to easily mitigate the problem, other than having spare systems available for unit replacement.
- Finally, we do have the occasional nonresponsive custodian. Fortunately, we can minimize our reliance on these folks but, on the other hand, many of our local helpers have provided outstanding assistance.

7 Summary

In support of the NASA THEMIS program, a network of 20 Ground Based Observatories have been designed, built, and deployed across the active auroral zone of the North American continent. Figures 23 and 24 provide a sample of the data available from the THEMIS network of ground observatories.

The GBO Team now assumes the tasks of data collection, monitoring, and maintenance of the remote sites. This is an ongoing effort that will continue throughout the THEMIS science operations.

Acknowledgements Development of the GBO hardware and the THEMIS project overall is funded by NASA, under contract NAS5-02099. Canadian efforts in the implementation phase of the GBO program were supported by funding from the Canadian Space Agency (CSA). We also acknowledge I. Mann for use of the Canadian GMAG data, and the CSA for support of the CARISMA network.

References

S.B. Mende, H.U. Frey, S.P. Geller, J.H. Doolittle, Multistation observations of auroras: Polar cap substorms. J. Geophys. Res. **104**, 2333–2342 (1999)

E. Donovan, T. Trondsen, L. Cogger, B. Jackel, All-sky imaging within the Canadian CANOPUS and NORSTAR programs, in *Proceedings of the 28th Annual European Meeting on Atmospheric Studies by Optical Methods*. Sodankyla Geophysical Observatory Publications, vol. 92 (2003), pp. 109–112

The THEMIS Fluxgate Magnetometer

H.U. Auster · K.H. Glassmeier · W. Magnes · O. Aydogar · W. Baumjohann · D. Constantinescu · D. Fischer · K.H. Fornacon · E. Georgescu · P. Harvey · O. Hillenmaier · R. Kroth · M. Ludlam · Y. Narita · R. Nakamura · K. Okrafka · F. Plaschke · I. Richter · H. Schwarzl · B. Stoll · A. Valavanoglou · M. Wiedemann

Originally published in the journal Space Science Reviews, Volume 141, Nos 1–4, 235–264.
DOI: 10.1007/s11214-008-9365-9

Abstract The THEMIS Fluxgate Magnetometer (FGM) measures the background magnetic field and its low frequency fluctuations (up to 64 Hz) in the near-Earth space. The FGM is capable of detecting variations of the magnetic field with amplitudes of 0.01 nT, and it is particularly designed to study abrupt reconfigurations of the Earth's magnetosphere during the substorm onset phase. The FGM uses an updated technology developed in Germany that digitizes the sensor signals directly and replaces the analog hardware by software. Use of the digital fluxgate technology results in lower mass of the instrument and improved robustness. The present paper gives a description of the FGM experimental design and the data products, the extended calibration tests made before spacecraft launch, and first results of its magnetic field measurements during the first half year in space. It is also shown that the FGM on board the five THEMIS spacecraft well meets and even exceeds the required conditions of the stability and the resolution for the magnetometer.

Keywords Plasma physics · Substorm · Fluxgate magnetometer · Calibration

---

H.U. Auster (✉) · K.H. Glassmeier · D. Constantinescu · K.H. Fornacon · Y. Narita · K. Okrafka · F. Plaschke · I. Richter · B. Stoll
Institut für Geophysik und extraterrestrische Physik der Technischen Universität Braunschweig, Mendelssohnstrasse 3, 38106 Braunschweig, Germany
e-mail: uli.auster@tu-bs.de

W. Magnes · O. Aydogar · W. Baumjohann · D. Fischer · R. Nakamura · A. Valavanoglou
Space Research Institute, Austrian Academy of Sciences, Schmiedlstrasse 6, 8042 Graz, Austria

O. Hillenmaier · R. Kroth · M. Wiedemann
Magson GmbH Berlin, Carl Scheele Strasse 14, 12489 Berlin, Germany

E. Georgescu
MPE Garching, Giessenbachstrasse, Postfach 1603, 85740 Garching, Germany

P. Harvey · M. Ludlam
SSL at UCB, 7 Gauss Way, Berkeley, CA 94720-7450, USA

H. Schwarzl
IGPP at UCLA, Los Angeles, CA 90095-1567, USA

1 Introduction

Magnetic fields are essential in characterizing different plasma regions in and around the Earth's magnetosphere. Accurate measurements of the magnetic field vector along the orbits of the Themis spacecraft (hereafter referred to as probes) is the objective of the FGM experiment. The Themis probes follow elliptical, equatorial orbits. In the transfer orbits (coast phase) the probes have a perigee of about 1 Earth radius (R_E) and an apogee of about 15 R_E. The apogees of the final orbits vary from 10 R_E for the inner to 30 R_E for the outer probe. Changes of the orbits from the costal to the final phase and the seasonal variation of the apogee due to the Earth's orbital motion provide for an opportunity make to perform measurements of the magnetic field at various conditions in space. The magnetometer is designed to cover measurements in the solar wind, magnetosheath, magnetotail, and outer magnetosphere up to the region dominated by the Earth's dipole field. To achieve this goal several technical challenges had to be solved.

Frequent crossing of the radiation belt requires a reasonable radiation tolerance of the electronics, the spacecraft spin imposes a condition on high precision of timing, and the necessity to use the magnetic field at perigee for attitude determination defines the maximum measurement range. Furthermore, measuring the magnetic field within the required precision instrument design, magnetic environmental conditions, and constraints due to limited spacecraft resources had to be balanced.

The instrument itself is based on the heritage of the participating magnetometer teams, dating back to the missions such as the German Helios mission in the seventieth and the Russian Phobos missions in the eighties. Experience from magnetometer experiments on more recent missions such as Freja (Zanetti et al. 1994), Equator-S (Fornacon et al. 1999), Cluster (Balogh et al. 2001), Cassini (Dougherty et al. 2004), Double Star (Carr et al. 2005), VenusExpress (Zhang et al. 2006), or Rosetta (Auster et al. 2007; Glassmeier et al. 2007a) largely contributed to the successful design, fabrication, and operation of the Themis magnetometers. The instruments actually operating are very similar to those currently in use on the European Space Agency's cometary mission Rosetta (Glassmeier et al. 2007b; Auster et al. 2007) and VenusExpress (Zhang et al. 2006). Capabilities of these instruments are tailored to the science objectives of the Themis mission.

FGM benefits from a close cooperation between several institutions lead by the Institute of Geophysics and extraterrestrial Physics (IGEP) group of the Technical University Braunschweig. The hardware was developed at IGEP (sensor) and Magson GmbH Berlin (electronics). The Space Research Institute of the Austrian Academy of Sciences (IWF) in Graz supported the instrument development. Part procurement, integration, and qualification as well as the development of the onboard software has been done by the Space Science Laboratory of the University of California at Berkeley (UCB). Tests and preflight calibrations were performed in Braunschweig, Berlin and Graz. IGEP, supported by the University of California at Los Angeles (UCLA) group, is responsible for the in-flight calibration. The software for ground data processing has been developed by UCB, UCLA and the Max-Planck-Institute for extraterrestrial Physics (MPE) in Garching. This large team stands for a high level of expertise and guarantees an efficient adaptation of the existing hardware, software and other tools to Themis specific requirements.

Two features are specific for the Themis magnetometer experiments: a single sensor on a 2 m boom and the compact integrated instrument concept (Harvey et al. 2008). Placing just one sensor on a 2 m short boom is a novelty compared to, for example, the Cluster mission where each spacecraft has two sensors mounted on a 5 m boom. Limitations due to magnetic environmental conditions, which depend on the boom length, the number of sensors, and the level of spacecraft magnetic contamination are to be expected. An extensive magnetic cleanliness program was necessary to limit spacecraft disturbances below 1 nT DC and 10 pT AC at the sensor position. With only one sensor, the possibility to detect and remove s/c disturbances by a difference analysis is not possible anymore. The magnetic cleanliness program had to ensure that interferences caused by magnetic materials or generated by onboard currents are below the threshold given by the scientific requirements. In Sect. 5.2 remaining interferences detected by FGM measurements during commissioning as well as the policy for its removal are discussed. A detailed report describing methods and results of the magnetic cleanliness program is given by Ludlam et al. (2008).

The other Themis specific feature is that the spacecraft have a compact integrated instrument concept. The electronics is part of an instrument package inside the common electronics box. Therefore EMC and integration constraints are more difficult to meet. The fluxgate experiment can not be seen as an autonomous experiment. It is not placed, as usually done for larger spacecraft, in a stand-alone electronics box with internal DC/DC converter, own processing capability and well defined EMC conditions. The FGM electronics share a standard board inside the common electronics box together with the Power Control Unit (PCU). The secondary voltages are provided by a central DC/DC converter. The processing capability was divided into an instrument related part integrated in the FGM FPGA (Field Programmable Gate Array circuit), and a higher level onboard software implemented in the Instrument Data Processing Unit (IDPU). The integrated design had two consequences: first, the EMC environment depends on the operation status of nearby boards, and second, all parameters which can be influenced by environmental conditions had to be verified during and after spacecraft integration. Test facilities, which guarantee measurements with the full precision were developed, to verify instrument parameters during the integration process. The test and calibration strategy is described in detail in Sect. 4.4, while Sect. 5.2 deals with conducted interferences.

2 Science Requirements

Themis is a multi-spacecraft mission allowing to separate spatial and temporal variations in the Earth magnetosphere. After the four-spacecraft Cluster mission it is the second mission of this kind. The prime objective is the study of the physical causes of substorm onsets in the magnetotail of the Earth. The major unresolved question is: Where does substorm onset occur, in a region closer to Earth or at a more distant location in the magnetotail. With the five Themis spacecraft the spatial propagation of the substorm associated magnetic field disturbance can be properly timed and its direction, tailward or Earthward, determined. Secondary science objectives are studies of magnetospheric processes such as the dynamic response of the magnetosphere to solar wind dynamic pressure variations, using the multi-spacecraft situation. This allows making use of special data analysis tools developed for the Cluster mission (e.g. Glassmeier et al. 2001).

The typical propagation speed of a substorm associated perturbation will be of the order of 1000 km/s and spatial scales of about 100 km are realistic. If a propagating structure with this scale and velocity passes a satellite it causes a temporal variation on a time scale of 0.1 s. Furthermore, in collisionless plasmas wave-particle interactions and thus also higher frequency plasma waves play an important role. In addition to the search coil magnetometer onboard the Themis spacecraft also the fluxgate instrument will provide important information about these waves. Baumjohann et al. (1999), for example, studied ELF waves in the frequency range 15–40 Hz using the fluxgate magnetometer onboard the Equator-S spacecraft. Amplitudes of the observed waves are of the order of 0.5 nT.

These and other science objectives enforce a couple of basic requirements to the magnetometer. First, the temporal resolution of the magnetometer should be at least 10 Hz and better. A second requirement needs to be imposed on the field resolution. Magnetic field changes associated with substorm processes will be as small as 1 nT. In order to trace the actual field variation of such small changes a resolution of at least 0.1 nT is required. Such a resolution is also suitable to observed higher ELF wave forms. A third requirement is imposed on the offset stability of the magnetometers. A key element of Themis measurements are coordinated observations at different locations within the magnetosphere. If, for example, substorm onset is triggered at a tail distance of 15 R_E and observed as close to the Earth as 5 R_E the time for the perturbation to travel this distance is of the order of minutes. During this time the offset should not change on the 0.1 nT level. This leads to a requirement for the offset stability of 0.2 nT/hour. A further requirement applies to the measurement range. Observations will also be taken close to Earth at fields levels of about 25,000 nT for attitude determination purposes. Thus, FGM needs to operate in a magnitude range between 0.1 and 25,000 nT.

3 Instrument Description

Fluxgate magnetometers are the most widely used magnetometers for space applications. The Themis fluxgate magnetometer FGM consists of a vector compensated three axis fluxgate sensor unit and a mainly digital electronics on a single printed circuit board. Magnetometer electronics and Power Control Unit share one of altogether five boards of the Instrument Data Processing Unit. Both, vector compensated sensor and sensor electronics, have flight heritage from magnetometers aboard the Rosetta Lander Philae (Auster et al. 2007) and VenusExpress (Zhang et al. 2006). The used ring cores—carrying the soft-magnetic material—are based on a 25 year-long continuous development phase carried out in Germany.

The special feature of the digital fluxgate electronics is the digitization of the AC output signal from the fluxgate sensor directly behind a preamplifier. It follows the general trend of a signal conversion from analog to the digital domain as close as possible to the sensor(s).

In this context, the replacement of analogue circuitry by digital processing in an FPGA improves the overall measurement stability, guarantees a precise timing of the field vectors relative to the system clock, independent from selected range and sampling rate, and furthermore reduces the susceptibility of the system to electro-magnetic interference. The feedback field in the fluxgate sensor is generated by two cascaded 12-bit Digital-to-Analog Converters (DACs). The field value is calculated by the sum of feedback field and measurement of the remaining field on the ring core position with a 14-bit Analog-to-Digital Converter (ADC). Both together provide field components with 24-bit resolution, which are transmitted to the Data Control Board (DCB).

The telemetry interface consists of two channels. The high telemetry channel (TMH) permanently provides 128 Hz samples and a low telemetry channel (TML) can be commanded to transmission rates between 4 and 128 Hz. The FGM output vectors are synchronized to a 1 Hz clock provided by the DCB. The DCB also contains the IDPU which shows responsible for all further processing of the FGM data like the generating of onboard data products as well as FGM controlling e.g. ranging. All secondary voltages (± 8 V analog, ± 5 V analog, $+5$ V digital and $+2.5$ V digital) required by FGM are provided by the Low Voltage Power Supply (LVPS) via the PCU.

Relevant housekeeping values are the temperatures of the sensor and the electronics as well as supply voltages and currents. Both temperatures sensor signals are conditioned on the magnetometer board and routed to the central housekeeping ADC as well as all power values.

The FGM resource requirements as well as its main instrument parameters are given in Table 1 and Table 2.

Table 1 Resources requirements

Mass	
Sensor	75 g
Harness	150 g (60 g/m)
Electronics	150 g
Dimensions	
Sensor	Diameter 70 mm, height 45 mm
Board	100 mm × 120 mm
Power consumption	800 mW
Data Interface to DCB	
TMH channel	128 Hz
TML channel	4–128 Hz; vector rate and filter mode are commandable
Data synchronization	Excitation frequency derived from IDPU clock; 128 Hz data centered to 1 Hz pulse

Table 2 Instrument parameters

Range	±25,000 nT
Resolution	3 pT (24bit)
Noise	10 pT/$\sqrt{\text{Hz}}$ at 1 Hz
Temperature range/calibrated	
Sensor	−100° C...60° C
Electronics	−55° C...80° C
Offset stability	
vs. time	<1 nT/year
vs. sensor temperature	<50 pT/°C
vs. electronics temperature	<50 pT/°C
Gain stability	
vs. sensor temperature	22 ppm/°C (copper)
vs. electronics temperature	15 ppm/°C
Axes alignment	
Mechanical tolerance	<1°
Knowledge of axes direction	<1 arcmin
Stability of axes direction	<1 arcmin

3.1 Fluxgate Sensor

The ring-cores used for Themis have been developed by Karl Heinz Fornacon in Germany for more than 20 years (Müller et al. 1998). The main design goals have always been low noise and offset stability over a wide temperature range and period of time. Material selection and preparation as well as a proper thermal treatment are the key steps to achieve the performance parameters required for the Themis mission. The applied soft-magnetic material, a 13Fe-81Ni-6Mo alloy, is rolled to a foil of 20 μm thickness. Ribbons with a width of 2 mm are cut and 7 turns of it are wound on a bobbin made from Inconel. One of the most important permalloy parameters is the grain size which increases with the annealing temperature. The best noise results are achieved when the grain size is considerably smaller than the ribbon thickness (Fig. 1).

The selection of the ring-cores relies on an extended test procedure. After winding the excitation coil directly onto the ring core bobbins the noise of each ring core is measured before and after a specific aging process which consists of ultra sonic treatment, vibration, and temperature cycling. The sensor noise at 1 Hz of a ring core with a diameter of 13 mm is typically less than 5 $\mathrm{pT}/\sqrt{\mathrm{Hz}}$ as shown in Fig. 2.

After a pre-selection of those ring-cores with the lowest noise around 1 Hz, a quasi-longterm registration follows over a time period which must be longer than 1 day (typically one weekend) in order to verify the sensor noise at lower frequencies. This stability check is performed in a ferromagnetic shielding can. Several sensors are operated in parallel to

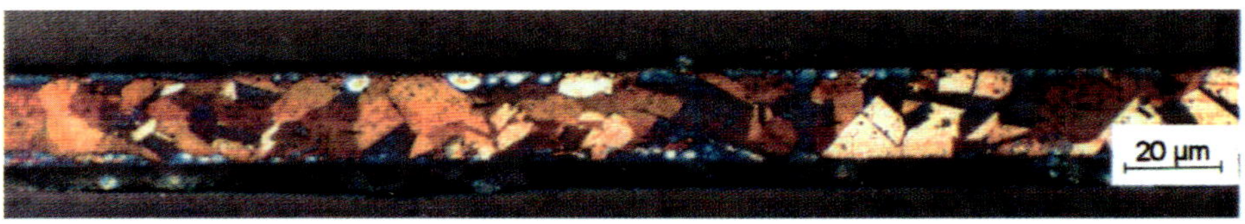

Fig. 1 Metallographic microstructure of the 13Fe-81-Ni-6Mo alloy annealed at 850°C (after Müller et al. 1998)

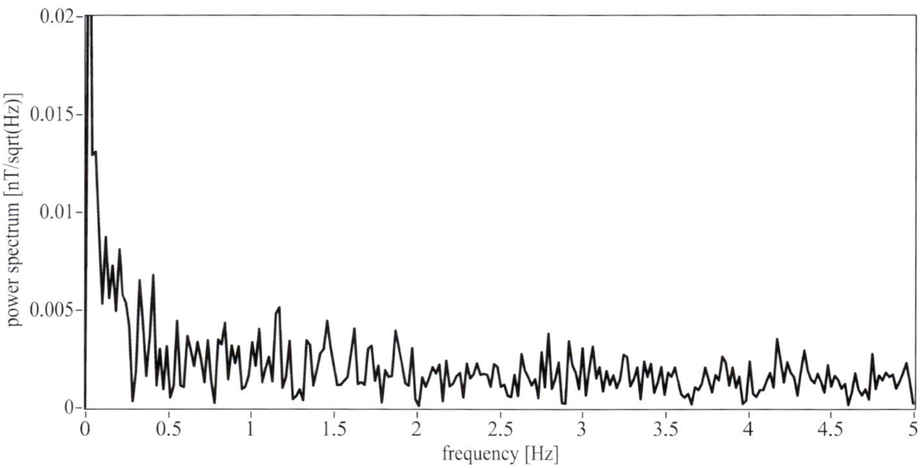

Fig. 2 Noise spectrum of a 13 mm ring-core as used for Themis

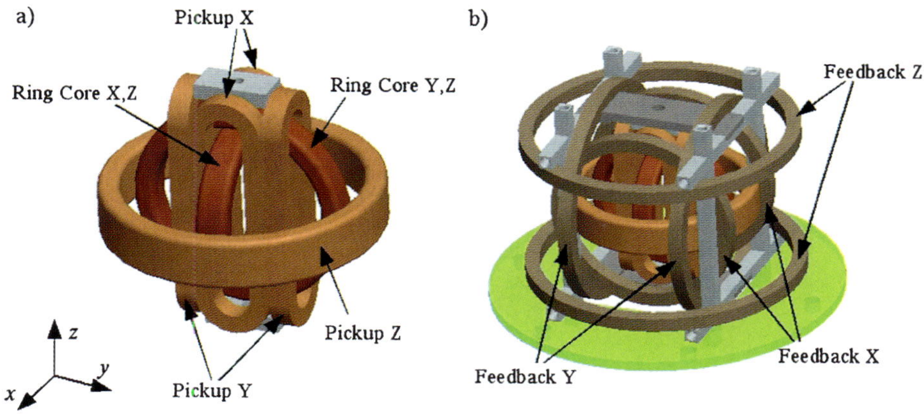

Fig. 3 3-D model of the FGM sensor with **a** ring cores and pick- up coil system and **b** fully functional sensor including the Helmholtz feedback system

separate the time and temperature dependence of the shielding can from ringcore related effects.

Two entwined ring-cores with a diameter of 13 and 18 mm are finally used to measure the magnetic field in three directions in the vector compensated sensor set-up. Via the smaller ringcore the magnetic field is measured in X and Z direction while the larger is used for Y and Z (see Fig. 3a). The ring-cores are equipped with two 3-D coil systems: an inner one to collect (pick-up) the magnetic field dependent second harmonic of the fundamental excitation frequency and an outer Helmholtz coil system to compensate the external field at the ringcore position. The pick-up coil system is attached as close as possible to the ring cores to increase the signal to noise ratio, in contrast to the comparably much larger Helmholtz coils which are used as feedback system to homogeneously compensate the magnetic field vector at the core position. The vector compensation keeps the sensitive sensor element in zero field. The single axis feedback design stabilized the scale value. The advantage of the vector compensation is the additional stabilization of the axis orientation. Thus both, scale value and axis direction depend only on the mechanically well stabilized feedback coil system. All coils are made from bond coated copper wire. By using this technology additional mechanical support, e.g. by ceramic rings, can be reduced to a minimum, the combination of materials with different thermal expansion coefficients can be avoided and mass can be saved. As a result, the mass of the sensor—excluding harness, mounting elements, protection cap and thermal hardware (see Fig. 3b)—could be reduced to less than 40 g for the type of sensor used for FGM.

3.2 Sensor Electronics

The block diagram of the FGM sensor electronics is shown in Fig. 4. An excitation AC current (excitation frequency at 8192 Hz, F0) drives the soft-magnetic core material of the two ring cores deep into positive and negative saturation. The external magnetic field distorts the symmetry of the magnetic flux and generates field proportional even harmonics of the excitation frequency in the pick-up coils.

In the digital fluxgate electronics design as used for FGM, analogue elements of traditional fluxgate magnetometers—such as filters and phase-sensitive integrators—are replaced

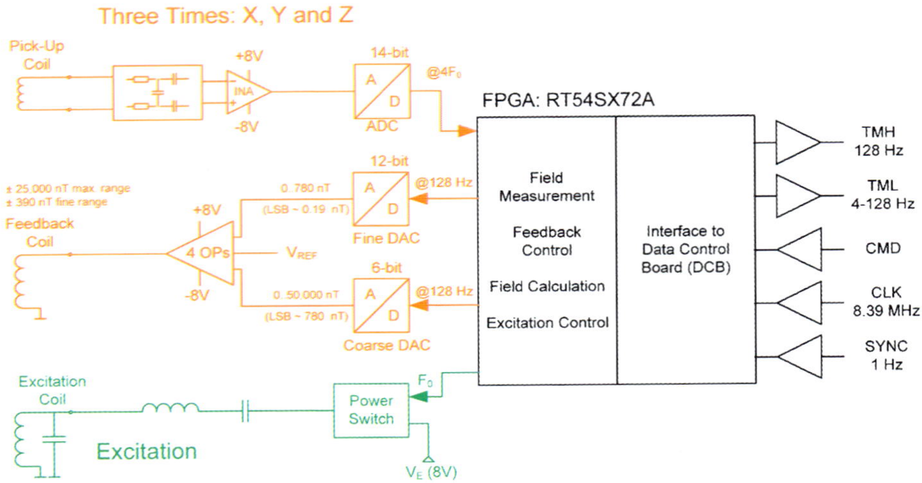

Fig. 4 Block diagram of the FGM sensor electronics

Table 3 Development steps of the digital magnetometer principle

Spacecraft mission	Control of ADC/DAC	Calculation of feedback	Calculation of magn. field	Type of FPGA
Rosetta/Lander	FPGA	DPU	DPU	RH 1280
VenusExpress	FPGA	FPGA	DPU	RT54SX32
Themis	FPGA	FPGA	FPGA	RT54SX72

by fast digitization of the sensor AC-signal and the subsequent data processing in FPGAs (Auster et al. 1995). Such a digital magnetometer was first development for the Rosetta Lander magnetometer followed by the magnetometer aboard the VenusExpress mission. From mission to mission the digital electronics has been further miniaturized as outlined in Table 3. In the ROMAP instrument, the near sensor FPGA mainly controls the converter components while the calculation of the feedback and the final output values are computed by a separate micro-processor (Auster et al. 2007). In the VEXMAG instrument aboard VenusExpress, the calculation of the feedback values is taken on by the FPGA (Zhang et al. 2006), and finally in the Themis FGM the complete digital processing is performed in a single near-sensor FPGA.

The replacement of analogue parts and the digitization on AC-level in general makes the sensed signal much more robust against changes of the environmental temperature and the supply voltage as well as insensitive to electro-magnetic interference, which are important features for the common E-box design of the Themis Instrument Data Processing Unit (IDPU).

The induced voltage in the pick-up coils is digitized behind the preamplifier at a sampling frequency of four times the excitation frequency. The accumulation of multiples of four consecutive data samples is necessary in order to eliminate all odd harmonics of the excitation signal, which couple from the excitation to the pick-up coil inductively. After processing the magnetic field digitally, the feedback settings are updated so that the field generated by the Helmholtz coil system compensates the external field almost completely.

The overall instrument performance is widely influenced by the sensor interface electronics. 14-bit ADCs (Maxwell 7872) and 12-bit DACs (Maxwell 8143) have been available with a radiation tolerant specification and reasonable power consumption for the Themis mission. The digital resolution of the 14-bit ADC at an input voltage range of ±5 V is 0.6 mV with a theoretically white quantization noise of 0.173 mV_{RMS}. Considering a ratio of 256 between sampling (4F0, 32,768 Hz) and maximum output frequency (128 Hz), the quantization noise in the signal bandwidth is 10.8 μV_{RMS}. With a nominal sensor sensitivity of 0.005 mV/nT and a pre-amplification of 40 dB—limited by the contents of odd harmonics in the pick-up signal—the amplitude of the digitization error is in the order of 21.6 pT_{RMS} for a signal bandwidth of 64 Hz which corresponds to a noise density of less than 3 $pT/\sqrt{Hz}$ assuming a white noise behavior. Thus, the digitization error does not exceed the design goal of 10 $pT/\sqrt{Hz}$ at 1 Hz. Nevertheless it is in the order of the sensor noise as shown in Fig. 2 and cannot be neglected completely.

More critical is the limited resolution of the DACs and here especially the non linearity which is in the order of half a Least Significant Bit (LSB). This corresponds to a non acceptable 6 nT error if one DAC is used for the whole measurement range of ±25,000 nT. Therefore two 12-bit DACs are cascaded (as shown in Fig. 4), a coarse one with a range of 50,000 nT (only the upper six bit active with the lower bits constantly set to binary 100000) and a fine one with a 780 nT range. The output voltages of the cascaded DACs are connected to a voltage to current conversion circuit. Using the fine DAC for the scientifically relevant low field range only, the maximum non linearity error for this range could be limited to <0.23 LSB and corresponding <43 pT by a pre-selection process of the best DACs. For fields above 400 nT (used for attitude determination) the linearity error of the coarse ADC has to be taken into account. During data post-processing on ground the non-linearity is partly corrected (MSB only). This is done before calibrating data inflight, because an uncorrected non-linearity of 2×10^{-4} would limit the accuracy of determination of scale value ratios and angle errors.

The core of the digital fluxgate electronics is an RT54SX72 FPGA from Actel. Its functionality can be divided into three sections: interface to sensor, interface to DCB and a 32-bit RISC processor module especially designed for the Themis magnetometer (see Fig. 5).

The sensor interface module enables the excitation; it starts the ADC sampling with programmable phase shift versus excitation clock at all three channels synchronously, averages (sign sensitive) a programmable number of ADC values and sends the results to the processor module. The processor calculates the magnetic field vector by adding the old DAC and new ADC values, both scaled by programmable conversion factors k_1 and k_2. Additionally, new feedback settings are calculated and passed to the sensor interface. High resolution 128 Hz data and low resolution low pass filtered or decimated data (4–128 Hz) are transferred via output register to the DCB interface. The DCB interface module receives commands for configuring hard- and software, synchronizes the data sampling to all other scientific instruments by a 1 Hz clock, and sends the serial data stream to the DCB containing the magnetic field vector (3 × 24 bit word) and a status word. In Table 4 all programmable configuration settings are listed.

Using these settings the instrument can be commanded into various modes. In the standard mode magnetic field values are calculated using commanded scaling factors. Lower time resolution data (TML) are calculated by filtering the raw data with a non-overlapping arithmetic averaging filter, by data decimation or a combination of both. For health checks,

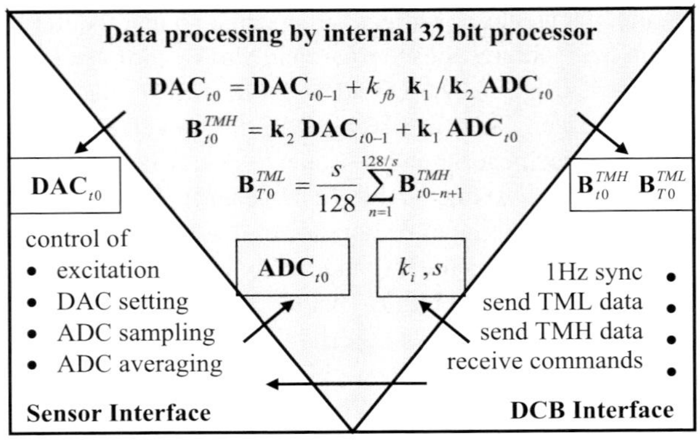

Fig. 5 Block diagram of the magnetometer FPGA. The magnetic field vector **B** is calculated by sum of the active ADC values and previous DAC settings. The factors k1 (ADC's) and k2 (DAC's) change converter units into magnetic field values with a basic resolution of 3 pT. The most significant bit of the DAC corresponds to a feedback field of 25,000 nT. The resulting field (24-bit) is transmitted via High Telemetry (TMH) and simultaneously after averaging via Low Telemetry (TML) to the Data Control Board (DCB)

Table 4 Summary of configurations settings

Hardware configurations	Excitation on/off Feedback on/off Relays on/off Type of filter
Software configurations	Sampling setup (phase, number) ADC/DAC scaling factors ki & offsets Fixed DAC values TML telemetry rate TML filter type

analysis of error sources and in the case of malfunctions in the feedback circuitry the feedback loop can be opened by software (open loop command) or hardware (relays). In this case all three k_2 values have to be set to zero.

Three calibration modes can be commanded by setting hard- and software options. In Cal-1 mode the instrument is operated in an open loop regime and the DAC values can be commanded manually. By this method the sensitivity of a sensor can be checked. Applying a constant calibration field and varying the phase between excitation and ADC sampling the balance of sensor and electronics input impedance can be checked and if necessary readjusted.

In Cal-2 mode the DAC setting are incremented automatically. This mode can be used to check the linearity of the sensor. Counting range as well as exposure time can be configured. If the sensor output is ignored ($k_1 = 0$) and the DAC values are not scaled ($k_2 = 1$), the count steps are transmitted directly. In this case the magnetometer generates independently from the external magnetic field a step function which can be used to check further data processing steps, telemetry quality and data timing. In Cal-3 mode ADC and DAC values are transmitted separately in TMH and TML channels. The mode is used to analyze the control behavior of the feedback loop.

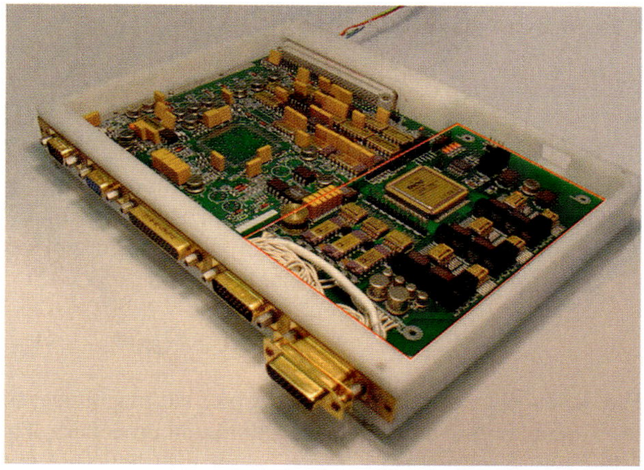

Fig. 6 Themis FGM electronics (*red square*) placed on a shared board

The electronics with the described functionality is placed on one side of a Themis standard board (see Fig. 6). The FGM board area is about 120 cm^2, the power consumption is 800 mW and its mass adds up to 150 g.

3.3 Onboard Data Processing at IDPU

The FGM electronics sends data over a serial interface to the processor board (DCB) inside the IDPU. Here the IDPU Flight Software (FSW) processes and packetizes the data. The 24 bit long vectors are shifted to select only 16 bits for telemetry. The selection of which 16 bits acts as a ranging function by selecting the widest range with the lowest resolution up to the smallest range with the highest resolution. As the samples are stored in memory, a header is written to the packet that includes the FGM message from the FGE board and the range and sample rate data (in the case of the variable rate packet). The packet timestamp is also added to this header when the packet is created and consists of time in seconds since January 1st 2001 as a 32 bit quantity and 16 bits of subseconds. Two separate telemetry streams are sent to the DCB board from the FGM. One is constant, 128 Samples/s data known as TMH and the other is variable rate data from 4–128 Samples/s known as TML. The FSW also takes the TMH stream and samples it to produce the attitude control packet that provides 8 Hz magnetometer data for spacecraft mission operations. This data is always in the widest least sensitive range. Two temperatures are sampled from the FGM thermistors, one on the FGE board and the other on the sensor. These, along with the FGM control word and message are reported in IDPU housekeeping.

The IDPU FSW also samples the FGM telemetry stream to process onboard spin fitted data. This is downlinked as a separate packet to the time series data. The software collects samples from the B-field vectors by taking 32 points at equal angles and fitting a sine wave least squares fit to the data. The best fit of the data is defined by the formula: $A + B \times \cos() + C \times \sin()$. The spin fit process calculates the least square fit and its standard deviation and then rejects the points that are far from the fit. The calculation is repeated until no more points are rejected. The fit can be chosen to be on the Bx or By data. Given a spin rate of 3 seconds, the use of 128 Hz data for spin fitting puts an apparent phase shift of $360/(3 \times 128)$ or roughly 0.9 degrees into the results. While this meets the 1.0 degree requirement, the phase shift correction can be determined on the ground using the spin pulse time data relative to the 1 Hz tick which is the basis of the 128 Hz data. In addition, the FSW averages the Z-axis data and provides it in the spin fit packet.

4 Instrument Calibration

4.1 Determination of Transfer Function

To measure the magnetic field vector correctly the magnetometer output ($\mathbf{B}_{\text{out}}$ in digital units) has to be scaled in nT, offset corrected and transferred into an orthogonal system. Assuming a diagonal matrix ($\mathbf{M}_{\text{gain}}$) to convert the digital units into nT, an offset vector ($\mathbf{O}_{\text{fgm}}$) and a matrix ($\mathbf{M}_{\text{ort}}$) to transform the measured components into an orthogonal system the calibrated field vector ($\mathbf{B}_{\text{fgs}}$) can be written as follows:

$$\mathbf{B}_{\text{fgs}} = \mathbf{M}_{\text{ort}}(\mathbf{M}_{\text{gain}}\mathbf{B}_{\text{out}} - \mathbf{O}_{\text{fgm}})$$

The dependence of these parameters on field magnitude, field dynamics, time and temperature shall be investigated by the calibration procedure. The offset is field independent per definition. If sensor and electronics are well balanced the offset should also not depend systematically on sensor and electronics temperature. The design goal is to keep the non systematic variation low, the goal of calibration is to record its behavior. To get a sufficient statistics, the offset was measured by sensor rotation in a weak field as often as possible, typically in the beginning and end of each calibration campaign. The determination of its temperature dependence was part of the test described in the following section.

The scale values in contrast are well defined by the feedback design. To investigate its temperature behavior, the expansion coefficients of the feedback coils and thermal coefficients of electrical parts have to be studied. Additionally its field and frequency dependency must be considered. The field non-linearity mainly caused by the DACs is discussed in Sect. 3.2, the frequency dependency later in this section. Due to the possibility to actualize the scale values by modification of k values, scaling ($\mathbf{B}_{\text{fs}} = \mathbf{M}_{\text{gain}}(k)\mathbf{B}_{\text{out}}$) can already be done onboard by the magnetometer software.

Cross coupling between magnetic axes caused by the electronics can be neglected due to the digital design. Therefore the misalignment is in contrast to sensitivity and offset a pure sensor property. Tests have been done to prove this assumption. If the orthogonality depends on the sensor only, arbitrary digital fluxgate electronics can be used for determining the orientation of magnetic sensor axes. To perform a scalar calibration the range of the qualification electronics has been extended to $\pm 50{,}000$ nT. The Earth field vector was measured at various sensor orientations and the calculated field magnitude has been compared to the field measured by a proton magnetometer. As derived by Auster et al. (2002) the motion about two sensor axes would be sufficient to provide the coefficients of a linear transfer function by this method. Measurements at arbitrary orientations, in practice at 24 sensor position which can by reached by 90° rotation of a cube, provide a sufficient redundancy. By this method the three angles of non-orthogonality were determined. In a second step the sensor was mounted in a fixture representing an orthogonal coordinate system with high precision (see Fig. 7).

The mechanical reference system of the fixture is defined by 6 center holes. By these holes the fixture can be pivoted along the three coordinate axes. If the rotation axis is oriented approximately perpendicular to the Earth field vector (e.g. in magnetic east-west direction), a misalignment of the true sensor axis to the reference axis causes a sinusoidal signal in the magnetic field measurement if the fixture is rotated about the reference axis. The sine amplitude normalized by the total Earth field and the phase versus Earth field direction provide the absolute misalignment of the true sensor axis. If the rotation is performed about all three axes of the reference system, all six angles of a transformation into the reference

Fig. 7 Sensor in fixture which defines the mechanical reference system with a precision of 10 arcsec

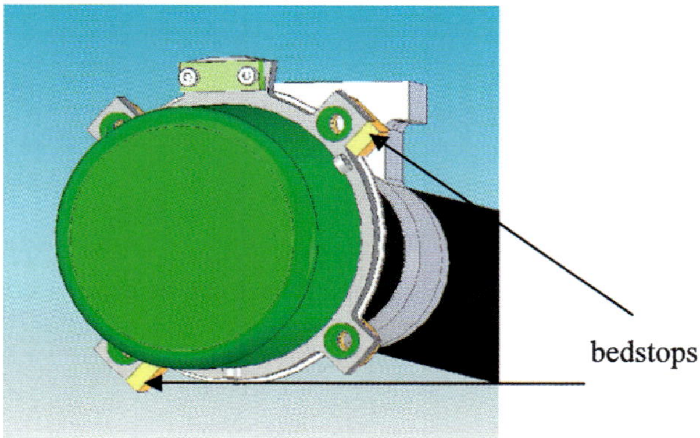

Fig. 8 Sensor fixation on boom for repeatable sensor mounting

system are determined. These six angles include the three angles of non orthogonality, which can be used to verify the first step, and additionally the rotation into the reference system.

Finally, the orientation of the sensor with respect to the probe has to be determined. The sensor interface is well defined by the mounting plane and two bedstops (see Fig. 8). This interface permits the repeatable mounting and demounting of sensors to the boom and in test facilities. The orientation of the sensor interface versus boom as well as the orientation of the boom vs. probe was measured by means of geodetic instruments in stowed and deployed boom configuration at the UCB workshop. This measurement completes the chain from raw data in a non orthogonal sensor system to a calibrated field vector in the probe system.

The only frequency dependent calibration quantity is the scale value. The sensor output signal is digitized exactly at the maximum and minimum of the second harmonic of the excitation signal with an sampling rate of 32,768 Hz. A certain number N of ADC samples are accumulated to one output value 128 times per second in order to produce the 128 Hz FGM raw data. To avoid the measurement during feedback updating, data sampling and feedback setting have to be done sequentially. Taking into account the time for the feedback calculation as well as the stabilization of the feedback current, only 232 samples of the maximum number of $N_{\max} = 256$ are accumulated. The frequency characteristic of the accumulated data is that of a standard average (boxcar) filter without overlapping. The fre-

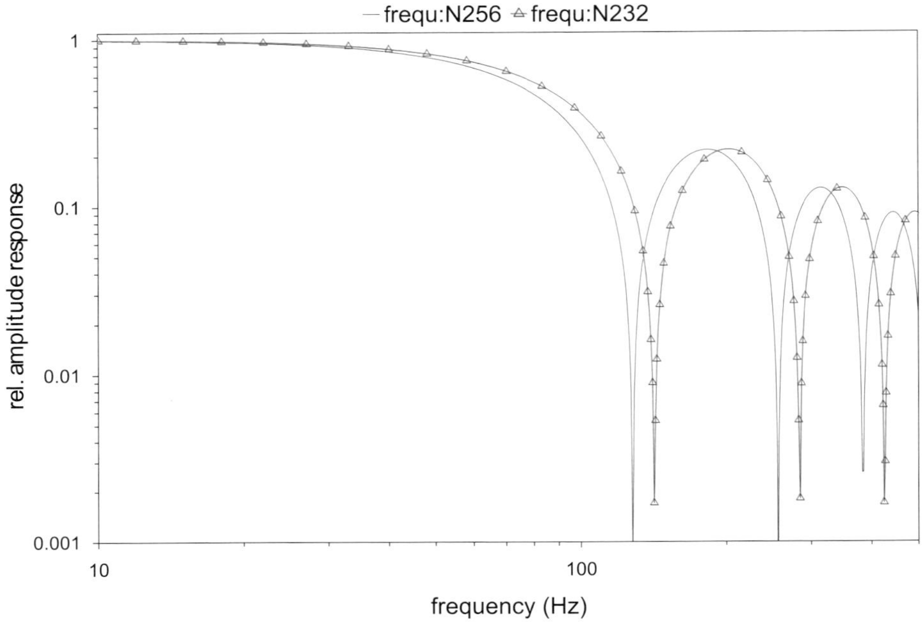

Fig. 9 Amplitude response of 128 Hz data

quency response of the averaging filter can be expressed analytically by amplitude $G(w)$ and phase $a(w)$ response:

$$G(\omega) = \frac{\sin(0.5N\omega T)}{N\sin(0.5N\omega T)}, \qquad \varphi(\omega) = -0.5N\omega T$$

where $\omega = 2\pi f$ denotes the angular frequency and T the sampling period (1 s/32,768).

Figure 9 shows the amplitude response for maximum ($N = 256$) and real ($N = 232$) samples accumulated. The filter characteristic of the sequential sampling mode is shifted by 13.24 Hz to higher frequencies. A verification of the frequency response has been done by measurements in Graz applying sine wave fields between 0.1 and 180 Hz generated in calibration coils. Amplitude and phase are measured with respect to the field generating current.

Low telemetry data are derived from 128 Hz raw data by averaging data using a non overlapping boxcar filter. Note that the DC field value is affected due to spin modulation by the filter characteristics. This has to be corrected during ground data processing.

4.2 Dependency on Electronics and Sensor Temperature

The test of the dependency of instrument parameter on electronics temperature was performed at TU-Braunschweig. The electronics boards were mounted inside a temperature chamber in which the temperature has been varied between -20 and $+60^{\circ}$C. The sensor was placed in the Themis sensor Control Unit (TCU), a ferromagnetic shield, in which the Earth field is suppressed by a factor of 10^4. The TCU is equipped with a coil system (see Fig. 10) to generate test fields and a rotation capability to check the sensor offset.

Fig. 10 TCU with coil system and sensor rotation capability

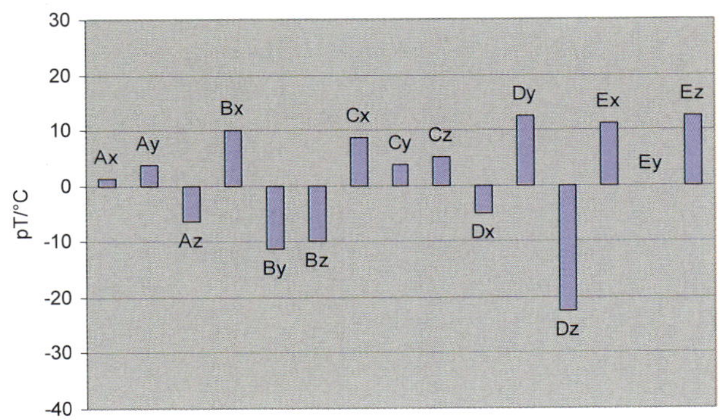

Fig. 11 Offset drift depending on electronics temperature of all sensor components (Probe A-E) caused by excitation and pick-up electronics

The dependency of the scale values on electronics temperature was tested by applying 20,000 nT in each vector direction. For each electronics channel a temperature sensitivity of less than 5 ppm/°C could be diagnosed. No measurable changes could be detected for linearity, noise, phase of second harmonics versus excitation clock and inrush current. The changes in power consumption are less than 5% in the tested temperature range.

Offsets are measured by rotating the sensor inside the screen. Due to the possibility to open the feedback, the sources of a changing offset could be separated into excitation and pick-up electronics (if feedback relays are open) and feedback current. Excitation and pick-up contribute to the temperature drift with less than ±20 pT/°C (see Fig. 11), the feedback

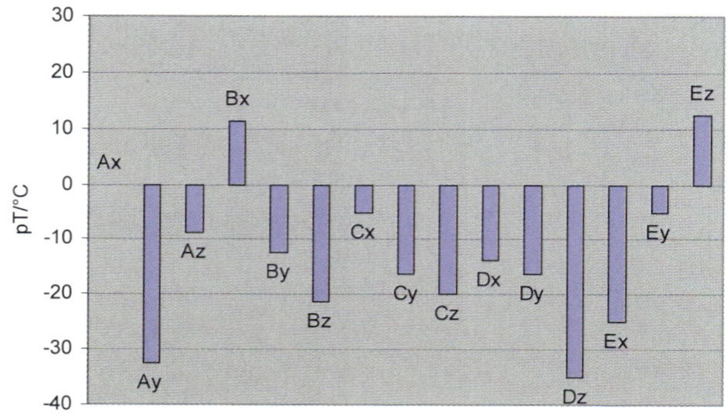

Fig. 12 Offset drift depending on electronics temperature of all sensor components caused by feedback circuitry

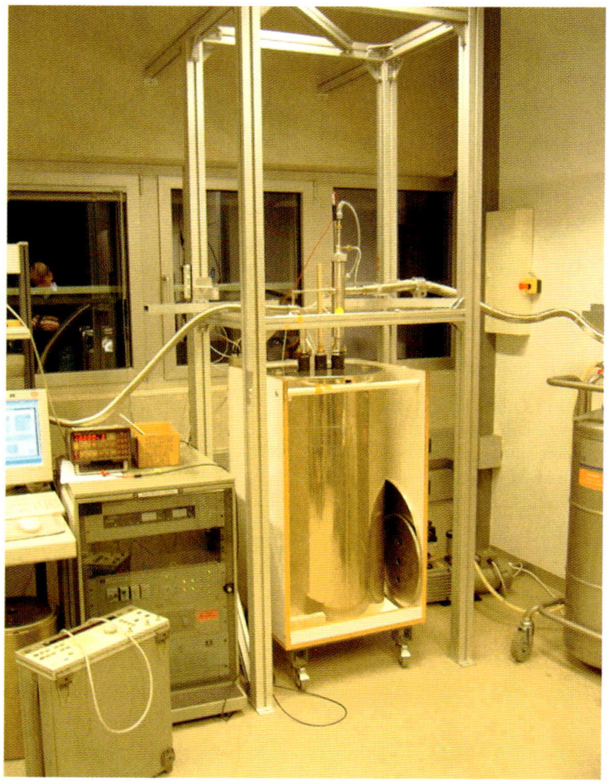

Fig. 13 Facility to measure the dependency of instrument parameters on sensor temperature

current instead shows an averaged negative temperature coefficient of −10 pT/°C with an error bar of ±20 pT/°C (see Fig. 12).

The dependency on sensor temperature was tested in a ferromagnetic shield equipped with a liquid nitrogen controlled temperature chamber (see Fig. 13) at IWF Graz in a temperature range between −100°C and +65°C. The tests showed that the noise levels (measured at 1 Hz) become higher at lower sensor temperatures. While FGM has a typical noise of 10 pT/$\sqrt{\text{Hz}}$ at temperatures between 0°C and 60°C, the noise increases from 15–20 pT/$\sqrt{\text{Hz}}$

at temperatures about -50°C up to 30 pT/$\sqrt{\text{Hz}}$ at -100°C which is still within the specifications. The expected sensor temperature in the Earth orbit is around 0°C.

Even as the temperature dependency of noise is an unintentional effect, properties like sensitivity or phase of sampling vs. excitation change inevitably with temperature. The permeability of the core and hence the inductivity of the pick up coil as well as its resistance are functions of the temperature and affect the balance conditions of the input circuitry. The advantage of the digital magnetometer is, that these effects can be determined and compensated by updating the instrument configuration (phase shifts, scaling factors). Figure 14 shows the input sensitivity versus ADC sampling phase at three temperatures. The phase shift changes due to the temperature dependent inductivity of the ringcore. The sensitivity at lower temperature increases due to the lower copper resistance.

Current sources are used to drive the feedback, therefore the sensitivity shall depend only on the thermal expansion coefficient and not on the resistance of the feedback coil system. Only materials with expansion coefficients of about 20 ppm/°C (aluminum, copper) are used. All sensitivity measurements confirm this temperature coefficient within an error bar of ± 3 ppm/°C. Due to the fact that combination of materials with different expansion coefficients are avoided (e.g. copper and ceramics) the temperature coefficient is constant over the whole temperature range.

Offsets are measured by sensor rotation at various temperatures. Also the offset dependency on sensor temperature is comparable to the one of the electronics temperature (<30 pT/°C). Figure 15 shows, that no systematic temperature behavior is noticeable.

4.3 Parameter Check under Well-Defined Field Conditions

Two tests were done to check the overall functionality and to verify the calibration parameters. First the magnetometer was tested by artificial fields generated in a coil system, and secondly by variations of the Earth field.

The sensors were mounted inside a thermal control box, which is placed in the coil center (see Fig. 16). After the setup measurements (standard mode, 4 Hz data rate, external field ± 20000 nT) the calibration was started with a test field sequence at 20°C. Then the temperature was increased to 60°C with a gradient of 0.3°C/min and the measurement sequences were repeated. The cooling down to about -70°C was performed using ceramic blocks (3.5 kg) which had been cooled in liquid nitrogen prior measurement. At all temperature levels sensitivity and orthogonality were checked.

The direct comparison between two instruments or if possible the comparison with an observatory magnetometer is an expedient method to verify the properties of the instrument. Two Themis sensors are respectively mounted on a pillar (see Fig. 17) and compared with a reference instrument. Unfortunately the earth field vector on ground cannot be measured by the Themis magnetometer because its range is adjusted for a perigee of more than 1000 km. Therefore, only the horizontal components are compared. The test was repeated with a sensor alignment rotated by 90°.

Sensors of observatory and Themis magnetometers are identical. The reference electronics, usually applied in geomagnetic observatories, can be used as standard because it is well tested and has no limitations due to the space restricted part assortment. The tests were performed in the Test Facilities of Magson GmbH in Jeserigerhuetten (Germany). First of all, irregularities like field jumps, data loss, timing problems etc. can be detected. Furthermore, the long term behavior, including stability of offsets, scale values and magnetic axes can be evaluated and finally, as shown in the extracted short term pulsation registration (see Fig. 18), it is an in situ test of measurements of field changes expected during substorm onsets.

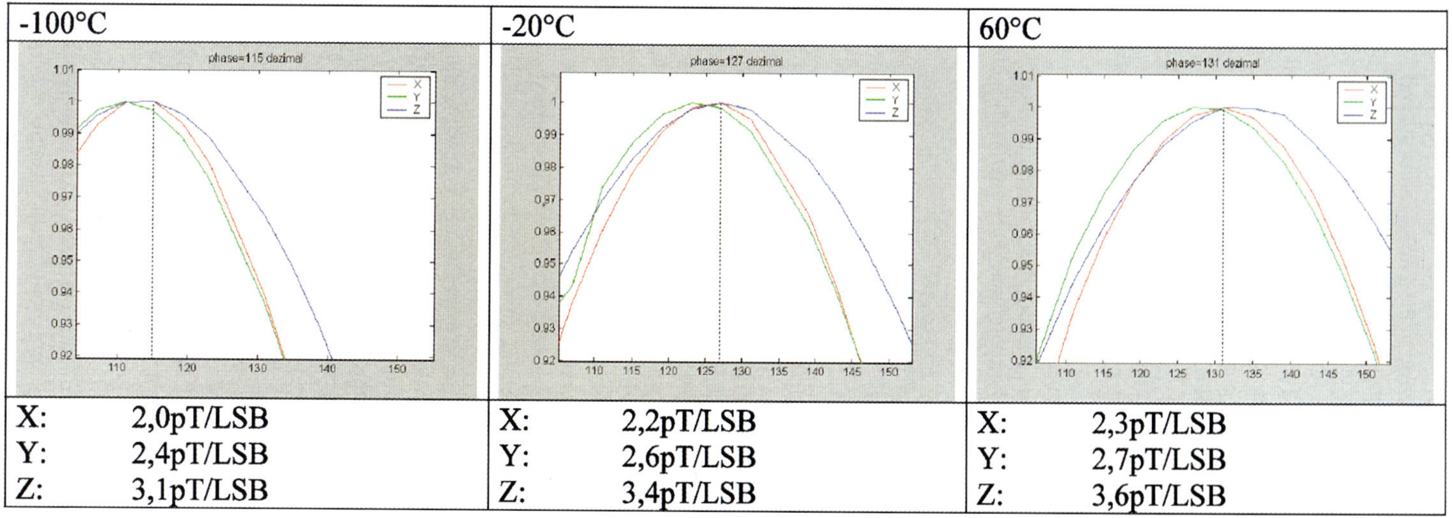

Fig. 14 Input sensitivity and phase of ADC sampling versus excitation clock in dependency on sensor temperature. *Upper panel* show the sensitivity versus phase angle, the numbers *below* provide the sensitivities absolutely

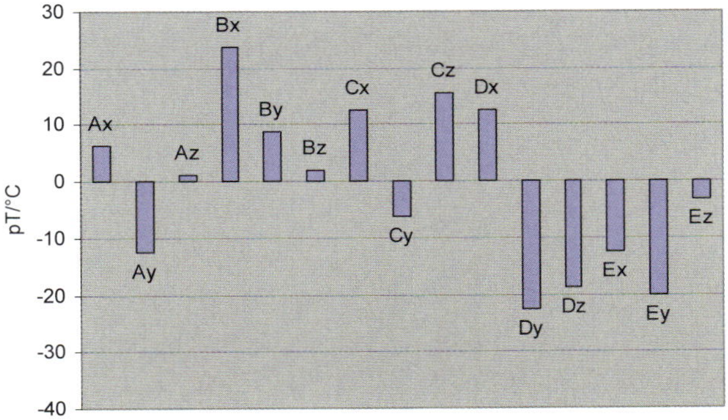

Fig. 15 Offset drift depending on sensor temperature of all sensor components

Fig. 16 Coil system with thermal box baseplate in Magnetsrode, Braunschweig

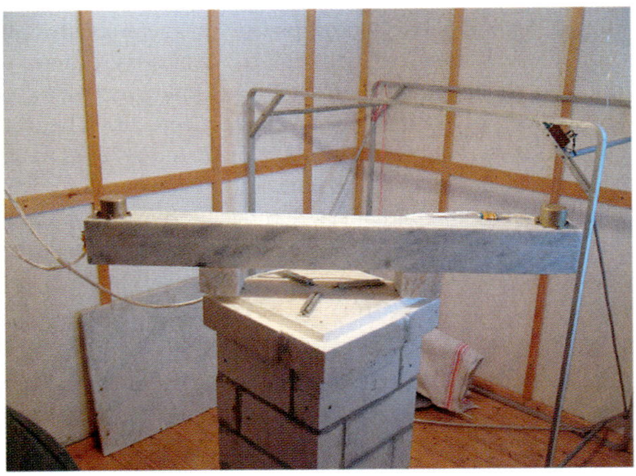

Fig. 17 Two Themis sensors mounted on a pillar to measure Earth field components for comparison with an observatory instrument

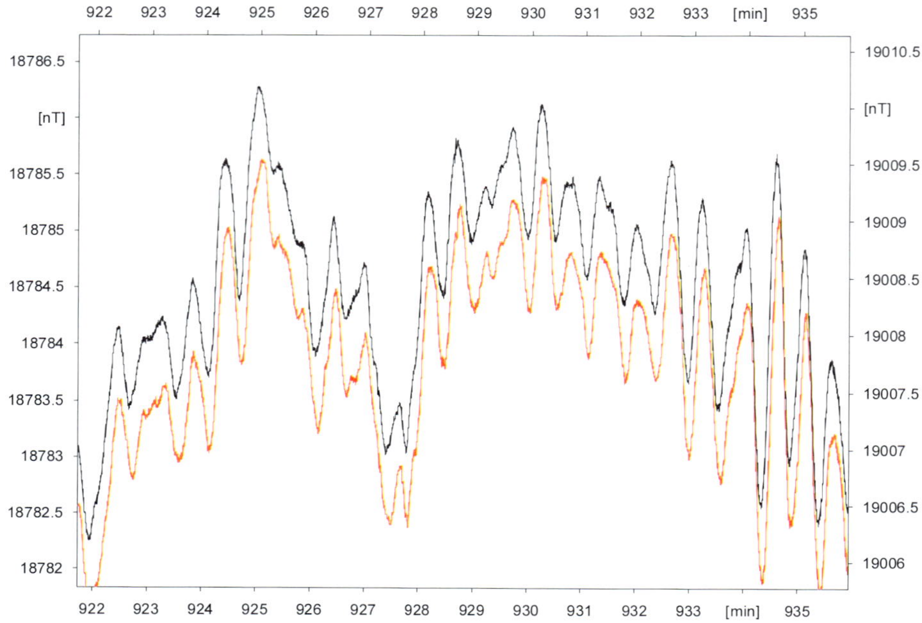

Fig. 18 Pulsation measurement of two Themis magnetometers (*black* THC, *red* THD) performed in north-south direction of the Earth field. Field variations measured with an independent observatory instrument confirm the result of both Themis magnetometers

4.4 Parameter Check during S/C Integration (SFT)

The integration took place in Berkeley. Effects on measurement quality had to be expected. The second half of the board (PCB) was powered for the first time together with the FGM electronics, the secondary voltages were provided for the first time by the original DC/DC converter, and finally the interface to DCB was established. This made the precise magnetic field measurement at the integration environment necessary. Especially, parameters like noise and offset had to be checked routinely before and after integration steps.

A test facility which protects the sensor from Earth and technical field variations and which is mobile enough to follow the magnetometer during its integration procedure was used. Three of the ferromagnetic shields which were already used to keep the sensor in a controlled environment during electronic temperature tests (see TCU description in Sect. 4.2) are installed in Graz, Braunschweig and Berkeley. The Berkeley unit was used for all tests before and after integration steps. During the tests the sensor was removed from the boom and placed inside the TCU, connected by an extension cable. The influence of the extension cable on calibration parameters has been tested and stated as negligible.

A Short Functional Test (SFT) procedure of 20 minutes duration, performed by the integration team, checks the overall functionality, offsets, scale values, noise, sensor-electronics balance and telemetry errors. Each instrument was tested during the s/c integration about 20 times by this procedure. As a result we found two errors—a sensor was replaced due to increased noise level, a cable short was detected and removed—and it provided statistics of the tested parameters covering more than one year. Although S/C induced disturbances were investigated by these tests, some interference could only be identified in space as shown in Sect. 5.2.

Table 5 Coordinate systems which are used to transform the magnetometer output data into a spin aligned sun oriented system as defined in detail in Angelopoulos (2008). Abbreviations are referred to the terms used for Cluster

Abbreviation	Description
FS	Non orthogonal sensor system
FGS	Orthogonal sensor system
UNIT	Boom aligned system
SPG	Spinning Probe Geometric
SSL	Spinning Sunsensor L-oriented
DSL	Despun Sun oriented L-oriented

4.5 Creation of Calibration Files

In the following sections calibration relevant coordinate systems are introduced and the creation of CalFiles is described. Elements of the calibration matrix are derived from many individual parameters which can be clearly related to instrument/spacecraft properties.

The magnetometer provides data in digital units in a non orthogonal coordinate system (FS). The digital units are pre-scaled by the magnetometer processor. The conversation factor of 2.98 pT/bit is specified by the ratio between dynamic range (±25,000 nT) and digital resolution (24 bit). The selection of the transmitted 16 bit is done by the so-called ranging in the IDPU. Range 8 stands for transmitting the lower 16 bits, range 0 for transmitting the upper ones. The range dependent conversion factor can be expressed by: $k_r = 50{,}000/2^{(15+\text{range})}$. The sensor offsets $\mathbf{O}_{\text{fgm}}$ have to be corrected in the FS system and the data has to be transformed by $\mathbf{M}_{\text{ort}}$ into an orthogonal sensor system (FGS):

$$\mathbf{B}_{\text{fgs}} = \mathbf{M}_{\text{ort}}(k_r \times \mathbf{B}_{\text{fs}} - \mathbf{O}_{\text{fgm}})$$

The orientation of the sensor coordinate system is defined by the mechanical interfaces of sensor and boom (see Fig. 19) as well as by the moment of inertia of the probe which determines the rotation axis. All angles of these three coordinate transformations are measured on ground. The determination of the sensor alignment versus boom interface ($\mathbf{M}_{\text{unit}}$) is part of the sensor calibration program, the boom alignment versus spacecraft ($\mathbf{M}_{\text{probe}}$) is measured during the boom verification procedure. Using these coordinate transformations, the magnetic field data can be rotated into the probe coordinate system:

$$\mathbf{B}_{\text{spg}} = \mathbf{M}_{\text{probe}}\,\mathbf{M}_{\text{unit}}\,\mathbf{B}_{\text{fgs}}$$

In the probe coordinate system errors caused by the magnetic properties of the spacecraft are considered. Spacecraft offsets $\mathbf{O}_{\text{sc}}$ are added. The influence of probe soft-magnetic material on the direction of the sensor axes can be neglected, its influence on the sensitivity is compensated by multiplying the magnetic field with $\mathbf{M}_{\text{scale}}$. To align the coordinate system with the spin axis and to align the x axis with the sun direction, the field vector has to be rotated by $\mathbf{M}_{\text{spin}}$ and $\mathbf{M}_{\text{phase}}$. The nominal spin axis and spin phase alignment are determined during the spin balance tests at JPL and the sun sensor integration. Additionally the delay and the spin dependent damping factor of the boxcar filter for TML data has to be compensated. This is done by $\mathbf{M}_{\text{filter}}$ which contains the rotation about the angle α_{delay} for the filter delay and the correction of the sensitivity in the spin plane of d_{filter}:

$$\alpha_{\text{delay}} = -\pi\,\frac{f_{\text{spin}}}{f_{\text{sample}}}; \quad \text{and} \quad d_{\text{filter}} = \frac{f_{\text{sample}}}{128}\,\frac{\sin(\frac{\pi}{128} f_{\text{spin}})}{\sin(\pi\,\frac{f_{\text{spin}}}{f_{\text{sample}}})}$$

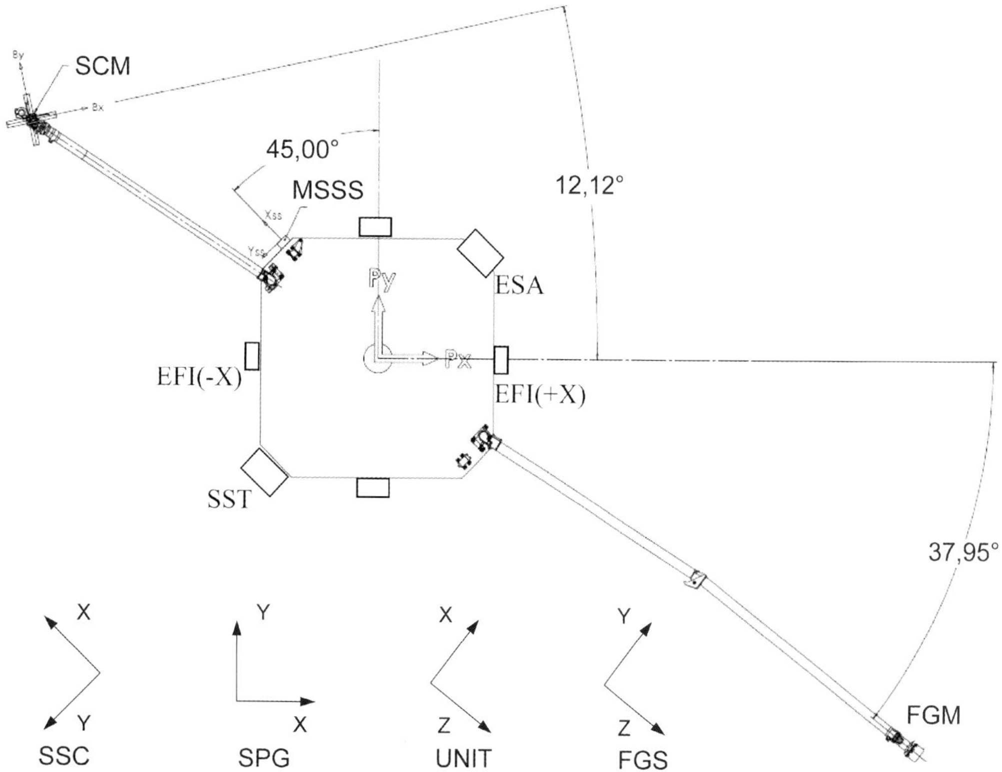

Fig. 19 Accommodation of boom mounted FGM sensor and orientation of instrument, boom, probe and sun sensor (MSSS) related coordinate systems

The magnetic field in the spin aligned sun oriented system can be calculated by:

$$\mathbf{B}_{\text{ssl}} = \mathbf{M}_{\text{filter}}\,\mathbf{M}_{\text{phase}}\,\mathbf{M}_{\text{spin}}\,\mathbf{M}_{\text{scale}}\,(\mathbf{B}_{\text{spg}} - \mathbf{O}_{\text{sc}})$$

The calibration File contains all corrections/transformation up to the SSL-system. Calibration matrix $\mathbf{M}_{\text{cal}}$ and offset $\mathbf{O}_{\text{cal}}$ are calculated by the single transformations as follows:

$$\begin{aligned}\mathbf{B}_{\text{ssl}} &= \mathbf{M}_{\text{filter}}(\mathbf{M}_{\text{cal}} k_{\text{r}} \times \mathbf{B}_{\text{fs}} - \mathbf{O}_{\text{cal}})\\ \mathbf{M}_{\text{cal}} &= \mathbf{M}_{\text{phase}}\,\mathbf{M}_{\text{spin}}\mathbf{M}_{\text{scale}}\,\mathbf{M}_{\text{probe}}\,\mathbf{M}_{\text{unit}}\,\mathbf{M}_{\text{ort}}\\ \mathbf{O}_{\text{cal}} &= \mathbf{M}_{\text{phase}}\,\mathbf{M}_{\text{spin}}\,\mathbf{M}_{\text{scale}}(\mathbf{M}_{\text{probe}}\,\mathbf{M}_{\text{unit}}\,\mathbf{M}_{\text{ort}}\,\mathbf{O}_{\text{fgm}} + \mathbf{O}_{\text{sc}})\end{aligned}$$

$\mathbf{M}_{\text{filter}}$, $\mathbf{M}_{\text{probe}}$, $\mathbf{M}_{\text{unit}}$ and $\mathbf{O}_{\text{fgm}}$ are assumed to be constant. Their values are determined by ground calibration. $\mathbf{M}_{\text{phase}}$, $\mathbf{M}_{\text{spin}}$, $\mathbf{M}_{\text{scale}}$, $\mathbf{M}_{\text{ort}}$ and $\mathbf{O}_{\text{sc}}$ are time dependent and therefore subjects to the inflight calibration procedure. Initial values are taken also from ground calibration.

5 First Results

5.1 Inflight Calibration Result

During commissioning all basic functions are tested by a procedure similar to the one applied for short functional tests on ground. Some modifications are necessary due to the rotation period of the probes. Sensor-electronics balance and sensitivity are unchanged compared to preflight tests, telemetry quality and onboard data processing are error free. The tests have been repeated after the successful deployment of all magnetometer booms. After deployment the total noise level of the magnetic field measurement was checked at apogee crossings. A statistic about the noise level of all 15 sensors is shown in Fig. 20.

The number of sensors was counted for certain noise levels. At 1 Hz the averaged noise level of all 15 components is about 12 pT/$\sqrt{\text{Hz}}$, which is less than half of the required level of 30 pT/$\sqrt{\text{Hz}}$. Based on the results of the preflight calibration (see Sect. 4.1) we can assume a linear transfer function between the magnetometer output in a non orthogonal sensor

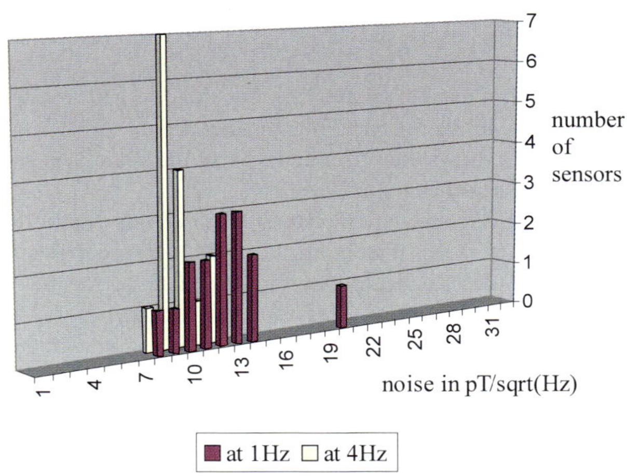

Fig. 20 Noise Statistic measured inflight: The overall noise was measured for each sensor at quiet field conditions. The sensors are sorted by noise levels at 1 Hz and 4 Hz. A noise level less than 30 pT/$\sqrt{\text{Hz}}$ at 1 Hz was required

system and the magnetic field vector in a spin axis aligned spacecraft system. Updating the initial elements of the transfer function at regular intervals is a task of the in-flight calibration. The result of the in-flight calibration is a calibration file (CalFile) which contains the 12 elements of the vector transformation, the spin period and the time of validity.

The elements of the transformation consist of scale values, non-orthogonality, sensor orientation and offsets. Deviations from nominal values are caused by many reasons, either constant in time (e.g. boom and sensor alignment) or time and temperature dependent (e.g. sensor and spacecraft generated offsets).

To determine the transfer function in flight we need a multitude of inputs. First the rotation of the spacecraft can be used. The fact that the spin frequency and its first harmonic have to be absent in the field magnitude provides 4 equations. Furthermore one axis is defined by the spin axis (2 equations). 8 of 12 elements are affected by the spacecraft rotation namely two spin plane offsets, the ratio between spin plane scale values, all three angles of non-linearity and the two angles of orientation versus spin axis. Using n times 6 equations for n different field conditions (variable in field direction and amplitude) these 8 elements can simply be determined by minimizing the spin tone frequencies in the field magnitude.

The remaining four elements—spin axis offset and scale value, scale value of spin plane components and spin phase—have to be determined by criteria derived from field properties (e.g. non compressible waves) and field models (e.g. IGRF). Special field conditions are required for this calibration. The determination of sensitivities and spin phase need the Earth fields which is known by models at the perigee at least with an accuracy of 0.1%. The spin axis offsets can be determined during solar wind passages in the first summer season and later more rarely at low field in the magnetosphere at selected intervals. Additionally the comparison of magnetic field measurements between the spacecraft can be used for calibration. At special field conditions it can be assumed that the field is homogeneous over the distance of the probes ($\mathbf{B}_1 = \mathbf{B}_n$), spatially linearly distributed and current free ($\mathrm{curl}\,\mathbf{B} = 0$), or only spatially linearly distributed ($\mathrm{div}\,\mathbf{B} = 0$). Themis constellations which fulfill these requirements are rare and, if available, e.g. in solar wind, the spin axis offsets can also be determined by single spacecraft analysis. Therefore the spacecraft comparison might be useful to check the in-flight calibration from time to time but cannot provide a significant input for the routine in-flight calibration. As described above, different field conditions are necessary for one in-flight calibration. It has to be assumed that the elements are constant over the whole calibration interval. Therefore the repeatability of the in-flight calibration (at least once per orbit) defines the requirements on the stability of the magnetometer. On the other hand the results of the in-flight calibration present a reality check of the instrument stability.

Calibration results are available for the first half a year of FGM operations. All angles and scale values were constant with an accuracy of 10^{-4}.

Figure 21 shows the offset behavior of the spin plane components of spacecraft A. Both offsets vary less than 0.2 nT over half a year.

In Fig. 22 the standard deviation of all offsets within this interval is plotted. The maximum variation is less than 0.3 nT/6 month. The required stability was 0.2 nT/12 hours.

The statistics has been done for spin-plane offsets only. Spin plane offsets are easy to determine, since the offset is a DC contribution to a signal that should have a spin-frequency variation. The few spin axis offsets we got from solar wind passages are variable in the same order of magnitude, so that we can assume that the stability presented for spin plane offsets is representative for all axes.

To consider the time dependency of the calibration parameters, CalFiles are updated each day, which is the orbital period of the inner spacecraft. Additionally high resolution

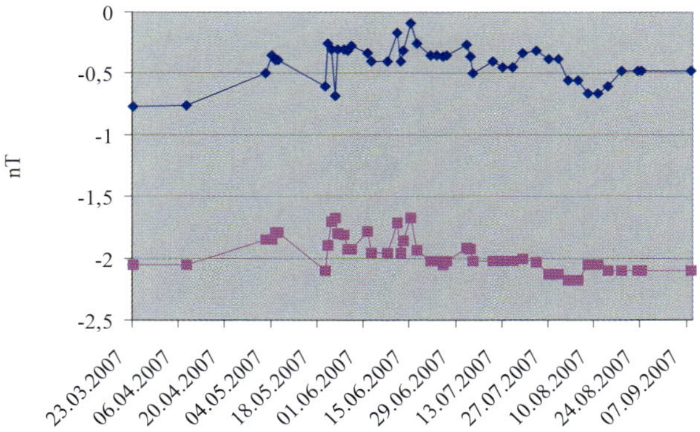

Fig. 21 Offset of Probe A spin plane components in the first half year of flight

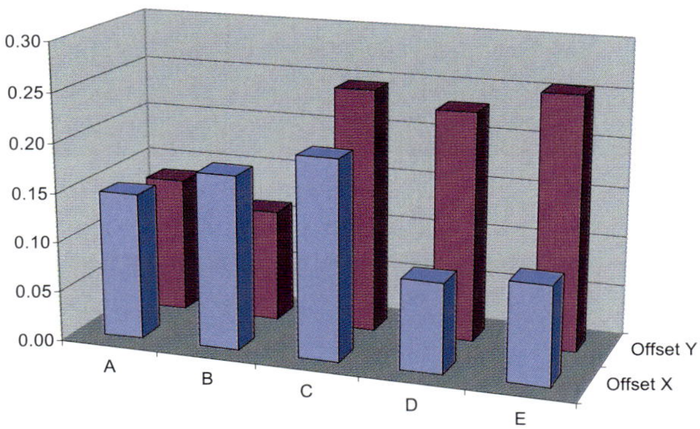

Fig. 22 Offset variation (standard deviation in nT) of spin plane components of all five Probes

CalFiles can be provided on request. These files are based on daily CalFiles with small adaptations of the two spin plane offsets, one scale value and the angle between the two spin plan components. This is not a calibration in the truest sense of the word, because in an underdetermined system simply the most prominent 4 elements are modified in order to minimize the spin tones in the field magnitude.

5.2 Spacecraft Interferences

Two types of interferences could be detected in space. Both have maximum amplitude of 0.3 nT peak to peak. The first one is related to the solar cells driven power management and therefore strongly spin synchronized. A model of this interference has been developed. After the spin axis was aligned precisely at high field conditions, the remaining content of spin frequency and its harmonics of the spin axis component at low field conditions has been used as input parameter for the model. The derived field wavelet was scaled for the spin plane components by the amplitude of the spin tone harmonics and subtracted from the raw data. Figure 23 shows the dynamic spectra of the spin axis component in SSL system before and after correction. The error in spin tone of 35 pT and double spin tone of 15 pT could be suppressed by a factor of four. The remaining periodic content of spin tone appearing in the corrected data can be interpreted as a non constant phase of the interference with respect to the sun pulse. This seems reasonable because the sun dependent power switch sequence is synchronized with a finite time resolution.

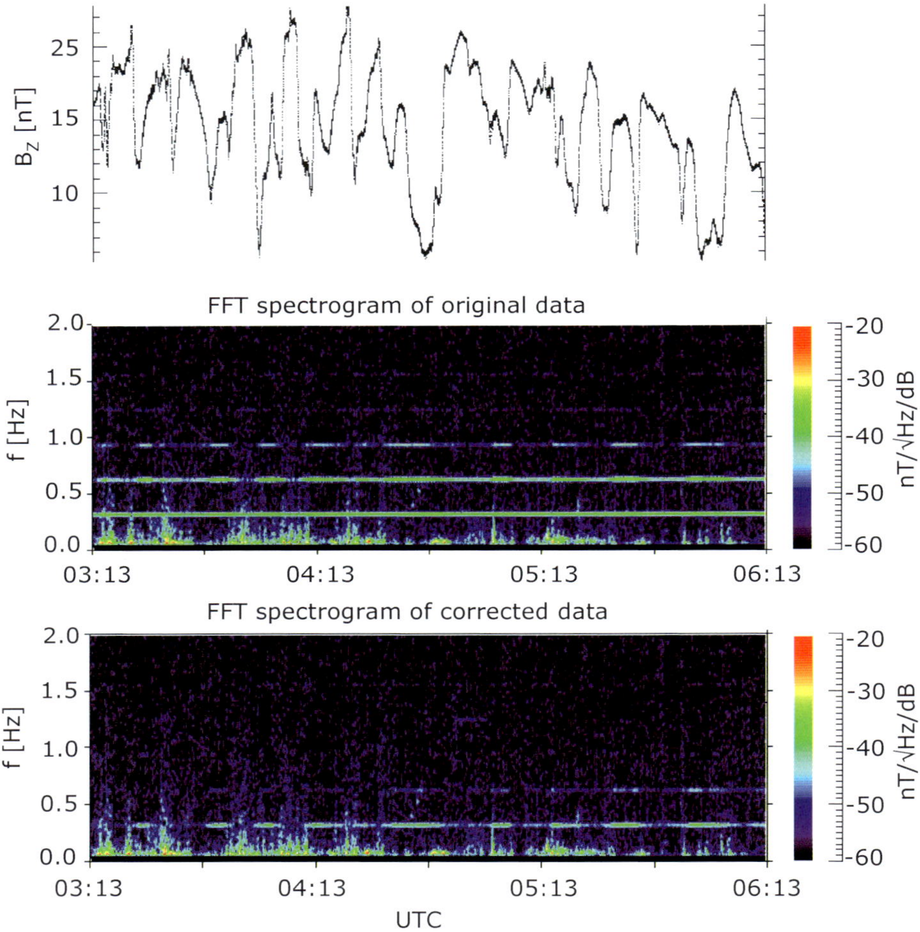

Fig. 23 Data before and after correction of interferences induced by the power supply. *Upper panel*: time plot of the spin axis component. *Central panel*: dynamic spectra of original spin axis data. *Bottom panel*: dynamic spectra of spin axis data after applying the correction using the sun pulse triggered interference model. The FFTs are calculated using 128 4 Hz samples

The second error is caused by sectoring of the particle instruments. The signatures measured by the magnetometer are certainly not generated by mode dependent magnetic moments of the particle instruments. The interference is conducted due to the power profile of the particle instruments. Facilities to detect the interferences in the magnetic field data (see Sect. 4.4) as well as grounding options to prevent the magnetic field measurement from conducted interferences were available. Due to the complex test assembly for such a test on bench level (sun simulation & operation of more than one experiment) the common operation has unfortunately never been tested on ground. The sectors are switched by the 32nd part of a spin period. This corresponds to a 11 Hz switch frequency. Also the sector switching is performed by a finite time resolution continuously synchronized by the sun pulse. This leads to a jitter in the switch frequency and therefore to a dilatation of the interference frequencies. The disturbance can be avoided by changing the flight software timing. It will

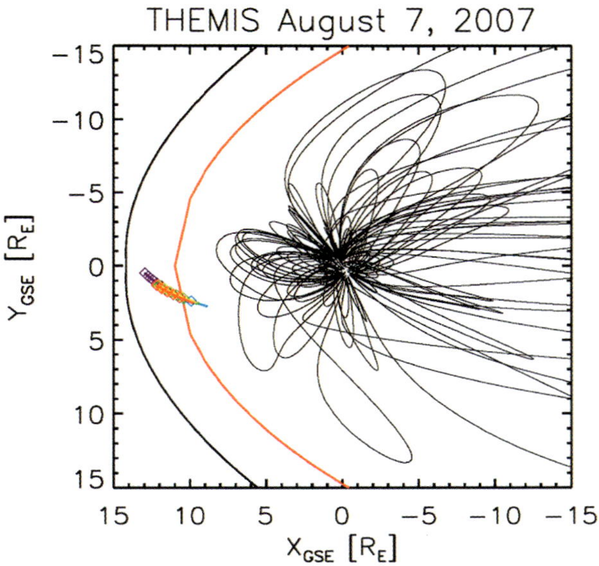

Fig. 24 Themis orbits on August 7 2007 between 09:00 and 11:30 UT. The magnetic field has been obtained using the Tsyganenko 96 model

be done only in a later phase of the mission because the amplitude of the interference is small (0.1 nT) and the affected frequency bands ($n \times 11$ Hz ± 2 Hz) are covered by SCM (Roux et al. 2008) too.

5.3 Magneto Pause (MP) Oscillations Visible in FGM Data

To demonstrate the FGM capabilities we study a magnetopause crossing which occurred on August 7, 2007 close to the sub-solar point. At this date, the spacecraft were still in the injection phase, sharing the same orbit with a 15.4 RE apogee (see Fig. 24). This "string of pearls" configuration is particularly well suited for timing analysis of the magnetopause position.

Figure 25 shows the magnetic field magnitude measured by all five probes between 09:00 and 11:30 UT as they move from the magnetosheath into the magnetosphere. Probe A, being the last in the string, does not reach the magnetopause during this time interval. The first to cross the magnetopause is probe B at around 09:25 UT. Probes C, D, and E follow five minutes later, one shortly after another. During the following 90 minutes all four leading probes experience multiple magnetopause crossings. Due to the fact that the spacecraft move along the same track we can draw a position-time diagram such as the one shown in Fig. 26. Here we plotted the distance along the orbit, from a common reference point to each spacecraft as a function of time. It can be seen that probe B leads the formation, at a distance of about 1 R_E from probes C, D, and E, which are grouped closer together. About 1.5 R_E away, Probe A closes the formation. A magnetopause crossing detected at a certain moment in time by one of the spacecraft is represented by a dot on the corresponding line.

From the slope of each crossing we can derive the speed of the magnetopause along the spacecraft orbit. The resulting mean values are 72 km/s for inward motion and −95 km/s for outward motion. These values are comparable with 67 km/s, which is the maximum speed of the magnetopause motion if we assume harmonic oscillations.

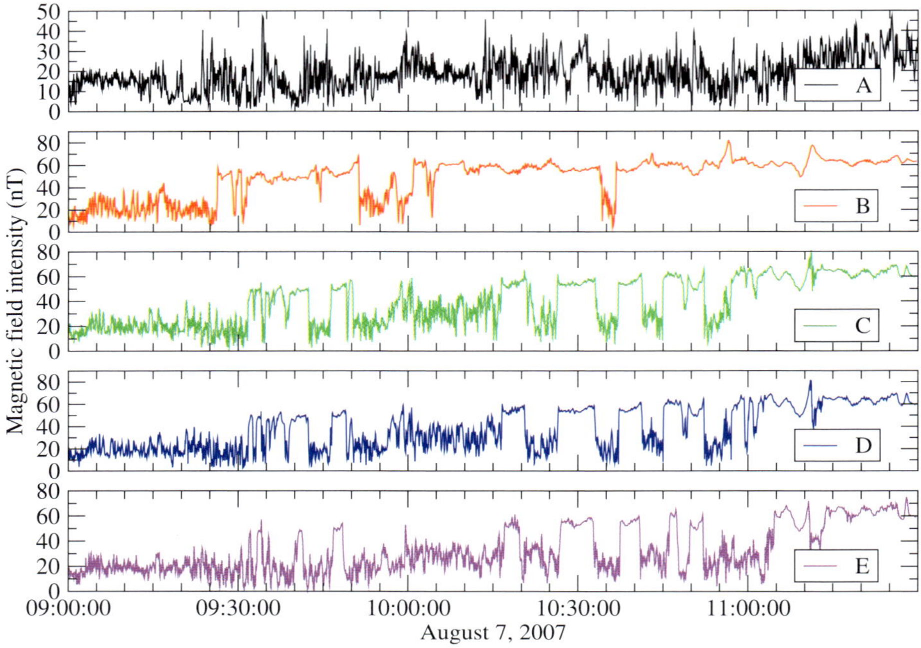

Fig. 25 The magnetic field magnitude measured by Themis magnetometers. All probes but Themis A exhibit multiple magnetopause crossings

In total we detect 81 single-spacecraft events which group themselves in 17 crossings. The motion of the magnetopause is visible in the position-time diagram as indicated by the curved line connecting the crossings between 10:10 and 11:00 UT. Roughly, we see an oscillation with an amplitude of about 2 R_E and a period close to 10 minutes.

6 Summary

The THEMIS FGM benefits from elaborate works for the development of ring cores and the sensor design, the technology of digital fluxgate magnetometers, and tests and calibrations in the high precision facilities developed for a number of previous missions. The general characteristics of FGM, calibration procedure and results are summarized in the present paper.

FGM provides accurate and stable magnetic field measurements in the near-Earth space. The stability was proven to be better than 0.5 nT during the first half year operation. Five point measurements lead to a number of data analysis methods. One example is presented from the magnetopause crossings and the speed of the magnetopause motion is estimated. This reconstruction of the time history of the magnetopause motion is a good example of a new analysis method which uses the specific Themis multi-point configuration.

Acknowledgement The THEMIS team is greatly indebted to many individuals who made the THEMIS mission possible and who contributed greatly to the success of developing, building, testing, and flying the FGM instrument. Special thanks are to Ernst Jelting and Sabine Filbrandt (IGEP Braunschweig) for carefully

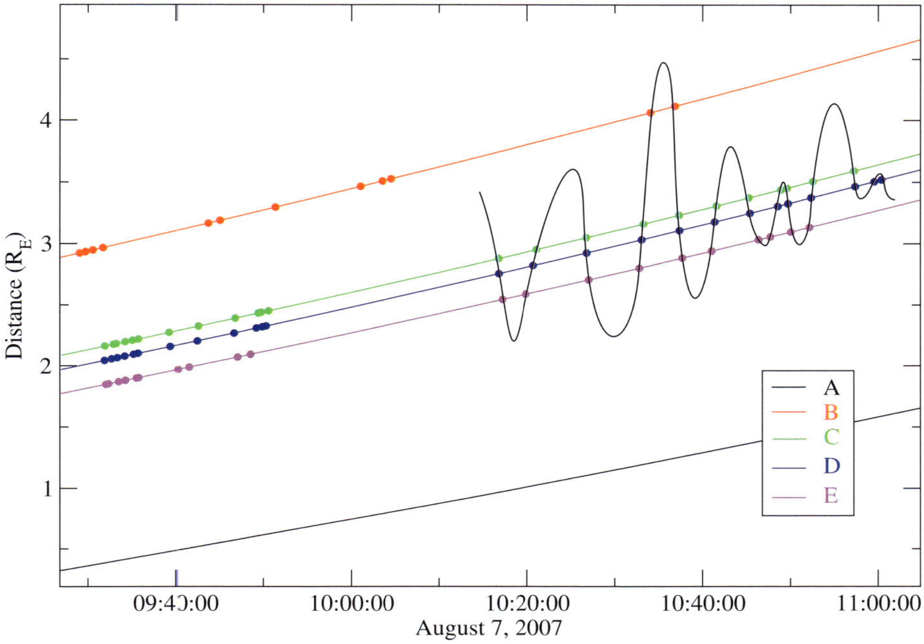

Fig. 26 Position-time diagram of the magnetopause crossings. The y-axis shows the distance along the orbit. For each spacecraft there is a position curve on which the magnetopause crossings are marked

handling the many technical and financial activities of the Lead Investigator group in Braunschweig. The Project Team at UCB has done an outstanding job in running the THEMIS project. Special thanks go to Peter Harvey, Vassilis Angelopolous, and Dave Sibeck. Financial support for the work of the FGM Lead Investigator Team at the Technical University of Braunschweig by the German Ministerium für Wirtschaft und Technologie and the Deutsches Zentrum für Luft- und Raumfahrt under grant 50QP0402 is acknowledged. Financial support of the Austrian Academy is also gratefully acknowledged. THEMIS was made possible by NASA, under contract NAS5-02099.

References

V. Angelopoulos, The Themis Mission, Space Sci. Rev. (2008, this issue). doi:10.1007/s11214-008-9336-1

H.U. Auster, A. Lichopoj, J. Rustenbach, H. Bitterlich, K.H. Fornacon, O. Hillenmaier, R. Krause, H.J. Schenk, V. Auster, Meas. Sci. Technol. **6**, 477–481 (1995). doi:10.1088/0957-0233/6/5/007

H.U. Auster, K.H. Fornacon, E. Georgescu, K.H. Glassmeier, U. Motschmann, Meas. Sci. Technol. **13**, 1124–1131 (2002). doi:10.1088/0957-0233/13/7/321

H.U. Auster, I. Apathy, G. Berghofer, A. Remizov, R. Roll, K.H. Fornacon, K.H. Glassmeier, G. Haerendel, I. Hejja, E. Kührt, W. Magnes, D. Moehlmann, U. Motschmann, I. Richter, H. Rosenbauer, C.T. Russell, J. Rustenbach, K. Sauer, K. Schwingenschuh, I. Szemerey, R. Waesch, Space Sci. Rev. **128**, 221–240 (2007). doi:10.1007/s11214-006-9033-x

A. Balogh, C.M. Carr, M.H. Acuna, M.W. Dunlop, T.J. Beek, P. Brown, K.-H. Fornacon, E. Georgescu, K.H. Glassmeier, J. Harris, G. Musmann, T. Oddy, K. Schwingenschuh, Ann. Geophys. **19**, 1207–1217 (2001)

W. Baumjohann, G. Haerendel, R.A. Treumann, T.M. Bauer, J. Rustenbach, E. Georgescu, U. Auster, K.H. Fornacon, K.H. Glassmeier, H. Lühr, J. Büchner, B. Nikutowski, A. Balogh, S.W.H. Cowley, Adv. Space Res. **24**, 77–80 (1999). doi:10.1016/S0273-1177(99)00428-7

C. Carr, P. Brown, T.L. Zhang, J. Gloag, T. Horbury, E. Lucek, W. Magnes, H. O'Brien, T. Oddy, H.U. Auster, P. Austin, O. Aydogar, A. Balogh, W. Baumjohann, T. Beek, H. Eichelberger, K.H. Fornacon, E. Georgescu, K.H. Glassmeier, M. Ludlam, R. Nakamura, I. Richter, Ann. Geophys. **23**, 2713–2732 (2005)

M.K. Dougherty, S. Kellock, D.J. Southwood, A. Balogh, E.J. Smith, B.T. Tsurutani, B. Gerlach, K.H. Glassmeier, F. Gliem, C.T.G. Erdos, F.M. Neubauer, S.W.H. Cowley, Space Sci. Rev. **114**, 331–383 (2004). doi:10.1007/s11214-004-1432-2

K.H. Fornacon, H.U. Auster, E. Georgescu, K.H. Glassmeier, G. Haerendel, J. Rustenbach, M. Dunlop, Ann. Geophys. **17**, 1521–1527 (1999). doi:10.1007/s00585-999-1521-3

K.H. Glassmeier, U. Motschmann, M. Dunlop, A. Balogh, M.H. Acuna, C. Carr, G. Musmann, K.H. Fornacon, K. Scheda, J. Vogt, E. Georgescu, S.J. Buchert, Ann. Geophys. **19**, 1439–1448 (2001)

K.H. Glassmeier, I. Richter, A. Diedrich, G. Musmann, U. Auster, U. Motschmann, A. Balogh, C. Carr, E. Cupido, A. Coates, M. Rother, K. Schwingenschuh, K. Szegö, B. Tsurutani, Space Sci. Rev. **128**, 649–670 (2007a). doi:10.1007/s11214-006-9114-x

K.H. Glassmeier, H. Boehnhardt, D. Koschny, E. Kührt, I. Richter, Space Sci. Rev. **128**, 1–21 (2007b)

P. Harvey, E. Taylor, R. Sterling, M. Cully, Space Sci. Rev. (2008, this issue)

M. Ludlam, V. Angelopoulos, E. Taylor, R.C. Snare, J.D. Means, Y. Ge, P. Narvaez, H.U. Auster, O. LeContel, D. Larson, T. Moreau, Space Sci. Rev. (2008, this issue)

M. Müller, T. Lederer, K.-H. Fornacon, R. Schäfer, J. Magn. Magn. Math. **177**, 231–232 (1998)

A. Roux, O. Le Contel, P. Robert, C. Coillot, A. Bouabdellah, B. la Porte, D. Alison, S. Ruocco, M.C. Vassal, Space Sci. Rev. (2008, this issue)

L. Zanetti, T. Potemra, R. Erlandson, Space Sci. Rev. **70**, 465–482 (1994). doi:10.1007/BF00756882

T.L. Zhang, W. Baumjohann, M. Delva, H.U. Auster, A. Balogh, C.T. Russell, S. Barabash, M. Balikhin, G. Berghofer, H.K. Biernat, H. Lammer, H.I.M. Lichtenegger, W. Magnes, R. Nakamura, T. Penz, K. Schwingenschuh, Z. Vörös, W. Zambelli, K.H. Fornacon, K.H. Glassmeier, I. Richter, C. Carr, K. Kudela, J.K. Shi, H. Zhao, U. Motschmann, J.-P. Lebreton, Planet. Space Sci. **54**, 1336–1343 (2006). doi:10.1016/j.pss.2006.04.018

The Search Coil Magnetometer for THEMIS

A. Roux · O. Le Contel · C. Coillot · A. Bouabdellah · B. de la Porte · D. Alison · S. Ruocco · M.C. Vassal

Originally published in the journal Space Science Reviews, Volume 141, Nos 1–4, 265–275.
DOI: 10.1007/s11214-008-9455-8 © Springer Science+Business Media B.V. 2008

Abstract THEMIS instruments incorporate a tri-axial Search Coil Magnetometer (SCM) designed to measure the magnetic components of waves associated with substorm breakup and expansion. The three search coil antennas cover the same frequency bandwidth, from 0.1 Hz to 4 kHz, in the ULF/ELF frequency range. They extend, with appropriate Noise Equivalent Magnetic Induction (NEMI) and sufficient overlap, the measurements of the flux-gate magnetometers. The NEMI of the searchcoil antennas and associated pre-amplifiers is smaller than 0.76 $\mathrm{pT}/\sqrt{\mathrm{Hz}}$ at 10 Hz. The analog signals produced by the searchcoils and associated preamplifiers are digitized and processed inside the Digital Field Box (DFB) and the Instrument Data Processing Unit (IDPU), together with data from the Electric Field Instrument (EFI). Searchcoil telemetry includes waveform transmission, FFT processed data, and data from a filter bank. The frequency range covered depends on the available telemetry. The searchcoils and their three axis structures have been precisely calibrated in a calibration facility, and the calibration of the transfer function is checked on board, usually once per orbit. The tri-axial searchcoils implemented on the five THEMIS spacecraft are working nominally.

Keywords Search-coil magnetometer · Magnetospheric mission · Substorm physics

1 Introduction

The primary thrust of the Time History and Macroscale Interaction during Substorms (THEMIS) mission is to establish where and when substorms start, and to determine the nature of the instability involved in this explosive process (Angelopoulos et al. 2008;

A. Roux (✉) · O. Le Contel · C. Coillot · A. Bouabdellah · B. de la Porte · D. Alison · S. Ruocco
Centre d'étude des Environnements Terrestre et Planétaires (CETP), 10-12 avenue de l'Europe, 78140 Vélizy, France
e-mail: alain.roux@cetp.ipsl.fr

M.C. Vassal
3D Plus, 641 rue Hélène Boucher, 78532 Buc, France
url: http://www.3d-plus.com

Harvey et al. 2008). In the magnetospheric plasma, where binary collisions are almost absent, plasma waves are expected to ensure the collisionless dissipation requested by some of the substorm models, and, for guided waves, to allow a remote sensing of the dynamics of active regions. The Electric Field Instrument (EFI, see Bonnell et al. 2008 for more details) and the Search Coil Magnetometer (SCM) instruments on THEMIS are tailored to investigating the possible role played by waves at substorm breakup and during the expansion phase. THEMIS SCM has a long heritage; earlier versions of the instrument have been built for GEOS 1 and 2, Ulysses, Galileo, Interball, and more recently for Cluster, and Cassini. A description of the design for Cluster can be found in Cornilleau-Wehrlin et al. (1997, 2003). Each instrument had specific characteristics (frequency range, NEMI and weight), tailored to the constrainsts of the missions listed above. Section 2 starts with a short description of science objectives and of the requirements that they imposed on THEMIS SCM. In Sect. 3 we describe the design of THEMIS SCM antennas and preamplifiers. Section 4 provides a description of the tests and calibrations applied to the various models (one qualification, five flight models, and one spare), and compare the performances of THEMIS SCM's to the specifications. Data from SCM are digitized and processed in the Digital Fields Box (DFB) and in the Instrument Data Processing Unit (IDPU); the corresponding modes are described in the IDPU paper (Taylor et al. 2008; Cully et al. 2008). Here we simply give a short description (in Table 6) of the main characteristics of various operation modes.

2 Measurement Requirements

2.1 Science Objectives

There are basically two types of models for substorm breakup (see for instance Lui 2001 and references therein). For the first type of model, magnetic reconnection (MR) occurs first and triggers substorms. In the second type of model, labelled "Current Disruption" (CD), the breakup is triggered by a reduction in the cross-tail current, associated with the development of an instability. In both cases ULF and ELF waves are believed to play a critical role. In the MR models whistler mode waves are expected to accelerate electrons up to large (super-Alfvénic) velocities (Mandt et al. 1994). Very thin current sheets can also be destabilized by HF tearings in the whistler mode (Bulanov et al. 1992). In CD models the cross tail current is disrupted by HF cross field instability (Lui et al. 1992), or undergo LF (ballooning modes) instabilities (Roux et al. 1991) which are coupled to higher frequency waves: ion cyclotron, lower hybrid drift and/or whistler modes. Thus wave observations provide a critical test to substorm scenarios. THEMIS SCM, and EFI (which measure electric fields in the same frequency range) are designed to identify waves associated with the breakup and to investigate their role in substorm dynamics. Furthermore, when the waves are guided, they can be used to help remotely track the active region where breakup starts. SCM is also needed to assess the nature of the waves; are they electrostatic (such as lower hybrid waves), or electromagnetic (such as whistler mode waves)?

2.2 Requirements

2.2.1 Frequency Range and NEMI

The science objectives briefly sketched above, and described in more details in a companion paper by Le Contel et al. (2008), have been used to specify the characteristics of THEMIS

SCM. We summarize the main instrument requirements below. Among the various types of wave modes listed above, whistler mode waves have the highest frequency cut-off: f_{ce} (the electron gyrofrequency), in the regions of interest. This cut-off has been used to fix the maximum frequency to be covered by the SCM. CD is expected to develop in the near-earth plasma sheet, typically around 8 R_{E} (Lui et al. 1992). At 8 R_{E}, $B \simeq 50$ nT is a typical value for the DC field, then $f_{ce} \simeq 1.4$ kHz. At the geostationary orbit (6.7 R_{E}), assuming a dipole field, one gets $f_{ce} \simeq 4$ kHz, while during a substorm growth phase, where the average equatorial magnetic field is reduced by the effect of the increasing cross tail current, one gets $f_{ce} \simeq 2.8$ kHz. At 5 R_{E}, the dipole field gives $f_{ce} \simeq 9$ kHz, which is about twice the maximum frequency covered by the SCM instrument. Since CD is not expected to occur inside of 5–6 R_{E}, and whistler mode waves are damped and/or unguided above $f_{ce}/2$, an upper cutoff of the SCM frequency bandwith at 4 kHz has been imposed. Given that the amplitude of waves measured in the plasma sheet by Cluster is typically on the order of 10–100 pT/$\sqrt{\mathrm{Hz}}$ at 10 Hz, the requirement for SCM NEMI is to be better than 1 pT/$\sqrt{\mathrm{Hz}}$ at 10 Hz.

2.2.2 *Other Requirements*

Given the constrainsts on spacecraft mass, the allocated mass budget was 800 g (600 g for the sensors, including mounting hardware, and 200 g for the preamplifiers), while the power allocation was fixed to be smaller than 100 mW, for each set of SCM preamplifiers (Harvey et al. 2008). In order to determine the polarization of the waves, the direction of the magnetic axis of the antennas has to be known. Thus the direction of each antenna axis has to be known to better than 1 degree. In Sect. 3 we show that these requirements are met by the five tri-axis SCMs deployed on THEMIS.

3 Description of the Instrument

3.1 Modeling

3.1.1 *Magnetic Amplification*

The instrument is based upon the combination of a high magnetic permeability material and a large number of turns which passively detect voltage induced by the changing external field (an AC-current measurement). The high magnetic permeability core amplifies the external magnetic field (Ripka 2001; Osborn 1945; Bozorth and Chapin 1942; Coillot et al. 2007). This core is located inside two types of windings. The main winding is the sensing element; it has a very large number of turns; 51 600 in our case, while the secondary winding is used to introduce a flux feedback in order to flatten the frequency response by removing the resonance associated with the main winding. The flux feedback stabilizes the phase response of the instrument, independent of temperature variations. Figure 1 shows a picture of a search coil before potting. The magnetic amplification, or apparent permeability μ_{app}, characterizes the ratio between the magnetic field at a given position in the core B_{core}, and the external magnetic field B_{ext}. In practice one must define a mean apparent permeability that describes the average field seen inside a winding of length L. The main winding has N turns wound around a core with a relative permeability μ_r. In these conditions, Lenz's law

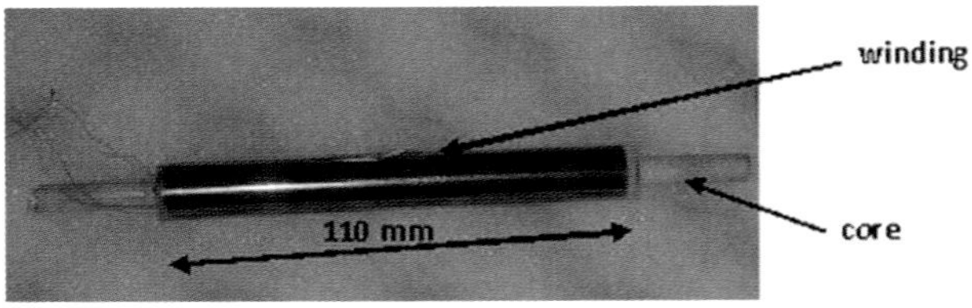

Fig. 1 Picture of a THEMIS search coil before potting

$e = -Nd/dt(\phi)$, where ϕ is the magnetic flux across one loop, gives the modulus of the induced voltage as

$$e = NS\left(\frac{1}{L}\int_0^L \mu_{app}(l)dl\right)B_{ext}\omega = NS < \mu_{app} > B_{ext}\omega. \tag{1}$$

S is the core section, assuming a sinusoidal magnetic perturbation ($B_{ext} \propto \exp(j\omega t)$) and the brackets denote an averaging along the winding. When a magnetic field is applied to ferromagnetic materials, they become magnetized. The intensity of this magnetization depends on the relative permeability of the magnetic material, and the shape of the sample via its demagnetizing coefficient N_z. The latter depends on the ratio m between the length and the diameter of the magnetic core. In the case of a cylinder we utilize the expression:

$$\mu_{app}(m) = \frac{B_{core}}{B_{ext}} = \frac{\mu_r}{[1 + (\mu_r - 1)Nz(m)]}. \tag{2}$$

For the magnetic material used for THEMIS, the relative permeability is very high, thus $(\mu_r - 1)Nz(m) \gg 1$ and the previous expression becomes: $\mu_{app}(m) = 1/Nz(m)$. Then, in order to increase the apparent permeability, one must decrease the coefficient of the demagnetizing field, increasing the ratio m. This can be done either by increasing the length L or decreasing the diameter d. In the first case, the size and mass of the sensor is increased, and in the second the intensity of the induced voltage is reduced, since the section of the core is decreased. The design of the THEMIS search coil results from a compromise between these two constrainsts; the dimensions of the magnetic core are: $L = 170$ mm and $d = 7$ mm, see Fig. 1.

3.1.2 Electrical Modeling

The frequency behaviour can be represented by an RLC circuit excited by a voltage source corresponding to the induced voltage collected by the main winding, as shown in Fig. 2. The transmittance (ratio between the output voltage V and the measured magnetic field B) of such a sensor can be easily put in the following form :

$$T(j\omega) = \frac{V}{B} = \frac{-j\omega NS\mu_{app}}{(1 - LC\omega^2) + jRC\omega}. \tag{3}$$

The sensor has a main resonance which limits its measurement bandwidth and reduces the dynamics. This is because a signal measured at a frequency close to the resonant frequency leads to a high transmittance value. To remove this effect, the output of the high gain preamplifier (from the primary winding) is used to generate a feedback flux inside the core via the secondary winding as shown in Fig. 3. In other words the feedback current flowing through the secondary winding (via a feedback resistor) generates a magnetic field along the search-coil axis proportional to the external magnetic field but in the opposite direction. Therefore

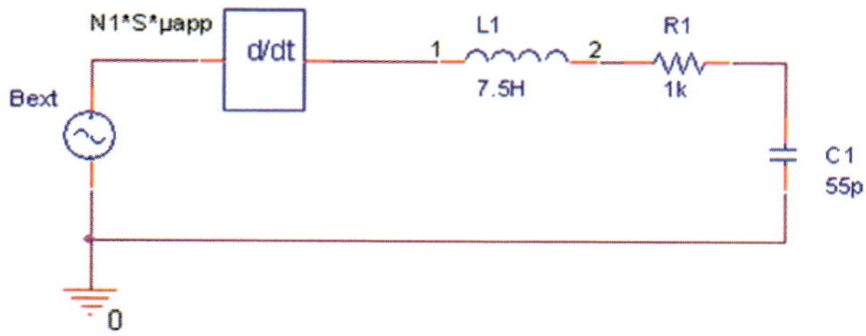

Fig. 2 Electrokinetic representation of a search coil sensor

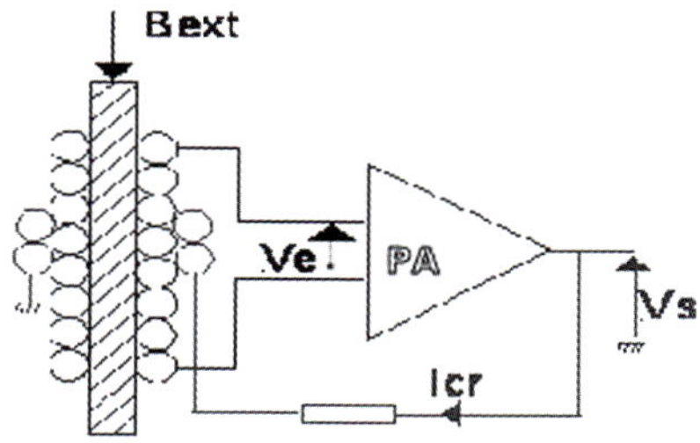

Fig. 3 Principle of fluxmeter. The output (from the primary winding) of the high gain preamplifier (PA) is used to inject a current to the secondary winding (via a feedback resistor) in order to generate a feedback flux inside the core. This feedback loop allows to flatten and stabilize the frequency response in amplitude and phase

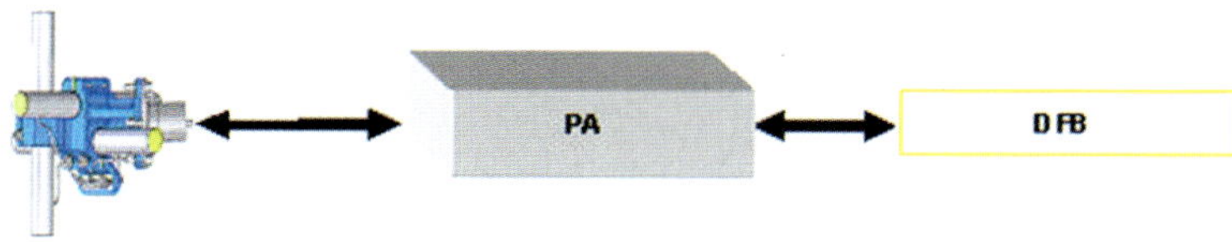

Fig. 4 Block diagram: the three analog signals are amplified by the preamplifier (PA) and then digitized by the digital field box (DFB)

this feedback flux, which is proportional to the flux generated by the external magnetic field, allows to flatten and stabilize the frequency response of the antenna. The value of the feedback resistor controls the strength of the feedback. The smaller the value of the resistor, the stronger the feedback is.

3.2 Design of THEMIS SCM

3.2.1 SCM Antennas

The Search Coil Magnetometer consists of three sensors and a preamplifier box connected to the DFB (Digital Field Board) as shown in Fig. 4. The sensors are mounted in a tri-axial configuration to measure the x, y, and z components of the magnetic field in the frequency range of 0.1 Hz–4 kHz. They are mounted on the tip of a one meter length rigid boom as shown in Fig. 5 (stowed position). Such a boom length allows for a reduction of the level of the electromagnetic noise coming from the spacecraft and detected by the antennas (Ludlam et al. 2008). The deployed configuration is displayed in a companion paper (Le Contel et al. 2008) and the axis definitions are described. The antenna structure is designed to allow a precise alignment of the sensors with respect to the spacecraft axis. Two sensors lie in the spin plane and the third one is parallel to the spin axis. The 3 SCM antennas are held orthogonally on a mechanical structure mounted on a one-probe diameter boom; they are thermally isolated and covered by a thermal blanket of Multi-Layer Insulation (MLI).

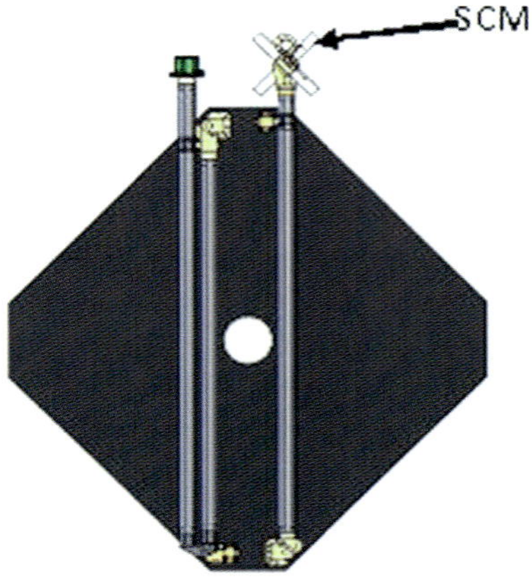

Fig. 5 SCM position on the probe (drawing from UCB)

The identical sensor (18 cm) and its kin (27 cm) have been previously flown by CETP on more than 7 earth-orbiting and interplanetary missions; most-recently it was flown as part of the Cluster/STAFF experiment. The mass of the SCM antennas and mounting structure is 568 g, compliant with THEMIS specifications (less than 600 g). Electrostatic shielding is implemented around the search coils to minimize their sensitivity to electric fields.

3.2.2 Preamplifier Design

Three identical low-power preamplifiers (one for each sensor signal) are mounted in an electrical unit, fixed on the IDPU box (Taylor et al. 2008). They have been developed using a new technology: multichip module vertical (MCM-V). The MCM-V technology allows a significant reduction in the mass and the volume of the preamplifiers. It is also well adapted to withstand a severe radiation environment. This technique consists of dividing the circuit in smaller functions. Each function is implemented using bare chips on thin flexible printed circuit boards (PCB), which are piled up to form a compact module (essentially a cube, see Fig. 6). Layers containing tantalum plates are inserted between the PCBs to protect the sensitive components against radiation. These plates are placed only on the top and bottom of these sensitive components. This spot shielding has the advantage of being lighter than more traditional techniques. These preamplifiers have low-noise input stages. Their dynamic range is about 100 dB, which allows weak signals to be measured in the presence of the large voltage signals induced by the rotation of the spacecraft in the DC magnetic field. They benefit from a specific power supply which is also realized in MCM-V technology and corresponds to a fourth cube. A specific SCM calibration signal is generated inside the preamplifier box. The use of this calibration is discussed in the next section. The mass and power consumption of SCM preamplifier are 200 g and 75 mW, respectively, which is compliant with specifications.

4 Calibrations and Tests

The experiment transfer function and NEMI are given in Figs. 7 and 8; they were measured on the ground in a calibration facility, at Chambon la Forêt, France. The NEMI of each flight model is shown in Tables 1 to 5. The largest NEMI is 0.76 pT/$\sqrt{\text{Hz}}$ at 10 Hz which complies

Fig. 6 Preamplifier box. The 3 upper cubes correspond to the 3 antennas preamplifiers, the fourth one is for power regulation. They are manufactured in 3D technology. The electronics for the in-flight calibration is also shown on the PCB. The preamplifier box size is 95 mm × 81 mm (109 with the mounting "ears") × 30 mm. The mass is 200 g

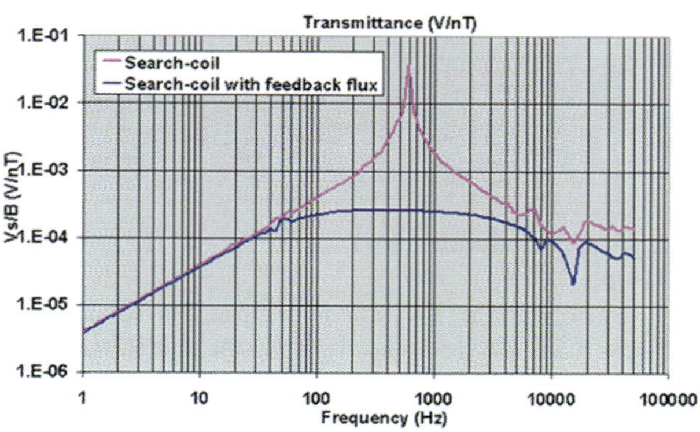

Fig. 7 Frequency response (transfer function) of the SCM with feedback (*blue*) and without (*pink*)

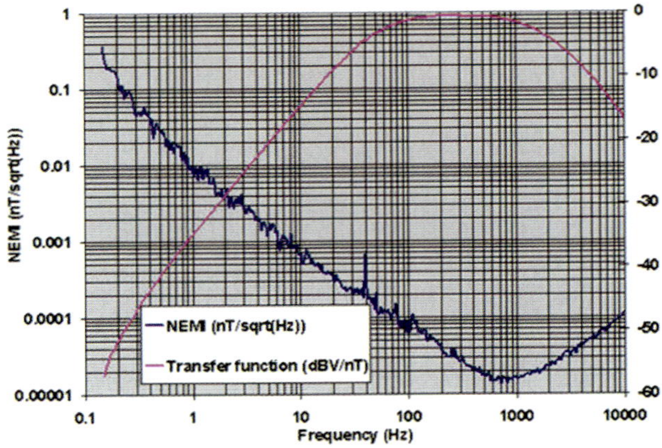

Fig. 8 Transfer function (*pink*) in dB V/nT and NEMI (in *blue*) in nT/$\sqrt{\text{Hz}}$ of THEMIS search coil measured in the calibration facility at Chambon la Forêt

Table 1 NEMI of FM1

FM1	10 Hz	100 Hz	1 kHz
X	0.69	0.070	0.022
Y	0.65	0.072	0.021
Z	0.64	0.077	0.021

Table 2 NEMI of FM2

FM2	10 Hz	100 Hz	1 kHz
X	0.64	0.0699	0.020
Y	0.65	0.0698	0.020
Z	0.645	0.076	0.022

Table 3 NEMI of FM3

FM3	10 Hz	100 Hz	1 kHz
X	0.61	0.070	0.0196
Y	0.66	0.067	0.016
Z	0.74	0.066	0.019

Table 4 NEMI of FM4

FM4	10 Hz	100 Hz	1 kHz
X	0.71	0.079	0.017
Y	0.76	0.08	0.019
Z	0.74	0.077	0.019

with the specified value (1 pT/$\sqrt{\text{Hz}}$). At 100 Hz the largest NEMI is 0.08 pT/$\sqrt{\text{Hz}}$, and at 1 kHz it is 0.022 pT/$\sqrt{\text{Hz}}$.

It can be seen that they are quite close. The transfer functions are also almost identical, with less than 1 dB difference between the models. In order to determine the polarization of the waves on each satellite and to accurately compare data from the five satellites, careful measurements of the angle between each magnetic axis and its corresponding mechanical

Table 5 NEMI of FM5

FM5	10 Hz	100 Hz	1 kHz
X	0.66	0.065	0.016
Y	0.70	0.074	0.016
Z	0.72	0.069	0.016

Table 6 Telemetry modes for SCM

APID	Processing	Characteristics and comments
440	Filter bank (fbk)	Mean value of the signal in 6 frequency bands kHz: [4–2]; [1–0.5]; Hz: [256–128]; [64–32]; [16–8]; [4–2] Sampling rate: 1/16 to 8 S/s There are only 2 inputs to be shared within 3 SCM and 12 EFI components
444	Fast survey (scf)	Waveform data for the 3 SCM components Sampling rate within 2 to 256 S/s, nominal value is 8
448	Particle burst (scp)	as 444 but nominal value is 128 S/s
44C	Wave burst (scw)	as 444 but nominal value is 8192 S/s
44D	Particle burst spectra (ffp)	Compressed FFT with 16, 32 or 64 frequency lines, nominal value is 32 Sampling rate from 1/4 to 8 Spectra/s, nominal value is 1 There are only 4 inputs to be shared within 20 possible signals including the 3 SCM components
44E	Wave burst spectra (ffw)	as 44D but nominal values are 64 frequency lines and 8 Spectra/s

axis have been made. These angles may be just a few degrees, but need to be known with precision. The estimated error in the knowledge of the magnetic axis is 0.5 degrees which is less than the 1 degree in specifications. For example the angle of sensor X of the flight model 3 (FM3) is 0.2 degrees. For each sensor the angle difference between mechanical and magnetic axes can be neglected. Thus the magnetic field can be accurately transformed into any required reference frame. A free running onboard oscillator generates a 9 Hz triangular wave used to check the transfer function inflight. A digital command is activated by the DFB (Cully et al. 2008) to connect this signal to the feedback winding of the sensors. This signal is detected by the primary winding, and seen in B_x, B_y, and B_z data. The default duration is 30 seconds with a maximum of 60 seconds. This signal contains many frequencies throughout the bandwidth, which are used for in flight calibration.

Further signal processing takes place inside the IDPU, on a single board, together with the EFI instrument. The analog signals output by SCM are filtered, sampled, and converted to digital in the DFB and the IDPU. Taking into account that SCM measurements are digitized with 16 bits and that it must comply with the NEMI requirements, the SCM measurement is likely to saturate for an AC magnetic field about 1000 nT. The spin of the spacecraft will cause the DC magnetic field to saturate the SCM at the spin frequency, at altitudes less than approximately 4 R_{E}. The different modes of data transmission are listed in Table 6 and described in the IDPU and DFB papers (Taylor et al. 2008; Cully et al. 2008). SCM data is transmitted in the following modes: Fast Survey (Application Identifier or APID number

444) at sampling rates between 2 Hz and 256 Hz, Particle Burst (APID 448) at 2–256 Hz, Wave Burst (APID 44C) at 512 Hz to 8 kHz, Spectra (APID 44D and 4E), and Filter Banks (APID 440) where the power in six frequency bands is calculated: 2–4 kHz, 0.5–1 kHz, 128–256 Hz, 32–64 Hz, 8–16 Hz, and 2–4 Hz (see Table 6 for more details).

5 Summary and Conclusions

The SCM instruments implemented on the five THEMIS probes measure the 3D magnetic field fluctuations in the frequency bandwidth from 0.1 Hz to 4 kHz. The weight and power consumption comply with THEMIS requirements. They extend with appropriate NEMI and sufficient overlap, the measurements of the FGM beyond the 1 Hz range. We have described the design of THEMIS SCM. Tests and calibrations carried out in a calibration facility, at Chambon la Forêt, demonstrate that the five sets of triaxial search coils cover the same frequency range, have the same frequency response, and the same NEMI. SCM instruments fulfill all mission requirements. The NEMI at 10 Hz is smaller than 0.76 $\mathrm{pT}/\sqrt{\mathrm{Hz}}$ for each instrument while the NEMI at 1 kHz is about 20 $\mathrm{fT}/\sqrt{\mathrm{Hz}}$. Onboard calibration is usually performed once per orbit in order to check the stability of the transfer function. The magnetic axes are known to better than 1 degree; as requested. The 0.1 Hz to 4 kHz frequency range is covered by EFI and SCM, in the same manner. SCM and EFI data are filtered, processed and transmitted to the ground via THEMIS DFB/IDPU. The tri-axes SCMs, on each of the five THEMIS spacecraft, are working nominally.

Acknowledgements The French involvment on THEMIS is supported by CNES and CNRS. Vacuum tests and integrations of the SCM were made at UCB. We are pleased to acknowledge the friendly collaboration and the help of our colleagues at UCB, in particular V. Angelopoulos, P. Harvey, R. Jackson, M. Ludlam, D. Meilhan, H. Richard, and E. Taylor. Work in the US was supported by NASA contract NAS5-02099.

References

V. Angelopoulos et al., The THEMIS mission. Space Sci. Rev. (2008, this issue). doi:10.1007/s11214-008-9336-1

J.W. Bonnell, F.S. Mozer, G.T. Delory, A.J. Hull, R.E. Ergun, C.M. Cully, V. Angelopoulos, The electric field instrument (EFI) for THEMIS. Space Sci. Rev. (2008, this issue)

R.M. Bozorth, D.M. Chapin, Demagnetizing factors of rods. J. Appl. Phys. **13**, 320–326 (1942)

S.V. Bulanov, F. Pegoraro, A.S. Sakharov, Magnetic reconnection in electron dynamics. Phys. Fluids B **4**, 2499–2508 (1992)

C. Coillot, J. Moutoussamy, P. Leroy, G. Chanteur, A. Roux, Improvements on the design of search coil magnetometer for space experiments. Sens. Lett. **5**, 1–4 (2007)

N. Cornilleau-Wehrlin, P. Chauveau, S. Louis, A. Meyer, J. Nappa, S. Perraut, L. Rezeau, P. Robert, A. Roux, C. de Villedary, Y. de Conchy, L. Friel, C.C. Harvey, D. Hubert, C. Lacombe, R. Manning, F. Wouters, F. Lefeuvre, M. Parrot, J.L. Pinçon, B. Poirier, W. Kofman, P. Louarn, The cluster Spatio-temporal Analysis of Field Fluctuations (STAFF) experiment. Space Sci. Rev. **79**, 107–136 (1997)

N. Cornilleau-Wehrlin, G. Chanteur, S. Perraut, L. Rezeau, P. Robert, A. Roux, C. de Villedary, P. Canu, M. Maksimovic, Y. de Conchy, D. Hubert, C. Lacombe, F. Lefeuvre, M. Parrot, J.-L. Pinçon, P.M.E. Décréau, C.C. Harvey, P. Louarn, O. Santolik, H.St. Alleyne, M. Roth, T. Chust, O. Le Contel, STAFF team, First results obtained by the Cluster STAFF experiment. Ann. Geophys. **21**, 437–456 (2003)

C.M. Cully, R.E. Ergun, K. Stevens, A. Nammari, J. Westfall, The THEMIS digital fields board. Space Sci. Rev. (2008, this issue). doi:10.1007/s11214-008-9417-1

P.R. Harvey et al., The THEMIS observatory. Space Sci. Rev. (2008, this issue)

O. Le Contel, A. Roux, P. Robert, C. Coillot, A. Bouabdellah, B. de la Porte, D. Alison, S. Ruocco, V. Angelopoulos, K. Bromund, C.C. Chaston, C. Cully, First results of THEMIS Search Coil Magnetometers (SCM). Space Sci. Rev. (2008, this issue). doi:10.1007/s11214-008-9371-y

M. Ludlam et al., The THEMIS magnetic cleanliness program. Space Sci. Rev. (2008, this issue). doi:10.1007/s11214-008-9423-3
A.T.Y. Lui, Current controversies in magnetospheric physics. Rev. Geophys. **39**(4), 535–563 (2001)
A.T.Y. Lui, R.E. Lopez, B.J. Anderson, K. Takahashi, L.J. Zanetti, R.W. McEntire, T.A. Potemra, D.M. Klumpar, E.M. Greene, R. Strangeway, Current disruption in the near-Earth neutral sheet region. J. Geophys. Res. **97**, 1461 (1992)
M.E. Mandt, R.E. Denton, J.F. Drake, Transition to whistler mediated magnetic reconnection. Geophys. Res. Lett. **21**(1), 73–77 (1994)
J.A. Osborn, Demagnetizing factors of the general ellipsoid. Phys. Rev. **67**(11–12), 351–357 (1945)
P. Ripka, Magnetic sensors and magnetometers. Artech house (2001)
A. Roux, S. Perraut, P. Robert, A. Morane, A. Pedersen, A. Korth, G. Kremser, B. Aparicio, D. Rodgers, R. Pellinen, Plasma sheet instability related to the westward traveling surge. J. Geophys. Res. **96**, 17,697 (1991)
E. Taylor et al., The THEMIS instrument data processing unit. Space Sci. Rev. (2008, this issue)

The THEMIS ESA Plasma Instrument and In-flight Calibration

J.P. McFadden · C.W. Carlson · D. Larson · M. Ludlam · R. Abiad · B. Elliott · P. Turin · M. Marckwordt · V. Angelopoulos

Originally published in the journal Space Science Reviews, Volume 141, Nos 1–4, 277–302.
DOI: 10.1007/s11214-008-9440-2

Abstract The THEMIS plasma instrument is designed to measure the ion and electron distribution functions over the energy range from a few eV up to 30 keV for electrons and 25 keV for ions. The instrument consists of a pair of "top hat" electrostatic analyzers with common $180° \times 6°$ fields-of-view that sweep out 4π steradians each 3 s spin period. Particles are detected by microchannel plate detectors and binned into six distributions whose energy, angle, and time resolution depend upon instrument mode. On-board moments are calculated, and processing includes corrections for spacecraft potential. This paper focuses on the ground and in-flight calibrations of the 10 sensors on five spacecraft. Cross-calibrations were facilitated by having all the plasma measurements available with the same resolution and format, along with spacecraft potential and magnetic field measurements in the same data set. Lessons learned from this effort should be useful for future multi-satellite missions.

Keywords THEMIS · Space plasma instrument · Calibrations · Electrostatic analyzer · In-flight calibrations

PACS 94.80.+g · 06.20.fb · 94.30.C- · 94.05.-a · 07.87.+v

1 Introduction

The THEMIS mission was designed to study fundamental processes concerning the nature of magnetospheric substorms, the explosive release of solar wind energy stored in the Earth's magnetotail (Sibeck and Angelopoulos 2008). To address the substorm problem, the THEMIS team built five identical spacecraft which were placed in highly-elliptical near-equatorial orbits with apogees of $\sim$11.8 Re for the three inner probes, and apogees of $\sim$19.6

J.P. McFadden (✉) · C.W. Carlson · D. Larson · M. Ludlam · R. Abiad · B. Elliott · P. Turin · M. Marckwordt · V. Angelopoulos
University of California, Berkeley, USA
e-mail: mcfadden@ssl.berkeley.edu

V. Angelopoulos
University of California, Los Angeles, USA

and ~31.6 Re for the outer probes (Angelopoulos 2008). Orbital periods for the probes are 1, 2 and 4 days allowing magnetotail alignment conjunctions once every 4 days as required for substorm timing analysis.

THEMIS was launched on February 17, 2007, into an initial insertion orbit with an apogee of 14.7 Re near ~ 21 MLT and shifting to the dayside with the Earth's orbital motion. This orbit required a 7 month coast phase, where the spacecraft were kept in a close configuration to keep orbital parameters optimized for final orbit insertion in the fall of 2007. This close proximity allowed accurate cross-calibration of the plasma instruments as described in Sect. 2 in preparation for the substorm campaign in early 2008. In addition, "First results" papers in this issue are primarily concerned with dayside science investigations. First results from the plasma sensors, along with other performance issues, can be found in the companion paper, McFadden et al. (2008).

Each THEMIS spacecraft (Harvey et al. 2008) contains a fluxgate magnetometer (Auster et al. 2008), a search-coil magnetometer (Roux et al. 2008), electric field instrument (Bonnell et al. 2008), solid state telescopes (Larson et al. 2008) and the ESA (Electro-Static Analyzer) plasma instrument described below. These core instruments provide a set of measurements needed to resolve the in-situ dynamics of substorms. The plasma instrument provides detailed ion and electron particle distribution function measurements along with on-board moment calculations. The overall mechanical and electrical design of the THEMIS ESA plasma instrument was directly derived from the FAST Plasma Instrument (Carlson et al. 2001). Below we present a description of the instrument technical design and data products, followed by an in depth discussion of the calibrations. Lessons learned from this calibration effort should be useful for future multi-satellite missions.

1.1 Sensor Description

The THEMIS plasma instrument consists of a pair of top-hat electrostatic analyzers (ESAs) (Carlson et al. 1983) that measure ion and electron energy per charge (E/q). Figure 1 shows a picture and cross-section of the sensors which are packaged together to provide a common field-of-view (FOV). The electron and ion analyzers have $\Delta R/R$ of 0.060 and 0.075, respectively, with inherent energy resolutions of about 15% and 19%. The sensors have

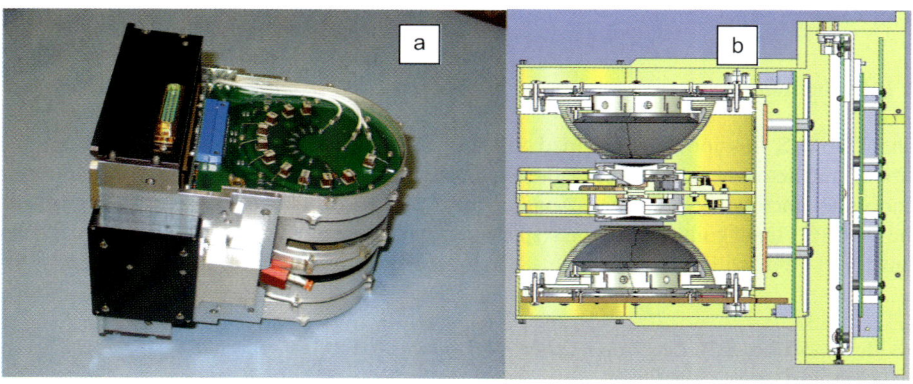

Fig. 1 The THEMIS ion and electron Electrostatic Analyzers (ESAs) are packaged together to provide a common field of view. (**a**) ESA with the anode cover removed reveals coupling capacitors. The LVPS (*black*) is mounted on the side and the nitrogen purge inlet (*red*) is exposed. (**b**) Analyzer cutaway shows the common aperture mechanism and electronics packaging

selectable energy sweeps (programmable starting energy with logarithmic or linear sweep steps) and are generally swept in energy (logarithmically) from ~32 keV for electrons and ~25 keV for ions, down to ~6–7 eV. Nominal operations have 32 sweeps per spin, with 31 energy samples per sweep, plus one sample energy retrace, resulting in a typical measurement resolution with $\Delta E/E \sim 32\%$. Particle events are registered by microchannel plate (MCP) detectors.

Both sensors have a 180° × 6° FOV (FWHM), with the 6° rotating with the spacecraft to provide 4π steradian coverage each spin. 32 sweeps per spin provides 11.25° resolution in rotation phase (ϕ). The electron sensor has 8 anodes giving 22.5° resolution in the polar angle (θ), while the ion sensor has 16 anodes with up to 5.625° resolution. The high angular resolution anodes in the ion sensor are concentrated at the spin plane to resolve solar wind ions. The ion sensor can also be operated in a double-sweep mode (64 sweeps/spin, 16 energies) to provide a similar 5.625° angular resolution in rotation phase.

Table 1 summarizes the electron and ion instrument specifications. Geometric factors are for 180° FOV sensors and include energy acceptance such that differential flux, $(\mathrm{s\,cm^2\,sr\,eV})^{-1}$, is given by count rate divided by geometric factor at the given energy "E". A typical count binning pixel has 1/8 of this geometric factor (22.5°/180°), with two sweeps (1/16 spin) contributing to a pixel, or 6 ms accumulation time per energy sample. The in-flight calibrations indicate the measured geometric factors are roughly equal to the predicted geometric factors. Predicted geometric factors are from simulations, combined with grid losses and assumed MCP efficiency. The discrepancies are likely due to a combination of internal scattering and leakage fields near the analyzer exit. Leakage fields through an exit grid and into the analyzer, caused by a bias voltage applied to the front of the MCP detectors, can increase the sensor's response at low energies as discussed in Sect. 2.2.

Figure 2 illustrates the ESA plasma instrument's modular design which allows subassemblies to be constructed and tested separately. Figure 2a shows the MCP (red) detectors and mounting hardware that attach to the anode. Spring fingers on both the inner and outer edges of the MCP annulus distribute the force providing uniform clamping that can withstand ~85 Gs of acceleration. This subassembly allows MCP testing and characterization prior to sensor assembly. The tab on the lower right corner of the anode is the interface to the purge tube that supplies filtered dry nitrogen during storage. Figure 2b reveals an analyzer design that maintains its concentricity to ~15 μm under normal assembly. Although the hemisphere alignment is controlled by three interfaces, the flexible insulator that supports the inner hemisphere is self-centering producing an alignment equivalent to the tolerance at the outer mounting plate interface.

The outer hemisphere was serrated and all internal surfaces were blackened with ebanol-C to reduce scattered sunlight from reaching the detectors. (Ebanol-C is found to reduce scattered sunlight by at least a factor of 10 over gold-black, which was used on the FAST satellite.) Figure 2c shows the combined anode, analyzer, outer aperture, and the top-hat. The top-hat is supported by a torsion spring and contains a conductive gasket that seals against the outer hemisphere during launch to prevent contamination. Figure 2d shows the release plate mechanism that pushes both ion and electron sensor top-hats into closed positions against their outer hemispheres. The reset-able, SMA-actuator release mechanism was developed for the THEMIS program to simplify sensor refurbishment during ground testing, replacing a melt-wire design used for the FAST mission. The release mechanism also acts as a poppet valve during nitrogen purge and rocket ascent, preventing over-pressure by venting gas at the top-hat.

The modular design also extends to the ESA electronics as illustrated in Fig. 3. The instrument simplifies assembly by eliminating most of the harnessing by coordinating the

Table 1 ESA instrument specifications

Parameter	Value	Comments
Electron electrostatic analyzer		Spherical top-hat
$\Delta R/R$	0.06	37.5 mm inner hemisphere radius
Analyzer constant	7.9	0 to 4 kV sweep supply
Energy range	2 eV to 32 keV	
Analyzer energy resolution	17% (15%)	$\Delta E/E$ measured (predicted)
Measurement energy resolution	32%	Typical 31 step log energy sweep
Energy sweep rate	32 per spin	~94 ms for nominal 3 s spin
Instantaneous field of view	180° × 6° FWHM	180° centered on spin plane
Anode angle resolution	22.5°	8 anodes
Field of view each spin	4π	steradians
Analyzer geometric factor	0.0075 cm^2 sr E	Predicted from simulations, analyzer only
	0.0047 cm^2 sr E	One 90% exit grid and 70% MCP efficiency
Sensor geometric factor	0.0066 cm^2 sr E	From in-flight calibration, includes grid losses, MCP efficiency @10 eV
Counter readout	1024 per spin	~3 ms
Full distribution array	32E×88Ω	32 or 128 spin cadence, continuous
Burst distribution array	32E×88Ω	1 spin cadence, 30–60 min/orbit
Reduced distribution array	32E×6Ω or	1 spin cadence, continuous, 6Ω in FS mode
	32E×1Ω	1Ω in SS mode
Ion electrostatic analyzer		Spherical top-hat
$\Delta R/R$	0.075	37.5 mm inner hemisphere radius
Analyzer constant	6.2	0 to 4 kV sweep supply
Energy range	1.6 eV to 25 keV	
Analyzer energy resolution	18% (19%)	$\Delta E/E$ measured (predicted)
Measurement energy resolution	32%	Typical 31 step log energy sweep
Energy sweep rate	32 or 64 per spin	64/spin in solar wind mode with 16 energies
Instantaneous field of view	180° × 6° FWHM	180° centered on spin plane
Anode angle resolution	5.6 ° to 22.5°	16 anodes
Field of view each spin	4π	steradians
Analyzer geometric factor	0.0181 cm^2 sr E	Predicted from simulations, analyzer only
	0.0073 cm^2 sr E	Two 90% grids and 50% MCP efficiency
Sensor geometric factor	0.0061 cm^2 sr E	From in-flight calibration, includes grid losses and MCP efficiency @500 eV
Counter readout	1024 per spin	~3 ms
Full distribution array	32E ×88Ω	32 or 128 spin cadence, continuous
Burst distribution array	32E ×88Ω	1 spin cadence, 30–60 min/orbit
Reduced distribution array	24E×50Ω or	1 spin cadence, continuous, 50Ω in FS mode
	32E×1Ω	1Ω in SS mode

mechanical and electrical designs. Figures 3a and 3b show the preamplifier board which contains 24 Amptek A121 preamplifiers for both sensors. This board uses a single FPGA to implement counters and a command decoder that controls its test pulser and the preamplifier

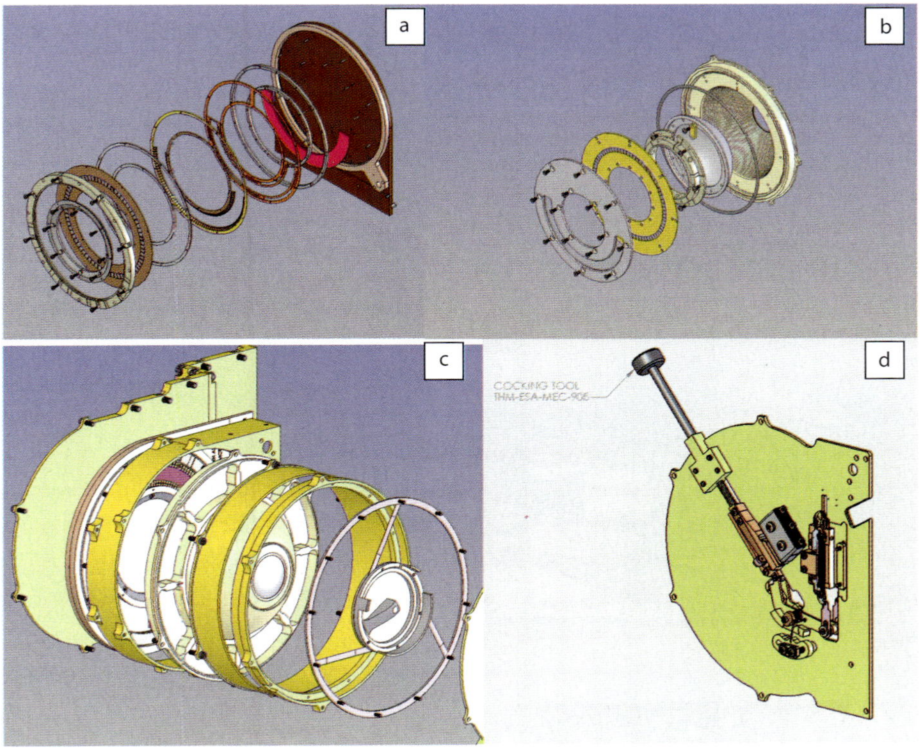

Fig. 2 (**a**) THEMIS detector-anode module, (**b**) analyzer module, (**c**) sensor subassembly, and (**d**) the aperture release plate mechanism with SMA actuator. The modular mechanical design allowed subsystem assembly and testing to proceed in parallel for the 10 ESA sensors

gains. Separate high voltage (HV) boards are used for electron and ion sensors, each board (Fig. 3c) containing a raw sweep supply and MCP supply for a single sensor. The HV boards mount to a common mechanical frame (Fig. 3d) that also supports the Interface-Sweep (IS) board (Fig. 3e) and a small mother board (Fig. 3d) that provides an electrical interface between the IS and HV boards. The IS board contains opto-coupler (Amptek HV601) circuits for analyzer HV sweeps, in addition to providing FPGA control voltages for the sweeps and the HV boards. The combined electron–ion instrument requires two +28 V power lines, one for low voltage (LV) and one for HV. Figure 3f shows the LV power supply which mounts to the side of the ESA instrument.

The ESA interfaces to the Instrument Data Processing Unit or IDPU (Taylor et al. 2008) which provides power, control, and data interfaces. Figure 4 shows a block diagram of the ESA. MCP detectors in chevron configuration are voltage biased at $\sim$2 kV to amplify input events to $\sim 2 \times 10^6$ e^-, approximately -320 fC. Amptek A121 preamplifiers are used to detect output charge pulses and have programmable gain to facilitate testing. Events are recorded by counters which are read out 1024 times per spin. The ESA electronics include a programmable test pulse generator to provide electronic stimulation when high voltage is off. The rate of this stimulation can be slaved to the analyzer sweep control to confirm internal timing. For nominal operation the preamplifier thresholds are set at -40 fC ($\sim$250,000 electrons), well above electronic noise in the system. For MCP gain testing, the

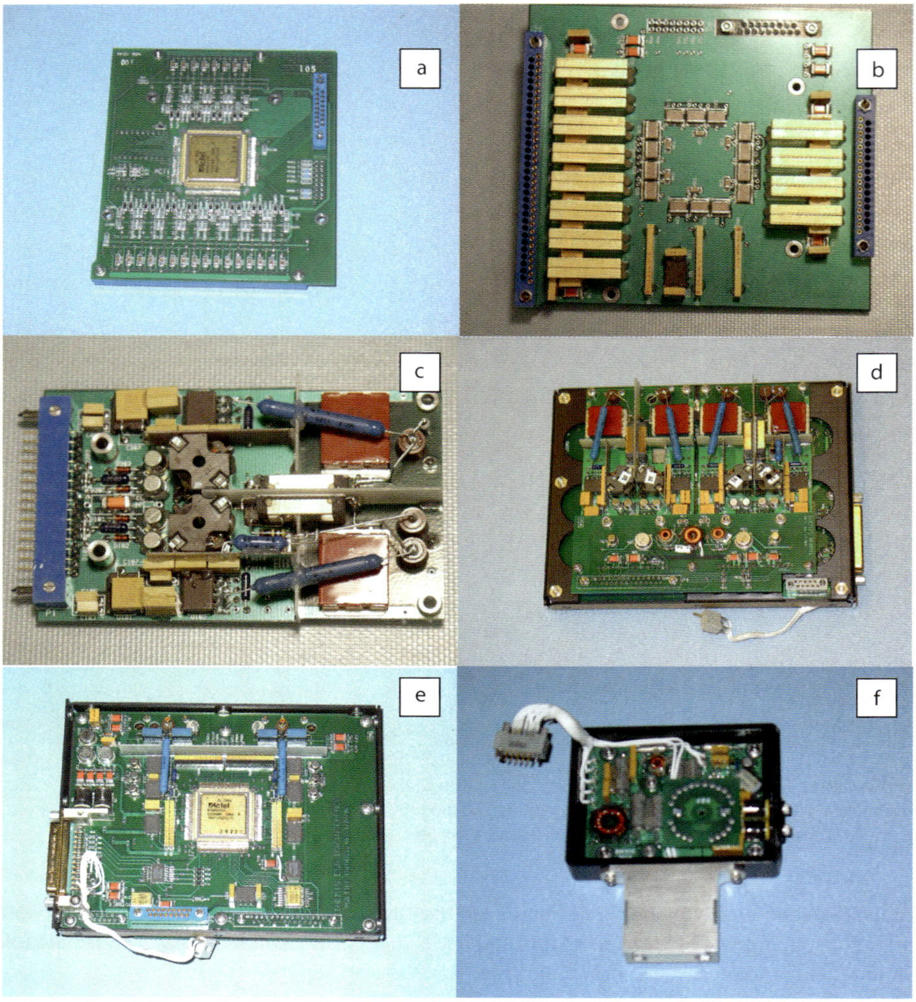

Fig. 3 The THEMIS ESA electronics shown here includes the preamplifier board, with front (**a**) and back (**b**) views, the paired sweep and MCP high voltage power supplies shown individually (**c**) or mounted to the back of the interface and sweep control board (**d**) along with the ESA mother board, the front of the interface and sweep control board (**e**), and the low voltage power supply (**f**). The modular mechanical-electrical design minimized harnessing, allowing quick assembly and disassembly of the instruments

preamplifier gain is toggled, switching the threshold from −40 fC to −330 fC several times. A factor of two change in count rate indicates the peak in the MCP's pulse height distribution is about −330 fC, the desired 2×10^6 gain.

As mentioned above, the ESA contains four separate high voltage power supplies (HVPS), two for MCPs and two for the ion and electron energy sweeps. All HVPS are separately controlled by the IDPU to allow independent operation of both sensors. The sweep supplies produce a 5 kV maximum output (−5 kV for ions), that is used as a raw input to an opto-coupler circuit that provides voltage to the inner hemispheres. The hemispheres are swept from high voltage to low voltage in ∼100 ms, with ∼1 ms required for the high volt-

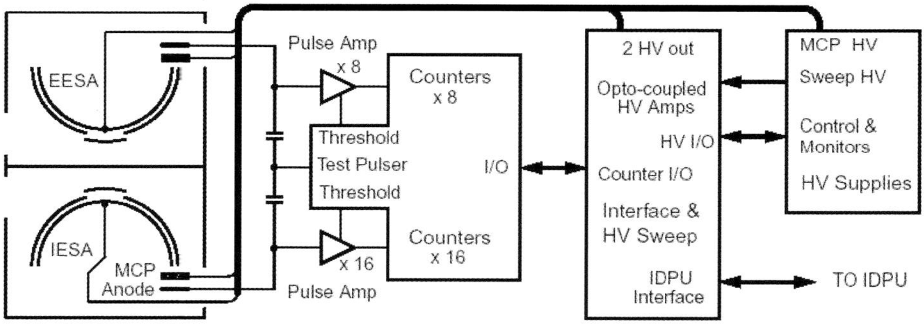

Fig. 4 THEMIS ESA block diagram

age to retrace to its starting value. The energy sweep is synchronized to the 1024 pulse/spin input clock that also controls counter readout.

1.2 ESA Modes and Data Products

ESA data are collected and formatted by the ETC board in the IDPU into seven data products: two full packets, two burst packets, two reduced packets, and moment packets. Each sensor has separate packets, except for moment packets which are made by combining data from both ESAs and the Solid State Telescopes (SSTs) (Larson et al. 2008). The format of each data product is programmable and may change with time depending on the "Spacecraft-Mode", the location within the magnetosphere, or with the development of new "instrument-modes" that can be uploaded to the satellites. Below we describe data products generated by the nominal "magnetospheric mode" used for the majority of the first nine months of the mission, and the "solar wind" mode which can be used to resolve the cold solar wind ions.

"Full packets" are a low-time-resolution data product useful for large scale surveys of THEMIS data. They generally maintain the "full" 32 energies sampled, and have an 88 solid-angle map as illustrated in Fig. 5a. Full distribution data are 1-spin snapshots of the plasma with a measurement cadence of either 32 spins (in "Fast-Survey Spacecraft-Mode") or 128 spins ("Slow-Survey Spacecraft-Mode"). These high energy-angle resolution measurements are the primary data product used for the in-flight calibration effort described below. Full packets are used to generate summary plots, to validate the on-board plasma moment computations, and to provide detailed distribution functions in all regions of the magnetosphere. Full distribution data are particularly helpful for the identification of unique features, such as counter-streaming field-aligned beams, that may not be easily identified from other data products such as moments. Different angle maps can be selected for full packets in different region of space. Figure 5b illustrates the angle map for a "solar wind" spacecraft mode which provides $\sim 5.6^\circ$ resolution for solar wind beams.

"Burst packets" contain high-resolution 3-D plasma distribution functions with spin-period time resolution. Due to telemetry limitations, burst packets are generally limited to several five-minute-intervals each orbit. Burst packet format is usually the same as full packet format, with 32 energies and 88 solid angles. The selected time intervals are chosen by ground command or by on-board triggers as discussed in Angelopoulos (2008). Burst data provide the high resolution measurements needed to resolve boundary crossings such as the magnetotail neutral sheet, plasma sheet boundary layer, magnetopause, and bow shock.

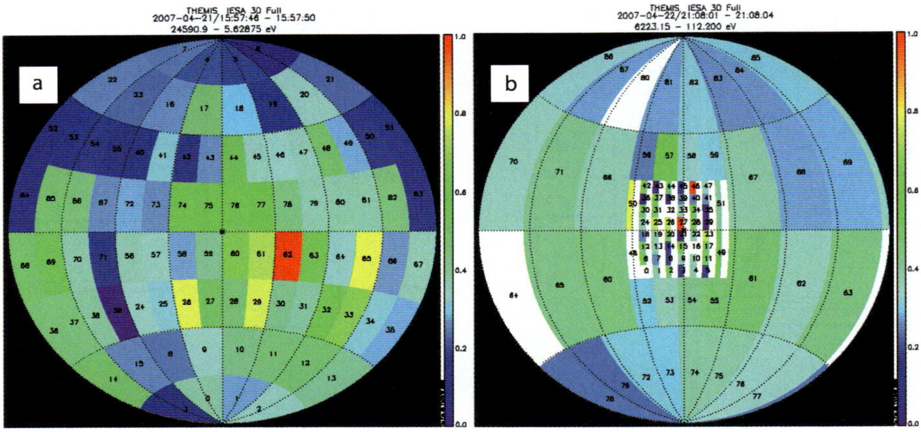

Fig. 5 (**a**) Typical 88 solid-angle map used for collecting both ion and electron ESA data into full and burst data packets in the magnetosphere. (**b**) Ion angle map used for ion solar wind mode packets, where only bins 0–87 are labeled. Bins 88–175 are shifted 5.6 degrees to the right of bins 0–87 and have bin numbers given by the adjacent bin plus 88. Angle maps are programmable allowing different anodes and sweeps to be combined to produce solid angle resolution that support a particular science goal

"Reduced packets" are 1-spin-resolution plasma distributions sampled continuously, but with limited solid-angle and/or energy coverage. When in Slow-Survey mode, reduced packets are generally composed of 32-energy, omni-directional spectra which allow energy-time spectrograms with the same cadence as on-board moment data. In Fast-Survey mode, the nominal ion reduced packet consists of a 24-energy, 50-solid-angle distribution, while the electron reduced packet is a 32-energy, 6-solid-angle distribution (see Fig. 6). These Fast-Survey mode formats were chosen to maintain enough angular information so that plasma moments could be computed and so features of the distribution, such as field-aligned beams, could be deduced. When combined with on-board spin-resolution moment data, reduced data allows high-time resolution science investigations to be conducted on data gathered throughout the orbit.

"Moment packets" contain spin-resolution on-board computations of the ion and electron ESA and SST (> 30 keV) moments. These moments include the plasma density, three components of flux, six components of the momentum tensor, and 3 components of energy flux, computed separately for the four instruments (iESA, eESA, iSST, eSST). These partial moments can be combined on the ground to get the total moments, or combined to calculate related quantities such as velocity and pressure. Combining ESA and SST data can be especially important for determining the total pressure in the plasma sheet. THEMIS on-board moment calculations are unique in that THEMIS is the first mission to include corrections for spacecraft charging. Spacecraft potential, as measured by the EFI instrument (Bonnell et al. 2008), is used to correctly shift the energies of particles in the moment computations. In particular this correction eliminates photo-electrons which often contaminate electron plasma measurements. Moment computations also include weighting factors to correct for energy and angle efficiency variations in the sensors.

1.3 Ground Testing and Calibrations

In addition to standard functional tests, ground testing of the ESA sensors included several subsystem optimizations prior to assembly and calibration. Before loading into flight

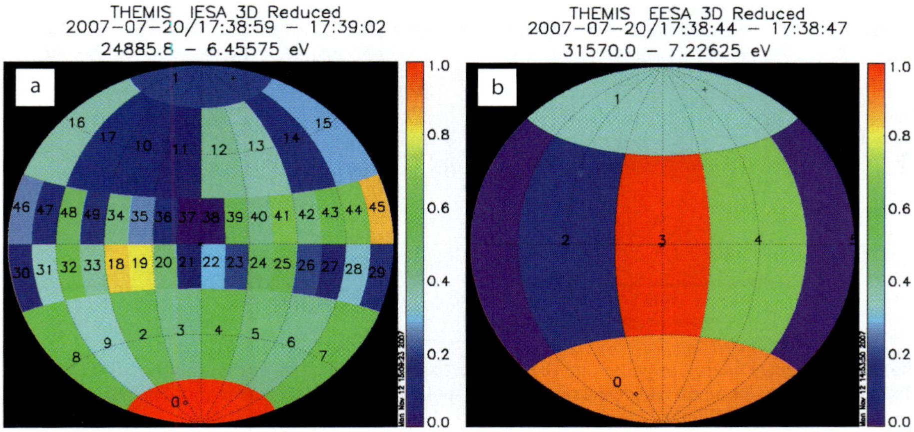

Fig. 6 Typical ion (*left*) and electron (*right*) solid angle maps showing angle bin number for reduced data packets during fast survey magnetospheric mode. Color coding is in normalized counts. Reduced data are produced at spin resolution and maintain 32 energies for electrons and 24 energies for ions. The 50 solid angles for the ions are adequate for accurate velocity moments except during narrow beams along the spin axis. The 6 solid angles for electrons are adequate to identify anisotropies such as counter-streaming electrons

boards, A121 preamplifiers were individually tested for response including threshold, output pulse length and dead time to assure nearly identical characteristics for all amplifiers. Opto-coupler high voltage sweep electronics were tuned to provide near zero offset at the lowest DAC control setting (actual voltage offsets were < 10 mV on a 4 kV output), to provide nearly identical ($\sim 1\%$) high voltage control gain, and to have rapid settling ($\sim$1–2 ms) with no overshoot during high voltage retrace. MCP detectors were matched for bias current and selected for low background rates and gain uniformity. After sensor assembly, calibrations were performed in high vacuum ($< 10^{-6}$ Torr) and included background noise sensitivity, MCP background rates, MCP pulse height distribution tests, energy-angle characterization, analyzer concentricity tests, relative sensitivity characterization, and sweep mode testing.

MCP tests within the analyzer assembly showed background rates 10 to 100 times higher then those observed during subassembly testing with MCPs exposed to the chamber. After several subsystem and system level checks, it was recognized that the higher background resulted from higher pressure at the MCPs which were buried deep in a newly assembled instrument. Internal instrument out gassing, and its associated higher pressure, was the source of higher background rates. By allowing more time in high vacuum prior to testing, and by scrubbing the MCPs prior to the beginning of calibration testing, background rates were reduced to relatively low levels that did not impact calibrations. Similar high background rates were observed on some sensors during thermal vacuum testing at the spacecraft level at UCB, with background seen to correlate with chamber pressure and to decrease with time. Spacecraft level Integration and Test (I&T) at JPL, with higher vacuum during the HV tests, showed expected background rates of $\sim 1/\mathrm{s\,cm}^2$. In-flight data have similarly low background rates.

Figure 7 shows the energy-angle calibration for the electron sensor on THC. The inner hemisphere voltage is kept constant while the beam energy and "out-of-plane" alpha angle are varied. The average of this response over alpha angle provides a characteristic energy curve and determines the analyzer energy constant (average energy divided by hemisphere voltage). When the response is averaged over energy, the test provides a measure of the out-

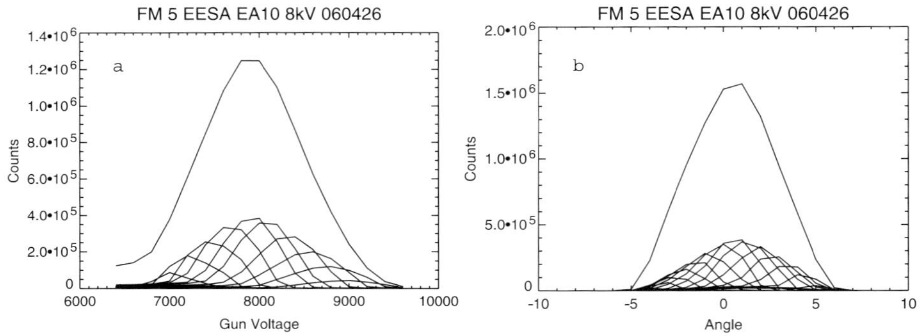

Fig. 7 (**a**) Average energy response and (**b**) average "out of plane" alpha angle response of the THEMIS electron ESA. The *lower curves* are energy (angle) response at a single angle (energy). The analyzer energy constant is determined from the *curve on the left*

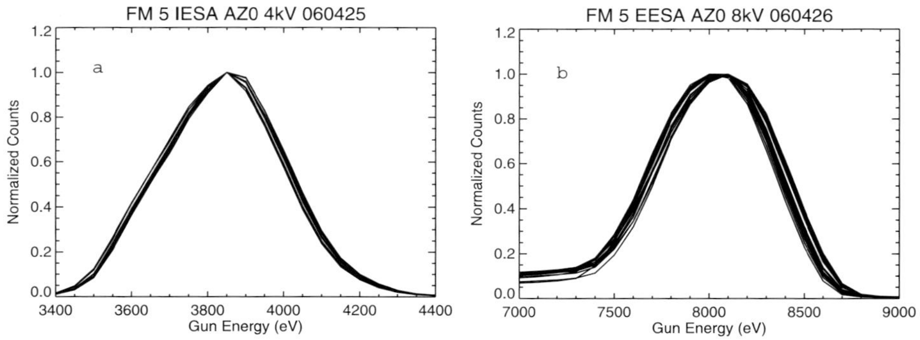

Fig. 8 Energy response curve at 0° alpha angle for the beam at 15° increments around the 180° field of view for ions (**a**) and electrons (**b**). Variations between curves indicate about 1% variation in the analyzer energy constant with look direction

of-plane angular acceptance of the instrument. These tests were compared with simulations to confirm proper analyzer operation. Energy-angle tests are performed at three different rotation angles and the instrument's energy calibration is determined from the average energy constant. No significant differences in the energy constant with look direction were found.

One of the most important tests performed was for concentricity of the analyzer's hemispheres. If the hemispheres are not concentric, the energy of measured particles will be a function of its 180° FOV, which will complicate data analysis and in-flight calibrations. Although the energy-angle test described above can often identify hemisphere misalignment, a faster test uses a single alpha angle for the beam and sweeps the beam energy at a dozen different look directions. Figure 8 shows examples of analyzer concentricity tests for the electron and ion sensors on THC. All THEMIS ESAs were found to have good concentricity with about 1% or less variation in energy with look direction. This accuracy corresponds to a misalignment of hemispheres of $\sim$15 μm.

Figure 9 shows results from relative sensitivity tests performed on the THEMIS ESAs. For this test the sensor and the beam are optimized for beam throughput and the analyzer is rotated about the symmetry axis. For a parallel beam, the response should be roughly

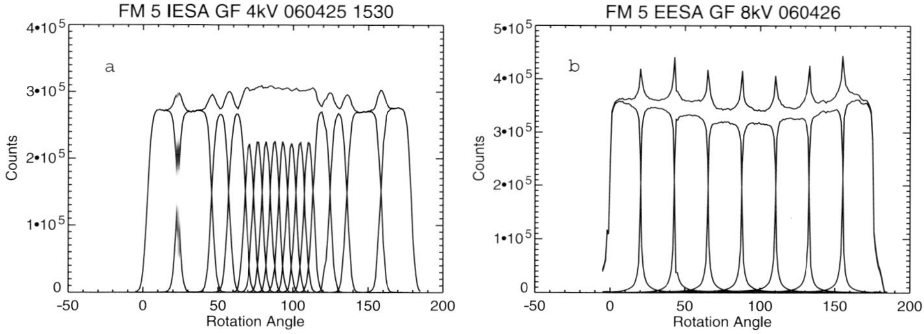

Fig. 9 Azimuthal response of the ion (**a**) and electron (**b**) sensors on THC to a parallel beam at 0° alpha angle reveal the 16 anode and 8 anode patterns for the two sensors. A small amount of particle double counting can be seen as enhanced response at the borders between anodes

flat, with deviations revealing any large asymmetry in the assembly. The $\sim 10\%$ variations in response with look direction are normal and result from a combination of MCP detector bias angle (Gao et al. 1984) and double counting of events at anode boundaries. This test was used early in the THEMIS calibrations to identify a problem with an aperture opening mechanism. A $\sim 40\%$ variation in sensitivity with rotation was observed on two sensors that simultaneously showed little variation in concentricity. This indicated the top-hat was not seating properly. Careful examination with backlighting revealed the seating problem which was barely visible to the eye. A clearance problem was identified and the problem was fixed on all sensors. Although these tests provide a preliminary estimate of the uniformity of response, relative sensitivity is more accurately quantified during in-flight calibrations discussed in Sect. 2.4.

Additional ground calibrations were performed to assure proper operation of the sensor in flight configuration. These included tests of nominal sweeping modes with a beam source and full spacecraft level testing of flight data packaging. Absolute sensitivity of the sensors and relative efficiency of the detectors with energy were not calibrated due to lack a beam monitor. However, the electron and ion count rates were very similar for different instruments with the same beam settings and same instrument geometry providing confidence that beam fluxes were stable over time and that the different sensors were nearly identical. Since an instrument's absolute efficiency will change on orbit as its detectors age, in-flight calibration procedures are required to determine absolute and relative calibrations. These tests are discussed in Sects. 2.2, 2.5 and 2.6.

2 In-flight ESA Calibrations

The THEMIS and Cluster missions are the only multi-spacecraft missions in which four or more satellites have been kept in close proximity, allowing detailed cross-calibration efforts between instruments. Unlike the Cluster mission where the electron and ion plasma measurements originated from separate groups and where data products differed in both time and angular resolution, THEMIS offers the advantage of having all the plasma measurements available with the same resolution and format. Furthermore, since all the THEMIS data are distributed as a single data set, spacecraft potential (Bonnell et al. 2008) and magnetic field (Auster et al. 2008) measurements are instantly available in raw or processed form for use in the cross-calibration efforts. This advantage has allowed a detailed cross-calibration effort to be performed in a rather short period of a few months early in the mission. Lessons learned from this effort should be useful for future multi-satellite missions.

The in-flight calibration of the THEMIS electron and ion plasma instruments required inputs from several other instruments including spacecraft potential measurements supplied by EFI, magnetic field measurements supplied by FGM, and spacecraft attitude and timing information supplied by the Mission Operations Center at Berkeley. These inputs were crucial for proper interpretation of the measurements and for confirmation that the sensor operation was nominal. In addition, ESA data were used for in-flight calibration of the EFI and SST instruments as described by Bonnell et al. (2008) and Larson et al. (2008), respectively.

The following sections describe the methodology used for the ESA in-flight calibrations, along with the basic rational for each analysis. Part of this process was the identification of data collection times where known properties of the plasma, such as charge neutrality and gyrotropy, could be used to identify small variations in the sensor response. In addition, the absolute sensitivity calibrations required the use of solar wind data from the Wind spacecraft. During this calibration effort, unexpected variations in the ratio Ni/Ne were discovered for several orbits. This led to an investigation of sensor efficiency as a function of energy and the discovery of an unexpected small variation in analyzer sensitivity due to leakage fields through the analyzer's exit grids. Lastly, the THEMIS ESA calibration effort is not a one-time process. Maintaining accurate calibrations will require continuous monitoring of the detector gain with regular bias voltage adjustments, along with repeated iterations of the techniques described below.

2.1 Spacecraft Potential Corrections

In order to cross-calibrate plasma sensors using density as a measure of sensitivity, one must include corrections for spacecraft charging (Pederson et al. 1998). This is especially important for the electron density calculation where inclusion of spacecraft photoelectrons can result in large errors to the density. The electric field experiment (Bonnell et al. 2008) provides a proxy for the spacecraft-to-plasma potential by measuring the potential of Langmuir sensors relative to the spacecraft. With a proper bias current to the Langmuir sensors (roughly 25% to 50% of the sensor photoemission current) these sensors should float within about a volt of the "local" plasma potential. There are several caveats here. First, the Langmuir sensor and spacecraft must be in sunlight to obtain good current balance between photoemission and plasma collection. Second, by "local" potential, we mean local plasma potential adjacent to the Langmuir sensor. This potential differs from the plasma potential we wish to use as a reference for the spacecraft since the spacecraft, the antenna, and their photoelectrons all perturb the "local" plasma environment. To account for this difference, we introduce a potential "scale factor" that corrects the measured "local" potential to actual plasma potential at large distances from the spacecraft. This "scale factor" is a function of spacecraft potential (or plasma and photoelectron distributions), complicating the calculation. Third, the difference potential between the Langmuir sensor and "local" plasma can vary from near zero in high density plasmas to about two volts in low density plasmas. This potential "offset" also depends on bias current applied to the Langmuir sensor. So, like the "scale factor", the "offset" potential varies with spacecraft potential (or with plasma and photoelectron distributions). Therefore, calculating the spacecraft potential to be applied to a particle distribution is non-trivial since the measured Langmuir-sensor-to-spacecraft potential must be corrected for changing parameters, such as Langmuir sensor bias current and the plasma distribution.

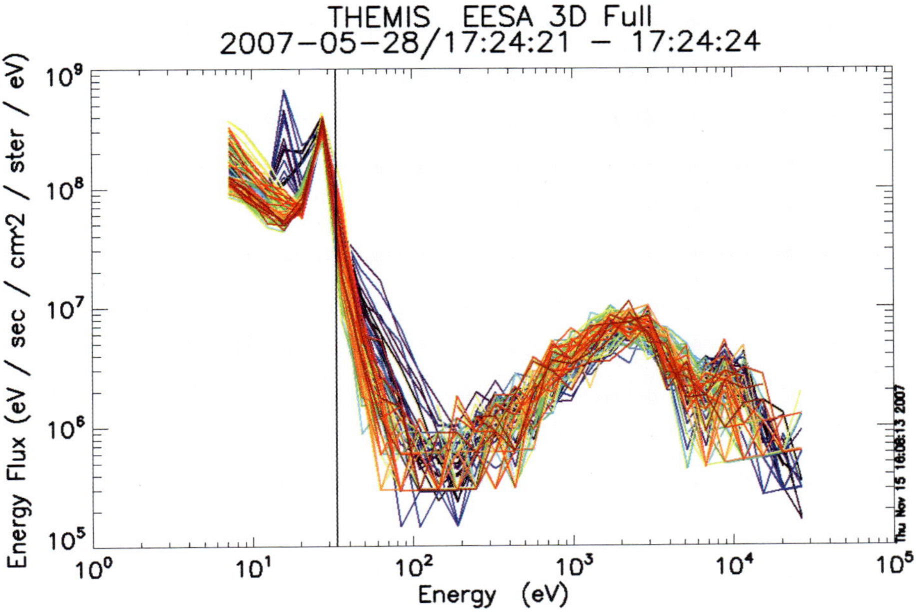

Fig. 10 Electron spectra from THC in a low density ($n_e \sim 0.2$) magnetospheric plasma. Each of the 88 solid angle bins are plotted separately. The line indicates the inferred spacecraft potential using (1). The peaks at 15 eV and 28 eV are due photoelectrons emitted from the axial and radial Langmuir sensors, respectively. Spacecraft photoelectrons dominate most solid angle bins below 21 eV. Field-aligned electrons are resolved between 40 eV and 150 eV (*blue-black*), and isotropic hot plasma is observed at higher energies

For THEMIS, the spacecraft potential, Φ_{sc}, is estimated with the following equation.

$$\Phi_{sc} = -A(\Phi_{sensor} + \Phi_{offset}) \tag{1}$$

where A is a near unity "scale factor", Φ_{sensor} is the average radial Langmuir-sensor-to-spacecraft potential, and Φ_{offset} is the potential "offset" discussed above. Axial Langmuir sensors are not used in the estimate. Current software sets Φ_{sensor} to the spin-averaged potential of two or four radial Langmuir sensors, and treats A and Φ_{offset} as constants over an interval with default values of $A = 1.15$ and $\Phi_{offset} = 1.0$ V, respectively. These values were determined empirically from spacecraft THC early in the calibration effort by comparisons of calculated electron and ion densities as a function of A and Φ_{offset}. This choice of parameters generally prevents spacecraft photoelectrons from being included in the electron density calculations. Both numbers are consistent with those estimated from previous missions (F. Mozer private communication) and from THEMIS EFI modeling (C. Cully private communication).

The need for a "scaling factor" is illustrated in Fig. 10. The two electron peaks at 15 eV and 28 eV are the result of photoelectrons emitted from the axial and radial Langmuir sensors, respectively. The vertical line indicates the spacecraft potential determined from (1) using the default values for the scale factor and offset. If no scaling factor were needed, the proximity of the spacecraft would not matter and both peaks would be at the same value. However, at the ~2 m distance of the axial Langmuir sensors, the local plasma potential is about half the spacecraft potential relative to the distant plasma. Since the axial Langmuir sensor comes to equilibrium with the local plasma, its photoelectrons only gain $\sim 0.5e\Phi_{sc}$ before reaching the plasma sensor. In contrast, the plasma near the radial Langmuir sensors (~ 20 and ~ 24 m distance) is at a potential much closer to the distant plasma potential resulting in a spectral peak that is much closer to the actual plasma potential.

In plasma regimes where the bulk of the electrons and ions have energies larger than $e\Phi_{sc}$, plasma moment calculations are not be very sensitive to small errors in the estimated Φ_{sc} as long as Φ_{sc} is large enough to eliminate spacecraft photoelectrons. However, within the magnetosphere there is often a cold electron component with a sizeable density. This is especially true within the plasmasphere or within plasmaspheric plumes where cold plasma dominates. These cold electrons appear at energies $\sim e\Phi_{sc}$ and are difficult to separate from spacecraft or Langmuir sensor photoelectrons. In addition, these cold electrons reduce the Debye length, resulting in "scale factors" close to one and Langmuir sensors that float very close to plasma potential. For THEMIS data, which generally have energy bins such that $\Delta E/E \sim 0.32$, it may be impossible to separate these cold plasma electrons from the Langmuir sensor photoelectrons. Data with these problems were avoided during the in-flight calibration effort.

THA and THB spacecraft did not have Langmuir sensors deployed during the first eight months of the mission and therefore made no in situ potential measurements. To perform cross-calibrations of plasma sensors between spacecraft, it was essential to correct for spacecraft charging. Therefore we developed an empirical method to estimate the spacecraft potential on THA and THB based on the potential of the near by THD. It was not sufficient to assume the same potential as THD since the bias currents applied to the Langmuir sensors have a significant affect on spacecraft potential. To develop an empirical relationship between spacecraft, we used data from the EFI bias sweeps (Bonnell et al. 2008) to estimate the change of spacecraft potential as a pair of Langmuir sensors were cycled on/off. A perpendicular Langmuir sensor pair was then used to measure the change in spacecraft potential. The measurements showed a linear relationship between spacecraft potential with and without bias current applied.

$$\Phi_{\text{no-bias}} = 0.7\Phi_{\text{bias}} + 0.5, \tag{2}$$

where Φ_{bias} and $\Phi_{\text{no-bias}}$ are difference potentials between the spacecraft and a perpendicular Langmuir sensor pair in volts, with bias current to a radial Langmuir sensor pair cycled on and off, respectively. In addition, the axial-booms were found to produce a ~ 0.3 V shift in spacecraft potential. Not enough samples were obtained to determine any dependence of this shift on Φ_{bias}. Combining the above measured differences with (1), we estimated the spacecraft-to-plasma potential on THA and THB as

$$\Phi_{\text{THA}} = \Phi_{\text{THB}} = 0.49\Phi_{\text{THD}} + 1.22, \tag{3}$$

where Φ_{THD} is determined from (1).

The above estimated potentials were used for some of intervals in the cross calibration effort described in Sect. 2.5. However, the EFI operating voltages on the usher and guard surfaces (see Bonnell et al. 2008 for the antenna geometry) were changed from -8 V to $+4$ V relative to the Langmuir sensor surface on June 22, 2007. This reduced spacecraft charging and changed the 0.49 scale factor determined above. To recover an equivalent formula to (3), we performed empirical comparisons between electron spectra and calculated densities, varying the scaling factor to get the best agreement. These comparisons resulted in (4) which should be valid between June 23 and September 10, 2007.

$$\Phi_{\text{THA}} = \Phi_{\text{THB}} = 0.8\Phi_{\text{THD}}, \tag{4}$$

For most of the in-flight calibration effort discussed below, we selected periods where errors in our estimated spacecraft potential would have minimal effect on the calibration result. During the course of this effort, we observed additional aspects of the dependence of spacecraft potential on the plasma that were not included in this in-flight calibration but are worth noting for future data analysis efforts. First, the Langmuir sensor-to-spacecraft potentials on different "identical" spacecraft in the same environment were not identical and could differ by $\sim 5\%$ at potentials of $\sim$ 6–8 V. Smaller differences were observed at lower potentials. Although this error is not large, it could cause $\sim 5\%$ difference in estimated electron plasma density in the solar wind. Second, the ratio of potentials on different spacecraft as the environment changed was not a constant, indicating this was not just a simple difference in a constant "scale factor". These results indicate that for very precise estimates of density, (1) will require a different "scale factor" and "offset potential" for each spacecraft and that these parameters likely have a weak functional dependence on potential. For missions such as MMS, where resolving small differences between spacecraft is essential, in-flight calibration efforts must be planned to quantify these small differences.

2.2 Energy-Dependent Efficiency Corrections

Microchannel plate (MCP) detectors are known to vary in efficiency with the incident particle energy (Goruganthu and Wilson 1984; Straub et al. 1999) and with the incident particle angle relative to the microchannel bias angle (Gao et al. 1984). For THEMIS ESAs, the bias angle of the plates was oriented so that average variations of incident angle around the MCPs was minimized, and any bias angle efficiency variations are included in "relative efficiency calibration with look direction" discussed later. For energy dependent efficiencies, we initially adopted values published in the literature for electron and ions as shown in Fig. 11a and 11b.

However, early in the calibration effort it was discovered that the calculated ratio of electron and ion density appeared to depend upon ion energy. In particular the presence of low energy (< 100 eV) ions seemed to increase the Ni/Ne ratio. Not knowing the source of this energy-dependent efficiency change, an empirical approach was adopted, testing various energy-dependent changes to the ion efficiency to determine the approximate variation required to improve the Ni/Ne ratio. These comparisons indicated the ion sensor was significantly (> 40%) more efficient at low energies (<100 eV) than at higher energies (>500 eV). Since ions are pre-accelerated to $\sim$2 keV before striking the MCP, it is unlikely that small variations in initial energy could cause large changes in detector efficiency. Instead we realized that the efficiency change must stem from leakage fields into the electrostatic analyzer

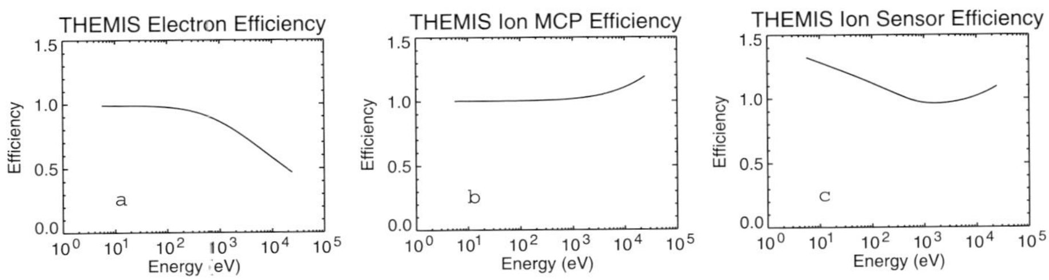

Fig. 11 (**a**) Electron MCP energy efficiency adapted from (Goruganthu and Wilson 1984), (**b**) ion MCP energy efficiency adapted from Funsten (private communication), and (**c**) ion sensor energy efficiency (analyzer+MCP) which includes leakage fields at the exit grids that increase analyzer geometric factor over the ideal analyzer. See text for discussion

from the -2 kV bias voltage on the front of the MCPs. Analyzer simulations had ignored this effect because an exit grid was used to shield the inside of the analyzer.

To quantify the effects of the leakage fields, we used a combination of analyzer simulations and grid transparency corrections. First, a 3-D model of the exit grid and nearby surfaces showed that 2–3% of the MCP bias voltage would penetrate the grid. Second, the impact of this leakage field was characterized with a 2-D analyzer simulation by replacing the "ideal" exit grid with a potential surface that varied linearly from 40 V at the edges to 60 V at the center. The analyzer geometric factor was found to increase by $\sim$30% at low energies due to this leakage field, with an e-fold drop in this additional sensitivity at $\sim$180 eV. Third, we recognized that a secondary effect of the leakage field would be to focus low energy ions away from the grid wires, increasing the 90% exit grid transmission to about 100% at low energy. This was also simulated to estimate the exit grid effective transmission as a function of energy. Fourth, a separate "MCP grid" was placed in front of the detector and biased at the MCP voltage to increase detector efficiency. Due to the alignment of the MCP grid with the exit grid, the transmission of the MCP grid could also vary with energy. Simulations showed a complicated transmission function for the MCP grid caused by particle focusing by the exit grid, with some increased transmission at lower energies. Full characterization of this dependence was not attempted and instead an increase from 90% to 95% at low energies was estimated from the simulations. Combining these leakage field effects we obtained a $\sim$45% increase in analyzer geometric factor at low energies. Figure 11c shows the final energy-dependent efficiency correction used for THEMIS ion ESAs, which includes both the leakage field and the MCP detector variations (see Fig. 11b).

We point out that a similar energy-dependent analyzer affect is likely present in the electron analyzers. It would manifest itself at lower energies since the voltage on the front of the MCPs is only $\sim$450 V. Based on ion sensor simulations, we might expect the enhanced sensitivity to have an e-folding decrease at $\sim$45 eV from the maximum transmission at $\sim$0 eV. In addition, there is a second effect exclusive to electron analyzers. Electrons at energies $>$50 eV are capable of producing secondary electrons when they strike the outer or inner hemisphere. These low energy secondary electrons will be accelerated to the detector by the leakage fields if the incident particle strikes the hemisphere near the analyzer exit. This additional enhancement in analyzer secondary production kicks in at about the same energy that the enhanced throughput falls off. If the effects are similar in magnitude, it may be that the combined response is relatively flat. To sort out these effects would require a complex analyzer simulation which is beyond current THEMIS data analysis plans. However, after many comparisons between ion and electron densities, we conclude there are no large changes in the electron sensor (detector+analyzer) energy efficiency curve at these energies indicating the combination of these two effects has a relatively flat dependence on energy. Therefore the energy efficiency curve shown in Fig. 11a was adopted and any energy-dependent sensitivity changes due to leakage fields are assume to be absorbed in the overall geometric factor of the electron sensors.

2.3 Instrument Dead Time Corrections

Corrections for instrument dead-time can be important in regions of high particle count rates such as a high-density magnetosheath. For plasma sensors, lost counts due to instrument dead-time result from a combination of electronic and detector dead-time. For THEMIS, electronic dead-time was measured as part of the calibration and determined to be 170 ± 10 ns for all Amptek A121 preamplifiers. Detector dead-time is more difficult to determine. For microchannel plate detectors (MCPs), the dead-time is caused by a decrease in gain at high count rate that results in some events dropping below the preamplifier threshold. The gain drop occurs when the microchannels are unable to completely replenish the charge lost after the previous firing. THEMIS ESAs were fitted with high current MCPs for fast recharging. "Back of the envelope" estimates of the detector dead time, assuming a 6 MHz broad angle flux that illuminates a 22.5° anode, suggest the effective detector dead time is only ~30 ns at this high rate. The calculation assumes the signal current is limited to 10% of the fractional MCP strip current for the anode (~4 μA), and that the detector pulse height distribution maintains a Gaussian shape with a FWHM equal to its peak. Since the estimated detector dead time is much smaller than the preamplifier dead time, detector dead-time is not expected to be important unless the particle flux manifests as an intense narrow beam that is focused on a small portion of the detector. The software assumes a nominal 170 ns dead-time for all detector-preamp combinations.

Nature allowed in-flight testing of our dead-time correction. Figure 12 shows ion and electron data collected during a period of high density magnetosheath plasma. The top panels show ion and electron spectrograms, while panel c shows the uncorrected ion and electron densities and panel d the ratio of these densities. The electrons require significant dead time corrections prior to 18:30 UT, resulting in a density ratio that exceeds one. Panels e and f show the dead-time corrected densities with an expected ratio of about 0.9 during this period, confirming that the preamplifier dead time correction accounts for the majority of high rate dead time correction. While interpreting this plot, the reader should ignore the period with un-shocked solar wind (19:30–21:20 UT) where ion densities are underestimated in 32 sweep/spin mode.

2.4 Calibration of Relative Efficiency with Look Direction

Themis ion and electron ESAs have 16 and 8 discrete anodes, respectively, each of which requires a "relative efficiency" calibration factor to account for small variations in sensitivity with look direction. These relative efficiency corrections are approximately unity and do not reflect changes in overall instrument sensitivity. Overall instrument sensitivity is accounted for in the "absolution efficiency" calibration factor discussed in the next section.

For THEMIS electron sensors, the "relative efficiencies" correspond directly to the eight 22.5° anodes that cover 180° of polar angle. For ion sensors, which have 16 anodes and up to 5.6° resolution, an ideal relative efficiency calibration would result in 16 efficiency values. However, THEMIS data collected in regions useful for calibrations only maintain 22.5° resolution. Therefore ion relative efficiencies are determined in the same manner as those for electrons. Anodes within a 22.5° sector are assumed to have the same relative efficiency as the sector.

THEMIS relative calibrations were accomplished by finding data intervals inside the magnetosphere that met the following criteria:

1. Flows determined by the ion sensor must be small (< 30 km/s) and random. This allows us treat the raw data as if it is collected in the plasma frame and ignore pitch angle asymmetries caused by flows.
2. The magnetic field must be relatively constant and make a significant angle (> 20°) with the spin axis. This assures that the same pitch angle is measured by more than one anode during a spacecraft rotation.
3. The average pitch angle distribution of 1–20 keV plasma must be relatively smooth and vary by no more than a factor of 3 (no large anisotropies). This allows us to use a small number of polynomials in the fit and avoid high order terms that tend to be affected by statistical fluctuations or beams much narrower than the angle bins. The high energy range (> 1 keV) is selected so spacecraft potential variations are unimportant.

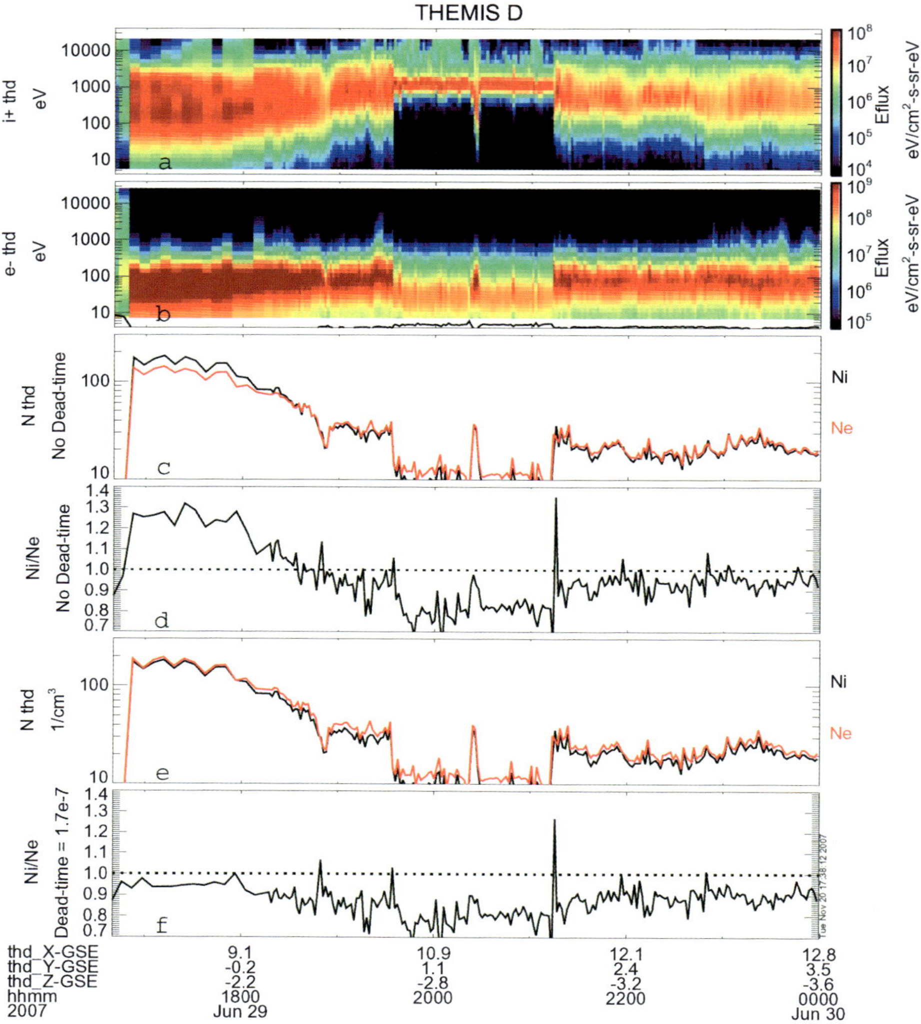

Fig. 12 The plot illustrates the importance of dead-time corrections during a day with high magnetosheath density (<19:00 UT). *Top panels* are ion (panel **a**) and electron (panel **b**) spectrograms. Panel **c** shows the ion (*black*) and electron (*red*) densities uncorrected for dead-time, and panel **d** illustrates how the ratio of these densities results in unphysical 1.0–1.3 values due to underestimation of the electron density. Density ratios should be ~0.9 since solar wind alpha content causes an underestimation of ion density (the calculation assumes protons only). Panel **e** shows the dead-time corrected densities have good agreement and panel **f** illustrates density ratios of ~0.9 as expected. In addition, within the solar wind (19:30–21:20 UT) the ion density is further underestimated due to the narrow beam width

4. Data intervals must contain 1–2 hours of Fast Survey data. This criterion assures that 40–75 spins of high angle resolution (88 solid angles) data are available for the fit.

Data between May 31 and August 13 were examined for these criteria, and 10 to 20 intervals were found for each spacecraft. Data from each interval was averaged, sorted by pitch angle, and fit to a 6th order symmetric polynomial $f = a + bx^2 + cx^4 + dx^6$, where

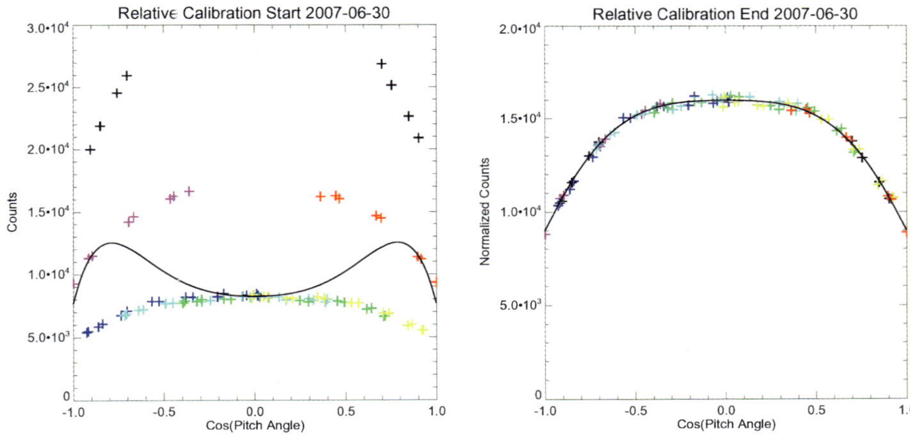

Fig. 13 Plots generated as part of the relative anode efficiency calibration algorithm. Different anodes are different colors and each "+" represents a different solid angle bin. (**a**) The *left plot* shows the raw counts in each solid angle bin as a function of pitch angle. *Black* colored bins have 4 times the counter accumulation time and *red/magenta* bins have twice the accumulation time of the other bins, accounting for the initial large variations. (**b**) The *right plot* shows normalized counts after anode efficiencies have been applied. The variation in normalized counts (9×10^3 to 1.6×10^4) is due to actual pitch angle anisotropy. Anode efficiencies are determined through minimum variance to a 6th order symmetric polynomial. The initial and final best fit polynomials are the *solid lines* in each plot. Different anodes had relative efficiencies within 10% of unity

$x = \cos\theta$ and θ is the pitch angle. Relative efficiencies were calculated by minimizing the variance in the least-squares fit and the fitting algorithm was repeated. This procedure was iterated until efficiency estimates converged to optimal values for each interval. Figure 13 illustrates the initial and final pitch angle distributions, with different anodes indicated by color and each point a different solid angle. After convergence to the right plot in the figure, the combined anode efficiencies and integration time effects are determined. Two anodes (red, magenta) have twice the integration time and two anodes (black) have four times the integration time, which accounts for the majority of the initial difference in the left plot.

Once efficiencies were calculated for the different intervals, they were averaged and incorporated into the code. Anode efficiencies were generally within about 10% of unity, with the largest variations associated with slightly smaller anodes in the pole channels (which were designed to prevent noise counts near the edge of the microchannel plate detectors). The standard deviation of the relative efficiencies over the intervals was $\sim 1.5\%$ for ions and $\sim 1\%$ for electrons indicating sound methodology and high precision. Since there were no systematic trends in these efficiency variations with time, we assume that these relative efficiencies are constant.

In using a symmetric polynomial to model the pitch angle distribution, we assumed that the distribution was stable and that sufficient bounce-averaging of plasma between magnetic poles had removed any asymmetry in the distribution. This is probably not the case and therefore this assumption may introduce an asymmetric, systematic error to the relative efficiencies. This asymmetry cannot be removed by simply including asymmetric terms in the pitch angle model. This is because the pitch angle overlap between anodes is generally limited to adjacent anodes so the fitting algorithm does not have a strong constraint on large-angle variations. Any large-angle asymmetry in the input distribution cannot be distinguished from an asymmetry in efficiencies. Therefore we are forced to use other methods to evaluate and correct for asymmetries in the response.

A simple estimate of the magnitude of any asymmetric response can be made by comparing the sensitivity of the two halves of the sensor. Since the sensors are symmetric, with hemispheric concentricity generally better than 1%, the average response of the 0° to 90° and 0° to −90° sensor halves are expected be very close. Indeed they are found to differ by only ∼0.5% for electrons, and by <2% for ions (except for one sensor that differed by ∼4%). Ion distributions during quiet periods in the magnetosphere were then checked for systematic flows in the z-direction and no statistically significant flows were found. Since small errors in electron ESA asymmetry can result in large flow errors along the spacecraft spin axis, we compared electron and ion flows in the magnetosheath during July 21–25, 2007. A first order ($\cos\theta$) asymmetry at the 1–3% level was then introduced to the electron efficiencies to obtain agreement between ion and electron flows. This correction was then tested for data from August 21–25, 2007 and found to provide consistent spin axis velocities between ions and electrons.

2.5 Cross-Calibration of the Sensors

The five electron and five ion ESAs on the THEMIS spacecraft each require a "sensor-level relative efficiency" calibration that accounts for overall variations in efficiency between instruments. These "near unity" efficiencies will change with time as detectors age and as detector bias voltages are increased to compensate for decreasing detector gains. We chose to separate this relative cross-calibration from an absolute calibration since we had all five satellite data sets in close proximity at the start of the mission and because we lacked a reference for absolute calibration. Since THEMIS does not have a high-frequency measurement of the plasma frequency, absolute calibration will require comparisons with upstream solar wind monitors, such as Wind or ACE, as discussed in the next section.

For this cross-calibration we used the already-determined relative anode efficiencies and energy-dependent detector efficiencies to calculate a sensor-level calibration factor that forces agreement between the separately determined densities. These calibrations cannot be performed in the magnetosphere since it often contains significant cold plasma that is unmeasured. Instead we focused on the magnetosheath where flows were large enough to assure nearly all ions were measured and where spacecraft potential corrections were relatively small. We emphasize that even though the spacecraft potential in the magnetosheath is rather low, typically 5–6 volts, inclusion of the spacecraft potential in the density calculation is critical. For spacecraft whose electric field sensors were not deployed (THA and THB), we used measured potentials from other spacecraft, with corrections as described in Sect. 2.1. Lastly, since the ion plasma sensor is not mass resolving, we must also account for differences in estimated density between ion and electron sensors due to the presence of alpha particles in the solar wind. These were accounted for by using upstream solar wind measurements from the Wind spacecraft.

Since THC was the first spacecraft with deployed electric field sensors, early cross-calibration efforts focused on inter-comparisons of its electron and ion sensors. Its observations were used to investigate two months of data (late-May to early-August, 2007) and select ten days where cross-calibrations would be possible. Selected days had measurements from all spacecraft while in the magnetosheath, and if possible were selected for low alpha content to minimize mass-dependent corrections. Since these are relative calibrations, the electron sensor on THC on June 28, 2007, was selected as a reference, its pre-launch estimated geometric factor used as a baseline, and its efficiency set to one. THC's ion sensor was then cross-calibrated, and its pre-launch geometric factor adjusted to give the same density as the electrons, and its efficiency was set to one. Following these baseline determinations, we set the geometric factors of all other electron and ion sensors to the same values determined above and calculated their sensor-level efficiencies to get agreement between densities. This was accomplished by first cross-calibrating ion and electron sensors on each spacecraft, then cross calibrating electron sensors on each spacecraft with THC's electron sensor.

Figure 14 shows an example of the type of data used to perform the cross-calibration. The top four panels are ion and electron spectrograms and allow quick determination of the various regions: magnetosphere (<12:30 UT), magnetosheath (12:30–17:00 UT) and solar wind (>18:00 UT). Panel e shows the density determined from the electron and ion sensors on THC and THD, after the calibration process. As discussed above, ion–electron sensor cross-calibrations on the same spacecraft are performed in the magnetosheath. Panels f and g show that the Ni/Ne ratios have been matched to ∼0.99 during the 12:30–17:00 UT period. (Upstream Wind-3dp plasma data indicated very low alpha content.) The inter-spacecraft cross-calibration is performed between electron sensors in the solar wind (>18:00 UT) as illustrated in panel h. The variance in the plots results from a combination of counting statistics and temporal variations less than the spin period.

The above calibration procedure was repeated for the 10 selected days. It was then assumed that relative efficiencies of sensors only decreased with time as the detectors aged unless detector bias voltage was increased. This assumption forced a renormalization of the sensor-level efficiencies on each day, except our reference day of June 28, so that each sensor's efficiency monotonically decreased in time. Over the 72 day interval starting 2 months after sensor turn-on, sensor efficiency degradation of 5% to 11% was estimated by this method. Table 2 shows the initial (07-05-15) and final (07-08-25) values for relative efficiency for the 10 sensors illustrating the variations in their loss of sensitivity. Sensor geometric factors can be calculated by combining the geometric factor from Table 1 with the relative sensitivity in Table 2 and the energy dependent efficiency in Fig. 11. In addition, the MCP detector voltage was raised on three of the sensors toward the end of this interval, with no measurable impact on detector efficiency. This suggests the drop in sensitivity may not be a detector gain issue but rather a front-end particle detection efficiency change. In any case, sensitivity degradation is assumed to come from MCP scrubbing, which is expected to stabilize after a few months of high counting in the magnetosheath. Since random errors in the determination of these detector efficiencies may result in an over estimation of this degradation, it may be necessary to correct these calibrations with an overall mission-level efficiency using upstream solar wind measurements.

We close this section by pointing out that the ESA MCP detectors are tested each month for gain to assure that the peak in their pulse height distribution is about a factor 8 above preamplifier threshold (see Sect. 1.1). During 16 months of operations, four of the five electron sensors, and three of the five ion sensors, have had their MCP voltage increased one or more times to maintain this gain. No noticeable change in sensitivity was observed during any of these increases indicating they were performed well before a significant number of events dropped below preamplifier threshold. It is not known whether the drop in efficiency discussed above was due to front-end changes in incident particle secondary electron production, or due to changes in gain that result in the loss of events below threshold. Continued monitoring and comparison of instrument sensitivity should provide information on MCP degradation and longevity. Lastly, absolute calibrations, described in the following section, should provide additional information to track long term changes in sensitivity.

2.6 Absolute Calibrations

The calibration efforts described above resulted in a consistent set of relative calibrations between measurements within a sensor, and consistent measurements between different

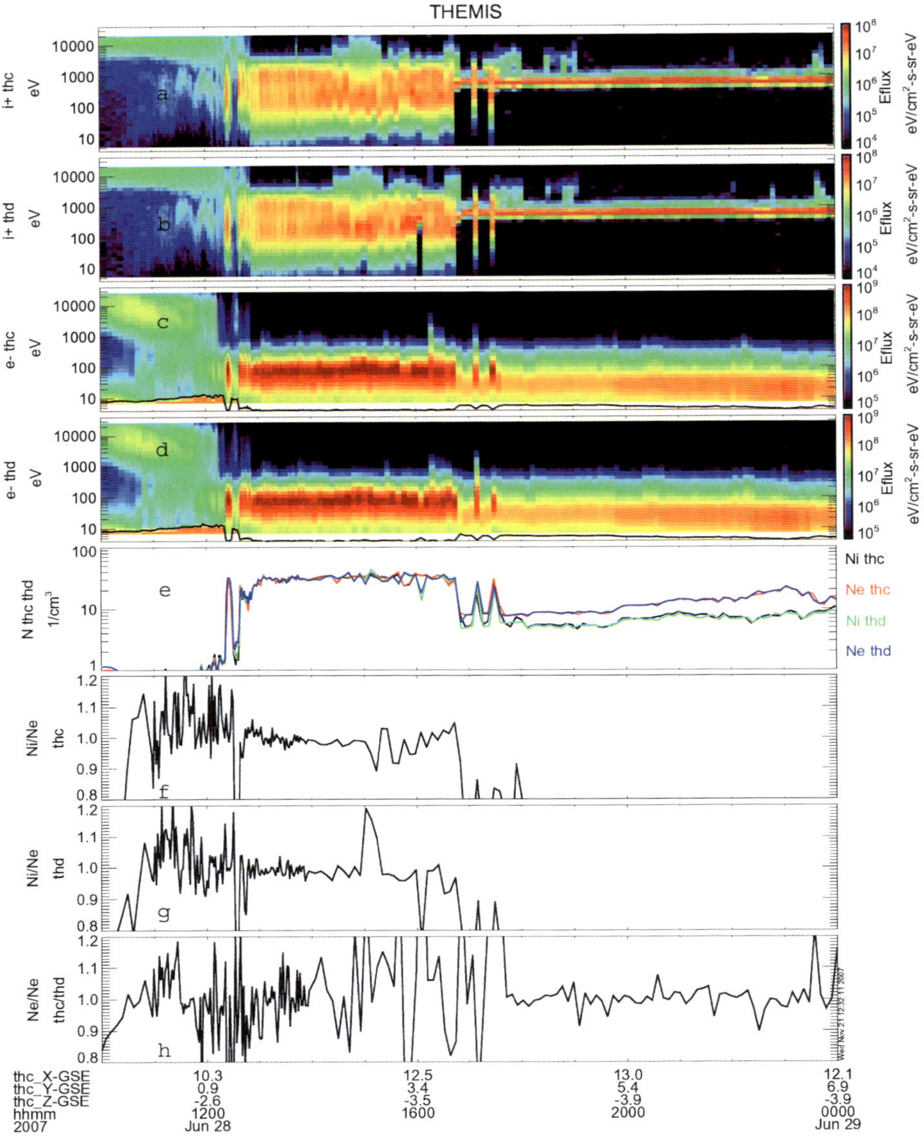

Fig. 14 Example of the cross-calibration analysis. Ion–electron sensor cross-calibrations use Ni/Ne ratios measured on the same spacecraft (panel **f**, **g**) within the magnetosheath (12:30–17:00) and match the result to the expected ratio based on upstream alpha content, ~0.99 in this case. Inter-spacecraft cross-calibrations match electron densities within the solar wind (>18:00 UT) as illustrated in panel **h**. Solar wind ion densities are underestimated due to the narrow solar wind beam and therefore not used in calibrations

sensors. However, there was no test in the above procedures that determined the absolute sensitivity of the sensors. All measurements of the even moments, such as density and pressure, or measurements of flux and energy flux, which are proportional to density, were likely incorrect by some scale factor. For these initial calibrations, the absolute sensitivity, or ab-

Table 2 Relative sensor efficiency

Spacecraft	eESA 07-05-15	eESA 07-08-25	iESA 07-05-15	iESA 07-08-25
THA	1.010	0.935	1.075	1.010
THB	1.000	0.890	1.100	1.050
THC	1.030	0.910	0.995	0.920
THD	0.945	0.845	1.015	0.970
THE	0.865	0.805	1.085	1.035

solute geometric factor, was estimated from the electron analyzer simulation combined with expected losses from an exit grid and an estimated 70% MCP detection efficiency. To determine the absolute calibration requires a comparison of the THEMIS ESA response to a known "standard candle" plasma parameter. Generally this standard candle is the plasma frequency which determines the electron density. However since THEMIS spacecraft do not have high frequency wave receivers, absolute calibration has to be determined through comparisons with other spacecraft. Since the THEMIS spacecraft are rarely near any other magnetospheric spacecraft for cross-calibrations and since magnetospheric plasma regimes vary dramatically in density and pressure, the most favorable location for consistent cross-calibrations is the solar wind.

Absolute sensitivity of the plasma sensors was tested through cross-calibration with the Wind-SWE instrument (Ogilvie et al. 1995). Electron densities measured by THEMIS THC and THD in the solar wind were compared with SWE proton densities, with appropriate corrections for time delays based on the location of the Wind spacecraft. Five intervals during a two month period were compared and a $\sim$0.7 correction to the THEMIS densities was required to give good agreement. Figure 15 shows an example of this cross-calibration. The top panels show the THC and THD electron spectrograms with nearly identical solar wind plasma. The black lines indicate spacecraft potential, $e\Phi_{sc}$. The third panel shows the IMF magnitude (Wind-black, THC-red, and THD-green) which assists in the determination of the temporal alignment and the verification of the suitability of the time interval. The bottom panel demonstrates that the Wind-SWE, THC and THD densities have good agreement after the correction factor is applied. Since $e\Phi_{sc}$ was below the lower energy cutoff of the THEMIS electron sensors, this comparison used a density calculation algorithm that extrapolated the distribution function to $e\Phi_{sc}$, assuming a Maxwellian distribution. For the data in Fig. 15, this algorithm introduced a 2% to 9% increase in the calculated the THEMIS electron density relative to a simple algorithm that just used the measured energy range.

The above cross-calibration indicates the THEMIS electron ESA pre-flight geometric factors were underestimated by $\sim$40%($\sim$1./0.7). Recall in Sect. 2.2, the ion sensor energy efficiency was adjusted to account for leakage fields through the exit grid. A similar low-energy efficiency-correction was not performed for electrons since high energy electrons produce secondary electrons at the analyzer exit which can also add to the combined analyzer-detector sensitivity. We assumed the combined effects of the leakage fields were flat in energy and would be determined when the overall geometric factor was calibrated. It now seems likely that the above 40% correction is at least partly due to these leakage fields. Part of the correction may also be due to an underestimate of the MCP detection efficiency (we assumed 70%). For the five intervals tested, the correction factor was relatively constant and showed no systematic change in time. Temporal variations between the measurements indicate that errors in this cross-calibration are at the $\sim$10% level. The 40% correction fac-

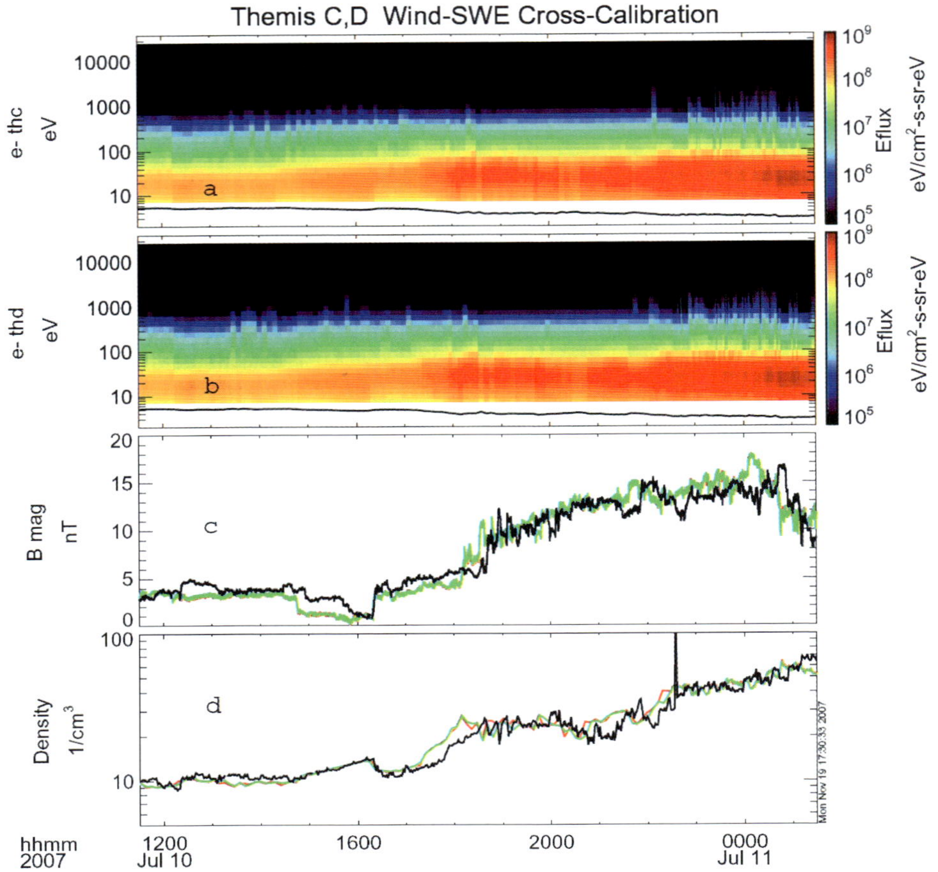

Fig. 15 Plot illustrates the cross calibration between Wind-SWE and THEMIS electron ESAs used to determine absolute sensitivity. The plot shows electron spectrograms from THC (panel **a**) and THD (panel **b**) in the solar wind. Panel **c** shows the IMF magnitude on Wind (*black*), THC (*red*) and THD (*green*). Panel **d** shows the agreement between ion density on Wind (*black*), and electron density on THC (*red*) and THD (*green*) after the cross calibration. Since THC and THD measurements of IMF and density are nearly identical, only the green curve (THD) shows in panels **c** and **d**

tor was combined with the pre-flight sensor geometric factors to obtain the absolute sensor geometric factors.

As a final test of the absolute calibration, magnetopause crossings were evaluated to check for pressure balance. Figure 16 shows an example that contains several magnetopause crossings in addition to flux transfer events. The bottom panel shows the electron (red), ion (green), and magnetic (blue) pressures, in addition to the combined pressure (black). The nearly constant total pressure during these crossings indicates accurate absolute calibrations.

3 Summary

The THEMIS ESA plasma instrument measures the 3-D distribution functions of electrons (up to 30 keV) and ions (up to 25 keV) using a pair of "top hat" electrostatic analyzers. Parti-

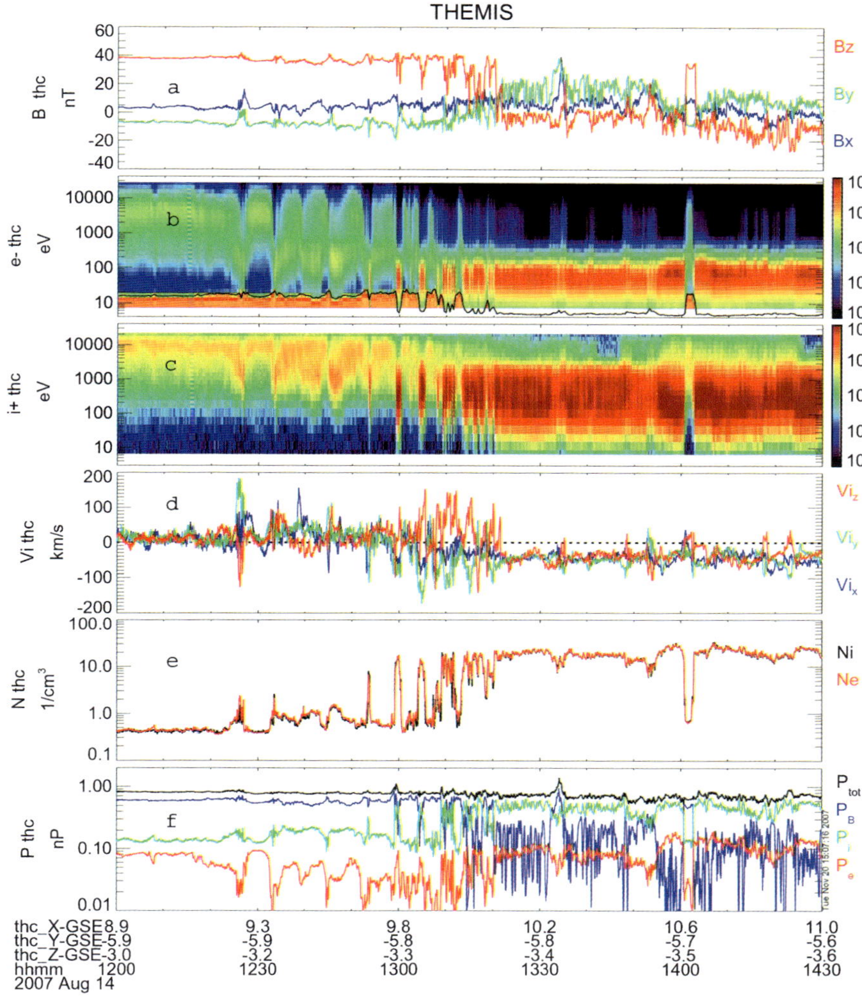

Fig. 16 THEMIS magnetopause crossing used to test calibrations. From top to bottom: magnetic field, electron spectrogram, ion spectrogram, ion velocity, density, and pressure. Good agreement of the electron (*red*) and ion (*black*) densities (panel **e**) indicates good cross-calibration. Nearly uniform pressure (panel **f**) across several magnetopause boundaries, on either side of flux transfer events, and within magnetosheath mirror modes, reveals accurate absolute calibrations

cle events identified by microchannel plate detectors are binned into six types of distributions whose energy, angle, and time resolution depend upon instrument mode. Omni-directional spectra or coarse-angle resolution distributions are continuously available at spin resolution (3 s). Higher energy-angle resolution distributions are available at a lower cadence or at spin resolution during burst data collection. In addition, on-board data processing generates plasma moments at spin resolution that include corrections for spacecraft charging.

The overall design of the THEMIS ESA plasma instrument was directly derived from the FAST Plasma Instrument (Carlson et al. 2001). This modular design simplified assembly and subsystem testing of the 10 flight ESAs. The primary changes from the FAST design were the development of a resetable closing mechanism that utilizes an SMA actuator to seal the detector from contamination, and the change from gold-black to ebanol-C blacking which reduces scattered sunlight from reaching the detector. Ground calibrations showed the five sensor pairs to be nearly identical in response and all ten sensors continue to perform optimally after 16 months.

The close proximity of the THEMIS satellites during the first 7 months of the mission allowed extremely accurate multi-satellite cross-calibrations of the plasma sensors. These calibrations were facilitated by having all the plasma measurements available with the same resolution and format, along with spacecraft potential and magnetic field measurements in the same data set. The methodology of the in-flight calibration effort has been outlined in this paper, and its precision demonstrated through comparisons with Wind-SWE and total pressure across the magnetopause. However, this calibration effort is not complete. The THEMIS plasma instruments will require monitoring throughout the mission to track and quantify degradation of the MCP detectors and to determine the adjustments to their bias voltages. Future tracking and calibration efforts will utilize the same methods employed above, but will be more difficult due to larger spacecraft separations. It is envisioned that additional cross-calibration efforts will rely on the few month period each summer when multiple satellites encounter the solar wind. Lessons learned from this effort should be useful for future multi-satellite missions.

Acknowledgements The analysis of THEMIS data was supported by NASA NAS5-02099. We thank J. Bonnell and F. Mozer for providing spacecraft potential, and K.-H. Glassmeier for providing magnetic field data. These data assisted the ESA in-flight calibration effort. Financial support for the work of the FGM Lead Investigator Team at the Technical University of Braunschweig by the German Ministerium für Wirtschaft und Technologie and the Deutsches Zentrum für Luft- und Raumfahrt under grant 50QP0402 is acknowledged.

References

V. Angelopoulos, Space Sci. Rev. (2008). doi:10.1007/s11214-008-9336-1

H.U. Auster, K.H. Glassmeier, W. Magnes, O. Aydogar, D. Constantinescu, D. Fischer, K.H. Fornacon, E. Georgescu, P. Harvey, O. Hillenmaier, R. Kroth, M. Ludlam, Y. Narita, K. Okrafka, F. Plaschke, I. Richter, H. Schwarzi, B. Stoll, A. Valavanoglu, M. Wiedemann Space Sci. Rev. (2008). doi:10.1007/s11214-008-9365-9

J.W. Bonnell, F.S. Mozer, G.T. Delory, A.J. Hull, R.E. Ergun, C.M. Cully, V. Angelopoulos, P.R. Harvey, Space Sci. Rev. (2008, this issue)

C.W. Carlson, D.W. Curtis, G. Paschmann, W. Michael, Adv. Space Res. **2**, 67–70 (1983)

C.W. Carlson, J.P. McFadden, P. Turin, D.W. Curtis, A. Magoncelli, Space Sci. Rev. **98**, 33–66 (2001)

R.R. Goruganthu, W.G. Wilson, Rev. Sci. Instrum. **55**(12), 2030 (1984)

R.S. Gao, P.S. Gibner, J.H. Newman, K.A. Smith, R.F. Stebbings, Rev. Sci. Instrum. **55**(11), 1756 (1984)

P.R. Harvey, E. Taylor, R. Sterling, M. Cully, Space Sci. Rev. (2008, this issue)

D. Larson, T. Moreau, R. Lee, R. Canario, R.P. Lin, Space Sci. Rev. (2008, this issue)

J.P. McFadden, C.W. Carlson, D. Larson, J. Bonnell, F. Mozer, V. Angelopolos, K.-H. Glassmeier, U. Auster, Space Sci. Rev. (2008, this issue)

K.W. Ogilvie, D.J. Chorney, R.J. Fitzenreiter, F. Hunsaker, J. Keller, J. Lobell, G. Miller, J.D. Scudder, E.C. Sittler Jr., R.B. Torbert, D. Bodet, G. Needell, A.J. Lazarus, J.T. Steinberg, J.H. Tappan, A. Mavretic, E. Gergin, Space Sci. Rev. **71**, 55–77 (1995)

A. Pederson, F. Mozer, G. Gustafsson, in *Measurement Techniques in Space Plasmas: Fields*. Geophysical Monograph, vol. 103, (1998), p. 1

A. Roux, O. Le Contel, C. Coillot, A. Bouabdellah, B. de la Porte, D. Alison, S. Ruocco, M.C. Vassal, Space Sci. Rev. (2008, this issue)

D.G. Sibeck, V. Angelopoulos, Space Sci. Rev. (2008). doi:10.1007/s11214-008-9393-5

H.C. Straub, M.A. Mangan, B.G. Lindsay, K.A. Smith, R.F. Stebbings, Rev. Sci. Instrum. **70**(11), 4238 (1999)

E. Taylor P. Harvey, M. Ludlam, P. Berg, R. Abiad, D. Gordon, Space Sci. Rev. (2008, this issue)

The Electric Field Instrument (EFI) for THEMIS

J.W. Bonnell · F.S. Mozer · G.T. Delory · A.J. Hull · R.E. Ergun · C.M. Cully · V. Angelopoulos · P.R. Harvey

Originally published in the journal Space Science Reviews, Volume 141, Nos 1–4, 303–341.
DOI: 10.1007/s11214-008-9469-2

Abstract The design, performance, and on-orbit operation of the three-axis electric field instrument (EFI) for the NASA THEMIS mission is described. The 20 radial wire boom and 10 axial stacer boom antenna systems making up the EFI sensors on the five THEMIS spacecraft, along with their supporting electronics have been deployed and are operating successfully on-orbit without any mechanical or electrical failures since early 2007. The EFI provides for waveform and spectral three-axis measurements of the ambient electric field from DC up to 8 kHz, with a single, integral broadband channel extending up to 400 kHz. Individual sensor potentials are also measured, providing for on-board and ground-based estimation of spacecraft floating potential and high-resolution plasma density measurements. Individual antenna baselines are 50- and 40-m in the spin plane, and 6.9-m along the spin axis.

The EFI has provided for critical observations supporting a clear and definitive understanding of the electrodynamics of both the boundaries of the terrestrial magnetosphere, as well as internal processes, such as relativistic particle acceleration and substorm dynamics. Such multi-point electric field observations are key for pushing forward the understanding of electrodynamics in space, in that without high-quality estimates of the electric field, the underlying electromagnetic processes involved in current sheets, reconnection, and wave-particle interactions may only be inferred, rather than measured, quantified, and used to discriminate between competing hypotheses regarding those processes.

Keywords Electric field instrumentation · Electrodynamics · Reconnection · Substorms

J.W. Bonnell (✉) · F.S. Mozer · G.T. Delory · A.J. Hull · P.R. Harvey
Space Sciences Laboratory, University of California, Berkeley, USA
e-mail: jbonnell@ssl.berkeley.edu

R.E. Ergun · C.M. Cully
Laboratory for Astrophysics and Space Physics, University of Colorado, Boulder, USA

C.M. Cully
Institute for Space Physics, Uppsala, Sweden

V. Angelopoulos
Institute for Geophysics and Planetary Physics, University of California, Los Angeles, USA

1 Introduction

The Electric Field Instrument (EFI) on the THEMIS spacecraft measures the three components of the ambient vector electric field. Waveform measurements cover DC up to 4 kHz (with AC-coupled differential measurements from $\sim$10 Hz up to 8 kHz), with on-board spectral measurements covering the same ranges, as well as providing an estimate of integrated power in the 100- to 400-kHz band. On-board spin-fit E-field estimates are provided for both the spin plane and axial measurements. On-board estimates of the spacecraft floating potential are produced for use in particle moment calculations and burst trigger evaluations.

The choice of capabilities for the THEMIS EFI were driven by both the primary scientific measurement requirements, as well as the relatively limited power and telemetry resources available to the THEMIS spacecraft. Nonetheless, the EFI waveform and spectral data allow the observation of DC filed phenomena throughout the various THEMIS mission phases, and resolution of wave phenomena up to the electron plasma and cyclotron frequencies at densities less than 0.8 cm^{-3} and distances greater than 4.7 R_E (290 nT). The lower hybrid and proton and heavy ion cyclotron frequencies are within the sampling interval of the waveform and spectral data throughout the entire orbit of the THEMIS spacecraft.

The orientation and location of the six EFI sensors and their booms are shown in Fig. 1. The individual sensors are identified by number (1–6), and when deployed, are closely aligned with the spacecraft coordinate system (indicated by the triad in the upper right portion of the diagram). Both individual (V_n, $n = 1, \ldots, 6$) and differential ($E_{mn} = V_m - V_n$, $m, n = [(1, 2), (3, 4), (5, 6)]$) sensor potentials are measured. The differential sensor potentials are used to estimate the ambient vector E-field. The individual sensor potentials are used to verify proper sensor operation, estimate the spacecraft floating potential, and measure the ambient plasma density with high (sub-spin-period) time resolution.

Below, we describe the measurements the EFI is required to make in the context of the THEMIS mission, present some examples of the sorts of measurements made by the EFI drawn from the first year of operation, describe the detailed mechanical, electrical, and operational design of the EFI, and detail the on-orbit operation and performance of the EFI.

1.1 Measurement Requirements

The measurement requirements for the THEMIS-EFI flow down from the need to make accurate, quantitative estimates of the electric fields associated with the flows, current systems, and electromagnetic fluctuations during substorm onset in the terrestrial magnetotail, energetic electron acceleration in the radiation belts, and crossings of the various dayside magnetospheric boundary layers (magnetopause, bow shock, low-latitude boundary layer, etc.). Formally, the top-level science requirements imposed upon EFI and the basic justification for those requirements are as follows:

- Determine the 2D spin plane electric field (to an accuracy of 1 mV/m or 10%) at times of substorm onset at 8–10 R_E radius in the magnetotail (plasma sheet and plasma sheet boundary layer (PSBL)). This allows one to estimate the electric fields associated with flows, flow diversions, and interchange-like macro-instabilities at the inner edge of the plasma sheet.
- Determine the dawn-dusk electric field (to an accuracy of 1 mV/m or 10%) at 18–30 R_E radius in the magnetotail. This allows one to estimate the electric fields associated with up- and down-tail flows during substorm onset.

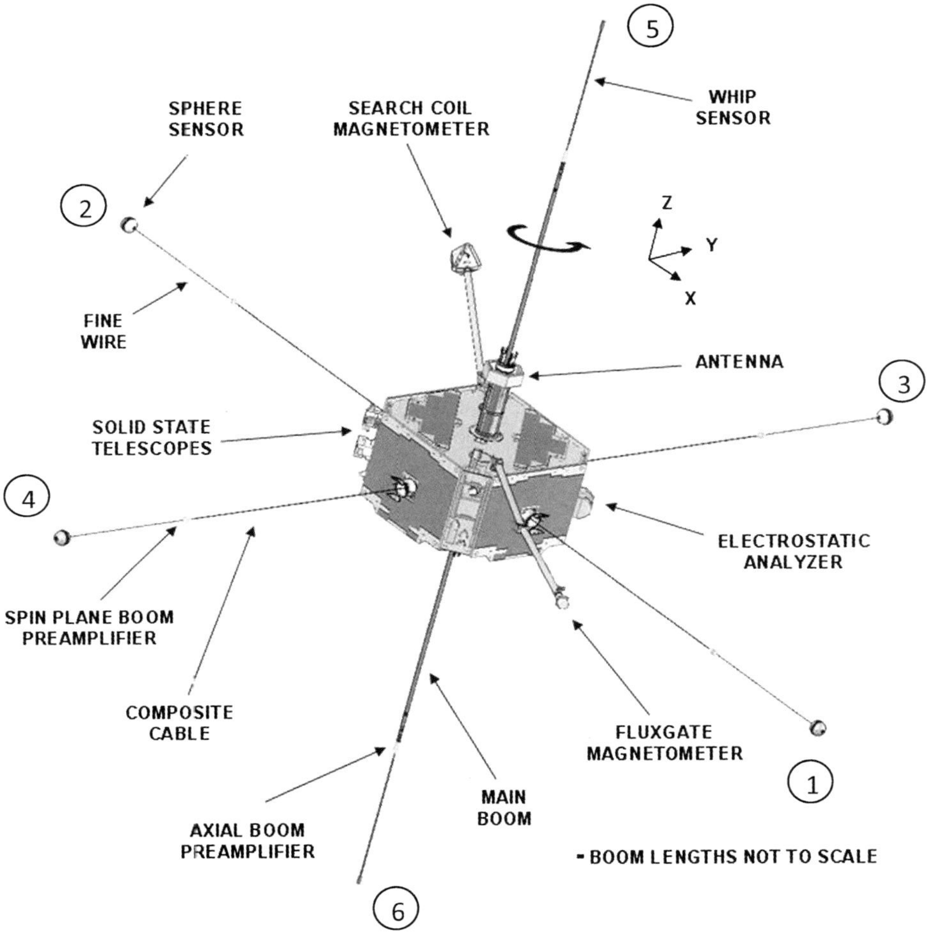

Fig. 1 Schematic diagram of THEMIS spacecraft, including body- and boom mounted sensors

- Measure the 3D wave electric field from 1–60 Hz at times of substorm onset at 8–10 R_E radius in the magnetotail. This allows one to estimate the electric fields associated with current disruption and interchange-like instabilities at the inner edge of the plasma sheet.
- Measure the 3D wave electric field at frequencies up to the local electron cyclotron frequency in the radiation belts. This allows one to measure the electric fields associated with the energization, scattering, and loss of energetic electrons in the outer radiation belts (e.g. whistler-mode hiss, chorus, or electron cyclotron fundamentals or harmonics).

In addition to these instrument-specific measurement requirements, the THEMIS-EFI was required to comply with the general environmental (radiation, thermal, shock, vibration, acoustic), resource (mass, power), and compatibility (EMI/EMC, DC Magnetics, Electrostatic Cleanliness) requirements imposed at the mission level on THEMIS. The EFI team itself imposed the Electrostatic Cleanliness requirement upon the mission, and that specification (design, implementation, and verification) is detailed in an appendix to this article.

2 Design

The EFI is in essence no more than a set of high-input-impedance, low-noise, broadband digital voltmeters. By measuring the potential difference between three pairs of electrodes separated along orthogonal axes, one can estimate the external vector electric field (Pedersen et al. 1998, and references therein).

Complications in the mechanical design arise because of the need to separate the electrodes of the individual sensors by significant distances from both the spacecraft and from each other in order to reduce systematic errors due to the influence of the spacecraft itself, as well as increase the signal level for a given external field strength. This signal is proportional to sensor separation through $V_n - V_m = -\mathbf{E} \cdot (\mathbf{X}_n - \mathbf{X}_m)$, where V_n and V_m are the potentials of the sensors located at the (vector) positions $\mathbf{X}_n$ and $\mathbf{X}_m$, and $\mathbf{E}$ is the external (vector) electric field.

Complications in the electrical design arise from the need to control the electrical properties of the plasma sheaths that form around the sensors, the emission and collection of photoelectrons by the sensors and surrounding surfaces on the booms and spacecraft, as well as to maintain stringent control over the DC offsets and AC noise levels present on each of the sensors.

2.1 Mechanical Design

The mechanical design of the THEMIS-EFI sensors and deployment systems draws on over 30 years of flight heritage, stretching back to the earliest successful DC magnetospheric electric field measurements (S3-3, ISEE), through to several recent missions in low-Earth or magnetospheric orbits (CRRES, FAST, Polar, Cluster-II). A schematic diagram showing the THEMIS-EFI in fully-deployed configuration, as well as smaller-scale mechanical details of the sensors is shown in Fig. 2. As can be seen in the figure, two distinct types of sensor deployment units are utilized on THEMIS: a wire boom system for the four spin-plane sensors, and a stacer boom system for the two spin-axis sensors. The final deployed lengths for each of the boom systems are collected in Table 1.

2.1.1 Spin Plane Booms

The Spin Plane Booms (SPBs) consist of a sensor and preamp assembly affixed to the end of a $\approx$22-m-long custom cable, along with the deployment mechanism itself. The sensor consists of an 8-cm diameter graphite-coated (Acheson Colloid DAG-213; graphite in baked-on epoxy-resin matrix) Al sphere with internal spring keyreel mechanism, along with a 3-m-long, 0.2-mm (0.009-inch) diameter 302/304 stress-relieved stainless steel fine wire (7×7 braided). The preamp enclosure is a 2.3-cm diameter by 3.7-cm long dual-tapered cylinder that contains the preamp electronic board, as well as acts as two of the photoelectron control surfaces (usher and guard). The custom cable (W.L. Gore and Associates) consists of an inner coax (AWG-36 single conductor with Al-mylar overwrap shield layer), surrounded by eight insulated single conductors (AWG-36), a kevlar load-bearing braid, and an outer braid consisting of a continuous, helically-wrapped aluminized mylar shield surrounded by a silver-plated copper braid, and is approximately 2.5 mm in diameter ($\sim$0.100 inch). The distal three meters of the outer braid is electrically-isolated from the proximal portion, and can be driven using the distal braid, or DBraid, circuits of the Boom Electronics Board (BEB). Total deployed length of each SPB (sensor plus cable) is up to 25 meters (spacecraft centerline to sphere center). On orbit, one pair of booms is deployed to a greater tip-to-tip

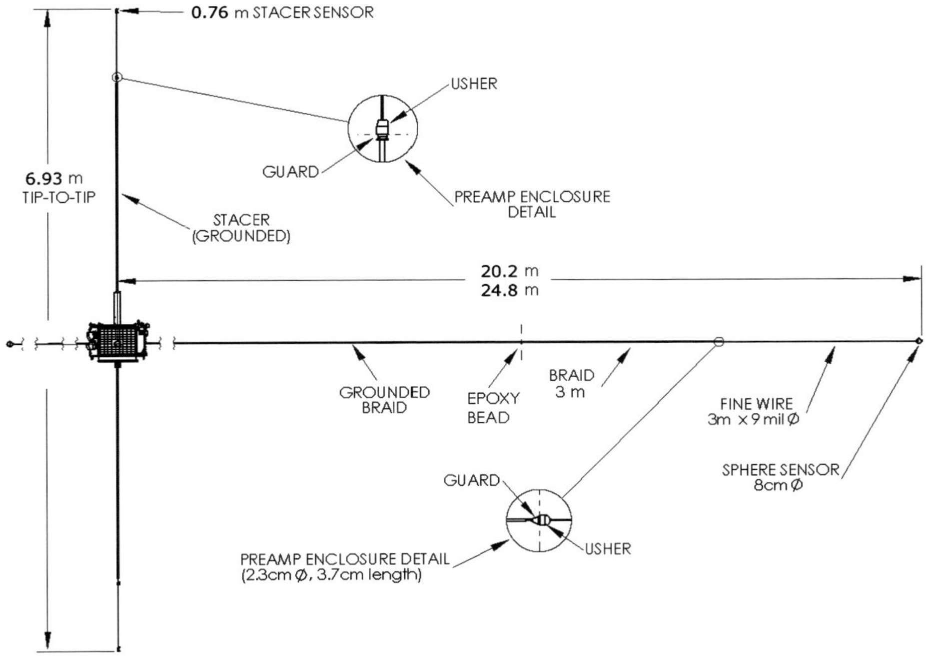

Fig. 2 Deployed configuration of the THEMIS-EFI sensor and boom systems

Table 1 Final deployed lengths of EFI boom systems

Axis	Deployed length	Notes
X (V1, V2, E12)	49.6-m tip-to-tip	None
Y (V3, V4, E34)	40.4-m tip-to-tip	None
Z (V5, V6, E56)	6.93-m tip-to-tip	0.76-m whip length; $\approx$6.2-m whip center-to-center

length than the other to allow for better determination of certain systematic errors in the DC and AC E-field measurements (see Table 1 for exact deployed lengths). Also note that the effective sensor separation distance on-orbit is less than the physical separation due to electrostatic effects. Estimation of this effective sensor separation is discussed at length in the Cross Calibration section below.

The deploy mechanism, or SPB proper, and a partially-deployed section of cable, preamp enclosure, and sphere is shown in side view in Fig. 3. The unit consists of a spool, upon which the cable is wound for stowing during launch; a motor and meter wheel assembly that pulls the cable from the spool at a controlled rate, measures the deployed length of cable through a calibrated cam and switch system (click counter) with a resolution of better than 4.8 cm, and provides for electrical (wire tension switch) and mechanical (shear pin) end-of-wire deploy stop controls; and a snout and doors, which cage the sphere and preamp for launch, and provide for electrical isolation of both the sphere and preamp from the SPB chassis, along with a low-impedance (few-hundred-ohm) contact to the sphere itself for self-test excitations. The SPB doors are held closed by a plunger and spring mechanism that is released through a custom TiNi-wire-driven actuator requiring $\approx$2 amps for 2–3 seconds

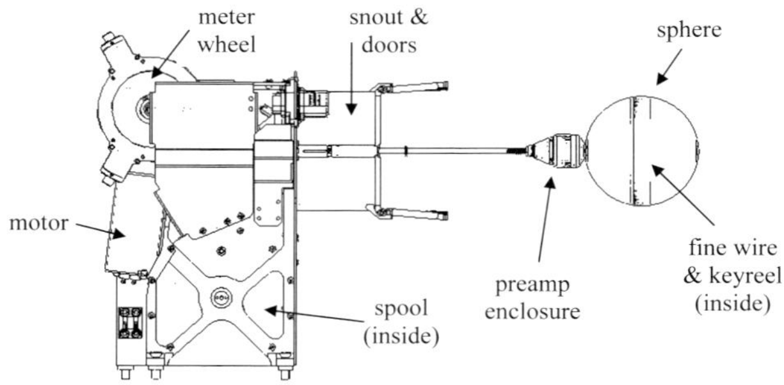

Fig. 3 Side view of SPB and partially-deployed cable, preamp enclosure, and sphere

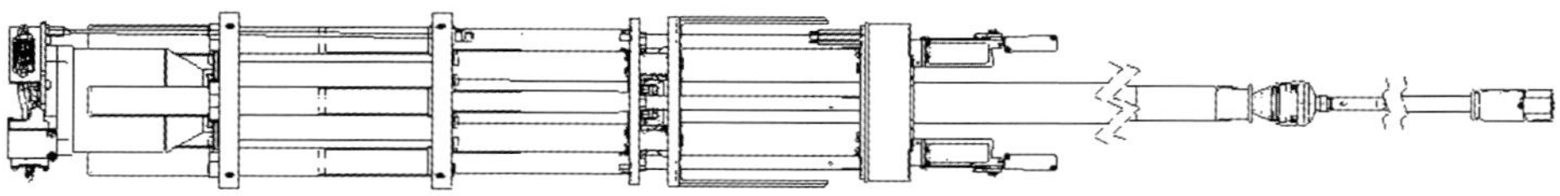

Fig. 4 Side view of deployed THEMIS-EFI AXB, showing (left-to-right) cable bobbin, main stacer can, double deploy assist device mechanism, main stacer, preamp enclosure, and axial sensor (whip stacer and can)

to fire under nominal on-orbit thermal conditions. The SPB motors are DC brush motors, requiring 0.1–0.2 A at 36 V to operate. The permanent magnets in the SPB motors are shielded with a three-layer mu-metal shield, reducing the dipole moment to typically less than 10 mA-m^2 and at most 20 mA-m^2 ($\approx$0.1 to 0.2 nT at FGM location; for details, see Ludlam et al. 2008).

2.1.2 Axial Booms

A diagram of one of the two THEMIS-EFI Axial Booms (AXBs) is shown in Fig. 4. Each boom and sensor system consists of a sensor and preamp assembly affixed to the end of a $\approx$2.5-m-long stacer boom assembly, as well as the stacer Deploy Assist Device (DAD). The sensor consists of a 0.75-m-long, tapered (4.8 to 7.0-mm) graphite-coated (DAG-213) Elgiloy whip stacer with 1.6-cm diameter, 4.6-cm long can at the outboard end. The preamp enclosure is similar to that found on the SPB, and contains the preamp electronic board, as well as acting as two of the photoelectron control surfaces (usher and guard). The preamp enclosure is mounted to the outboard end of the $\approx$2.5-m-long, $\approx$2-cm diameter graphite-coated (DAG-214) Elgiloy main stacer. The two-stage DAD is spring-loaded, and serves to initiate the main stacer deploy after its release from the Frangibolt actuator, to form the stacer during deploy through two sets of roller nozzles, and to provide for lateral stability of the main stacer when fully-deployed. The preamp enclosure and sensor (whip stacer) are caged for launch by two sets of doors located on the top of the outer DAD plate. The rollers on the two doors engage with a groove around the inboard end of the whip stacer can, and provide both the DC grounding path for the AXB sensor as well as the contact for the ACTEST line. The two AXB assemblies are mounted back-to-back inside a 102-mm diameter (four-inch) graphite-composite tube that serves as the central strength member of the THEMIS spacecraft bus and the mounting point for the six-panel S-band telemetry antenna and elements of the reaction control system.

2.2 Electrical Design

A functional block diagram of the THEMIS-EFI is shown in Fig. 5. The EFI follows a basic instrumentation amplifier design for each of the three axes. One branch of one axis is shown in the block diagram, running from the sensor, through the preamp and cable system, to the Digital Fields Board (DFB), Boom Electronics Board (BEB), and Low-Voltage Power Supply (LVPS) located in the central Instrument Data Processing Unit (IDPU).

Each sensor electrode is connected to the input of high-input-impedance ($\sim 10^{12}$ ohm), low-noise unity-gain preamplifier circuit. The preamp circuit drives the buffered sensor potential down the cable to the spacecraft into the BEB and DFB boards of the IDPU, as well as serves as the point where the sensor bias current is injected. Each preamp operates off of a separate floating power supply. The floating ground potential of each supply (FGND) is derived from a low-pass-filtered version of the preamp output, and is generated on the BEB. The FGND has a dynamic range of ± 100 volts relative to spacecraft ground, and the preamp power supplies provide ± 10-V with respect to that floating ground potential. This floating power supply topology allows for the use of low-noise, low-input-bias-current, rad-hard (100-kRad(Si)) op amps in the preamp circuit (OP-15, in this case, with input bias currents in the few pA range), while still allowing for the large dynamic range of sensor to spacecraft potentials (tens of volts) that can occur due to varying plasma conditions and current biasing of the EFI sensors.

The signal from the preamp is shared between the DFB and BEB. All on-board scientific signal processing (analog and digital) is performed by the DFB, while the voltage and current sources required to support the operation and biasing of the EFI sensor and surrounding voltage and photoelectron control surfaces are supported on the BEB. Each of these functional elements is discussed below.

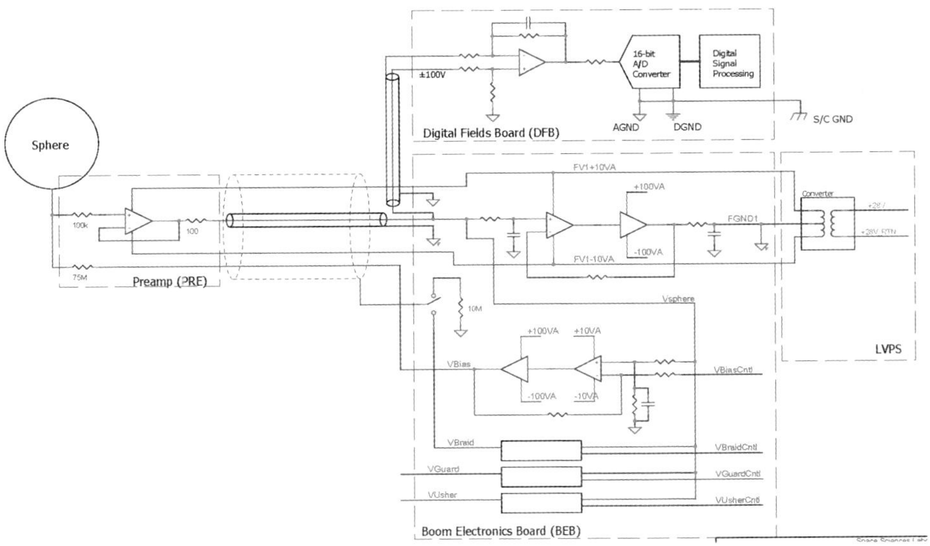

Fig. 5 Functional Block Diagram of THEMIS EFI, showing preamp, BEB, floating supply, and DFB connections for a single sensor

2.2.1 Sensors and Preamps

The EFI preamp design is a simplification over previous designs flown for magnetospheric applications. Previous designs have included circuitry supporting a low-input-impedance current-measurement mode, also known as Langmuir or Density mode that allowed for direct measurement of ambient electron fluxes and derivation of total electron density via voltage bias sweeps of the sensor potential. The measurement requirements of THEMIS did not necessitate such a precision measurement, and mass and volume resources did not support inclusion of the circuitry supporting such a mode, and so a simple, high-input-impedance design was chosen to support the required E-field measurements. However, the absolute and relative ambient density can be estimated with modest precision (at least 50% accuracy, typically) from the measurement of the relative floating potentials between the EFI sensors and spacecraft ground using well-established modeling and analysis techniques (e.g. Pedersen et al. 1998).

As shown in Fig. 6, the EFI preamp consists of a single op amp (OP-15) in a unity-gain follower configuration, powered via a floating-ground ± 10-V supply, with FGND roll off at ≈ 100 Hz. A 100-kΩ resistor in parallel with a 10-pF capacitor provides for input ESD protection and response compensation, while the 100-Ω output resistor mitigates the attenuation and stability issues that arise when driving the significant (≈ 2 nF) capacitive load due to the SPB cable coax.

Figure 7 shows the predicted frequency response (voltage gain and phase) for both the spin-plane and spin-axis EFI sensors in sunlight, as well as the lumped electrical parameters used in the response analysis. The total response is denoted by the black dashed line, while the contributions to the response due to circuit elements upstream (sheath and effective input impedances) and downstream (output impedances, cable and analog electronics loads) of the follower circuit (red and blue lines). No on-board compensation for the variable gain of the EFI sensor system with frequency is applies. Corrections for the response are applied in either the frequency domain (for spectral data products) or time domain (for waveform data products) by ground data processing software.

The relative phase and timing of the EFI and SCM signals collected by the DFB are known from ground testing to better than 1 degree, both in terms of absolute knowledge, and variation from channel to channel and spacecraft to spacecraft. The exact coupling im-

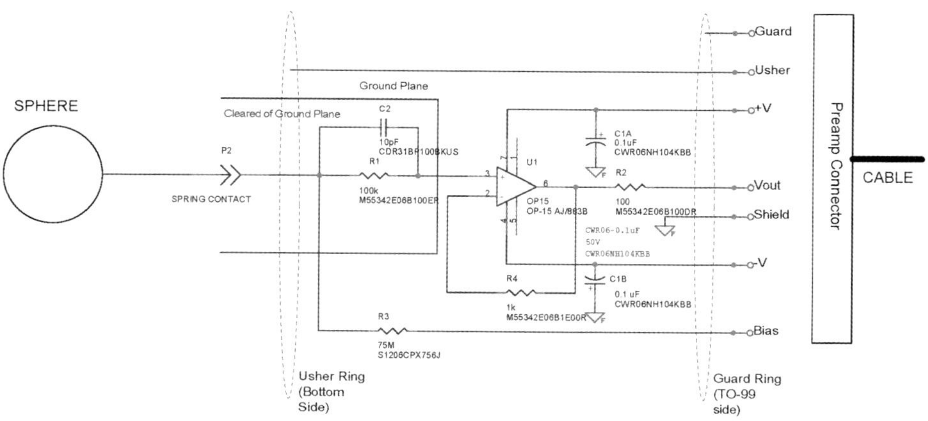

Fig. 6 Schematic diagram of THEMIS-EFI preamp

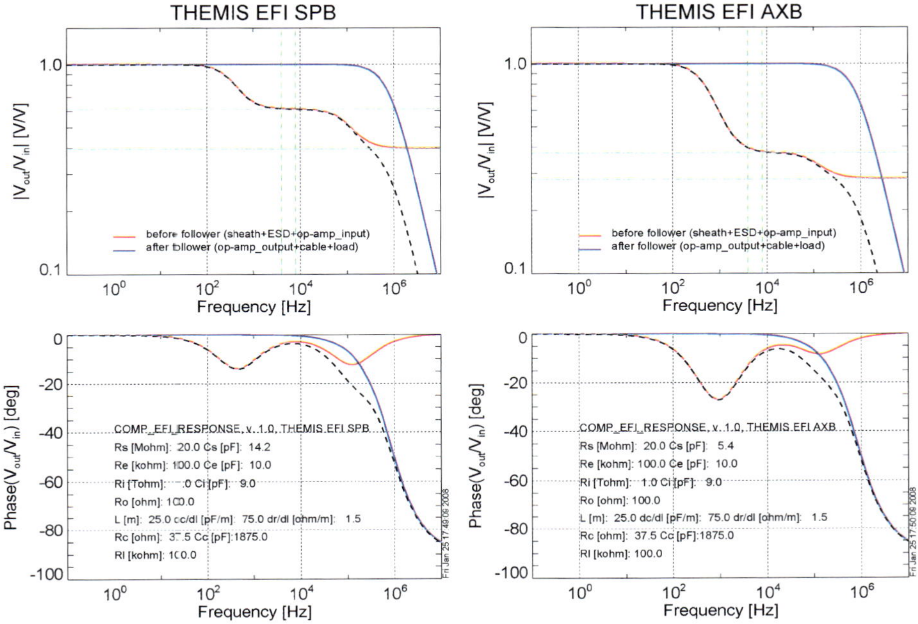

Fig. 7 Predicted frequency response of THEMIS-EFI spin-plane and spin-axis antennas

pedance of the EFI sensors to the plasma is unknown; the exact value of effective sheath resistance and capacitance are not known precisely. Figure 8 shows the variation in predicted spin plane and axial sensor response for the expected range of plasma conditions and bias currents. From these curves, one can see that the uncertainty one can expect are on the order of 20% in gain and 5–10 degrees in phase, sufficient to allow for accurate determination of polarization, Poynting flux, and mode.

Wide temperature excursions of the preamp and enclosure were predicted to occur during the three-hour eclipses that occur during some portions of the main THEMIS science season (down to $\approx -120^\circ$C in eclipse, up to $\approx 70^\circ$C in full sunlight). These temperature excursions, coupled with the tight mechanical tolerances of the preamp and preamp enclosure, and lack of electronic parts qualified to that temperature range, led to the need for a formal, vigorous, and extensive thermal qualification and acceptance program for both the electrical and mechanical design of the preamp. The preamp design had passed informal thermal survival testing of the sort performed on previous missions (powered-down emersion in liquid nitrogen), but such a test was deemed unrepresentative, both in temperature range, and level of thermal shock, to properly qualify the design for use on THEMIS.

The formal thermal qualification and acceptance program for the preamps was developed and executed at UC Berkeley, with concurrence from the THEMIS Parts Control Board at Goddard Space Flight Center. This program involved an initial vacuum thermal balance test to validate the analytical modeling of the preamp thermal behavior during eclipse and sunlight, followed by mechanical and electrical qualification testing of the design and components to be used (24 hot–cold–hot thermal-vacuum test cycles on eight ETU preamps and enclosures), with six-cycle acceptance testing of each flight preamp (30 units total; five flight sets, plus one flight spare set). Extensive automation of both the data acquisition during the qualification and acceptance runs, as well as the comparison of the pre- and post-test data

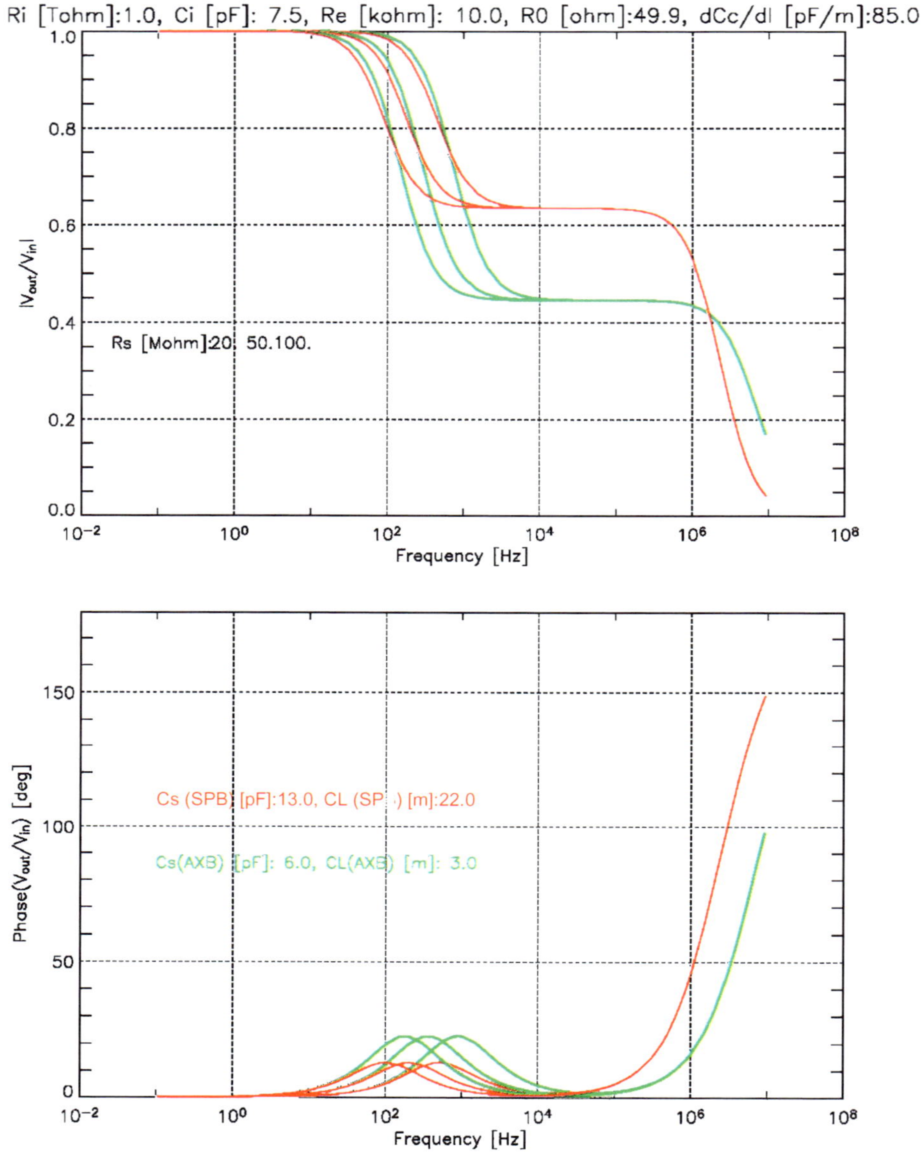

Fig. 8 Predicted SPB (*red*) and AXB (*green*) sensor response as a function of frequency (*top panel*: gain; *bottom panel*: phase) for three IBIAS settings ($R_{\text{sheath}} = 20$, 50, and 100 MΩ)

(input bias current and voltage; frequency response; power consumption; etc.) allowed for the stringent level of testing to occur, even under the aggressive schedule pressure during flight deliveries on THEMIS (<1 month between flight set deliveries to Instrument Suite I&T). No units failed the thermal testing, and the change in all relevant component values and performance specifications was less than 5% in all cases, indicating the extreme robustness of the mechanical and electrical design of the preamp.

In addition to the shielding mass provided by the outer shell of the preamp enclosure, the preamp PWB includes a fully-enclosing (better than 98% coverage) tantalum radiation shield around the OP-15 that provides for a total shielding mass equivalent to 7 mm of Al, reducing the expected total dose on the OP-15 to approximately 44 kRad(Si) (RDM = 2) for the two-year THEMIS mission. This design feature was driven by a requirement to maintain the low nominal input bias current over the entire nominal mission duration; while the OP-15 is qualified rad hard (to 100-kRad(Si) total dose), increases in DC input bias current and the fluctuations thereof begin to appear at the 40–50 kRad(Si) total dose level (Sahu and Kniffen 1998).

2.2.2 Boom Electronics Board (BEB)

The THEMIS-EFI utilizes a single, centralized Boom Electronics Board (BEB), mounted in the THEMIS Instrument Data Processing Unit (IDPU), with power services provided by the IDPU-Low Voltage Power Supply (LVPS; providing both fixed and floating ground power supplies). This design choice is in contrast to the individual BEB and LVPS utilized on previous missions (FAST, Polar, Cluster-II), and was driven by the tight mass and volume budgets for THEMIS.

As is shown in Fig. 5, the BEB is responsible for generating and controlling the floating grounds (FGND), current- (BIAS) and voltage- (USHER, GUARD, DBRAID) bias signals for each of the six EFI sensors, as well as DBRAID reference selection. Current and voltage biasing is controlled by 16-bit digital-to-analog converters (AD5544) with a dynamic range of ±528 nA/sensor (BIAS), and ±40 V relative to sensor potential for USHER, GUARD, and DBRAID for sensor floating potentials in the ±60-V range. BIAS, USHER, GUARD, and DBRAID signals are all referenced to a low-pass-filtered (passive one-pole at ≈400-Hz) version of the sensor potential to allow for stable DC biasing of the various surfaces in the presence of the low-frequency, up to tens-of-volts excursions of sensor-to-spacecraft potential that occur due to changes in ambient plasma conditions. Only 11–12 bits of the total DAC resolution is required to implement the 0.1% precision needed for bias matching on-orbit, with the remaining 4–5 bits acting as margin in case of DAC degradation over time.

Typical bias settings for each of the surfaces are collected in Table 2, and have been determined using the on-orbit Sensor Diagnostic Tests described in the On-Orbit Performance and Operation section below.

As noted above, the outer braid of the SPB cables is divided into two electrically-isolated surfaces, the Distal and Proximal Braids. The Proximal Braids (PBRAIDs) are individually connected to analog ground (AGND) via 330-kΩ resistors. All four Distal Braid (DBRAID) surfaces are controlled together, and can either be connected to AGND via 1-MΩ resistors,

Table 2 Typical current and voltage bias settings for THEMIS-EFI

Surface	Axis	
	SPB	AXB
Sensor	≈180 nA (≈120 nA initially)	≈180 nA (≈120 nA initially)
Usher	≈+4 V	≈+4 V
Guard	≈+4 V	≈+4 V
DBraid	≈0V, V1 Reference or Grounded	n/a (no DBRAID)

or voltage-biased with a ±40-V dynamic range with respect to either AGND, or low-pass-filtered versions of the sensor potentials V1 or V3. This biasable surface was included in the design to allow for control of the wire boom length at spacecraft potential, and to reduce the magnitude of boom shorting and electrostatic wake effects. This aspect of the design is akin to the "bootstrapped braid" design feature in the Polar-EFI and Cluster-II-EFW designs (c.f. Fig. 3, Pedersen et al. 1998).

Switching of the various power services (EFI_X, _Y, _Z; EFI_BOARDS) is handled through the IDPU-PCB (Power Control Board). The floating supplies serving each of the preamps are grouped in pairs (by axis), and may be switched separately. This allows for graceful response to anomaly or failure of one or more of the preamps or BEB channels in an axis-by-axis fashion.

Switching for the various actuator power supplies (SPB doors; SPB motors; AXB FrangiBolt), as well as click counter read backs from the SPBs are handled through harnessing connected to the IDPU-Power Controller Board (PCB). Each actuator line is carried in shielded twisted pair in the individual branches of the EFI harness, with the shield terminated to actuator power ground at the PCB. The preamp output signals are carried via coax (shield connected to floating ground; FGND), with all other preamp power (FGND, ±10-VF) and sensor biasing (BIAS, USHER, GUARD, DBRAID) signals carried on single-wire lines. The entire harness is overwrapped with a shield layer, terminated at BEB end of the harness to AGND.

The exposed portions of the SPBs and AXBs were connected to AGND via the EFI harness, while being electrically-isolated from the nearby spacecraft deck in order to properly implement the THEMIS ESC Specification. Similarly, all exposed thermal blankets, tapes, and coatings were connected to AGND, either explicitly via ground wires for larger blanket sections, or through proper tabbing and interconnects for smaller sections. Because of the significant radiative heat leak due to the open SPB doors after deploy of the EFI SPB sensors, the SPB chassis are thermally-isolated from the bottom deck of the THEMIS spacecraft by G-10 fiberglass spacers, as well as covered by Indium-Tin-Oxide-coated MLI thermal blankets on all inward facing surfaces.

The BEB also provides the resources for a self-test capability when the EFI sensors are in their stowed configuration. When stowed, the individual sensors are connected to analog ground via grounding resistors ($\approx$10 MΩ for SPBs, $\approx$7 MΩ for AXBs), as well as being connected to an ACTEST line. The BEB can switch a fixed-amplitude, 5 V_{pp}, 128-Hz square wave signal onto any of the individual ACTEST lines. A simple DC Functional Test with a dynamic range of approximately ±5 volts can be run by commanding the individual sensor BIAS lines through a set of values while collecting waveform telemetry data. A similar simple AC Functional Test can be run by commanding the individual sensor ACTEST lines on and off while collecting waveform and spectral telemetry data. This self-test capability was used extensively during Instrument Suite and Mission Integration & Test (I&T), eliminating the need for external electrical GSE to support each performance test, and standardizing the instrument configuration for each test run.

As shown in Fig. 5, each individual sensor potential is sent in parallel to both the BEB circuits, as well as the analog conditioning circuits on the IDPU-Digital Fields Board (DFB; Cully et al. 2008b). Connection from the BEB to the DFB is provided by short coax lines and SMA connectors on the front panel of the IDPU.

2.3 Data Quantities

The set of data quantities available from the EFI, as well as their dynamic ranges, precision, and other properties are determined by the analog and digital signal processing performed

Table 3 Data products available from EFI

Data product	Range (not adjusted for frequency-dependent gain)	Bits	Resolution	Available sampling rates
V1, ..., V6	±105 V (±100 V supply limited)	16	3.2 mV/ADC	2-8192 samp/s
EDC12 (49.6-m)	±300 mV/m	16	(9.2 μV/m)/ADC	2-8192 samp/s (16384 samp/s for AC)
EAC12 (49.6-m)	±51 mV/m		(1.6 μV/m)/ADC	
EDC34 (40.4-m)	±370 mV/m	16	(11 μV/m)/ADC	2-8192 samp/s (16384 samp/s for AC)
EAC34 (40.4-m)	±63 mV/m		(1.9 μV/m)/ADC	
EDC56 (6.2-m)	±2.7 V/m	16	(81 μV/m)/ADC	2-8192 samp/s (16384 samp/s for AC)
EAC56 (6.2-m)	±450 mV/m		(14 μV/m)/ADC	
HF	4 μV/m to 12 mV/m	8	0.01 decade of amplitude/ADC	2-8192 samp/s
Spin fit E_{xy}	Same as EDC12 or EDC34, depending upon source setting	16-bit floating point	Same as EDC12 or EDC34, depending upon source setting. ≈1 deg. In angle	1 vector/spin (typ. 3-s period)
Spin avg E_z	Same as EDC56	16-bit floating point	Same as EDC56	1/spin (typ. 3-s period)
Spacecraft potential		16		1/spin (typ. 3-s period)
Filter bank and FFT spectral products	See Digital Fields Board description (Cully et al. 2008b) for details			

by the DFB. The design of the DFB is described in detail in Cully et al. (2008b). The data quantities available from the EFI are shown in Table 3.

The single-ended V channels, along with the DC-coupled differential EDC channels have the same frequency response as the combined sensor-preamp-cable response shown in Fig. 7, combined with the four-pole Bessel anti-aliasing filters on the DFB (Cully et al. 2008b). The AC-coupled differential EAC channels incorporate a six-fold increase in gain, as well as a passive one-pole RC, 10-Hz high-pass filter in addition to the usual four-pole Bessel anti-aliasing filter on the DFB. In order to minimize the impact of spacecraft potential fluctuations on the differential E-field estimates, the DC-coupled channels have a

DC CMRR of better than 80 dB (better than 0.1 mV/V common mode input), and the AC-coupled channels have a AC CMRR of better than 40 dB (10 mV/V common mode input).

Note that displacement of the electrical center of the spacecraft from the center of the EFI antenna array is an additional source common-mode error in the EFI measurement. On-orbit testing shows that while this effect is insignificant for the measurements made by the spin plane sensors, it is quite significant for the axial sensor measurements (see Axial Performance section below).

The HF channel is a special-purpose, broadband filter channel covering the auroral kilometric radiation emission band (100–400 kHz) with peak response at 130 kHz, and half-max points at 60 and 300 kHz (see Fig. 5 of Cully et al. 2008b). Response is logarithmic in amplitude, covering approximately 3–1/2 decades in RMS amplitude integrated across the channel response. The HF channel is designed to both monitor the integrated power in the AKR spectral band during substorms, as well as act as a possible trigger for on-board high-rate burst data.

The spin-fit E_{xy} and spin-averaged E_z data products are computed on-board in the IDPU. For the spin-fit E_{xy}, 128-sample/s data from either the DC-coupled E12 or E34 channel is fit to a model of the form: $A + B \cdot \sin(\psi) + C \cdot \cos(\psi)$, where ψ is spin phase relative to the Sun pulse, using a standard least-squares method with iterative outlier subtraction. For the spin-averaged E_z, 128-sample/s data from the DC-coupled E56 channel is fit to a constant with iterative outlier subtraction. The estimated standard error and final number of points for each fit or average is returned via telemetry as well.

An on-board estimate of the spacecraft floating potential is computed in the IDPU using either the average of several samples of one pair of spin plane sensors (V1 and V2, or V3 and V4), or a spot-sample of the same. This quantity is then scaled and offset through two adjustable parameters, and then provided to the on-board particle moments calculation for purposes of correcting the measured energy of particles for the effect of the spacecraft floating potential. This quantity is also telemetered to ground as part of the particle moments data.

Each of the basic waveform (V, EDC, EAC) EFI data quantities may be used as input the Filter Bank and FFT spectral products processing on the DFB. See Cully et al. (2008a) for details of those spectral products (frequency resolution, dynamic range and sensitivity, and cadence). Note that the spectral products provided by the DFB only cover the range from DC to 4 kHz (8 kHz for AC-coupled E-field products), and that power present at higher frequencies is integrated into the single HF channel described above. This feature of the design was driven by the maximum frequencies required to achieve the science requirements, and the tight power and telemetry resources available for the THEMIS mission.

Table 4 presents a typical on-orbit configuration for EFI data collection in terms of waveform sampling rates, spectral cadences and resolutions, and gain states in the four modes of operation. Also shown (in bold italics) are the data types that represent each data quantity in the Level-1 and Level-2 CDF data files, as well as the processed versions available through either the THEMIS Data Analysis Software (TDAS) IDL libraries, or the Science Data Tool (SDT) stand-alone browser. These tags (vaf, efp, etc.) indicate both the data type (va = sensor voltages; ef = E-field estimates) and acquisition mode (f, p, w = fast survey, particle burst, wave burst) for each data type. For more detailed information, the interested reader should reference the THEMIS Fields Variables Names documentation (apollo.ssl.berkeley.edu/pub/THEMIS/3%20Ground%20Systems/3.2%20Science%20Operations/Science%20Operations%20Documents/thm_soc_105_FIELDS_VARNAMES_20060929.pdf).

The EFI waveform data are available in a variety of operationally and geophysically relevant coordinate systems. These are described in detail in the, "THEMIS Coordinate

Table 4 EFI on-orbit data quantities

	Slow survey	Fast survey	Particle burst	Wave burst
Duration	Up to full orbit (24 hours maximum)	Up to 12 hours	Variable, few to tens of minutes, up to 8/orbit	Variable, few to tens of seconds, up to 8/P burst
Data quantity				
V_n $(n = 1, \ldots, 6)$	0, 2, or 4 samp/s ***vaf***	2 samp/s ***vaf***	16 samp/s ***vap***	8192 samp/s ***vaw***
E_{nm} $(nm = 12, 34, 56)$	0, 2, or 4 samp/s DC-coupled, ***eff***	4 samp/s DC-coupled, ***eff***	128 samp/s DC-coupled, ***efp***	8192 samp/s DC-coupled, ***efw***
Spin fit E_{xy}, Spin avg E_z	1/spin (1/3 s typ.) ***efs***			
Filter Banks, HFF	EDC12, SCM1 1 spec/4 s ***fb_*, fbh***			
FFT spectra	–	–	16 or 64 bins. 4 or ¼ spec/s. ***ffp_16_*, ffp_64_****	16 or 64 bins. 4 or ¼ spec/s. ***ffw_16_*, ffw_64_****

Systems," document (apollo.ssl.berkeley.edu/pub/THEMIS/3GroundSystems/3.2Science Operations/ScienceOperationsDocuments/thm_soc_110_COORDINATES_20060929.pdf. At Level-2, as well as through special processing of Level-1 data, the so-called "_0" and "_dot0" versions of the 3D E-field waveform data are available. These two data types address the radical difference in accuracy between the spin plane and spin axis estimates of the E-field (see discussion in Axial Performance section below).

The "_0" data quantities replace the axial E-field estimate with zero prior to transforming the E-field into despun and/or geophysically relevant coordinate systems. They are useful for estimating the impact of the axial field component upon the full 3D E-field estimate.

The "_dot0" data quantities replace the axial E-field estimate with a value computed from the spin plane E-field estimates and the ambient B-field under the assumption that $E_{\|} = 0$, or equivalently $\mathbf{E} \cdot \mathbf{B} = 0$; in other words, one replaces E_{axial} by $E_{\text{axial}} = -((B_x/B_z) \cdot E_x + (B_y/B_z) \cdot E_y)$. It is often the case, especially at the macroscopic scales of interest for THEMIS, that the dominant contribution to the electrodynamics arises from the perpendicular component of the electric field, and this technique allows one to estimate that component, and then transform it into other coordinate systems without having to remove the large systematic errors in the axial E-field estimate. If the B-field is too close to the spin plane (i.e. B_z is too small relative to B_x and B_y), then the error associated with this method grows, leading to a ten-to-one increase in error for angles smaller than ~6 degrees.

In addition to the science quantities, a complete set of EFI housekeeping quantities are collected by the IDPU and telemetered as part of the IDPU State-Of-Health packets (APID

Table 5 EFI housekeeping quantities

Quantity	Description	Resolution (bits)	Measurement range
IMON_EFI_BOARD	Aggregate primary-side current drawn by DFB and BEB fixed-voltage supplies (analog and digital)	8	0 to 150 mA. Typ. 100–130 mA
IMON_EFI_X, Y, Z	Primary-side current drawn by EFI floating supplies (axis-by-axis)	8	0 to 130 mA. Typ. 55–70 mA in sunlight, up to 80 mA in eclipse (cold, saturated)
IEFI_IBIAS1, ..., 6	Readback of BIAS offset voltage DAC output	16	−528 to 528 nA
IEFI_USHER1, ..., 6, GUARD1, ..., 6, BRAID	Readback of USHER1, ..., 6, GUARD1, ..., 6, and BRAID offset voltage DAC output	8	−40 to 40 V
ISPB_TEMP, IAXB_TEMP	Preamp board temperature (connector)	8	−258 to 357°C. Typ. 20 to 30°C (SPB), 30 to 40°C (AXB). −135°C (3-hr eclipse)
IBEB_TEMP	BEB board temperature (BEB FPGA)	8	−76 to 176°C. Typ. 20 to 30°C

0x404 and 0x406). These housekeeping quantities are summarized in Table 5. The current monitors are sampled once per second, while the other housekeeping channels that monitor signals on the BEB are sampled at either variable rates between once per 256 s to once per 2^{16} s ($\approx$18 hours), or on demand, when a bias update command is sent to the BEB. The relatively slow sampling rate for the BEB-derived housekeeping channels is driven by both the relatively slow variations of the quantities involved (either fixed for many orbits (bias settings, for example), or varying on minutes to hours time scales), as well as a modest, but noticeable level of noise injection associated with the housekeeping sampling seen on the EDC and EAC channels (unipolar signals of amplitudes $\approx$ few tenths of mV/m).

3 First Results

Below, we present several examples of EFI data taken on-orbit during the first year of operations on the five THEMIS satellites. These examples demonstrate the capabilities of the EFI instrument in a variety of geophysical contexts, including regions of whistler energization of relativistic electrons in the radiation belts, and reconnection and electrodynamics at the dayside magnetopause.

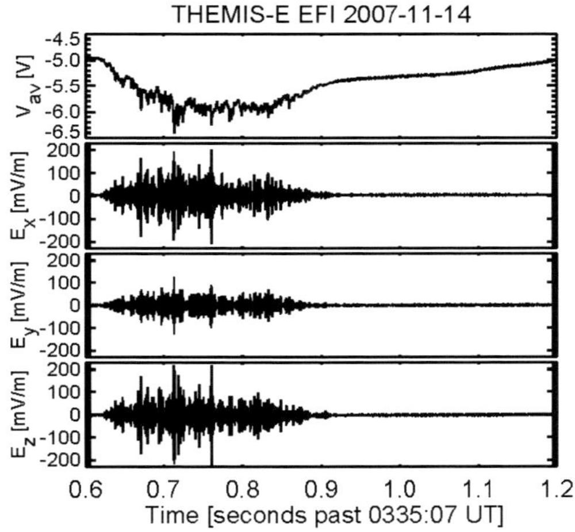

Fig. 9 Full bandwidth (8192 samp/s) estimates of spacecraft floating potential and despun (GSE) E-field showing large amplitude (200–400 mV/m, pk-to-pk) whistler mode waves embedded in an ambient density depletion (THEMIS-E, 14 Nov 2007, 0335:07 UT) (figure courtesy AGU; Cully et al. (2008a), GRL, doi:10.1029/2008GL033643)

3.1 Large-Amplitude Whistler Wave Observations in the Radiation Belt

An example demonstrating the capabilities of the full bandwidth (8192 samp/s), large dynamic range measurements of the EFI is shown in Fig. 9 (taken from Cully et al. 2008a). Whistler-mode fluctuations (bottom three panels) of up to 400 mV/m pk-to-pk (corrected for AC gain) are observed embedded within a significant (10–20%) variation in ambient density, as registered by the $\approx$1 volt change in the spacecraft floating potential computed from the average of the four spin plane sensor potentials (top panel). As can be seen in Fig. 10, the enhancement is fairly broadband, and extends at least up to the Nyquist frequency for the burst waveform measurements in this mode (4096 Hz). When these burst waveform observations are combined with the survey mode spectral data products (filter banks and FFT spectra), it can be shown that the upper- and lower-band hiss in the inner magnetosphere consists primarily of spatially or temporally-localized (filling factor or duty cycle of $\approx$1%) bursts of these large-amplitude whistler-mode fluctuations, rather than a more uniform bath of such fluctuations as has been assumed based on previous, purely spectral observations, leading to a rather different model for relativistic electron acceleration and scattering in the presence of such fluctuations (Cattell et al. 2008; Cully et al. 2008a).

3.2 Hall Electric Field Observations at the Magnetopause

Figure 11 presents data collected by the THEMIS C, D, and E probes during an inbound-outbound magnetopause crossing between 1736–1743 UT on 20 July 2007 (Mozer et al. 2008). Each spacecraft's passage is represented by two panels, the first (panels a, c, and e) being the measured ion and electron density from the THEMIS ESA instrument (McFadden et al. 2008b), and the second (panels b, d, and f) being the GSE X (sunward) component of the perpendicular electric field computed from spin fits of the long (E12) and short (E34) spin plane EFI sensors (black and red traces), and the analogous component of the ideal MHD E-field, $-\mathbf{V}_i \times \mathbf{B}$ (green). The close agreement (better than a few tenths of a mV/m) between the E-field estimates in the high-density magnetosheath regions at the beginning and end of the interval, as well as through the bulk of the magnetospheric region, clearly indicate that EFI is measuring geophysical E-fields, rather than fields due to cold plasma wakes (for example). The discrepancies between the measured $\mathbf{E}$ and $-\mathbf{V}_i \times \mathbf{B}$ at the magnetopause crossings are coincident with gradients in $\mathbf{B}$ (finite $\mathbf{J}$), and has be shown to be

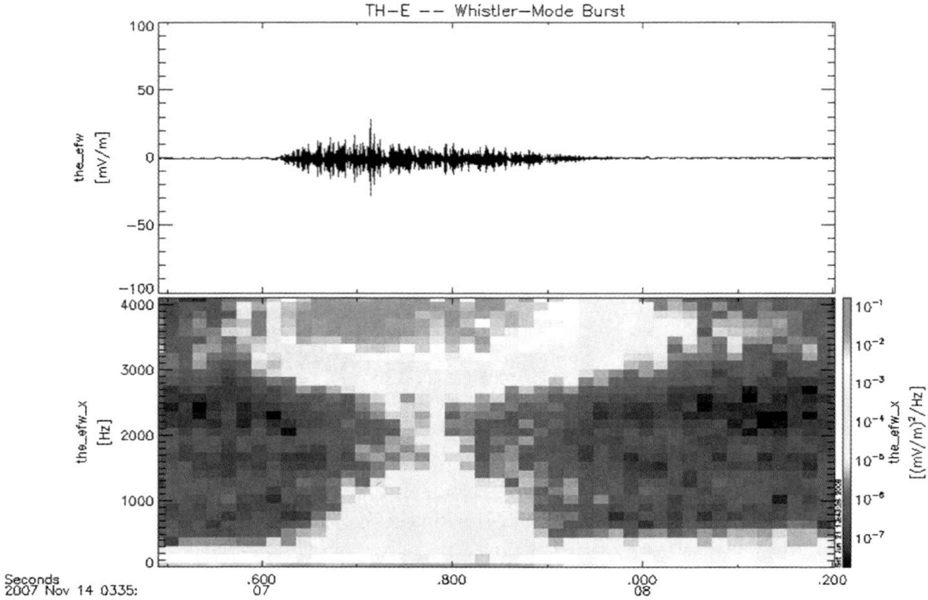

Fig. 10 Spectrogram of single E-field component in spinning spacecraft frame (THE, 2007-11-14, 0335:07.06) (note: AC attenuation factor of ~0.6 not compensated for in this plot)

consistent with the predictions of the Hall term of the generalized Ohm's Law (Mozer et al. 2008):

$$E + V_i \times B = j \times B/en - \nabla \circ P_e + \left(m_e/ne^2\right)(\partial j/\partial t) + \eta j.$$

Figure 12 demonstrates this agreement at the level of fractions of a mV/m out of a several to 10 mV/m E field for two crossings by THD. The top panels show the variation in the GSE-Z component of **B** across the magnetopause, while the bottom panels show the estimated Hall term computed from $\mathbf{E} + \mathbf{V}_i \times \mathbf{B}$ and $\mathbf{j} \times \mathbf{B}/en$ (**j** is computed from the temporal variation in B combined with the estimated magnetopause velocity computed from multi-satellite timing). In each case, the two estimates agree to better than a mV/m over the bulk of the magnetopause crossing, demonstrating the dominance and importance of the Hall field for magnetopause electrodynamics during these crossings. In addition, the Hall field occurs on the magnetospheric (low-density, high-**B**) side of the magnetopause, unlike the symmetric quadripolar Hall field distributions observed around magnetotail reconnection sites (e.g. Oieroset et al. 2001). This asymmetric distribution of the Hall E-field has be shown to be consistent with the predictions of the generalized Ohm's law (see Mozer et al. 2008 for a detailed analysis).

3.3 Electron Diffusion Region-Scale E-Field Structures at the Dayside Magnetopause

The electron diffusion region associated with magnetic field reconnection is a site whose size is $\sim c/\omega_{pe}$ (~4 km for a plasma density ~2 cm^{-3}), where ω_{pe} is the electron plasma frequency, and that contains a parallel electric field such that electrons are not frozen to magnetic field lines within it. Electron diffusion regions have been observed by both the Polar (Mozer 2005 and references therein) and Cluster satellites (Mozer et al. 2005), but the

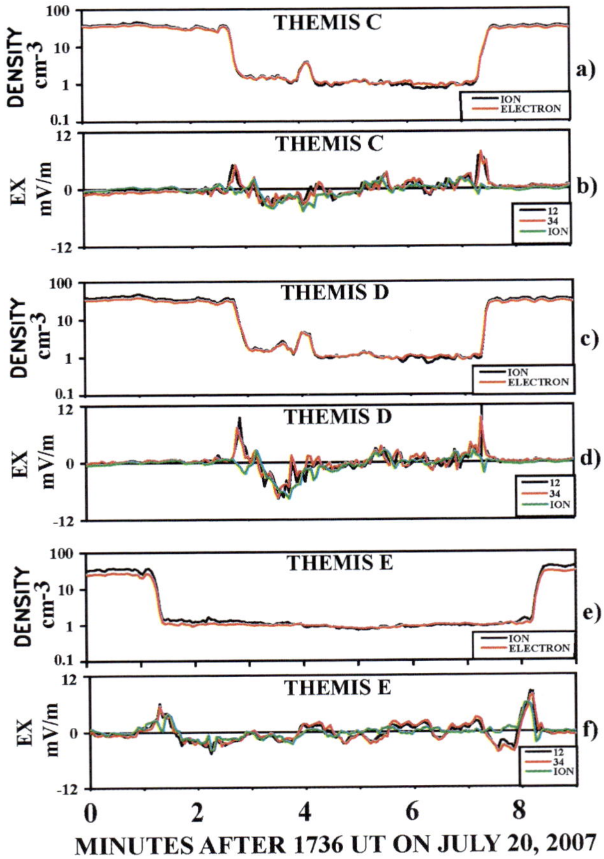

Fig. 11 Density and electric field estimates during magnetopause crossings by THEMIS-C, D, and E on 20 July 2007 (courtesy AGU)

plasma data did not have adequate time resolution for quantitative studies of the physics in such regions. THEMIS, with its high speed bursts, offers the possibility of such studies, so it is important to verify that THEMIS sees the electric fields in candidate electron diffusion regions.

Figure 13 presents one example of a candidate electron diffusion region observed by THEMIS-D at 0041:39 UT on 24 June 2007. At this time, THEMIS-D was located at a geocentric distance of 11.8 R_E, and a GSE local time and latitude of 1500 and -15^0 respectively. The top three panels of this figure present one second of the estimated perpendicular component of the electric field, E_{perp}, in the GSE coordinate system, calculated from the spin plane wire boom measurements using the $\mathbf{E} \cdot \mathbf{B} = 0$ assumption described above. At this time, the spacecraft was passing from the magnetosphere into the magnetosheath as is evidenced by the increasing plasma density (estimated from spacecraft potential) in the middle plot and the changes of the magnetic field components of the bottom three plots. The key feature of this interval is the spike in E_{perp} occurring at 0.7 s into the interval, indicated by the red rectangle. Here, the electric field reached an amplitude of nearly 100 mV/m over a region that lasted about 30 ms in the spacecraft frame. For a magnetopause velocity across the spacecraft of 10–20 km/s, as estimated from the differences in magnetopause crossing times of the multiple THEMIS spacecraft, the size of this region is less than 1 km, or a fraction of an electron skin depth for these plasma conditions. Thus, this event is a candidate

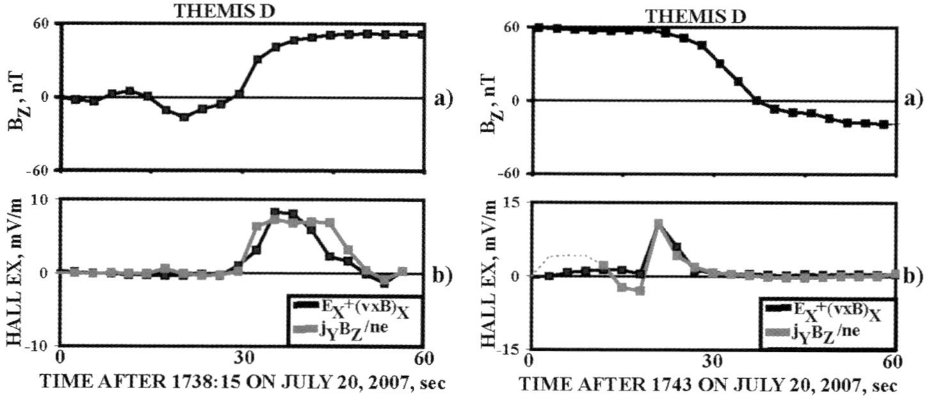

Fig. 12 Comparison of Hall E-field estimated during inbound and outbound magnetopause crossings of THEMIS-D

Fig. 13 Candidate electron diffusion region at the dayside magnetopause observed by THEMIS-D

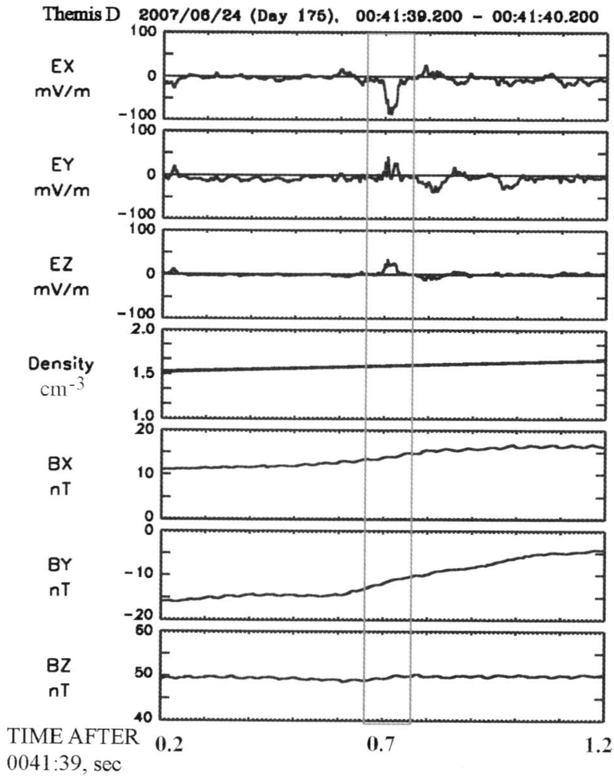

electron diffusion region and it remains an exciting challenge to analyze such events that are found when high time resolution plasma data is also available.

4 On-Orbit Performance and Operation

Ground testing and calibration of the EFI sensors, BEB, and DFB demonstrated that gains, offsets, and other relevant operational parameters were within specification and within better than 0.1% of each other, both taken in pairs (for a single axis), as well as across the entire instrument complement. This allows for a significant simplification in operation of the EFI across the constellation, as very little in the way of individual instrument character needs to be taken into account when determining the optimal bias configurations.

Once on orbit, a set of optimization and calibration procedures are run in order to determine the parameters required to convert the potential differences measured by the EFI sensors into E-field estimates, as well as to minimize the systematic errors in those estimates. This process for the spin plane sensors is described in some detail below, and is followed by a discussion of the relative level of confidence and systematic error between the spin plane and axial EFI sensors.

4.1 On-Orbit Bias Optimization

In order to control the magnitude and impact of the sources of systematic error present in the EFI measurement in the tenuous plasma of the magnetosphere, it is necessary to bias both the sensor itself, as well as the surrounding conductive surfaces. The sensor is current-biased in order to minimize both the DC offset voltage of the sensor with respect to the surrounding plasma, and the small-signal impedance of the plasma sheath around the sensor. Figure 14 shows a model current–voltage characteristic for the sphere portion of one of the EFI SPB sensors in sunlight in the tenuous plasma of the central plasma sheet (density of 0.3 cm^{-3}, T_e of 600 eV, T_i of 4.2 keV (assumed protons), sunlit conditions). The model includes ambient electron and ion fluxes, as well as photoelectron fluxes from the sensor surface.

The top two panels show the current collected by the sensor as a function of potential difference between the sensor and the local plasma potential. The topmost panel shows the contributions from the photoelectrons, and ambient electrons and protons (dashed, solid, and dotted lines, respectively). The third panel shows the small-signal resistance $(1/(dI/dV))$ of the sheath, and is a direct measure of the sensitivity of the sensor potential to variations in current due to either changes in the ambient plasma conditions (density or temperature variations), or differences in the collection of photoelectron currents from nearby or distance sources. The fourth panel shows this sheath resistance as a function of bias current to the sensor.

Minimization of the sheath resistance is very important for making accurate DC E-field measurements due to the relative magnitudes of the external fields and error sources on the measurements. For example, a 1 mV/m field associated with a 100 km/s flow in the central plasma sheet (CPS) produces a potential difference of at most 50 mV between the 50-m spin plane antenna sensors (The actual potential difference is less due to the partial "shorting" of the external field due to the grounded portion of the EFI booms). Typical photoelectron current sources to the EFI sensors are on the order of 1–100 nA, with differences in the 1–10 percent range ($\approx$10 nA maximum). As can be seen in Fig. 14, an unbiased sensor would have a sheath resistance on the order of 10^9 Ω, leading to error voltages on the order of 10–100 mV, and very poor signal-to-noise ratio. By biasing the sensor, the sheath resistance is reduced by a factor of 100–1000, into the 10^7 Ω range, leading to error voltages on the order of fractions of a mV, dramatically improving the accuracy of the measurement.

As noted in Laasko et al. (1995), there is a broad optimum (minimum) in the sheath resistance for a sunlit sensor in a tenuous plasma centered around approximately half the

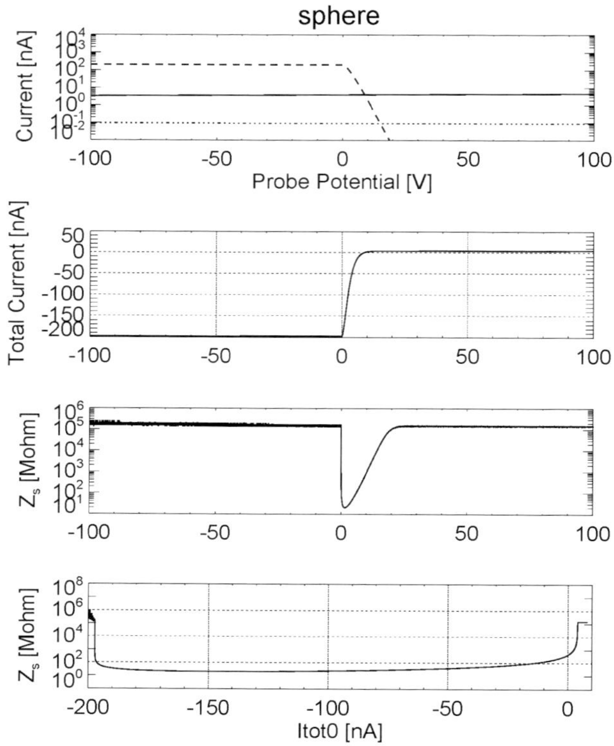

Fig. 14 Current–voltage characteristic and small-signal sheath resistance for spherical sensor

total photoelectron emission current (in this case, ≈120 nA/sensor for the assumed 4 nA/cm^2 photoelectron emission current density), giving a typical sheath resistance of several tens of MΩ for optimal bias. This can be readily seen in the bottom panels of Fig. 15 where the sheath resistance is plotted as a function of IBIAS. The relative insensitivity of the sheath resistance to bias current around the minimum is important for operation of the THEMIS EFI sensors, due to the relatively small collection area of the spacecraft body and other grounded appendages.

Figures 15 and 16 demonstrates this issue, and shows the expected floating potentials as a function of the bias current per sensor of both the central spacecraft (red triangles) and the individual EFI sensors (blue cross and plus sign; and green diamond, corresponding to SPB sphere+fine wire, SPB sphere only, and AXB whip) for nominal central plasma sheet and plasma sheet boundary layer conditions (i.e. nominal operational conditions for EFI in the magnetotail).

In order to supply the desired bias current to the EFI sensors (electrons flowing from spacecraft ground to the sensors and out in order to balance a significant fraction of the photoelectron current), the spacecraft will charge positive with respect to the ambient plasma in order to collect sufficient ambient electrons and spacecraft photoelectrons to balance the current to the EFI sensors. This can only occur up to a potential difference between the EFI sensors and spacecraft ground of up to the dynamic range of the EFI floating ground system, in this case approximately ±80 volts when the details of the bias current driver system are taken into account. At that point, the EFI sensors saturate at a fixed voltage, and no longer directly measure external field variations (note that the individual sensor and spacecraft potentials relative to the ambient plasma shown in Figs. 15 and 16 saturate at the extremes

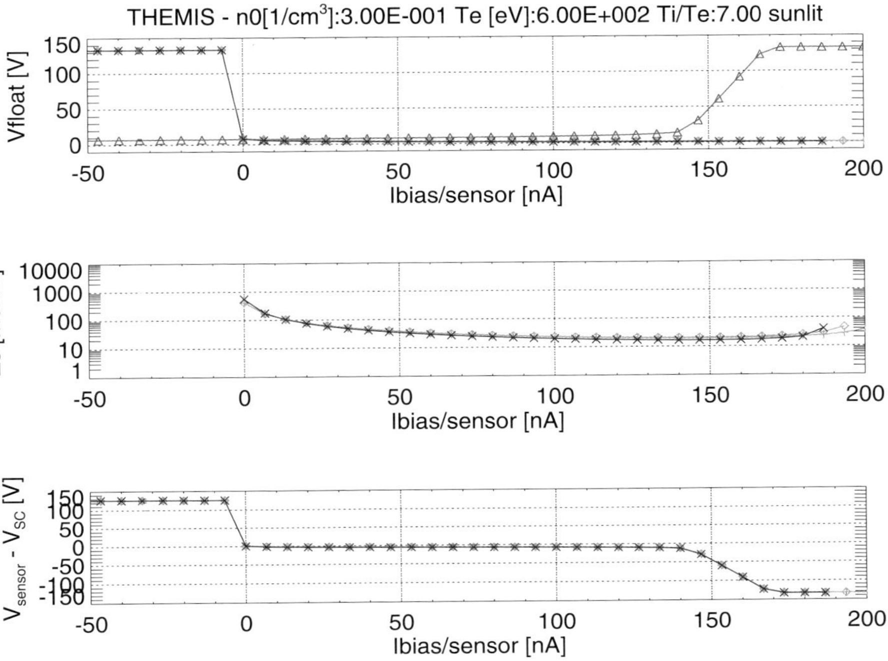

Fig. 15 EFI sensor and spacecraft floating potentials (*top panel*) and sheath resistances (*middle panel*), and sensor potentials relative to spacecraft (*bottom panel*) as a function of sensor bias current for nominal central plasma sheet conditions

of the bias current scale due to the limited range of potentials used in the current–voltage curve computations, rather than due to an actual physical mechanism).

In addition, the potential of the spacecraft directly affects the ability of body-mounted particle detectors (such as the ESA or SST) to collect or accurately measure low-energy particle populations, and so moderating the excursions in the spacecraft potential as ambient conditions change figures into the choice for the EFI sensor bias current as well.

Comparing the potential curves for the plasma sheet and (lower-density) PSBL, one can see that the range of sensor bias currents that maintain the spacecraft potential in the few volts to tens of volts range is significantly less in the PSBL than in the CPS ($\approx$20 to 30 nA/sensor, rather than the optimal 120 nA/sensor, effectively doubling the sheath resistance from 20 MΩ to 75 MΩ). On orbit, the optimal range of bias current is less restricted than one would expect from the model shown above due to the recollection of sensor photoelectrons by the central spacecraft, allowing for bias currents up to $\approx$180 nA/sensor. However, because the recollection process is difficult to model accurately, the EFI was designed with the more stringent operational requirement of relatively low bias currents in place.

In addition to optimization of the sensor's operating point, the use of voltage-biased surfaces to control the emission and collection of photoelectrons by the sensor and nearby surfaces is an important aspect of EFI design and operation. While the approximate current densities and characteristic energies (4 nA/cm^2 and few eV, respectively) of the photoelectrons emitted by the neighboring surfaces are known with some accuracy (Whipple 1981), the potential distributions surrounding the EFI sensors, booms, and the central spacecraft are

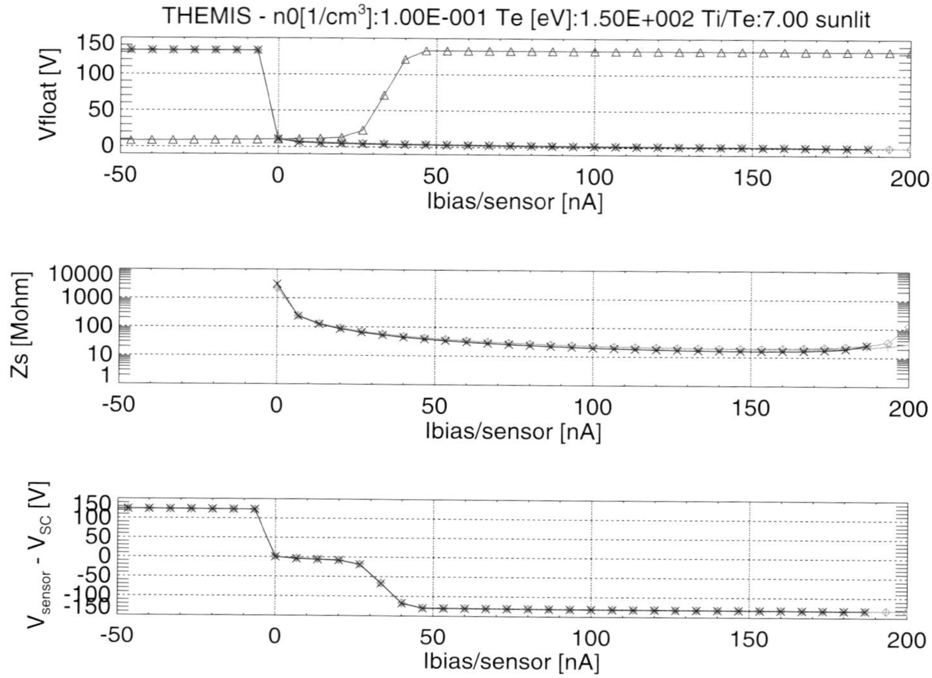

Fig. 16 Same as Fig. 15, but for nominal plasma sheet boundary layer conditions

not trivial to model (e.g. Cully et al. 2006), and the estimation of photoelectron exchange between surfaces in these complex potential structures are difficult to model accurately enough to predict on the ground the exact bias conditions required for optimal EFI operations.

Thus, the voltage-biased surfaces near the sensor (usher, guard, DBraid) were designed with sufficient dynamic range (±40 volts relative to sensor potential) to allow those surfaces sufficient leverage over the photoelectron fluxes, both in terms of the voltage relative to the characteristic energies of the photoelectrons, and the expected falloff in space of the potential structures around the surfaces due to the relatively small sizes (the usher and guard surfaces are a few cm in each dimension; the DBraid is three meters long by a fraction of a cm in diameter). The exact biasing of both the sensor and surrounding surfaces was determined on-orbit by stepping through a likely set of bias currents and voltages, and monitoring the response of the individual EFI sensor potentials and measured E-field estimates through Sensor Diagnostic Tests (SDTs), as well as examining any interference effects on the measurements made by other instruments (the low-energy plasma measurements made by the ESA in particular).

4.2 Sensor Diagnostic Tests

Figure 18 shows one portion of such a SDT run, taken from THEMIS-A on 15 Jan 2008, 0230–0313 UT. The left panel shows the entire SDT run on the X-axis sensors, while the right panel zooms in on one IBIAS sweep to show the variation in sensor potential and sensitivity to stray photoelectron currents with bias current.

The current and voltage sweeps used in this SDT are shown in the bottom three panels of the plot (IBIAS, USHER, and GUARD). This SDT run used five steps on each of the

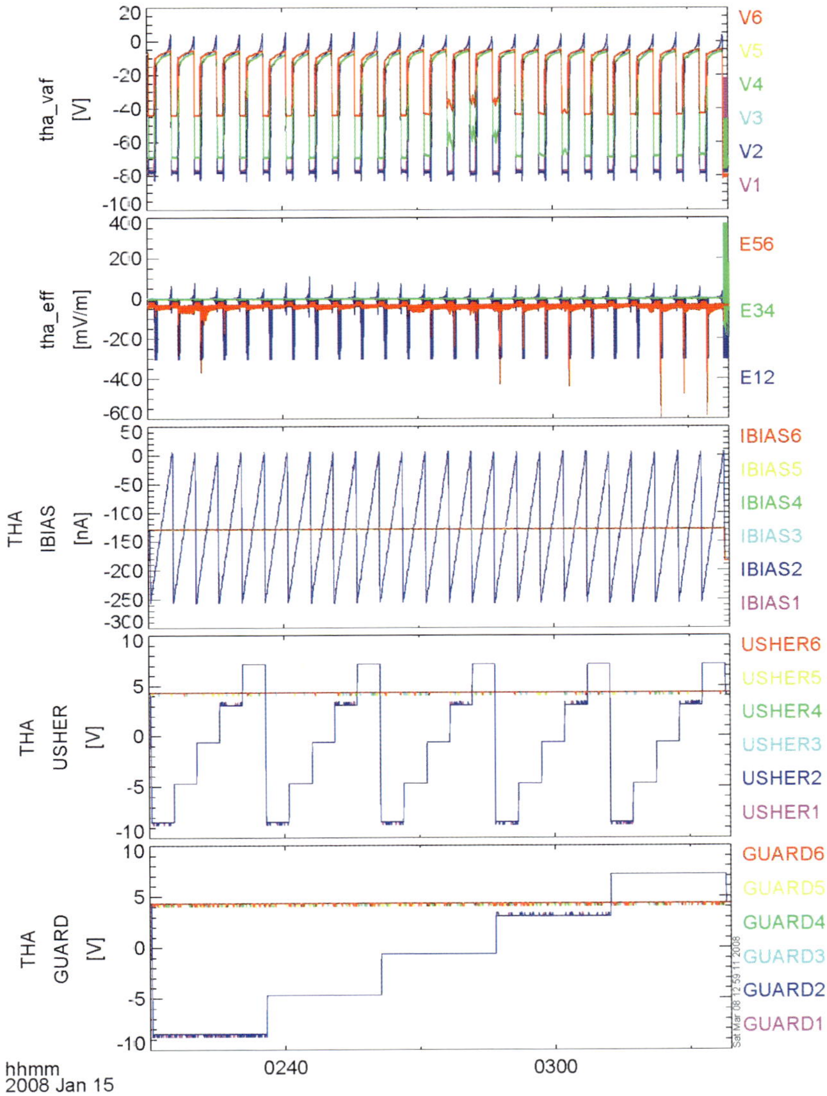

Fig. 17 Example data taken during an X-axis SDT run on THEMIS-A, 15 Jan 2008, 0230–0313 UT; panels described in text

Usher and Guard voltages from ≈ -8 to $+8$ volts relative to the sensor, and 16 steps on the bias current, from -260 to $+10$ nA/sensor. The bias currents and voltages are kept constant for a single spin of the spacecraft (3 s). The bias currents and voltages on the Y and Z axis sensors are kept constant during the sweeps on the X axis so as to use those sensors to monitor the ambient field for comparison against the measured field on the X axis sensors and variations in spacecraft potential driven by the changing bias current. The top two panels show the measured sensor voltages and potential differences in the body-fixed, spinning coordinate system aligned with the spacecraft geometric and EFI boom axes ("spinning

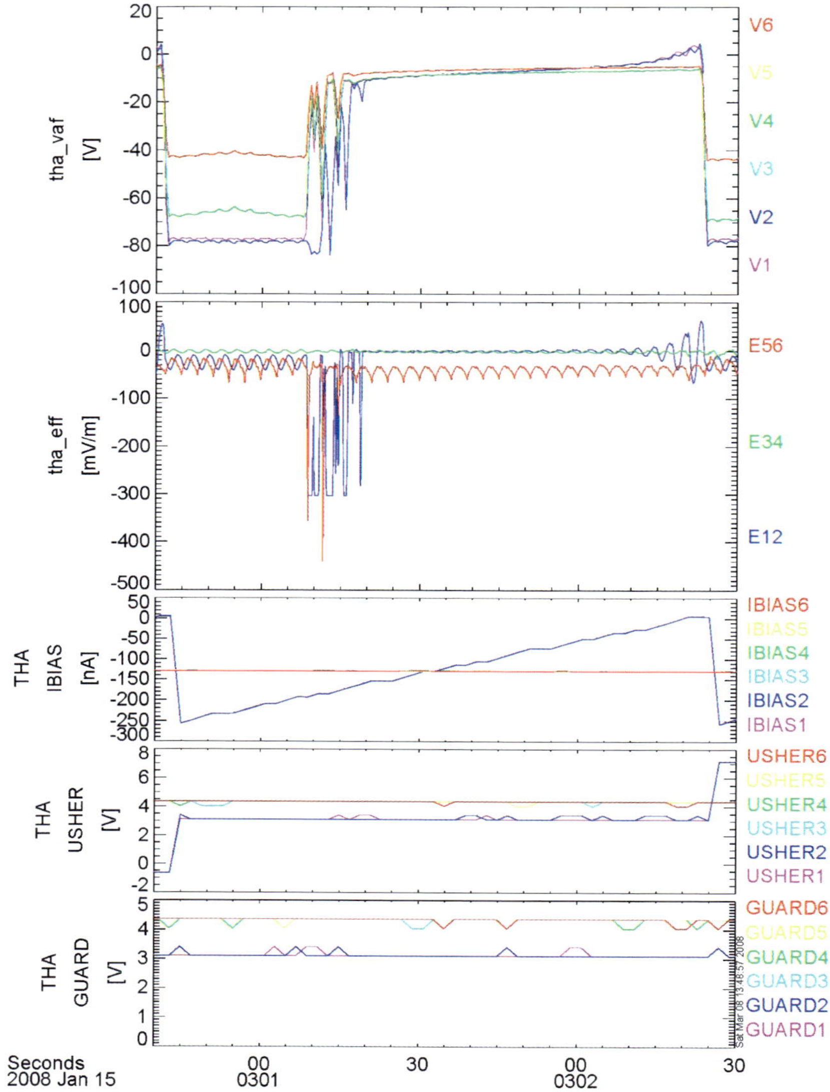

Fig. 18 Expanded view of single bias current sweep from X-axis SDT on THEMIS-A, 15 Jan 2008, ≈0300:30–0302:30 UT

probe geometric", or SPG). The typical quasi-DC E-field amplitudes due to geophysical processes in the region where this SDT was run are on the order of a few mV/m, and so one can clearly see the need for optimization of the biases, with measured potential differences between the probes corresponding to >200 mV/m over most of the parameter space.

The bias current sweep corresponding to the usher and guard potentials of ≈+4 volts is shown in more detail in the right panel of Fig. 18. Initially, the −250 nA of current to the sensor is greater than the saturation photoelectron current, driving the X-axis sensors into saturation at −80 V (blue and magenta traces in the top panel), as well as driving the spacecraft floating potential significantly positive (sensors negative relative to spacecraft ground) in order to collect enough electrons to satisfy the current bias as can be seen in the Y- and Z-axis sensor traces (cyan and green; yellow and red, respectively). The marked difference between the Y- and Z-axis potentials is due to the significant difference in distance of those sensors from the spacecraft ($\approx$20 m vs. $\approx$2.5 m), and is due to the radial falloff in potential around the spacecraft.

The X-axis sensors come out of saturation at a bias current between 190 and 150 nA/sensor. This indicates that the saturation photoemission current is in this range. The predicted photocurrent is 200 nA; eventually, the sensors achieve this photoemission after some time ($\approx$ one month) on orbit, based on both previous missions and data from earlier deploys on THEMIS. As the magnitude of the bias current is reduced, there is a slow variation in the magnitude of the E12 signal, increasing from a few mV/m up to 120 mV/m peak-to-peak as the bias current drops to zero and goes positive. The positive bias currents have a very modest effect on the spacecraft floating potentials, although the X-axis sensors themselves do go positive with respect to the spacecraft in order to collect electrons themselves, as predicted from the biased I–V curves (c.f. Fig. 15 for negative bias currents (current sense definition is opposite to this discussion)). The variation of the magnitude of E12 with IBIAS is consistent with the predicted variation of sheath resistance with bias current, assuming that the differences in photoelectron emission and collection from nearby sources are roughly constant with respect to IBIAS.

SDTs were run on each of the three EFI sensor axes during the boom deploy sequence, as well as several times afterward in the fully-deployed configuration in order to determine an optimal set of bias currents and voltages for the primary regions of interest for the THEMIS mission. The IDPU allows for storage of up to four different bias configurations on-board, and switching between those setups (sunlit/eclipse, low-/high-density conditions, for example), but that flexibility has not been exploited thus far in the first year of operations on THEMIS.

The optimal sensor current biases, the positive voltage biases on the usher and guard surfaces, as well as the zero-offset, V1-driven DBraid configuration that is the nominal flight configuration resulted from a campaign held on THEMIS-C, D, and E in the dayside magnetosphere and magnetosheath during June–July 2007. During this campaign it was found that negative biasing of the usher and guard led to relatively intense, nearly mono-energetic, low-energy (few tens of eV) fluxes of electrons impinging upon the ESA detector at energies above the nominal cutoff in returning spacecraft photoelectron fluxes. These fluxes were several orders of magnitude higher than the fluxes due to ambient electrons, and constituted a significant source of error on the ESA measurements and for the on-board electron moment calculations. Such intense return fluxes for negative guard biases have not been observed on previous missions even though the guard surface (or its analog) is routinely run at a negative bias voltage so as to block the access of spacecraft photoelectrons to the region around the sensor. The relatively compact dimensions of the THEMIS satellite and its boom systems may lead to a greater susceptibility to this effect than on other spacecraft, or the effect may not have been noticed in the low-energy electron data on previous missions.

4.3 Spacecraft Potential Variations

Once an optimal set of current and voltage biases are determined, there are still significant (few tenths of volt) variations in spacecraft and sensor potentials as a function of the aspect angle of the sensors and booms with respect to the Sun. A representative example of these variations is shown in Fig. 19, where five seconds (just under two full spin periods) of data

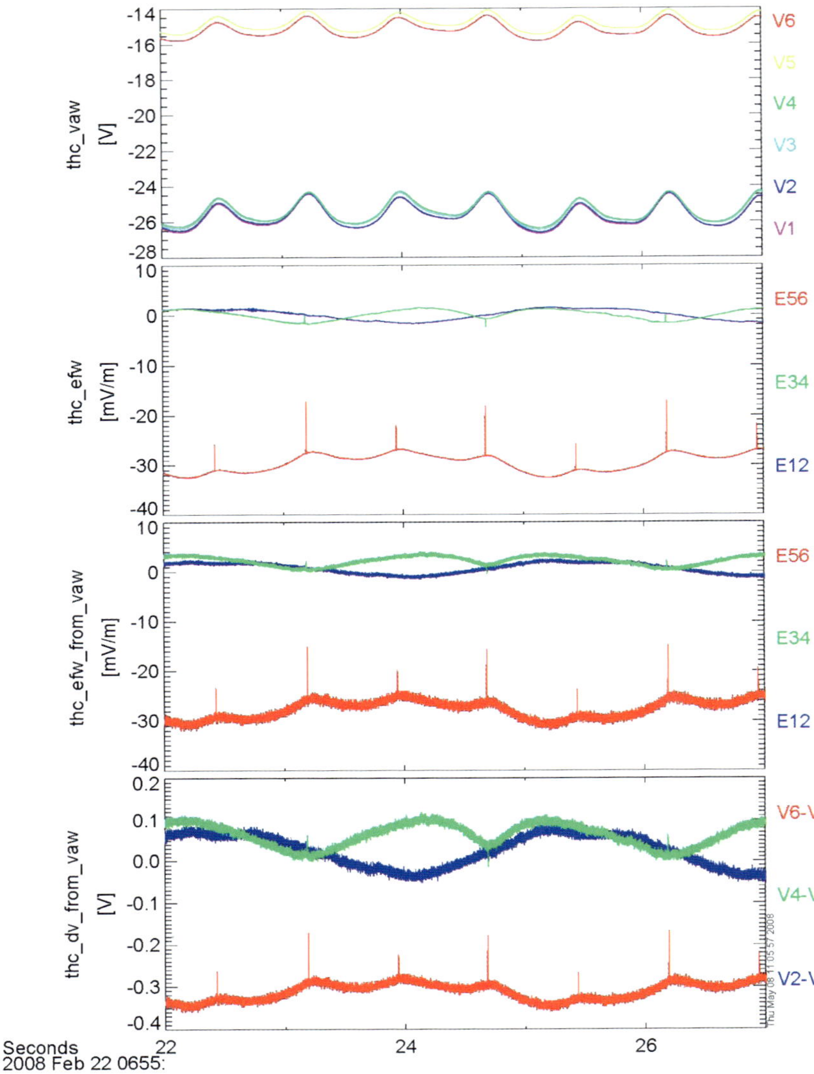

Fig. 19 Representative spin-periodic variations in sensor potentials, on-board differential channels, and ground-computed differential channels (TH-C, 22 Feb 2008, ≈0655 UT)

from THEMIS-C taken on 22 Feb 2008, at ≈0655 UT are shown. The top panel shows the six individual sensor potentials relative to spacecraft ground, the middle panel the three on-board DC-coupled differential channels, and the bottom panel the equivalent differential measurements computed on the ground from the individual sensor potentials. The highest rate (8192 samp/s wave burst data; vaw and efw) data products are used in this comparison to bring out the finest scale features of the potential variations.

The primary fluctuation one sees in the sensor potentials is a four-per-spin variation in the potential with a magnitude (in these plasma conditions) of ≈1 V. This variation represents a spin-periodic fluctuation in the spacecraft floating potential, and arises because of the

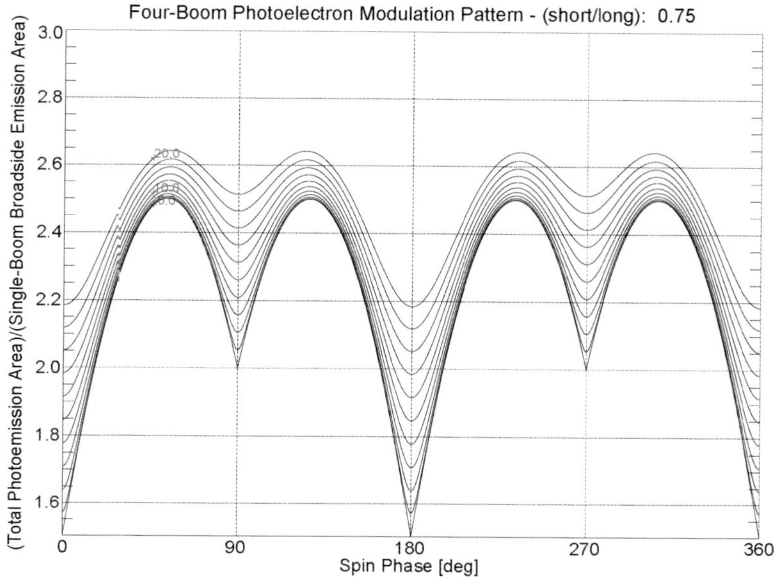

Fig. 20 EFI wire boom photoemission area (*ordinate*) as a function of spin phase (*abscissa*) and sun angle (multiple curves; *red labels*) for fully-deployed configuration

relatively large photoemission and collection area of the PBraid portion of the SPB wire booms as compared to the exposed area of the central spacecraft (≈25–30%). Figure 20 shows the expected variation of this emission area as a function of spin phase and Sun angle (minimum angle of Sun with respect to spin plane), which matches quite well with the observed variation.

The several volt difference in potential difference between the spin plane (V1, ..., V4) and axial sensor potentials (V5, ..., 6) arises due to the radically different separations of those sensors from the central spacecraft; essentially, the axial sensors are ≈10 times closer to the center of potential, and see a smaller floating potential because of that difference. The much smaller (few tenths of a volt) difference between the long (V1, ..., V2) and short (V3, ..., V4) sensor potentials is also consistent with this hypothesis. The four-per-spin spikes in the axial measurements arise from the momentary shadowing of an axial sensor by one of the four wire booms during this particular season on this spacecraft.

4.4 Cross-Calibration and Determination of Boom Shorting Factors

From the example shown in the Sensor Diagnostic Test section above, one can see that when properly and symmetrically biased, systematic differential errors can be minimized, and the large-amplitude common-mode signal due to the difference between the spacecraft and sensor floating potentials can be subtracted from opposing sensors, leaving a signal that is linearly related to the ambient electric field.

If the external electric, magnetic, and plasma flow fields are uniform and quasi-static, and the fluids themselves are collisionless, then the fields and flows are related by the standard ideal MHD approximation to the generalized Ohm's law, $\mathbf{E} = -\mathbf{V}_i \times \mathbf{B}$, where $\mathbf{V}_i$ is the ion bulk flow velocity. Thus, by finding intervals of time and regions of space away from strong current systems (leading to uniformity of $\mathbf{B}$) where the flow and electromagnetic fields are slowly varying, one can perform a straightforward correlation analysis between $\mathbf{E}and -\mathbf{V}_i \times \mathbf{B}$ in order to determine important EFI calibration factors, such as the so-called boom shorting factor and offsets arising from spin-dependent differential photoelectron emission and collection (the well-known sunward offset observed on Polar-EFI and Cluster-II-EFW). Once having done that, one can then use the calibration factors to accurately determine the electric field in regions where $\mathbf{E}$ does not equal $-\mathbf{V}_i \times \mathbf{B}$ due to current systems, pressure gradients, flow shears and the like, and thus get at the details of the electrodynamics in such regions (The observations of Hall electric fields associated with asymmetric reconnection sites on the dayside magnetopause noted earlier from Mozer et al. (2008) is a prime example of this methodology).

The boom shorting effect is a well known (Mozer et al. 1974; Pedersen et al. 1998; Cully et al. 2006), but difficult to accurately predict, electrostatic effect that arises from the redistribution of charge on the spacecraft and EFI boom systems in response to a given external E-field. Even though the EFI sensors are electrically isolated from the spacecraft, the presence of the conducting booms and spacecraft between the two EFI sensors making up a given axis of measurement reduces the potential difference between those two sensors from that would be present in the absence of the boom system, effectively shorting out a portion of the external field. Comparison between the observed magnitude of $\mathbf{E}$ measured by EFI ($\mathbf{E}_{\mathrm{EFI}}$) to that predicted by B-field and plasma measurements (typically FGM $\mathbf{B}$ ($\mathbf{B}_{\mathrm{FGM}}$) and ESA ion vector ion velocity ($\mathbf{V}_{\mathrm{i,ESA}}$)) allows one to estimate the magnitude and variation of the boom shorting factor.

Note that in the magnetosphere and magnetosheath regions the ambient plasma flow velocities are almost always much larger than the spacecraft orbital velocity (tens to hundreds of km/s versus a few to less than 1 km/s). In the inner magnetosphere, e.g. the plasmasphere, this situation is reversed, and the contribution of the spacecraft orbital velocity dominates that of the plasma flow itself, so that if the ion velocity measurement is not available or is too inaccurate, the spacecraft orbital velocity derived from ephemeredes data can be substituted for $\mathbf{V}_{\mathrm{i}}$ and a similar calibration procedure performed.

The effects of spin-dependent photoelectron emission and collection, as well as any large-scale space charge asymmetries surrounding the spacecraft and its appendages (spacecraft photoelectron cloud; electrostatic structures (wakes) arising from ambient cold plasma interactions; etc.) are also detected and quantified through the comparison of $\mathbf{E}_{\mathrm{EFI}}$ and $-\mathbf{V}_{\mathrm{i,ESA}} \times \mathbf{B}_{\mathrm{FGM}}$.

An example of this cross-calibration process is shown in Fig. 21 which shows the correlation analysis for the sunward (left panel) and dawn-dusk (right panel) components of the spin plane spin-fit $\mathbf{E}_{\mathrm{EFI}}$ and spin-resolution $-\mathbf{V}_{\mathrm{i,ESA}} \times \mathbf{B}_{\mathrm{FGM}}$ for data taken on THEMIS-C, on 21 July 2007, 0825–1320 UT for intervals when the spacecraft was in the magnetosheath, well away from regions where significant non-uniformities could exist (e.g. the magnetopause). The individual data points are shown as crosses, a one-to-one relationship between $\mathbf{E}_{\mathrm{EFI}}$ (ordinate) and $-\mathbf{V}_{\mathrm{i,ESA}} \times \mathbf{B}_{\mathrm{FGM}}$ (abscissa) is shown by the dashed red line, and the best-fit linear model of each component of $\mathbf{E}_{\mathrm{EFI}}$ versus $-\mathbf{V}_{\mathrm{i,ESA}} \times \mathbf{B}_{\mathrm{FGM}}$ by the solid blue line.

One can clearly see that the EFI underestimates the magnitude of both the sunward and duskward components of the electric field relative to the prediction based on ESA and FGM data by a factor of 0.69, corresponding to a boom shorting factor of $1/(0.69) \approx 1.45$. One can also see the clear sunward offset in the EFI data of $\approx$0.8 mV/m, and the negligible offset in the dawn–dusk E-field component. Typical values of 2.5 mV/m for the sunward offset field, 0 mV/m for the dawn–dusk offset field, and boom shorting factors between 1.3–1.6 were observed in the calibration campaign performed using data from the THEMIS-C spacecraft during July–Aug 2007. These values have been adopted as the nominal calibration

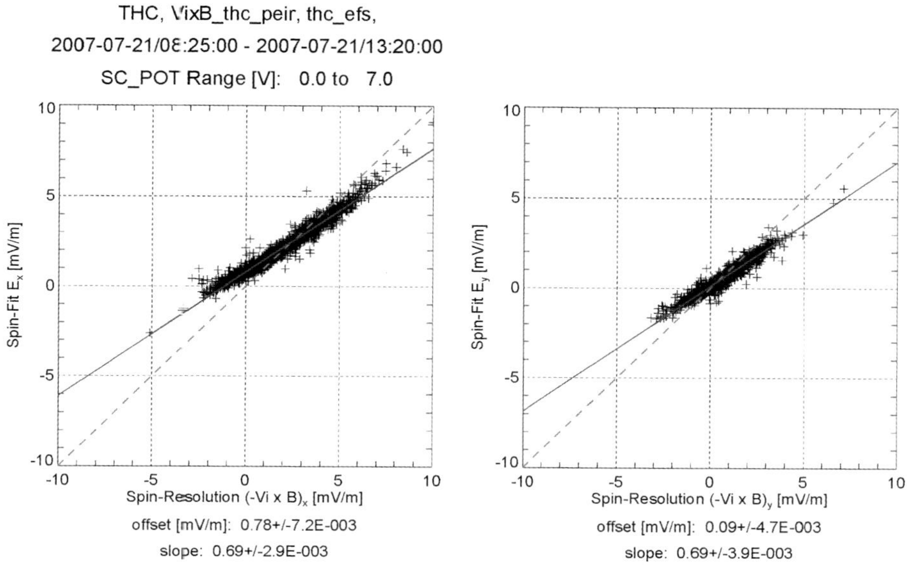

Fig. 21 Correlation analysis between EFI, iESA, and FGM data in the magnetosheath

parameters for data derived from the spin plane EFI sensors. For comparison, the boom shorting factors for the Polar-EFI have been measured to be on the order of 1.2 to 1.4 (F.S. Mozer, private communication, 2008); note that while the boom shorting factor is predicted to approach 1 as the boom length increases, one analytical theory (U. Fahleson, private communication, 1974) predicts a smaller boom shorting factor for THEMIS than Polar, given the detailed differences between the EFI sensor geometries on Polar and THEMIS. This points out the vital importance of this form on on-orbit calibration of the DC E-field data for accurate, quantitative work.

At a given value of $-\mathbf{V}_{i,ESA} \times \mathbf{B}_{FGM}$ one can see a spread in $\mathbf{E}_{EFI}$ of $\approx$2 mV/m. This spread represents both the level of accuracy in the EFI measurement, as well as the intrinsic differences between the EFI, ESA, and FGM measurements—while the EFI and FGM data are spin-fit, the estimates of the ion velocity come from spin-resolution 3D particle measurements, and are subject to a different set of errors than EFI and FGM (counting statistics; aliasing of time-variable flows), and so the accuracy of the DC E-field estimate provided by EFI is on the order of 1 mV/m over long intervals of time. More detailed calibrations may be made over shorter time intervals, and accuracies of better than a few tenths of mV/m can be achieved in those cases (e.g. Mozer et al. 2008).

4.5 Electrostatic Wake Effects

The presence of significant densities of cold ($<$ tens of eV) plasma in the magnetosphere is well-documented, and the impact of that cold plasma on double-probe E-field measurements through electrostatic wake formation has been quantified and modeled using Cluster-II observations (Puhl-Quinn et al. 2008; Engwall 2004, internal report). The impact of such electrostatic wake effects on the THEMIS EFI measurements is significant, as the occurrence of significant cold plasma densities throughout the magnetosphere (dayside in particular) is far higher than previous observations may have suggested (McFadden et al. 2008a).

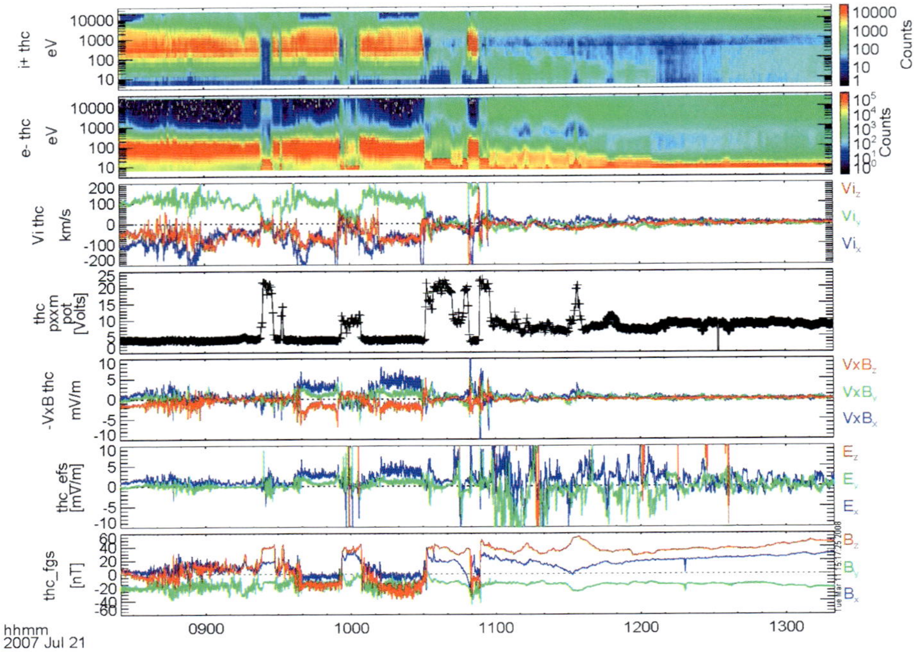

Fig. 22 Plasma and fields data from THEMIS-C, 21 July 2007, 0825–1320 UT

Figure 22 shows the impact of cold plasma wake fields in EFI measurements during a series of magnetopause crossings observed by THEMIS-C on 21 July 2007. The top two panels show the ion and electron count rates (proportional to differential energy flux) as a function of energy. The third panel shows the ion velocity in despun spacecraft coordinates (DSL, close to GSE). The fourth panel shows the estimated spacecraft floating potential. The fifth and sixth panels show the estimated E-field from spin-resolution $-\mathbf{V}_{\mathrm{i,ESA}} \times \mathbf{B}_{\mathrm{FGM}}$ and $\mathbf{E}_{\mathrm{EFI}}$, and the seventh panel shows the spin-fit $\mathbf{B}$. Comparison of the blue and green traces of the two E-field estimates reveals intervals of quite good agreement (typically magnetosheath, prior to 1030 UT) mixed with intervals of profound disagreement (typically plasma sheet, after 1100 UT). The presence of a significant cold ion population can be seen in the 10–100 eV energy range below the peak in flux due to the main plasma sheet ion population, and is well-correlated with significant (tens of mV/m) discrepancies between the two E-field estimates.

While the spurious fields generated by electrostatic wakes are large, the multiple boom lengths available on THEMIS allow for the discrimination of these spurious fields whenever waveform data are available. This strategy is similar to that used on Polar-EFI (Harvey et al. 1995), although the longer boom lengths on Polar (130-m/100-m) mean that the impact of wake fields is much less than on THEMIS (theoretical predictions and observations suggest that the impact of ES wake fields scales with boom length, L, as $1/L^2$ or $1/L^3$, depending upon the details of spacecraft geometry and boom grounding schemes). The Cluster-II/EFW has booms of equal length (88 m), and so must rely upon either comparison with electron-drift-based E-field estimates or careful comparison between plasma densities provided by the on-board plasma instruments and that provided by spacecraft floating potential estimates to infer the presence of such spurious E-fields.

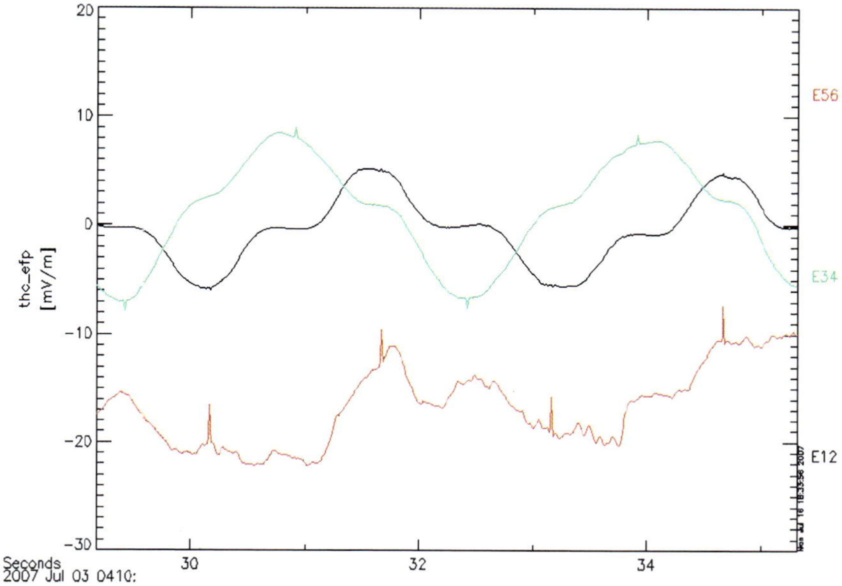

Fig. 23 Spin plane (*black, green*) and axial (*red*) E-field measurements during cold plasma wake event (THEMIS-C, 3 Jul 2007, 0410 UT)

Figure 23 shows an example of the body-frame E-field estimates during a instance of significant ES wake formation. The dark blue and green traces are the E-field estimates from the long (E12) and short (E34) spin plane booms. Unlike the signal from a uniform external E-field, the signals on the short boom antenna are significantly larger in amplitude than that on the long booms, and both signals show significant deviation from sinusoidal behavior. The deviation from sinusoidal behavior is reminiscent of that found in the ES wake models for the Cluster-II polar wind results (Engwald 2004). By comparing spin-period estimates of the variance of the two spin-plane E-field estimates, one can readily detect this effect and flag the data as arising from local electrostatic wake effects rather than external geophysical processes.

4.6 Axial Boom Performance

The E-field estimates from the relatively short axial EFI booms on THEMIS are strongly affected by variations in the spacecraft floating potential. Such effects scale with roughly the inverse third power of boom length for the displacement of the true charge center of the spacecraft from the center of the boom pair in question ($E_{\text{err}} \approx V_{\text{sc}} \cdot (2ad/L^3)$, where V_{sc} is the spacecraft potential, a is the effective radius of the spacecraft, d is the displacement of the charge center along the boom axis, and L is the length of one of the booms of the axis), and so one would expect this effect to be roughly a factor of 1000 larger on the axial booms than the spin plane booms. This can be seen in Fig. 24, where E-field in the despun spacecraft coordinate system (top three panels) and sensor potentials (V1, V3, and V5; bottom three panels) are shown. The marked correlation between the axial (DSC_Z) component of E and the sensor potentials demonstrates a common-mode (spacecraft potential) sensitivity of $\approx$4 (mV/m)/V, corresponding to a displacement of $\approx$6 cm of the charge center along the axial boom direction.

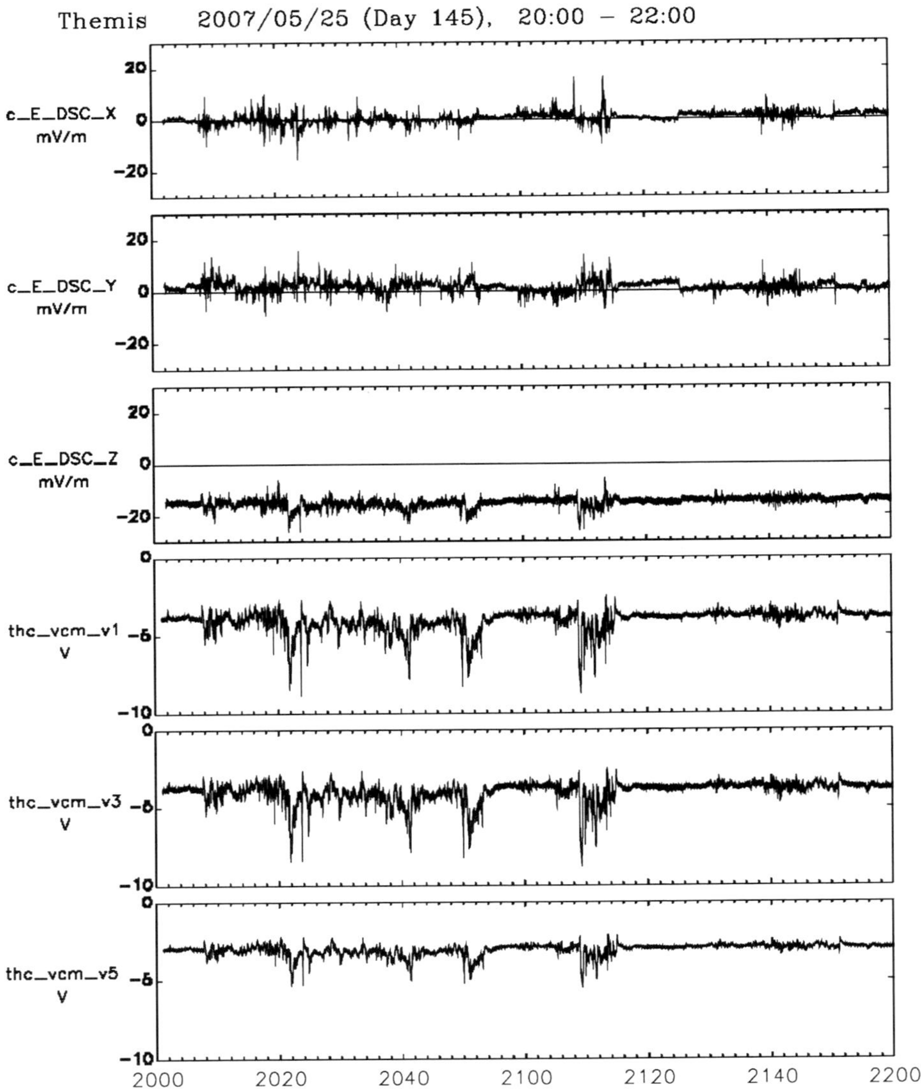

Fig. 24 Despun electric field estimates and sensor floating potentials observed on THEMIS-C on 25 May 2007, 2000–2200 UT, showing the correlation between the axial E-field estimate and spacecraft floating potential

Some displacement of the electrical center of the spacecraft from its geometrical center was expected, based on the $\approx$0.5-m antenna mast that protrudes from the top deck of the spacecraft. The actual displacement of the electrical center was expected to be less than this dimension, however, due to redistribution of charge on the surfaces of the grounded axial boom stacers. Thus, the axial booms were designed to deploy with the $+Z$ axis sensor being 4 cm further from the geometric center of the spacecraft body in an attempt to compensate for this charge center displacement effect, based on electrostatic modeling of the central body, antenna mast, and grounded portions of the axial booms themselves. The magnetometer booms were not included in this model, nor possible asymmetries in the photoelectron cloud between the top and bottom decks of the spacecraft, and either of these effects may be the source of the discrepancy between the predicted and actual position of the charge center.

Figures 25 and 26 demonstrate more quantitatively the significant issues involved with extracting a quasi-DC E-field estimate from the relatively short THEMIS axial boom system using data acquired by THEMIS-C during an inbound crossing of the magnetopause on 8 Aug 2007 between 00:04:00 and 00:005:30 UT. The top three panels of Fig. 25 shows the spin plane (efc_12 and efc_34) and axial (efc_56) E-field estimates in the spinning

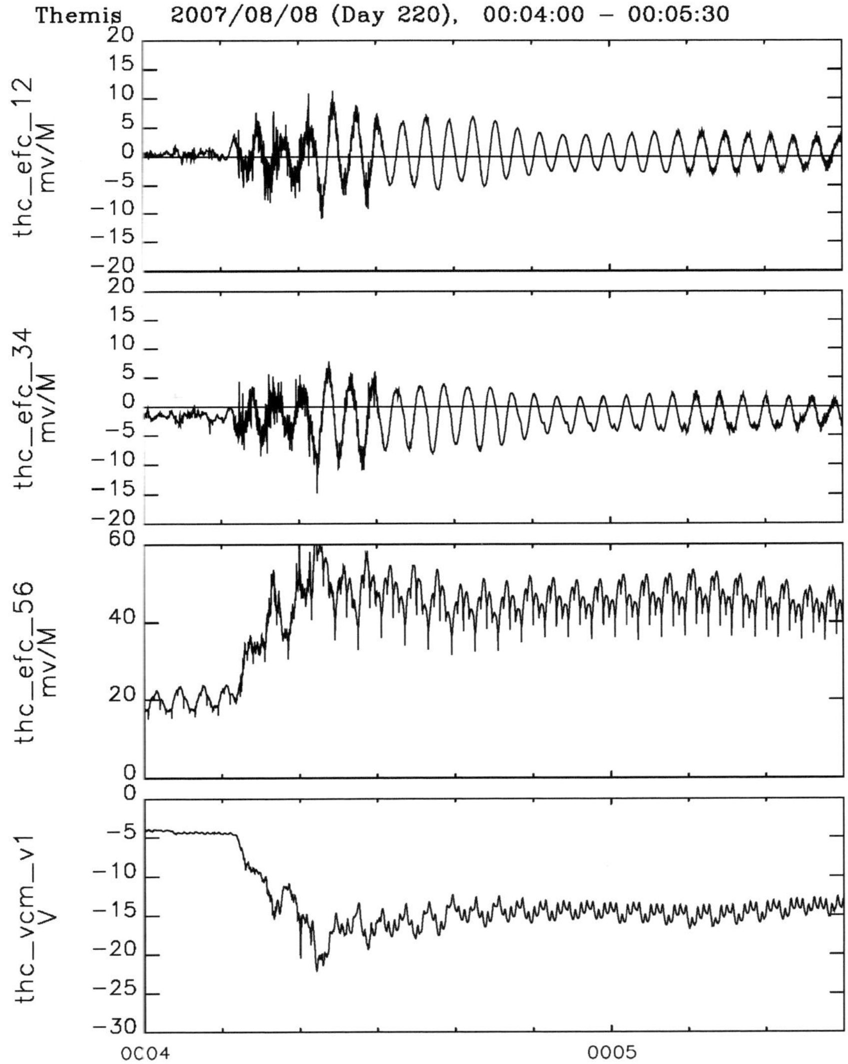

Fig. 25 Electric field and sensor potential measurements during an inbound crossing of the magnetopause (THC, 8 Aug. 2007, 00:04:00–00:05:30 UT). Panels described in text

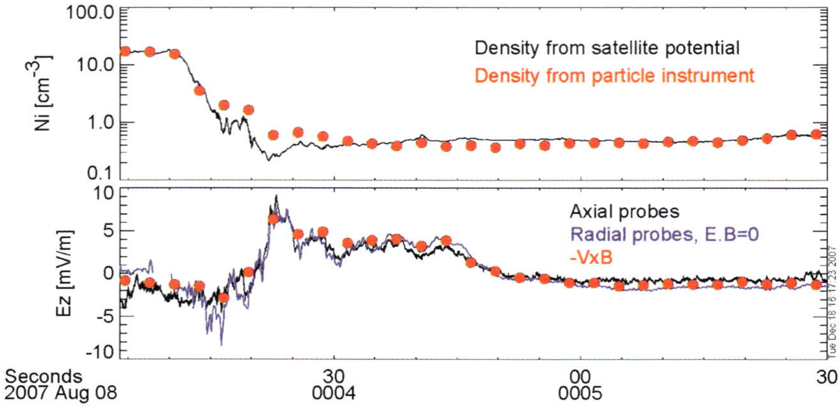

Fig. 26 Ambient ion density estimate (*top*) and axial E-field component (*bottom*) from particle and EFI instruments

spacecraft coordinate system. The potential of sensor 1 (spin plane, 25-m separation from spacecraft) is shown in the fourth panel as a proxy for the spacecraft floating potential. Similar to the example shown in Fig. 24, one can see the relatively clean signals measured by the 40- and 50-m long spin plane booms, and the marked correlation between the fluctuations in the $\approx$6-m axial measurement and the spacecraft potential, in this case with a proportionality factor of $\approx$2.7 (mV/m)/V (rather than the 4 (mV/m)/m observed in that event. In addition, one can see the change in the amplitude of the spin-dependent fluctuations in the axial signal, increasing from roughly 6 mV/m peak-to-peak on the high-density (smaller spacecraft potential) magnetosheath side of the boundary up to roughly 20 mV/m peak-to-peak on the low-density (larger spacecraft potential) magnetospheric side of the boundary. Most of this increase in the spin-dependent variation is due to the increase in the amplitude of the spin-dependent variations in spacecraft potential from the magnetosheath to magnetosphere (fractions of a volt to roughly 3 volts peak-to-peak), but the constant of proportionality changes from one side to the other, indicating some significant environment-dependent effects on the proportionality factor.

Careful modeling and removal of the common mode signal due to spacecraft potential variations from the axial E-field estimate, as well as filtering of the spin-periodic signals due to spacecraft potential variations and the few-degree tilt of the axial sensor from the spin axis, one can get quite good agreement between the axial E-field estimate and that predicted from $-\mathbf{V}_{\mathrm{i,ESA}} \times \mathbf{B}_{\mathrm{FGM}}$ and $\mathbf{E} \cdot \mathbf{B} = 0$, as can be seen in Fig. 26. One can also see that the axial component of $\mathbf{E}$ is of similar, although lesser, magnitude to the spin plane components of $\mathbf{E}$ in this example (5 mV/m versus 10 to 20 mV/m, based on the peak-to-peak amplitudes of the spin plane E-field estimates in FIG).

However, this agreement is achieved in the presence of large (tens of mV/m), event-dependent offset subtractions, opening the question of whether the small ($\sim$1 mV/m) excursions of the corrected axial signal from $-\mathbf{V} \times \mathbf{B}$ represent artifacts of the calibration process, or actual geophysical E-fields. When the relative difficulty (many parameters, event-dependent vs. few parameters, relatively constant) and accuracy of the calibration of the DC axial measurement is contrasted with that required for the spin plane measurements, the value of longer E-field booms in this context becomes clear. Note that while the short axial booms do lead to significant issues with the DC field measurement, the AC field measurement (above a few hundred Hz) does not suffer from these failings, and has been used to make clean, three-dimensional E-field estimates successfully (e.g. Cully et al. 2008a).

5 Summary

The THEMIS-EFI provides high-quality estimates of the near-ecliptic components of the DC electric field, allowing for accurate (better than 1 mV/m) estimation of the perpendicular components of E in the relevant plasma environments to support the THEMIS magnetotail observations. The DC measurement is susceptible to contamination by local electrostatic fields arising from cold plasma wakes, but the spin plane boom length differential allows for routine detection of this effect and monitoring of data quality. The instrument supports 3D E-field measurements at frequencies up to 4 kHz, and provides significant supporting data for studies of large-scale electrodynamics as well as smaller-scale wave phenomena.

Acknowledgements This work supported by NASA Contract NAS5-02099.

The authors would like to acknowledge all the members of the THEMIS EFI, IDPU, Science, Ops and Management teams without whose efforts the daunting task of completing more than six spacecraft of sensors and electronics for EFI (prototypes, ETU, flights, and spares!) within the aggressive schedule and tight resource allocations of the mission would not have been possible, or nearly as much fun:

P. Berg, M. Bester, R. Canario, N. Castillo, C. Chen, D. Cosgrove, L. Croton, B. Dalen, G. Dalton, B. Donakowski, R. Duck, M. Eckert, J. Fischer, S. Frey, D. Glasser, W. Greer, C. Grimmer, R. Gupta, S. Harris, S. Heavner, Y. Irwin, R. Jackson, S. Jelinsky, D. Larsen, M. Larsen, J.W. Lewis, M. Lewis, M. Ludlam, J. McCauley, J. McDonald, J.P. McFadden, K. McKee, S. Marker, S. Martin, D. Mielhan, T. Moreau, A. Nammari, D. Pankow, H. Richard, B. Roberts, D. Rummel, D. Schickele, C. Scholz, C. Smith, R. Sterling, K. Stevens, E.R. Taylor, P. Turin, J. Westfall, H. Yuan.

Appendix: The THEMIS Electrostatic Cleanliness (ESC) Specification

The understanding that careful consideration of surface charging effects and mitigation of possible sources of strong differential charging in tenuous plasma conditions is part of robust spacecraft design has deepened over the past decade (Garrett and Whittlesey 2000). However, the THEMIS science objectives for DC E-field and low-energy plasma measurements impose more stringent requirements on the magnitude of charging of spacecraft surfaces (few volts to few tens of volts in sunlight) as well as differential charging (<1 V differential between any exposed surfaces, with a goal of <0.1 V) between those surfaces than is typical for low-earth, geosynchronous, or high-earth orbit spacecraft that are not required to make accurate in situ E-field and particle measurements.

Realization of this fact led to an early and vigorous effort on the part of the instrument and spacecraft engineering teams towards providing an electrostatically clean spacecraft. This effort brought together the EFI science and engineering teams, and the lead mechanical, electrical, thermal, and systems engineers from the instrument and spacecraft teams in order to: specify the requirements for ESC on THEMIS; determine any problem areas on the spacecraft early and decide what, if any forms of mitigation could be used; select and verify appropriate materials and designs for exposed surfaces and the grounding thereof; and finally validate the final ESC implementation throughout the Instrument Suite and Spacecraft I&T flows.

While stringent ESC programs have been imposed on previous missions (FAST, Polar, Cluster-II, etc.), the aggressive schedule and limited budget of THEMIS meant that the implementation had to be streamlined and relatively foolproof in order to avoid both costs incurred to develop the implementation (large-scale electrostatic modeling, for example), as well as insure that the risk of finding an ESC issue late in the I&T flow was minimized. This goal of the management process led to the development of a modest library of surfaces and grounding schemes (round and square patches, grounded at corners or edges; surfaces embedded in apertures; insulating epoxy bond lines; etc.) that allowed for the specification of maximum levels of surface resistivity (for materials selection), as well as determination of the maximum allowed resistance to ground from any point on such a surface once installed (allowing for simple verification using an ohmmeter, rather than a surface resistance measurement). This is in contrast to the typical industry standard specification of a global maximum allowed surface resistivity for all surfaces on a given spacecraft, which tends to be stricter (and thus harder to comply with) as it can not take into account the inverse relationship between allowed maximum resistivity and exposed area of any given sub-assembly.

For THEMIS, this relationship was that the allowed resistance to ground of any exposed surface was at most on the order of $125\ \mathrm{M\Omega\text{-}cm^2}/A$, where A was the exposed area in cm^2, under the assumption that the primary source of differential current collection/emission was photoelectrons at a current density of 8 $\mathrm{nA/cm^2}$ (twice expected). The exact geometry of the exposed area and grounding contacts reduced or increased this value for any particular surface.

Verification of proper surface resistance and grounding was performed as early as possible in the I&T flow, and was an integral part of the close out and Quality Assurance process of each instrument, spacecraft sub-system, and completed vehicle. Specialized tools and techniques were developed and tested for use on potentially fragile surfaces, such as the indium-tin-oxide coated regions of certain thermal tapes and coatings, in order to provide a robust measurement of the resistances involved (few kΩ to tens of MΩ) while allowing for a broad range of personnel to use the equipment (thermal treatment technicians, I&T engineering staff, even instrument scientists).

The close correlation between predicted and observed levels of charging, as well as the relatively low biases seen in the EFI data suggest that this streamlined ESC implementation and verification plan was very successful, and should provide a model of such processes for future magnetospheric missions.

References

C. Cattell, J.R. Wygant, K. Goetz, K. Kersten, P.J. Kellogg, T. von Rosenvinge, S.D. Bale, I. Roth, M. Temerin, M.K. Hudson, R.A. Mewaldt, M. Wiedenbeck, M. Maksimovic, R. Ergun, M. Acuna, C.T. Russell, Discovery of very large amplitude whistler-mode waves in Earth's radiation belts. Geophys. Res. Lett. **35**, L01105 (2008). doi:10.1029/2007GL032009

C.M. Cully, R.E. Ergun, A.I. Eriksson, Electrostatic structure around spacecraft in tenuous plasmas, J. Geophys. Res. (2006)

Cully et al., THEMIS DFB, Space Sci. Rev. (2008a, this issue)

C.M. Cully, J.W. Bonnell, R.E. Ergun, THEMIS observations of long-lived regions of large-amplitude whistler waves in the inner magnetosphere. Geophys. Res. Lett. **35**, L17S16 (2008b). doi:10.1029/2008GL033643

E. Engwald, Numerical studies of spacecraft-plasma interaction: simulation of wake effects on the Cluster electric field instrument EFW. IRF Scientific Report 284, Feb 2004

H.B. Garrett, A.C. Whittlesey, Spacecraft charging, an update. IEEE Trans. Plasma Sci. **28**(6), 2017–2028 (2000). doi:10.1109/27.902229

P.R. Harvey, F.S. Mozer, D. Pankow, J. Wygant, N.C. Maynard, H. Singer, W. Sulivan, P.B. Anderson, R. Pfaff, T. Aggson, A. Pederson, C.-G. Falthammar, P. Tanskannen, The electric-field instrument on the Polar satellite. Space Sci. Rev. **71**(1–4), 583–596 (1995)

H. Laasko, T.L. Aggson, R.F. Pfaff, Plasma gradient effects on double-probe measurements in the magnetosphere. Ann. Geophys. **13**, 130–146 (1995)

M. Ludlam et al., THEMIS magnetic cleanliness specification. Space Sci. Rev. (2008, this issue)

J.P. McFadden, C.W. Carlson, D. Larson, J. Bonnell, F.S. Mozer, V. Angelopoulos, K.-H. Glassmeier, U. Auster, Structure of plasmaspheric plumes and their participation in magnetopause reconnection: first results from THEMIS. Geophys. Res. Lett. **35**, L17S10 (2008a). doi:10.1029/2008GL033677

J.P. McFadden, C.W. Carlson, D. Larson, V. Angelopoulos, M. Ludlam, R. Abiad, B. Elliot, P. Turin, M. Marckwordt, The THEMIS ESA plasma instrument and in-flight calibration. Space Sci. Rev. (2008b, in press)

F.S. Mozer, Criteria for and statistics of electron diffusion regions associated with sub-solar magnetic field reconnection. J. Geophys. Res. **110**(A12), A12222 (2005)

F.S. Mozer, J.-J. Berthelier, U.V. Fahleson, C.-G. Fälthammar, A proposal to measure the quasi-static vector electric field on the low altitude and the elliptic orbiting electrodynamics explorer satellites. Research Proposal to the National Aeronautics and Space Administration, UCBSSL No. 552/75, 1974

F.S. Mozer, S.D. Bale, J.P. McFadden, R.B. Torbert, New features of electron diffusion regions observed at subsolar magnetic field reconnection sites. Geophys. Res. Lett. **32**(24), L24102 (2005)

F.S. Mozer, V. Angelopoulos, J. Bonnell, K.H. Glassmeier, J.P. McFadden, THEMIS observations of modified Hall fields in asymmetric magnetic field reconnection. Geophys. Res. Lett. **35**, L17S04 (2008)

M. Oieroset, T.D. Phan, M. Fujimoto, R.P. Lin, R.P. Lepping, In situ detection of collisionless reconnection in the Earth's magnetotail. Nature **412**, 414 (2001)

A. Pedersen, F. Mozer, G. Gustafsson, Electric field measurements in a tenuous plasma w0ith spherical double probes, in *Measurement Techniques in Space Plasmas: Fields*. Geophysical. Monograph, vol. 103 (AGU, 1998)

P.A. Puhl-Quinn, H. Matsui, V.K. Jordanova, Y. Khotyaintsev, P.A. Lindqvist, An effort to derive an empirically based, inner-magnetospheric electric field model: Merging Cluster EDI and EFW data. J. Atmos. Sol.-Terr. Phys. **70**(2), 564–573 (2008)

K. Sahu, S. Kniffen, Technical memo, PPM-98-008, Radiation report on OP15 (Analog Devices) (LDC9722A), Unisys Corporation—Federal Systems Division, 23 April 1998

E.C. Whipple, Potentials of surfaces in space. Rep. Prog. Phys. **44**, 1197–1250 (1981)

The THEMIS Digital Fields Board

C.M. Cully · R.E. Ergun · K. Stevens · A. Nammari · J. Westfall

Originally published in the journal Space Science Reviews, Volume 141, Nos 1–4, 343–355.
DOI: 10.1007/s11214-008-9417-1

Abstract The Digital Fields Board (DFB) performs the data acquisition and signal processing for the Electric Fields Instrument and Search Coil Magnetometer on each of the THEMIS (Time History of Events and Macroscale Interactions during Substorms) satellites. The processing is highly flexible and low-power ($\sim$1.1 watt orbit-averaged). The primary data products are time series waveforms and wave power spectra of the electric and magnetic fields. The power spectra can be computed either on the raw signals (i.e. in a system co-rotating with the spacecraft) or in a coordinate system aligned with the local DC magnetic field. Other data products include spectral power from multiple passbands (filter banks) and electric power in the 30–500 kHz band. The DFBs on all five spacecraft have been turned on and checked out in-flight, and are functioning as designed.

Keywords THEMIS · Signal processing · Electric field instrument · Search-coil magnetometer

PACS 07.50.Qx · 07.87.+v · 94.80.+g · 95.55.-n · 84.40.Ua

1 Introduction

Despite many advances in our understanding of the substorm process, there remain some fundamental open questions. Where and when does the onset occur? What physical processes are primarily responsible? How do the individual components of a substorm interact? What is the mechanism through which a substorm drives the aurora?

The THEMIS (Time History of Events and Macroscale Interactions during Substorms) mission has been built to help answer these and other questions. It is a five-satellite NASA medium-class explorer mission launched in February 2007 in conjunction with an extensive ground-based instrument array. A detailed description of the THEMIS mission can be found elsewhere in this issue (Angelopoulos 2008).

C.M. Cully (✉) · R.E. Ergun · K. Stevens · A. Nammari · J. Westfall
Laboratory for Atmospheric and Space Physics, University of Colorado, Boulder, CO 80303, USA
e-mail: chris@irfu.se

Table 1 Analog signals available for sampling

Mnemonic	Description
SCMX, SCMY, SCMZ	3-axis magnetic field from SCM
V1 to V6	Probe-spacecraft voltage for all 6 EFI sensors
E12, E34, E56	DC-coupled electric field from EFI
E12AC, E34AC, E56AC	AC-coupled electric field from EFI
E12HF	High frequency electric field from EFI

Some of the most intriguing open questions concern the role of local processes in the current sheet. For example, which plasma instabilities disrupt the cross-tail current? Which instabilities lead to reconnection? How are they related? Resolving these questions will require good electric and magnetic field measurements at frequencies well above DC.

Electric and magnetic fields are observed by three types of sensors on THEMIS. The Flux-Gate Magnetometer (FGM) (Auster et al. 2008) and Search-Coil Magnetometer (SCM) (Roux et al. 2008) measure the magnetic fields, while the Electric Fields Instrument (EFI) (Bonnell et al. 2008) measures electric fields. The DFB is responsible for acquiring and processing all of the signals from the EFI and SCM, including multi-rate digital filtering and spectral processing. A unique processing function allows the calculation of spectra in field-aligned coordinates.

To reduce power consumption, all of the digital processing is done by three Field Programmable Gate Arrays (FPGAs) without the aid of a dedicated digital signal processor. The orbit-averaged DFB power consumption is roughly 1.1 watt, of which roughly 390 mW is devoted to the analog electronics.

This paper discusses the DFB in detail. The next section presents an overview of the DFB functionality. Sections 3 and 4 discuss the analog and digital electronics, while Sects. 5, 6 and 7 present the three main digital functions: digital filtering, power spectrum calculation and rotation to field-aligned coordinates. Section 8 lists all of the available data products.

2 Overview

Figure 1 gives a top-level overview of the DFB. There are sixteen analog signals available for sampling, as listed in Table 1. The three SCM signals are filtered to remove aliasing to give the 3 signals we call SCMX, SCMY and SCMZ, while the six EFI signals similarly yield signals V1 to V6. The EFI signals from opposing sides of the spacecraft are also paired to give differential electric field measurements, which can be selected as either DC or AC coupled (E12, E34, E56, E12AC, E34AC and E56AC). Finally, there is one channel devoted to the high-frequency (50–200 kHz) electric field (E12HF). These analog signals are converted and processed by the DFB, and the resulting digital products are sent to the Digital Control Board of the Instrument Data Processing Unit (IDPU) (Taylor et al. 2008).

To maximize the scientific output given the telemetry, the THEMIS mission makes extensive use of burst modes (Angelopoulos 2008). The lowest level of data collection, termed slow survey, covers the entire orbit at low resolution. The second level, termed fast survey, covers a relatively large fraction of the orbit at medium resolution. "Particle burst" covers a small fraction of the orbit with high resolution particle and wave data, while the final "wave burst" level applies only to the EFI and SCM data, giving very high resolution data for short intervals. Data from higher-resolution burst levels do not interrupt the data flow from lower-resolution levels; for example, slow survey data is still collected when the satellite is in wave

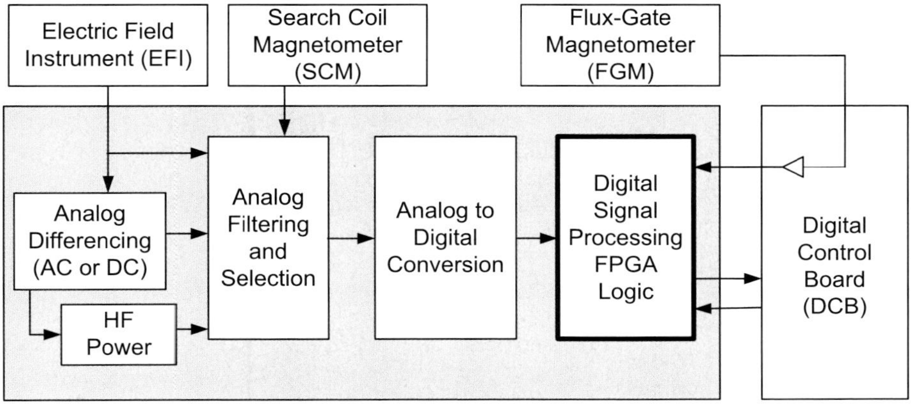

Fig. 1 Top-level schematic of DFB operation

burst mode. Consequently, the DFB must produce a large number of data products tailored for the various burst levels.

3 Analog Electronics

Nine signals arrive at the DFB input: 3 from the SCM and 6 from the EFI. All nine of these signals are available for sampling, as well as six differential signals for the EFI (Exx, ExxAC). Additionally, the power in the 50–200 kHz band is measured using a pseudo-logarithmic rectifying amplifier. Figure 2 shows the analog functionality.

Since the spacecraft to plasma potential is much greater than potential differences due to the ambient electric field, the common mode rejection ratio for the differential amplifiers must be large. The DC common mode rejection ratio was tested on each board and is better than −80 dB. By precision matching capacitors, an AC common mode rejection ratio of better than −40 dB is attained over the full temperature range on all boards.

All signals are filtered to avoid aliasing. The EFI signals are filtered using 4-pole Bessel filters at either 4.0 kHz (for the 9 DC channels) or 8.0 kHz (for the 3 AC channels). The SCM signals are filtered using 4-pole Butterworth filters at 4.0 kHz. Bessel filters were chosen for the EFI instrument to optimize for time-domain data analysis (constant group delay), while the SCM filters are optimized for frequency-domain analysis (flat frequency response).

To measure the power in the high-frequency electric field (E12HF), we use a 6-stage rectifying pseudo-logarithmic amplifier. Each stage is composed of a high-pass filter, a pair of clipping diodes and an amplifier with a gain of 5. The total amplification is $5^6 = 15\,625$, so that later amplifier stages easily (and intentionally) saturate. For larger input values, an increasing number of stages saturate, which gives the pseudo-logarithmic response seen in Fig. 3. The output from each of the six amplifier stages is half-wave rectified and summed before being low-pass filtered (1 pole, 500 Hz) to give the output signal. The frequency range matches the frequency range of auroral kilometric radiation (AKR) frequently seen at substorm onset.

All 16 analog signals are routed through a series of multiplexers to two redundant Linear Technologies LTC1604 analog to digital converters (ADCs). To save power, only signals

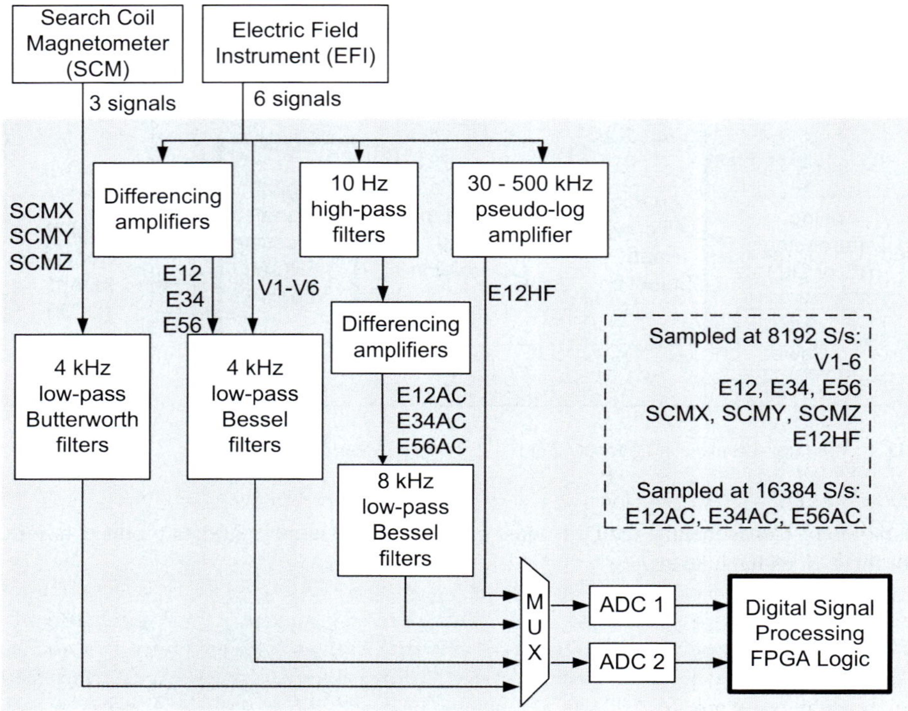

Fig. 2 Analog functionality

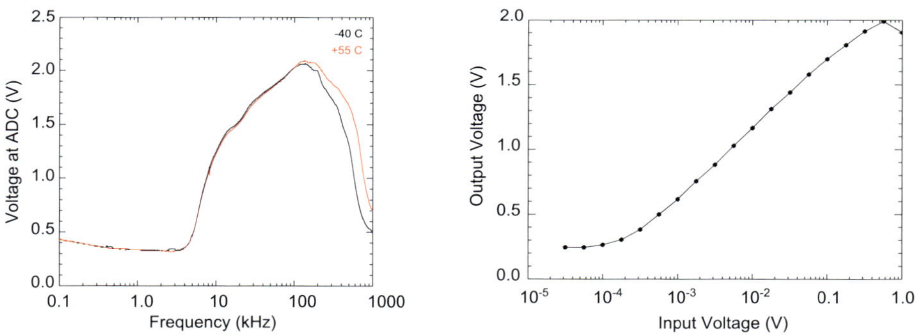

Fig. 3 E12HF response for board F6 (S/N 009); other units are similar. *Left*: output voltage as a function of frequency. *Right*: output voltage as a function of input voltage

that are actually needed for the telemetry are sampled, and the ADCs are put into a power-conserving nap mode whenever possible between samples. The worst-case crosstalk between channels is −81 dB, with most combinations considerably less than that. The signals are sampled sequentially according to a fixed schedule, with different schedules for normal operation and for backup operation in case of an ADC failure. The maximum time delay between any two similar signals (e.g. V1 and V6) is 38 microseconds for signals sampled at 8 kS/s, and 15 microseconds for signals sampled at 16 kS/s.

The total harmonic distortion of the DFB waveforms is minimal. With a monochromatic input signal, the output is free from spurious harmonics down to at least the −60 dB distortion floor of our test equipment. However, a more meaningful figure for the scientific data is the complete end-to-end distortion of the instrument, and not just the distortion from the DFB. The complete EFI electronics, including the DFB, have second and third harmonic amplitudes below −40 dB. For large signals in its operational environment, harmonics may also be generated by differences in the plasma sheaths surrounding opposing EFI probes (Boehm et al. 1994; Bonnell et al. 1997) in addition to the distortion in the electronics.

4 Digital Electronics

The main component of the digital electronics is a set of three Actel RT54SX72S FPGAs that together function as an application-specific processor. These FPGAs are triple-module redundant antifuse parts specifically made for space flight. We designed the FPGAs using the VHDL hardware description language, which allowed us to simulate the design in software during development, with full access to all internal signals.

One FPGA acts as the master controller. It is responsible for all external communications, runs the analog to digital converters and performs the digital filtering. The other two FPGAs are co-processing devices that calculate spectra and derived quantities. The clock to these two FPGAs can be shut down by the master FPGA to conserve power whenever their functionality is not required. All DFB functions are synchronized to a 1-Hz synchronization pulse provided by the IDPU in addition to the 8.4 MHz clock.

Internal bitwidth is carefully controlled to avoid roundoff or truncation errors. Any operations that could be subject to out-of-range numerical errors have been built to saturate cleanly and avoid wrap-around.

The three FPGAs share a common bank of 512 kB of RAM arranged on a 32-bit bus. The RAM is radiation-hard (Honeywell HX6228TBRT), and memory access conflicts are handled by a fixed time-division multiplexed scheme.

5 Digital Filters

There are two main digital filtering functions: the bandpass filter banks and the lowpass waveform filtering. Both of these filtering operations are performed using the common filter bank architecture shown in Fig. 4.

The left half of Fig. 4 shows the first two levels of the low-pass filtering scheme. In the first level, the signal passes through two Finite Impulse Response (FIR) digital filters. The first FIR filter uses the function

$$y_n = a_1 x_n + a_2 x_{n-1} + a_3 x_{n-2} + a_4 x_{n-3} + a_5 x_{n-4} + a_6 x_{n-5} + a_7 x_{n-6} \tag{1}$$

$$a_{1,\dots,7} = \{-8, 0, 72, 128, 72, 0, -8\}/256 \tag{2}$$

where x_m is the input sequence and y_n is the output sequence. This filter operates as a low-pass filter at roughly 40% of the Nyquist frequency. As a FIR filter, the phase response is perfectly linear (i.e. equal group delay at all frequencies).

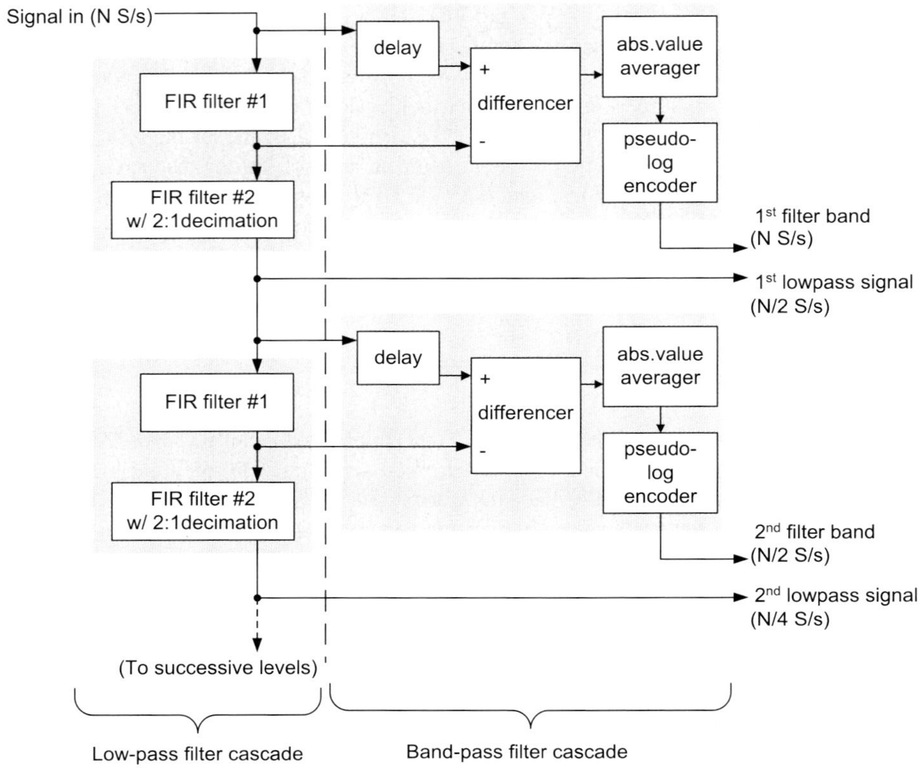

Fig. 4 Filter architecture

The filter coefficients are chosen both to give a sharp filter cutoff and also to minimize resources in the FPGA. Since there are no built-in multipliers in the FPGAs, each multiply operation must be physically built up from the basic logic cells. By using coefficients that have only one or two non-zero bits in a binary representation, the multiplier can be constructed from simple shift operations, which greatly reduces complexity and logic usage.

The second filter is a decimating FIR filter. It uses the function

$$y_n = 0.25x_n + 0.5x_{n-1} + 0.25x_{n-2} \tag{3}$$

but only keeps every second point. Consequently, the output rate is only half the input rate. The output of this filter forms the first level in the low-pass filter cascade.

This same filtering process is cascaded 12 times for most signals, and 13 times for the ExxAC signals (which are sampled at 16384 S/s rather than 8192 S/s). The result is a set of signals at every rate 2^n from 2 S/s up to 8192 S/s (16384 S/s for ExxAC). Since the rate decreases by a factor of 2 at each step, the computational resources required also decrease by a factor of 2 at each step. The full 12-step cascade therefore only requires about twice as much computation as the first step.

The left side of Fig. 5 shows the frequency response of the low-pass filter cascade. The filter response is -13 dB at the Nyquist frequency, and falls at roughly 80 dB/octave from

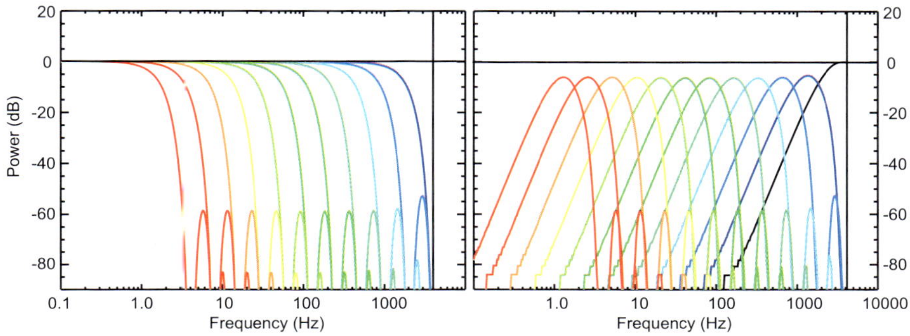

Fig. 5 Frequency response for the low-pass filter cascade (*left*) and the bandpass filter cascade (*right*)

there. A side lobe occurs at roughly 3 times the Nyquist frequency, with a maximum amplitude of -59 dB.

For the bandpass signals, the filter cascade calculates the difference between the signal at each level before and after it passes through the first FIR filter in the lowpass cascade (see right hand side of Fig. 4). As a symmetric 7-tap filter, the FIR filter has a group delay of 3 points. Consequently, the unfiltered signal must be delayed by 3 points to avoid a phase mismatch. After differencing the two signals, the amplitude in the bandpass-filtered signal is estimated by taking the average of the absolute value. The amplitude is then compressed to 8 bits by a pseudo-logarithmic encoder.

The predicted frequency response of the bandpass-filtered signals is shown in the right hand panel of Fig. 5. The highest-frequency signal is larger by a factor of 2, and similar sidelobes can be seen as in the lowpass filters. There is some quantization error visible in the small-amplitude response (below -70 dB), which is caused by the pseudo-logarithmic 8-bit encoding.

6 Derived Quantities

From a scientific perspective, electric and magnetic field spectra are more informative in a magnetic-field-aligned coordinate system. Thus, the DFB has the ability to rotate the observed 3D electric field magnetic fields signals into a field-aligned system before calculating the power spectra.

The parallel (to B) direction is uniquely specified. We take advantage of the fact that the perpendicular direction is not unique by selecting the signal that is both orthogonal to B and also in the spin plane. In this way, the perpendicular electric field signal is entirely derived from the spin-plane wire booms which have much lower noise and a well-defined gain. In the rare case that the spin axis is exactly parallel to B, a default direction is used.

On the ground, one could calculate the parallel and perpendicular components as

$$X_{\|} = \frac{\vec{X} \cdot \vec{B}}{|\vec{B}|} = \frac{X_x B_x + X_y B_y + X_z B_z}{\sqrt{B_x^2 + B_y^2 + B_z^2}} \tag{4}$$

$$X_{\perp} = \frac{\vec{X} \times \vec{B}}{|\vec{B}|} = \frac{X_x B_y - X_y B_x}{\sqrt{B_x^2 + B_y^2 + B_z^2}} \tag{5}$$

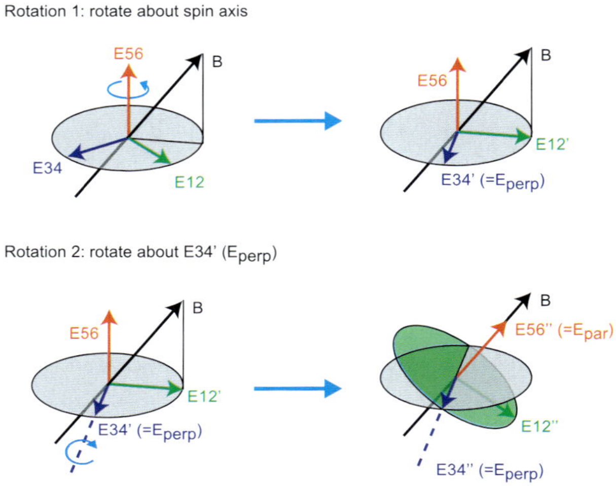

Fig. 6 Calculation of the derived quantities $E_{||}$ and $E_{\perp}$ using two rotations of the coordinate system. The first rotation is about the spin axis, and rotates the E12 axis to align with the projection of the FGM magnetic field vector onto the spin plane. This rotation places E34′ in the $E_{\perp}$ direction. The second rotation is about the $E_{\perp}$ axis, and rotates E56 to the $E_{||}$ direction (i.e. parallel to the FGM magnetic field vector). SCM quantities $\delta B_{||}$ and $\delta B_{\perp}$ are rotated similarly

where $\vec{X}$ is either the electric field from the EFI or the AC magnetic field from the SCM, and $\vec{B}$ is the DC magnetic field from the FGM. Although these operations could be performed by the FPGA, they are expensive in terms of FPGA logic (especially the square root), and we have devised a more efficient implementation.

The implementation relies on a method known as CORDIC (COordinate Rotation for DIgital Computers) (Volder 1959; Andraka 1998) that can arbitrarily rotate 2-dimensional vectors using minimal hardware resources. The DFB uses CORDIC both to rotate vectors for the derived quantities and also to calculate $\cos(\theta)$ for the Fast Fourier Transform calculations (see Sect. 7).

The implementation we use in the DFB is sketched in Fig. 6. Two CORDIC rotations are performed on the time-series waveforms after digitization but before any conversion to the frequency domain (Sect. 7). The first rotation is about the z axis (E56 or SCMZ), which yields the perpendicular component $X_{\perp}$. Note that the perpendicular component is thus constrained to the spin plane, so that $E_{\perp}$ uses data only from E12 and E34. The second rotation is about this perpendicular axis $X_{\perp}$, and yields the parallel component. See Fig. 6 for a sketch of the rotations.

The data is calibrated before the rotations occur. Variable offsets and gains are applied separately to the FGM data, the EFI data and the SCM data. These gains need to be computed on the ground and uploaded. For the EFI data, the offset can alternatively be calculated on-board using a 10-minute average. The FGM data are also rotated to align with the EFI or SCM coordinate system. The rotation operation is combined with the FGM gain into two 9-element rotation matrices: one for the EFI system and one for the SCM system.

The FGM data arrive at the DFB in 24-bit precision at a rate of 128 S/s. To ensure synchronization, the EFI and/or SCM data are delayed to account for the full delay through the FGM chain to the DFB. The calculated angles θ and ϕ from the FGM vectors are then linearly interpolated in the DFB to avoid jumps in the coordinate system which would create noise at harmonics of 128 Hz.

As discussed in Sect. 3, the ADC samples the data sequentially, so that a small time delay exists between measurements taken along different axes. At frequencies approaching the Nyquist frequency, this delay introduces an error in the measured direction of the wave. The worst-case error for a pathological signal (narrow-band signal near the Nyquist frequency, FGM vector in the spin plane at 45 degrees to the measurement axes, ADCs in backup mode)

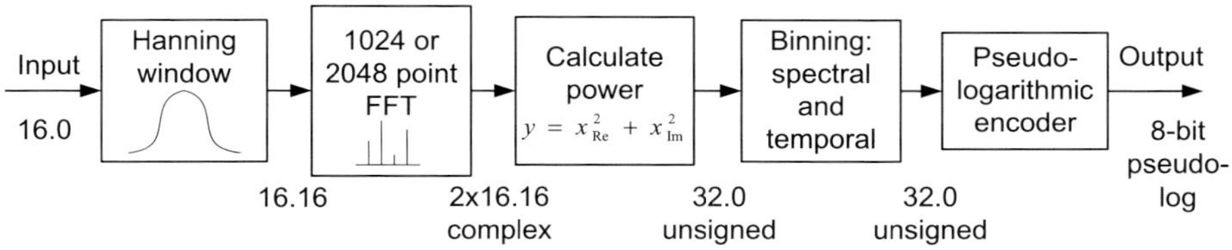

Fig. 7 Top-level flow diagram for FFT. Internal bitwidths are shown at each stage

is 5 degrees for the ExxAC data and 3 degrees for the SCM data. The accuracy using more representative signals is ± 1 degree.

The result of this processing is a set of four signals that we refer to as the "derived quantities". These four signals are the magnetic components $\delta B_{||}$ and $\delta B_{\perp}$ and the electric components $E_{||}$ and $E_{\perp}$. The electric components can be either AC-coupled (from ExxAC) or DC-coupled (Exx).

7 Calculation of Spectra

Power spectra are available in the particle burst and wave burst modes, and provide coverage for the higher frequency range. The spectra are computed by Fast Fourier Transforms (FFTs) implemented in the FPGAs. Figure 7 is a top-level flow diagram.

The DFB can calculate continuous FFTs of any four of 19 available signals: the 15 analog quantities in table 1 not including E12HF, plus the 4 derived signals discussed in section 6. 1024-point FFTs are used for all signals at 8192 S/s, while 2048-point FFTs are used for signals at 16384 S/s (ExxAC and any derived quantities using ExxAC). The resulting spectra are available at up to 8 spectra per second.

In order to optimize the use of FPGA resources, all operations are performed in a fixed-point format (not floating-point). The bitwidths of all internal signals were analyzed to ensure signal integrity; the resulting widths are shown in Fig. 7. The notation "16.16" means a 32-bit wide signed number, with 16 bits to the left of the decimal place and 16 bits to the right. By using a 32-bit internal representation, the resulting powers are identical with double-precision calculations to the level of ~ -95 dB.

We use a standard in-place decimation in time (Cooley-Tukey) algorithm for the FFT, with a Hanning window to reduce side lobes. To increase efficiency, the DFB packs two N-point real vectors into one N-point complex vector, processes the FFT for that one complex vector, and then unpacks the results appropriately (Press et al. 1992). A CORDIC algorithm is used for the sine and cosine calculations.

In a fixed-point representation with no roundoff error, every multiplication doubles the bitwidth. In an FFT operation, this is not acceptable, since the bitwidth would geometrically grow with each butterfly operation. To avoid this, we clip the result of each butterfly pass to a 16.16 representation for each component (real and imaginary) and accept the small roundoff error. In the event of a rail-to-rail signal on the input, the maximum possible error with a pathological signal is -87 dB (3 bits). Using more reasonable signals, the maximum error is below the 1-bit floor (i.e. the calculated power is exact to the limit of its bitwidth).

After calculating the power from the complex FFT result, the DFB bins the spectral power in frequency and time according to the commanded mode. Binning in time is accomplished by averaging together spectra to reduce the cadence from the natural 8 spectra per

Table 2 Spectral binning for 1024-point FFT results

64-point spectra		32-point spectra		16-point spectra	
Bins	Bandwidth	Bins	Bandwidth	Bins	Bandwidth
0–31	8 Hz	0–15	16 Hz	0–7	32 Hz
32–39	32 Hz	16–19	64 Hz	8–9	128 Hz
40–47	64 Hz	20–23	128 Hz	10–11	256 Hz
48–55	128 Hz	24–27	256 Hz	12–13	512 Hz
56–63	256 Hz	28–31	512 Hz	14–15	1024 Hz

Table 3 Spectral binning for 2048-point FFT results

64-point spectra		32-point spectra		16-point spectra	
Bins	Bandwidth	Bins	Bandwidth	Bins	Bandwidth
0–15	8 Hz	0–7	16 Hz	0–3	32 Hz
16–23	16 Hz	8–11	32 Hz	4–5	64 Hz
24–31	32 Hz	12–15	64 Hz	6–7	128 Hz
32–39	64 Hz	16–19	128 Hz	8–9	256 Hz
40–47	128 Hz	20–23	256 Hz	10–11	512 Hz
48–55	256 Hz	24–27	512 Hz	12–13	1024 Hz
56–63	512 Hz	28–31	1024 Hz	14–15	2048 Hz

second to as low as 1 spectra per 16 seconds. The number of combined spectra are 2^n with integral n. Frequency binning reduces the spectrum to either 16, 32 or 64 bins. Bin spacing is approximately logarithmic, as shown in Tables 2 and 3. The lower frequency of each bin is equal to the sum of the bandwidths below it, starting from zero.

8 Data Products and Data Rates

There are three basic types of data provided by the DFB: waveforms, filter bank data and spectra. Table 4 lists the available waveform data products, grouped by data ID and burst mode. All waveform data are 16 bits wide.

For slow survey, seven waveform signals (V1 through V4, E12, E34 and E56) at 128 S/s each are passed to the IDPU for spin fitting. The other data streams (IDs) can contain up to six different quantities at a single user-selectable rate of 2^n S/s (for integer n). Quantity and speed selection in the streams is entirely independent; for example, E12, E56 and E34AC could be grouped together on data ID 443 (hexadecimal), with E12 and E34 on data ID 447.

Filter bank data are 8 bits wide, and use a pseudo-logarithmic compression to attain a 19-bit range. The filter bank data types are listed in Table 5. There are two output streams from the filter bank: one for slow survey data and one for input to the burst-mode triggering algorithm. The slow-survey filter bank data consist of 2 banks of 6 spectral bands each, covering the range from ~2 Hz to ~4 kHz with logarithmic spacing. Each bank may be set to any of the 16 input quantities except E12HF. The trigger data are the same as the slow-survey data, except that 11 spectral bands are reported (instead of 6), and the rate is fixed at 16 S/s. Two additional quantities are included along with the filter bank quantities: the peak and average levels of the E12HF signal in the time period since the last filter bank data point.

Table 4 Waveform data products (data IDs in hexadecimal representation)

Data ID	Description	Burst mode	Quantities	Rate
450	Spin-fit data	slow survey	V1–4, E12, E34, E56	128 S/s
441	Voltage group A	fast survey	any V1–6	2-8192 S/s
442	Voltage group B	fast survey	any V1–6	2-8192 S/s
443	Electric field	fast survey	any Exx or ExxAC	2-16384 S/s
444	SCM data	fast survey	any SCM (X, Y, Z)	2-8192 S/s
445	Voltage group A	particle burst	any V1–6	2-8192 S/s
446	Voltage group B	particle burst	any V1–6	2-8192 S/s
447	Electric field	particle burst	any Exx or ExxAC	2-16384 S/s
448	SCM data	particle burst	any SCM (X, Y, Z)	2-8192 S/s
449	Voltage group A	wave burst	any V1–6	2-8192 S/s
44A	Voltage group B	wave burst	any V1–6	2-8192 S/s
44B	Electric field	wave burst	any Exx or ExxAC	2-16384 S/s
44C	SCM data	wave burst	any SCM (X, Y, Z)	2-8192 S/s

Table 5 Filter bank data products

ID	Burst mode	Quantities	Bands	Rate
440	slow survey	any 2 of: V1-6, Exx, ExxAC, SCM	6	1/16–18 S/s
440	slow survey	E12HF: peak and average	–	same as above
451	trigger	same 2 quantities as ID 440	11	16 S/s
451	trigger	E12HF: peak and average	–	16 S/s

The −6 dB frequencies defining the filter passbands for 8192 Hz signals (i.e. all signals except ExxAC) are listed in Table 6 as $f_{\min}$ and $f_{\max}$. The appropriate frequencies for the 16 kS/s ExxAC signals are exactly twice these values. The frequency at which the filter response is maximized (f_{center}) and the total bandwidth ($f_{\max} - f_{\min}$) are also listed. The highest-frequency band is limited at high frequency only by the analog filters, and hence has an anomalously large bandwidth. In slow survey, filter levels 0, 2, 4, 6, 8 and 10 are included in the telemetry for 8 kS/s signals, and levels 1, 3, 5, 7, 9 and 11 are included for 16 kS/s signals, so that the passband frequencies remain similar.

Finally, the DFB computes the power spectral densities of any four signals from the set of the 15 analog quantities (Table 1, not including E12HF) and the four derived quantities. Table 7 lists the spectral data available in particle and wave burst modes. The telemetry consists of either 16, 32 or 64 spectral power bands at a rate selectable from 1/16 S/s to 8 S/s. The bands cover the range up to 4 kHz in a roughly logarithmic fashion. For ExxAC, the range is extended to 8 kHz.

9 Conclusion

The Digital Fields Board is responsible for the digital signal processing for the THEMIS waves package. The processing is performed by field programmable gate arrays instead of

Table 6 Filter bank passbands for 8 kS/s signals (double frequencies for 16 kS/s)

Level	f_{min} (Hz)	f_{center} (Hz)	f_{max} (Hz)	Bandwidth (Hz)
0	1390	2689	5994	4604
1	631	1149	1836	1204
2	316	572	904	587
3	159	287	453	293
4	80.2	144.2	227.4	147.2
5	40.2	72.3	113.9	73.7
6	20.1	36.2	57.0	36.9
7	10.1	18.1	28.5	18.4
8	5.04	9.05	14.26	9.23
9	2.52	4.53	7.13	4.61
10	1.26	2.26	3.57	2.31
11	0.63	1.13	1.78	1.15

Table 7 Spectral data products

ID	Burst mode	Quantities	Bins	Rate
44D	particle burst	any 4 signals from: V1-6, Exx, ExxAC, SCM, derived	16–64	1/16–18 S/s
44E	wave burst	same 4 as above	16–64	1/16–18 S/s

using a dedicated processor, and consumes very little power. Many functions that have previously been implemented in analog electronics are performed digitally, resulting in significant power savings.

The data products produced by the DFB are extremely flexible. By using cascading filter banks, the same signal can be transmitted at a number of different rates in low-pass or band-pass filtered form, and with minimal processing. This strategy is efficient for satellites like THEMIS that have many different burst modes.

Spectral data are calculated using a Fast Fourier Transform. One of the unique features of the DFB is its ability to transform to a field-aligned coordinate system prior to calculating the wave spectra. This allows us to separate parallel and perpendicular components in the wave spectra.

The Digital Fields Board provides the digital processing necessary for observing waves in the electric and magnetic fields. All five DFBs have been turned on and checked out in-flight, and are functioning perfectly. We look forward to new and exciting science that can come from the DFB and all of the instruments on THEMIS.

Acknowledgements The THEMIS project is made possible by the contributions of a large number of scientists, engineers and supporting personnel. We particularly thank V. Angelopoulos, J. Bonnell, P. Harvey, E. Taylor., M. Ludlam and H. Richard. This work was supported under NASA contract NAS5-02099.

References

R. Andraka, A survey of CORDIC algorithms for FPGAs, in *Proceedings of the 1998 ACM/SIGDA Sixth International Symposium on Field Programmable Gate Arrays*, 22–24 Feb. 1998, Monterey, CA (1998)

V. Angelopoulos, The THEMIS mission. Space Sci. Rev. (2008, this issue). doi:10.1007/s11214-008-9336-1

H.U. Auster, K.H. Glassmeier, W. Magnes, O. Aydogar, W. Baumjohann, D. Constantinescu, D. Fischer, K.H. Fornacon, E. Georgescu, P. Harvey, O. Hillenmaier, R. Kroth, M. Ludlam, Y. Narita, R. Nakamura, K. Okrafka, F. Plaschke, I. Richter, H. Schwarz, B. Stoll, A. Valavanoglou, M. Wiedemann, The THEMIS fluxgate magnetometer. Space Sci. Rev. (2008, this issue). doi:10.1007/s11214-008-9365-9

M. Boehm, C. Carlson, J. McFadden, J. Clemmons, R. Ergun, F. Mozer, Wave rectification in plasma sheaths surrounding electric field antennas. J. Geophys. Res. (1994)

J. Bonnell, P. Kintner, J.E. Wahlund, J. Holtet, Modulated langmuir waves: observations from freja and SCIFER. J. Geophys. Res. (1997)

J.W. Bonnell, F.S. Mozer, G.T. Delory, A.J. Hull, R.E. Ergun, C.M. Cully, V. Angelopoulos, P.R. Harvey, The electric field instrument (EFI) for THEMIS. Space Sci. Rev. (2008, this issue)

W.H. Press, S.A. Teukolsky, W.T. Vetterling, B.P. Flannery, *Numerical Recipes in C*, 2nd edn. (Cambridge University Press, Cambridge, 1992)

A. Roux, O. Le Contel, C. Coillot, A. Boubdellah, B. de la Porte, D. Alison, S. Ruocco, M.C. Vassal, The search coil magnetometer for THEMIS. Space Sci. Rev. (2008, this issue)

E. Taylor et al., The instrument data processing unit for THEMIS. Space Sci. Rev. (2008, this issue)

J.E. Volder, The CORDIC trigonometric computing technique. IRE Trans. Electron. Comput. (1959)

The THEMIS Array of Ground-based Observatories for the Study of Auroral Substorms

S.B. Mende · S.E. Harris · H.U. Frey · V. Angelopoulos · C.T. Russell · E. Donovan · B. Jackel · M. Greffen · L.M. Peticolas

Originally published in the journal Space Science Reviews, Volume 141, Nos 1–4, 357–387.
DOI: 10.1007/s11214-008-9380-x

Abstract The NASA Time History of Events and Macroscale Interactions during Substorms (THEMIS) project is intended to investigate magnetospheric substorm phenomena, which are the manifestations of a basic instability of the magnetosphere and a dominant mechanism of plasma transport and explosive energy release. The major controversy in substorm science is the uncertainty as to whether the instability is initiated near the Earth, or in the more distant >20 Re magnetic tail. THEMIS will discriminate between the two possibilities by using five in-situ satellites and ground-based all-sky imagers and magnetometers, and inferring the propagation direction by timing the observation of the substorm initiation at multiple locations in the magnetosphere. An array of stations, consisting of 20 all-sky imagers (ASIs) and 30-plus magnetometers, has been developed and deployed in the North American continent, from Alaska to Labrador, for the broad coverage of the nightside magnetosphere. Each ground-based observatory (GBO) contains a white light imager that takes auroral images at a 3-second repetition rate ("cadence") and a magnetometer that records the 3 axis variation of the magnetic field at 2 Hz frequency. The stations return compressed images, "thumbnails," to two central databases: one located at UC Berkeley and the other at the University of Calgary, Canada. The full images are recorded at each station on hard drives, and these devices are physically returned to the two data centers for data copying. All data are made available for public use by scientists in "browse products," accessible by using internet browsers or in the form of downloadable CDF data files (the "browse products" are described in detail in a later section). Twenty all-sky imager stations are installed and running at the time of this publication. An example of a substorm was observed on the 23rd of December 2006, and from the THEMIS GBO data, we found that the substorm onset brightening of the equatorward arc was a gradual process (>27 seconds), with minimal

S.B. Mende (✉) · S.E. Harris · H.U. Frey · V. Angelopoulos · L.M. Peticolas
Space Science Laboratory, University of California, Berkeley, CA 94720, USA
e-mail: mende@ssl.berkeley.edu

E. Donovan · B. Jackel · M. Greffen
University of Calgary, Calgary, Canada

V. Angelopoulos · C.T. Russell
University of California, Los Angeles, CA 90095, USA

morphology changes until the arc breaks up. The breakup was timed to the nearest frame (<3 s) and located to the nearest latitude degree at about ±3°E in longitude. The data also showed that a similar breakup occurred in Alaska ∼10 minutes later, highlighting the need for an array to distinguish prime onset.

Keywords Auroral substorms · Magnetospheric instability · Ground-based observatories · Auroral imagers · Magnetometer array · All sky camera · Substorm onset

1 Introduction

The NASA Time History of Events and Macroscale Interactions during Substorms (THEMIS) project is intended to answer fundamental questions regarding magnetospheric substorms (Angelopoulos 2008). THEMIS will distinguish between the two most likely magnetotail processes responsible for initiating substorms—local disruption of the near-tail plasma sheet current at <10 Re, or magnetic interaction with the rapid influx of plasma ejected from lobe flux annihilation at >20 Re (Lui 1991; Baker et al. 1996). Correlative observations from five identical probes located at strategic positions will document the phenomena and the timing of their occurrences, and will infer the direction of propagation of energy in the substorm process. If the process were to initiate with local current disruption at <10 Re, then the phenomena would be expected to propagate outward along the tail. Whereas if the initiation point were at the tail reconnection region at >20 Re, then the phenomena would be expected to propagate inward towards the earth and precede the auroral signature that is visible from the ground. It is therefore crucial to locate and time the onset of the auroral signature at substorm onset. The five identical probes (satellites) measure particles and fields on orbits which permit the alignment of the satellites in the tail while North America is in the night side. The satellite observations can be supplemented by a set of ground observatories in North America which will time the auroral breakup onset. There are three inner probes at ∼10 Re that monitor current disruption onset, while two outer probes, one at 20 and one at 30 Re, monitor plasma acceleration due to magnetic reconfiguration, such as tail flux dissipation and/or field dipolarization. THEMIS answers many critical questions in radiation belt physics and solar wind-magnetosphere energy coupling, in addition to addressing its primary objective. Detailed design of the THEMIS mission, the satellites, and their instrumentation is described in accompanying articles. This paper will discuss the requirements that led to the design of the Ground Based Observatories, describe the instruments, the array concept and the analysis techniques, as well as data formats and data products. The technical implementation and deployment of the GBOs are discussed by Harris et al. (2008).

Magnetospheric substorms are impulsive changes in the energy balance of the magnetosphere. They were discovered during studies of auroral images taken from an array of simultaneously operating ground-based all-sky cameras (ASCAs) (Akasofu 1977). Substorms are highly evident in optical auroral observations as a sudden dynamic activity and brightening in a pre-existent quasi-steady arc structure, and subsequent rapid poleward and local time propagation of the auroral brightening. During the International Geophysical Year (IGY: July, 1957 to December 31, 1958) and the following International Geophysical Collaborations (the year 1959), auroral observations greatly improved. Globally, there were about 120 all-sky cameras, divided between the Northern and Southern Hemispheres in the approximate ratio of 3:1, but never more than half of the Arctic polar sky was under observation at any one time; this was due to land and sea distribution, and clouds (Akasofu 1963). During

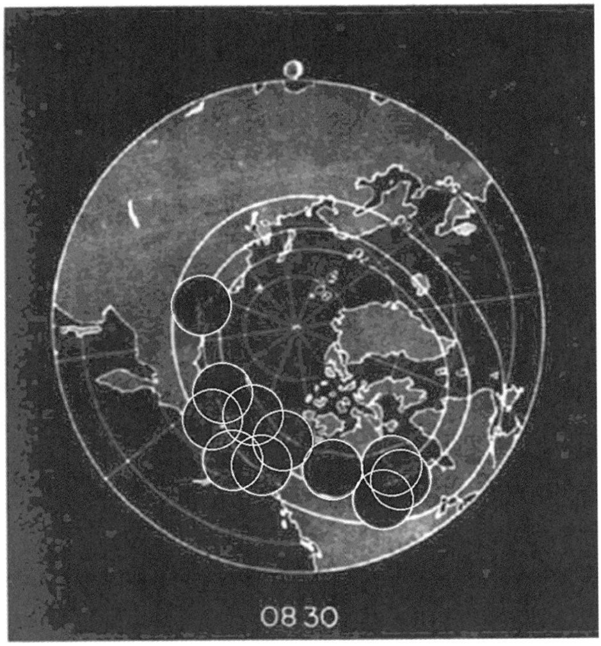

Fig. 1 Group of ground-based all-sky cameras (ASCA) in the Northern Hemisphere, employed by Akasofu (1963) during the IGY

the IGY, ASCA photographs were taken at most stations at 1-minute intervals during dark periods in fine weather; the exposure was 55 seconds followed by a 5-second period to move the film. The program produced a great volume of records, which has shed much insight into the morphology of a typical auroral substorm (Akasofu 1964, 1965, 1968; Davis 1966).

Since the IGY. several other attempts were made at fielding all-sky imagers and related instruments to study magnetospheric substorms. Magnetospheric substorms are a global phenomenon, therefore a single all-sky camera station is inadequate to study them. For this reason, several attempts have been made to field and operate chains of coordinated optical observatories. More recent ground-based arrays were/are the Canopus/Norstar array in Canada (Donovan et al. 2003), the United States AGO network in Antarctica (Mende et al. 1999; Rosenberg 2000) and the Magnetometers–Ionospheric Radars–All-sky Cameras Large Experiment (MIRACLE) network (Syrjäsuo et al. 2002) in Northern Fennoscandia and Svalbard for monitoring auroral ionospheric dynamics. The MIRACLE includes five newly implemented digital all-sky cameras of the type used in the THEMIS GBOs.

High-altitude satellite-based imaging has been the most productive way of documenting substorms on a global scale and for time periods relevant to substorm phases—substorms generally last from one to several hours. To make useful observations, satellites must have orbits that permit continuous observation for long periods of time. Thus low-altitude earth orbit (LEO) satellites are fundamentally unsuited for observations of substorm phases. DE-1 was the first high-altitude satellite that permitted extended duration of global auroral viewing (Frank et al. 1981; Frank and Craven 1988). The far ultraviolet wavelength band is strongly favored for space-based observations because it permits observation of the aurora in daytime sunlit conditions. Most importantly, no special effort is required to extend the dynamic range of the instrument to accommodate the relatively faint aurora in the presence of bright sunlit earth in the field-of-view. Thus many authors successfully pursued space-based substorm studies by using the ultraviolet imagers on POLAR (Torr et al. 1995) and IMAGE

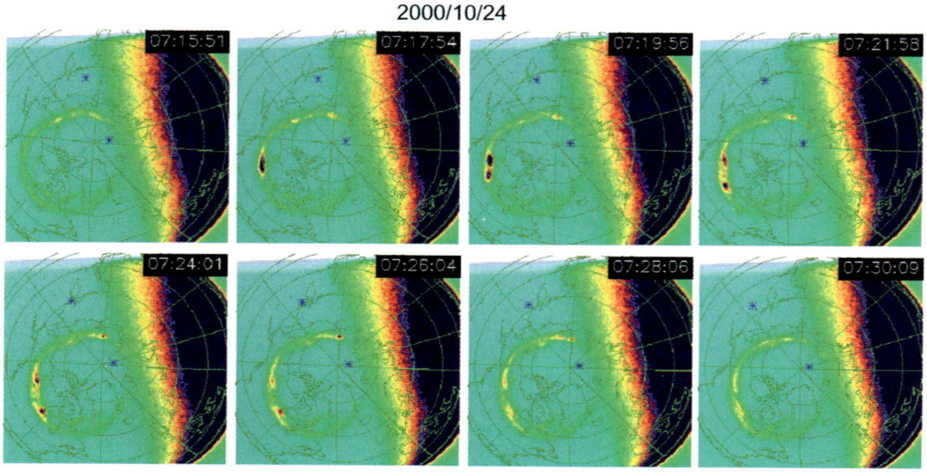

Fig. 2 Global imaging of substorms. Substorm observed on the 24th of October 2000 by the Wideband Imaging Camera (WIC) on the NASA IMAGE satellite. Substorm onset occurred between 07:15:51 and 07:17:54. Although each exposure is only a 10-second integration, the 2 minute rotation period of the IMAGE satellite restricted the time resolution to about 2 minutes. THEMIS ground-based observatories need to make quasi-global images at a much higher repetition rate

(Mende et al. 2000) (Fig. 2). These latter instruments had adequate wavelength resolution to distinguish between auroras caused by low- and by high-energy electrons. In addition, the IMAGE FUV instrument had sufficient high spectral resolution to sense Doppler-shifted auroral hydrogen emission and distinguish proton-induced aurora from the optically thick and greatly intense geocorona.

It would have been very difficult to coordinate the orbit of a separate high-altitude and high-inclination imaging satellite to that of the five THEMIS spacecrafts, and cover the relevant part of the polar regions conjugate to the THEMIS satellites. A 63-degree inclination orbit satellite in a 12-hour orbit could have been synchronized to observe the northern Canadian arctic in the winter at midnight—however, such an arrangement would require launching a separate satellite platform into a specific orbit and would have been prohibitively expensive. It was decided early in the program that a GBO network would be a more cost-effective way of satisfying the needs of THEMIS to determine substorm onset location and timing. An additional drawback to high-altitude space-based imaging is the limitation on spatial resolution. It is difficult to satisfy the competing technical requirements arising from the need for global coverage and fine spatial resolution with a single space-based imager.

Substorm onsets can be determined from their characteristic magnetic signatures. A large westward Hall current is initiated at substorm onset, and magnetometers located near the onset point pick up the magnetic field change due to the ionospheric and magnetospheric currents overhead. Figure 3, upper box, illustrates the signature of a substorm onset taken with the GBO magnetometer located in Athabasca. The overhead westward current produces a "negative bay" as observed in the BH (magnetic north-south) component. The strong negative bay in the B_z (vertical) component signifies that the bulk of the current was flowing northward of the station and the latitude of the onset was most probably located poleward of Athabasca. The two lower panels show a magnified view of the magnetic signature of a typical onset and recovery of a substorm.

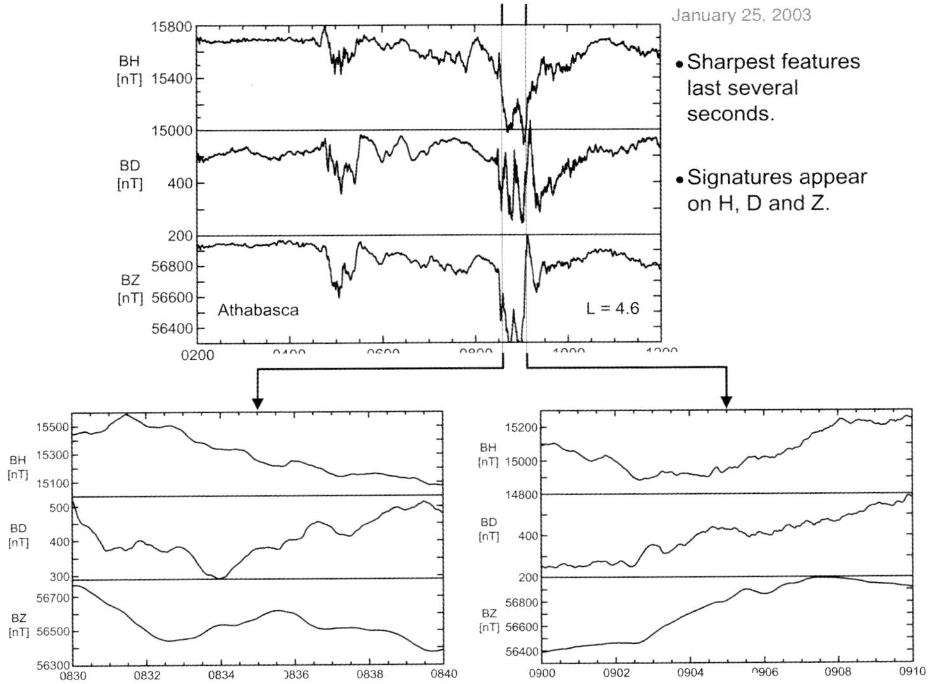

Fig. 3 Magnetometer data illustrating typical substorm signatures

2 Requirement Definition

The primary, Level 1 THEMIS mission requirement is to be able to determine the time and location of the initial auroral intensification in magnetic local time (MLT) and latitude coordinates. This requirement is met with an array of all-sky imagers and ground magnetometers distributed over an 8-h geographic local time sector in northern Canada and Alaska. There are two ground magnetometers (GMAG) and at least one auroral zone ASI per MLT hour. The ASIs have a time resolution of <10 s and a spatial resolution $<1^{\circ}$ of latitude.

Figure 4 shows substorm onset locations as observed by the IMAGE FUV experiment, with the originally proposed fields of views of 20 ASIs shown (Frey et al. 2004). The coverage of the all-sky imaging array would have been sufficient to capture the greatest majority of substorm onsets (asterisks), had all the GBOs been up and working at the time IMAGE was taking the data. It also shows that station coverage is quite comprehensive in latitude and that it covers a longitude region from 190 to approximately 310 degrees (east longitude). This is equivalent to local time coverage of 8 hours. Extensive coverage is desirable to cover substorms in which the onset occurs at various local times and latitudes.

The latitude resolution requirement of one degree of latitude is equivalent to a dipole L value change ΔL of 0.2 at 60 degrees. From Fig. 4, it can be seen that substorm onset locations are seldom higher than 70 degrees magnetic. Thus the one degree of latitude resolution translates to $\Delta L = 1$, which is approximately equivalent to 1 Re distance in the near tail region.

With regard to time resolution, THEMIS needs to determine the substorm onset time corresponding to the time it takes for the plasma effects to propagate about 1 Re in the near

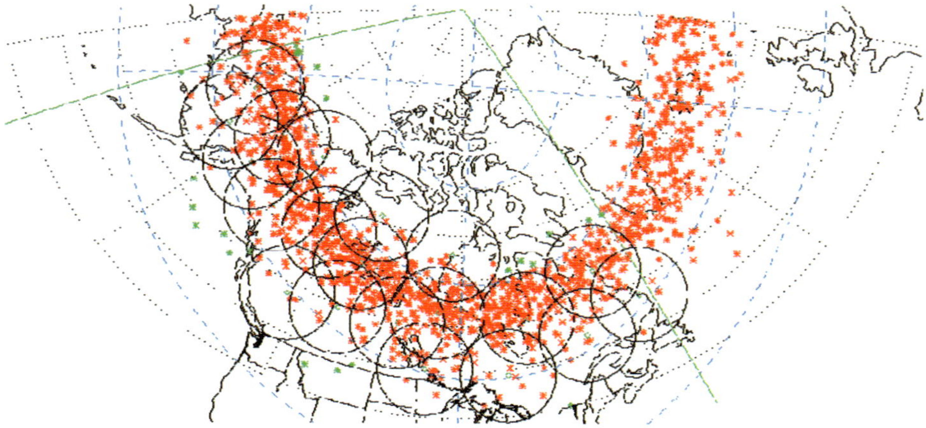

Fig. 4 Substorm onset occurrences as observed by the IMAGE FUV experiment (*asterisks*) and the coverage of the THEMIS GBO stations

equatorial magnetosphere. The propagation speed is the Alfven speed,

$$V_a = B/(4\pi nm)^{1/2}.$$

In the equatorial region where the onsets occur the magnetic field B, is of the order of $\sim$50 nT (5×10^{-4} gauss) or less. If we assume a lower limit of 1 particle per cm^3 of hydrogen ($m = 1.67240 \times 10^{-24}$ g) for the plasma density, the largest Alfven speed is $V_a = 1.09 \times 10^3$ km s. Thus one L value propagation takes about 6 s and the substorm phenomena travel from regions >20 Re to <10 Re, taking about a minute. The waves would travel more slowly in weaker magnetic fields or in the presence of greater plasma densities. The THEMIS requirement of timing accuracy is 10 seconds. The practical limit of our hardware is an exposure cadence of 3 seconds, which satisfies the THEMIS time resolution requirement and provides definitive determination of substorm onset time.

A sample rate of 2 Hz was specified for the GMAG instrument to take advantage of the accurate timing and location of substorm onsets from Pi1 pulsations (1–40 s period). These pulsations are expected to provide more precise timing than the more widely-used Pi2 pulsations (40–150 s period) (Posch et al. 2004).

The coverage of a single ASI is approximately circular in geographic space, with a radial distance equivalent to about 4.5 degrees of latitude. At high latitudes (60 and above) this is equal to about twice as many degrees in longitude; thus each station is providing between 16 and 20 degrees of longitude, i.e., larger than 15 degree = 1 hour local time coverage (see Fig. 4). Therefore, 1 hour local time per ASI is adequate for contiguous coverage.

Using similar arguments, it can be shown that two GMAG instrument per one hour local time spacing provides sufficient resolution to determine the location of onset of the substorm current.

Substorm auroras are relatively bright. A threshold sensitivity of 10 kR would be quite adequate to satisfy THEMIS Level 1 requirements. As we will see, THEMIS ASIs have more than an order of magnitude higher sensitivity. Similarly, the detection of magnetic bays associated with substorms do not require very high sensitivity—1 nT sensitivity would be satisfactory. This is surpassed by the THEMIS GMAGs, whose sensitivity is in the 0.1 nT range.

In summary, the THEMIS GBOs are monitoring the auroral light and ionospheric currents across the North American continent in order to record the time, location, and evolution of the auroral signatures of substorms.

In Table 1, we have summarized the requirements of the GBO stations.

3 Observatory Chain Design

The most difficult THEMIS requirement demands comprehensive coverage of the North American sector of the auroral oval. In order to satisfy this requirement, 20 stations were needed. The station locations are given in Table 2. Figure 5 illustrates the station locations, with circles representing the fields-of-view of the all-sky cameras when taking a 160-degree total field projected to 110 km altitude. The diameter of each circle is about 9 latitude degrees. The station array provides fairly complete coverage, with the exception of a small gap between stations GILL–FSMI, GILL–KAPU and SNKQ–GBAY. Even with these small "holes," the system fulfills the THEMIS requirements. It is estimated that the array will cover over 90% of substorms occurring in the North American local time sector (Fig. 4).

When locating the stations it was necessary to consider the resource requirements at each prospective site. The stations require clear, unobstructed field of view (without lights), power, internet connection and some minimal custodian attention. Historically, it has been difficult to find sites that have power but only minimal exterior lighting. Large cities or population centers are generally unsuitable, due to light pollution. The requirement for quick data retrieval and "real time" programming/commanding capability demanded a high speed internet connection, further restricting the choices. Fortunately there are geosynchronous satellite-based internet providers and their service can be accessed even from the higher latitude stations. They can provide sufficient baud rates to allow the retrieval of the compressed Near Real Time (NRT) thumbnail ASI data, and full resolution GMAG data.

The technique for observing the global aurora from an array of stations is limited by the inherent distortion of the observing geometry, as illustrated in Fig. 6. For equal areas on the "sky" at auroral altitudes ($\sim$110 km) represented by dx, the corresponding angular distance $d\theta$ for a ground observer is compressed near the horizons, with increasing distance from the station. Conversely, outer pixels in the circular image represent larger regions of auroral precipitation than pixels near the middle.

Figures 7a and b illustrates an image from Rankin Inlet. Figure 7a represents the original image as it is read out and digitized from the CCD. A rayed arc is seen on the right half of the image and a thin, long east-west extended arc is seen near the bottom, in the vicinity of the horizon. Lights from a nearby settlement can be seen at the very bottom right of the image. The crosses represent the magnetically projected track of the NASA FAST auroral satellite that was passing through the field of view. Figure 7b is a representation of the same all-sky image projected on an imaginary layer at 110 km altitude. The distortions produced by this treatment are quite evident—the regions near the horizons are greatly extended. The method to create this image is called "backward projection" because this picture was produced by starting from the final latitude/longitude pixel matrix. This matrix represents the "bins" on the sky at an assumed 110 km altitude. These bins on the final image matrix are mapped to the appropriate pixels in the all-sky camera image. The pixel intensity in the all sky image pixel is simply copied as the output intensity in the latitude/longitude bin. This method is very simple and it overcomes the complexities of the mismatch in the area between the pixels in the all sky camera image and the corresponding latitude/longitude bins. In cases when the latitude/longitude bins spans multiple pixels in the all-sky camera

Table 1 GBO requirement summary

	Requirement driver	Requirement	Unit
		Imager	
Field of view	Large spatial coverage 9 lat deg. circular	170	° full angle
Spectral passband	Capture visible aurora	400–700 (with IR filter)	nm
Sensitivity	Record substorm aurora	<1 (@ 5:1 S/N)	kR
Spatial resolution number of pixels	Locate substorm 1° of latitude (~100 km)	32 (diameter all-sky-image)	pixel
Exposure duration	High cadence/ high sensitivity	Programmable, 1	s
Cadence	Resolve substorm onset with 10 s accuracy	3	s
		Fluxgate magnetometer	
Dynamic range	Cover the earth background field	±72,000@0.01 nT resolution (~23 bits)	nT
DAC offset system		256	Possible ranges per axis
Data rate	Record Pi2 waves	2 (3 component vectors)	Per second
Power	N/A	<4	W
Size	Simplicity of installation	22 × 13 × 5	cm × cm × cm
Sensor design	Survivability in arctic conditions	Ruggedized all weather	
Data retrieval and firmware upload	Flexibility of operation	USB interface	
		Enclosure	
External operating temperature	Operate during winter cold. Survive full sunlit summer days	−50° to +40°C	°C
Maintained internal temperature	Commercial electronics operating temperature	20° ± 10°C	°C
Heating power	Minimize utility costs	<150	W
Environmental protection	Survive/operate extreme weather conditions	Hermetically sealed unit w/ nitrogen purge and sealed electrical connectors	
Optical I/F with ambient	Good optical transmission	Polycarbonate/acrylic dome	
Mounting	Optic axis to be vertical within 1°	Flexible mounting with in field leveling adjustment	
Thumbnail data rate (out)	Retrieve thumbnails	1.23/ continuous (2.7 kbits/s)	Mbytes/h
Data rate (in)	Accept s/w uploads	50	kbit/s
Full image data rate (out)	20 full images per minute	2.5	Mbits/min
Storage	On site image archive	1.2	Gigabit/day
		GPS receiver	
Configuration	Timing and geolocation	Integrated antenna and electronics package	
Remoteness from host computer	Flexible for installation	>30	m
Time accuracy	Substorm timing <1 s	1 (NTP compatible)	ms

Table 2 GBO stations. Columns 1–6 are self explanatory. Column 7 is the UT time at local magnetic midnight, Column 8 is the serial number of the imager, Col. 9 is the serial number of the GBO electronics, Col. 10 the type of magnetometer and Col. 11 the station installation date

1. No.	2. Site	3. Abbrev.	4. Latitude	5. Longitude	6. Mag. latitude	7. Mag. longitude	8. Mag. midnight (UT)	9. ASI#	8. GBO#	10. GMAG type	11. Deploy date
20	Goose Bay	GBAY	53.316 N	299.540 E	60.73 N	23.08 E	3:37	03	GBO-14	GMAG-6 (10532/5009)	Feb-06
18	Kuujjuaq	KUUJ	58.155 N	291.468 E	66.89 N	13.23 E	4:15	13	GBO-13	GMAG-8 (10547/5011)	Nov-07
19	Chibougamau	CHBG	49.814 N	285.581 E	59.57 N	3.62	4:49	16	GBO-17	GMAG-9 (10546/5013)	Sep-06
16	Sanikiluaq	SNKQ	56.536 N	280.769 E	66.45 N	356.99 E	5:12	09	GBO-22	NRCan w/ GPS-9	Oct-06
17	Kapuskasing	KAPU	49.392 N	277.680 E	59.76 N	351.95	5:29	21	GBO-15	GMAG-7 (10545/5012)	May-06
10	Rankin Inlet	RANK	62.828 N	267.887 E	72.41 N	335.74	6:24	12	GBO-09	CGSM w/ GPS-4 (10528)	Sep-05
13	Gillam	GILL	56.354 N	265.344 E	66.18 N	332.78 E	6:34	19	GBO-19	CGSM w/ GPS-7 (10516)	May-06
15	Pinawa (LdB)	PINA	50.163 N	263.934 E	60.08 N	331.46 E	6:38	18	GBO-16	CGSM w/ GPS-8	May-06
14	The Pas	TPAS	53.994 N	259.059 E	63.27 N	323.80 E	7:05	08	GBO-06	GMAG-1 (10505/4001)	May-05
11	Fort Smith	FSMI	59.984 N	248.158 E	67.38 N	306.64 E	8:06	10	GBO-10	CGSM w/ GPS-3 (10527)	Jul-05
12	Athabasca	ATHA	54.714 N	246.686 E	61.98 N	307.76 E	8:07	02	GBO-02	NRCan w/ GPS-0	Aug-04

Table 2 (*Continued*)

1. No.	2. Site	3. Abbrev.	4. Latitude	5. Longitude	6. Mag. latitude	7. Mag. longitude	8. Mag. midnight (UT)	9. ASI#	8. GBO#	10. GMAG type	11. Deploy date
7	Ekati	EKAT	64.717 N	250.667 E	72.28 N	307.66 E	8:02	04	GBO-04	GMAG-3 (10503/4003)	Dec-04
9	Prince George	PGEO	53.815 N	237.172 E	59.13 N	295.67 E	8:52	15	GBO-03	GMAG-2 (10501/4002)	Sep-04
8	Fort Simpson	FSIM	61.762 N	238.779 E	67.30 N	293.85 E	8:57	05	GBO-21	CGSM w/ GPS-6 (10539)	Nov-06
6	White Horse	WHIT	61.010 N	224.777 E	63.66 N	278.14 E	10:01	07	GBO-07	GMAG-4 (10533/4015)	Jul-05
5	Inuvik	INUV	68.413 N	226.230 E	71.23 N	275.09 E	10:17	17	GBO-08	GMAG-11 (10550/5017)	Jun-05
1	Gakona	GAKO	62.407 N	214.842 E	63.06 N	269.02 E	10:48	20	GBO-18	GI w/ GPS-10	Aug-06
2	Fort Yukon	FYKN	66.560 N	214.786 E	67.24 N	266.14 E	11:00	14	GBO-12	GI w/ GPS-5 (10529)	Oct-05
3	Mcgrath	MCGR	62.953 N	204.404 E	61.72 N	259.84 E	11:32	11	GBO-11	GMAG-5 (10525/4016)	Aug-05
4	Kiana	KIAN	66.971 N	199.562 E	65.13 N	253.47 E	12:02	22	GBO-20	GMAG-10 (10554/4009)	Sep-06
Spare	Berkeley	BERK	37.881 N	237.756 E	43.19 N	301.21 E	8:38	01	GBO-05	GMAG-0 (proto s/n 1)	Mar-05

Table 3 The name, location, abbreviation, IP address, altitude, longitude and latitude of the Education and Public Outreach (EPO) magnetometers. These instruments—besides providing high quality magnetic data—are used by teachers and students in high school science courses

City	State	Acronym.	Latitude	Longitude	Mag. latitude	Mag. longitude
Bay Mills	MI	BMLS	46.24	−84.34	56.760 N	11.15 W
Carson City	NV	CCNV	39.19	−119.8	45.071 N	57.006 W
Derby	VT	DRBY	44.95	−72.13	55.039 N	6.015 E
Fort Yates	ND	FYTS	46.09	−100.68	55.756 N	35.281 W
Hot Springs	MT	HOTS	47.61	−114.67	54.730 N	53.498 W
Loysburg	PA	LOYS	40.18	−78.38	51.063 N	3.367 W
Pine Ridge	SD	PINE	43.08	−102.59	52.439 N	37.396 W
Petersburg	AK	PTRS	56.83	−133.16	59.910 N	76.67 W
Remus	MI	RMUS	43.66	−85.14	54.647 N	12.992 W
Shawano	WI	SWNO	44.78	−88.60	55.31 N	17.400 W
Ukiah	OR	UKIA	45.14	−118.93	51.317 N	57.711 W
San Gabriel	CA	SGD1	34.20	−117.85	40.376 N	53.632 W
Table Mountain	CA	TBLE	34.38	−117.68	40.596 N	53.493 W

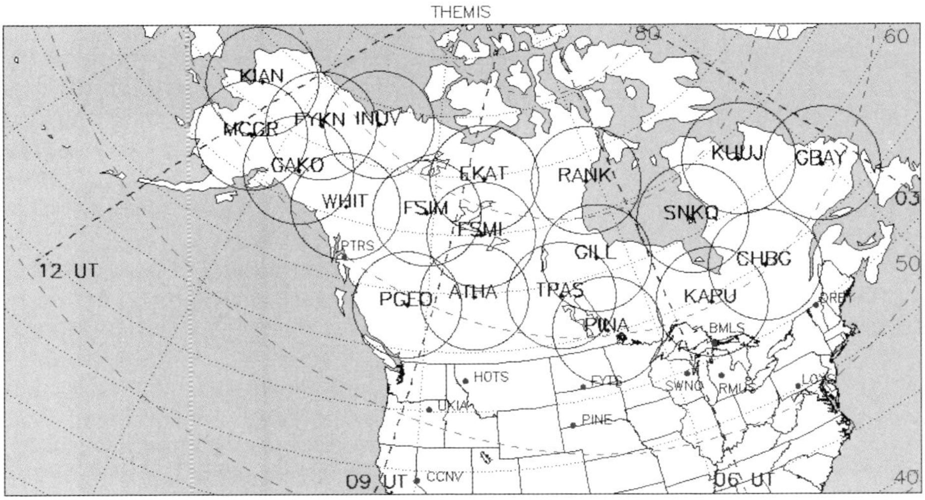

Fig. 5 Map of North America with the GBO station names, locations and their approximate fields of view of the all-sky cameras. Magnetic latitudes are shown with *dashed lines*. Meridians of local magnetic midnight at 03, 06, 09, and 12 UT are also indicated

image, the intensity of only one of the pixels is used. This procedure does not make use of the ultimate SNR obtainable from the co-adding of multiple pixels, however, in general there is no point in enhancing the SNR in small selected region of the image. In cases when several latitude/longitude bins correspond to a single all sky pixel the same intensity is used repeatedly for all the bins.

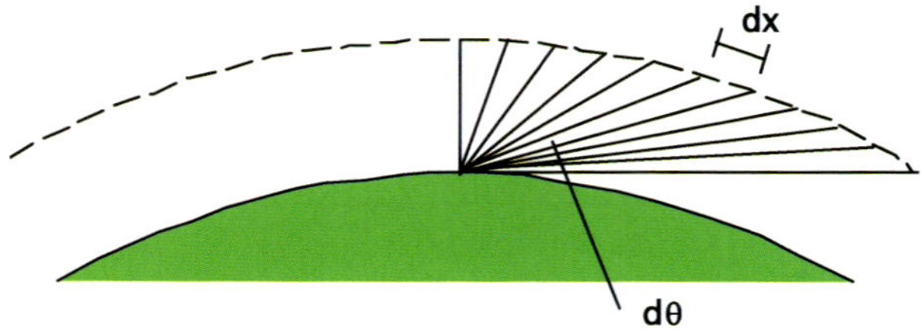

Fig. 6 Schematic showing the cross-sectional view of the all-sky imager, illustrating that equal distance on the "sky" at auroral altitude, dx, represents progressively smaller angles, $d\theta$, towards the horizon

Fig. 7a All-sky image of arcs near the poleward edge of the aurora oval taken at a time when the FAST satellite passed through. The image was taken at Rankin Inlet Station. The *small white crosses* represent the FAST positions at each full minute starting at the top at 03:20 UT

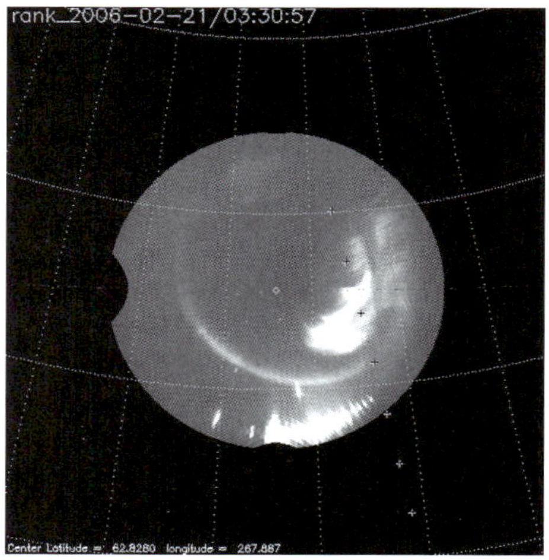

Fig. 7b The same all-sky image backward-projected and shown on a geographic latitude longitude grid

Mathematically if the latitude/longitude bins are represented by coordinate system x_0 and y_0 and the all sky image pixels by x_i and y_i then the intensity in the output bin:

$$I(x_0, y_0) = I(x_i, y_i) = I(f(x_0, y_0), g(x_0, y_0))$$

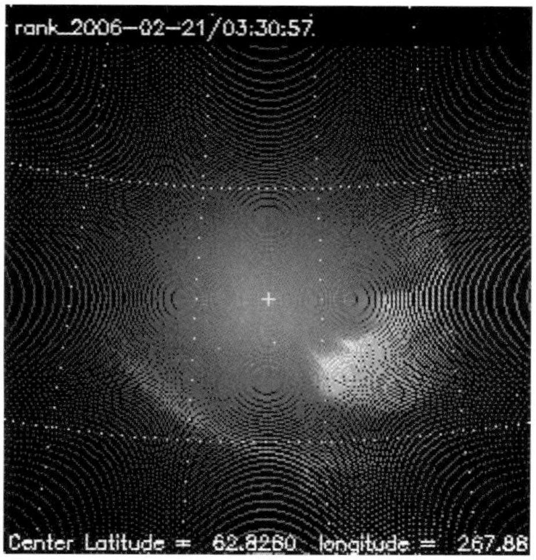

Fig. 8a A forward projected all-sky image demonstrating that, although near its center there is complete coverage, towards the edges only some pixels are illuminated. This is because each pixel in the output image represents larger and larger areas of the "sky"

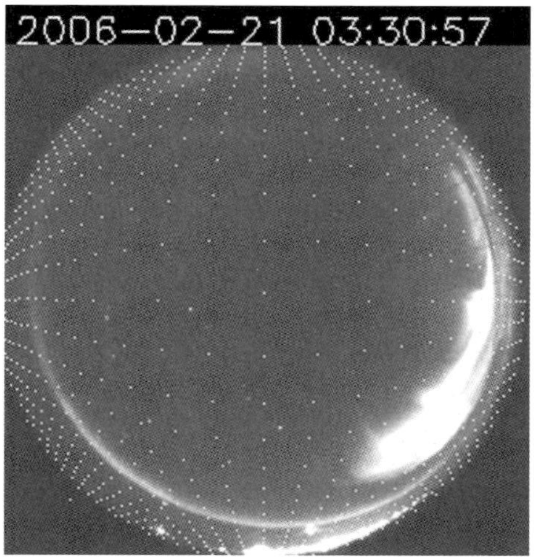

Fig. 8b If the image were divided into 1024 approximately equal area bins, on the sky the *white dots* would signify the corners of the projection of the bins on the original image

where the functions $x_i = f(x_0, y_0)$ and $y_i = g(x_0, y_0)$ define the image pixel coordinates in terms of the corresponding output bin coordinates.

Figure 8a presents the same image in a forward-projected transformation, in which pixels in the original image (see Fig. 7a) are transformed to the 110 km altitude region and then projected onto a latitude/longitude coordinate system. On a 256 × 256 output image matrix, not all pixels receive information due to the coarseness of the grid. Upon close inspection, this image is sparsely populated with light, especially near the outer regions.

Figure 8b illustrates how a matrix of similar size bins of approximately 30 km × 30 km projected on the sky would translate to various size bins on the original image. Each white point on the image represents the corners of approximately equal area bins at 110 km altitude on the "sky."

The minimization of the data volume for NRT transmission to the home base was a high priority. We use a data compression scheme that takes advantage of the fact that the all-sky cameras over sample the central region of the image. In this scheme, a 1024-element vector is generated from the all-sky images, where each element represents the intensity of an approximately equal area regions of the aurora. These vectors are transmitted through the internet.

Upon receiving the data. the 1024-element vectors are converted into reduced resolution thumbnails. These thumbnails satisfy the Level 1 requirements of the THEMIS GBO program because they retain the spatial and temporal resolution necessary to locate and time the substorm onsets. The thumbnail images permit the construction of quasi global auroral images, so-called mosaics. These mosaics are produced as soon as the NRT data are received through the internet—usually with less than one day delay.

4 GBO Instrumentation: The All-Sky Imager (ASI)

There were two primary considerations in the imager design. The first was to satisfy the THEMIS scientific requirements. The second was to keep the cost of each camera well under $10,000 so the program could afford the large number (20) of them. This reduction in cost was made possible by using a white light (panchromatic imager). Because the THEMIS science does not require accurate quantitative measurements of the auroral light in the various different wavelength bands, it was possible to use a white light imager.

In the past decades, filtered all-sky imagers based on the telecentric scheme of Mende et al. (1977) became accepted as the standard. With the use of a filter, an auroral spectral feature can be isolated. The measured photon flux can be meaningfully interpreted in terms of the auroral brightness in absolute units such as Rayleighs (Chamberlain 1961). Unfortunately, when filtering is applied, the resultant light from the aurora usually becomes so faint that the systems operate in a photon-starved mode and un-intensified CCDs are generally unsuitable detectors. To illustrate this, consider a CCD with a measured readout noise of 10 electrons r.m.s. (root mean square), which means that it will need 30 signal electrons to have a signal-to-noise ratio (SNR) of 3. In the detection of auroras, filtered auroral images may contain no more than 2 to 3 photo electrons per pixel, which would be undetectable by an unintensified CCD. However, the same signal would be quite visible with an intensified system which has intensification gain in front of the CCD. Thus the introduction of a narrow band filter in auroral observations almost always requires the use of some kind of intensification in front of the CCD. Unfiltered auroral imagers, like the THEMIS ASIs, collect light in the entire visible range and the signal is almost 10 times the intensity of any single auroral spectral feature. Thus the 2 to 3 photo electrons of a single spectral feature would be equivalent to 20 to 30 electrons in white light—just about detectable by an un-intensified CCD that has 10 electron r.m.s. noise. To minimize system complexity in the THEMIS camera design, we were able to stay away from intensification without compromising the instrument performance.

By using the white light system, do we abandon all possibility of quantitative interpretation? To answer this question we have modeled the spectral profile of auroral emissions due to different energy electrons (Lummerzheim and Lilenstein 1994; Chaston et al. 2005) and simulated the collection efficiency of a typical white light video camera of the type described by Maggs and Davis (1986). We found it quite remarkable that in most cases the camera signal was closely proportional to the observed auroral precipitated energy.

Figure 9 shows the results of the modeling for this type of camera. In general, the response is close to unity for all energies higher than 3 keV, indicating that in this energy range a white light camera provides data that is a good measure of the total precipitated energy. The legend shows the different assumptions that were made for magnetic activity,

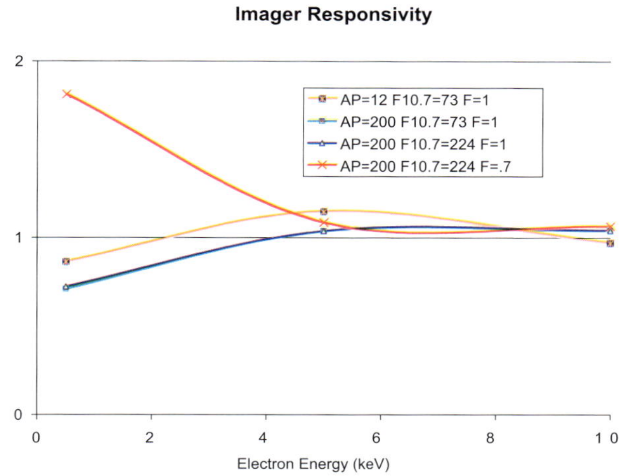

Fig. 9 This shows the modeled camera white light response in equivalent kR, based on the properties of a camera described by Maggs and Davis (1986) for different atmospheric conditions. By modeling three different types of auroral electron precipitation spectra—a 0.5 Maxwellian, a 5 keV monoenergetic and a 10 keV monoenergetic—three points were generated and the points were connected by lines

such as A_p and $F/10.7$ indices used in the modeling. The F factor in the atmospheric model (MSIS-90) is used to investigate the dependence on the atmospheric O/N_2 ratios when they are different from the nominal atmospheric model. The O scale-height in the model was reduced to 70%. This was an unrealistically drastic reduction showing that the response was significantly enhanced at low energies. Presumably this is because low energy electrons are more efficient in producing white light when interacting with N_2 rather than O atoms. In most studies the dependence of auroral emission efficiency on the atmospheric composition is rarely accounted for (Hecht et al. 2006).

Using white light imaging gave us the following major advantages, simplifications and corresponding cost savings:

(1) No expensive filters were needed.
(2) No mechanisms were required for changing filters (e.g., filter wheels) or complex wedge prisms in order to split the light between various wavelengths.
(3) The large bandwidth provided large gains in photon collection and allowed the use of short (1 s) exposure times with very high SNR, even with faint aurora.
(4) Since the system was not "photon-starved," there was no requirement for image intensification with such complexities as a high voltage system and daylight protection.
(5) The white light images had good SNR for many stars, permitting the geometric and intensity calibration and monitoring of the weather conditions above the imagers.

In summary, a simple wide field-of-view optical system, black and white CCD camera operating with a 1-second duration exposure satisfied all the requirements. We built a prototype all-sky imager and tested it. In our first prototype, there were no moving parts, just the passive optics, the camera and a computer.

The basic design of the THEMIS all-sky camera (Mende et al. 1977) was originally developed to record narrow-band evenly-filtered images of the whole wide all-sky camera field. Although filtering was not a requirement for THEMIS, other considerations still favored this same design. For example, it allows use of the inexpensive mass-produced fish-eye lenses commonly used with 35 mm format cameras. These lenses were available with moderate speeds ($F/3.5$) and could be obtained readily. The design permits attachment of such lenses to a wide range of different size CCD cameras with inexpensive coupling optics. The Peleng $F/3.5$ 8 mm focal length lens was selected. Its large format image was

Fig. 10a A schematic cross-section of the all-sky imager (ASI)

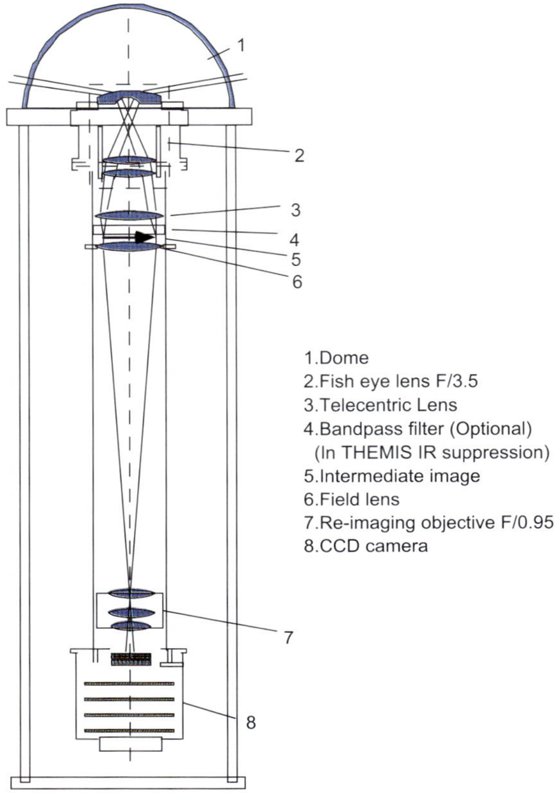

de-magnified and projected on the CCD by the combination of a closeup lens and a Canon $F/0.95$ 25 mm Soligor lens. This arrangement resulted in an all-sky system that uses only inexpensive mass-produced lenses, while having an overall system speed of $F/0.95$, coupled with excellent resolution. Our estimates show that the camera should produce about 100 electrons per pixel per kilo Rayleigh during the selected 1-second exposure. Twenty-two of these relatively inexpensive camera systems were built. They had sufficient sensitivity to detect less than 1 kR aurora with an exposure duration of 1 second. These cameras were about fifty times more sensitive than the IGY cameras, which required long (55 s) exposures. The sensitivity and resolution of the system is adequate to record fairly deep starfields for calibration.

In the THEMIS cameras, the telecentric filter space was used to include a heat-reducing infrared (IR) suppression filter. The camera is illustrated in Fig. 10a. The camera is under the dome (1) in a hermetically sealed aluminum chamber with internal thermal insulation. The Peleng fish eye lens (2) is followed by a simple condenser lens, which also serves as a telecentric lens (3) ensuring that the rays coming from the Peleng lens exit pupil are approximately parallel with the optic axis while passing through the filter. This lens moves the apparent center of the Peleng lens exit pupil far away to "infinity," hence the name "telecentric." An intermediate image (5) is formed near the filter. Another simple field lens (6) directs all the rays into the entrance pupil of the $F/0.95$ re-imaging lens (7). There is a 2-diopter closeup lens (not shown) in front of the re-imaging lens (7) so that the lens operates nearer to infinity conjugate to where its operation was optimized. The CCD camera is a Starlight

Fig. 10b Photograph of the actual camera unit assembled

Fig. 10c As installed on the roof

Express camera containing a Sony black and white CCD, with a USB output connection to the computer.

Heaters, temperature sensors and other auxiliary items are packaged in the camera housing. The sun exclusion shutter is the only moving part of the system. The first prototype imager was installed and operated without a shutter in Athabasca for a whole year as a trial operation. Some of the coatings and the interior paint of the lens deteriorated noticeably during this first year. It was thought that the lifetime of the cameras could be greatly extended if they were protected from continuous year-round direct daytime sun exposure. Furthermore, the CCD has a plastic microlens array in front of the sensitive area and the plastic may be subject to UV degradation. Although none of these deleterious affects made any noticeable change in the camera operation for the first trial year, a mechanical clamshell shutter was designed and installed on each camera as a precautionary measure. The ASI array should be in operation for several years and these shutters should prevent the cumulative effects that might be more significant over several years. The design was "fail-safe" because the shutter has to be powered to close, and it is most likely to fail in the open position.

5 The THEMIS GBO Magnetometers

The GBO fluxgate magnetometer system consists of the sensor—normally buried in the ground—and the auxiliary electronics box located in the electronics enclosure. The sensor is connected to the support electronics via an extended cable threaded inside a garden hose, which provides effective protection of the cable at a modest expense. The sensor is protected by waterproof housing made of plastic pipe.

The magnetometer (GMAG) electronics contain the GPS receiver, which is used to generate time stamps for the GBO data products. The magnetometer auxiliary electronics are illustrated schematically in Fig. 11. The sensor cable connector (on the right) provides the drive signal from the generator to the fluxgates. The 3 sense axes are first amplified and then

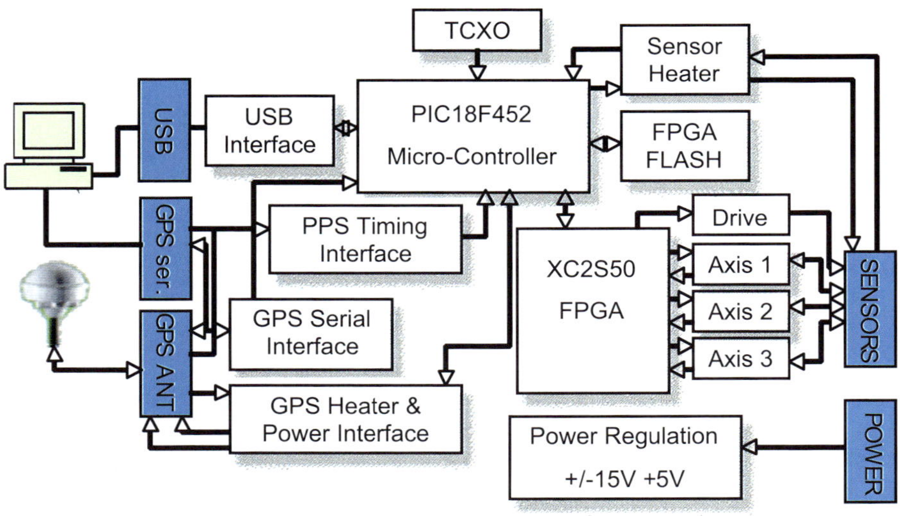

Fig. 11 GBO magnetometer system diagram

digitized in the Field Programmable Gate Array (FPGA). The data are transmitted to the GBO station processor by the micro controller, through a USB interface. The GPS signals are processed and sent to the GBO computer via a serial line. The whole system was built and calibrated by UCLA.

6 The THEMIS GBO Station Design

Figure 12 illustrates the station design. The science detectors are the All-Sky Imager (ASI) and the magnetometer (GMAG). They produce the science data stream, which is transmitted to the GBO processor. The GPS receiver generates more accurate timing than the THEMIS requirements demand (~1 s). The internet serves as the data retrieval and command channel at any station where wired high speed internet is available. At other stations, where high speed internet is still not available, a synchronous satellite-based internet service system was installed. This required a local transmitting and receiving station with an associated dish antenna (Telesat dish illustrated in Fig. 12). At some sites, an additional Iridium satellite transmitter/receiver was also installed to send commands or receive low-rate and housekeeping data.

A typical installation is shown in Fig. 13. The photograph was taken in Athabasca (ATHA). The station at Athabasca has several optical instruments, and the bigger dome on the roof is not related to THEMIS—the THEMIS GBO ASI is the small white unit to the right of the big dome and has a small dome on it. In front of the building, the white box is the computer and auxiliary electronics environmental housing. Most of the magnetometer sensor head is buried except its top, which is clearly visible in the foreground.

At each site, the installation configurations can be different. The computer hardware is situated where a conveniently-located building exists. The camera is in its own cylindrical environmental housing, either on the roof of the building or somewhere nearby mounted on a pole or a tower. The camera housing is hermetically sealed to keep out dust and humidity.

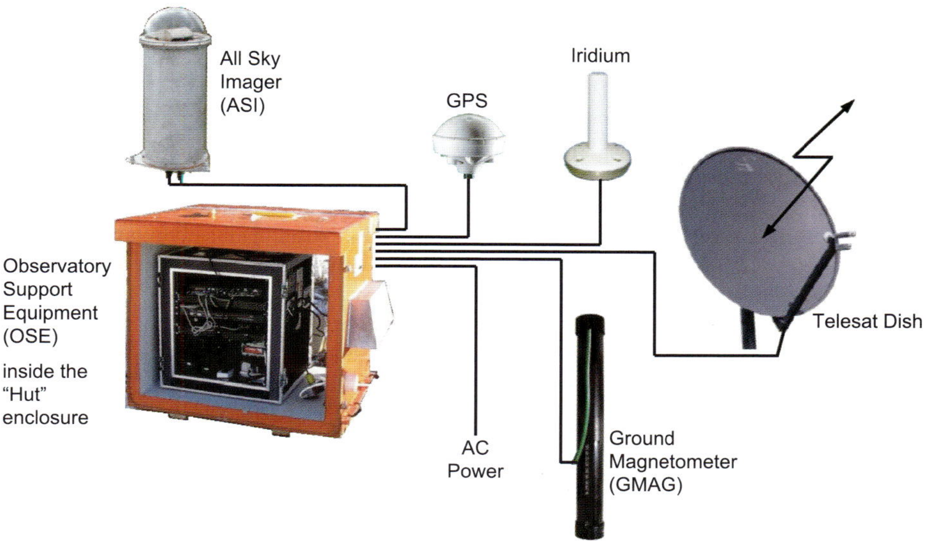

Fig. 12 GBO station schematic

Fig. 13 GBO station as installed in Athabasca, Canada

It is made of aluminum with a layer of insulation inside and the dome is made of clear acrylic.

The availability of a suitable shelter determined the enclosure selection for the computer and auxiliary electronics at each site. If a building was available, then the GBO electronics were installed in their portable 19-inch rack (Fig. 15) without any additional protection. In this configuration, the ambient room temperature took care of the thermal requirements. In cases where no shelter was readily available, a special environmental fiberglass enclosure was installed to protect the GBO auxiliary electronics. It was relatively simple to add heat and keep a reasonable temperature inside the environmental enclosure during cold weather, using power available at each site from the local power grid. A much greater challenge was thermal conditioning the enclosure during the summer hot weather. The heat generated by the electronics, including the computer, needed to be conducted outside of the enclosure through the insulating shelter walls. A solid state thermal electric cooler was installed in one of the walls of the environmental enclosure to satisfy this requirement. This cooler had sufficient power to handle the heat generated by electronics inside the shelter.

The electronics enclosure had an active temperature control system to regulate the heaters and the thermal electric cooler. In Fig. 14, we show a schematic of the auxiliary electronics of the GBOs. The ASI and the GMAG are connected to the system computer via USB interface. The GPS data are provided through a serial input into the computer. The heart of

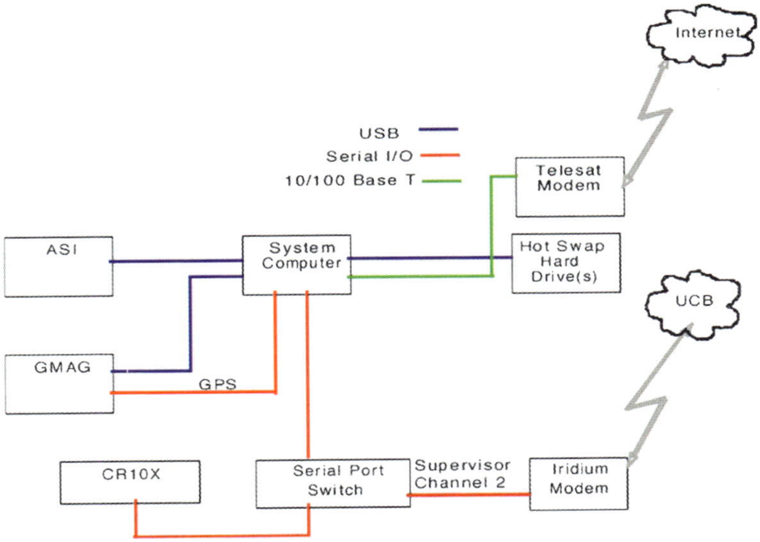

Fig. 14 Schematic of GBO support electronics

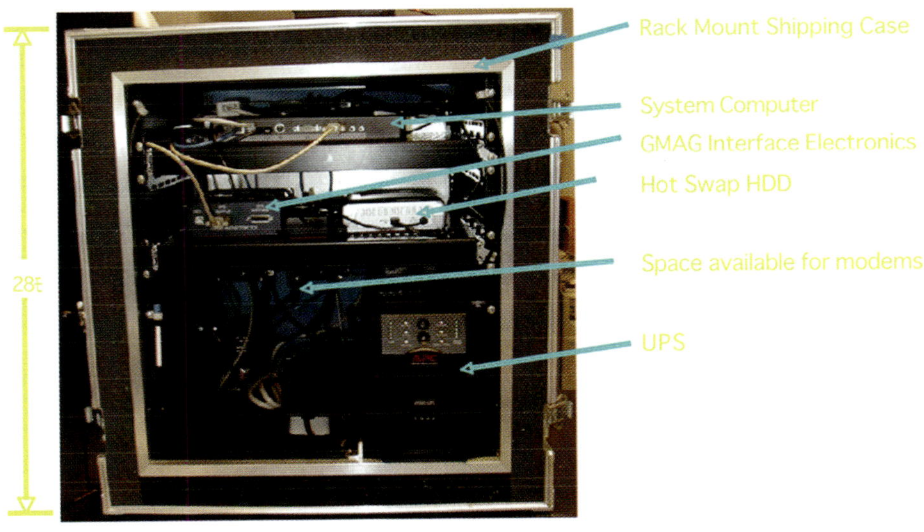

Fig. 15 Photograph of GBO support electronics

the environmental control system is the CR10X micro-controller. This is a programmable controller, it has temperature sensor inputs and can turn on and off heaters and other equipment. This device has been qualified over a wide operating-temperature range and uses minimal power. It has its own battery backup system, which provides uninterruptible operation for several weeks. A major advantage of the system is that it can be re-programmed remotely through one of the communication systems—either the internet or the Iridium system (Fig. 14). For example, the temperature setpoints, where heaters and coolers turn on or off, can be changed without visiting the site. In addition, the CR10X permits turning off the system computer if the system overheats. The system computer uses substantial power because it has to process and compress the image data into thumbnails in real time for transmission via the internet. One of the peripherals of the system computer is the Hot Swap hard drive. When this drive is full, the custodian physically disconnects it and sends it back to the University of Calgary. The rack mount enclosure (Fig. 15) is part of every GBO electronics subsystem.

7 The THEMIS GBO Data

The time resolution requirements dictate the retrieval of large data sets. The GBO ASI data system supports a fast, three-second cadence with each image having 256 × 256 pixels. Twelve to fourteen bits intensity resolution is required to cover the large dynamic range of the aurora. The full image data collected every three seconds result in a ~2.5 MB/min continuous data stream per station. For 20 stations, this represents ~50 MB/min of combined data rate. It would not be possible to reliably retrieve these data from our remote sites via the internet. Therefore, we adopted two separate data retrieval methods. All images were stored at the site, for later retrieval by physically mailing back the hard drives. For NRT data transmission through the internet, a compressed version was generated, consisting of 1024 vector elements per image. The vector data stream is produced from the CCD images via address lookup tables at each site. This data stream provides adequate spatial resolution to satisfy the THEMIS Level 1 requirement and publish the data in near real time.

The THEMIS data retrieval system is schematically illustrated in Fig. 16. Some of the data are retrieved via the internet, as shown by solid black arrows. The full resolution images cannot be retrieved this way and they are recorded on-site on external USB hard drives. The various data products are shown schematically at the bottom of Fig. 16. The global mosaics are produced at Calgary from the thumbnails and later on from the full images. Several "browse products" will be produced from both types of retrieved data and will be available online through a standard internet browser. The team is also committed to producing CDF

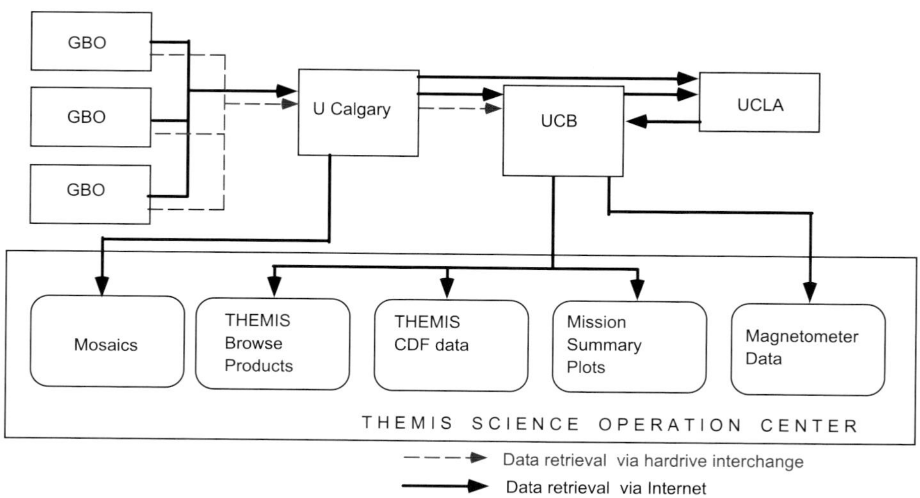

Fig. 16 THEMIS data flow and summary of GBO data types

data files from both the thumbnails and the full images. The mission summary plots contain GBO-derived keogram data in addition to the key satellite data displays. The magnetometer data are available both as data plots and as sets of downloadable CDF files.

We distinguish the two types of data: (1) the near real time thumbnail image data, real time keograms produced at the sites and hourly images and magnetometer data to be retrieved electronically via internet on a daily basis and (2) full resolution image data that will be copied from hot swap hard drives mailed back from the sites.

The NRT data will be transmitted daily to UCB by internet via a main node at the University of Calgary. Those data consist of thumbnails images reconstructed from the 1024 vectors, keograms and hourly full resolution images. The latter two are jpeg images, which are also useful as engineering data with which to monitor station health. These are not converted to CDF format. They are available for viewing on the web in their native jpeg format. The daily and hourly keograms (32 or 256 pixel vertical scans at 1-min or 6-s cadence) and the thumbnails (at 3-s or 6-s cadence) are the highest temporal resolution data possible given the available bandwidth used in the retrieval.

The high-resolution data stored at the site is retrieved by the local custodians periodically (1–3 months). The drives from the stations are sent to the University of Calgary and subsequently to UCB by mail, where they are copied and archived on servers for retrieval. The ASI hard drive data consist of high spatial resolution 256 × 256 pixels images at the highest time resolution (3 s).

Both data types are processed to serve as THEMIS GBO data products accessible on the web.

8 Description of the THEMIS-GBO Web-based "Browse Products"

The thumbnail ASI and GMAG data from all stations are collected by the University of Calgary through the internet. The magnetometer data will be passed to UCLA for validation. The magnetometer data are included in the THEMIS GBO data products. The first six browse data products are created from the NRT data.

Data Product 1 Hourly full average and jpeg compressed 1-minute images. Once an hour, a full jpeg image is transmitted from each GBO site. This format is intended for data quality evaluation, including assessment of the sky clarity over the station. Once a minute, a jpeg compressed full image is also retrieved.

Data Product 2 Clickable KEOGRAMS. A set of Keograms are produced at each GBO station from the full resolution images. These are transmitted as separate files. An example of the available Keograms from all stations is shown in Fig. 17 in collage form. On the THEMIS science operation center (SOC) web display, these Keograms are accessible directly. Clicking at any UT time, displayed as the horizontal axis on the Keogram, will open one of the GBO summary images (Data Product 3) for the appropriate time and station from the THEMIS gifs produced from the images.

Data Product 3 GBO summary images. This is a set of gif images produced first from the thumbnails that contain 1024 pixel vectors. As the full data sets become available from the mailed hard drives, the thumbnail gif images are replaced with gifs produced from the full fidelity images. On the Keogram in Data Product 2, in the very rightmost column, the presence of the "*T*" indicates that the gif images still represent the thumbnail data. A click on the Keogram will reveal the appropriate hour data as a collage of gif images taken on the even minute. Clicking on any one image will expand it to the full time resolution (3-s cadence) or 20 gif images per minute.

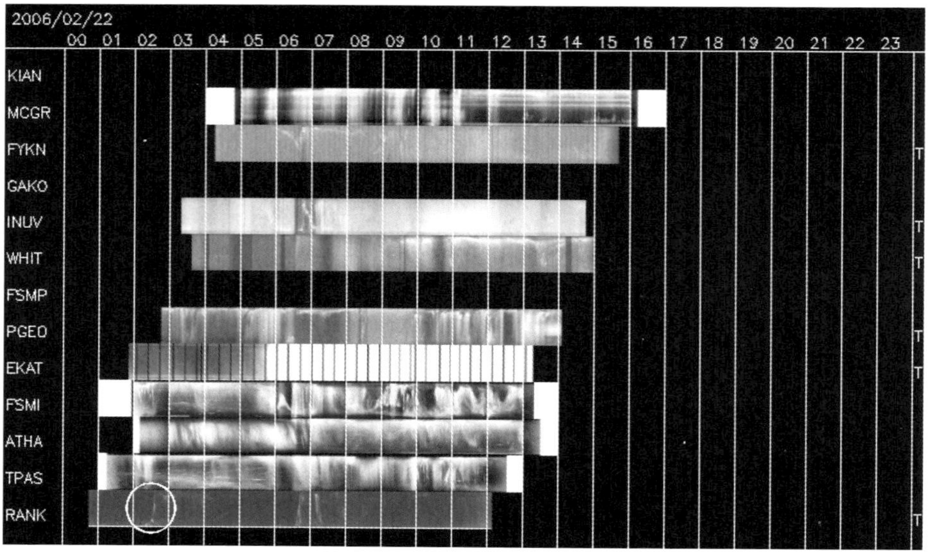

Fig. 17 Keograms from the available stations for February 21, 2006. Clicking on the Keogram derived from Rankin Inlet Data (RANK) hour 2–3 UT (marked with *black circle*) will produce the full resolution collage of Fig. 18a

Data Product 4 Magnetometer data. Individual magnetometer station data X, Y, and Z components are presented as a function of time.

Data Product 5 Mosaic. Full mosaics are produced for the entire array from the thumbnail images or, when available, from full images. These data products are available about 2–4 days after data collection.

8.1 THEMIS Downloadable CDF Files

In addition to the "browse products" described here, the THEMIS science operation center will produce downloadable CDF files. If the science requirement is simply to view images, then the browse products will suffice in most cases. However, if further processing of the GMAG or the ASI data are required, then the CDF files described below will be most useful.

The THEMIS data Level 1 (L1) will be in CDF file format. The THEMIS team provides the community with additional calibration files. The team also provides software tools that read these files and plot the data in a scientifically useful way, producing plots of physical quantities. The software can support integrated analysis and combined plotting with the rest of THEMIS data products. Derivative of the L1 data files, the L2 files are created. These are used in NSSDC data center, facilitating further distribution and plotting through CDAWeb capabilities.

All CDF files are processed at UC Berkeley and University of Calgary independently, but using common processing code, in order to adhere to the same format and file structure described herein. The same structure is intended for use by future ancillary data sets, as they become available. All ASI data will be transformed into 16 bits before they are written to L1 data files. The three types of CDF L1 data products are shown in Table 4.

The first L1 data product is the high time resolution Keogram produced from the 3-s cadence full spatial resolution data. The second data product is comprised of the thumbnail

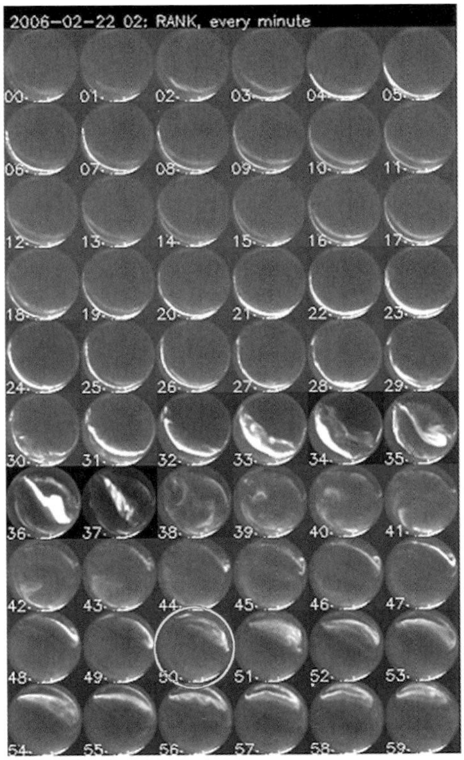

Fig. 18a One hourly image collage from the full resolution images. Clicking on minute 50 (shown with *circular frame*) will open the full time resolution collage, which is illustrated as Fig. 18b

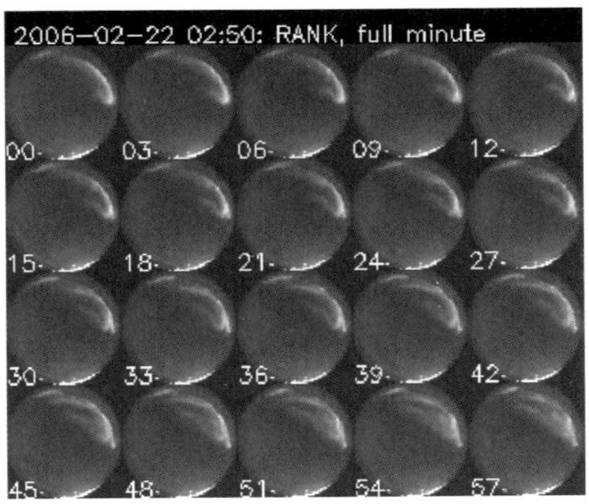

Fig. 18b Full time resolution 3-s cadence, 20 images per minute

images at full time resolution. Some of the stations have reduced transmission bandwidth and the near real time data will only support 6-second cadence or 10 exposures per minute. When the full resolution data are retrieved via the mailed hard drives, then the missing alternate frames can be filled in and full time resolution can be restored with 20 frames per minute. The last L1 CDF set is the full images after they have been copied off the returned hard drives.

Table 4 ASI Level 1 data types and their file names

	L1 VARNAME	Width (pixel)	Frame height (pixel)	Size (bytes)	Cadence (frames/minute)	Sites (per file)	Hour (Mega bytes)	Day Ave (Mega bytes)	Max (Mega bytes)	Year 20 sites (GB)
High T-Res Keograms	thg_ask	1	256	512	20	20	12.3	98.3	295	700
High T-Res Thumbnails	thg_ast_ssss	32	32	2048	10	1	1.23	9.8	29.5	1391
High T-Res Full frames	thg_asf_ssss	256	256	131072	20	1	150	1200	3600	8554

	L1 VARNAME	L1 FILENAME
High Time Res Keograms	thg_ask	thg_l1_ask_yyyymmdd_vnn.cdf
High Time Res Thumbnails	thg_ast_ssss	thg_l1_ast_ssss_yyyymmdd_vnn.cdf
High Time Res Full frames	thg_asf_ssss	thg_l1_asf_ssss_yyyymmdd_hh_vnn.cdf

8.2 Example of THEMIS-GBO Observation Data Set Taken on the 23rd of December 2006

The THEMIS GBOs provide a new view of the auroral regions, and allow the unambiguous recognition of the temporal onset from ground-based data. Mende et al. (2007) describes a substorm that occurred on the 23rd of December 2006 that was captured using the THEMIS GBOs and discusses the accuracy of determining the onset location and timing. Mosaics were made from the station images where the sky was clear. To produce the mosaics, the aurora at each station was projected on the "sky" at 110 km altitude. The resulting image collage consisted of a 1024×512 pixel matrix. Where pixels from image regions of adjacent stations overlapped, the mean of the intensities was taken. This mapping process was only performed once, and a lookup table was generated, allowing subsequent rapid mapping of the entire station array for a selected time period.

For comparison with the optical aurora, the horizontal magnetic vector data were superimposed on the images (Fig. 19). These are the magnetic deviation components produced by subtracting data from a quiet day (December 28, 2006) from each measurement. The arrows represent the horizontal components (B_x in meridional and B_y in east-west [zonal] direction) by the red vectors (Fig. 19). A substorm mosaic movie was made from 06:17:00 to 06:30:00 UT for December 23, 2006, using optical data from six stations (SNKQ, GILL, FSMI, WHI, INUV and FYKN). This can be viewed at: http://www.agu.org/journals/gl/gl0717/2007GL030850/supplement.shtml or ftp://sprite.ssl.berkeley.edu/pub/mende/GBO_movies/12_23_06_movie_quicktime.mov

Prior to onset, the aurora was relatively stationary, with an extended east-west arc located relatively near the zenith at stations SNKQ, GILL, and FSMI. The sky clarity at WHIT was not good enough to assess the situation between FSMI and the Alaskan sector. The latter is clearly seen from the INUV and FYKN data. The arcs were also visible in the Alaskan sector. Another poleward arc system was most visible at FSMI. The magnetic field variations were minimal at most stations, except those that were near the field-of-view of

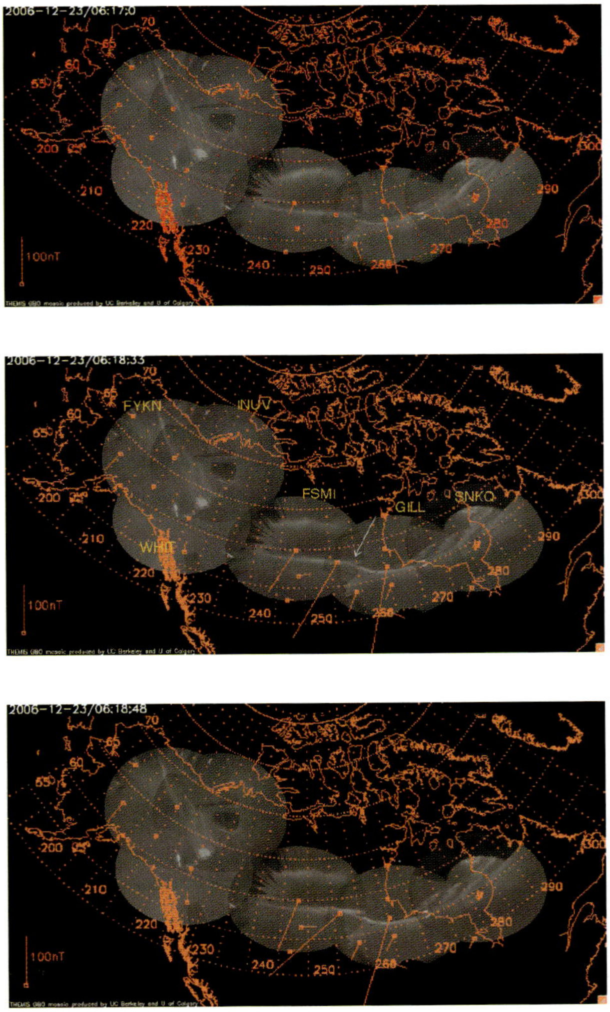

Fig. 19 Mosaic presentation of six auroral images from the THEMIS GBO chain. The stations are: SNKQ, GILL, FSMI, WHIT, INUV and FYKN. The magnetic field vectors were superimposed in the manner described in the text. Image times are indicated on the *top left*. 6:17:00—stationery arc structure with low level of magnetic variation. The westward current is somewhat larger at the sector where the breakup will occur. Some enhancement of the westward current is seen (6:18:33), followed by bifurcation of the arc (6:18:48) and the first frame showing the structured arc prior to breakup (6:19:36). The poleward arc feature is still present westward of the surge (6:22:18) and it disappears by 6:21:42. Breakup is in progress and field vectors are east-west and west-east because of strong FAC east and west of breakup. At (6:26:57) apparent substorm onset intensification is seen in Alaska

GILL and FSMI where the breakup subsequently occurred (6:17:00). The magnitude was slowly increasing prior to and reaching ~200 nT at breakup. The deviation was mainly southward, signifying a westward ionospheric Hall current. There was no eastward current anywhere in the region, therefore the location of the Harang discontinuity could not be established.

The first sign of substorm onset was that the equatorward arc began brightening at 6:18:21 UT with simultaneous increase in the westward current. The first morphological change occurred at 6:18:33 UT (Fig. 19), when the arc appeared bifurcated at the eastern sector of the FSMI frame—here indicated by an arrow. Otherwise the morphology did not change significantly until the arc breakup at 6:18:48 UT, unlike the onset observation of Donovan et al. (2006) who found that the auroral brightening was preceded by east-west structuring. Note that Donovan et al. had only single-station ASI observations. They were unable to rule out a non-local onset and the subsequent propagation of the disturbance from

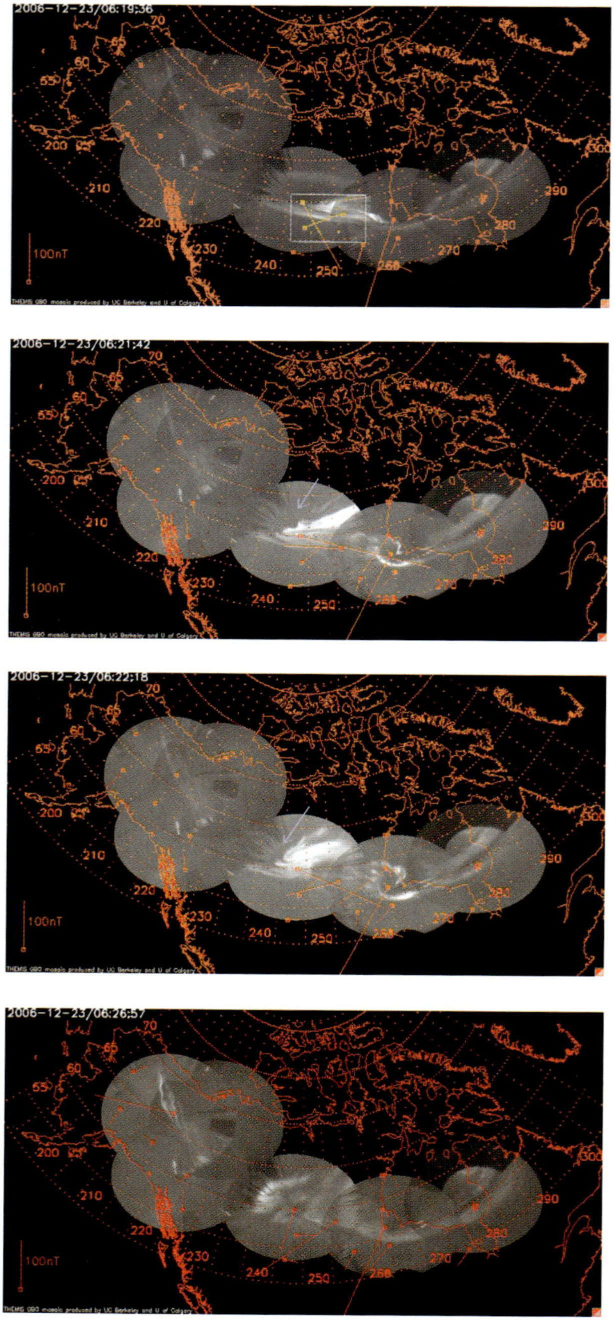

Fig. 19 (*Continued*)

outside of their field-of-view. In our case, the breakup occurred at latitude 58°N and longitude 256°E (67°N magnetic and 22.1 hours MLT) near the region of the bifurcation, and 27 s later than the first discernable brightening. Because the aurora is an east-west elongated arc, the onset determination is accurate to the nearest degree in latitude, but the longitude could be in error of 2 to 3 degrees.

The 6:19:36 UT image (Fig. 1) shows a significant poleward auroral surge. Superimposed on the image is a rectangle showing oppositely-directed magnetic field vectors at two adjacent stations. Counterflowing (north-south) ionospheric currents could cause such a configuration. However, in this case the scenario is consistent with the superposition of the magnetic field from a vertical field aligned current (FAC) due to the substorm current wedge (Akasofu 1972; McPherron et al. 1973). In fact, the overall distribution of the magnetic variations (06:19:36 UT Fig. 1) is consistent with a model of a current system made up of a westward flowing electrojet current, a vertical upward flowing 500 A FAC centered to the west (61°N, 250°W), and a similar magnitude downgoing FAC to the east (60°N, 268°W) of the onset point.

As discussed earlier, at onset there was a clearly visible arc poleward of the breakup arc. The most poleward arc is often associated with the closed field line region poleward (tailward) boundary, therefore it maps to the nightside steady state reconnection line (e.g. Mende et al. 2003). In our case, this arc is present east and westward of the surge and it becomes fainter on the west side, becoming barely visible by 6:21:42 (indicated with an arrow) at FSMI west of the surge. By 6:22:18 (36 seconds later), this feature disappears, and after that the surge arc is the most poleward feature. It has been found in prior observations involving satellite in-situ and imaging data that, prior to substorm onset, the bright onset aurora is not the most poleward feature. However, later, when the substorm is in progress, the surge arc becomes the most poleward feature. The significance of this is that the most poleward arc is often associated with the last closed field line, therefore it maps to the nightside re-connection line. The disappearance of the poleward arc can be interpreted as the opening of the previously closed magnetic flux. And we interpret the surge arc as the new poleward (tailward) boundary of the closed magnetic flux. This would be topologically consistent with a plasmoid formation and its disappearance downtail (Hones 1972). The subsequent poleward expansion of the surge would be also consistent with dipolarization.

The image at 6:26:57 UT shows a situation where the aurora displays the pre-onset arc in Alaska just prior to its undergoing breakup five minutes later. Subsequent to this image, the aurora underwent a similar breakup in Alaska, with electrojet current intensification and a westward turning of field signature. This illustrates that the large local-time coverage of the multiple GBO stations was necessary in order to correctly identify that the prior onset location was in Canada. In general, the widespread westward turnings of the horizontal magnetic field vectors after onset can be associated with an upward FAC poleward of the stations. This shows that the propagating breakup produced a similar upward-flowing current in the Alaska sector, but some minutes later.

9 Summary

A set of ground-based observatories were designed, constructed and fielded as an integral part of the NASA THEMIS satellite program so that the substorm onsets could be timed and located. The instrumentation was designed to cover a large region of the northern arctic in the Canadian and Alaskan sectors. The instruments have a demonstrated high light sensitivity, permitting short exposure and high time resolution. They were designed to incur relatively modest construction costs. The stations were also designed to operate reliably in the Arctic, requiring minimum maintenance. Operation of the chain in the winter of 2005–06 and 2006–07 validated the station design concepts and produced data set that are scientifically useful in their own right. An example of the data set taken during the substorm of December 23, 2006 was presented in this paper. In this example, the onset brightening took 27 seconds, which is too long for an effective time marker to satisfy the THEMIS timing requirements. However, the auroral arc breakup was relatively instantaneous and could be timed to the nearest 3-second, located to the nearest latitude degree and to about ± three degrees in longitude. During the initial arc brightening, the intensification of the magnetic field variations was also quite slow. The largest intensification took place after breakup. Significant magnetic impulses in the Pi2 frequency range occurred, but with a significant ($\sim$40 s) delay. The long periods associated with these type pulsations limited their timing accuracy. In summary, the THEMIS GBOs satisfies the timing and location requirements dictated by the THEMIS program. The coordinated operation of the ground based observatories and the THEMIS satellites will create unprecedented opportunities to study auroral magnetospheric physics.

References

S.-I. Akasofu, The dynamical morphology of the aurora polaris. J. Geophys. Res. **68**, 1667 (1963)

S.-I. Akasofu, The development of the auroral substorm. Planet. Space Sci. **12**, 273 (1964). doi:10.1016/0032-0633(64)90151-5

S.-I. Akasofu, Dynamic morphology of auroras. Space Sci. Rev. **4**, 498 (1965). doi:10.1007/BF00177092

S.-I. Akasofu, *Polar and Magnetospheric Substorms* (Reidel, Dordrecht, 1968)

S.-I. Akasofu, Magnetospheric substorms: A model, in *The Magnetosphere: Part III of Solar-terrestrial Physics/1970* (Reidel, Dordrecht, 1972), pp. 131–151

S.-I. Akasofu, *Physics of Magnetospheric Substorms* (Reidel, Dordrecht, 1977), p. 358

V. Angelopoulos, The THEMIS mission. Space Sci. Rev. (2008, this issue)

D.N. Baker, T.I. Pulkkinen, V. Angelopoulos, W. Baumjohann, R.L. McPherron, Neutral line model of substorms: past results and present view. J. Geophys. Res. **101**, 12975–13010 (1996)

J.W. Chamberlain, *Physics of the Aurora and Airglow*. International Geophysics Series (Academic Press, San Diego, 1961)

C.C. Chaston, L.M. Peticolas, C.W. Carlson, J.P. McFadden, F. Mozer, M. Wilber, G.K. Parks, A. Hull, R.E. Ergun, R.J. Strangeway, M. Andre, Y. Khotyaintsev, M.L. Goldstein, M. Acuña, E.J. Lund, H. Reme, I. Dandouras, A.N. Fazakerley, A. Balogh, Energy deposition by Alfvén waves into the dayside auroral oval: Cluster and FAST observations. J. Geophys. Res. **110**(A2), A02211 (2005)

T.N. Davis, The application of image orthicon techniques to auroral observation. Space Sci. Rev. **6**, 222 (1966)

E.F. Donovan, S. Trond, L.L.C. Trondsen, B.J. Jackel, All-sky imaging within the Canadian CANOPUS and NORSTAR. Sodankyla Geophys. Observatory Publ. **92**, 109–112 (2003)

E.F. Donovan, S. Mende, B. Jackel, H. Frey, M. Syrjäsuo, I. Voronkov, T. Trondsen, L. Peticolas, V. Angelopoulos, S. Harris, M. Greffen, M. Connors, The THEMIS all-sky imaging array—system design and initial results from the prototype imager. J. Atmos. Terr. Phys. **68**, 1472–1487 (2006)

L.A. Frank, J.D. Craven, K.L. Ackerson, M.R. English, R.H. Eather, R.L. Carovillano, Global auroral imaging instrumentation for the Dynamics Explorer mission. Space Sci. Instrum. **5**, 369–393 (1981)

L.A. Frank, J.D. Craven, Imaging results from Dynamics Explorer 1. Rev. Geophys. **26**, 249–283 (1988). doi:10.1029/RG026i002p00249

H.U. Frey, S.B. Mende, V. Angelopoulos, E.F. Donovan, Substorm onset observations by IMAGE-FUV. J. Geophys. Res. **109**, A10304 (2004). doi:10.1029/2004JA010607

S.E. Harris, S.B. Mende, V. Angelopoulos, W. Rachelson, E. Donovan, B. Jackel, M. Greffen, C.T. Russell, D.R. Pierce, D.J. Dearborn, K. Rowe, M. Connors, THEMIS, Ground based observatory system design. Space Sci. Rev. Online First (SSRv Homepage) (2007). doi:10.1007/s11214-007-9294-z

J.H. Hecht, D.J. Strickland, M.G. Conde, The application of ground-based optical techniques for inferring electron energy deposition and composition change during auroral precipitation events. J. Atmos. Solar-Terr. Phys. **68**(13), 1502–1519 (2006)

E.W. Hones Jr., Plasma sheet variations during substorms. Planet. Space Sci. **20**, 1409 (1972). doi:10.1016/0032-0633(72)90048-7

A.T.Y. Lui, A synthesis of magnetospheric substorm models. J. Geophys. Res. **96**, 1849–1856 (1991). doi:10.1029/90JA02430

D. Lummerzheim, J. Lilenstein, Electron transport and energy degradation in the ionnosphere: Evaluation of the numerical solution, comparison with laboratory experiments and auroral observations. Ann. Geophys. **12**, 1039–1051 (1994)
E. Maggs, T.N. Davis, Measurements of the thicknesses of auroral structures. Planet. Space Sci. **16**, 205 (1986)
R.L. McPherron, C.T. Russell, M.P. Aubry, Satellite studies of magnetospheric substorms on August 15, 1968. J. Geophys. Res. **78**, 3131–3149 (1973). doi:10.1029/JA078i016p03131
S.B. Mende, R.H. Eather, E.K. Aamodt, Instrument for the monochromatic observation of all-sky auroral images. Appl. Opt. **16**, 1691–1700 (1977)
S.B. Mende, H.U. Frey, S.P. Geller, J.H. Doolittle, Multistation observations of auroras: Polar cap substorms. J. Geophys. Res. **104**, 2333–2342 (1999). doi:10.1029/1998JA900084
S.B. Mende, H. Heetderks, H.U. Frey, M. Lampton, S.P. Geller, S. Habraken et al., Far ultraviolet imaging from the IMAGE spacecraft. 1. System design. Space Sci. Rev. **91**, 243–270 (2000). doi:10.1023/A:1005271728567
S.B. Mende, C.W. Carlson, H.U. Frey, L.M. Peticolas, N. Østgaard, FAST and IMAGE-FUV observations of a substorm onset. J. Geophys. Res. **108**, 1344 (2003). doi:10.1029/2002JA009787
S.B. Mende, V. Angelopoulos, H.U. Frey, S. Harris, E. Donovan, B. Jackel et al., Determination of substorm onset timing and location using the THEMIS ground based observatories. Geophys. Res. Lett. **34**, L17108 (2007). doi:10.1029/2007GL030850
J.L. Posch, M.J. Engebretson, S.B. Mende, H.U. Frey, R.L. Arnoldy, M.R. Lessard, A comparison of Antarctic Pi1 signatures and substorm onsets recorded by the WIC imager on the IMAGE satellite. American Geophysical Union, Spring Meeting 2004, abstract #SM53B-05 (2004)
T.J. Rosenberg, Recent results from correlative ionosphere and magnetosphere studies incorporating antarctic observations. Adv. Space Res. **25**(7–8), 1357–1366 (2000)
M.T. Syrjäsuo, T.I. Pulkkinen, P. Janhunen, A. Viljanen, R.J. Pellinen, K. Kauristie, S. Wallman, P. Eglitis, P. Karlsson, O. Amm, E. Nielsen, C. Thomas et al., Observations of substorm electrodynamics using the MIRACLE network, in *Proceedings of the ICS-4* (2002)
M.R. Torr, D.G. Torr, M. Zukic, R.B. Johnson, J. Ajello, P. Banks et al., A far ultraviolet imager for the international solar-terrestrial physics mission. Space Sci. Rev. **71**, 329–383 (1995). doi:10.1007/BF00751335

THEMIS Ground-Based Magnetometers

C.T. Russell · P.J. Chi · D.J. Dearborn · Y.S. Ge · B. Kuo-Tiong · J.D. Means · D.R. Pierce · K.M. Rowe · R.C. Snare

Originally published in the journal Space Science Reviews, Volume 141, Nos 1–4, 389–412.
DOI: 10.1007/s11214-008-9337-0

Abstract The THEMIS mission includes a comprehensive ground-based measurement network that adds two additional dimensions to the information gained in the night magnetosphere by the five THEMIS spacecraft. This network provides necessary correlative data on the strength and extent of events, enables their onsets to be accurately timed, and provides an educational component in which students have an active participation in the program. This paper describes the magnetometers installed to obtain these ground-based North American magnetic measurements, including the magnetometers installed as part of the educational effort, and the support electronics provided by UCLA for the ground-based observatories. These magnetometers measure the Earth's magnetic field with high resolution, and with precise timing provided by the Global Positioning System. They represent UCLA's next generation of low-cost, ground-based magnetometers using an inexpensive personal computer for data collection, storage and distribution. These systems can be used in a stand-alone mode requiring only AC power. If there is Internet connectivity, they can be configured to provide near real-time data over the web. These data are provided at full resolution to the entire scientific community over the web with minimal delay.

Keywords THEMIS · Magnetometers · Ground-based magnetometer

1 Introduction

The underlying thesis of the THEMIS mission is that a combination of spatial information on the location and timing of events in the tail during substorms with complementary two-dimensional ground-based data at the feet of these field lines, will enable the location and time of onset of substorms to be identified, and the evolution of the disturbance to be

C.T. Russell (✉) · P.J. Chi · D.J. Dearborn · Y.S. Ge · B. Kuo-Tiong · J.D. Means · D.R. Pierce · K.M. Rowe · R.C. Snare
Institute of Geophysics and Planetary Physics, University of California, Los Angeles, 90095-1567 CA, USA
e-mail: ctrussel@igpp.ucla.edu

followed in space and time. This information will then be used to distinguish between competing models of substorms and other dynamical phenomena in the magnetotail. The auroral ionosphere is essentially a giant TV screen for THEMIS, a plasma screen if you will, that gives high definition pictures complementing the very coarse pixels available in space.

Both magnetometers and all-sky imagers (ASI) monitor this plasma screen. The imagers have the largest field of view outlining where electrons are accelerated down into the upper atmosphere. Twenty stations can provide nearly complete coverage over Canada and Alaska. Magnetometers sense the currents flowing in the auroral ionosphere at an altitude of close to 100 km. Thus magnetometers sense a region above them of only about 100 km in radius and not 500 km as the all-sky imagers do. Ideally it would be desirable to have instruments every 100 km. This would necessitate having the number of magnetometers be 25 times larger than the number of ASI's. However, a 500-site array is not affordable and some compromises must be made. Thus the magnetometer array is collocated with the all-sky-imager array to give coarse latitudinal and longitudinal data. It is crossed with several more finely spaced latitudinal chains, most notably CARISMA and its extension, McMAC and the Alberta Chain and its extension to the south. With these stations current intensifications should be resolvable to about 1° in latitude and 10° in longitude.

Accurate timing and high temporal resolution are very important for understanding the substorm onset. One needs to determine the time of the first acceleration, and one needs to be able to resolve rapidly oscillating phenomena such as Pi1B waves near 1 Hz. Thus GPS timing is used both for the imagers and the magnetometers and the magnetic field data are returned with a 2 Hz cadence.

Of the 50 states, only Alaska lies under the quiet-time auroral zone. However, many of the northern tier of the 48 contiguous states lie close enough to the auroral zone that under disturbed conditions both aurora and significant magnetic disturbances will be seen. Thus an education program was developed around the installation of magnetometers at 11 U.S. high schools. This allows the students to make a meaningful contribution to the science of THEMIS and participate with the THEMIS scientists in the excitement of the program.

In the sections that follow we describe the sensor that measures the magnetic signal; the electronics unit that powers the sensors and handles the data; the GPS timing circuit; the installation of the magnetometers; the data returned; and the plans for archiving the data.

2 Magnetometer Sensor

The mechanical design of the sensor used in the ground-based magnetometers for the THEMIS mission is based on the successful design for the earlier Sino Magnetic Array at Low Latitudes (SMALL) terrestrial vector fluxgate magnetometer (Gao et al. 2000). Three single axis sensors are mounted on a printed circuit board in an orthogonal arrangement. Each of these three sensors is constructed, as shown in Fig. 1, on a one-inch diameter ring core wrapped with a multi-layer toroidal winding. This core is then slipped inside an outer solenoidal winding. The sense axis is aligned with the axis of the solenoid. Precise machining and winding of the solenoid and mounting fixture allow for orthogonality within 0.1°.

The printed circuit board fixture holding the set of three orthogonal sensors is reinforced with side stiffeners to ensure mechanical stability as shown in Fig. 2. It is assembled with a hermetically sealed 1.5-inch diameter white polyvinyl chloride (PVC) tube filled with paraffin oil to provide a stable thermal environment. Paraffin oil is chosen for its non-reactivity with the materials used in the construction of the fluxgate sensor as well as for its high thermal capacity. The right side of Fig. 2 shows the end cap of the white PVC tube containing a

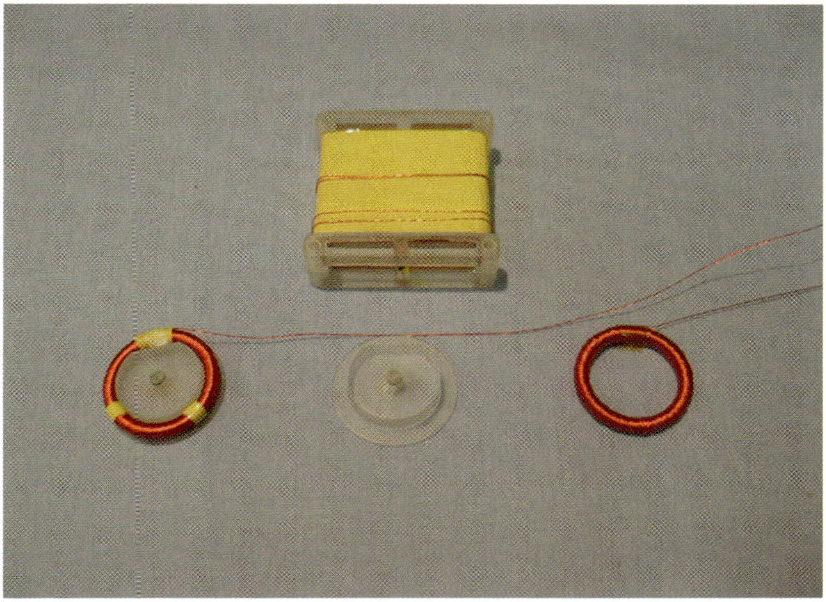

Fig. 1 THEMIS ground-based magnetometer sensor showing (*right*) toroidal winding on core and (*upper middle*) outer solenoidal winding into which core slips

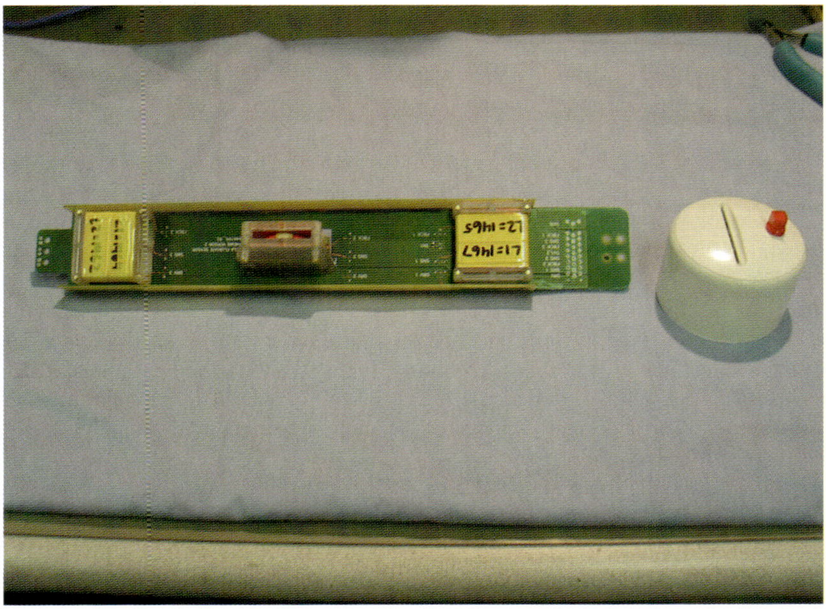

Fig. 2 Printed circuit board fixture holding three orthogonal sensors (*right*). End cap of surrounding PVC tube has slit to allow end of circuit board to interface to cable

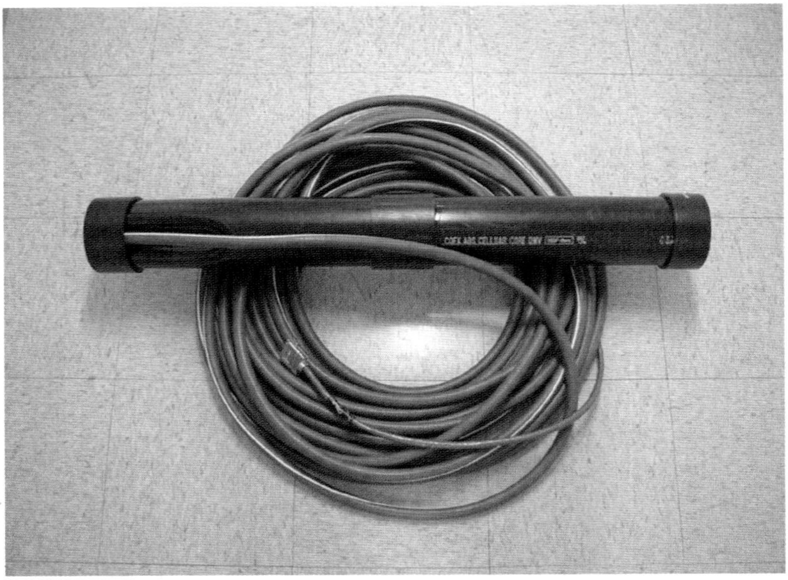

Fig. 3 Outer casing of magnetometer with 30-m cable covered by garden hose attached

slot through which the end of the printed circuit board extends. Electrical connections to the sensor triad are made directly to this portion of the printed circuit board. The board is then fastened with epoxy to the PVC tube end cap. A slotted disk (not shown) is fastened, again by epoxy into one end of the PVC tube. The sensor fixture has a protrusion that fits into the slotted disk ensuring proper axial alignment of the sensor in the tube. The fit between the slotted disk and the sensor fixture is left unglued to allow for slippage during any thermal expansion or contraction of the sensor fixture. In this way, the alignment of the sensor is maintained over a broad range of ambient temperatures.

A 30-meter cable, Belden type 8774 with nine shielded twisted pairs connects the sensor to the electronics chassis. Prior to assembly of the magnetometer, the 30-meter cable is encased in a 3/4 inch diameter common garden hose. This hose provides sufficient protection to the cable to allow it to be buried in the ground without the need for a special conduit.

The white PVC tube is then placed inside a larger four-inch diameter, 1.1-meter-long black PVC tube. Closed cell foam is installed between the walls of the white and the black PVC tubes for additional insulation. The black outer PVC tube is 0.5 meters longer than the inside white tube, allowing the sensor to be easily buried to greater depth. This provides both greater mechanical and greater thermal stability. Figure 3 shows the final assembled sensor ready to be installed. When installed, the sensor assembly is buried vertically in the ground with the top extending sufficiently above the ground to ensure that any standing water does not breach the top cap of the sensor.

A typical site installation is shown in Fig. 4. The PVC tube is vertically oriented and the cable buried. Once it is properly installed, the sensor assembly's thermal design ensures a stable environment for the three orthogonal fluxgate sensors. Temperature variations at the sensor are less than 20°C seasonally resulting in less than 2 nT seasonal offset error. Diurnal temperature changes are attenuated by 95%. Thus a site with a 20°C diurnal change external to the magnetometer would see a diurnal change of only 1°C at the sensor or less than 0.1 nT offset error over the diurnal cycle.

Fig. 4 Completed installation showing buried sensor housing and buried cable

3 Electronics and GPS Circuits

The THEMIS ground-based magnetometer has been designed to provide a cost-effective, high-resolution (10 pT), low-noise ($\pm$25 pTrms) vector measurement of the geomagnetic field at 2 Hz with a GPS-driven timing accuracy of better than 1 msec while being suitable for field deployment requiring a minimum of on-site attention. To achieve this level of performance, several design improvements have been made to UCLA's (SMALL) terrestrial vector fluxgate magnetometer. These include the incorporation of an innovative sigma-delta data processing technique eliminating the need for a precision high-resolution analog-to-digital converter, the addition of a programmable offsetting system capable of fully offsetting a 72k nT field in all three axis independently, and the combined use of a state-of-the-art GPS receiver and temperature-stabilized, crystal, local oscillator capable of assuring timing accuracy in the event of the loss of GPS signal. These elements are illustrated in Fig. 5.

Traditionally, the fluxgate magnetometer electronics has been implemented using amplification, a bandpass filter, synchronous demodulator, integrator, feedback, and analog to digital converter (ADC) as shown in Fig. 6. In this design, a magnetic field is generated in the sense winding that counters the ambient field resulting in a zero-field environment within the sensor along the measurement axis. By operating the fluxgate sensor core in a zero-feedback field, a high degree of linearity is achieved. Two drawbacks of this design, however, are the need for a multi-pole bandpass filter and a precision ADC. Instability in the filter and the limited resolution of conventional ADC's limits the long-term stability and accuracy of the magnetometer in the geomagnetic field.

The use of the sigma-delta technique eliminates the need for these two components while preserving the performance of the fluxgate sensor. The architecture chosen is that of a second order sigma-delta modulator, illustrated in Fig. 7. Optimal performance is achieved by using the field canceling action of the fluxgate sensor in place of the input differencing element of the sigma-delta modulator input as shown in Fig. 8. The output of the modulator, shown

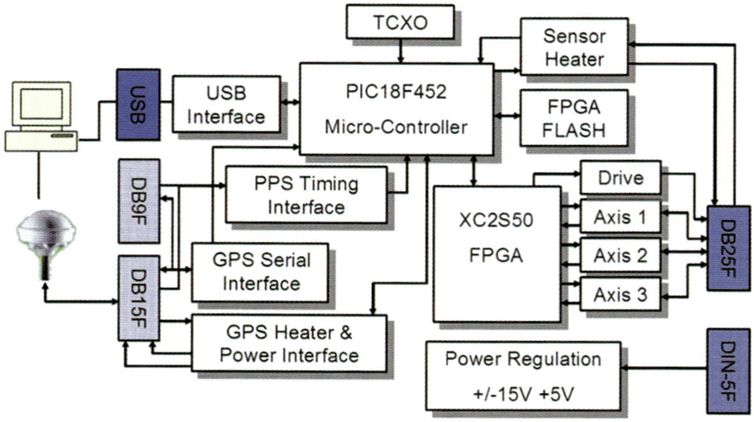

Fig. 5 Functional block diagram of the fluxgate magnetometer

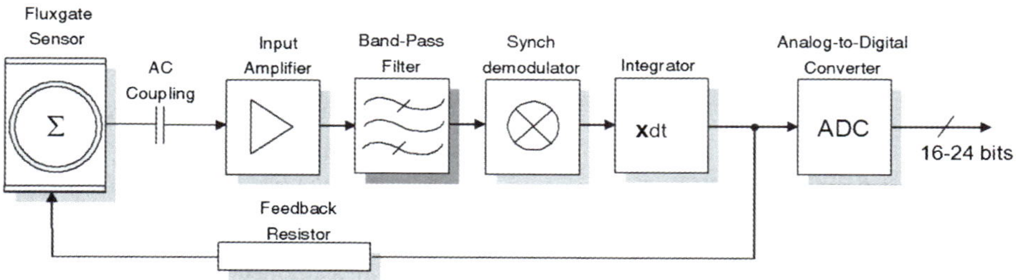

Fig. 6 Feedback circuit of traditional fluxgate magnetometer

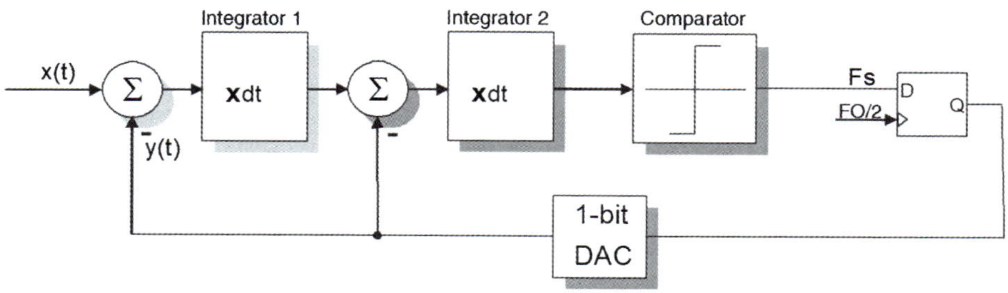

Fig. 7 Functional block diagram of a second-order sigma-delta modulator used for analog-to-digital conversion

as Dout, is a serial bit stream. This bit stream has the characteristic that the density of logic high versus logic low is an accurate representation of the magnetic field. Furthermore, a desirable feature of the sigma-delta modulator is its digitization noise shaping characteristic (Magnes et al. 2003).

The serial bit stream is filtered by use of a comb filter comprised of a set of 4 integrators connected in series followed by a set of 4 differentiators in series as is standard practice for sigma-delta modulators for analog-to-digital converters. The output of this comb filter

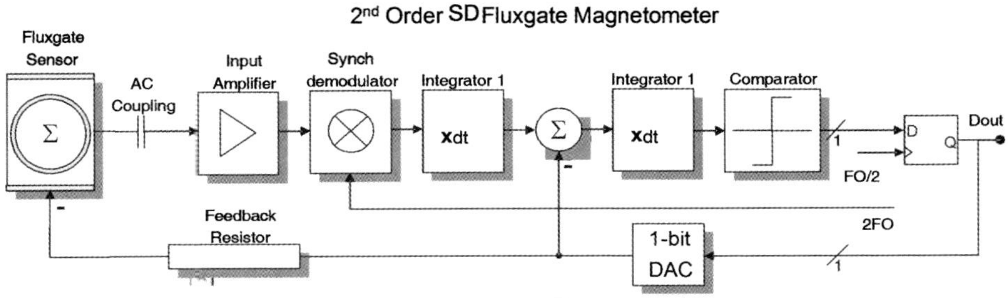

Fig. 8 Feedback circuit of a second-order sigma-delta fluxgate magnetometer

is decimated by a factor of 32, resulting in a 17-bit value at a data rate of 100 Hz. This 17-bit data stream is further filtered and decimated by 50 to reject the high-frequency digitization noise resulting in a 2 Hz data stream with 20-bit resolution and 1 Hz low-pass corner frequency. Figure 9 shows the time series and spectral output of the magnetometer in the presence of a 6 nT 0.2 Hz field applied to the vertical axis.

To achieve a resolution of 10 pT with 20 bits, the dynamic range of the sigma-delta modulator is limited to $\pm$4,000 nT. In order to operate the magnetometer in the geomagnetic field of 72k nT, an offset system was employed. For ease of installation, a three-axis bipolar offset system has been implemented. This allows for each of the three axes of the magnetometer to offset the full geomagnetic field in either polarity. This feature facilitates installation in the southern hemisphere. To achieve this, a 3-channel DAC under the control of the data collection system has been configured to provide 128 offset field values in each polarity. Each step in field is approximately 550 nT in size. Under program control, the magnetometer can be commanded to automatically select the optimal offset field setting to provide maximum dynamic range within the sigma-delta loop of each axis. This need only be done during initial system installation. The value of the offset for each of the three magnetometer axis is recorded within the data stream so that the value of the geomagnetic field for each axis can be reconstructed.

Timing accuracy is obtained by combining the time of day accuracy of the Acutime-2000 GPS receiver with the long-term stability of a Vectron Series TC-400 temperature compensated crystal oscillator (TCXO). The combination of these two time sources allows for synchronization of data collection between magnetometers anywhere in the world. The Acutime 2000 GPS provides a 1 Hz strobe that is synchronized to occur at the zero millisecond boundary of every 1 second tic. When sufficient GPS satellite signals are present, this 1-second strobe is used to time tag the data processing logic within the magnetometer. The local oscillator derived from the Vectron TCXO is also periodically resynchronized to the GPS 1-second strobe. Each time this occurs, a measurement is obtained of the drift rate of the TCXO. These data are collected as a part of the overall GMAC data log stream. In the event that the GPS satellite signal is lost, the TCXO will free run allowing seamless data collection for an extended period of time. Without GPS satellite signal, the TCXO will be expected to drift at a rate of less than 0.2 seconds per day. This drift rate limits the duration in which synchronization of data sampling between stations can be maintained. The drift rate of the TCXO for each station is however well-known, due to the continual measurement of its drift relative to the GPS 1-second strobe. Therefore, time knowledge can be corrected to an uncertainty of less than 10 msec per day.

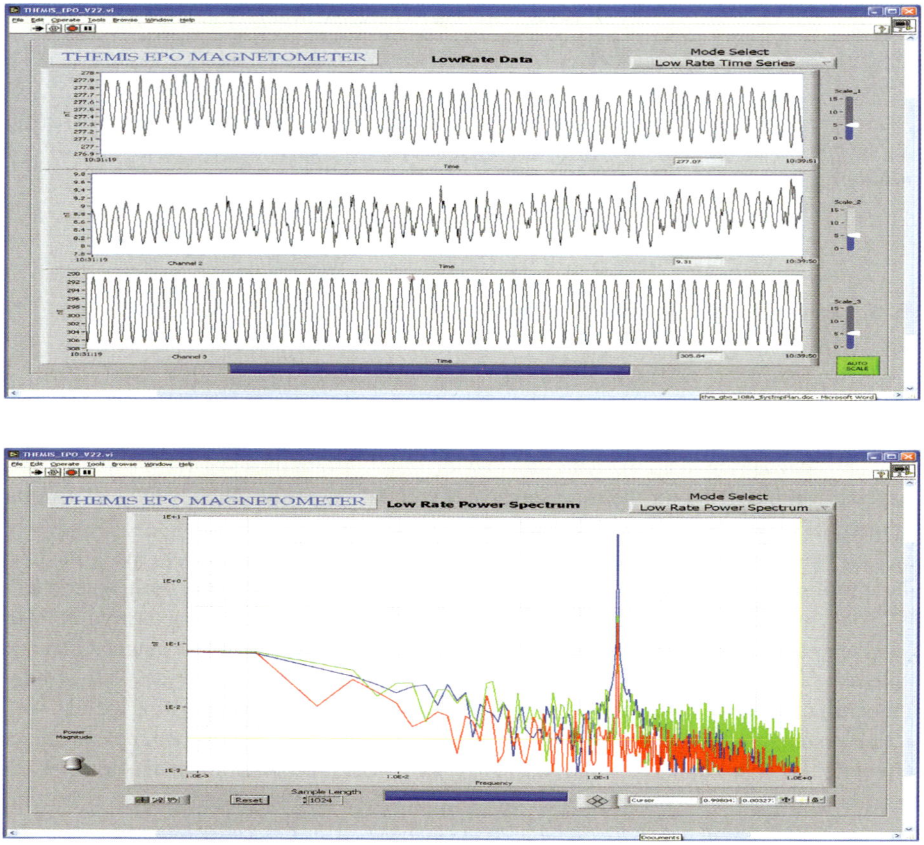

Fig. 9 Response of magnetometer to an applied 6 nT amplitude, 0.2 Hz frequency sinusoidal input. (*Top*) Eight minutes of time series in each of the three components. (*Bottom*) Amplitude spectrum on log–log scale from 10^{-3} to 1 Hz showing spectral peak at 0.2 Hz

4 Installation

The final configuration of the THEMIS ground magnetometer array consists of ten UCLA magnetometers installed with all-sky imagers at Canadian and Alaskan ground-based observatory (GBO) locations. Ten other Canadian all-sky camera sites already had magnetometers installed. Eleven further magnetometers in Alaska and in the contiguous states were installed in schools across the northern portion of the country. Only the Alaskan, Education and Public Outreach (EPO), site could be considered to be in the auroral zone but all were sufficiently sensitive to subauroral geomagnetic activity that they were useful both pedagogically and scientifically. The GBO magnetometers were installed by University of Calgary personnel. These sites have been described by Harris et al. (2008), and Mende et al. (2008). The EPO sites were installed by UCLA personnel. The THEMIS Education and Public Outreach program has been described by Peticolas et al. (2008).

The magnetometer system installed at the schools is shown disassembled in Fig. 10. This diagram shows the outer black PVC tube and its extension section on the right and left sides of the picture. The closed cell foam, the white PVC tube, the printed circuit board with

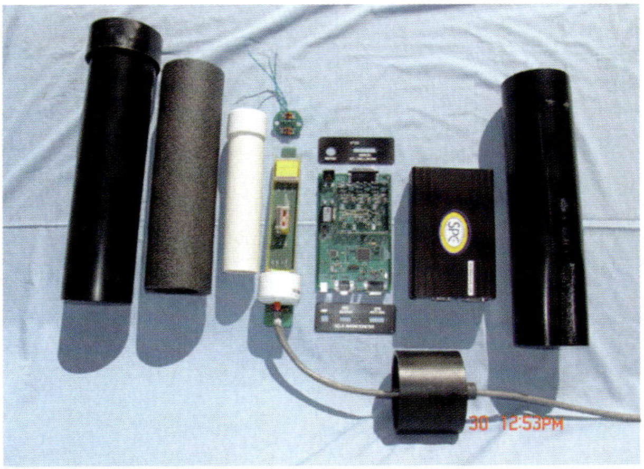

Fig. 10 Ground-based magnetometer sensor and electronics disassembled. On the *left* and *right* are segments of the exterior black PVC tube. The *grey* cylinder is the closed cell foam. The white PVC tube contains the sensors mounted on the printed circuit board. Next are the magnetometer electronics unit and the GPS electronics

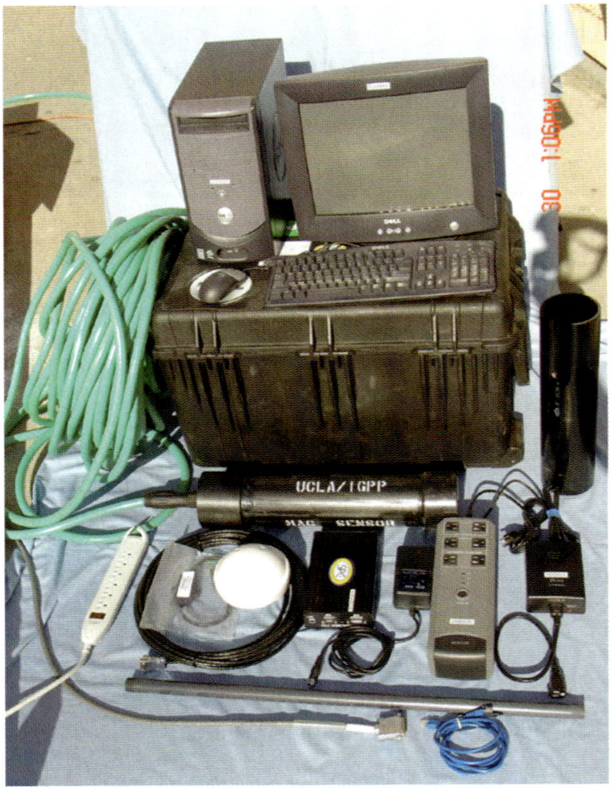

Fig. 11 Magnetometer kit shipped to each EPO site. In *back row*, personal computer tower and monitor sit on shipping box. *To the left* is the cable encased in a garden hose. *In front* are the sensor, GPS system, UPS system, cords and cables

sensors, the electronics board and its housing are arrayed left to right across the picture. Completing the installation require the 30 m cable in a garden hose, a global positioning system (GPS), an uninterruptible power supply (UPS), a computer monitor and modem, and a shipping box as shown in Fig. 11. As shown in Fig. 12, the teachers and students assist

Fig. 12 Teacher and students at Remus assist with the installation of the magnetometer

in the installation of the magnetometers by helping dig the holes and test for aliveness of the magnetometers. The GPS unit provides precise timing for the magnetometers so that all units can be intercompared precisely. These units are generally installed on the roof of the facility as shown in Fig. 13.

The orbits of the THEMIS satellites are phased so that they come into alignment in the geomagnetic tail over North America. The longitudinal extent of auroral processes is large, typically 90° to 180° in extent. Thus it is important to have not only good latitudinal coverage, but good longitudinal coverage as well. Figure 14 shows the location of both the GBO and EPO magnetometers, and Table 1 lists the geographic location and magnetic coordinates of each station. While the data from these 21 sites are collected daily and provided to the THEMIS community, many more sites are providing high cadence magnetic field measurements that will ultimately be available for use by the community, many available already. The full array of sites is shown in Fig. 15.

5 Data Management and Dissemination

The UCLA magnetic observatory consists of four components, the sensor, the GPS receiver, the magnetometer electronics unit and the PC based suite of software. The PC software is designed to facilitate the setup of the magnetometer, monitor the performance and health of the instrument, collect, archive and distribute the magnetometer data products and provide for remote access to the observatory over the Internet.

In most of the installations the software is designed to operate in a Microsoft Windows OS (XP Home) environment. The primary software component is implemented in LabView.

Fig. 13 Antenna for Global Positioning System signal acquisition installed on roof of facility in which magnetometer electronics is housed

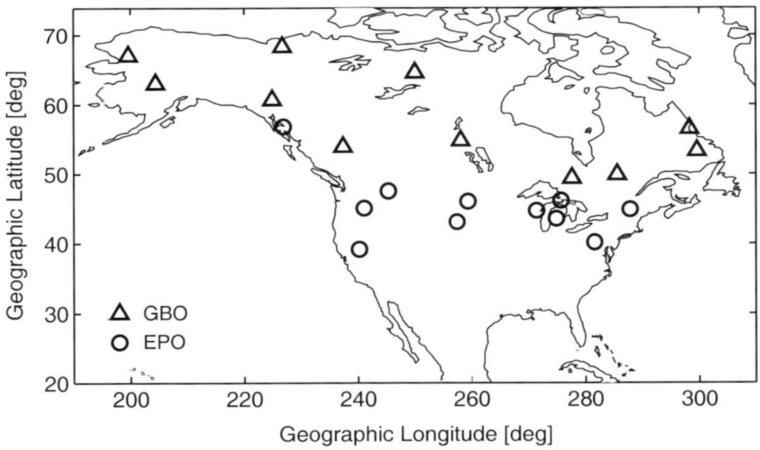

Fig. 14 Locations of GBO and EPO magnetometers

This application allows for setup of the magnetometer and the GPS unit, monitoring the real time data from the magnetometer both in time series and spectral density formats, monitoring the temperatures of the GPS receiver and the magnetometer electronics and sensor, and control of the data products.

There are two types of data products produced by the software, data files stored on the PC and real time data packets transmitted over the Internet. There are three types of data files created and stored on the PC, 'RMD', 'HKP' and 'LOD' files. The 'RMD' files contain the

Table 1 Location of the GBO and EPO magnetometers

Station (Code)	Geogr. lat. (deg)	Geogr. long. (deg)	MAG lat. (deg)	MAG long. (deg)	L	Inv. lat. (deg)	CGM long. (deg)
GBO							
Chibougamau (CHBG)	49.9	285.6	59.9	356.7	3.92	59.7	3.7
Fliti Flon (TPAS)	54.8	258.1	63.0	320.3	5.18	63.9	322.2
Goose Bay (GBAY)	53.3	299.6	63.0	15.2	4.18	60.7	23.2
Inuvik (INUV)	68.3	226.7	70.9	271.9	9.64	71.2	275.7
Kapuskasing (KAPU)	49.4	277.6	59.2	346.5	3.95	59.8	351.8
Kiana (KIAN)	67.0	199.6	65.1	248.8	5.67	65.2	253.5
Kuujjuarapik (KUUJ)	55.3	282.3	65.2	352.0	5.66	65.1	359.2
Lac de Gras (EKAT)	64.6	250.0	71.4	303.2	10.5	72.1	306.8
McGrath (MCGR)	63.0	204.4	62.2	256.6	4.47	61.8	259.8
Prince George (PGEO)	53.9	237.4	59.3	296.3	3.83	59.3	295.9
White Horse (WHIT)	60.7	224.9	63.7	278.3	4.98	63.4	279.4
EPO							
Bay Mills (BMLS)	46.24	275.66	55.92	344.48	3.33	56.8	348.9
Carson City (CCNV)	39.19	240.22	45.35	304.84	2.00	45.0	303.5
Derby (DRBY)	44.95	287.87	54.93	359.67	2.97	54.5	6.4
Fort Yates (FYTS)	46.09	259.35	54.54	324.71	3.12	55.5	325.5
Hot Springs (HOTS)	47.59	245.34	54.31	307.96	2.98	54.6	307.2
Loysburg (LOYS)	40.17	281.62	50.08	352.21	2.48	50.6	357.2
Petersburg (PTRS)	56.83	226.84	60.33	283.11	3.98	59.9	283.3
Pine Ridge (PINE)	43.11	257.40	51.39	323.14	2.67	52.3	323.3
Remus (RMUS)	43.60	274.84	53.25	343.78	2.92	52.2	347.6
Shawano (SWNO)	44.78	271.40	54.24	339.46	3.09	55.3	342.6
Ukiah (UKIA)	45.13	241.07	51.30	304.02	2.55	51.2	302.9

Raw Magnetometer Data. Each 'RMD' file contains one hour's worth of raw, unprocessed data from the electronics unit along with timing (accurate to 1ms) and station identification information. Each file takes up $\sim$ 104 kilobytes disk space. The 'HKP' files contain temperature, offset DAC values and a status byte from the GPS receiver, one record per minute for an entire day. Each 'HKP' file takes up $\sim$345 kilobytes. The 'LOD' files contain log data recording magnetometer configuration data, GPS status changes, and 'RMD' file creation and closing. Each 'LOD' file contains log data for one day and takes up $\sim$ 103 kilobytes of disk space. The sum of the disk space required for 'RMD', 'HKP' and 'LOD' files over one year is slightly greater than 1 Gbyte. A typical PC used for the installation will have approximately 40 Gbytes free space on hard drive at time of installation. The PC's are equipped with DVD R/W optical drives for local extraction of data archives.

The second type of data product produced by the software are the 'real time' data packet datagrams transmitted over the Internet. The protocol used is 'User Datagram Protocol' (UDP). Unlike TCP/IP protocol, UDP is connectionless, i.e. no response from the client is required or expected. This frees the server software from waiting for a 'packet received'

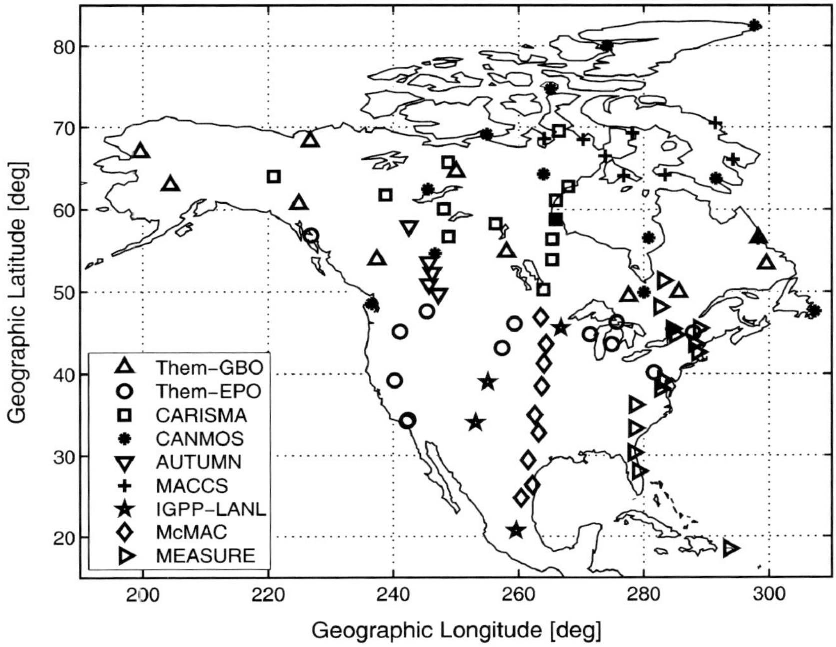

Fig. 15 Locations of ancillary magnetometers in North America together with two THEMIS arrays

acknowledgment from the client. Loss of Internet connection or a bottle neck slowing down transmission will have no effect on the server software. To improve chances of reception, all packets are kept to <512 bytes to preclude the packets being broken up when transmitted over media unable to accommodate packets >512 bytes.

Each mag data packet contains project and site identification strings (ProjID & SiteID), a datagram count number (DGC—incrementing by one for each packet transmitted to a client), temperatures of the magnetometer, sensor and GPS unit, GPS status, data timing information accurate to 1 ms, and up to 50 vectors of 2sps magnetic field data. These packets are transmitted every 5 seconds with a 20 second overlap. Under normal circumstances most data is received with no more than 5 seconds latency resulting in the client being able to present an almost 'real time' time series plot. The 20 second overlap means that most of the time, four consecutive datagrams would have to be lost in order to have a loss of data.

Experiments with a magnetometer installation at a remote site transmitting to a client at UCLA showed that over a 24 hour period $\sim$80% percent coverage of the data was achieved using this method. Running the same test with the client at a remote site resulted in $\sim$98% coverage, implying that most of the Internet bottle necks occur at UCLA. The GMAG software is capable of duplicating the UDP transmission to up to three different clients. By having the clients then retransmit the datagrams received to the other two clients who then discard datagrams received with identical SiteID, ProjID and DGC, almost 100% coverage can be achieved. This is accomplished with $\sim$3 packets of <512 bytes every 5 seconds resulting in a bandwidth requirement of $\sim$2400 bps.

In addition to the magnetometer data packets, log packets containing Project ID, site ID, DGC, time, date, magnetometer offset and scale values, transformation matrix, temperature conversion coefficients, observatory latitude and longitude and any new information added to the LOD file at the observatory. These packets are also <512 bytes long and are transmitted every 5 minutes, whenever the log buffer is full or the configuration information changes. Under normal circumstances, no more that one log packet is transmitted every 5 minutes having little effect on bandwidth requirements.

The primary purpose of this UDP data product is real time data presentation both as a teaching tool at the EPO sites (a teacher can have a magnetometer UDP client running on classroom PC) and as a quick check by UCLA personnel on the health of all observatories currently operated by UCLA. A secondary purpose of the UDP packets is to duplicate the 'RMD', 'HKP' and 'LOD' files at UCLA in 'real time'. While a secure shell link (SSH) is setup on each observatory computer for downloading via RSYNC (a public domain database synchronization system), RSYNC will only transmit those portions of the database that are different. If 'RMD,' etc. files can be created 'real time' at the server, then once a day when RSYNC is scheduled to run very little data will need to be retransmitted, further reducing the observatory bandwidth requirements.

Each observatory has CYGWIN (a public domain LINUX 'look alike' that runs under Windows) installed with the SSH option. This allows engineering personnel to access the observatories remotely while keeping the holes in the local (to the observatory) firewall to a minimum. Remote access includes RSYNC, mentioned above, and VNC (public domain software implementing a virtual terminal) to allow the engineer to remotely upgrade software, change configuration parameters and diagnose and correct problems.

Each site also has an iBoot device that monitors a heart beat signal sent out by the magnetometer software over the local area network (LAN). The heartbeat signal should be sent once every 5 seconds. If the iBoot has not received a heartbeat in 10 minutes, it will power cycle the computer, magnetometer and electronics GPS. The iBoot has a password secured website built in that can be accessed by engineers at UCLA for configuration and manual power cycling of the PC, etc.

Each site also has an uninterruptible power supply (UPS) capable of running the observatory for ~1 hour during AC power outages. The UPS interface software on the computer monitors the condition of the AC power and the status of the UPS batteries and automatically notifies the engineering staff at UCLA of changes.

Data products from the THEMIS EPO sites are updated daily, using RSYNC and are published at UCLA. Data Products from the THEMIS GBO sites are updated daily by UCB and Calgary. All magnetometer data, 'RMD', 'HKP' and 'LOD' files, both EPO and GBO are then 'mirrored' at UCLA, UCB and Calgary, again using RSYNC, insuring that complete databases are available at all three institutions. In addition, at UCLA, a website is available where magnetometer data is published in corrected ASCII format upon request.

The GBO observatories, in addition to the magnetometers also support the all-sky camera developed and built at UCB. The needs of this system dictated that a version of the magnetometer software be developed to run under Linux with a minimal impact on computer resources. The data product capabilities of the LabView software were duplicated in a software package written in 'C'. The 'RMD', 'HKP' and 'LOD' files created and UPD packets transmitted are the same as those produced by the LabView software, but without the graphical user interface provided by LabView. The LabView software is still used to setup and configure the magnetometer, then the configuration data is saved in an 'ini' file and the 'C' version of the magnetometer software is started as background task that runs with very little impact on the UCB provided software. UCB maintains and operates the GBO sites and collects the magnetometer and camera data.

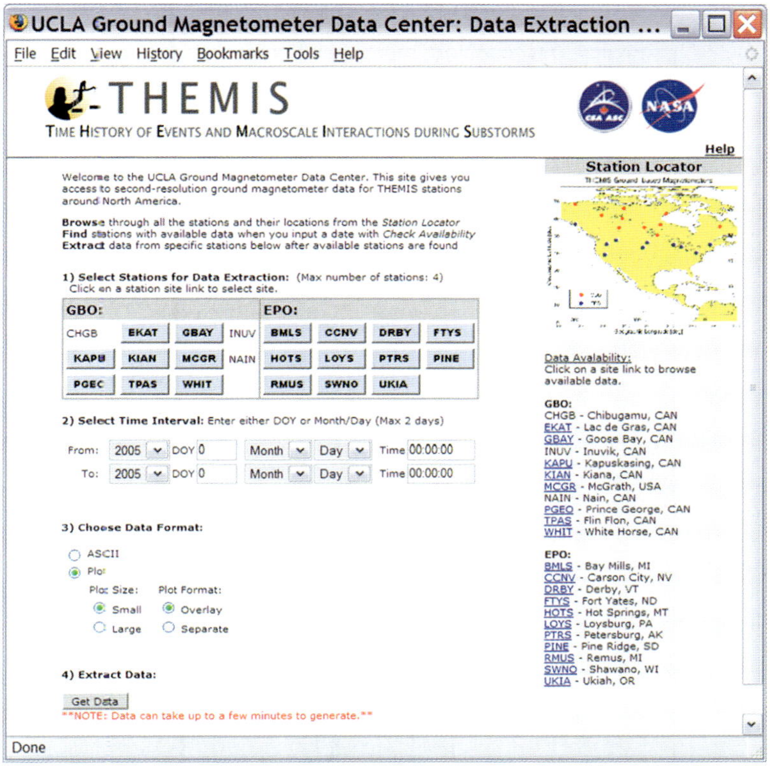

Fig. 16 THEMIS ground magnetometer data center at UCLA (URL: http://www-ssc.igpp.ucla.edu/uclamag/themis_center/)

6 Web-Based Data Server

The THEMIS ground magnetometer data are available to the public through a dedicated online server at http://www-ssc.igpp.ucla.edu/uclamag/themis_center/. The front page of the data server is shown in Fig. 16. Users can either plot or download the data in the ASCII format in the selected time interval. When the plot function is selected, one plot can show data from up to four stations at the same time. When the ASCII data mode is chosen, the data listing includes a header that shows the essential attributes of the data, such as the time interval and calibration values of the magnetometer. The data server has a simple, straightforward user interface. A help file is also linked with the front page, explaining each step of the data extraction in detail.

7 Scientific Return

The THEMIS ground-based magnetometers have been providing high-quality data for scientific research since their installation in the years leading up to launch. Thus they were ready for action as soon as the instruments on THEMIS began taking data. For substorm studies, the prime objective of THEMIS, high-resolution geomagnetic field measurements provide

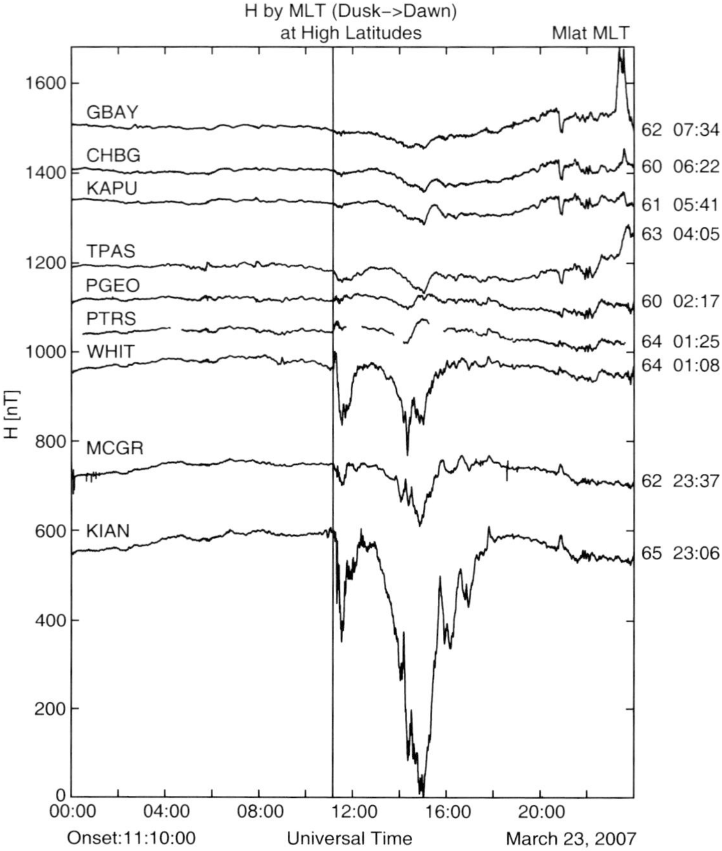

Fig. 17 *H*-component magnetograms for high latitude stations during the first THEMIS substorm event, on March 23, 2007

accurate information on substorm timing. At the expansion onset of substorms, the substorm current wedge (SCW) forms. The initial generation of this 3-D current system including the ionospheric electrojet and field-aligned currents (FACs) can be observed by ground-based magnetometers. The auroral stations (high-latitude stations) provide measurements which detect the variations of ionospheric currents, especially the westward electrojet at the expansion phase of substorms, while the magnetic field perturbations caused by FACs then are detected by the mid-latitude magnetometers. Below we show an example of how the magnetometers document the changing ionospheric currents and their locations during substorms, and the timing of substorm events with Pi1 and Pi2 waves.

Figure 17 shows the *H* components of geomagnetic field for high-latitude stations during the first THEMIS substorm event with conjugate observations and which has been studied comprehensively. The substorm onset is near 11:10 UT (the vertical line in Fig. 17) on March 23, 2007 (McPherron et al. 2008), when the THEMIS spacecraft constellation is located in the dusk–premidnight sector and all spacecraft are within the near-tail region at this time. Dipolarizations of the near-Earth tail magnetic field have been observed by all THEMIS spacecraft. At the same time, the images of aurora during this substorm are recorded by the

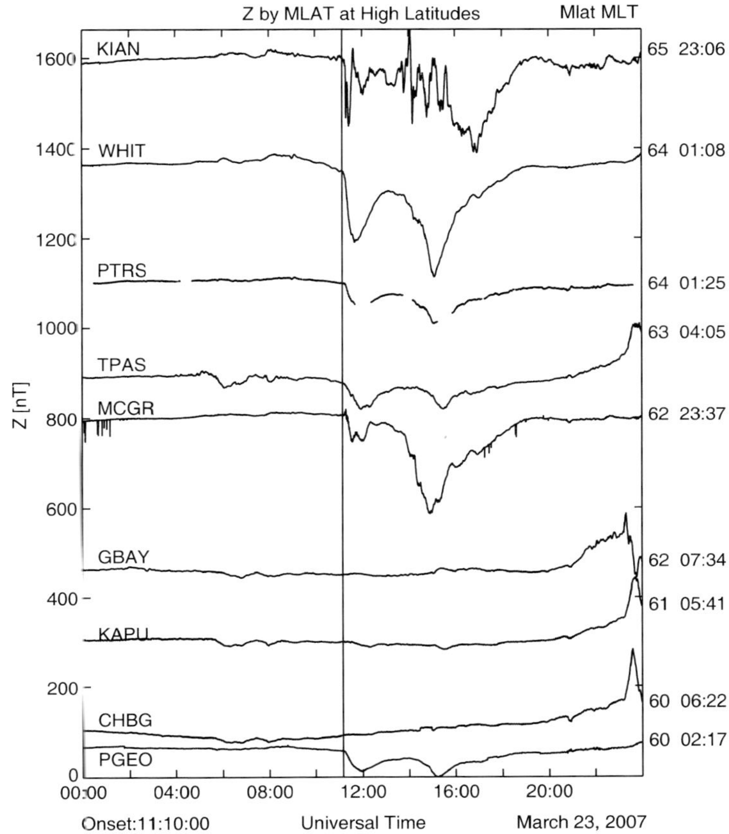

Fig. 18 Z-component magnetograms for high latitude stations during the March 23, 2007 substorm

Polar spacecraft in the southern hemisphere and also by the ground all-sky imagers at Fort Yukon and Kiana (KIAN). In Fig. 17, sharp decreases of H components appear on Whitehorse (WHIT) and KIAN at the substorm onset, showing that the westward electrojet forms over these stations close to the magnetic local time (MLT) of 2300. The magnitude of the largest decrease of H component at this onset is about 250 nT at KIAN (2306 MLT). Weaker negative perturbations appear on the further east stations. The H component perturbations indicate that this substorm is a moderate substorm comparing the following one which has the maximum perturbation of 500 nT at around 1255 UT. The westward electrojet is located at the premidnight sector and close or above the latitudes of KIAN (65°) and WHIT (64°). The latitude of the westward electrojet can be inferred from the perturbations of Z components at the onset which is shown in Fig. 18. The stations are ordered by their magnetic latitudes in this figure from high to lower latitudes. All stations north of Goosebay (magnetic latitude 62°) record negative perturbations in Z components, suggesting that the westward electrojet forms at a higher latitude than this magnetometer network during this substorm.

The middle-latitude stations (the EPO chain) record the perturbations generated by the FACs of the substorm current wedge (SCW) at the substorm onset. During the substorm, the H components of mid-latitude stations are usually enhanced when the stations are located within the sectors of the SCW, which is called a 'mid-latitude positive bay.' In this

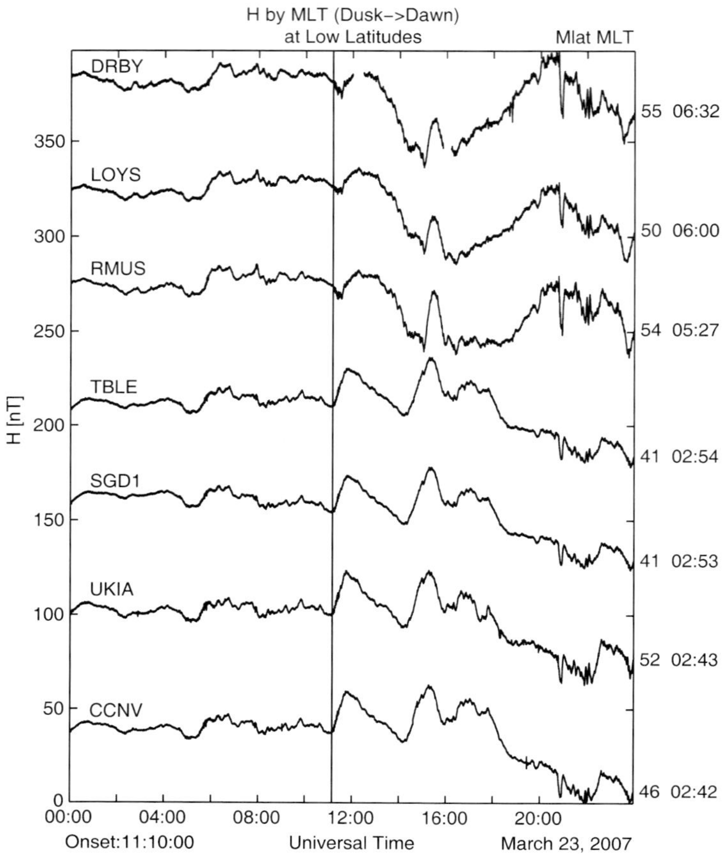

Fig. 19 *H*-component magnetograms from the EPO stations during the March 23, 2007 substorm

event, clear positive *H* bays are shown in Fig. 19 at the onset. The azimuthal expansion of SCW is also signaled by the shifting of positive bays on the stations further east. The *D* component perturbations in Fig. 20 indicate that all THEMIS EPO stations are located east of the SCW central meridian because all perturbations are negative (westward), which is caused by the downward FAC in space. Thus, high-latitude and mid-latitude stations detect the formation and expansion of the SCW at the onset. They also find that the westward electrojet is at a latitude higher than 65° and that on the onset, the SCW central meridian is west of Carson City (CCNV) at 0242 MLT.

For the timing of substorm onsets, ground Pi2 pulsations are generally used, but in this frequency band (40–150 seconds), the uncertainty of the timing is around half minutes (a quarter of typical period). Thus the higher-frequency wave-like pulsations in Pi1 band have been proposed to provide the more precise timing. The association of Pi1 bursty pulsations (Pi1 B) with substorm onsets was proposed in some early studies (e.g., Boesinger et al. 1981) and has been statistically investigated by Ge et al. (2007). It has been found that the Pi1 B pulsations generally appear at substorm onsets but can be detected only in a limited local time range, which can make them less generally useful than Pi2 pulsations (Ge et al. 2007).

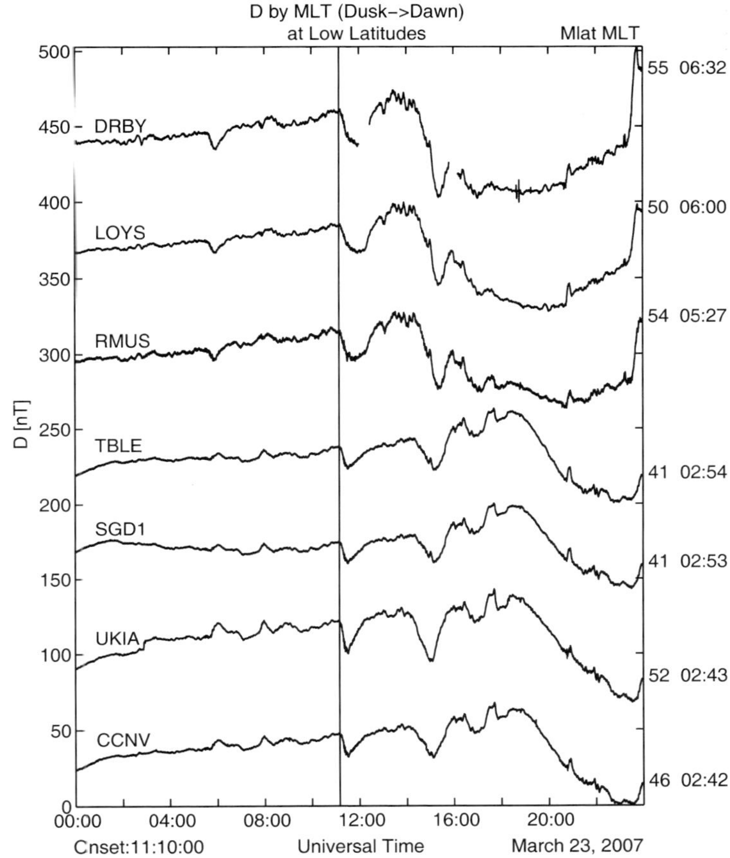

Fig. 20 *D*-component magnetograms from EPO stations during the March 23, 2007 substorms

The band-pass filtered *H* components in Pi1 and Pi2 bands are respectively shown in Figs. 21a and 21b. The Pi2 pulsation onsets are seen at all high-latitude stations but Pi1 B onsets are clear at only three stations (KIAN, McGrath, WHIT), demonstrating the localization of Pi1 B pulsations. The filtered *H* components for the mid-latitude EPO stations are shown in Figs. 22a and 22b. No clear Pi1 B onset is found on any mid-latitude stations while all stations record pulsations in Pi2 band Ge et al. (2007) shows that there is a considerable number of events in which Pi1 B pulsations appear also on mid-latitude stations. The authors believe that the absence of Pi1 B at mid-latitude stations in this event is due to the localization of Pi1 B because the MLT of the station (CCNV) which is the closest EPO station to the SCW central meridian is still 0242 at the substorm onset.

The apparent time delay among high-latitude stations for both Pi2 and Pi1 B pulsation onsets (Fig. 22) indicates that these pulsations are associated with the currents generated at the substorm onset and propagating azimuthally as the SCW expands. However, the Pi2 onsets on the mid-latitude EPO stations appear quite close to each other and are apparently later than the Pi2 onsets at high-latitude stations, suggesting that in the magnetosphere, Pi pulsations propagating inward from the tail, which gives the earlier onsets at high-latitude

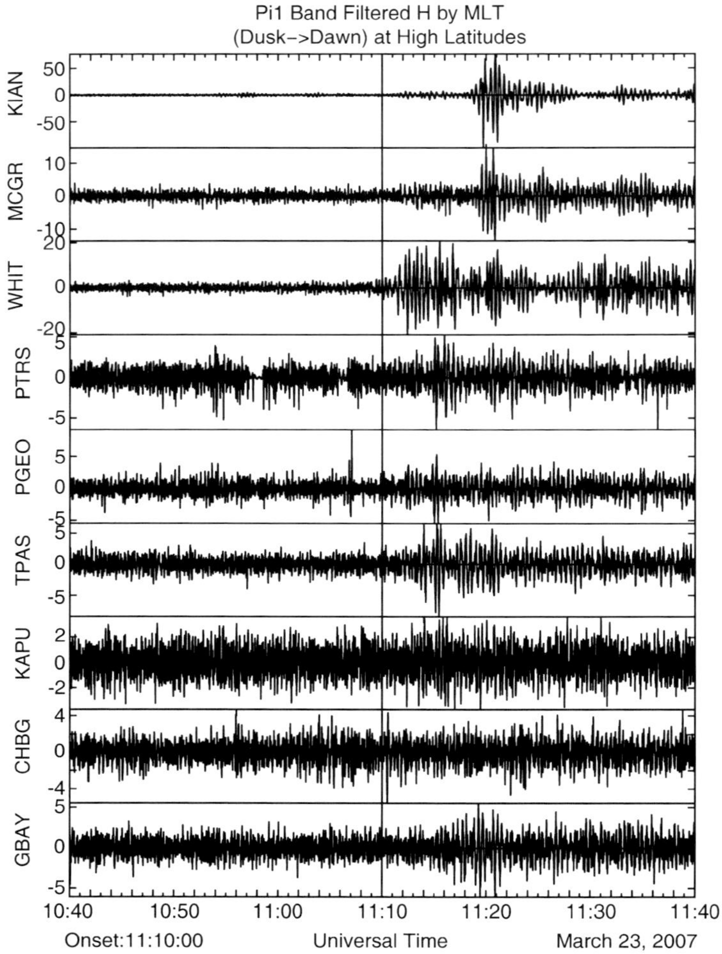

Fig. 21a Bandpass-filtered high-latitude magnetograms covering the Pi1 frequency band

stations. As they propagate closer to the Earth, they more likely become cavity-mode wave to give the almost simultaneous Pi2 onsets at mid-latitude stations.

The earliest onsets both in Pi1 and Pi2 bands are found at WHIT and close to the substorm onset time determined by McPherron et al. (2008) with more ground stations. The Pi2 onsets is more and more delayed as the station is further away from WHIT showing the clear propagation of signals as the expansion of SCW. According the Pi2-band filtered data on WHIT, the onset can be determined between the minima before the vertical line 11:09:00 and the later peak 11:10:20. The uncertainty of timing using Pi2 pulsations is inevitably from half minutes to even one minute. However, the determination from Pi1 B pulsations is more precise. Similar to the method for Pi2, the times for the minima and peak of the wave cycle closest to the Pi1 B onset are found at 11:10:58 and at 11:11:12. The uncertainty then comes down to about 7 seconds, much smaller than the uncertainty from the Pi2 timing. We note that the Pi1 B onset time in this event is about 30 seconds to one minute later than the Pi2 onset time and close to the mid-latitude Pi2 onset time, consistent with the statistical

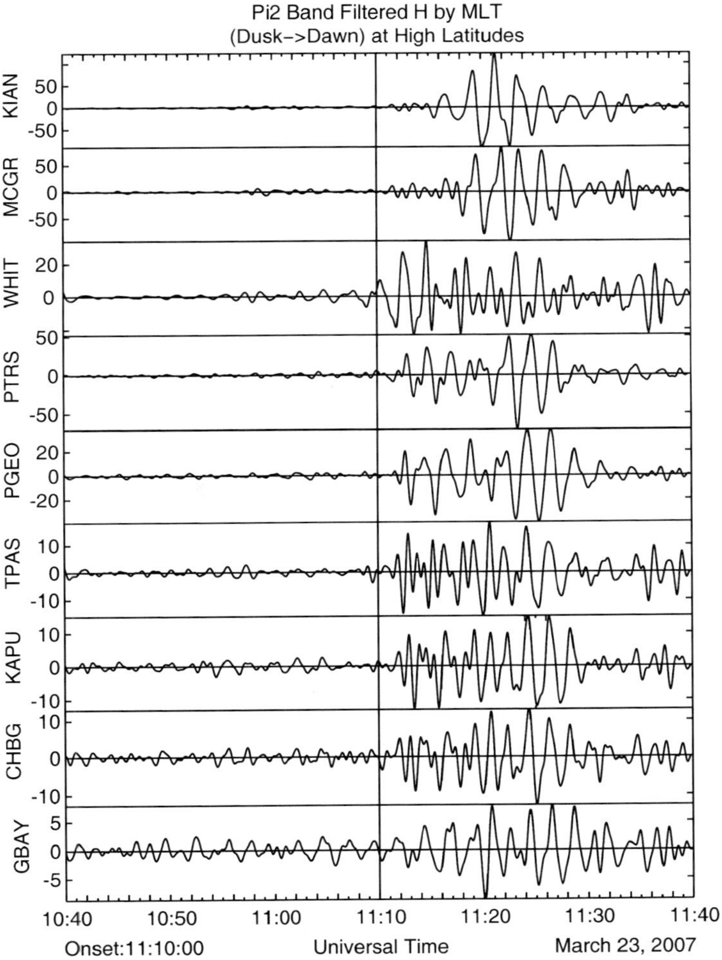

Fig. 21b Bandpass-filtered high-latitude magnetograms covering the Pi2 frequency band

results by Ge et al. (2007). Thus the Pi1B and the Pi2 waves are responding to different parts of the substorm onset and the relative accuracy of the two techniques may be a moot point.

8 Summary

The THEMIS ground-based magnetometers are installed and operating. Their design and operation represent an advance in the state-of-the-art scientific magnetometry. Costs have been minimized to enable the installation of extensive networks. Their operation has been kept simple so they can be operated remotely and reliably for long periods. They are quite appropriate to simultaneously act as a teaching device to introduce students to science and at the same time provide high-quality scientific data. Already the magnetometers are proving their worth, pointing accurately to both the location and the time of substorm onsets.

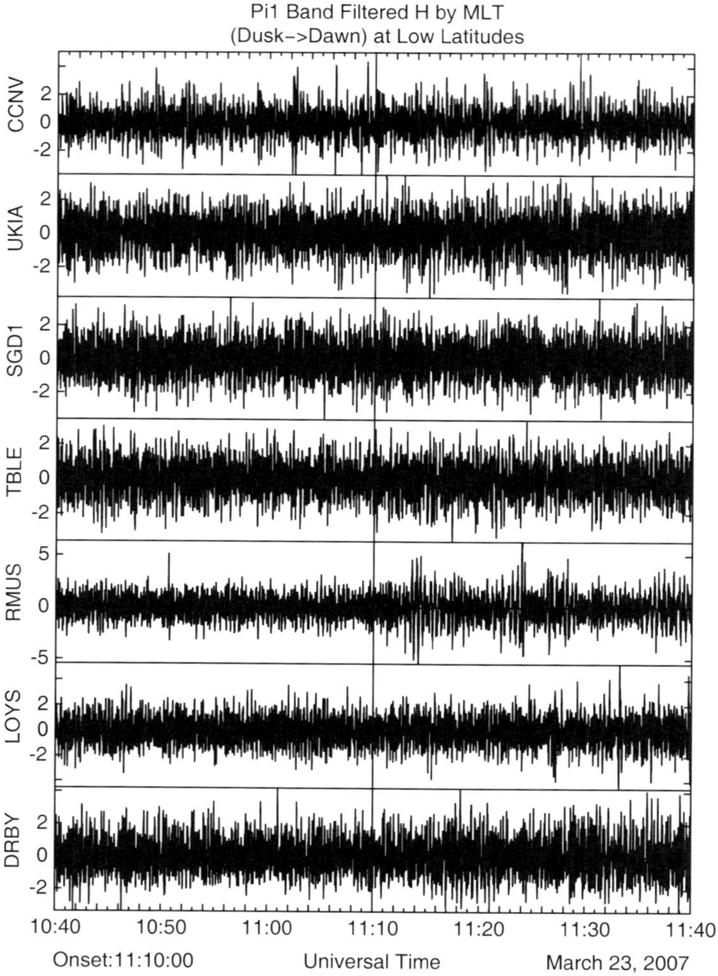

Fig. 22a Bandpass-filtered H-component magnetograms in the Pi1 range for EPO stations

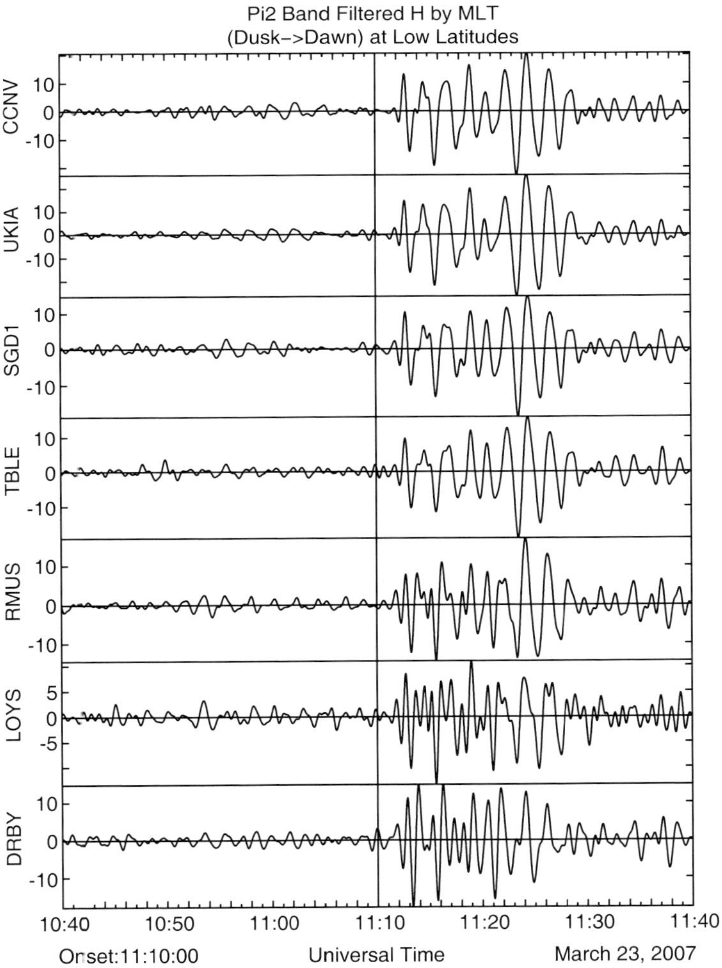

Fig. 22b Bandpass-filtered H-component magnetograms in the Pi2 range for EPO stations

Acknowledgements We would like to express our appreciation to the many educators who have offered to host our EPO sites and have engaged their students in the process. We also thank the many UCLA students who worked in the laboratory, especially Carrie Selsky, who helped immensely in this project. This project was supported by the National Aeronautics and Space Administration under grant NAS5-02099.

References

T. Boesinger, K. Alanko, J. Kangas, W. Baumjohann, H. Opgenoorth, Correlations between PiB type magnetic micropulsations, auroras and equivalent current structures during two isolated substorms. J. Atmospheric Terr. Phys. **43**, 993–945 (1981)

Y.F. Gao, P.J. Chi et al., Sino-Magnetic Array at Low Latitudes (SMALL) including initial results from the sister sites in the United States. Adv. Space Res. **25**(7/8), 1343–1351 (2000)

Y.S. Ge, C.T. Russell, T.-S. Hsu, S. Mende, V. Angelopoulos, J.I. Rae, R.L. McPherron, Investigation on Pi1 B pulsations using THEMIS ground-based magnetometers. American Geophysical Union, Fall Meeting, 2007

S.E. Harris, S.B. Mende, V. Angelopoulos, W. Rachelson, E. Donovan, B. Jackel, M. Greffen, C.T. Russell, D. Pierce, D. Dearborn, THEMIS ground-based observatory system design. Space Sci. Rev. (2008, this issue)

W. Magnes, D. Pierce et al., A sigma-delta fluxgate mangnetometer for space applications. Meas. Sci. Technol. **14**, 1003–1012 (2003)

R.L. McPherron, V. Angelopoulos et al., Pi 2 timing of substorm expansion onset using THEMIS ground based observations and other data. Geophys. Rev. Lett. (2008, submitted)

S.B. Mende, S.E. Harris, H.U. Frey, V. Angelopoulos, E. Donovan, B. Jackel, M. Greffen, C.T. Russell, L.M. Peticolas, Space Sci. Rev. (2008, this issue)

L.M. Peticolas, N. Craig et al., The Time History of Events and Macroscale Interactions during Substorms (THEMIS) education and public outreach program. Space Sc. Rev. (2008, this issue)

The Upgraded CARISMA Magnetometer Array in the THEMIS Era

I.R. Mann · D.K. Milling · I.J. Rae · L.G. Ozeke · A. Kale · Z.C. Kale · K.R. Murphy · A. Parent · M. Usanova · D.M. Pahud · E.-A. Lee · V. Amalraj · D.D. Wallis · V. Angelopoulos · K.-H. Glassmeier · C.T. Russell · H.-U. Auster · H.J. Singer

Originally published in the journal Space Science Reviews, Volume 141, Nos 1–4, 413–451.

Abstract This review describes the infrastructure and capabilities of the expanded and upgraded Canadian Array for Realtime InvestigationS of Magnetic Activity (CARISMA) magnetometer array in the era of the THEMIS mission. Formerly operated as the Canadian Auroral Network for the OPEN Program Unified Study (CANOPUS) magnetometer array until 2003, CARISMA capabilities have been extended with the deployment of additional fluxgate magnetometer stations (to a total of 28), the upgrading of the fluxgate magnetometer cadence to a standard data product of 1 sample/s (raw sampled 8 samples/s data stream available on request), and the deployment of a new network of 8 pairs of induction coils (100 samples per second). CARISMA data, GPS-timed and backed up at remote field stations, is collected using Very Small Aperture Terminal (VSAT) satellite internet in real-time providing a real-time monitor for magnetic activity on a continent-wide scale. Operating under the magnetic footprint of the THEMIS probes, data from 5 CARISMA stations at 29–30 samples/s also forms part of the formal THEMIS ground-based observatory (GBO) datastream. In addition to technical details, in this review we also outline some of the scientific capabilities of the CARISMA array for addressing all three of the scientific objectives of the THEMIS mission, namely: 1. Onset and evolution of the macroscale *substorm instability*,

I.R. Mann (✉) · D.K. Milling · I.J. Rae · L.G. Ozeke · A. Kale · Z.C. Kale · K.R. Murphy · A. Parent · M. Usanova · D.M. Pahud · E.-A. Lee · V. Amalraj
Dept. of Physics, University of Alberta, Edmonton, Alberta, Canada, T6G 2G7
e-mail: imann@phys.ualberta.ca

D.D. Wallis
Magnametrics, Ottawa, Ontario, Canada

V. Angelopoulos · C.T. Russell
Department of Earth and Space Sciences, University of California at Los Angeles, Los Angeles, USA

K.-H. Glassmeier · H.-U. Auster
Institut fur Geophysik und Extraterrestrische Physik, Technische Universitat Braunschweig, Braunschweig, Germany

H.J. Singer
NOAA Space Weather Prediction Center, Boulder, CO, USA

2. Production of storm-time MeV electrons, and 3. Control of the solar wind-magnetosphere coupling by the bow shock, magnetosheath, and magnetopause. We further discuss some of the compelling questions related to these three THEMIS mission science objectives which can be addressed with CARISMA.

Keywords Magnetosphere · Magnetometry · Ionospheric currents · Remote-sensing · Substorms · ULF waves · Radiation belts · Plasmasphere · Cross-phase · Discrete wavelet transform · Field line resonance

1 Introduction

The Canadian Array for Realtime InvestigationS of Magnetic Activity (CARISMA) magnetometer array is a network of ground-based magnetometers with stations deployed across the North American continent. The CARISMA array monitors the 3-D vector magnetic field and its fluctuations at the surface of the Earth, and hence can monitor the magnetic field perturbations driven in the magnetosphere by coupling to the solar wind. The global, meso-scale, and local magnetic effects from electrical current systems and waves can be remote-sensed using this network of sensitive instruments. These instruments monitor the magnetic perturbations arising from currents flowing in the magnetosphere, the magnetic plasma bubble carved out in the solar wind by the Earth's magnetic field, or in the ionosphere, a region of Earth's atmosphere above around 110 km altitude which is itself perturbed by currents and energetic particles from space. These magnetic perturbations provide the capability to remote-sense energy transfer and track disturbances driven in near-Earth space by the sun.

The CARISMA array operates as an integral part of the Canadian Geospace Monitoring (CGSM) program, a multi-instrument program funded by the Canadian Space Agency (CSA) whose goal is to "understand the transport of mass and energy across multiple scales throughout the solar-terrestrial system". CGSM has five grand challenge science themes, namely to address the processes that are responsible for:

- Driving magnetospheric convection and controlling energy injection into the global magnetosphere.
- The triggering and development of magnetotail instabilities and flows.
- The generation, modulation, and multi-scale structure of auroral arcs and auroral particle acceleration.
- The role of wave–particle interactions in the acceleration and loss of energetic particles in the magnetosphere.
- Cold plasma injection, transport, and loss in the global magnetosphere.

CARISMA data contributes to scientific examinations of the process and causes for all of these five CGSM grand challenge science themes.

Data from selected CARISMA stations also represent a formal element of the data set for the NASA Time History of Events and Macroscale Interactions during Substorms (THEMIS) mission (Sibeck and Angelopoulos 2008). CARISMA data will contribute directly to each of the science objectives of the THEMIS mission (see Angelopoulos 2008) which can be summarized as:

- Onset and evolution of the macroscale substorm instability.
- Production of storm-time MeV electrons.
- Control of the solar wind-magnetosphere coupling by the bow shock, magnetosheath, and magnetopause.

There are additional scientific foci for the CARISMA array within the framework of the CGSM grand challenge science themes, as well as support for the science objectives of the International Living with a Star (ILWS) program. However, in this paper we concentrate upon the THEMIS-related science capabilities of the CARISMA array.

In the following sections we describe the details of the CARISMA magnetometer array, including instrument characteristics and station locations, and then outline some of the key areas where the CARISMA magnetometer data will make a crucial contribution to reaching closure on the science objectives for the THEMIS mission. Section 3 briefly discusses the three THEMIS mission science objectives, and Sects. 4, 5 and 6 provide some case study examples and detailed discussion of CARISMA capabilities to address each objective. Section 7 then provides some conclusions.

2 The CARISMA Magnetometer Array

2.1 The Array

The CARISMA magnetometer array is the successor to the CANOPUS (Canadian Auroral Network for the OPEN Program Unified Study—see Rostoker et al. 1995 for details) magnetometer array. The CANOPUS magnetometer array operated as an integral part of the CANOPUS program, ran from 1986 to 2005, and made available magnetometer data from 13 stations at 5 s sampling resolution. The CARISMA project officially started on 1st April 2005, and data from CARISMA is available as a standard 1 s cadence data product and at the raw instrument sampling rate of 8 samples/s on request. CARISMA is operated by the University of Alberta as part of the CGSM program, and is funded by the CSA.

The thirteen original CANOPUS sites have undergone a program of upgrades to site infrastructure within the CSA CGSM program, supporting the continued real-time collection of scientific data from the magnetometers, riometers, meridian scanning photometers and all sky imagers deployed across the collective CGSM array. In the CANOPUS era the science instruments at each site were tightly coupled into a central processor, which built combined data packets for transmission over the Skyswitch satellite link. This has been replaced by autonomous science instrument data loggers which are linked to the internet via the Information Technology Infrastructure (ITI) firewall computer and associated hardware. The ITI provides a Ka-band satellite internet link, uninterruptible Power Supply (UPS), network switchable power outlets and GPS disciplined Network Time Protocol (NTP) timing for the science instruments. Each of the CGSM instrument arrays is operated under a separate contract from the CSA, and each instrument array has been developed further within CGSM under the leadership of the individual instrument array PI.

The CARISMA array now benefits from a significant upgrade as compared to the system which supported operation within the old CANOPUS magnetometer array. Most significantly, the new CARISMA magnetometer infrastructure addresses the most significant weakness of the previous system—the fact that there was no local data storage such that when the real-time satellite data transfer from the site to the central data archive was interrupted the data dropouts in fact represented a permanent loss of data. All CARISMA data is now stored at the remote site on the CARISMA data loggers' RAID disks, and data can be retrieved after any period of network downtime. An increase in bandwidth also allows the collection, retrieval and archiving of magnetometer data at the full 8 samples/s sampling rate, as compared to the previous CANOPUS array which could only provide a 5 second cadence data stream. GPS timing ensures that the CARISMA data time stamps have high

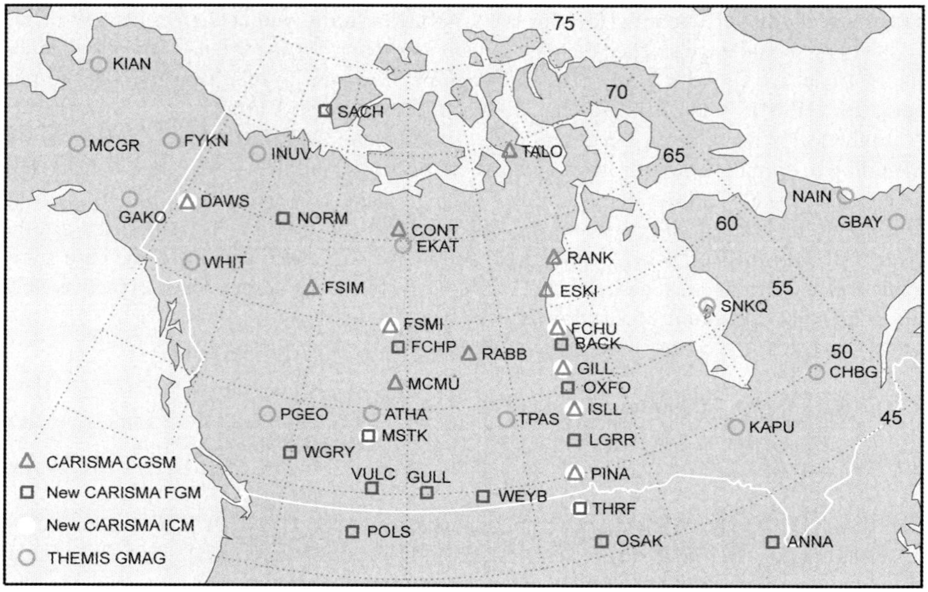

Fig. 1 The locations of the current CARISMA and THEMIS GMAG fluxgate magnetometers, together with the future CARISMA fluxgate (FGM) and search coil (SCM) magnetometers

accuracy, to within around 1 ms. Overall this creates an excellent high-quality data stream which is collected continuously from the CARISMA sites in near-real time over the Ka-band internet link.

The CARISMA network is also currently undergoing a significant expansion, funded by the Canadian Foundation for Innovation (CFI), with the deployment of 15 new fluxgate magnetometer sites underway and due to be finished by March 2009. With a total of 28 stations, CARISMA will constitute one of the foremost magnetometer arrays in the world. Additional stations will be added at strategic locations to improve the scientific capabilities of the array to: i) Provide coverage at mid-latitudes for plasmasphere, radiation belt, and sub-auroral polarization stream science; ii) Decrease Churchill line latitudinal spacing for cross-phase monitoring of the Alfvén continuum and total plasma mass density profiles; iii) Constitute a second meridional line in Alberta to enable some spatial (LT) and temporal (UT) ambiguities to be resolved, iv) Create a grid of stations at mid-latitudes enabling the location of the substorm current wedge (SCW), and Pi2, Pi1 and Pi1B timing, during substorm onset; and v) Provide additional coverage within the field of view of the western Canadian Super-DARN radars. Elements of the enhanced science capabilities afforded through operation of the new array are discussed further below.

In addition, induction coil magnetometers consisting of two crossed coils have been deployed at 8 of the expanded CARISMA array sites. The expanded array will be known as the CARISMA Magnetometer Network (CMN). The proposed CMN station locations are shown in Fig. 1. Site coordinates are given in Table 1.

The CMN covers over 25° in Geomagnetic Latitude (from $L \sim 2.8$ into the polar cap) and 5 hours in magnetic local time. The array is designed around 2 meridional chains, the Churchill line (333° magnetic longitude) and the Alberta line (308° magnetic longitude). These are connected by 2 chains at constant latitude, one in the auroral zone ($L \sim 6.6$) and

Table 1 CMN site coordinates CGM coordinates for 2008 at 100 km calculated using http://nssdc.gsfc.nasa.gov/space/cgm/cgm.html

Site code	Site name	Site type	Geodetic lat. (N)	Geodetic long. (E)	CGM lat. (N)	CGM long. (E)	L
CONT	Contwoyto	CGSM	65.75	248.75	72.9	304.6	11.7
DAWS	Dawson	CGSM	64.05	220.89	65.9	273.7	6.1
ESKI	Eskimo Point	CGSM	61.11	265.95	70.6	333.0	9.2
FCHU	Fort Churchill	CGSM	58.76	265.91	68.4	333.4	7.5
FSIM	Fort Simpson	CGSM	61.76	238.77	67.3	294.1	6.8
FSMI	Fort Smith	CGSM	60.03	248.07	67.3	306.7	6.8
GILL	Gillam	CGSM	56.38	265.36	66.1	332.9	6.2
ISLL	Island Lake	CGSM	53.86	265.34	63.7	333.2	5.2
MCMU	Fort McMurray	CGSM	56.66	248.79	64.2	309.0	5.4
PINA	Pinawa	CGSM	50.20	263.96	60.0	331.6	4.1
RABB	Rabbit lake	CGSM	58.22	256.32	66.9	319.0	6.6
RANK	Rankin Inlet	CGSM	62.82	267.89	72.3	335.8	11.0
SACH	Sachs Harbour	CGSM	71.99	234.74	76.2	280.0	NA
TALO	Taloyoak	CGSM	69.54	266.45	78.4	330.7	NA
BACK	Back	CFI	57.72	265.83	67.4	333.4	6.9
WGRY	Wells Gray	CFI	51.88	239.97	57.8	299.8	3.6
VULC	Vulcan	CFI	50.37	247.02	57.7	308.7	3.6
FCHP	Fort Chipewyan	CFI	58.77	248.89	66.3	308.4	6.3
GULL	Gull Lake	CFI	50.06	251.74	58.2	314.8	3.7
LGRR	Little Grand Rapids	CFI	52.03	264.54	61.9	332.3	4.6
MSTK	Ministik Lake	CFI	53.35	247.03	60.7	307.9	4.2
NORM	Norman Wells	CFI	65.26	233.31	69.6	285.5	8.3
POLS	Polson	CFI	47.66	245.79	54.7	307.9	3.1
OSAK	Osakis	CFI	45.87	264.92	55.9	333.4	3.2
OXFO	Oxford House	CFI	54.96	264.47	64.7	331.8	5.6
THRF	Thief River Falls	CFI	48.03	263.64	57.9	331.4	3.6
WEYB	Weyburn	CFI	49.69	256.20	58.6	320.8	3.7
ANNA	Ann Arbor	CFI	42.42	276.10	53.0	349.4	2.8

the other at mid-magnetic latitudes ($L \sim 3.6$). A small number of other sites are deployed to extend the coverage in both latitude and longitude. The highest concentration of sites will be deployed along the Churchill line to increase the latitudinal spatial resolution to a density where the average station separation is ~ 170 km. This is designed to be particularly useful for density diagnosis by the cross-phase technique (discussed further below). The inter-station separation is less dense along the Alberta line but will still yield very useful information for resolving UT/LT spatio-temporal ambiguities. The majority of the new stations have been added at mid-latitudes which will enable the array to be used for substorm location studies, for instance using the substorm current wedge modeling technique described in Sect. 3.1, or for substorm timing with waves in the Pi2 or Pi1 bands (e.g. Jacobs et al. 1964). The additional station coverage will also provide the capability to monitor the dynamics and erosion of the plasmapause and plasmasphere, using the cross phase technique, as well as the characteristics of ultra-low frequency (ULF) waves which propagate or are excited in the mid-latitude magnetosphere in the region magnetically conjugate to the outer Van Allen radiation belt.

2.2 Instrumentation

2.2.1 Fluxgate Magnetometers

The ringcore fluxgate magnetometers used in the CANOPUS array were designed and built by Narod Geophysics Ltd (NGL). They continue to operate reliably and have yielded high quality data since 1989. An upgraded version of this design has been supplied by NGL for the expansion of the CARISMA array. The instruments are very similar except where end-of-line electronic components needed to be replaced. The specifications of the fluxgate magnetometers are given in Table 2.

Note that the noise level of the new instrument has actually increased slightly as compared to the earlier CANOPUS instruments due to the lack of availability of the previous generation's ringcore material. However, the noise floor is still less than the amplitude resolution of the instrument and so in practice the performance of the instruments remains high and comparable to that obtained in the previous generation.

2.2.2 Induction-Coil Magnetometers

The induction-coil magnetometers (also known as search coil magnetometers) deployed in the expanded CARISMA array are LEMI-30 sensors designed and built by the Lviv Centre of Institute of Space Research, Ukraine. The specifications are given in Table 3.

Table 2 NGL magnetometer specifications

Characteristic	Specification
Dynamic range	±70000 nT
Resolution	0.025 nT
Temperature stability	<0.1 nT/°C
Drift	<0.01 nT/day
RMS noise	S100: <7 pT/$\sqrt{}$Hz at 1 Hz
	CFI: <20 pT/$\sqrt{}$Hz at 1 Hz
Sampling rate	8 Hz
Low-pass cutoff	2 Hz
Power	<1.3 W average

Table 3 Induction coil magnetometer specifications

Bandwidth	0.01 to 30 Hz	
Sensitivity	Channel 1:	Channel 2:
	20 mV/nT (1 to 30 Hz)	200 mV/nT (1 to 30 Hz)
	20 * f mV/nT (0.01 to 1 Hz)	200 * f mV/nT (0.01 to 1 Hz)
Sensitivity error	<3 dB	
Magnetic noise level	<0.2 pT/$\sqrt{}$Hz @ 1 Hz	
Noise rejection	>60 dB at 60 Hz	
Operating temperature range	−10 to +50°C	

2.3 Data Products and Data Access

The CARISMA data is distributed by the Canadian Space Science Data Portal (CSSDP) via the website http://www.cssdp.ca. The raw 8 sample/s data is filtered and decimated to yield 1 sample/s day-files which are the primary data product available on the CSSDP. The CSSDP functionality also allows the browsing of summary plots and provides an interface to a user-defined plotting tool which allows time ranges and filtering to be applied to data which is retrieved and displayed in plots within CSSDP. The full 8 samples/s resolution data is available on request to the CARISMA PI.

An additional data product is also produced at 2 samples/s cadence from five of the CARISMA sites at Fort Simpson, Fort Smith, Rankin Inlet, Gillam, and Pinawa. To form this data set, the data is decimated to 2 samples/s and rotated from the measured geographic coordinate system into local geomagnetic coordinates. This data stream is provided on a next day basis to the THEMIS Science Operations Center (SOC) at the University of California, Berkeley. This 2 samples/s CARISMA data forms an integral and formal element of the THEMIS Ground-Based Observatory (GBO; see also Mende et al. 2008) magnetometer data set, as per a NASA-CSA formal letter of agreement. Provision of the remaining CARISMA data in a 2 samples/s data product is also planned. The search-coil magnetometer data will be made routinely available at 20 samples/s (and on request at 100 samples/s) resolution, and in addition we will publish daily dynamic spectrograms from each site.

The CARISMA team are also founding members of the Ultra Large Terrestrial Magnetometer Array (ULTIMA; http://www.serc.kyushu-u.ac.jp/ultima/ultima.html) a collaboration and formal partnership between operators of international magnetometer networks, promoting scientific cooperation, collaboration, and mutual exchange of scientific data from world-wide arrays.

3 THEMIS Science with CARISMA

The upgraded and expanded CARISMA array can provide data which represents a crucial contribution toward science closure on the three THEMIS mission science objectives. The GBO network which forms part of the THEMIS mission provides global scale instrument coverage in the North American sector, both optically and magnetically. In the prime mission phase, the THEMIS orbits are designed to make repeated magnetic conjunctions to the Canadian sector. The ground-based data from the GBOs and from additional programs such as CARISMA provides an unprecedented framework in which to interpret the scientific data from the five THEMIS probes. In the following subsections, we briefly outline some of the scientific capabilities of the CARISMA array in the context of the three THEMIS science objectives. In Sects. 4, 5 and 6 we illustrate the CARISMA array capabilities with scientific examples from the THEMIS mission thus far.

3.1 Onset and Evolution of the Macroscale Substorm Instability

One of the most important outstanding questions in Space Physics concerns understanding the explosive dynamics of the magnetotail during substorm expansion phase onset. The location of, and relative timing between, the physical processes which constitute the substorm expansion phase remain controversial. The onset of the expansion phase of the magnetospheric substorm is marked by a rapid topological change in the magnetic field configuration in the nightside magnetosphere, resulting in a rapid transfer of energy from the magnetotail into the ionosphere, and triggering vibrant and dynamic auroral displays. There are two competing models proposed to explain substorm onset. Both models agree on the importance of magnetic reconnection at a near-Earth neutral line (NENL) at around 20–25 R_E to power the substorm, but disagree on the causal sequence of events. In the so-called current disruption model, onset initiates closer to the Earth (around 10–15 R_E) and is followed later by NENL reconnection once disturbances from the near-Earth onset reach the mid-tail. In the NENL model, it is reconnection at the NENL which begins the expansion phase and nearer-Earth disturbances follow (see e.g. Lui et al. 1991; Angelopoulos 2008).

The study of the "onset and evolution of the macroscale substorm instability" is the primary objective of the THEMIS mission. Specifically, the THEMIS mission targeted measurements to address the following topics:

- Establish when and where substorms start.
- Determine how the individual substorm components interact macroscopically.
- Determine how substorms power the aurora.
- Identify how the substorm instability couples dynamically to local current disruption modes.

The coverage from the ground arrays, including CARISMA, is key to tackling these problems. Especially important is the need to establish the location of the THEMIS probes in the context of the ionospheric auroral and magnetic signatures of the substorm onset process(es). Despite uncertainties in the mapping from the ionosphere to the near-Earth tail, the ground-based context of the disturbances and the sequence of events leading to the in-situ perturbations measured by the probes is crucial if the causality and time sequence of events are to be correctly identified. In Sect. 4 below, we illustrate how CARISMA and supporting magnetometer network data can be used to constrain the substorm process. We concentrate on a case study example from the 7th March 2007. This event followed the launch of the THEMIS probes on February 17th 2007, but was sufficiently early in the commissioning phase that only the fluxgate magnetometers (FGM; see Auster et al. 2008) were operating following the magnetometer boom deployment on all probes.

3.2 Control of the Solar Wind–Magnetosphere Coupling by the Bow Shock, Magnetosheath, and Magnetopause

The orbit of the THEMIS probes generates a configuration whereby the apogees of the probes align over the Canadian continent in their prime science phase configuration. This allows the constellation to sample regions of the upstream solar wind, the bow shock region, magnetosheath, magnetopause and even inside the magnetosphere, at the same time producing the unique capability to monitor the energy flow and processing of upstream solar wind disturbances by the bow shock and magnetopause boundaries. As with the substorm studies described in Sect. 3.1 above, the continent-scale ground-based magnetic monitoring provided by CARISMA has the capability to diagnose the magnetic signatures of this coupling on the ground. Scientific targets which can be addressed include the waves, current systems and convection driven by (see also Angelopoulos 2008):

- Pc3–4 upstream waves and IMF conditions (e.g. Le and Russell 1996);
- Sudden Impulses ($SI^{+/-}$ e.g. Araki 1994; Takeuchi et al. 2000);
- Traveling Convection Vortices (TCVs; e.g. Glassmeier 1992; Kivelson and Southwood 1991; Kataoka et al. 2003; Murr and Hughes 2003);
- Hot Flow Anomalies (HFAs; e.g. Sibeck et al. 1999);

- Fast solar wind streams and magnetopause Kelvin-Helmholtz instability (e.g. McKenzie 1970; Pu and Kivelson 1983; Miura 1992; Mann et al. 1999; Fairfield et al. 2007, and references therein);
- Solar wind dynamic pressure fluctuations (e.g. Kepko et al. 2002; Mathie and Mann 2000c);

In this review we concentrate including the effects of solar wind pressure pulses and KH shear flow instabilities excited at the magnetopause, and these are outlined in Sect. 5.

3.3 Production of Storm-Time MeV Electrons

One of the most interesting and important questions in current solar-terrestrial physics research concerns the acceleration of electrons to relativistic speeds in the Earth's Van Allen radiation belts. The fundamental mechanisms proposed to explain the dynamics, energization and loss of these particles are numerous, and which processes are dominant in response to different solar wind forcing conditions remains largely unknown (see e.g. the review by Friedel et al. 2002). Likely the most influential acceleration mechanisms are resonance with VLF lower band chorus, which operates through violation of the first adiabatic invariant (e.g. Meredith et al. 2003; Chen et al. 2007, and resonance with ULF waves which typically operates through violation of the third (e.g. Fälthammer 1966; Schulz and Lanzerotti 1974; Elkington et al. 2002). CARISMA has an excellent capability for studies of ULF wave related radiation belt acceleration and loss processes, including the drift resonant interaction with ULF waves as well as studies of the potential role of electromagnetic ion cyclotron (EMIC) waves for scattering MeV energy electrons into the loss cone and hence into the atmosphere (e.g., Horne and Thorne 1998; Friedel et al. 2002; Meredith et al. 2003; Summers and Thorne 2003).

A very surprising recent observation is the correlation of the inner edge of the radiation belt with the plasmapause (e.g. Tverskaya et al. 1986; O'Brien and Moldwin 2003; Li et al. 2006). One suggestion links this to the operation of VLF acceleration just outside the plasmasphere, with internal loss such as that occurring due to resonance with plasmaspheric hiss scattering the radiation belt particles internal to the plasmasphere into the atmosphere (e.g. Meredith et al. 2007). Alternatively, if a large element of radiation belt morphology is determined by inwards (and outwards; e.g. Shprits et al. 2004, 2005) diffusion then perhaps the effects of plasma density in controlling the penetration of ULF wave power to low-L such as that described by Loto'aniu et al. (2006) plays an important role. Comparing CARISMA observed ULF power, and energetic particle flux measured in-situ on-board THEMIS (with the solid state telescope (SST) for energies up to 900 keV) and other satellites, allows the role of ULF waves in producing the observed correlation to be tested.

Since plasma density can also influence the growth rates of EMIC waves (e.g. Kozyra et al. 1984), and therefore the efficiency of EMIC scattering of MeV electrons into the loss cone (e.g. Meredith et al. 2003), the plasmasphere and plasmapause morphology and location can be expected to influence radiation belt loss. A powerful remote-sensing capability of the CARISMA array involves the use of the properties of the waves supported by the plasma to determine the natural standing Alfvén wave eigenfrequencies of entire closed field lines. Through the solution of this inversion problem, the mass density in the equatorial plane can be determined in an assumed magnetic field. This so-called "cross-phase" technique (e.g. Baransky et al. 1985; Waters et al. 1991, 1995, 2002; Menk et al. 2000; Dent et al. 2003, 2006) enables the CARISMA array to monitor the equatorial mass density of field lines conjugate to CARISMA stations. This capability delivers a powerful means of studying the structure of plasmasphere itself, as well as supporting studies of the influence of mass density on radiation belt dynamics. Examples illustrating the use of CARISMA data in support of radiation belt studies are presented in Sect. 6.

4 Substorm Science Capabilities

In the following subsections, we detail the analysis that can be performed with ground magnetometry to examine the characteristics of the processes and causal sequence of events associated with substorm onset. We utilize an early cruise-phase "THEMIS substorm" on the 7th March 2007 to present a case study which illustrates the science capabilities of the CARISMA array in support of substorm science. There were three substorms during the period 03-09 UT on the 7th March 2007 and we study the second substorm during this interval whose onset occurred at ~0600 UT, where the CARISMA "Churchill Line" was situated close to magnetic midnight. During the period 0300-0700 UT, the ionospheric footprint of the THEMIS constellation first mapped to the Canadian sector and then towards the end of the interval to the Alaskan sector. The GOES geosynchronous spacecraft also provide a valuable in-situ resource for substorm studies, the CARISMA magnetometer array being particularly well-suited since GOES East (GOES-12 at this epoch) and GOES west (GOES-11 at this epoch) span the range of longitudes covered by the ~geosynchronous chain of CARISMA magnetometers (DAWS-FSIM-FSMI-RABB-GILL). In addition to THEMIS magnetic conjunctions, the GOES satellites provide the capability for contemporaneous ground-spacecraft conjugate studies of geophysical phenomena on the same flux tube including an examination of substorm depolarization and studies of onset related Pi2 waves (see e.g., the review by Olson 1999), the highest resolution GOES magnetic field data (0.512 s resolution) also allowing studies of higher frequency Pi1 (cf. Lessard et al. 2006) and EMIC waves.

4.1 In-situ THEMIS and GOES Measurements

During this substorm the THEMIS probes were in the cruise-phase whereby their orbits were constrained to a "string-of-pearls" configuration. On the 7th March 2007, the CARISMA ground magnetometers were hence not directly magnetically conjugate to the THEMIS probes during the substorm onset at ~06 UT. The THEMIS probes were separated from the CARISMA Churchill line (330° magnetic meridian) by ~2 hours of local time at 08 UT and by the end of this period at 08 UT were ~5 hours to the west. During this time the THEMIS probes were out-bound around the dusk flank.

Figure 2 shows the THEMIS FGM data for the interval 0500-0700 UT. The THEMIS spacecraft are traversing the dusk-side magnetosphere close to apogee at radial distances between ~11–14 R_E during this interval. From top to bottom, Fig. 2 shows FGM data from TH-A, TH-B, TH-C, TH-D and TH-E in the GSM x (blue), y (green) and z (red) directions. There is a small dipolarization evident in B_z in 3 of the 5 spacecraft around 0545 UT, which is curiously not observed in TH-C and TH-E, and a B_y variation around the same time observed in all 5 spacecraft, perhaps indicative of nightside depolarization on the flanks. Note that there is also a dipolarization and associated B_y perturbation around 0610 UT.

In this interval, the ionospheric footprint of GOES-11 is situated close to FSIM, and the ionospheric footprint of GOES-12 is less than an hour of local time to the east of GILL, close to the CANMOS magnetometer and THEMIS ASI site at SNKQ. Figure 3 shows the high-resolution (0.512 s) GOES-11 (west) and -12 (east) magnetic field data in local field-aligned coordinates for the interval 0500-0700 UT. In the co-ordinate system used here p is

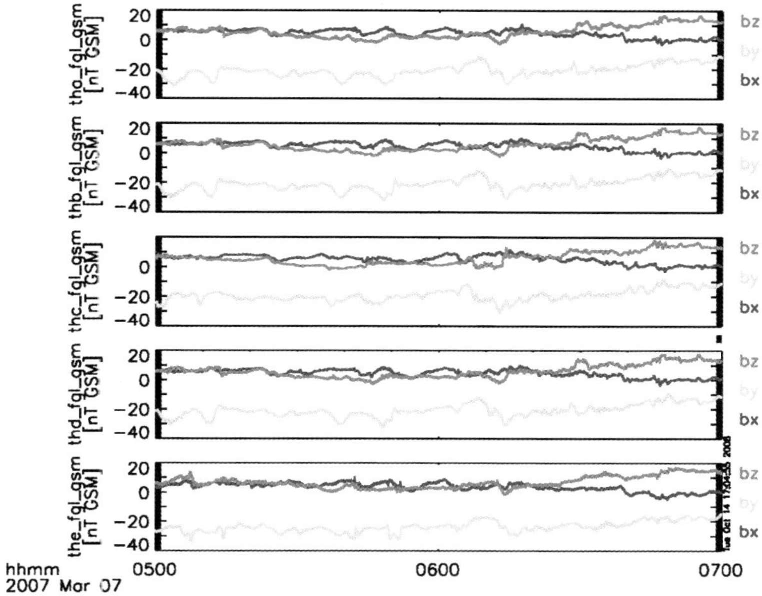

Fig. 2 Fluxgate magnetometer data from the THEMIS probes (A to E in the *top* to *bottom* panels, respectively) in GSM co-ordinates from 0500-0700 UT on the 7th March 2007

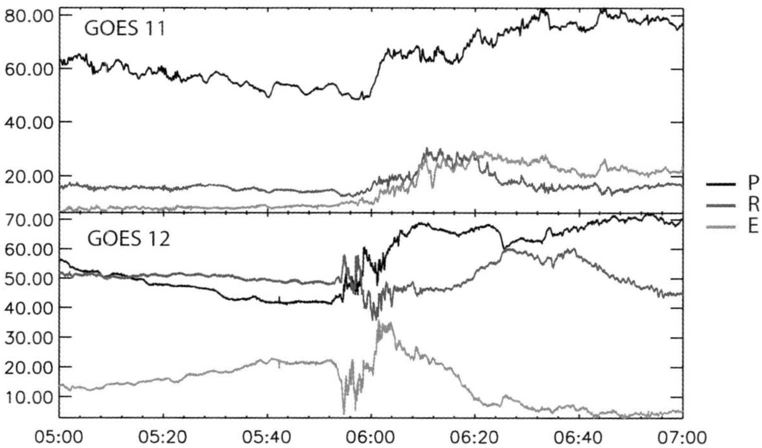

Fig. 3 GOES-11 and -12 magnetic field data from 0500-0700 UT on 7th March 2007 in p, r, and e coordinates (for details see text)

northward and perpendicular to the satellite orbit (parallel to the Earth's spin axis in a 0° inclination orbit, e is perpendicular to p and is directed earthward and is hence approximately radial, and n completes the triad, is directed eastward and is approximately azimuthal. Clear from Fig. 3 is the dipolarization in Hp, evident in both spacecraft at around ~0559 UT and ~0554 UT for GOES-11 and -12, respectively. In general, in order to interpret in-situ point measurements such as these, the broader continent scale monitoring provided by CARISMA as well as the THEMIS GBOs is crucial.

4.2 Substorm Timing and Location with CARISMA

4.2.1 Substorm Bays

Figure 4 shows the (Fig. 4a) H- and (Fig. 4b) D-component ground magnetograms from the Canadian sector from the CARISMA and THEMIS GMAG arrays for the same interval 0500-0700 UT. Figure 4 is arranged in latitude and longitude order such that the magnetic traces at the top (bottom) of the figures are the most eastern (western) stations. Clear in Fig. 4a is the presence of a substorm, identified as a large bay, for example, in the FCHU-GILL traces at ~0550 UT. The bays develop in response to the substorm current wedge (SCW) which is believed to be established through a diversion of the cross tail magnetospheric current into the ionosphere (e.g. McPherron et al. 1973). In this model, a downward field-aligned current (FAC) is established to the east, closes along the electrojet latitude in the ionosphere, and returns to connect to the cross tail current in a upward FAC which flows back to the tail to complete the circuit known as the SCW.

The magnitude and sign of the magnetic bays which develop at stations deployed across a latitudinal and longitudinal grid depends on the location of the stations with respect to the SCW and each of the H, D and Z components on the ground are affected (Cramoysan et al. 1995; Smith et al. 1999). We discuss this further below. At auroral latitudes, the Alfvén waves which must propagate to establish the SCW FAC elements are believed to bounce between the plasmasheet and the ionosphere, multiple reflections creating the Pi2 pulsations which "ride" on the back of the SCW bays as the FAC elements are established.

4.2.2 Substorm Pi2 Waves

A traditional method for timing substorm onset is to filter the ground magnetic perturbations in the Pi2 (40–150 s period) band. Figure 5 shows the Pi2 filtered H-component magnetometer data, together with similarly Pi2 filtered GOES-11 and GOES-12 p-component magnetometer data. Clear in Fig. 5 is that stations poleward of the electrojet (e.g., RANK) observe the Pi2 onset at later times than those closer to the onset latitude (e.g., TPAS). Using the Pi2 pulsations in Fig. 5, we can coarsely estimate substorm onset to be ~0554 UT. However, since the Pi2 pulsation has a period of ~2 minutes, it is virtually impossible to calculate the time at which the signal rises out of the noise to within a time accuracy that is shorter than the period of the wave. By looking at shorter period ULF waves and using improved time series analysis techniques, this uncertainty can be improved.

4.2.3 Substorm Pi1 Waves

Figure 6 shows the Pi1 (1–40 s) filtered H-component magnetometer data for the same stations and time period, together with similarly filtered GOES-11 and GOES-12 p-component data. In Fig. 6, the Pi1 signal rises out of the noise clearly at an earlier time than is obvious in Fig. 5, the most obvious onsets being at GILL, ISLL, PINA and MSTK prior to 0554 UT. Also notable in Fig. 6 is that there appears to be a pattern in the onset times of the Pi1 pulsations from station to station depending on their relative locations. For example, the onset of the large amplitude band-pass filtered Pi1 pulsations appears to be around 0554 UT at TPAS, but similar large amplitude Pi1 wavepackets only appear at a much later time shortly

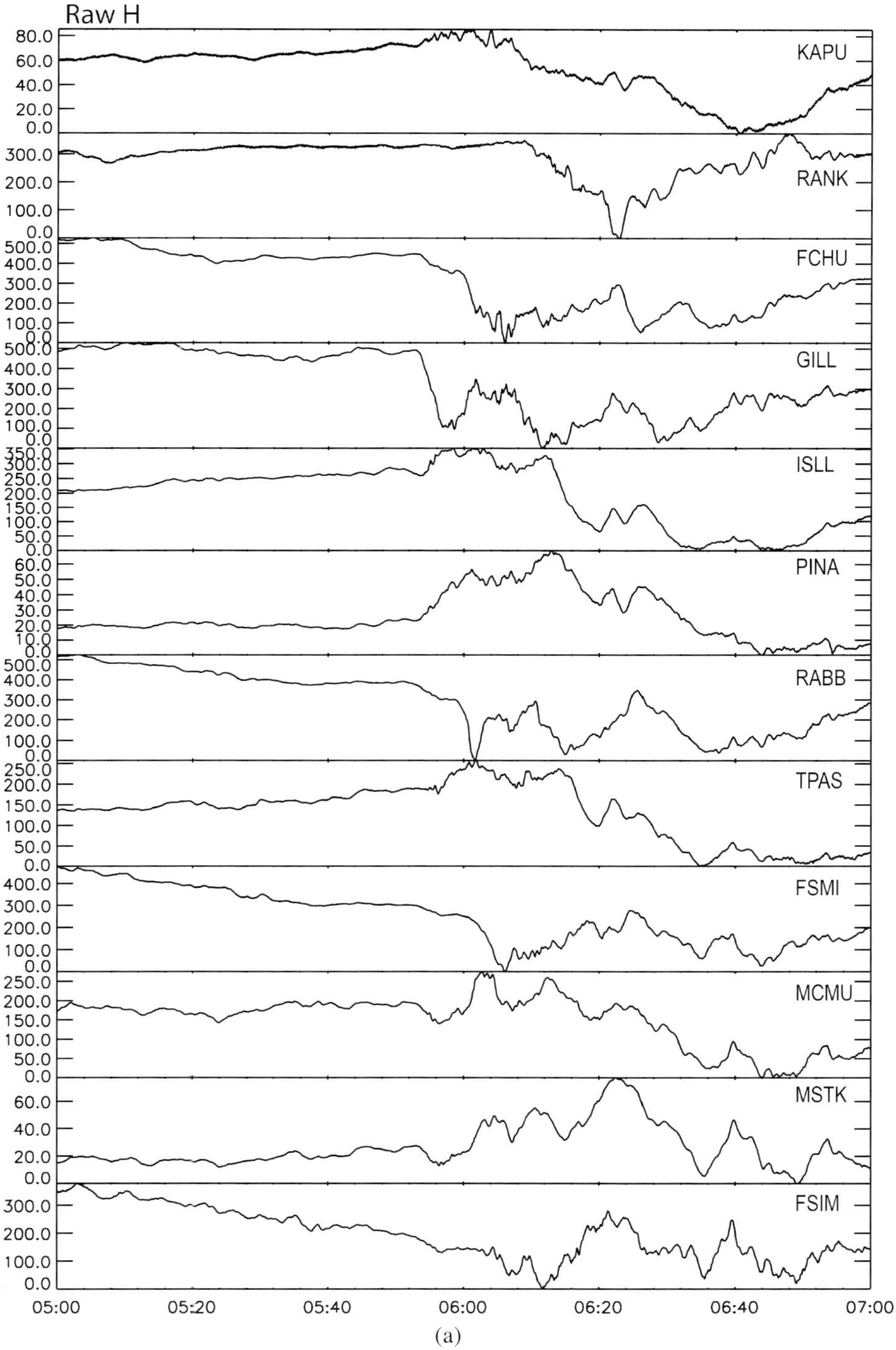

(a)

Fig. 4a Selected CARISMA, THEMIS and CANMOS H-component magnetometers from 0500-0700 UT on the 7th March 2007

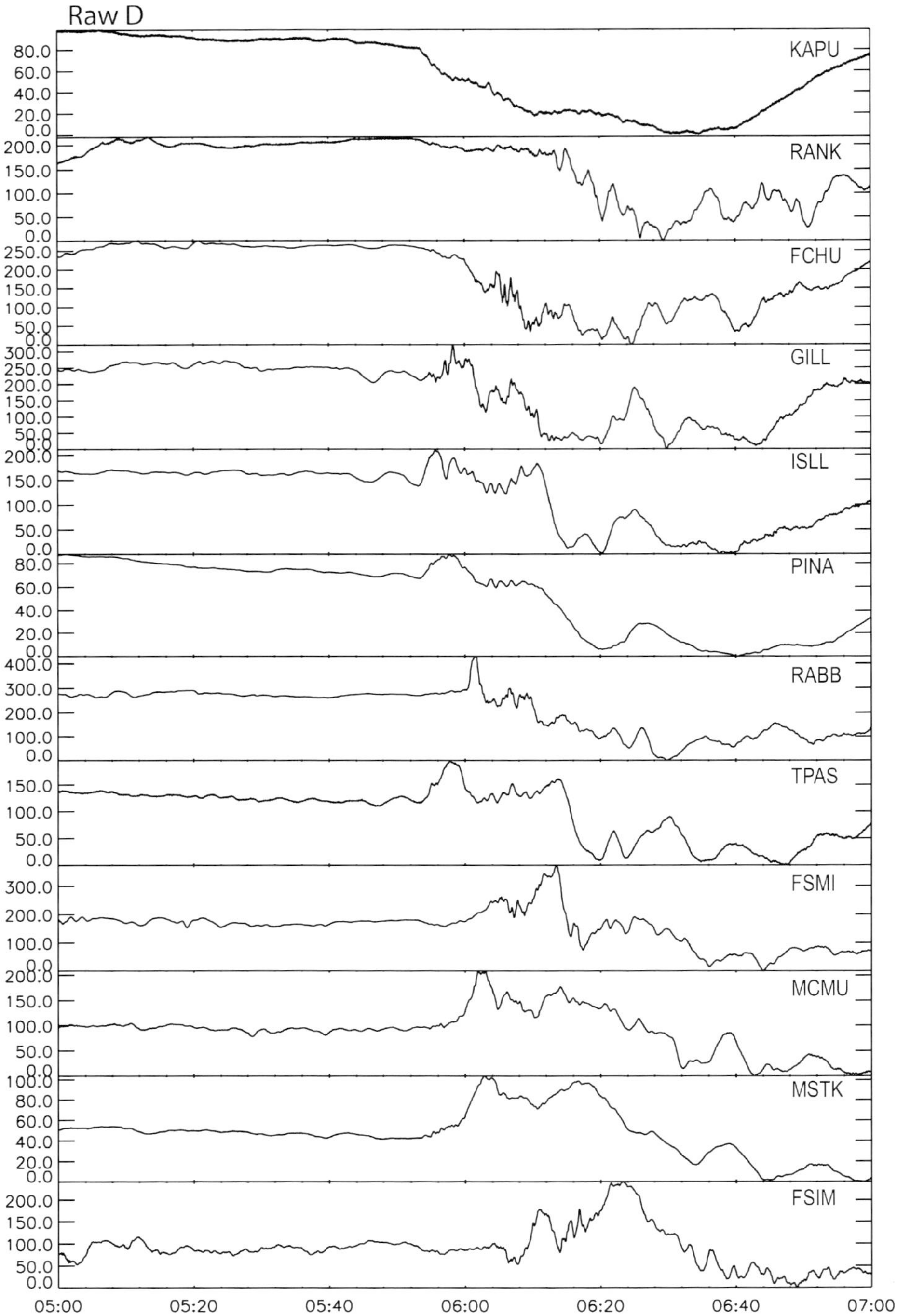

(b)

Fig. 4b Selected CARISMA, THEMIS and CANMOS D-component magnetometers from 0500-0700 UT on the 7th March 2007

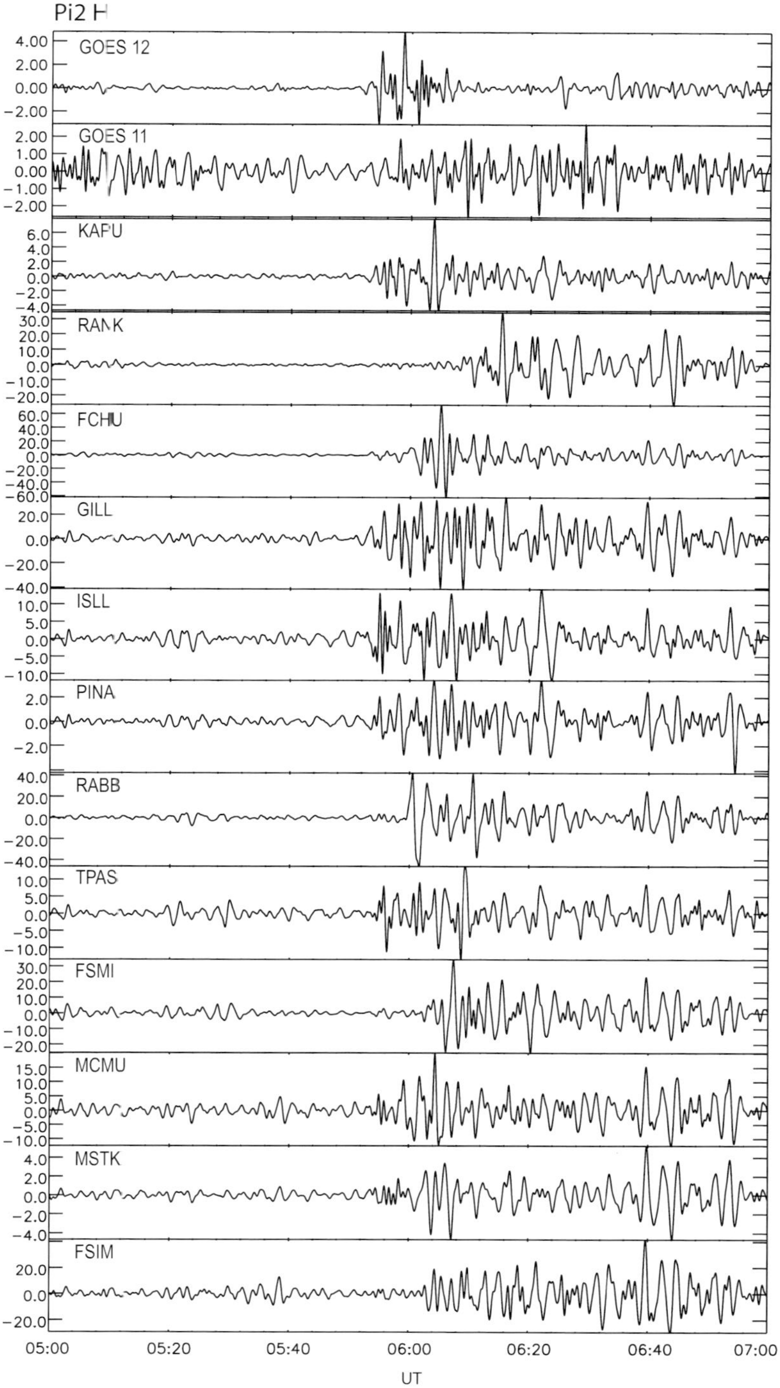

Fig. 5 Selected CARISMA, THEMIS and CANMOS Pi2 filtered (40–200 s) H-component magnetometers from 0500-0700 UT on the 7th March 2007

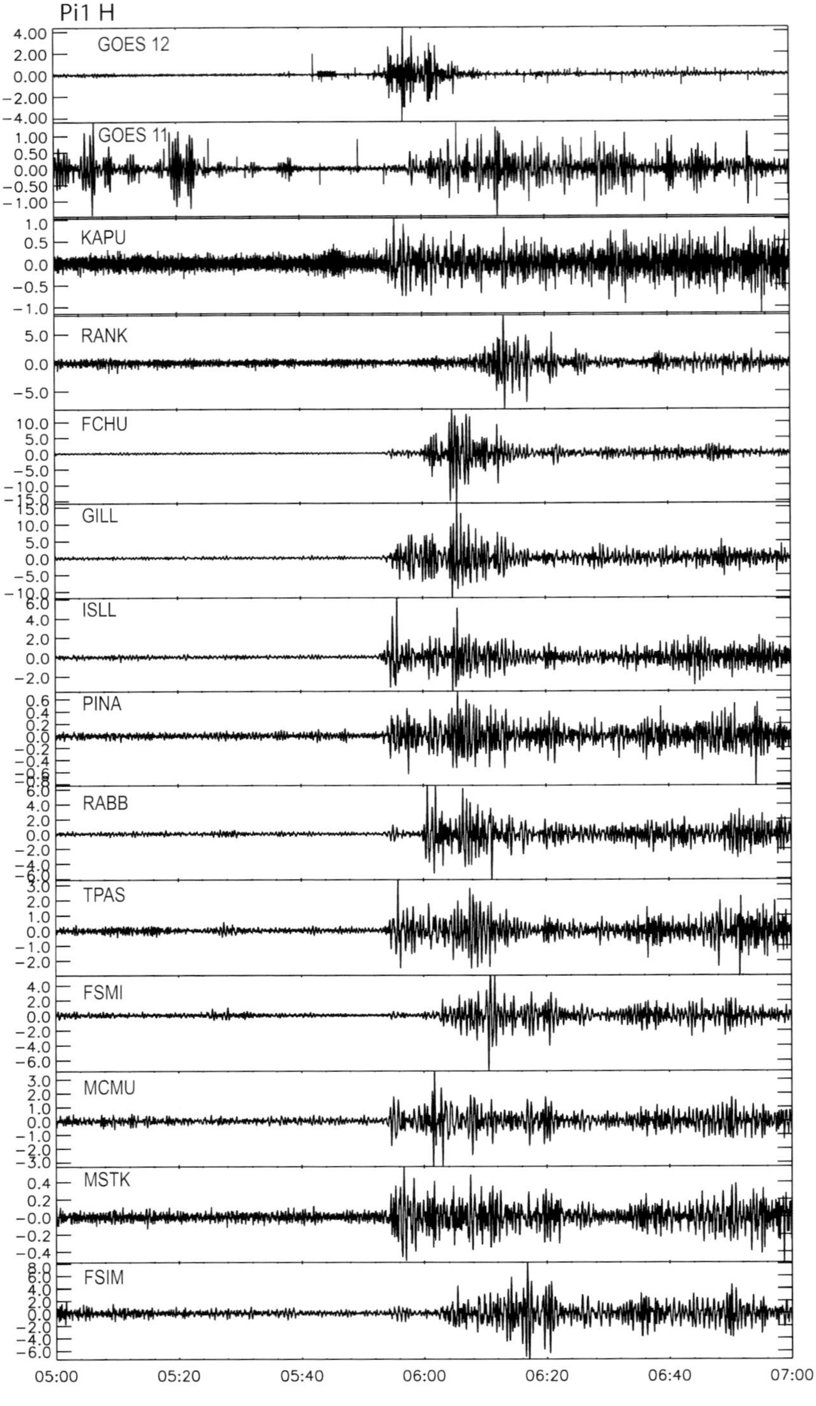

Fig. 6 Selected CARISMA, THEMIS and CANMOS Pi1 filtered (1–40 s) D-component magnetometers from 0500-0700 UT on the 7th March 2007

after 0600 UT at RABB, and even later (~0602 UT) at FSMI. Figure 6 illustrates an element which requires caution when using filtered time-series to identify pulsation onset times and propagation. Even though the largest amplitude Pi1 wavepackets appear to be delayed from, for example, TPAS to RABB and FSMI, there is a much smaller amplitude wavepacket which appears earlier at FSMI, and especially clearly at FSIM and RABB, which is more contemporaneous with the TPAS magnetic Pi1 onset. Similar features were seen for example in the study of Arnoldy et al. (1998) in relation to Pi1B waves, these authors suggesting the large amplitude Pi1B propagation polewards from a mid-latitude onset region followed the poleward motion of the optical auroral surge at substorm onset, whilst the lower amplitude more contemporaneous Pi1B wave propagation might be linked to ionospheric ducting. Using more advanced wavelet time-series analysis techniques, we can use formal definitions of threshold powers to definitively determine the timing of the onset of the ULF pulsations, and Pi1 pulsations in particular. We can therefore mathematically determine propagation delays across the array by using the time at which the signal rises above the preceding noise, providing a systematic timing which improves upon the estimates that are currently often obtained by eye.

4.2.4 Wavelet Substorm Onset Timing and Location

The Discrete Wavelet Transform (DWT) is a method to decompose a time series into basis functions (wavelets) that are localized in both frequency and time. In this study, we utilize the Meyer wavelet outlined and applied to the Pi2 waveform by Nose et al. (1998). The Meyer wavelet is band-limited in frequency, and therefore minimizes overlap between adjacent wavelet bins. Wavelet coefficients with large J have a high time resolution but a low frequency resolution and vice versa. An important aspect of the Meyer wavelet is the resemblance of the waveform to an impulsive ULF wave such as the Pi1 and Pi2. Table 4 shows the frequency range and time resolution of the Meyer wavelet coefficients for a time series of 512 points. For details of the analysis techniques and detailed analysis of an isolated substorm, we refer the reader to Milling et al. (2008) and to Murphy et al. (2008) and Rae et al. (2008a, 2008b) for further details of the technique.

The DWT enables the waves in both the Pi1 and Pi2 bands to be analyzed at the same time using different wavelet coefficients, J. An onset time for each station for each J can be determined by first finding a quiet period and performing the DWT on this data. The mean and standard deviation are then determined for each J. The onset for any coefficient

Table 4 Frequency and time resolution for the Meyer wavelet functions on the 512-point time series used in this study

Band J	Frequency range (mHz)		Period range (seconds)		Resolution (seconds)
9	166.67	666.67	1.5	6	2
8	83.33	333.33	3	12	4
7	41.67	166.67	6	24	8
6	20.83	83.33	12	48	16
5	10.42	41.67	24	96	32
4	5.21	20.83	48	192	64
3	2.60	10.42	96	384	128
2	1.30	5.21	192	768	256

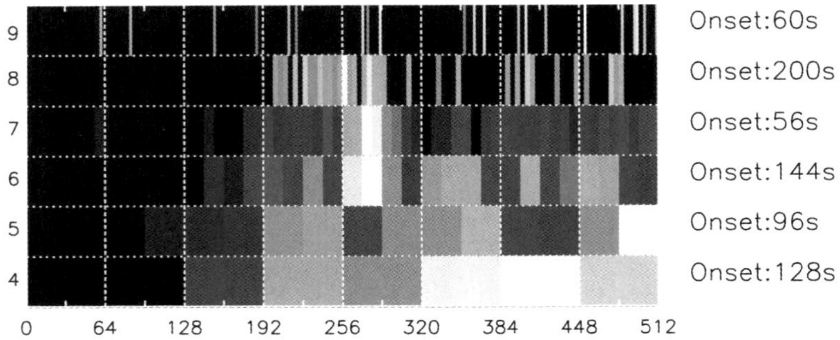

Fig. 7 (Colour online) Meyer DWT analysis of the SNKQ ground magnetometer station. The y-axis represents the power in different wavelet J bands (higher J represents higher frequency), and the x-axis denotes seconds since 0550 UT. Colour represents normalized DWT coefficient power, where *blue* is low power, and *yellow-red* is high DWT coefficient power. The onset times listed represent the first time that power in that particular J band rises above a threshold of two standard deviations from the mean defined in terms of earlier noise, indicating a change in the characteristics of the power at that time (see text for details)

J is then the initial time when the power in that coefficient rises above the mean plus two standard deviations of the power during the quiet period. Using the technique outlined above for the determination of a noise threshold means that there is a 98% chance that the signal observed is not noise of the same kind as observed during the preceding interval. The onset time for a given magnetometer station is then defined as the earliest onset time from all frequency bands, J. The power in any J band is calculated using the vector sum of the power from the H- and D-components such that we estimate the time when any transverse perturbation in the same frequency band rises above the defined threshold.

Figure 7 shows the results of the Meyer DWT analysis outlined above on the SNKQ magnetometer data, and starting from epoch time zero at 0551 UT. Figure 7 is a plot of the power in particular J bands. Since each J wavelet has a different length in the time series, dependent on the frequency being analyzed, the temporal resolution changes for each J. The power is colour coded such that purple/blue is low power, and white is the highest power, normalized to each frequency band. For example, only two wavelet coefficients are needed to describe character of the $J = 2$ band in a 512 point time series, each coefficient representing 256 points from that time series. For the $J = 9$ frequency band, 256 coefficients are required, which in this particular time series (1 sample/s magnetometer data) appears to be predominantly continuous noise. The Pi2 frequency band is denoted by $J = 3$–5, and the Pi1 frequency range by $J = 5$–9. Note that Table 4 denotes that there is significant (but minimized) overlap between the Pi2 and Pi1 frequency bands. For example, the $J = 5$ band represents the longer period Pi1 waves as well as the shorter period Pi2 pulsations. Clear in Fig. 7 is the J dependence of the time when power appears above the noise for this substorm onset. In the $J = 5$ frequency band the power appears above the noise at 05:51:48 UT $\pm$ 16 s (i.e., 112 s past 0550 UT). Note that the error quoted is an estimate given the size of the temporal resolution; the signal is simply observed during that window, and so we quote the middle value. Using this method, we can establish the onset time at which power in a given J rises above the pre-onset noise, as well as determine the frequency band of first signal arrival, for stations across the entire CARISMA and THEMIS GMAG array, as well as in the geosynchronous GOES magnetometer data. From Fig. 3 we apply the Meyer DWT analysis described above, and find that the onset of the Pi1/2 ULF waves in GOES-11 and -12 are clear also (see also the top two panels in Figs. 5 and 6). GOES-11 is situated closer to

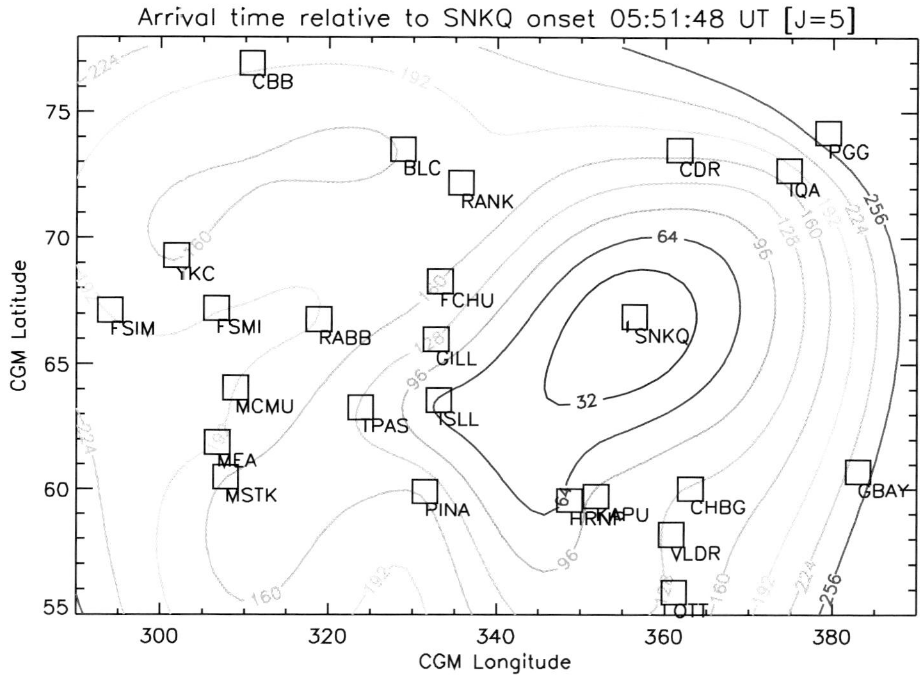

Fig. 8 Contour plot of the spherical harmonic fit to the $J = 5$ DWT first magnetic disturbance arrival time for 25 magnetometer stations across Canada as a function of CGM co-ordinates

the dusk flank (as is the THEMIS constellation) and the onset of the $J = 5$ band is 0556:52 UT $\pm$ 16 s, whereas the onset as defined by GOES-12 is at 0552:04 UT $\pm$ 16 s.

It is remarkable that the signals at SNKQ and observed in-situ with GOES-12 are observed within experimental error of each other. This contemporaneous onset of Pi1/2 signal on the ground and at a closely conjugate satellite in space appears to be too fast to be explained by bouncing of Alfvén wavepackets along the field to establish the magnetosphere-ionosphere coupling; rather, it points to the intriguing possibility that energetic electrons may be their rapid communication mechanism (e.g. Watt et al. 2005).

Figure 8 represents a contour plot of the arrival time of the $J = 5$ wavelet band for the 25 magnetometers used in this study. The contours of equal time are created from a two-dimensional spherical harmonic fit to the $J = 5$ onset times from these stations. Each coloured contour is 32 s apart, and times are shown as seconds since 0551:48 UT, from the initial onset time observed at SNKQ. There is a clear and coherent onset pattern to these $J = 5$ Pi1/2 ULF waves as seen in the ground magnetometer data; the onset of the $J = 5$ Pi1/2 ULF waves occurring first at SNKQ (0551:48 UT), and 64 s later at ISLL and later at surrounding stations (96 s at GILL, 160 s at FCHU, etc.). Since there is a definitive onset location and propagation of the onset of the Pi1/2 signal, the implication is that there must be a physical mechanism which this onset location and propagation pattern is associated with. This will be the subject of future studies, but the initial conclusion may be that this is a signature of the evolution of the magnetospheric source of these pulsations. Note that these propagation times are much faster than the propagation of the maximum Pi1B wave power which Arnoldy et al. (1998) associated with the propagation of the optical auroral substorm surge. Arnoldy et al. (1998) (see also Posch et al. 2007) also present a by-eye analysis from approximately one hour spectrograms showing a faster "prompt" propagation of lower amplitude Pi1B which starts from a lower latitude station with a large amplitude Pi1B and "promply" propagates, albeit at low amplitude, to higher latitude stations. We believe that the expansion of the onset of Pi1 power in the $J = 5$ Pi1/2 ULF waves presented in Fig. 8 indicates that even the "prompt" Pi1B propagation described by Arnoldy et al. (1998) in fact clearly propagates, both polewards and azimuthally, from a localized low latitude epicenter.

4.2.5 Substorm Current Wedge Location

Evaluating the location of the substorm current wedge is an integral part of the analysis and interpretation of the ground magnetic signals surrounding substorm onset needed to place the location of the THEMIS satellites in context. Figure 9 shows the ground magnetic perturbations obtained when the magnetic bays are estimated using a Biot-Savart law integration of an imposed line current model for the substorm current wedge in an assumed dipolar magnetic field (e.g. Cramoysan et al. 1995). Such a model can be used to locate the elements of the SCW using the magnitude and sign of the bays which are seen across a grid of magnetometer stations. From top to bottom, Fig. 9 shows the H-, D- and Z-component deflections which are expected to be observed by ground magnetometers in response to the model SCW. Superimposed upon these three figures are dashed lines indicating the upward and downward field-aligned current elements (vertical) and electrojet latitude (horizontal) which are assumed in the model. By careful analysis of the initial bay disturbance deflection on the ground across an array of stations, the location of the SCW elements in relation top these stations can be deduced. Note that although this model assumes a dipolar field,

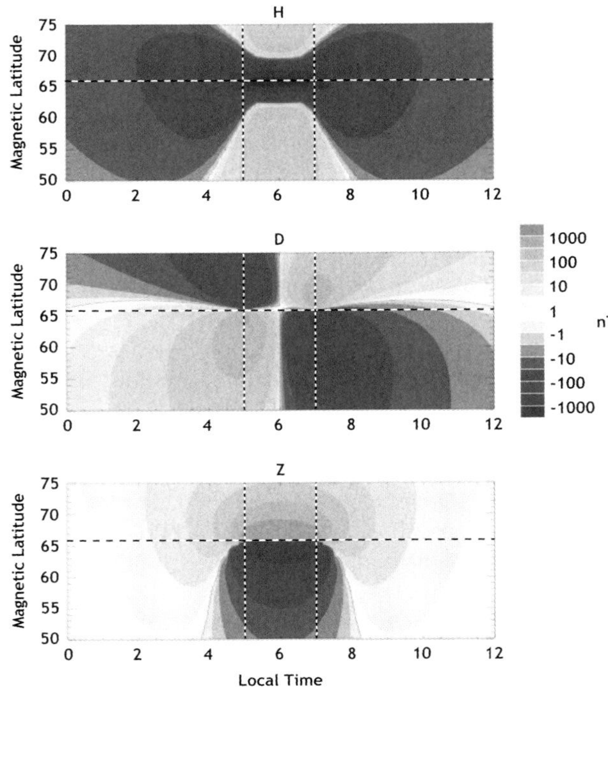

Fig. 9 (Colour online) Magnetic perturbations from the Cramoysan et al. (1995) substorm current wedge model in the (*top*) H-, (*middle*) D- and (*bottom*) Z-component magnetic field deflections. *Red* denotes a positive deflection, whilst *blue* represents a negative deflection. The *horizontal dashed line* denotes the electrojet latitude, and the *vertical lines* denote the (*left*) upward and (*right*) downward field-aligned current elements

the largest magnetic contributions to the ground signal arising from the Biot-Savart integration occur from the sections of the FAC closest to the ionosphere. Since this region is also likely to be the most dipolar, the assumed dipolar field geometry should not generate any significant errors in the location tool, especially on the scale of the spatial separations of the stations within the array used to locate the SCW. Through best-fit to the observed bays, CARISMA can provide diagnosis of the locations of the latitude of the auroral electrojet, and the longitudes of the upward and downward FAC elements. Such analysis can be completed by the CARISMA team on a collaborative basis in future for the second THEMIS tail season. Locating the SCW elements using this technique across the combined magnetometer arrays shown in Fig. 8 places the meridian of the downward FAC between KAPU and VLDR in the same meridian as SNKQ, the upward FAC element between TPAS and ISLL, and the electrojet latitude between GILL and ISLL. The location of the downward FAC therefore appears to be coincident with the magnetic onset location at SNKQ, suggesting that the downward FAC and the onset initiation process may be intimately linked.

4.3 The Ionospheric Alfvén Resonator

The ionospheric Alfvén resonator (IAR) resonant cavity develops in the topside ionosphere as shear Alfvén waves become trapped in a standing wave pattern between regions of large Alfvén velocity gradients at the cavity boundaries. The induction coil component of the CARISMA expansion will enhance studies of the physics and morphology of the IAR in the Canada–US sector. Interest in the Alfvén resonator (Lysak 1991, 1994; Belyaev et al. 1990) has grown in recent years due to increasing indications that the resonator's effects on plasma dynamics and energy transport through the coupled magnetosphere–ionosphere–thermosphere system are significant.

The excitation of the IAR can be observed as multiple harmonic resonance bands in the 0.1 to 10 Hz range that rise and fall with diurnal changes in upper ionospheric parameters (Hebden et al. 2005; Yahnin et al. 2003). Magnetic fingerprints of the stimulated IAR above Athabasca station ($L = 4.61$) on September 25, 2005, are shown in Fig. 10. Recently, magnetic signatures of the IAR occurring at sites spread across a range of L-shell

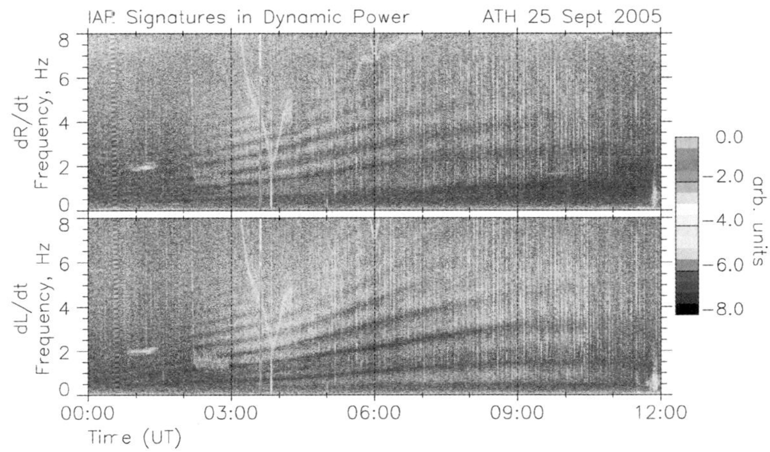

Fig. 10 (Colour online) Magnetic signatures of the IAR recorded in right- and left-hand circularly polarized dynamic power at Athabasca Geophysical Observatory on 25 Sept. 2005. The *color scale* indicates logarithmic power normalized to peak power in each spectrogram

and magnetic local time in the Canada–US sector and Russia have been studied (Parent et al. 2007). However, the excellent time resolution and strategic two-meridian coverage of the new CARISMA induction coil network will allow long-term statistical studies of IAR signatures at multiple stations across Canada.

The relationship between IAR signatures and topside ionospheric parameters can lead to the use of IAR observations as a diagnostic of F-region dynamics, including the ionospheric response during substorms. Particle precipitation during substorm events can disrupt and change the ionospheric plasma density profile in localized areas, thus directly affecting observed signatures of the IAR (Parent et al. 2008, in preparation). CARISMA induction coil observations of the ionospheric Alfvén resonator, along with CGSM riometers and optics, can nicely complement the study of substorm dynamics in the THEMIS era.

4.4 Discussion

In Sect. 4.2 above, we described some of the capabilities of the CARISMA array for substorm studies. These include timing and locating the first evidence of substorm onset waves in the ionosphere, as well as locating the regions of first wave arrival and SCW location. Since the capability to resolve the magnetic signals of substorm onset is not affected by cloud, as is the case for ground-based optics, these capabilities provide a powerful tool for substorm science especially when combined with data from the in-situ THEMIS probes. Specifically, the ionospheric signatures can be used to constrain the competing NENL and current disruption models of substorm onset since these models must ultimately be able to explain the magnetic timing and location observed in relation to the time sequence of events at substorm onset. CARISMA magnetic monitoring can also be used to do substorm science in its own right.

For example, the nature of the drivers of the Pi2 and Pi1 waves which are seen at onset is not fully understood. In general, the nightside magnetosphere during the substorm expansion phase is awash with geomagnetic activity, and it has long since been established that nightside ULF fluctuations are an integral part of the substorm (Saito 1961). However, there are debates as to the generation mechanism and drivers of many of these classes of ULF pulsations. For example, Rostoker et al. (1980) identified that the Pi2 pulsation (40–150 s period) was integral to the substorm process and substorm onset, and could be used to time substorm onset to ~minute timescales. The Pi2 ULF pulsations are thought to occur in the near-Earth plasmasheet, and field-aligned currents that establish the substorm current wedge (SCW) are established by the field-aligned propagation and ionospheric reflection of the Alfven waves in the Pi2 wavetrain. The characteristic decaying waveform of the Pi2 can therefore be explained by this reflection process (e.g., Baumjohann and Glassmeier 1984), and is observed in ground magnetograms as a series of Pi2 pulsations "riding on" the magnetic bays associated with the currents in the substorm current wedge. However, the Pi2 has since also been attributed to direct driving arising from the impact of earthward propagating of quasi-periodic Bursty Bulk Flows (BBFs; e.g. Kepko and Kivelson 1999; Kepko et al. 2001) or alternatively as a result of natural resonance frequencies in the nightside magnetosphere (e.g. Rae et al. 2007a). The frequencies of the Pi2 must be determined both or one of either the natural frequency content of the CPS disturbance, or the natural frequencies within the near-Earth CPS. Studies of the polarization of the Pi2 (e.g. Lester et al. 1983) have also shown them to be an excellent indication of the location of the substorm current wedge, in terms of location of both the upward and downward field-aligned current elements, and the centre of the electrojet. Furthermore, this can be verified with substorm bay analysis using a simple model for SCW location (e.g., Cramoysan et al. 1995;

Sect. 4.2). Finally, CPS disturbances have been shown to generate compressional fast mode waves, which may impact the plasmasphere and set up compressional plasmaspheric cavity modes (e.g. Allan et al. 1996). We direct the reader to Olson (1999) for a comprehensive review of the Pi2 pulsation. Overall, either the NENL model or the current disruption model must be able to explain both the relationship of these waves to the physical drivers arising from the processes during the expansion phase, as well as being able to explain their timing in relation to the sequence of events at onset.

The increase in temporal resolution and spatial coverage of the ground magnetometry has led to the discovery of higher frequency ULF waves associated with substorm onset. The Pi1 (1–40 s period) class of ULF wave is comparatively much less studied than those waves in the Pi2 band, and their relationship to onset processes much less well understood. If Pi1 waves are consistently observed before or even at the same time that the Pi2 is established, then it should be possible to time substorm onset with ground magnetometry to an increased accuracy due to the smaller wave periods in the Pi1 band. It is often stated that the first signature of substorm onset is the brightening of the auroral signature in optical measurements (e.g. Mende et al. 2007). However, the possibility of using Pi1 techniques may allow the relationship of Pi1s to the onset process to be determined. Moreover such timing could be done during cloudy conditions (e.g., Milling et al. 2008; Murphy et al. 2008; Rae et al. 2008a, 2008b). There is already some evidence that Pi1 signatures might be more local than Pi2s (e.g. Posch et al. 2007). The initial results from the DWT analysis also suggests that this might be the case, the results in this review already indicating the likely utility of Pi1s for both local substorm onset location and indeed for increased accuracy timing. Interestingly, broadband Pi1 signals in the 0.1–10 s period range known as Pi1Bs have also recently been studied in space and on the ground (e.g. Lessard et al. 2006). These signals may also provide a new window on the onset process, especially since one hypothesis suggest that Pi1Bs might be generated by FAC instabilities (see e.g. Lessard et al. 2006 and references therein). The long-period Pi1 waves which appear to provide a coherent substorm onset timing (cf. our Figs. 7 and 8) appear to be related to the location of the downward FAC (Milling et al. 2008). However, further studies are needed to establish their causal driver and their relation to the spatial and temporal development of process(es) operating during expansion phase onset.

5 Dayside Science Capabilities

One of the excellent scientific capabilities of the in-situ THEMIS orbits is their design to return to a meridional telescopic alignment once every four days. In the nightside this capability is designed for substorm science, but on the dayside it provides the capability to monitor the upstream solar wind as well as energy and processing of solar wind disturbances by the bow shock and magnetopause, and their ultimate role in energy transport into the dayside magnetosphere. As discussed in Sect. 3.2, this provides a unique capability for answering dayside science questions (the THEMIS tertiary objective), especially since in the prime mission phase the orbits of the THEMIS probes come into telescopic alignment over the Canadian sector both during tail (northern hemisphere winter) and dayside (northern summer) observing seasons, as well as on the magnetospheric flanks during the periods in-between.

The excitation of global scale ULF waves and field line resonances (FLRs) (see e.g. the review by Wright and Mann 2006) is increasingly recognized as an energetically significant component of solar-terrestrial energy transport (cf. Greenwald and Walker 1980). For example, at times the total energy deposited via ionospheric Joule heating due to Pc5 ULF waves may reach up to ~30% of the energy deposited during a substorm cycle (see also Rae et al. 2007b). Dayside long period Pc3–5 ULF wave modes can be classified by their external or internal excitation mechanisms into groups with high or low azimuthal wavenumber (m) (see e.g. the review by Hughes 1994). Nightside ULF waves, although critical for energy transport and diagnosing tail dynamics, are particularly poorly understood. Due to their energetic significance, and their role in transporting energy and coupling different energy plasma particle populations, ULF waves are an important research focus for CARISMA.

Below, we concentrate on highlighting the capabilities of CARISMA for completing studies of the solar wind excitation of dayside ULF waves, especially by solar wind impulses and through the development of magnetopause Kelvin-Helmholtz instabilities, as well as studies of the internal excitation of ULF pulsations by energetic ions. By analyzing the longitudinal phase change of a ULF wavepacket, the waves' azimuthal wave number, m, can be determined and hence these two populations of waves can be separated using ground-based magnetometer data (see e.g. Chisham and Mann 1999).

5.1 Excitation and Propagation of Low-m ULF Pulsations

The role of discrete frequency cavity/waveguide modes in the injection of low-m ULF energy into field line resonances (FLRs) in the magnetosphere is now on a strong theoretical (e.g. Samson et al. 1971; Kivelson et al. 1984; Kivelson and Southwood 1985; Wright 1994; Mann et al. 1999; Walker 2000) and experimental footing (e.g. Samson et al. 1992; Walker et al. 1992; Ruohoniemi et al. 1991; Mann and Wright 1999; Mathie et al. 1999a, 1999b; Mathie and Mann 2000a, 2000b, 2000c). The amplitude of the wave peaks at the location of the FLR and the phase of the wave changes with latitude by 180° across the FLR. Figure 11 illustrates these FLR features (cf. Samson et al. 1971) measured on the ground by the Churchill latitudinal array of CARISMA magnetometers (see Rae et al. 2005 for more details).

Solar wind discontinuities, pressure pulses, buffeting, as well as the Kelvin-Helmholtz instability (KHI) at the magnetopause (see the schematic in Fig. 12) have all been proposed as drivers for cavity/waveguide modes (e.g. Allan et al. 1986; Mann et al. 1999, 2002; Mann and Wright 1999; Mathie and Mann 2001; Rae et al. 2005). Pc5 ULF power, especially in the dawn sector, is also strongly correlated with solar wind speed (e.g. O'Brien and McPherron 2003; O'Brien et al. 2003; Mathie and Mann 2001; Mann et al. 2004). However, detailed studies are required to establish the dominant mechanisms of long-period ULF wave excitation in the dayside magnetosphere. Indeed, the detailed ULF wave response in the magnetosphere to fast solar wind streams, co-rotating interaction regions (CIRs), interplanetary coronal mass ejections (ICMEs), are not well-known.

The KHI is most likely to excite ULF waves during fast solar wind (>700 km/s) intervals, perhaps preferentially in the morning sector where the statistically the Parker spiral interplanetary magnetic field is approximately perpendicular to the Earth's magnetopause boundary. Figure 13 shows L-MLT maps of the 1–10 mHz integrated ULF amplitude for low (<300 km/s), medium (500–600 km/s) and high (>700 km/s) solar wind conditions using 10 years of CARISMA data from 1996 to 2006 for the H- and D-components observed on the ground, and mapped to the equatorial plane assuming a dipole field. The results presented in Fig. 13 clearly illustrate that ULF wave power increases with solar wind speed and that there is much more ground-based ULF wave power in the dawn sector compared to the dusk sector. The enhanced and localized H-component ULF wave power in the dawn sector is most likely caused by guided toroidal FLRs resulting from compressional waveguide modes excited by the KHI at the magnetopause boundary under fast solar wind conditions. The

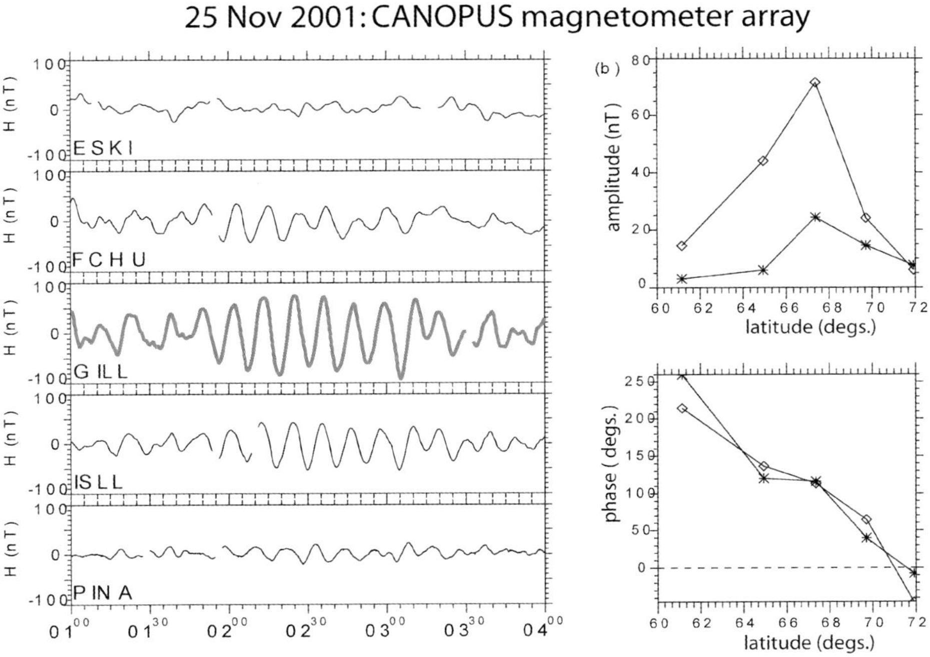

Fig. 11 (**a**) Unfiltered H-component ground magnetograms from the "Churchill line" of CARISMA magnetometer array between 0100 and 0400 UT on 25 November 2001. (**b**) Complex demodulation of the H-(*diamonds*) and D-(*stars*) components of the dominant spectral peak (i.e., 1.5 mHz) taken at 0235 UT. The *top* and *bottom panels* of (**b**) represent the amplitude and phase along the "Churchill line" of magnetometers. Adapted from Rae et al. (2005)

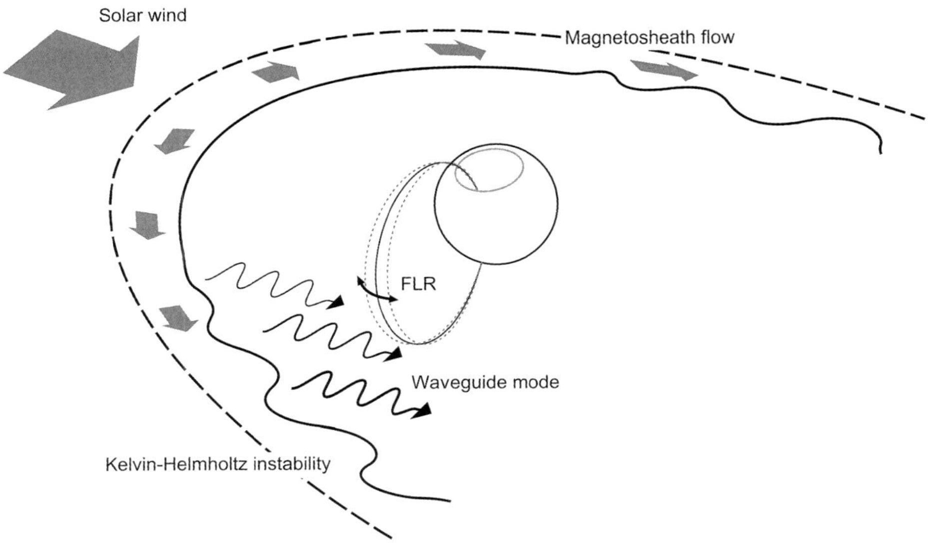

Fig. 12 Schematic showing the excitation of a field line resonance by compressional waves driven by magnetopause Kelvin-Helmholtz instability on the flanks of the magnetosphere. Taken from Rae et al. (2007b)

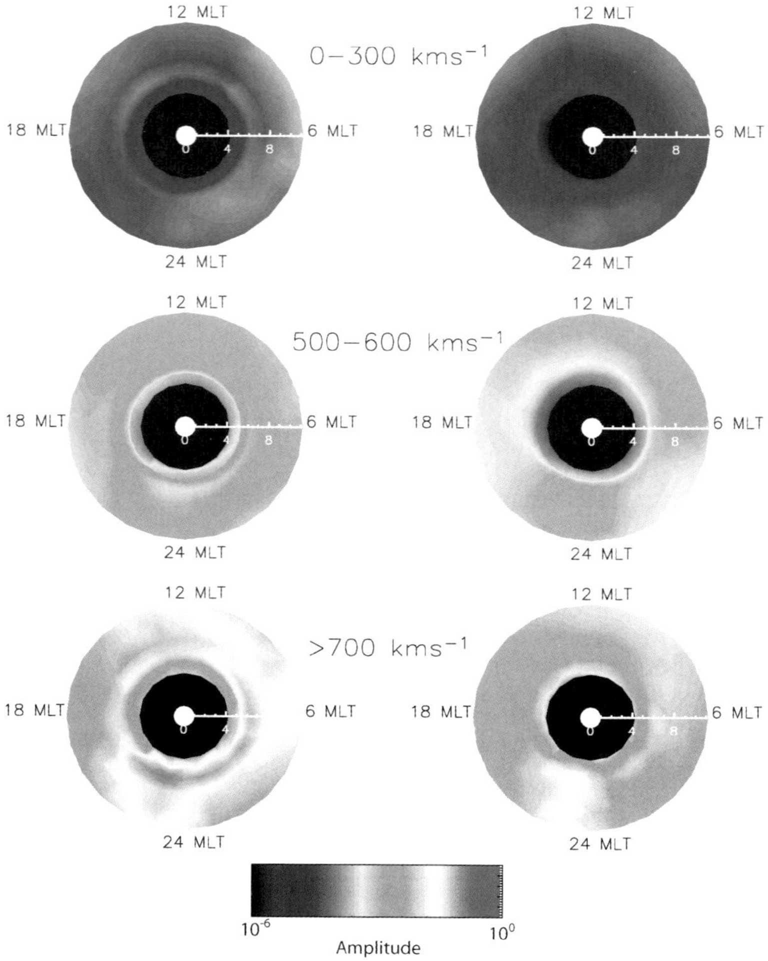

Fig. 13 Amplitude maps of 1–10 mHz frequency integrated ULF amplitude spectra for low (<300 km/s), medium (500–600 km/s) and high (>700 km/s) solar wind conditions, for the H-component (*left column*) and D-component (*right column*) from the Churchill line magnetometers at PINA, ISLL, GILL, FHCU, and RANK. The plots were produced using 10 years of CARISMA data from 1996 to 2006

field line resonance signature is strongest in for stations with L-values in the auroral zone in the H-component, and has a significantly lower amplitude at the lowest L-value station in the plot (PINA). A similar field line resonance signature can be seen in the D-component, but at much smaller amplitude. It is also interesting to note that this auroral zone FLR amplitude enhancement is strong in the morning sector, but extends around the nightside into the pre-midnight sector. There is also evidence for a pre-midnight H-component power enhancement which crosses all L-shells and is especially clear at high solar wind speeds. We believe that his may be evidence for tail waveguide modes (see e.g. Wright and Mann 2006, and references therein). Finally, in the pre-dawn sector there is evidence for an MLT localized H-component power feature which is confined predominantly from the pre-noon to the dawn sector, and whose amplitude decreases with decreasing L. We believe that this is likely a signature of the waveguide modes, or of magnetopause KHI surface waves, driving the observed FLR response. Future work combining the observed ground signatures with data from the in-situ THEMIS probes will be very useful for establishing the nature of the in-situ disturbances corresponding to these features on the ground.

The dominance of a field line resonance response in the morning sector over the afternoon, as seen by ground-based magnetometers, remains a puzzle. Previous research has suggested that potential explanations include stabilizing magnetic field tension at dusk (e.g. Miura 1992) or the comparative lack of seed perturbations for the KHI downstream of the quasi-perpendicular shock which for Parker spiral IMF orientation is on the dusk side (see e.g. Lee and Olson 1980). More recently, Glassmeier and Stellmacher (2000) suggested that the local time asymmetry of the radial gradients in plasma density around the dayside magnetosphere due to refilling might also preferentially screen Pc5 FLRs from the ground in the dusk sector. Finally, Rostoker and Sullivan (1987) suggested that different field line resonant responses might be generated either side of local noon because of the MLT dependence of the characteristics of driving solar wind disturbances. Specifically, Rostoker and Sullivan (1987) suggested that since solar wind dynamic pressure pulses impact the early afternoon magnetopause, these authors finding that the afternoon-side ULF wave response was more closely associated with solar wind impulses than that on the morning-side. However, more studies combining large scale latitudinal and longitudinal FLR characterization together with conjugate multi-point in-situ measurements such as those which are available with the THEMIS probes provide a capability to begin to solve this puzzle.

Another enigma is that there is often very little in-situ observational evidence of compressional waveguide modes of significant amplitude (e.g. Anderson and Engebretson 1995; Waters et al. 2002). Multiple satellites have seen the downtail propagation of waveguide modes (e.g. Mann et al. 1998), however, the number of such examples is limited. Further simulation studies have also suggested that due to dispersion down the magnetospheric waveguide, waveguide mode harmonics may not display quasi-sinusoidal signatures in the time domain at a single location, making them potentially difficult to identify in satellite time-series data (Rickard and Wright 1995). Multi-point studies with THEMIS can examine the waves in the magnetosphere at the times of well-defined CARISMA observed FLRs, as well as the nature of fluctuations at the magnetopause (e.g. Mann et al. 2002; Rae et al. 2005) and even in the sheath and solar wind, especially on the flanks. THEMIS conjunctions to CARISMA offer the ideal capabilities for these studies, especially during the early mission "string-of-pearls" configuration. Multi-point satellite studies could establish causality from studies characterizing directions of energy flow, magnetopause thickness, and magnetopause oscillation amplitudes. These could encompass studies of the role of the IMF in stabilizing the KHI, the role of direct coherent driving (cf. Kepko and Kivelson 1999) versus natural waveguide mode harmonic resonances (cf. the simulations of Wright and Rickard 1995), the role of the magnetopause over-reflection mechanism (e.g. Mann et al. 1999), and the role of non-linear KH vortex development (cf. Hasegawa et al. 2004; Fairfield et al. 2007), and the potential role of seed magnetosheath fluctuations downstream of the bow shock (e.g. Miura 1992).

5.2 Excitation and Propagation of High-m ULF Pulsations

Ring current ions injected into the magnetosphere naturally evolve as the ions drift and bounce through the inner magnetosphere. The energy and pitch angle dependence of the drift trajectories, including the effects of convection electric fields as well as magnetic gradient and curvature drifts, can generate spatial and energy gradients leading to a fundamental plasma instability where ion energy is transferred into guided high-m poloidal ULF waves in the Pc4–5 range especially through the drift-bounce resonance mechanism (e.g., Hughes et al. 1978; Southwood et al. 1969; Southwood 1976; Southwood and Kivelson 1982; Ozeke and Mann 2001). High-m waves are seen extensively in the local afternoon sector along ion injection paths from the tail during the main phase of geomagnetic storms (e.g. Cao et al. 1994, and Anderson et al. 1990), however, there are relatively few ground-satellite conjunction studies showing definitive evidence for free energy transfer to the waves.

Theory suggests that the drift-bounce resonance instability, and related diamagnetic drift, drift-mirror, and drift-Alfven ballooning modes (e.g. Chen and Hasegawa 1991; Vetoulis and Chen 1994) could explain the high-m poloidal waves observed. Observations capable of distinguishing between proposed excitation mechanisms are rare. Some statistical studies of the free energy in bump-on-tail distributions have been completed (e.g. Baddeley et al. 2004), yet debate remains about whether this loss process from the ring current is energetically significant (e.g. Wilson et al. 2006). Even for the simpler case of giant pulsation excitation, which typically occurs during quiet times likely also by drift-bounce resonance, debate continues (e.g. Chisham et al. 1992, 1997). The equatorial multi-point configuration of THEMIS, conjugate to CARISMA, offers the ideal vehicle with which to determine the physical mechanisms exciting these waves, and their role in ring current loss. Supporting measurements from other CGSM ground-based arrays, and partner satellite energetic particle observations especially resolving distribution function evolution along ion drift path, will aid science closure.

6 Radiation Belt Science Capabilities

One of the most interesting and important questions in current solar-terrestrial physics research concerns the acceleration of electrons to relativistic speeds. The fundamental mechanisms proposed to explain the dynamics, energization and loss of these particles are numerous, and which dominate remains largely unknown (see e.g. the review by Friedel et al. 2002). Likely the most influential acceleration mechanisms are resonance with VLF lower band chorus, which operates through violation of the first adiabatic invariant (e.g. Meredith et al. 2003; Chen et al. 2007, and resonance with ULF waves which typically operates through violation of the third (e.g. Fälthammer 1966; Schulz and Lanzerotti 1974; Elkington et al. 2002). Together with in-situ THEMIS measurements of both ULF wave electric and magnetic fields as well as energetic electron flux up to 900 keV with the solid state telescope (SST; see McFadden et al. 2008), CARISMA has an excellent capability for supporting studies of ULF wave related radiation belt acceleration and loss processes and hence to the secondary objective of the THEMIS mission. Indeed, having 5 probes provides an excellent capability to resolve the spatial and temporal structure of the evolution of energetic electron flux and phase space density during magnetic storms.

6.1 Pc5 ULF Wave Drift-Resonant Acceleration

Recent theoretical and observational developments have highlighted the possibility that Pc5 ULF waves might accelerate electrons to MeV energies in the outer radiation belt through drift-resonance (e.g. Rostoker et al. 1998; Elkington et al. 2002, 2003; Hudson et al. 2000; Mathie and Mann 2000c, 2001; O'Brien et al. 2001; Mann et al. 2004). Mathie and Mann (2001), showed clear correlations between daily dawn-side Pc5 ULF power and both solar wind speed and >2 MeV electron flux at geosynchronous orbit (GEO) for 6 months of the declining phase of the solar cycle in 1995.

Evidence in favor of longer timescale stochastic ULF wave radial diffusion also comes from ULF wave correlated MeV electron flux having even being shown to statistically propagate radially inwards from $L = 6.6$ to $L = 4.5$ (Mann et al. 2004). Historically, collisionless radial diffusion coefficients (e.g. Fälthammer 1966; Schulz and Lanzerotti 1974) have been quantified as a function of Kp from observations such as from CRRES (e.g. Brautigam and Albert 2000). Despite their simplicity, these have been able to generate models which reproduce many of the global morphological features of the radiation belts (e.g. Shprits et al. 2005; Mai Mai Lam and Richard Horne, Personal Communication 2007). The diffusion coefficient formalism developed by Brizard and Chan (1999, 2001, 2004) which derives the diffusion coefficients as a function of wave power at frequency $\omega = m\ \omega_d$, for 90° pitch angle particles in an uncompressed dipole field, can be used with the CARISMA magnetic field power to produce data driven energy dependent radial diffusion coefficients, once mapping from the ground measurements to the equatorial magnetosphere is computed (e.g. Ozeke et al. 2008). Such diffusion model studies can also look for regions where additional sources such as VLF acceleration are required.

The unusual MeV electron penetration into the slot region during the first day of the Halloween 2003 storms (e.g. Baker et al. 2004) was shown by Loto'aniu et al. (2006) to be consistent with enhanced ULF wave radial diffusion occurring in response to ULF wave penetration to anomalously low-L. On the 29th October 2003, a rapid decrease in eigenfrequency was observed using the cross-phase technique (see Sect. 6.3), most likely due to the injection of O^+ ions from the ionosphere, enabling ULF wave energy to penetrate much more deeply than usual (Loto'aniu et al. 2006; Kale et al., in preparation). Tantalisingly, this suggests that cold (eV energy) plasma might play a critical role in the dynamics of the apparently totally separate MeV energy radiation belt particle population, 6 orders of magnitude away in energy, via the intermediary of ULF waves. Recent studies have also suggested that eastward propagating moderate azimuthal mode waves ($m \sim 20$–40) (cf. Sect. 5.2), driven by drift-bounce resonance with ~few 100 keV O^+ ions outside a depleted plasmapause, can also energize MeV electrons via drift resonance at $L \sim 4$ (Ozeke and Mann 2008). Given there is an ample supply of energy in the ring current, such a mechanism is attractive for radiation belt electron acceleration. More case and statistical studies are required to validate these important concepts.

Recent studies completed by the CARISMA team have also shown the strong time domain coherence between ULF wave oscillations seen on the ground and modulation of energetic particle flux in a specific ULF wave packet (e.g. Mann et al. 2007). Simulations using the code described by Degeling et al. (2007) were able to reproduce the observed several hundred to ~1 MeV flux modulation, including a phase change as a function of energy we believe indicates a resonant response. One element of importance will be a careful consideration of the effects of mode polarization on any radiation belt response since poloidally polarized Alfvén modes should be dominated by azimuthal electric fields in the magnetosphere which can strongly interact with the azimuthal drift motion of radiation belt electrons. Conversely, toroidally polarized modes which are usually larger amplitude are expected to display a weaker interaction. Further studies are required including not only polarization, but also more realistic wave models including local time dependence. ULF wave data from CARISMA can be compared in time domain case studies to energetic particle data from THEMIS SST (up to 900 keV), as well as to those available from Polar, Cluster, HEO, SAMPEX, LANL and GOES satellites. Predictions from ULF wave-particle models which utilize tracing of Liouville trajectories in ULF fields (e.g. Degeling et al. 2007) also enable the results to be observationally tested in detail.

6.2 Electromagnetic Ion Cyclotron (EMIC) Waves

Continuous magnetic field fluctuations in the frequency range from 0.2 to 5 Hz are classified as Pc1 pulsations (Jacobs et al. 1964). Waves in the Pc1 band can be driven by the electromagnetic ion cyclotron (EMIC) instability, where free energy is provided by hot equatorial ions with temperature anisotropy ($T_{\text{perp}} > T_{\text{par}}$). Depending on ion composition, EMIC waves occur in three bands below the hydrogen, helium and oxygen ion gyrofrequencies. Despite satellite and ground observations since the 1960s, EMIC waves are still somewhat poorly understood; multi-point THEMIS and continent scale CARISMA observations provide the capability to examine the excitation processes, and the space-ground propagation characteristics, including the effects of ducting in the Earth-ionosphere waveguide (e.g. Fraser 1976). In the inner magnetosphere, EMIC waves are believed to be preferentially excited in a spatially localized zone along the high density dusk-side plasmapause, occur most frequently and are the most intense during magnetic storms (Horne and Thorne 1993; Kozyra et al. 1997; Jordanova et al. 2001), but in the outer magnetosphere have an occurrence rate which increases with L towards the magnetopause and are present even during very quiet geomagnetic conditions (see Anderson et al. 2002; Engebretson et al. 2002).

On the ground, EMIC waves are often observed as structured pulsations, which appear as sequence of discrete dispersive wave packets with repetition period of a few minutes (e.g. Mursula et al. 1997), while in space EMIC waves are typically unstructured: only a few cases of satellite observations of structured Pc1 pulsations have been reported so far. Of particular interest is debate about the mechanisms which generate structured Pc1 pulsations. Traditionally, structured Pc1's have been explained by a bouncing wavepacket (BWP) model (e.g. Jacobs and Watanabe 1964), in which a wavepacket travels along the magnetic field line between the conjugate hemispheres and compensates energy losses at the equator. However, observations of the Poynting flux of EMIC waves with the CRRES satellite (Loto'aniu et al. 2005) show that the Poynting flux propagates unidirectionally away from the equatorial plane, contradicting the BWP theory. Figure 14 illustrates the structure of EMIC waves observed simultaneously in space by the THEMIS E satellite and on the ground by the MCMU magnetometer located at $\sim 30^\circ$ west of the spacecraft's footprint. According to the BWP model, the ground EMIC waves should have a periodicity in space which is half that seen on the ground. Figure 14 shows an example where the relationship between wavepacket repetition period predicted by the BWP hypothesis is clearly not seen, and the observed wavepacket periodicities appear to be very similar both on the ground and in space (see Usanova et al. 2008 for detailed analysis).

Theoretical investigations have suggested that a Doppler shifted gyroresonant interaction between EMIC waves and MeV energy outer radiation belt electrons can lead to pitch-angle scattering and radiation belt electron loss into the atmosphere (e.g. Summers and Thorne 2003). Outer radiation belt MeV electrons typically drift around the Earth on time-scales of the order 5–10 min and spend only a small fraction of each orbit within the region of EMIC activity. While typical EMIC amplitudes usually exceed the level required for strong diffusion, because of the limited extent of the region of enhanced EMIC activity significant electron loss only occurs over many drift orbits. Under certain conditions (electron plasma frequency/electron gyrofrequency ≥ 10), MeV electrons can be removed from the outer radiation belt over a time-scale of several hours to a day (Summers and Thorne 2003). Such conditions are satisfied within the region of high plasma density and low magnetic field, such as the duskside plasmasphere or detached plasma regions at high L-values.

Despite the potential importance of EMIC waves for radiation belt loss, there are relatively few studies which examine this relationship. Meredith et al. (2003) performed a

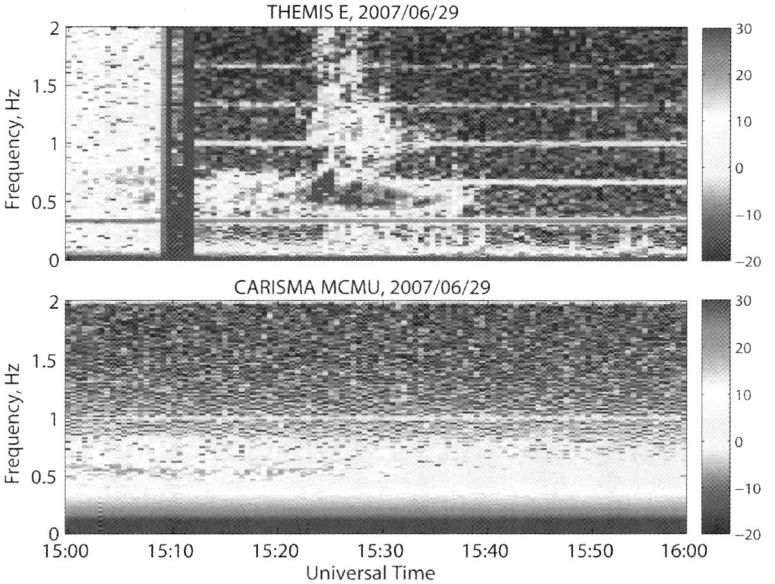

Fig. 14 Fourier spectrograms showing structured Pc1 pulsations detected in space by THEMIS E panel (**a**) and on the ground at the CARISMA MCMU magnetometer station ($L = 5.23$) panel (**b**)

statistical analysis of over 800 EMIC wave events observed on board the CRRES satellite, to establish whether the resonant scattering can occur at energies ≤ 2 MeV. The results of their analysis are consistent with the theoretical study of Summers and Thorne (2003). It has also been inferred from balloon observations of X-ray emissions in the dusk sector that EMIC waves can cause precipitation of MeV energy electrons (Foat et al. 1998; Millan et al. 2002) usually during storm recovery phase.

Further studies to establish the localization of EMIC waves in space, their propagation to the ground, and their potential role in MeV electron loss are required. Partner measurements of MeV electron precipitation loss from low altitude satellites such as NOAA, from balloons (e.g. the NASA funded BARREL project), or inferred from ground VLF networks (e.g. AARDVARK and SID/AWESOME) or even from riometer networks (e.g. from CGSM Norstar riometers) may be important. Ultimately, a characterization of EMIC waves as a function of L, MLT, and geomagnetic activity or storm phase could be used as an empirical input to loss modules within global models for radiation belt dynamics. At present there is a peculiarity that most MeV electron loss in the radiation belts occurs during main phase, yet on the ground EMIC power typically only appears during the recovery phase (e.g. Engebretson et al. 2008; Bortnik et al. 2008 and references therein). Whether this is due to internal reflection of EMIC waves in the magnetosphere (e.g. Rauch and Roux 1982), absorption, perhaps in the presence of heavy ions during the main phase (Horne and Thorne 1994), due to changes in reflection and transmission characteristics of a perturbed ionosphere (Mursula et al. 2000) or represents a lack of waves in the magnetosphere at the times when most MeV electron loss is observed is not clear. Waves below the He^+ gyrofrequency are found to be the most efficient for MeV electron scattering (Summers and Thorne 2003; Meredith et al. 2003), and a superposed statistical survey of EMIC wave power and radiation belt energetic electron flux as a function of L during storms would be valuable.

6.3 Role of the Plasmapause and Plasmasphere

The plasmasphere consists of cold ions and electrons with energies $\sim$1 eV, which corotate with the Earth. Interestingly, the outer boundary of the plasmasphere, the plasmapause, is believed to have a strong influence on the dynamics of the outer radiation belt's MeV electrons. Li et al. (2006) showed that the lowest L-shell which the outer radiation belt electrons can penetrate to is closely related to the minimum plasmapause location. Several different mechanisms may explain why the plasmapause location affects the penetration of the radiation belt electrons:

- The location of the plasmapause may affect how close to the Earth compressional fast mode waves can propagate inward. If these waves transport the electrons inward via ULF wave driven diffusion or coherent drift-resonance transport (Degeling et al. 2007), then these radial transport mechanisms would be affected by the location of the plasmapause.
- The radiation belt electrons may also be accelerated inward of a drift-resonance region by a guided poloidal FLR, which in turn is generated by an unstable ion distribution. These guided poloidal waves, which are able to resonantly accelerate the radiation belt electrons, are most likely to occur close to a depleted plasmapause (see Ozeke and Mann 2008).
- The electrons in the radiation belt may be locally energized by VLF waves which are most likely to occur in a region just outside the plasmapause (see Horne et al. 2005 and Shprits et al. 2006).

The explanation for the recent observation that the inner edge of the radiation belt is correlated with the plasmapause (e.g. O'Brien and Moldwin 2003; Li et al. 2006; see also Tverskaya et al. 1986) is hence still an enigma. Perhaps the explanation is that VLF acceleration operates just outside the plasmasphere, and losses internal to the plasmasphere remove any particles which diffuse inward across the plasmapause. Alternatively, if a large element of radiation belt morphology is determined by inward (and outward, e.g. Shprits et al. 2004, 2005) diffusion, then perhaps the penetration of ULF wave power such as that described by Loto'aniu et al. (2006) plays an important role. Comparing CARISMA observed ULF power, the cross-phase determined plasmapause location, and energetic particle flux from the satellites described above, would allow the role of ULF waves in producing the observed correlation to be investigated.

Recent observations (Fraser et al., personal communication 2007) using high resolution GOES magnetometer data have also demonstrated a link between EMIC waves and the extension of the plasmasphere into a dayside drainage plume. Other studies have also suggested a link between EMIC waves excited in dense plasma regions and sub-auroral red (SAR) arcs (e.g. Spasojević et al. 2004). Using the CGSM optical and precipitation infrastructure, the generation of SAR arcs could also be addressed using the combination of THEMIS probe and CARISMA infrastructure.

6.3.1 Diagnosing the Plasmapause and Plasmasphere with CARISMA

The plasmasphere was first discovered using ground-based VLF measurements of plasmaspheric ducts (Carpenter 1963). Despite several decades of study from the ground, and onboard satellites (e.g. Boskova et al. 1993; Park 1974) the basic rates of plasmaspheric refilling and the processes responsible are not well-understood, the problem being compounded by the fact that often satellite low energy ion detectors are "blind" to very cold ions due to spacecraft charging. The CARISMA research team and collaborators have been instrumental in developing cross-phase and related techniques for remote-sensing the distribution

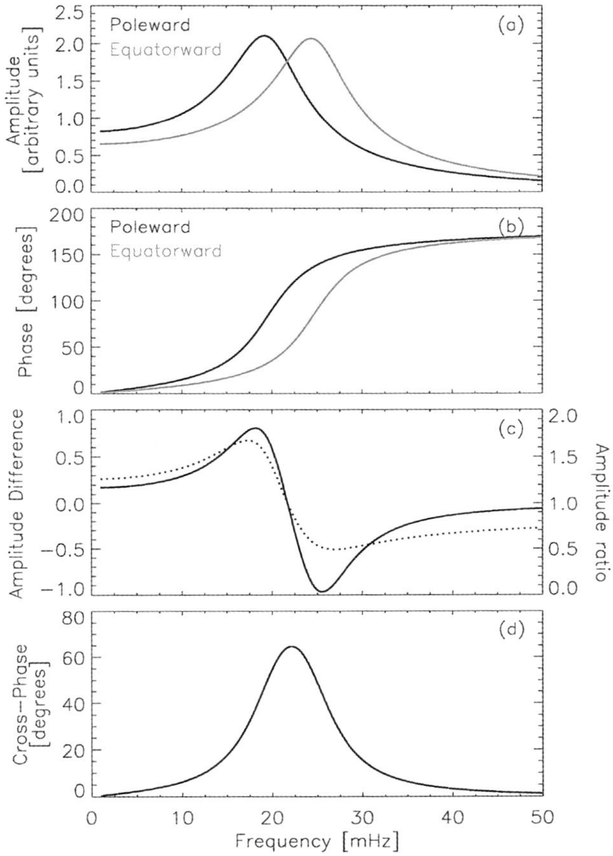

Fig. 15 Schemetic illustration of the cross-phase technique. Panels (**a**) and (**b**) show the amplitude and phase response of two latitudinally separated magnetometers. Panels (**c**) and (**d**) show the amplitude and phase difference between the two magnetometers

and dynamics of cold plasma using networks of ground-based magnetometers (e.g. Menk et al. 1999, 2000, 2004; Milling et al. 2001; Dent et al. 2003, 2006; Kale et al. 2007). By employing data from ground-based magnetometer networks, it is possible to determine both the location of the plasmapause and the equatorial density profile as a function of L-shell, via detection of local field line eigenfrequencies with the cross-phase technique.

The cross-phase technique examines the H-component amplitude and phase spectra from two latitudinally separated ground-based magnetometers in order to determine the eigenfrequency of a field-line with a foot-point assumed to be near the latitudinal and longitudinal midpoint between those two magnetometers (Waters et al. 1991). Gough and Orr (1984) explained that driven Alfvén waves may be treated as forced, damped simple harmonic oscillators. Figure 15 is based on Fig. 1 of Waters et al. (1991) and presents simple calculations of the response of forced, damped simple harmonic oscillators with eigenfrequencies of 20 mHz and 25 mHz, which represent the field lines at the locations of two latitudinally separated magnetometer stations. The top two panels show the amplitude and phase response as a function of frequency for each of the field-lines. These show the amplitude peak and 180° phase change, which are expected as the frequency passes through resonance. The resonance frequency of the field-line with a foot-point midway between the foot-points of the two field-lines being modeled is identified where the amplitude difference $= 0$, and am-

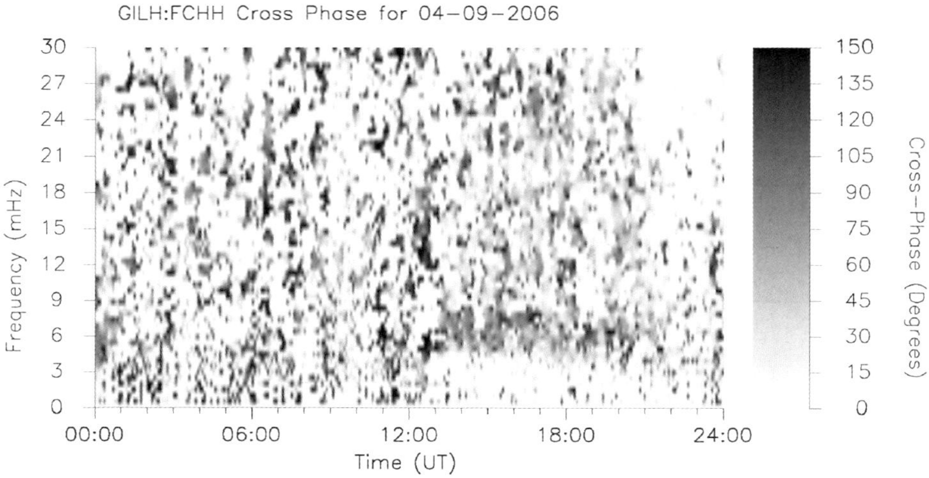

Fig. 16 Dynamic cross-phase spectra for the GILL-FCHU station pair, which belong to the CARISMA array

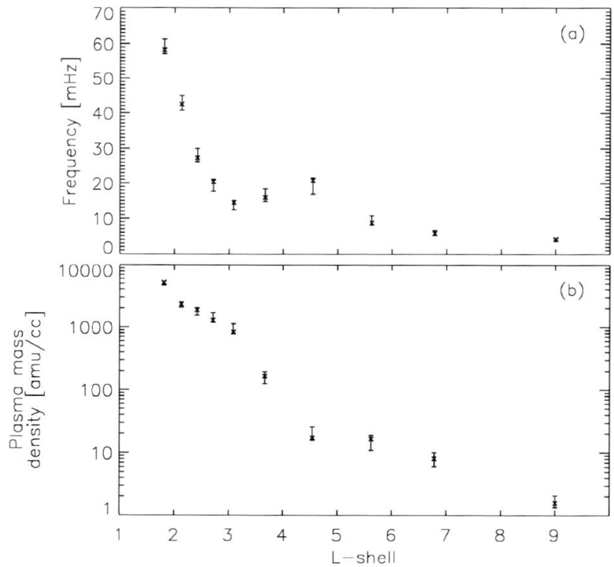

Fig. 17 (**a**) Fundamental mode field line eigenfrequencies as a function of L-shell, determined using the cross phase technique. (**b**) Equatorial plasma mass densities inferred from the eigenfrequencies presented in panel (**a**)

plitude ratio = 1 (both with negative gradient), and the cross-phase shows a local maxima at a value >0 (i.e., the phase difference maximizes with a positive value).

Figure 16 shows an example dynamic cross-phase spectrogram, which is plotting the cross-phase (i.e. phase difference) between two magnetometer data sets as a function of frequency and Universal Time. The cross-phase peak, representing the field-line eigenfrequency of the mid-point field-line, is shown by the dark band between ~5 and 8 mHz, between ~13 and 20 UT.

An example of the capability of ground-based magnetometers to diagnose eigenfrequency and density profiles in the North American sector is shown in Fig. 17. Data from magnetometers belonging to the CARISMA (4.54 < L_midpoint < 9.00) and McMAC (Mid-Continent Magnetoseismic Chain; 1.50 < L_midpoint < 3.09) ground-based magnetometer arrays have been employed. This example shows data from 4 September 2006, between 1700 and 1800 UT. The plasma mass density values have been determined assuming a dipolar magnetic field geometry, and a radial density variation $\propto r^{-1}$ along the field-lines (described in more detail in Dent et al. 2006). The error bars represent the uncertainty in determining the eigenfrequency values from the data, and the corresponding density uncertainty. The location of the plasmapause is shown to reside between $L = 3.09$ and $L = 4.54$, across which the resonance frequency rises with increasing L-shell, and the density drops from ~800 amu/cc to ~25 amu/cc. This region has limited coverage, and is part of the CARISMA expansion region. The new CARISMA expansion sites will provide improved spatial resolution along eigenfrequency and density profiles in the plasmapause region.

Through comparison to partner in-situ measurements of the electron density, such as from THEMIS spacecraft potential, the dynamical variations of heavy ion populations may also be determined (e.g. Fraser et al. 2005; Dent et al. 2006). Recent studies have shown that the "archetypal" plasmapause is rarely observed, and structure at the edge of the plasmasphere is created by competition between dynamical erosion and L-dependent refilling (e.g. Dent et al. 2003, 2006). Moreover, care needs to be taken in defining "the" plasmapause location, since different ion species, and electrons, can all indicate sharp gradients at different locations. A very steep plasmapause can create a local, and MLT (and likely time) limited feature of a reversed cross-phase peak (Kale et al. 2007). This corresponds to a local turning point in the Alfven continuum due to the rapid L-variation in density at the sharp plasmapause (Kale et al. 2007). At times, these "negative cross-phase peaks" demonstrate complex structure as a function of frequency, with both positive and negative peaks observed at different frequencies at the same time. Further work is required in order to establish the explanations for these cross-phase features.

Future ground-THEMIS satellite correlative studies can be used to examine the processes leading to dynamical structure of the plasmasphere, and which lead to plasmaspheric refilling. Work is certainly needed to establish whether the negative cross-phase peak results observed by Kale et al. (2007) apply universally to "steep plasmapause" profiles. Given the importance of the plasmasphere and plumes for radiation belt dynamics, THEMIS-CARISMA conjunctions also offer the basis for studies of the role of plasmaspheric drainage plumes in inner magnetosphere wave-particle interactions. Given that the IMAGE satellite is no longer operational following its failure in late 2005, ground-based magnetometers may offer a unique method for monitoring plume dynamics (e.g. Kale et al. 2008, in preparation). Ion outflow data from low Earth orbit, such as from the Canadian enhanced Polar Outflow Probe (e-POP) satellite (e.g. Yau et al. 2006) where spacecraft charging is less of a problem, can also be used to inform refilling studies. Additionally GPS TEC available from the Canadian High Arctic Ionospheric Network (CHAIN) and from the CHAIN GPS receiver deployed at the MSTK CARISMA station may also be employed in order to examine the ionospheric density variations associated with plasmaspheric depletion, refilling and dynamical plumes.

7 Conclusions

In this review, we have outlined some of the capabilities of the expanded CARISMA array for completing solar–terrestrial science. As part of the CSA funded Canadian Geospace Monitoring (CGSM) ground-based network, CARISMA data from an expanded array of fluxgate magnetometers and new induction coil magnetometers provides a powerful infrastructure to address scientific questions at the forefront of international efforts. With new real-time satellite data collection infrastructure, local data loggers, and increased cadence the collective array enables the high resolution characterization of magnetic activity on a continent scale. This magnetic activity results from waves and currents driven in the magnetosphere by solar forcing, providing a magnetic window on the solar–terrestrial interaction. CARISMA magnetometers are the only Canadian instrumentation which provides data to the formal THEMIS dataset. When combined with in-situ data from the THEMIS probes, which are regularly magnetically conjugate to the CARISMA array, the collective ground-satellite data set represents a powerful tool which can be used to address all three of the THEMIS mission's scientific objectives.

Acknowledgements CARISMA is operated by the University of Alberta and is funded by the Canadian Space Agency. THEMIS is funded by NASA contract NAS5-02099. GIMA data is provided by the Geophysical Institute of the University of Alaska Fairbanks. The Canadian Magnetic Observatory System (CANMOS) network is maintained and operated by the Geological Survey of Canada, and provided data used in this study. AP would like to acknowledge Kazuo Shiokawa and Athabasca Geophysical Observatory for data plotted in Fig. 10. Z.C.K. wishes to acknowledge C.L. Waters and F.W. Menk for providing cross-phase analysis programs.

References

W. Allan et al., Geophys. Res. Lett. **23**(7), 765 (1996)
W. Allan, S.P. White, E.M. Poulter, Planet. Space. Sci. **34**(4), 371 (1986)
B.J. Anderson, M.J. Engebretson, J. Geophys. Res. **100**(A6), 9591 (1995)
B.J. Anderson et al., J. Geophys. Res. **95**(A7), 10495 (1990)
B.J. Anderson et al., J. Geophys. Res. **107**(A6), 1079 (2002)
V. Angelopoulos, Space Sci. Rev. (2008). doi:10.1007/s11214-008-9336-1
T. Araki, in *Solar Wind Sources of Ultra Low Frequency Wave Pulsations*, ed. by M. Engebretson, K.T. Takahashi, M. Scholer (Am. Geophys. Union, Washington, 1994), p. 183
R.L. Arnoldy et al., J. Geophys. Res. **103**, 23581 (1998)
H.U. Auster et al., Space Sci. Rev. (2008). doi:10.1007/s11214-008-9365-9
L.J. Baddeley et al., Ann. Geophys. **22**, 4229 (2004)
D.N. Baker et al., Nature (2004). doi:10.1038/nature03116
L.N. Baransky et al., Planet. Space Sci. **38**(12), 1573 (1985)
W. Baumjohann, K.-H. Glassmeier, Planet. Space Sci. **32**(11), 1361 (1984)
P.P. Belyaev et al., J. Atmos. Terr. Phys. **52**(9), 781 (1990)
J. Bortnik et al., J. Geophys. Res. **113**, A04201 (2008). doi:10.1029/2007JA012867
J. Boskova et al., J. Atmos. Terr. Phys. **55**(11–12), 1595 (1993)
D.H. Brautigam, J.M. Albert, J. Geophys. Res. **105**(A1), 291 (2000)
A.J. Brizard, A.A. Chan, Phys. Plasmas **6**(12), 4548 (1999)
A.J. Brizard, A.A. Chan, Phys. Plasmas **8**(11), 4762 (2001)
A.J. Brizard, A.A. Chan, Phys. Plasmas (2004). doi:10.1063/1.1773554
M. Cao et al., J. Geophys. Res. **99**(A5), 8731 (1994)
D.L. Carpenter, J. Geophys. Res. **68**, 1675 (1963)
L. Chen, A. Hasegawa, J. Geophys. Res. **96**(A2), 1503 (1991)
Y. Chen et al., J. Geophys. Res. (2007). doi:10.1029/2007JA012314
G. Chisham et al., Planet. Space Sci. **40**(7), 953 (1992)
G. Chisham et al., J. Geophys. Res. **102**(A5), 9619 (1997)
G. Chisham, I.R. Mann, J. Geophys. Res. **104**(A7), 14717 (1999)
M. Cramoysan et al., Ann. Geophys. **13**, 583 (1995)
A.W. Degeling et al., Planet. Space Sci. (2007). doi:10.1016/j.pss.2006.04.039
Z.C. Dent et al., Geophys. Res. Lett. (2003). doi:10.1029/2003GL0169146
Z.C. Dent et al., J. Geophys. Res. (2006). doi:10.1029/2005JA011046
S.R. Elkington et al., J. Atmos. Sol. Terr. Phys. **64**(5–6), 607 (2002)
S.R. Elkington et al., J. Geophys. Res. (2003). doi:10.1029/2001JA009202
M.J. Engebretson et al., J. Geophys. Res. (2002). doi:10.1029/2001JA000198
M.J. Engebretson et al., J. Geophys. Res. 113, (2008). doi:10.1029/2007JA012362
D.H. Fairfield et al., J. Geophys. Res. (2007). doi:10.1029/2006JA012052

C.G. Fälthammer, J. Geophys. Res. **71**(5), 1487 (1966)
J.E. Foat et al., Geophys. Res. Lett. **25**(22), 4109 (1998)
B.J. Fraser, J. Atmos. Terr. Phys. **38**, 1141 (1976)
B.J. Fraser et al., Geophys. Res. Lett. (2005). doi:10.1029/2004/GL21315
R.H. Friedel et al., J. Atmos. Sol. Terr. Phys. **64**(2), 265 (2002)
K.-H. Glassmeier, Ann. Geophys. **10**(547), 547 (1992)
K.-H. Glassmeier, M. Stellmacher, J. Geophys. Res. **105**(18), 847 (2000)
H. Gough, D. Orr, Planet. Space Sci. **32**(5), 619 (1984)
R.A. Greenwald, A.D.M. Walker, Geophys. Res. Lett. **7**, 745 (1980)
H. Hasegawa et al., Nature **430**(7001), 755 (2004)
S.R. Hebden et al., Ann. Geophys. **23**(5), 1711 (2005)
R.B. Horne, R.M. Thorne, J. Geophys. Res. **98**(A6), 9233 (1993)
R.B. Horne, R.M. Thorne, J. Geophys. Res. **99**(A9), 17259 (1994)
R.B. Horne, R.M. Thorne, Geophys. Res. Lett. **25**(15), 3011 (1998)
R.B. Horne et al., J. Geophys. Res. (2005). doi:10.1029/2004JA010811
M.K. Hudson et al., Adv. Space Res. **25**(12), 2327 (2000)
W.J. Hughes et al., Nature **275**(5675), 43 (1978)
W.J. Hughes, in *Solar Wind Sources of Magnetospheric Ultra-Low-Frequency Waves*, ed. by M.J. Engebretson, K. Takahashi, M. Scholer (American Geophysical Union, Washington, 1994), p. 1
J.A. Jacobs et al., J. Geophys. Res. **69**(1), 180 (1964)
J.A. Jacobs, T. Watanabe, J. Atmos. Sol. Terr. Phys. **26**(8), 825–829 (1964)
V.K. Jordanova et al., J. Geophys. Res. **106**(A1), 7 (2001)
Z.C. Kale et al., J. Geophys. Res. **112** (2007). doi:10.1029/2007JA012367
R. Kataoka, H. Fukunishi, L.J. Lanzerotti, J. Geophys Res. **108**, 1436 (2003). doi:10.1029/2003JA010202
L. Kepko, M.G. Kivelson, J. Geophys. Res. **104**(A11), 25021 (1999)
L. Kepko, M.G. Kivelson, K. Yumoto, J. Geophys. Res. **106**(A2), 1903 (2001)
L. Kepko, H.E. Spence, H.J. Singer, Geophys. Res. Lett. (2002). doi:10.1029/2001GL014405
M.G. Kivelson, D.J. Southwood, Geophys. Res. Lett. **12**(1), 49 (1985)
M.G. Kivelson, J. Etcheto, J.G. Trotignon, J. Geophys. Res. **89**(A11), 9851 (1984)
M.G. Kivelson, D.J. Southwood, J. Geophys. Res. **96**(A2), 1661 (1991)
J.U. Kozyra et al., J. Geophys. Res. **89**(A4), 2217–2233 (1984)
J.U. Kozyra et al., in *Geophys. Monogr. Ser.*, vol. 98, ed. by B.T. Tsurutani et al. (AGU, Washington, 1997), p. 187
G. Le, C.T. Russell, J. Geophys. Res. **101**(A2), 2571 (1996)
L.C. Lee, J.V. Olson Geophys. Res. Lett. **7**, 777 (1980)
M.R. Lessard et al., Geophys. Res. Lett. (2006). doi:10.1029/2006GL026411
M. Lester, W.J. Hughes, H.J. Singer, J. Geophys. Res. **88**(A10), 7958 (1983)
X. Li et al., Geophys. Res. Lett. (2006). doi:10.1029/2006GL026294
T.M. Loto'aniu, B.J. Fraser, C.L. Waters, J. Geophys. Res. (2005). doi:10.1029/2004JA010816
T.M. Loto'aniu et al., J. Geophys. Res. (2006). doi:10.1029/2005JA011355
A.T.Y. Lui et al., J. Geophys. Res. **96**(A7), 11389 (1991)
R.L. Lysak, Feedback instability of the ionospheric resonant cavity. J. Geophys. Res. **96**(A2), 1553–1568 (1991)
R.L. Lysak, in *Auroral Plasma Dynamics*, ed. by R.L. Lysak (American Geophysical Union, Washington, 1994), p. 121
I.R. Mann, A.N. Wright, Geophys. Res. Lett. **26**(16), 2609 (1999)
I.R. Mann, G. Chisham, S. Bale, J. Geophys. Res. **103**(A3), 4657 (1998)
I.R. Mann et al., J. Geophys. Res. **104**(A1), 333 (1999)
I.R. Mann et al., Ann. Geophys. **20**(4), 405 (2002)
I.R. Mann et al., J. Atmos. Sol. Terr. Phys. **66**(2), 187 (2004)
I.R. Mann et al., EOS Transact. AGU, Fall Meet. Suppl. **88**(52), Abstract SM14A-04 (2007)
R.A. Mathie, I.R. Mann, Geophys. Res. Lett. **27**(20), 3261 (2000a)
R.A. Mathie, I.R. Mann, Geophys. Res. Lett. **27**(24), 4017 (2000b)
R.A. Mathie, I.R. Mann, J. Geophys. Res. **105**(A5), 10713 (2000c)
R.A. Mathie, I.R. Mann, J. Geophys. Res. **106**(A12), 29783 (2001)
R.A. Mathie et al., Geophys. Res. Lett. **26**(6), 659 (1999a)
R.A. Mathie et al., J. Geophys. Res. **104**(A4), 7025 (1999b)
J.P. McFadden et al., Space Sci. Rev. (2008, this issue)
J.F. McKenzie, Planet. Space Sci. **18**(1), 1 (1970)
R.L. McPherron et al., J. Geophys. Res. **78**(16), 3131 (1973)
S.B. Mende et al., Geophys. Res. Lett. (2007). doi:10.1029/2007GL030850

S.B. Mende et al., Space Sci. Rev. (2008, this issue)
F.W. Menk et al., J. Geophys. Res. **104**(A9), 19955 (1999)
F.W. Menk et al., J. Geophys. Res. **105**(A4), 7747 (2000)
F.W. Menk et al., J. Geophys. Res. (2004). doi:10.1029/2003JA010097
N.P. Meredith et al., J. Geophys. Res. (2003). doi:10.1029/2002JA009700
N.P. Meredith et al., J. Geophys. Res. (2007). doi:10.1029/2007JA012413
R.H. Millan et al., Geophys. Res. Lett. (2002). doi:10.1029/2002GL015922
D.K. Milling et al., Geophys. Res. Lett. **28**(1), 115 (2001)
D.K. Milling et al., Geophys. Res. Lett. **35**, 0 (2008). doi:10.1029/2008GL033672
A. Miura, J. Geophys. Res. **97**(A7), 10655 (1992)
K.R. Murphy et al., J. Geophys. Res. (2008). doi:10.1029/2008JA013548
D.L. Murr, W.J. Hughes, Geophys. Res. Lett. **30**, 1354 (2003). doi:10.1029/2002GL015498
K. Mursula et al., J. Geophys. Res. **102**(A8), 17611 (1997)
K. Mursula et al., J. Atmos. Sol. Terr. Phys. **62**(4), 299 (2000)
M. Nose et al., Earth Planets Space **50**(9), 773 (1998)
T.P. O'Brien, R.L. McPherron, J. Geophys. Res. (2003). doi:10.1029/2002JA009324
T.P. O'Brien et al., J. Geophys. Res. (2003). doi:10.1029/2002JA009784
T.P. O'Brien, M.B. Moldwin, Geophys. Res. Lett. (2003). doi:10.1029/2002GL016007
T.P. O'Brien et al., J. Geophys. Res. **106**(A8), 15533 (2001)
J.V. Olson, J. Geophys. Res. **104**(A8), 17499 (1999)
L.G. Ozeke, I.R. Mann, J. Geophys. Res. **106**(A8), 15583 (2001)
L.G. Ozeke, I.R. Mann, J. Geophys. Res. (2008). doi:10.1029/2007JA012468
L. Ozeke, I.R. Mann, I.J. Rae, J. Geophys. Res. (2008). doi:10.1029/2008JA013041
A. Parent et al., in *Proceedings of the 8th International Conference on Substorms*, ed. by M. Syrjäsuo, E. Donovan (University of Calgary, Canada, 2007), p. 225
C.G. Park, J. Geophys. Res. **79**(1), 169 (1974)
J.L. Posch et al., J. Atmos. Sol. Terr. Phys. **69**, 1775 (2007). doi:10.1016/j.jastp.2007.07.015
Z.-Y. Pu, M. Kivelson, J. Geophys. Res. **88**(2), 841 (1983)
I.J. Rae et al., J. Geophys. Res. (2005). doi:10.1029/2005JA011007
I.J. Rae et al., Planet. Space Sci. (2007a). doi:10.1016/j.pss.2006.02.009
I.J. Rae et al., Ann. Geophys. **25**(12), 2529 (2007b)
I.J. Rae et al., J. Geophys. Res. (2008a). doi:10.1029/2008JA013559
I.J. Rae et al., J. Geophys. Res. (2008b). doi:10.1029/JA2008JA013771
J. Rauch, A. Roux, J. Geophys. Res. **87**(A10), 8191 (1982)
G.J. Rickard, A.N. Wright, J. Geophys. Res. **100**, 3531 (1995)
G. Rostoker, S.-I. Akasofu, J. Foster, R.A. Greenwald, Y. Kamide, K. Kawasaki, A.T.Y. Lui, R.L. McPherron, C.T. Russell, Magnetospheric substorms—definition and signatures. J. Geophys. Res. **85**(A4), 1663–1668 (1980)
G. Rostoker, B.T. Sullivan, Planet. Space Sci. **35**, 429 (1987)
G. Rostoker et al., Space Sci. Rev. **71**(1–4), 743 (1995)
G. Rostoker, S. Skone, D.N. Baker, Geophys. Res. Lett. **25**(19), 3701 (1998)
J.M. Ruohoniemi et al., J. Geophys. Res. **96**(A9), 15697 (1991)
T. Saito, Sci. Rep. Tohoku Univ., 5, Geophys. **13** 53 (1961)
J.C. Samson, J.A. Jacobs, G. Rostoker, J. Geophys. Res. **76**(16), 3675 (1971)
J.C. Samson et al., Geophys. Res. Lett. **19**(5), 441 (1992)
M. Schulz, L.J. Lanzerotti, *Particle Diffusion in the Radiation Belts* (Springer, New York, 1974)
Y.Y. Shprits et al., Geophys. Res. Lett. (2004). doi:10.1029/2004GL019591
Y.Y. Shprits et al., Ann. Geophys. **23**(4), 1467 (2005)
Y.Y. Shprits et al., Geophys. Res. Lett. (2006). doi:10.1029/2005GL024256
D.G. Sibeck, V. Angelopoulos, THEMIS science objectives and mission phases. Space Sci. Rev. (2008). doi:10.1007/s11214-008-9393-5
D.G. Sibeck et al., J. Geophys. Res. (1999). doi:10.1029/1998JA900021
A.J. Smith et al., J. Geophys. Res. **104**(A5), 12351 (1999)
D.J. Southwood, J. Geophys. Res. **81**(19), 3340 (1976)
D.J. Southwood, J.W. Dungey, R.J. Etherington, Planet. Space Sci. 17 (1969)
D.J. Southwood, M.G. Kivelson, J. Geophys. Res. **87**(NA3), 1707 (1982)
M. Spasojević et al., Geophys. Res. Lett. (2004). doi:10.1029/2003GL018389
D. Summers, R.M. Thorne, J. Geophys. Res. (2003). doi:10.1029/2002JA009489
T. Takeuchi et al., J. Geophys. Res. **105**(A8), 18835 (2000)
L.V. Tverskaya et al., Geomagn. Aeron. **26**(5), 864 (1986)
M. Usanova et al., Geophys. Res. Lett. (2008). doi:10.1029/2008GL034458

G. Vetoulis, L. Chen, J. Geophys. Res. **21**(19), 2091 (1994)
A.D.M. Walker, J. Atmos. Sol. Terr. Phys. **62**(9), 799 (2000)
A.D.M. Walker et al., J. Geophys. Res. **97**(A8), 12187 (1992)
C.L. Waters, F.W. Menk, B.J. Fraser, Geophys. Res. Lett. **18**(12), 2293 (1991)
C.L. Waters, J.C. Samson, E.F. Donovan, J. Geophys. Res. **100**(A5), 7987 (1995)
C.L. Waters et al., J. Geophys. Res. **107**, 1284 (2002). doi:10.1029/2001JA000224
C.E.J. Watt et al., J. Geophys. Res. (2005). doi:10.1029/2004JA010877
M.E. Wilson et al., Ann. Geophys. **24**, 3027 (2006)
A.N. Wright, J. Geophys. Res. **99**(A1), 159 (1994)
A.N. Wright, G.J. Rickard, J. Geophys. Res. **100**(A12), 23703 (1995)
A.N. Wright, I.R. Mann, in *Magnetospheric ULF Waves: Synthesis and New Directions*, ed. by K. Takahashi et al.. AGU Monograph Series, vol. 169 (Am. Geophys. Union, Washington, 2006)
A.G. Yahnin et al., Ann. Geophys. **21**(3), 779 (2003)
A.W. Yau, H.G. James, W. Liu, Adv. Space Res. (2006). doi:10.1016/j.asr.2005.01.058

First Results from the THEMIS Mission

V. Angelopoulos · D. Sibeck · C.W. Carlson · J.P. McFadden · D. Larson · R.P. Lin · J.W. Bonnell · F.S. Mozer · R. Ergun · C. Cully · K.H. Glassmeier · U. Auster · A. Roux · O. LeContel · S. Frey · T. Phan · S. Mende · H. Frey · E. Donovan · C.T. Russell · R. Strangeway · J. Liu · I. Mann · J. Rae · J. Raeder · X. Li · W. Liu · H.J. Singer · V.A. Sergeev · S. Apatenkov · G. Parks · M. Fillingim · J. Sigwarth

Originally published in the journal Space Science Reviews, Volume 141, Nos 1–4, 453–476.
DOI: 10.1007/s11214-008-9378-4

Abstract THEMIS was launched on February 17, 2007 to determine the trigger and large-scale evolution of substorms. During the first seven months of the mission the five satel-

V. Angelopoulos (✉) · C.T. Russell · R. Strangeway · J. Liu
IGPP/ESS UCLA, Los Angeles, CA 90095-1567, USA
e-mail: vassilis@ssl.berkeley.edu

D. Sibeck · J. Sigwarth
Code 674, NASA/GSFC, Greenbelt, MD 20771, USA

C.W. Carlson · J.P. McFadden · D. Larson · R.P. Lin · J.W. Bonnell · F.S. Mozer · S. Frey · T. Phan · S. Mende · H. Frey · G. Parks · M. Fillingim
Space Sciences Laboratory, UCB, Berkeley, CA 94720-7450, USA

R. Ergun · X. Li · W. Liu
LASP, University of Colorado, Boulder, CO 80303, USA

C. Cully
Swedish Institute of Space Physics, Upsala, SE 751 21, Sweden

K.H. Glassmeier · U. Auster
TUBS, Braunschweig, 38106, Germany

A. Roux · O. LeContel
CETP/IPSL, 10-12 Avenue de l'Europe, 78140 Velizy, France

E. Donovan
Dept. of Physics and Astronomy, University of Calgary, Calgary AB T2N 1N4, Canada

I. Mann · J. Rae
Dept. of Physics, University of Alberta, Edmonton AB T6G 2J1, Canada

J. Raeder
Space Science Center, University of New Hampshire, Durham, NH 03824, USA

H.J. Singer
NOAA/Space Environment Laboratory, Boulder, CO 80303, USA

lites coasted near their injection orbit to avoid differential precession in anticipation of orbit placement, which started in September 2007 and led to a commencement of the baseline mission in December 2007. During the coast phase the probes were put into a string-of-pearls configuration at 100 s of km to 2 R_E along-track separations, which provided a unique view of the magnetosphere and enabled an unprecedented dataset in anticipation of the first tail season. In this paper we describe the first THEMIS substorm observations, captured during instrument commissioning on March 23, 2007. THEMIS measured the rapid expansion of the plasma sheet at a speed that is commensurate with the simultaneous expansion of the auroras on the ground. These are the first unequivocal observations of the rapid westward expansion process in space and on the ground. Aided by the remote sensing technique at energetic particle boundaries and combined with ancillary measurements and MHD simulations, they allow determination and mapping of space currents. These measurements show the power of the THEMIS instrumentation in the tail and the radiation belts. We also present THEMIS Flux Transfer Events (FTE) observations at the magnetopause, which demonstrate the importance of multi-point observations there and the quality of the THEMIS instrumentation in that region of space.

Keywords THEMIS · Magnetosphere · Substorms · Radiation belts · Magnetopause

PACS 94.30.-d · 94.30.cl · 94.30.cb · 94.30.ch · 94.30.cj · 94.30.C- · 94.30.cp · 94.30.Lr · 94.30.Va · 94.30.Xy · 96.50.Fm

1 Introduction

After a successful launch on February 17, 2007, and instrument commissioning by mid-March 2007, THEMIS's in-flight compatibility testing took place over a period of several weeks. Probe orbits were assigned constellation positions on March 27, followed by EFI boom deployments on probe C by May 16 and on probes D and E by June 7. During the first seven months of the mission, the five satellites coasted very near their injection orbit to avoid differential precession, in anticipation of a 3 month-long orbit placement maneuver period that would take the probes in their final position to the first baseline tail season. Probe placement maneuvers commenced on September 1st and ended on December 4th. Probe B's EFI booms were deployed after the main orbit maneuvers finished on November 17, and Probe A's EFI booms are expected to be deployed by January 12, 2008, following the final positioning of the replacement probe, TH-A (P5).

In the period between instrument commissioning and final orbit placement, the THEMIS probes were in a string-of-pearls configuration. In the first three months the string-of-pearls separations came about due to the natural orbit dispersions from the launch vehicle, with TH-C leading and E trailing at distances of $\sim$2 R_E from each other at apogee, while D, B, and A were at $\sim$1000 km separations between each other in the middle of the constellation. In the ensuing three months the probe positions were re-organized along their tracks, such that three middle probes were the ones equipped with EFI—deployed instruments (C, D, E) at $\sim$100 km separation, whereas B and A were leading and trailing respectively at separations of $\sim$2 R_E from each other. The orbit configuration during the coast phase is shown

V.A. Sergeev · S. Apatenkov
Institute of Physics, University of St. Petersburg, St. Petersburg, 198904, Russia

in Angelopoulos (2008), and it is further detailed in Sibeck et al. (2008) and in Frey et al. (2008).

Before leaving the tail on March 23, 2007 between 11:10 and 15:10 UT, all THEMIS instruments were ON, and data collection was commanded into "Fast Survey" mode, which enables storage and transmission of high time resolution ion distributions and fields waveforms. Two substorms were captured from a unique vantage point at the dusk sector. Simultaneous observations from the POLAR satellite captured the evolution of the first substorm in global imaging, which complemented the ground-based imaging from the THEMIS GBOs. MHD simulations of the event, using actual solar wind input, performed mapping that differs from statistical models and provides the best agreement with the data. The mapping enables interpretation of THEMIS in the context of the substorm current wedge as derived from modeling of ground-based magnetometer data. Timing using Pi2s and PiBs is consistent with that derived from imaging, and provides onset and intensification times to within a fraction of a minute. This paper provides the global context of the event and reports on a major finding from the event, namely the first simultaneous and commensurate observations of the expansion of the westward traveling surge in space and on the ground.

The optimal string-of-pearls configuration, at along-track separations of 100 s of km between probes on which EFIs were deployed and on the order of 1 R_E between the leading and trailing probes, provided an unprecedented dataset in anticipation of the first tail season. In this paper we also present THEMIS measurements at the magnetopause, which demonstrate the efficacy of multi-point observations from such a configuration in the dayside, and bespeak the quality of the THEMIS instrumentation.

2 THEMIS First Light: March 23, 2007, ~11:00 UT Substorm

At substorm onset, auroras expand poleward and westward (Akasofu 1976). In space the substorm current wedge forms at the meridian of the substorm activation (Atkinson 1967; McPherron et al. 1973). The westward portion of the auroral expansion is associated with the most intense field-aligned currents, as observed by low-altitude satellites (Hoffman et al. 1994). The field-aligned currents of the current wedge are expected to feed into and out of the expanding aurora. Specifically, the west portion of the substorm current wedge is expected to be responsible for the most intense current above the aurora, while the westward expansion of the two processes should match. These have never been observed to match each other with simultaneous measurements on the ground and in space. The expansion of the substorm current wedge in space has been studied statistically and with fortuitous conjunctions of satellites. Nagai (1982, 1991) used GOES satellite magnetometer data from geosynchronous altitudes and showed that the average current wedge expansion speed varies from <0.5 MLT h/min when the satellites are on average >1 MLT hours away from the substorm meridian, on the order of 1 MLT h/min if the satellites are located close to the substorm meridian. These expansion speeds were not well constrained near the substorm meridian due to the longitudinal separation of the GOES satellites (2 MLT h). They also indicate infinite speed at the substorm meridian due to the reversal in the expansion direction from duskward to eastward at the substorm meridian. The azimuthal expansion of the current wedge has also been studied using fortuitous conjunctions between near-Earth satellites (e.g. Lopez and Lui 1990) and found to take place in abrupt steps, comprised of localized dipolarization events of 1–2 R_E scale size.

The heated plasma within the dipolarized plasma sheet originates in a small (Ohtani et al. 1991) equatorial area ($\sim 1\ R_E^2$) and has been observed to expand radially both outward

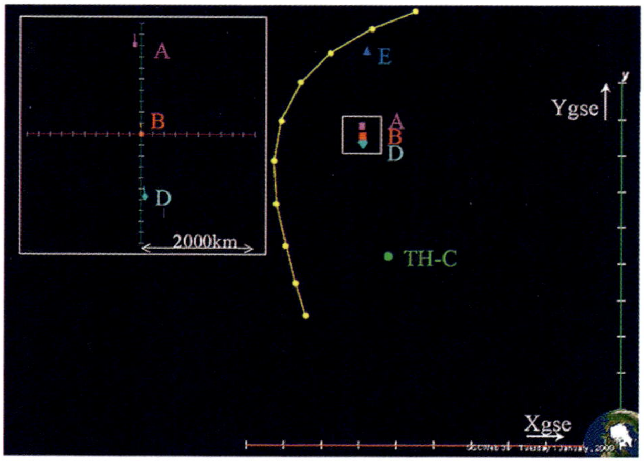

Fig. 1 THEMIS probe locations [GSE, R_E] on Mar. 23, 2007, 11:20 UT. Insert is an expanded view of the inner probes A, B, D, at a different scale, shown by the arrow within. (This product is available at http://sscweb.gsfc.nasa.gov/tipsod.) Note the typical color scheme for THEMIS probes is as shown in the insert. The *yellow curve* is an equatorial projection of an ionospheric trace on a constant (66 deg) magnetic longitude line, at 5 deg magnetic longitude increments from 285 deg to 330 deg using the Tsyganenko (1989) model

Satellite	Color	X	Y	Z
THEMIS-A (P5)		-6.947	8.808	-2.66
THEMIS-B (P1)		-6.928	8.544	-2.607
THEMIS-C (P2)		-6.227	5.202	-1.874
THEMIS-D (P3)		-6.918	8.357	-2.571
THEMIS-E (P4)		-6.817	10.89	-3.034

(Jacquey et al. 1991; Ohtani et al. 1992) and inward (Ohtani 1998) at ∼200 km/s. It has also been observed, on occasion, accompanied by near-Earth fast Earthward plasma flows (Fairfield et al. 1998; Angelopoulos et al. 1999). The relevance of these flows in the buildup of the magnetic flux, which comprises the substorm current wedge, differs in two prominent models of substorms (Lui 1996; Baker et al. 1996). However, the presence of such localized, transport-efficient fast flows in the near-Earth tail, especially further downtail of 10 R_E is well documented (Angelopoulos et al. 1994, 1997; Sergeev et al. 1996b). Those fast Earthward plasma flows are most often quite localized (e.g. Nakamura et al. 2004) and therefore the question arises how such localized flows participate in replenishing the magnetic flux eroded from the day side during the course of the substorm, and how they partake in the azimuthal expansion of the near-Earth dipolarization which extends over many hours of local time.

One attempt to explain the localized (1–2 R_E) nature of the fast flows and the extended nature of the substorm current wedge was made in Angelopoulos et al. (1996), in which the incoming flux transport is gradually diverted around the dipolar region, and further fast flows are layered on the outer edges of the substorm current wedge. In that picture, a westward expansion of the substorm current wedge would be accompanied by Earthward and duskward flow, piling additional magnetic flux further to the west. While consistent with single satellite observations available prior to THEMIS, this picture has not been verified by multi-satellite observations. THEMIS's string-of-pearls configuration available from the coast phase of the mission enables such observations for the first time. In this paper, we will demonstrate that the expanding westward traveling surge maps in space to a westward propagating dipolarization, which is accompanied by fast Earthward flows near its westward edge.

On March 23, 2007, THEMIS was in an ideal position to observe the westward propagation of a substorm (Fig. 1). The probes were on the inbound leg of their orbit, in the pre-midnight sector. The upstream solar wind conditions measured by Cluster, as well as by WIND and ACE when time-shifted to the magnetopause location, show that the interplanetary field was southward for an hour prior to 11:08 UT ($B_z \sim -5$ nT). The IMF turned momentarily northward at around 11:08 UT at which time the solar wind density, previously around $\sim 10/cc$, abruptly increased by 80%, though the solar wind speed remained steady at ~350 km/s. An overview of the conditions in the nightside magnetosphere is shown in Fig. 2. That figure shows data from TH-B in the standard "overview" plot of the THEMIS data distribution. (Such plots are available routinely on the web at 6 h and 24 h intervals at http://themis.ssl.berkeley.edu, and are automatically produced for arbitrary times using the machine-independent Graphical User Interface of the THEMIS software distribution.)

As is evidenced in the THEMIS pseudo-AE index shown in Fig. 2, the THEMIS GBOs recorded two main substorms around 11:10 and 14:10 UT. Probe TH-B observed an intensification of the energetic particle fluxes in both instances. THEMIS probes were commanded into time-based burst collection during this early period of the mission (i.e., not using on-board triggers); they were thus bursting simultaneously but not necessarily related to local activity. However, all probes were in fast survey mode, during which spin resolution particle data in high angular resolution were available. This allowed computation of accurate high-time resolution moments of the ion distributions.

During this period, the POLAR satellite collected images of the southern auroral oval and observed the entire sequence of the first substorm. Figure 3 shows a compendium of images for this event from the UVI (Torr et al. 1995) and the VIS (Frank et al. 1995) imagers on-board that satellite. Activations occurred at ~10:54 UT, ~11:10 UT and 11:18 UT, occurring at 21:00 MLT, 01:00 MLT and 23:00 MLT respectively. It is evident from the MLT-UT spectrogram of Fig. 3 that during the 11:10 UT onset the pre-cursor activation site, at 21:00 MLT, also intensified. Major substorm intensification expanded poleward very rapidly. By taking the difference between consecutive images from the POLAR UVI instrument, specifically at 11:18:26 UT and 11:19:03 UT (see Fig. 3), we determine the longitudinal expansion speed to be approximately 1 MLT/min. By 19:40 UT the intensification had expanded to ~21:30 MLT, and did not progress as much or as fast in the ensuing 3.5 minutes as evidenced by the last image in the series, taken at 11:23:21 UT. Although UVI's field of view did not extend much to the west of the 21:00 MLT meridian, the VIS global images at 11:18:45 and 11:23:21 UT do confirm the presence of a continuous auroral oval in the early pre-midnight sector (in fact, continuing all the way through dusk to the day side) and are in agreement with UVI regarding the longitude of the westernmost expansion of the substorm aurora. Since western Alaska midnight is at approximately 11:30 UT, the primary 11:10 UT onset took place 1 hour to the East of Alaska, within the field of view of station Inuvik, and the major intensification over the west coast of Alaska.

On the ground, the THEMIS GBOs over Alaska and western Canada are able to provide a partial view of the auroral intensification. Figure 4 shows a mosaic from the THEMIS All Sky Imager data in Alaska and western Canada. The main onset over Inuvik is evident in that imager, but the activity did not expand from it over Alaska. Smoke from a nearby factory at the Kiana site (since corrected), obscures the field of view at times over western Alaska, but does not prevent a clear identification of a newly formed arc which moved from over the ocean on the west to over western Alaska at 11:18:42 UT. Based on POLAR images, the onset occurred very near the west coast of Alaska and thus its timing was captured quite accurately (within seconds) by the Kiana GBO.

A comprehensive summary of the timing signatures on this event, using Pi2 pulsations from THEMIS and ancillary ground stations is presented by Keiling et al. (2008). Observations of field-aligned currents at the ionosphere and their relation to the THEMIS observations in space is presented in Strangeway (2008).

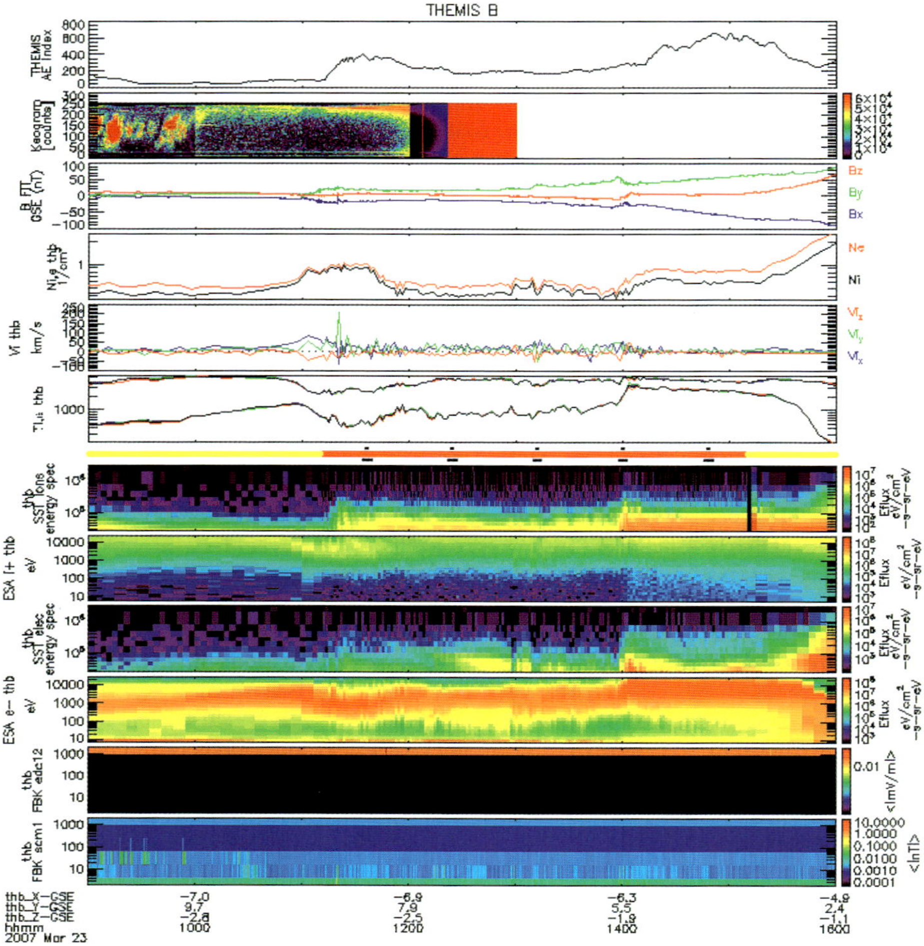

Fig. 2 Standard overview plot from THEMIS probe B, on Mar. 23, 2007, 09-16 UT. (**a**) THEMIS pseudo-AE index, generated from the THEMIS GBOs (Mende et al. 2008), indicating a substorm onset around 11:10 UT and another onset at around 14:10 UT; (**b**) Keogram (north-south stripe of an All Sky Imager camera) from Athabasca GBO (Harris et al. 2008); (**c**) Magnetic field (Auster et al. 2008) in GSE coordinates, measured on TH-B; (**d**) Ion and electron partial densities from the ESA instrument (Carlson et al. 2008; McFadden et al. 2008) which measures the thermal ions and electrons. Note that the electron density computation is affected by the presence of photoelectrons (artificial); this can be removed by special processing, or routinely after the EFI measurements of the spacecraft potential became available; (**e**) Ion partial flow velocity from the ESA instrument; (**f**) Ion and electron partial temperatures, from the ESA instrument (shown are the three components of the temperature matrix trace, with the lower trace being the electron temperature); (**g**) Bar indicating the data collection mode. *Yellow*: Slow Survey, *Red*: Fast Survey, *Black line underneath*: Particle Burst, *Black line above*: Wave Burst; (**h**) Ion omni-directional spectrum from the SST instrument (Larson et al. 2008); (**i**) Ion omni-directional spectrum from the ESA instrument; (**j**) and (**k**) electron spectra from the SST and ESA instruments; (**j**) on-board computed wave power spectrum at discrete filter-banks from the EFI instrument (Bonnell et al. 2008; Cully et al. 2008)—with the EFI booms stowed this quantity shows no real data on that day; (**k**) on-board computed wave power spectrum from the SCM instrument (Roux et al. 2008). Satellite position in GSE coordinates is shown at the *bottom*

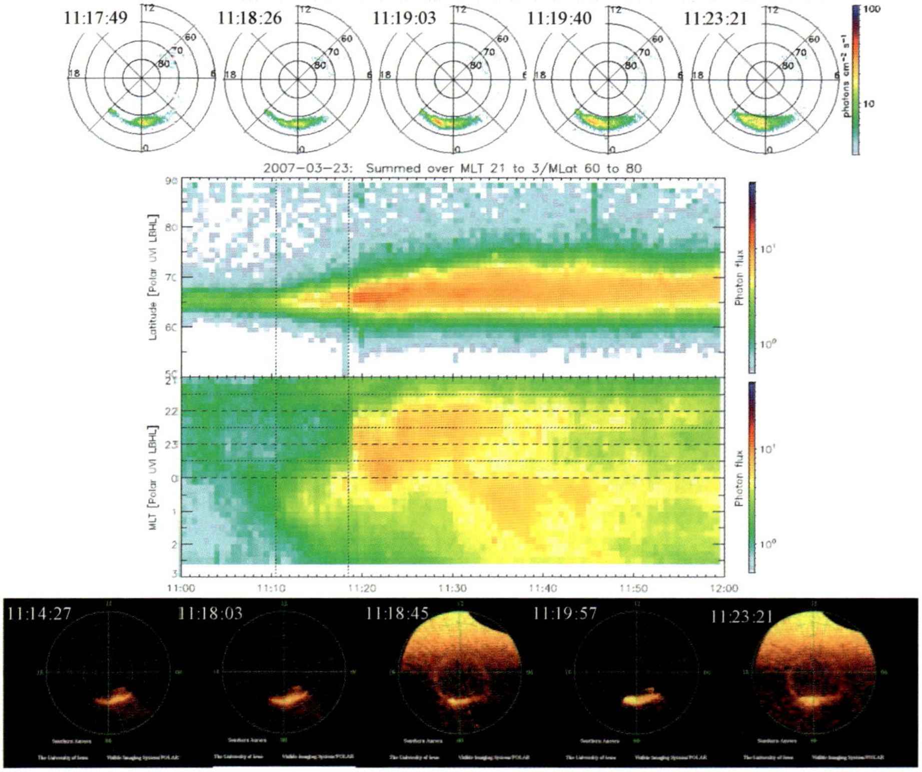

Fig. 3 POLAR UVI images (*top row*) and VIS images (*bottom row*) from select times during the course of the first substorm on March 23, 2007, projected onto a geomagnetic longitude/latitude map of the Northern Hemisphere. The *middle panel top row* shows a latitude-UT spectrogram of the intensity integrated over the pre-midnight sector, while the *middle panel bottom row* shows an MLT-UT spectrogram of the intensity integrated over auroral latitudes. The times shown next to the images correspond to the beginning times of the image collection. It is evident from the *middle panels* that a small auroral activation at 21:00 MLT at 10:54 UT preceded the major substorm activity. It was followed by a main onset at 01:00 MLT at 11:10 UT, and a major intensification at ∼23:00 MLT, at 11:18:45 UT

Mapping the THEMIS satellites with standard (average magnetic field) models presents a problem that is depicted in Fig. 5. The left portion of the image shows standard mapping using the Tsyganenko 2001 (Tsyganenko 2002a, 2002b) model, by varying the level of geomagnetic activity (Dst, AE) to determine the range of the footpoint locations and hence get an estimate of the mapping uncertainty. According to this mapping, all THEMIS satellites have footpoints about 2 h of MLT to the west of Alaska, and are well outside the westernmost location of the auroral activation at 21.5 MLT. Considering that all satellites, such as TH-B in Fig. 2, saw significant energetic particle flux increases, dipolarization, and flows, indicating that the spacecraft were within the meridian of the activity, the above mapping is suspect. One reason for this is the presence of a significant East-West interplanetary magnetic field that can penetrate inside the magnetosphere. Another reason is the distortion of the average magnetosphere during the late growth phase, which may be missing from the above average models. To represent those effects properly we relied on mapping using GGCM simulation model fields just prior to the onset occurring in the simulation. This is

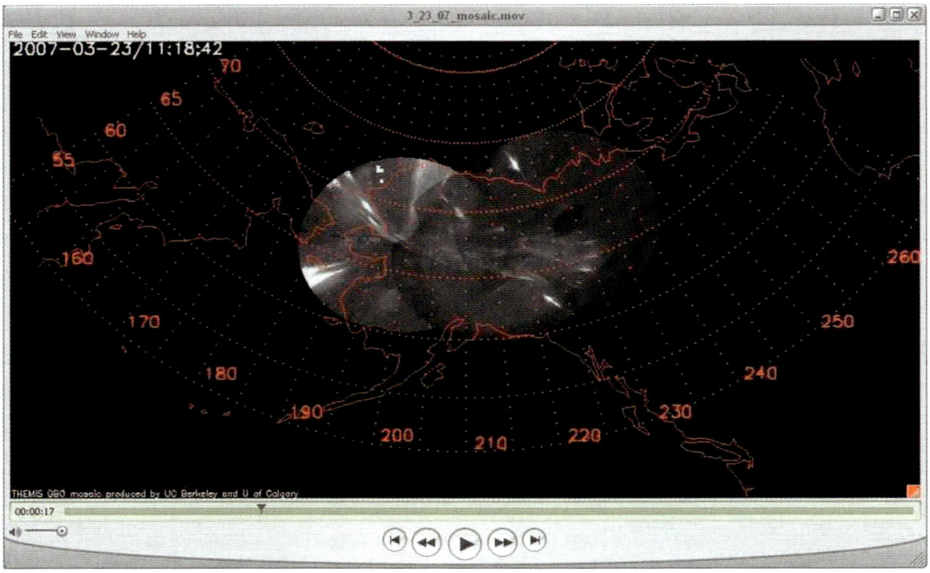

Fig. 4 THEMIS GBO/ASI mosaic of auroral images from stations: Kiana, Fort Yukon and Inuvik, at 11:18:42 UT, i.e. at the time of the substorm main intensification, as determined by the Kiana ASI. *Circles* indicate non-geophysical emissions, mainly due to smoke from a factory near the Kiana station. (The smoke contamination of the station has since been fixed.) The first indication of an auroral brightening over Kiana is shown with the *arrow*

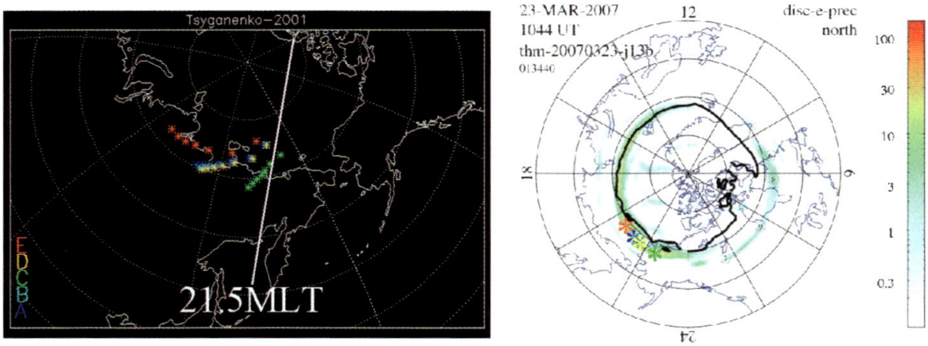

Fig. 5 Field line mapping of the THEMIS satellites C, D, B, A, E (from east to west, shown in *Green*, *Yellow*, *Ciel*, *Blue* and *Red*). *Left*: Mapping using the Tsyganenko 2001 model under various activity levels. *Right*: Mapping using realistic fields from GGCM simulation. The *right model* does a better job explaining the good relationship between the THEMIS measurements in space and the observed activity on the ground

accomplished as described in Raeder et al. (2008). Specifically, in response to the realistic solar wind magnetic field, the simulation onset took place at 11:00 UT, and thus fields just prior to it were used to map the THEMIS satellites. This mapping resulted in footpoints that were just to the west of Alaska, i.e., very near the substorm meridian, and is thus in better agreement with observations.

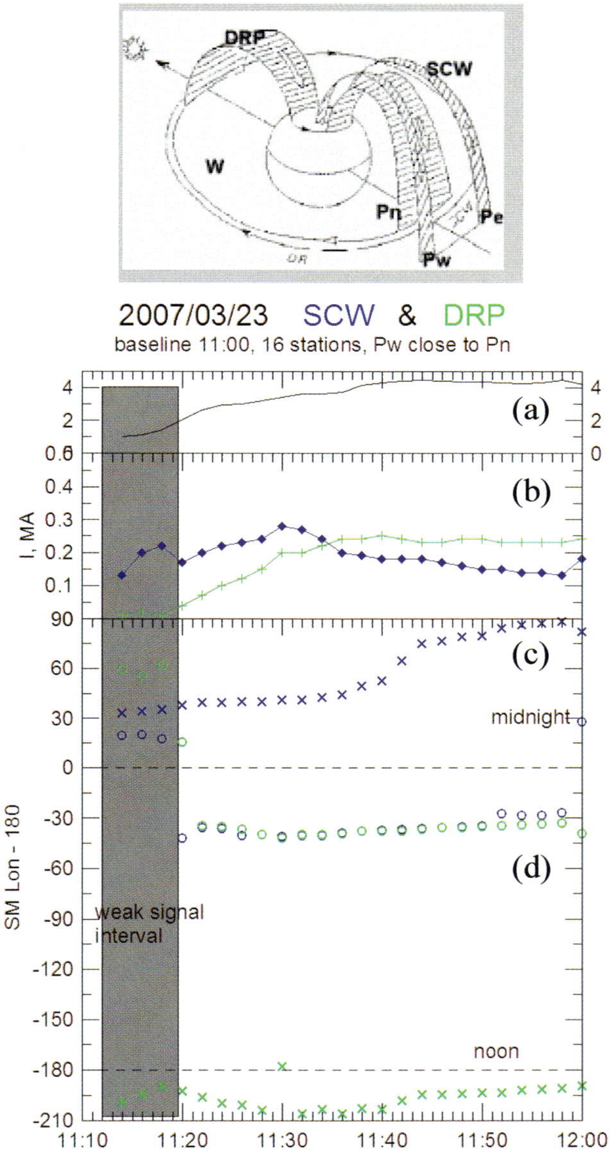

Fig. 6 Modeling mid-latitude ground magnetometer signatures to determine the central meridian, width and intensity of near-Earth space currents. *Top*: Pictorial representation of the model used (composed of line currents or series of line currents). *Bottom*: Modeled intensity (MA) and longitude (degrees) of the current system components as a function of time past 11:10 UT, specifically: (**a**) standard deviation of observations versus model disturbances; (**b**) SCW (*blue*) and DRP (*green*) current intensity; (**c**) Longitude of downward (*crosses*) and upward (*circles*) field-aligned currents, in SM-180 coordinates (i.e., Solar Magnetospheric longitude minus 180 deg)

Further evidence for a rapid evolution of the substorm current wedge comes from modeling of the ground magnetometer signals. Mid-latitude data from the intermagnet network were used in Fig. 6 to fit model parameters of the ring current disturbance (DR), the partial ring current disturbance (DRP) and the substorm current wedge (SCW) to determine their magnitudes and locations as a function of time (Horning et al. 1974; Sergeev et al. 1996b). The results show that starting around 11:14 UT and until 11:18 UT the substorm current wedge was localized around 01:30 MLT and had a width of 1.5 h of MLT. This is more or less consistent with the images from POLAR/UVI, which showed onset at

01:00 MLT. At 11:18 UT the westward portion of the substorm current wedge expanded rapidly to −45 deg longitude within 2 minutes.

As we have learned from the UVI images and the GBO/ASI images, the 11:18:45 UT intensification was an intensification distinct from the 11:10 UT onset, and appeared approximately at 23:00 MLT, i.e. at −15 deg longitude, and expanded both eastward and westward. The above technique cannot distinguish between the longitudes of the main onset and its intensification. If, however, we assume the location of the new activation to be at −15 deg, then the westward expansion speed of the SCW from the above analysis is about 2 MLT h/2 min, consistent with the expansion observed in the UVI images.

When mapped geometrically in space at the location of THEMIS, the aforementioned westward expansion speed on the ground translates to a 280 km/s azimuthal expansion at the location of TH-B (at a radius of 11.25 R_E). Due to the flaring of the magnetic field towards the flanks, standard Tsyganenko model projections relate a pure azimuthal ionospheric motion (along a constant magnetic latitude), to a westward motion of an equatorial footpoint in space, with a non-trivial radial component of a variable direction. The radial component of the velocity depends on mapping model details and activity level. Figure 1 shows an example of a Tsyganenko (1989) model equatorial mapping points starting at 100 km in the ionosphere at 66 deg magnetic latitude, and every 5 deg in longitude, i.e., starting from 22 MLT to the west, until 19 MLT. Such an exercise suggests that a speed of 1 MLT/min in the ionosphere corresponds to a propagation speed of 300 km/s, 350 km/s and 200 km/s predominantly in the Y_{GSM} direction, and is around 250 km/s between the location of TH-D, B, A and E satellites.

To determine if the projected speed matches the observed propagation speed in space we turn our attention to the THEMIS satellite observations. An expanded view of the data from TH-B around the time of the intensification is shown in Fig. 7. It is evident that TH-B, south of the equator, measured a sharp dipolarization at around 19:30 UT, accompanied by a reduction in field magnitude (which we interpret as entry into the high beta plasma sheet, closer to the neutral sheet) and a bipolar By signature superimposed on a permanent change in By which is evidence of a field-aligned current pair superimposed on a uni-directional current. These are classical signatures of plasma sheet recovery typically observed at late substorm expansion or substorm recovery (e.g. Pytte et al. 1976 JGR); however, in our case the satellites were already in the outer layers of a dense and hot plasma sheet.

One of the salient features important for understanding the buildup of the SCW is the fast Earthward flow pulse, seen almost simultaneously with the magnetic field depolarization, heating and slight density ramp-up and ramp-down. These are classical signatures of flow bursts, i.e., flow pulses observed within bursty bulk flow events (Angelopoulos et al. 1994; Sergeev et al. 1996a, 1996b). The flow burst is Earthward and slightly duskward and is superimposed on a long period flow wave (2.5 min period). It is preceded by an energetic particle flux enhancement as seen in panels (d) and (e). Spin phase is equivalent to GSE longitude in panel (d). Particles moving at 120 deg longitude are field-aligned (0 deg pitch angle) and particles moving at −60 deg are field-opposed (180 deg pitch angle). Particles at +30 deg have a 90 deg pitch angle and a gyro-velocity that is duskward. The flux enhancement is seen first in the duskward moving particles. It is followed, about a half a minute later, by enhancements in the dawnward particle fluxes, which denote complete immersion of the satellite into the heated plasma sheet plasma (100 keV gyro-centers on both sides of the satellite). It is evident from this panel that the hot plasma indeed expanded outward, i.e., southward, or lobe-ward over the satellite.

With the availability of more than one look directions and several energies it is possible to employ the technique of remote sensing (e.g. Daly et al. 1984; Kettmann and Daly 1988)

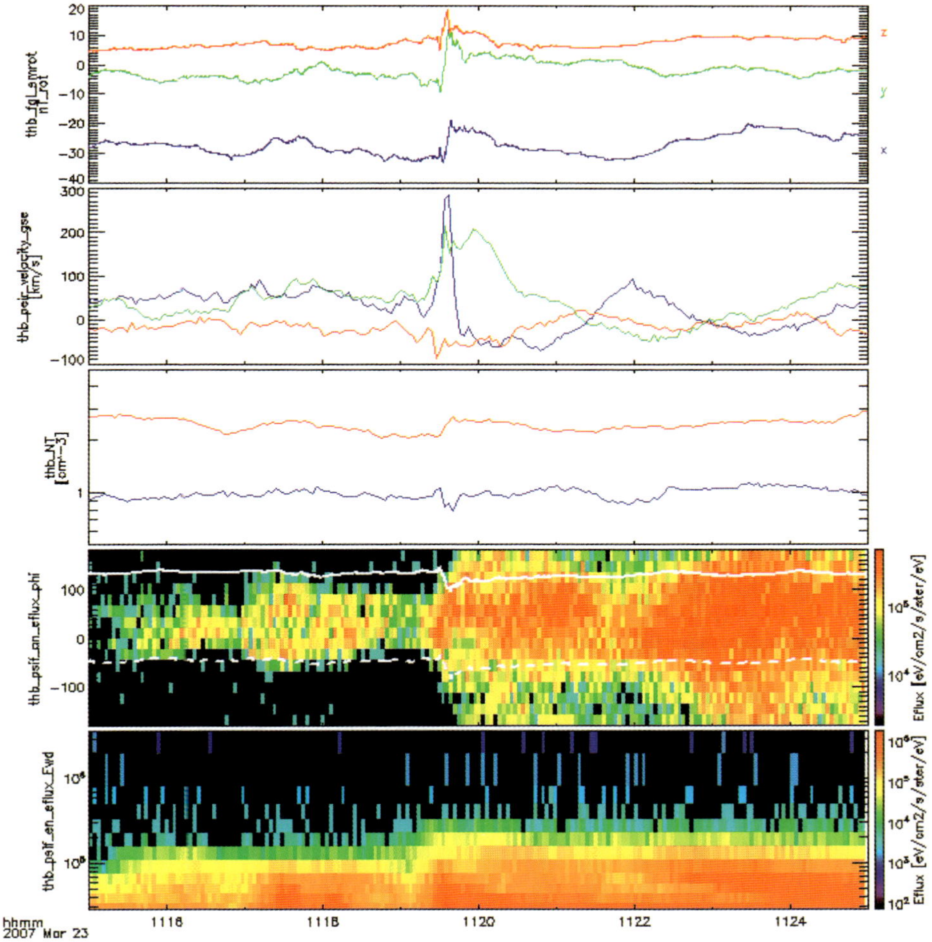

Fig. 7 Detailed signatures of the substorm intensification on TH-B. (**a**) FGM measurements, $4S/s$, SM coordinates, rotated about Z_{SM} such that X points towards Earth, and Y azimuthally westward (X, Y, Z correspond to *blue*, *green* and *red*); (**b**) ESA ion flow velocity in GSE coordinates; (**c**) Ion partial average temperature (*red*) and density (*blue*); (**d**) azimuth spectrogram of 100 keV SST ion differential energy flux, with the magnetic field azimuth in the parallel and anti-parallel directions to the field line (*solid* and *dashed white traces* respectively); (**e**) energy spectrogram of Earthward SST ion differential energy flux (±45 deg in azimuth and ±43 deg in elevation). The satellite spins with spin axis along ecliptic north; the magnetic field, pointing outward from Earth, is mostly on the spin plane (within 15 deg) and therefore the particle azimuths relative to the magnetic field azimuth in panel (**d**) are a good approximation of a directional pitch angle—a positive or negative direction relative to the magnetic field, corresponding to the particle motion in gyrophase

to determine the orientation and speed of the approaching hot energetic particle boundary. The technique relies on the timing of the appearance of the hot fluxes at various gyro-center locations.

We looked at arrival times of particles at four specific energies (40 keV, 100 keV, 150 keV and 300 keV). For simplicity we used only 90 deg pitch angles (i.e. approximately −60 deg and +30 deg particle azimuths). Those particles are measured in eight gyro-velocity directions, obtained from the four SST detector mounting locations (±52 deg, ±25 deg relative to the spin plane) times the two look directions possible normal to the magnetic field. The resultant gyro-velocity directions are $\delta = \pm 155, \pm 128, \pm 52, \pm 25$ measured from the Y-axis of an orthogonal right-handed coordinate system with Z that contains the spin direction and X along the field direction. Therefore, X is on the spin plane along the magnetic field azimuth, Z is along the spin axis along the ecliptic normal, and Y is 90 deg away from X, about 30 degrees away from Eastward. The resultant arrival times were modeled assuming a surface motion normal to the magnetic field of arbitrary orientation, ε, relative to the Y-axis, and the variance chi-square (χ^2) of the data was plotted as a function of arrival times, to determine the orientation that minimizes that variance. The velocity of the boundary is then determined for that specific orientation.

We used the singular value decomposition method (Press et al. 1989) for the least squares fit and obtained χ^2 values shown, for probe B, in Fig. 8. It is evident that the boundary orientation is well determined, since a deep minimum in chi squared is seen. The SST detector resolution in near-ecliptic elevations (±25 deg and ±52 deg) bespeaks a boundary orientation accuracy that is on the order of 10 deg or so.

The variance is minimized for a boundary angle of 280 deg, which corresponds to a boundary velocity of ~70 km/s, as is also shown in Fig. 8. Similar results were obtained from probes B and A and are tabulated in Table 1. They show that the motion of the active region at the outer layers of the plasma sheet where probes D, B and A were located, was lobe-ward and (possibly) slightly Eastward, in agreement with the convex shape of the plasma sheet surface near the flanks.

The differences between the arrival times of the boundary at the various spacecraft were $t_{\rm DB} = 4.2$ s and $t_{\rm BA} = 6.6$ s. The distances of the satellites on a coordinate system that has X along the magnetic field and Z containing the spin axis is approximately $dZ_{\rm DB} = 420$ km and $dZ_{\rm BA} = 600$ km, whereas if the magnetic field inclination is not accounted for in that distance (i.e. if the X-axis is along the magnetic field azimuth but on the spin plane) then $dZ_{\rm DB} = 220$ km and $dZ_{\rm BA} = 330$ km. The resultant expansion speeds are consistent between the two pairs and vary with the above assumption between: $V_{\rm DB} \sim V_{\rm BA} = 98$ km/s and $V_{\rm DB} \sim V_{\rm BA} = 49$ km/s. These speeds are consistent with the expansion speed determined from remote sensing.

Timing of the arrivals of the other signatures at the inner three spacecraft is also consistent with the expansion of the heated plasma from one probe to another. Figure 9 shows a summary of the time series flow velocity, magnetic field and energetic particle flux measurements on TH-D, B, A and E. The timing from the remote sensing technique is denoted in dashed lines. The results are consolidated in Table 2.

TH-E, further to the west, was further away from the nominal neutral sheet, but very close to the *actual* neutral sheet, as evidenced by the sign changes in Bx prior to the occurrence of the fast flow. Due to TH-E's proximity to the neutral sheet, TH-E maps onto the equatorial plane locally, contrary to TH-D, B, and A, whose equatorial footpoints likely map much further downtail. The energetic particles, which were measured at the location of TH-D, B and A at 11:19:30 UT and were likely injected near the neutral sheet at an earlier time, travel westward and were seen as a flux enhancement at 11:19:20 UT at TH-E, about 30 seconds prior to the arrival of the flow burst and the dipolarization there. The flux at all energies (not shown) increased gradually at TH-E, with no sharp injection, preventing the opportunity to apply remote sensing of an expanding boundary there. At the same time By perturbations prior to the dipolarization at TH-B complicate detection of a sharp field-aligned current boundary there. However, both dipolarization and fast flow were detected at TH-E and their timing is included in Table 2.

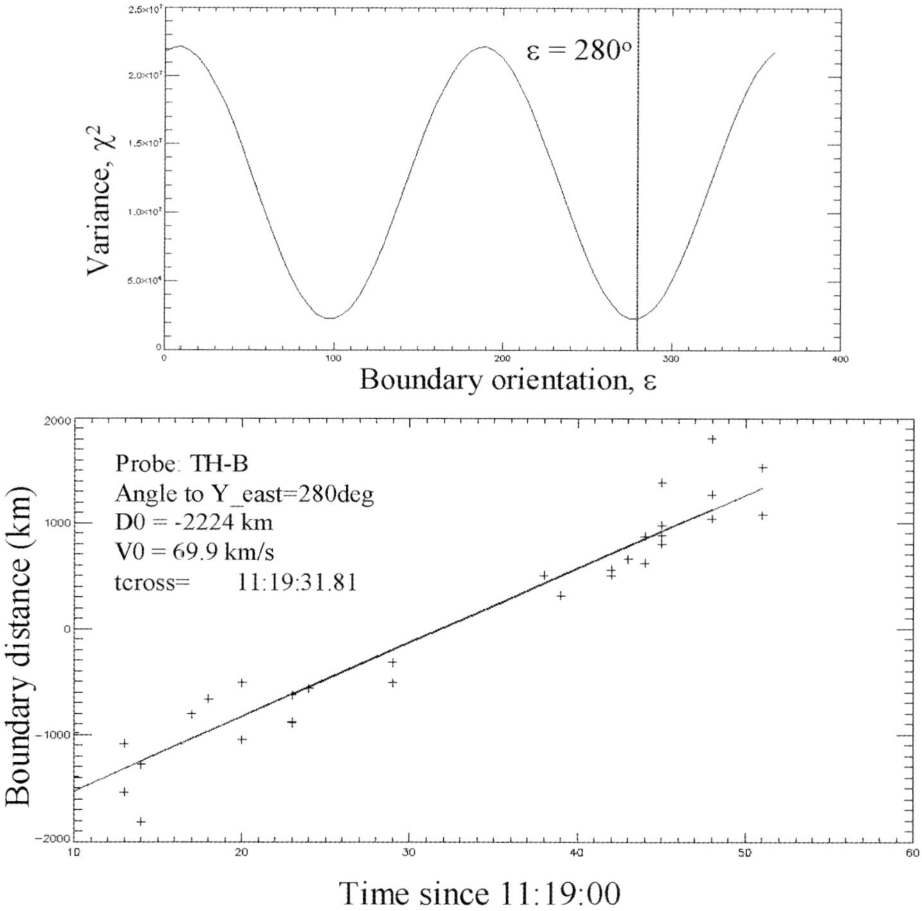

Fig. 8 Remote sensing of the approaching plasma sheet hot plasma during the 11:18:42 UT substorm onset on THEMIS satellite TH-B. The flux enhancement times of the 90° pitch angle particles of various gyro-phases and energies was modeled assuming an approaching boundary at arbitrary orientation relative to the Eastward direction, to determine the orientation of minimum variance. That orientation was then used to determine the optimal boundary speed and boundary crossing time. *Top*: variance of the arrival times for various boundary orientations. *Bottom*: boundary distances corresponding to the timing measurements for appropriate gyro-radius, gyro-velocity direction and optimal boundary direction versus time (in seconds relative to 11:19:00 UT). Negative or positive boundary distances correspond to an approaching or receding boundary respectively, assuming a positive speed

Table 1 Results of remote sensing analysis on the inner probes

	t_{cross}	V [km/s]	ε [deg]
D	11:19:27.6	75	270
B	11:19:31.8	70	280
A	11:19:38.4	80	275

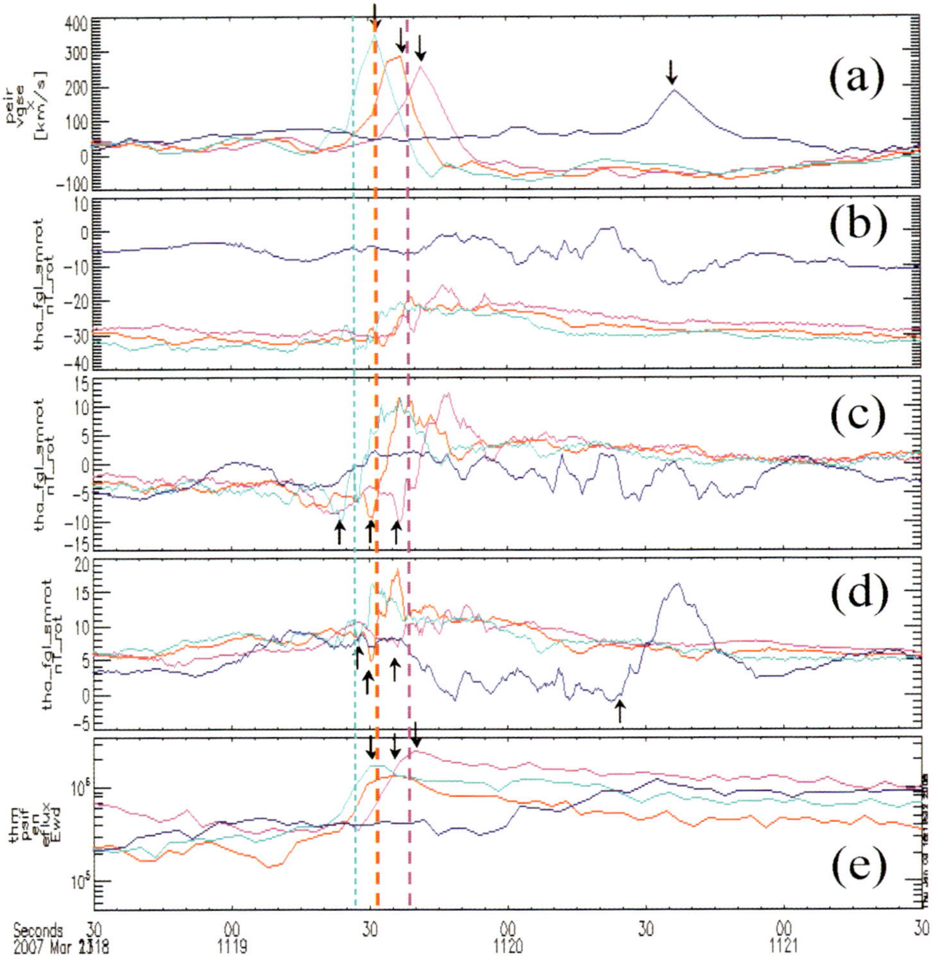

Fig. 9 Timing of particles and fields signatures at the various spacecraft. THD, B, A and E are shown in *Ciel*, *Red*, *Magenta* and *Blue*, respectively, per the standard THEMIS color scheme, as in Fig. 1. The times are also tabulated in Table 2. (**a**) X_{GSE} component of the flow velocity from the ESA instrument; (**b**), (**c**) and (**d**) X, Y, Z components of the magnetic field, in 'rotated' Solar Magnetospheric coordinates, where the X and Y axes have been rotated about Z_{SM} coordinate, such that X points towards Earth and Y is azimuthally westward; (**e**) the differential energy flux from the 40 keV channel of the SST ions. TH-E was near the neutral sheet at the time, and saw no sharp signature in B_y and in energetic particles, but did experience the fast flow and the associated dipolarization

Table 2 Summary of timing results (mm:ss) on probes TH-D, B, A and E

	D	B	A	E	t_{DB} [sec]	t_{BA} [sec]	t_{DE} [sec]
Vx	19:31	19:36	19:41	20:36	5	5	65
By	19:24	19:30	19:36	N/A	6	6	N/A
Bz	19:22	19:30	19:36	20:23	8	6	61
SST_{eflux}	19:29	19:35	19:42	N/A	6	7	N/A
RS	19:27.6	19:31.8	19:38.4	N/A	4.2	6.6	N/A

Timing of all signatures on TH-D, B and A consistently points toward a picture of outward local plasma sheet expansion, encompassing flows, depolarization and field-aligned current sheets. The activity boundary is layered horizontally. However, timing of the flow and dipolarization arrival to TH-E suggests a different process. The D-E spacecraft separation along Y is $\sim$2.5 R_E and the propagation delay using the timing tabulated in Table 2 is $\sim$250 km/s (we used TH-D since of the three, TH-D, B and A, TH-D is closest to the neutral sheet). This speed is much faster than the speed of the lobeward expansion of the boundary. Our interpretation of the rapid Westward evolution of the Earthward flow burst and dipolarization is that these phenomena are mapped to the westward traveling surge observed expanding westward on the ground. The observations of TH-D, B, A demonstrate the link between the evolving Earthward flow channel, the field-aligned currents and particle injections. The phenomena may have started earlier near the neutral sheet, but not seen at its outer layers where the D, B, A probes were located. Thus the time delay between TH-D and TH-E may be an upper bound on the expansion velocity. Nevertheless, that velocity is comparable to the expected expansion velocity from ground observations and our simple mapping arguments.

In summary, using THEMIS observations of a substorm on March 23, 2007, at 11:10 UT, we have shown that the westward traveling surge propagates on the ground at a speed that is commensurate with the expansion of the substorm current wedge in space. The speed of the expansion was observed to be $\sim$1 MLT/min, both on the ground and in space. This speed is much faster than could have been substantiated using the widely separated GOES satellites, but consistent with expectations from statistical results by Nagai (1982, 1991) that the expansion speed can be quite large when observed close to the activation meridian. To our knowledge these are the first clear simultaneous observations of the expansion of the westward traveling surge on the ground and in space. They show that the westward expansion of the substorm current wedge is due to the impulsive buildup of dipolarized flux, accompanied by Earthward flow bursts.

Further analysis of the timing on this event can be found in Liu et al. (2008). Further analysis of the low frequency compressional oscillations and particle perturbations can be found in Keiling et al. (2008). The variability of the inner edge of the plasma sheet is further expanded upon in Runov et al. (2008).

When the plasma sheet is observed further from the neutral sheet, close to or at the plasma sheet boundary, the flow burst, magnetic dipolarization and the heated plasma are observed to move lobe-ward, consistent with a local expansion of the plasma sheet. A single satellite is sufficient to determine the outward expansion speed and boundary orientation of the heated plasma. This is proposed for the primary phase of the THEMIS mission, when the inner satellites (P3, 4, 5) will be located at lower inclinations within the plasma sheet, i.e., nominally within $<$2 R_E from the neutral sheet. The lateral motion of a planar surface can then be modeled as a field-aligned current, and the spatial extent, latitudinal width and magnitude of that current can be inferred. When independent measurements of the field-aligned current are made using magnetometer data at appropriately positioned spacecraft separations, the planar assumption can be further validated, and—using high time resolution magnetometer measurements—the microstructure of these currents, including their spatial scales, can be determined and placed in a global context. This will be tested out at 1 R_E separation scales during the second-year nominal mission by taking advantage of the cross-sheet separation of P5 and P3/P4 satellites. It is well known that the westward traveling surge is composed of multiple current systems that map to scales smaller than 1 R_E to the equatorial magnetosphere (Hoffman et al. 1994). Plans for a more comprehensive study of smaller scale current filaments within the substorm current wedge will have to await an extended mission, with a tighter clustering of the inner THEMIS probes.

3 THEMIS Observations at the Day Side

THEMIS traversed the day side magnetosphere in a string-of-pearls configuration, which allowed it to determine the extent and motion of flux transfer events (FTEs, Russell and Elphic 1978, 1979) and dissect their structure. Figure 10 shows the spacecraft locations for such an encounter on July 12, 2007. During this inbound crossing, TH-A, the trailing probe, was in the magnetopause. Its data are shown in Fig. 11.

The magnetic field and velocity are in the magnetopause coordinates (LMN), where N is normal to a model magnetopause (Shue et al. 1998), with solar wind parameters obtained from WIND; L is normal to N containing the model field inside the magnetosphere (roughly Northward), and M is orthogonal to the other two directions; in our case predominantly duskward. The bipolar signature of a transient encounter with an FTE is evident at 07:04 UT in B_N (red), whereas B_L shows that the core of the FTE was pointing mostly southward.

As the FTE was approaching the spacecraft, the boundary moved outward and then inward, as evidenced in the bipolar signature in V_N. The ESA ion and electron spectra show a

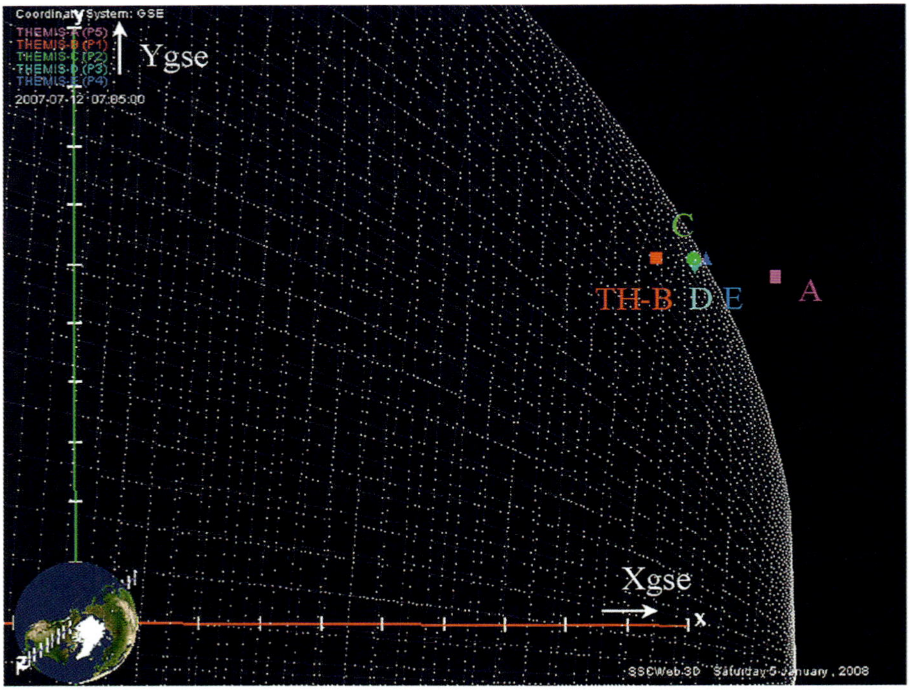

Satellite	Color	X	Y	Z
THEMIS-A (P5)		11.114	5.582	-3.39
THEMIS-B (P1)		9.214	5.928	-2.94
THEMIS-C (P2)		9.803	5.899	-3.101
THEMIS-D (P3)		9.827	5.849	-3.093
THEMIS-E (P4)		10.011	5.907	-3.174

Fig. 10 THEMIS probe locations [GSE, R_E] on July 12, 2007, 07:05 UT. Satellite locations are also tabulated in GSE coordinates. Magnetopause location for a solar wind dynamic pressure of 1.3 nPa is also shown, consistent with the observed location

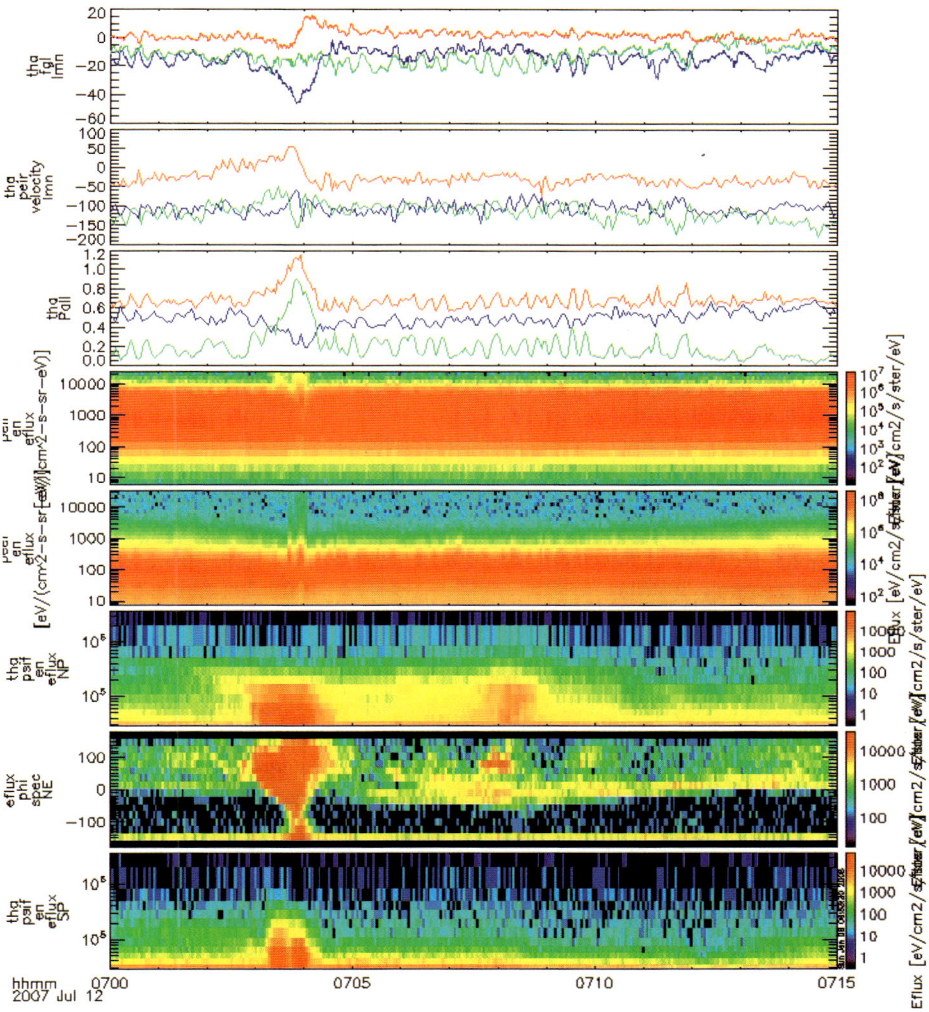

Fig. 11 TH-A observations July 12, 2007, 07:00–07:15 UT. The *dashed vertical lines* indicate FTE observations on TH-C (in Fig. 13). (**a**) Magnetic field (nT) in LMN coordinates: L, M, N are *blue*, *green* and *red lines*; (**b**) Partial velocity (km/s) from the ESA instrument in LMN coordinates; (**c**) Magnetic (*blue*), plasma (*green*) and total (*red*) pressure (nPa); (**d**) ESA ion omnidirectional differential energy flux (eV/(cm^2 s str eV)) (a.k.a. eflux); (**e**) Same as (**d**) but for ESA electrons; (**f**) energy spectrogram of the SST ion eflux from the North-Polar (NP) detector, looking at 52 deg above the spin plane; (**g**) angular (azimuth) spectrogram of SST ion eflux from the North Equatorial detector, looking at 25 deg above the spin plane; (**h**) energy spectrogram of the SST ion eflux from the South-Polar (SP) detector, looking at 52 deg below the spin plane. It is evident that the bipolar signature on TH-A (this plot) was centered slightly earlier (about 10 s) than the *dashed line* at 07:04 UT. The dashed line at 07:09 UT was an FTE observed on the other probes but only remotely sensed in TH-A energetic particles due to their finite gyroradius

transient crossing of the boundary layer, lasting about 20 s. The total plasma and magnetic pressure show a peak at the center of the FTE, which is evidence of the tension forces within the FTE due to the field line curvature there.

At the time prior to the FTE encounter, the magnetic field angle was 60 deg below the spin plane and +120 deg in azimuth, i.e., anti-Sunward and roughly tangent to the magnetopause. After the FTE passage the field direction retained the same azimuth, but the elevation changed to about 20 deg below the spin plane. In either case, the energetic particle data from the SST particles moving in the North Polar (NP) direction (i.e., measured by the detector looking in the south-polar direction). The particle velocities, centered at 52 deg above the spin plane, have roughly 90° pitch angles, with gyro-centers that were on the Earthward side of the spacecraft. The energy spectra of the NP particles show clearly the arrival of the FTE ahead of its magnetic signature, remotely sensing its arrival due to the finite gyroradius effect of the energetic particles. Higher energies were detected first, as they correspond to larger gyro-radii. The dispersion is clear in the case of the 07:04 UT FTE. Similar energy dispersion signatures are seen at 07:09 UT, but there is no accompanying magnetic signature of an FTE at that time at TH-A.

The SST particles moving in the North-Equatorial direction correspond to a larger solid angle portion of the nearly field-aligned and nearly field-opposed particles, and thus a wider range of pitch-angles at that time. That direction was partially affected by sunlight at the time, as evidenced by a horizontal bar at nearly anti-sunward particle directions; this was expected because the open detector is looking straight into the sun during a portion of the spin. The quick recovery of the electronics, which salvages most look directions, bespeaks a nominally operating instrument. The angular spectrogram of that detector is shown in Fig. 11, panel g. It is evident that for several minutes prior to the FTE arrival, the SST detector was measuring higher flux of 90° spin phase particles, coming from the magnetosphere. Just prior to the arrival of the FTE the 90 deg pitch angles increased in flux, and isotropized only inside the FTE. The SST particles moving in the South-Polar direction, 52 deg below the spin plane, correspond to either field-aligned or 90° pitch angles of gyro-centers that are predominantly on the sunward side of the spacecraft. The energy spectrum of the South-Polar moving particles is shown in the panel below. It is evident that the particle fluxes there are much reduced relative to North-Polar moving particles.

From the time delay of 55 s in this isotropization and from the gyroradius of 100 keV particles in the observed 28 nT field, we compute a speed of 50 km/s for this FTE boundary entry. This is smaller than the observed flow speed, but is expected since the encounter of the FTE happens at an oblique angle to the flow. The opposite dispersion is observed, on a similar time scale, upon exit from the FTE energetic particle boundary. Similar remote sensing signatures are evident in the angular spectrogram of (g) on either side of the dashed line centered at 07:09 UT, indicating similar effects on the second, remotely sensed, FTE.

The data from probe TH-E, TH-C and TH-B are shown in Figs. 12, 13 and 14, in that order, which represent their decreasing distance from Earth. TH-D was very close to TH-C and observed similar signatures as TH-C (at spin-resolution), though interesting differences appear in high-resolution fields data.

TH-E and TH-C were in the magnetosheath proper at 07:00 UT and entered the magnetopause boundary layer a couple of minutes later as evidenced by the electron and ion spectra which show enhancement in the 10–20 keV particles. TH-E and TH-C crossed of the magnetopause current layer about a minute later. Both boundary layer and magnetopause transitions were seen first on TH-C, closer to the magnetosphere, and then on TH-E, as expected for an outward motion of the magnetopause boundary. TH-E exited the current layer at 07:09:30 UT, followed by TH-C at 07:12:30 UT, i.e., in reverse order. After that the field at the two probes was in the same direction and magnitude as in the sheath (as monitored on TH-A, in Fig. 11). The flow within the magnetopause boundary layer was in the same average direction as in the sheath, but much more variable.

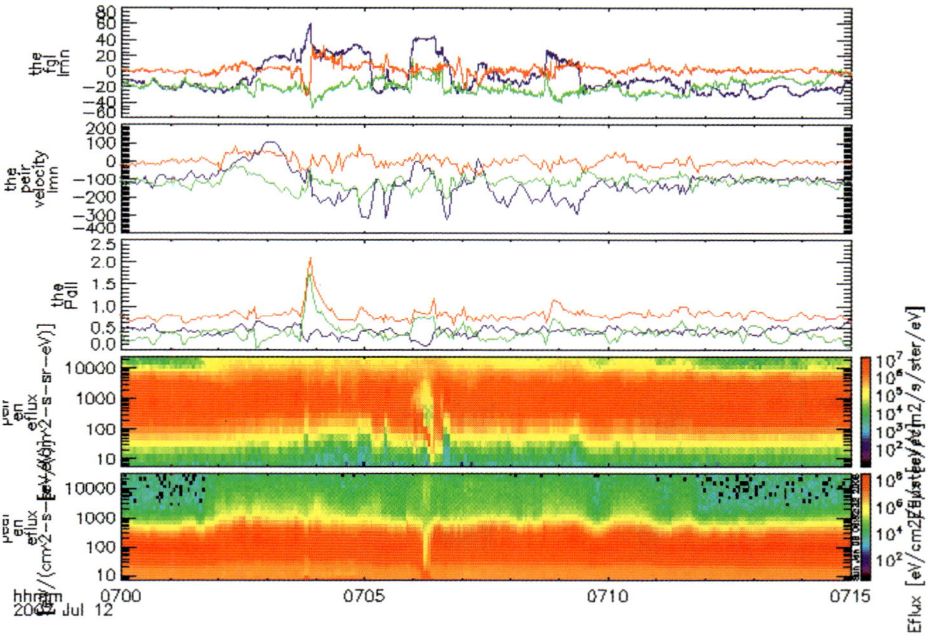

Fig. 12 TH-E data in identical format as Fig. 11. (**a**) Magnetic field (nT, *LMN*: *blue*, *green*, *red*); (**b**) Partial ESA velocity (km/s, *LMN*); (**c**) Magnetic (*blue*), plasma (*green*) and total (*red*) pressure (nPa); (**d**) and (**e**) ESA ion and electron omnidirectional e flux (eV/(cm^2 s str eV)). *Dashed lines* denote, again, timing of FTE centers on TH-C, observed slightly earlier on TH-E

TH-C observed clear bipolar signatures of two FTEs, as determined by the bipolar signatures in B_N. The FTEs were centered at 07:04 UT and 07:09 UT.

TH-B was in the magnetosphere at 07:00 UT and encountered the magnetopause boundary layer between 07:04:20 and 07:05:40 UT, as evidenced by the higher flux particle spectra there, with ions peaking in energy at 1 keV and electrons at around 100 eV. TH-B entered and exited the FTE current layer at 07:04:15–07:04:30 UT (as evidenced by the B_N, B_M values which match the data during the FTE encounters on TH-E and C), without traversing the magnetopause current layers, as shown by the fact that B_M never became negative, i.e., the field never attained a southward direction as in the magnetosheath.

Based on ion and electron spectra we can determine that TH-E remained in the boundary layer throughout most of the rest of the interval surrounding the FTEs, except for a short interval at 07:06–07:06:30 UT. Thus TH-E encountered the two FTEs from within the magnetopause boundary layer, though from the magnetospheric side of the current layer. TH-C entered the magnetosphere at 07:02:30 UT and again at 07:05:30 UT; in both cases it encountered the FTEs starting from the magnetospheric side, and entered/exited the FTE magnetic structure and surrounding magnetopause boundary layer particles due to the passage of the FTEs. Like TH-C, TH-B was at the magnetospheric side prior to the FTE encounter, and observed the first FTE particle and current layers, but did not observe the second FTE particle or current layer. Thus TH-C only remotely sensed the 07:09 UT FTE signatures from the magnetospheric side proper.

The low energy, cold particle signatures observed in the ion spectra inside the magnetopause (TH-C, B) are cold plasmaspheric plume ions, which are typically observed in the

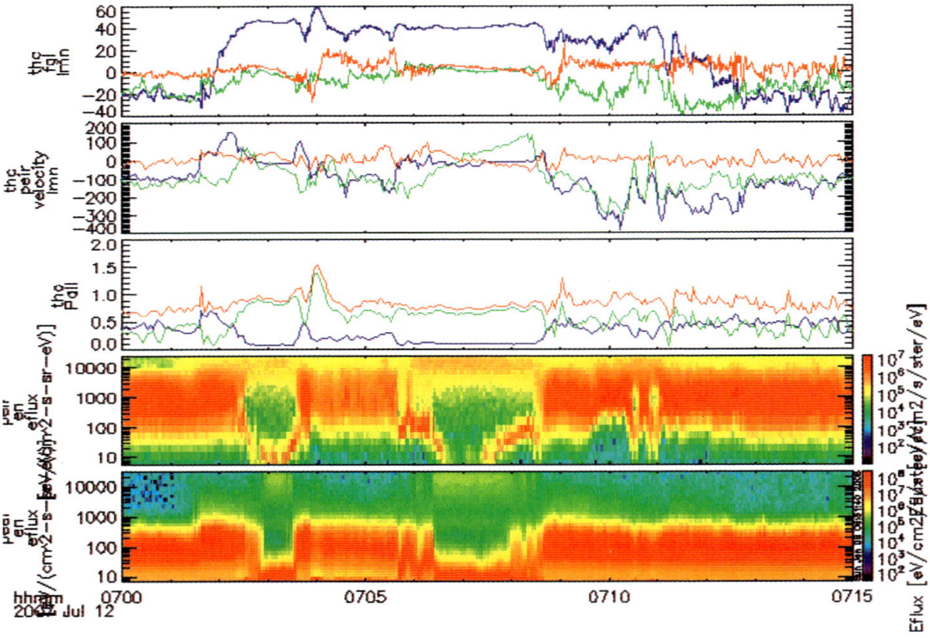

Fig. 13 TH-C data, in identical format as Fig. 12. *Dashed lines* indicate the center of the FTE (center of bipolar signature in B_N)

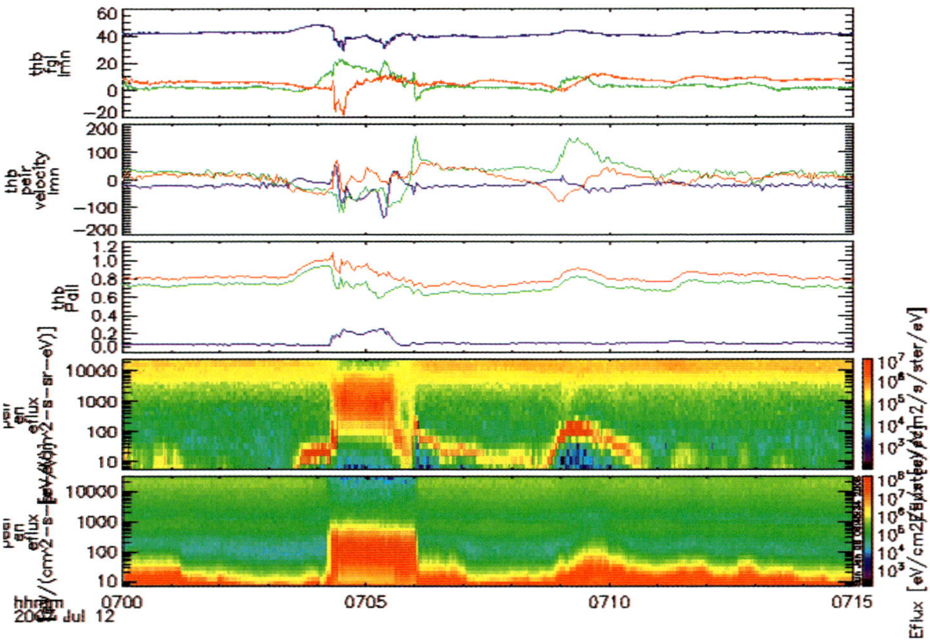

Fig. 14 TH-B data, in identical format as Fig. 12. *Dashed lines* indicate the center of the FTE on TH-C (center of bipolar signature in B_N). On TH-B, both FTEs were encountered slightly later than on TH-C

post-noon sector on THEMIS and are described in greater detail in McFadden et al. (2008). Their characteristic rise and fall in energy is related to increases in the plasma flow velocity, which enables this cold and otherwise nearly stationary population to ExB drift with the rest of the plasma populations into an energy where it is measurable by the ESA instrument (i.e., above the lower energy cut-off at ~5 eV).

The time delays of the FTE observations between TH-A, C and E are consistent with a lateral motion of the FTE past those spacecraft. However the time delays must be obtained after careful modeling to determine FTE orientation and structure, which is beyond the scope of this paper. Differences between the FTE motion and the magnetopause boundary layer motion are important in determining whether the FTE is a structure advected passively by the ambient flow, or an evolving magnetic structure that is moving relative to the ambient boundary layer plasma, subject to active reconnection forces.

That this event is an excellent "in vivo" FTE event is evident in the flow velocities, which are seen (on all probes) to deviate considerably from the magnetosheath flow around the FTE encounter. Monitoring of the steady magnetosheath flow on TH-A reassures us that these variations are not solar-wind related. Specifically, bipolar V_M and V_L flows are seen on TH-E and TH-D to reach or exceed 300 km/s, compared to a magnetosheath flow of 180 km/s.

Observations of the boundary normal flow velocity, V_M, reveal that the plasma motion is consistent with an approaching or retreating bulge in the magnetopause on both sides. This signature is cleanest when the satellite is outside the magnetopause layer prior to the encounter. For example, on the 07:04 UT FTE, TH-A observed an away ($V_N > 0$) then towards ($V_N < 0$) motion, while TH-E observed a towards ($V_N < 0$) then away ($V_N > 0$) motion. This shows that the FTE is not an undulation of the magnetopause but rather a bulge at the magnetopause. By integrating the normal velocity over the period prior to the FTE encounter at probes TH-A and TH-E we determined that the bulge extended +1700 km and −2200 km from the spacecraft on the magnetosheath or magnetospheric side respectively. Adding that on either side to the TH-A–TH-E spacecraft separation along the magnetopause normal (~9,900 km) we obtain an FTE thickness of 13,800 km or about 2.16 R_E. This is the first time such observations can be made definitively, and allow us to determine FTE shape, compute total flux content and determine the importance of the presence of the FTE in overall magnetopause energy coupling. Similar, unequivocal observations of an isolated FTE structure are reported by Sibeck et al. (2008). In that case, the observations are made possible due to a fortuitous crossing an isolated FTE by the five THEMIS spacecraft on either side, normal to its motion.

A salient feature of the observations of TH-E is that the spacecraft measured the flow signatures of the approaching FTE and a local magnetospheric pressure enhancement from the time-dependent perturbations of the nearby plasma, even though it never encountered the FTE current or particle populations. This is evidence of remote sensing of an FTE by a satellite on the magnetospheric side, and a counterpart of TH-A's remote sensing of an FTE on the magnetosheath side. These results are possible due to the simultaneous observations of the FTE by TH-E and TH-C closer to the magnetopause and thus able to observe the FTEs that generate these perturbations. The results herein are common in the string-of-pearls THEMIS dataset, and are substantiated using a different event in Liu et al. (2008).

In summary, analysis of the event on July 12, 2007 using the THEMIS probes from its string-of-pearls configuration shows that (i) FTEs can be remotely sensed using finite gyroradius techniques from the magnetosheath side even in the absence of a crossing of the FTE magnetopause boundary or current layer; (ii) the FTEs can be remotely sensed on the magnetospheric side from observations of the flow and plasma pressure, (iii) the flux rope structure can be revealed using multi-point measurements at separations commensurate to the size of the flux rope, and can be aided by plasma and remote sensing measurements to determine the rope's motion and shape. The scale size of the structures analyzed herein are on the order of 1 R_E but smaller scale sizes can be analyzed with probes TH-C, D and E, which are appropriately positioned in the middle. Although these inner satellites are not optimally separated to study details of the magnetopause current layer, current filaments, or wave particle interactions within FTEs, optimal separations at scales smaller than 1 R_E is the objective of an extended mission.

4 Summary

The unique, string-of-pearls configuration during its coast phase enables THEMIS to address questions on substorm and magnetosphere-ionosphere coupling which would not have been addressed otherwise. With a goal to exhibit the power of the THEMIS constellation and the quality of the data, we have presented initial results on two topics in substorm and magnetopause research. The first result, pertaining to the evolution of the substorm expansion phase in space, was obtained on March 23, 2007, when the major substorm intensification occurred at the local time sector of the THEMIS satellites. The probes observed in sequence the substorm dipolarization and Earthward flow bursts, and thus were able to determine that the propagation speed of these current wedge phenomena in space matches the propagation speed of the optical signatures of the westward traveling surge on the ground. Using the finite gyroradius technique of energetic particles we were able to distinguish between the outward expansion of the plasma sheet observed away from the neutral sheet and the westward motion of the dipolarization and Earthward flows. The second result concerns the structure and evolution of FTEs. THEMIS's multiple satellite observations were able to dissect the FTE structure at different distances from the FTE center and ascertain the lateral and radial motion of the FTE, as well as determine its extent and differentiate it from a boundary undulation. Using the remote sensing capabilities from the THEMIS SST detector we were able to show that an FTE can be remotely sensed further from the magnetopause, and using the ion data from the ESA instrument we were able to show that an FTE has remote signatures inside the magnetosphere in the plasma pressure and flow velocity even though the FTE boundary is not crossed. Similar analysis techniques employed in the baseline mission started in December of 2007 are expected to enable the resolution of THEMIS's primary objectives. The power of a string-of-pearls configuration and the capabilities which may arise from a tight but better-controlled clustering of the inner THEMIS probes will both be revisited in a possible extended phase of the mission in the years 2009–2012.

Acknowledgements This research was funded by NASA contract NAS5-02099.

References

S.-I. Akasofu, *Physics of Magnetospheric Substorms* (Reidel, Dordrecht, 1976)

V. Angelopoulos et al., Statistical characteristics of bursty bulk flow events. J. Geophys. Res. **99**, 21257 (1994)

V. Angelopoulos et al., Magnetotail flow bursts: association to global magnetospheric circulation, relationship to ionospheric activity and direct evidence for localization. Geophys. Res. Lett. **24**, 2271 (1997)

V. Angelopoulos et al., Multipoint analysis of a bursty bulk flow event on April 11, 1985. J. Geophys. Res. **101**, 4967 (1996)

V. Angelopoulos et al., On the relationship between bursty flows, current disruption and substorms. Geophys. Res. Lett. **26**, 2841 (1999)

V. Angelopoulos, The THEMIS mission. Space Sci. Rev. (2008, this issue). doi:10.1007/s11214-008-9336-1
G. Atkinson, The current system of geomagnetic bays. J. Geophys. Res. **23**, 6063 (1967)
U. Auster et al., Space Sci. Rev. (2008, this issue). doi:10.1007/s11214-008-9365-9
D.N. Baker et al., Neutral line model of substorms: Past results and present view. J. Geophys. Res. **101**, 12975 (1996)
J. Bonnell et al., Space Sci. Rev. (2008, this issue)
C.W. Carlson et al., Space Sci. Rev. (2008, this issue)
C. Cully et al., Space Sci. Rev. (2008, this issue)
P.W. Daly et al., Sounding of the plasma sheet in the deep geomagnetic tail using energetic particles. Geophys. Res. Lett. **11**, 1070 (1984)
D.H. Fairfield et al., Geotail observations of substorm onset in the inner magnetotail. J. Geophys. Res. **103**, 103 (1998)
L.A. Frank et al., The Visible Imaging System for the POLAR Spacecraft. Space Sci. Rev. **71**, 297 (1995)
S. Frey et al., Space Sci. Rev. (2008, this issue)
S. Harris et al., Space Sci. Rev. (2008, this issue). doi:10.1007/s11214-007-9294-z
R.A. Hoffman et al., Characteristics of the field-aligned current system in the nighttime sector during auroral substorms. J. Geophys. Res. **99**(A11), 21303 (1994)
B.L. Horning, R.L. McPherron, D.D. Jackson, Application of linear inverse theory to a line current model of substorm current systems. J. Geophys. Res. **79**(34), 5202–5210 (1974)
C. Jacquey et al., Location and propagation of the magnetotail current disruption during substorm expansion: analysis and simulation of an ISEE multi-onset event. Geophys. Res. Lett. **3**, 389 (1991)
A. Keiling, V.E. Angelopoulos, D. Larson, R. Lin, J. McFadden, C. Carlson, J.W. Bonnell, F.S. Mozer, K.-H. Glassmeier, U. Auster, W. Magnes, S.B. Mende, H. Frey, A. Roux, O. LeContel, S. Frey, T. Phan, E. Donovan, C.T. Russell, I. Mann, W. Liu, X. Li, M. Fillingham, G. Parks, K. Shiokawa, J. Raeder, Correlation of substorm injections, auroral modulations, and ground Pi2. Geophys. Res. Lett. (2008). doi:10.1029/2008GL033969
G. Kettmann, P.W. Daly, Detailed determination of the orientation and motion of the plasma sheet boundary layer using energetic protons on ISEE 1 and 2: waves, curves, and flapping. J. Geophys. Res. **93**, 7376 (1988)
Larson et al., Space Sci. Rev. (2008, this issue)
J. Liu, V. Angelopoulos, D. Sibeck, T. Phan, J. McFadden, K.H. Glassmeier, U. Auster, Z.-Y. Pu, THEMIS observations of the dayside traveling compression region and flows surrounding flux transfer events. Geophys. Res. Lett. (2008). doi:10.1029/2008GL033673
R.E. Lopez, A.T.Y. Lui, A multisatellite case study of the expansion of a substorm current wedge in the near-earth magnetotail. J. Geophys. Res. **95**, 8009 (1990)
A.T.Y. Lui, Current disruption in the Earth's magnetosphere: Observations and models. J. Geophys. Res. **101**, 13067 (1996)
J.P. McFadden et al., Space Sci. Rev. (2008, this issue)
R. McPherron et al., Satellite studies of magnetospheric substorms on Aug 15th, 1968. J. Geophys. Res. **78**, 3131 (1973)
S. Mende et al., Space Sci. Rev. (2008, this issue). doi:10.1007/s11214-008-9380-x
T. Nagai, Geophys. Monogr. Series (1991)
T. Nagai, Observed magnetic substorm signatures at synchronous altitudes. J. Geophys. Res. **87**, 4405 (1982)
R. Nakamura et al., Spatial scale of high-speed flows in the plasma sheet observed by Cluster. Geophys. Res. Lett. **31**, 9804 (2004)
S.-I. Ohtani et al., Tail current disruption in the geosynchronous region, in *Magnetospheric Substorms*. AGU Monogr. Ser., vol. 64 (AGU, Washington, 1991), p. 131
S.-I. Ohtani et al., Radial expansion of the tail current disruption during substorms: A new approach to the substorm onset region. J. Geophys. Res. **97**, 3129 (1992)
S.-I. Ohtani, Earthward expansion of tail current disruption: dual-satellite study. J. Geophys. Res. **103**, 6815 (1998)
W.H. Press et al., *Numerical Recipies: The Art of Scientific Computing* (Cambridge University Press, Cambridge, 1989)
T. Pytte, R.L. McPherron, S. Kokubun, The ground signatures of the expansion phase during multiple onset substorms. Planet. Space Sci. **24**, 1115–1132 (1976)
J. Raeder, D. Larson, W. Li, L. Kepko, T. Fuller-Rowell, Open GGCM simulations for the THEMIS mission. Space Sci. Rev. (2008)
A. Roux, O. Le Contel, C. Coillot, A. Bouabdellah, B. de la Porte, D. Alison, S. Ruocco, M.C. Vassal, The search coil magentometer for THEMIS. Space Sci. Rev. (2008)
Runov et al., Geophys. Res. Lett. (2008, submitted)

C.T. Russell, R.C. Elphic, Initial ISEE magnetometer results: magnetopause observations. Space Sci. Rev. **22**, 681 (1978)

C.T. Russell, R.C. Elphic, ISEE observations of Flux Transfer Events at the dayside magnetopause. Geophys. Res. Lett. **6**, 33 (1979)

V.A. Sergeev et al., Detection of localized, plasma-depleted flux tubes or bubbles in the midtail plasma sheet. J. Geophys. Res. **101**, 10817 (1996a)

V.A. Sergeev et al., Comparison of UV optical signatures with the substorm current wedge as predicted by an inversion algorithm. J. Geophys. Res. **101**(A2), 2615–2627 (1996b)

J.-H. Shue et al., Magnetopause location under extreme solar wind conditions. J. Geophys. Res. **103**, 17691 (1998)

D.G. Sibeck et al., Space Sci. Rev. (2008, this issue)

O. Strangeway, Geophys. Res. Lett. (2008). Manuscript in preparation

O. Torr et al., A far ultraviolet imager for the international solar terrestrial physics mission. Space Sci. Rev. **71**, 329–383 (1995)

N.A. Tsyganenko, A magnetospheric magnetic field model with a warped tail current sheet. Planet. Space Sci. **37**, 5 (1989)

N.A. Tsyganenko, A model of the near magnetosphere with a dawn-dusk asymmetry: 1. Mathematical structure. J. Geophys. Res. **107**(A8), 1179 (2002a)

N.A. Tsyganenko, A model of the near magnetosphere with a dawn-dusk asymmetry: 2. Parameterization and fitting to observations. J. Geophys. Res. **107**(A8), 1176 (2002b)

THEMIS ESA First Science Results and Performance Issues

J.P. McFadden · C.W. Carlson · D. Larson · J. Bonnell · F. Mozer · V. Angelopoulos · K.-H. Glassmeier · U. Auster

Originally published in the journal Space Science Reviews, Volume 141, Nos 1–4, 477–508.
DOI: 10.1007/s11214-008-9433-1

Abstract Early observations by the THEMIS ESA plasma instrument have revealed new details of the dayside magnetosphere. As an introduction to THEMIS plasma data, this paper presents observations of plasmaspheric plumes, ionospheric ion outflows, field line resonances, structure at the low latitude boundary layer, flux transfer events at the magnetopause, and wave and particle interactions at the bow shock. These observations demonstrate the capabilities of the plasma sensors and the synergy of its measurements with the other THEMIS experiments. In addition, the paper includes discussions of various performance issues with the ESA instrument such as sources of sensor background, measurement limitations, and data formatting problems. These initial results demonstrate successful achievement of all measurement objectives for the plasma instrument.

Keywords THEMIS · Magnetosphere · Magnetopause · Bow shock · Instrument performance

PACS 94.80.+g · 06.20.Fb · 94.30.C- · 94.05.-a · 07.87.+v

1 Introduction

The THEMIS mission provides the first multi-satellite measurements of the dayside magnetosphere, magnetopause and bow shock utilizing a string of pearls orbit near the ecliptic plane (Angelopoulos 2008). During a 7 month coast phase, the instruments were commissioned and cross-calibrated while spacecraft separations were adjusted from a few hundred

J.P. McFadden (✉) · C.W. Carlson · D. Larson · J. Bonnell · F. Mozer · V. Angelopoulos
Space Sciences Laboratory, University of California, Berkeley, 7 Gauss Way, Berkeley, CA 94720, USA
e-mail: mcfadden@ssl.berkeley.edu

V. Angelopoulos
University of California, Los Angeles, USA

K.-H. Glassmeier · U. Auster
Technical University of Braunschweig, Braunschweig, Germany

to ~10,000 kilometers. Most high-rate data capture was focused on the boundary crossings (magnetopause and bow shock), with inter-spacecraft spacing generally optimized to resolve these structures. Throughout this period the science team focused on dayside physics:

- The bow shock and foreshock (shock motion, particle acceleration, wave-particle interactions, and hot flow anomalies).
- The magnetosheath (shocked plasma relaxation, mirror modes, and reconnection).
- The magnetopause (reconnection, magnetopause structure, and flux transfer events).
- The dayside magnetosphere (dayside return flows, erosion of the plasmasphere, formation of the low latitude boundary layer, and field line resonances).
- The inner magnetosphere (radiation belt and ring current formation, plasmasphere).

In this paper we highlight the measurement capabilities of the ESA plasma sensors by presenting "first results" from several of these regions.

Each THEMIS spacecraft includes a fluxgate magnetometer (Auster et al. 2008), a search coil magnetometer (Roux et al. 2008), a 3-axis electric field instrument (Bonnell et al. 2008), solid state telescopes (SST) for energetic (>30 keV) ions and electrons (Larson et al. 2008), and electrostatic analyzers (ESAs) for electron and ion plasma (<30 keV) measurements (McFadden et al. 2008a). These instruments not only provide the information needed to perform substorm timing analysis during the prime mission, but also provide a core set of measurements needed to resolve most magnetospheric dynamics. Although plasma measurements will be the focus of these first results, observations from the other sensors are included to illustrate the synergy of these measurements and to demonstrate the ability of THEMIS satellites to resolve the basic plasma features of these regions. Throughout the paper we refer to the individual satellites by their abbreviated call letters—THA, THB, THC, THD and THE.

The THEMIS plasma instruments measure the 3-D plasma distribution function with ~3 s resolution. Although the highest-time, highest-phase-space resolution measurements are only available during bursts which are generally limited to ~30–60 minutes per orbit, coarser 3-D distributions at spin resolution are available for ~12 hours each orbit. These data have adequate angular resolution to allow accurate ground computation of moments and identification of beams. Even during periods where data collection is limited, the ESA data products include on-board calculated moments and omni-directional energy spectra at spin resolution. These on-board moments include corrections for spacecraft charging that generally provide accurate electron moment computations that eliminate photoelectrons. Spin-resolution energy spectra provide the additional information needed to interpret variations in the on-board moments and to correctly identify the dynamics associated with boundaries or changes in the multi-component plasma.

As demonstrated in the accompanying paper (McFadden et al. 2008a), the close proximity of the five THEMIS spacecraft during the early mission allowed very accurate in-flight calibration of the ESA sensors. The relative sensitivities of the ten sensors are believed to be determined to better than 5%, and the absolute sensitivity corrected to ~10% through cross calibration with Wind-SWE. The primary uncertainties in this calibration effort resulted from estimation errors of the proton to alpha ratio of the solar wind, from uncertainties in calculating the spacecraft-to-plasma potential from the measured spacecraft-to-Langmuir-probe potential, and by relying on the literature to correct for energy dependent efficiencies of the microchannel plate detectors. However, pressure balance checks across the dayside magnetopause (Fig. 16, McFadden et al. 2008a), performed independent of the calibration effort, provide additional confidence in our techniques. These accurate calibrations allow the combined electron and ion data to be used to deduce additional features about the plasma including mass composition and the presence of unmeasured cold plasma.

During the first 7 months of the THEMIS mission, the five spacecraft sampled the dusk, sub-solar and dawn regions of the magnetosphere. Section 2 presents first results from the encounters with these regions illustrating the abilities of the instruments to resolve small scale features and separate space and time. Although the THEMIS ESAs provide a data set of well calibrated observations, there are still several instrumental limitations to the data. During the course of the presentation, we will point out measurement limitations and uncertainties in the observations that can affect the accuracy of computed products such as moments. Some of these performance issues involve missing information, such as composition, while others are associated with the instrument's dynamic range.

THEMIS has an open data policy that strives for an immediate data release to the community. While data quality flags will be inserted into high level processed data, much of the data analysis effort will utilize unprocessed data. Therefore scientists need a reference where performance issues are identified, such as non-geophysical background counts, or where the impact of missing information, such as composition, is discussed. In Sect. 3 we provide a summary of all known performance issues with the ESA sensor including sources of background, non-ideal response of the instrument, limitations due to missing information, and telemetry formatting problems. Understanding and correcting for these performance issues will allow full use of the THEMIS measurement capabilities while avoiding any misinterpretation of the observations.

2 Multi-Point Observations by THEMIS

The multi-point measurements afforded by the five spacecraft provide the most important advantage of the THEMIS data set. THEMIS observes both the spatial and temporal variations in the structure of the magnetosphere, allowing detailed studies of time varying phenomena. This capability was most clear during the early mission when the spacecraft were organized in a string of pearls orbit that sampled the low latitude dayside magnetosphere. During this period, the spacecraft were ordered THB, THD, THC, THE, and THA, with an apogee of $\sim$14.5 R_e and perigee of $\sim$1.13 R_e. For magnetopause crossings, the inner three spacecraft were more closely bunched ($\sim$1000 km separations) while the lead and trailing satellites were generally separated by much larger ($\sim$4000–10 000 km) distances from the inner probes. This organization allowed sampling of the magnetopause and bow shock over multiple scales.

Figure 1 shows data from the fluxgate magnetometers and ion ESA sensors on the three inner probes during an outbound magnetopause crossing at $\sim$1330 LT on June 10, 2007. For this crossing, THD was leading THC by $\sim$500 km and THE was trailing THC by $\sim$1350 km. The panels are ordered top-to-bottom as the spacecraft are ordered along the orbit. Ion spectrograms provide clear identifications of changes in the plasma while the magnetometer provides identification of the magnetopause current. Ion velocity, plotted in LMN coordinates, is used to identify features of the boundary layer and distinguish between sheath flows, reconnection jets, and stagnant sheath plasma. The June 11 crossing was during a period of steady southward interplanetary magnetic field (IMF $B_z \sim -15$ nT at the magnetopause from 2130–2204 UT as determined by THB). Although external conditions were relatively constant, the magnetopause shows significant changes in structure on three crossings separated by less than 10 minutes.

The multi-satellite observations by THEMIS provide a time history of the evolution of the magnetopause during low-latitude reconnection. Beginning at the left side of Fig. 1, low

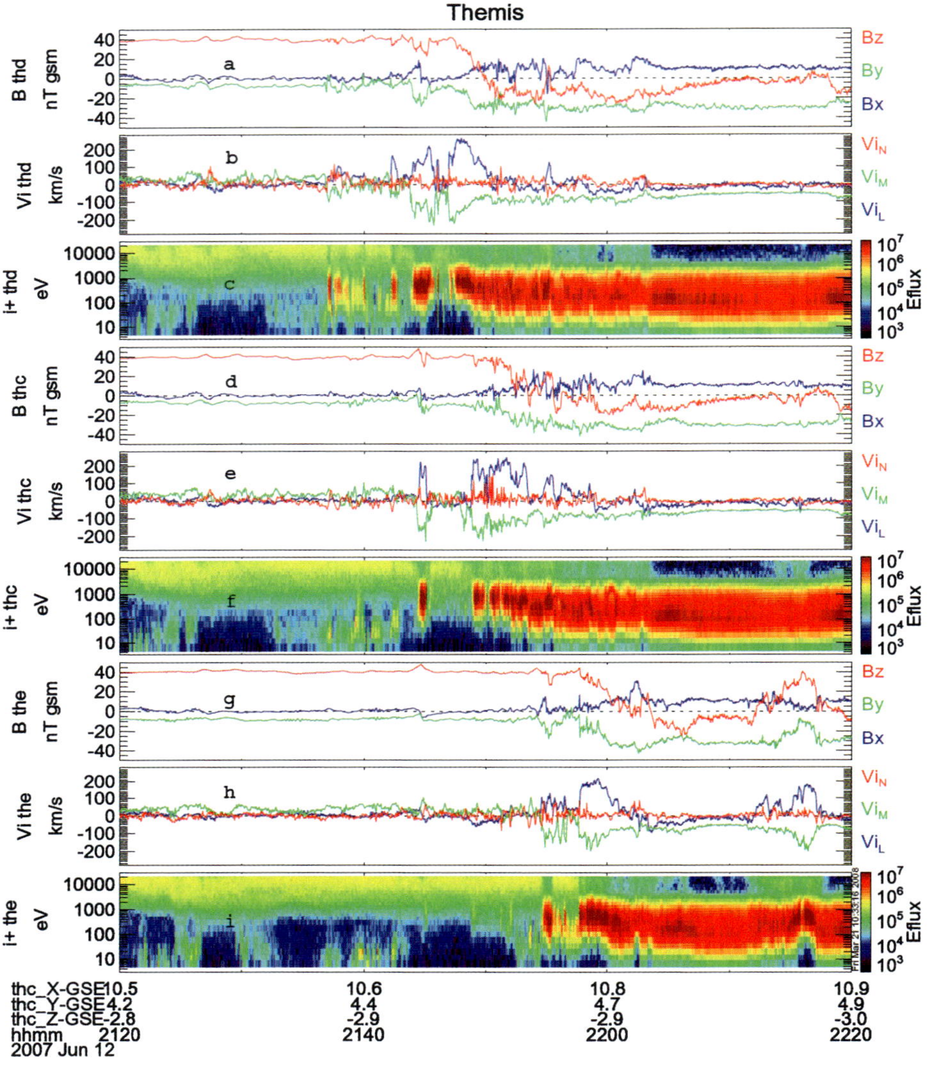

Fig. 1 Magnetic field in GSM, ion velocity in LMN, and ion spectrograms for the THD (panels **a–c**), THC (panels **d–f**) and THE (panels **g–i**) spacecraft. The plot shows three closely spaced crossing of the magnetopause during steady southward IMF illustrating the multi-point capabilities of THEMIS spin resolution observations

energy (<100 eV) ions are observed prior to 2137 UT in panel c (<2144 UT in panel f, <2155 UT in panel i). These are cold ions, generally repelled by spacecraft charging, which are revealed by magnetopause motion. Section 2.1 provides a closer look at this cold plasma component and McFadden et al. (2008b) provides a detailed look at cold plasma structure at the magnetopause. In addition to magnetopause motion inferred from cold plasma, multiple undulations of the magnetopause are clearly observed in the THD crossing (panels a–c) between 2137 and 2148 UT. Some of this motion is caused by the passage of flux transfer events (FTE), such as the FTE seen by all three spacecraft at 2145 UT. In Sect. 2.5 we take a closer look at FTEs to illustrate the plasma structure that can be resolved by THEMIS ESAs.

The additional magnetopause crossings by THC (panels d–f) and THE (panels g–i) show a similar overall structure, but differ in details from the THD crossing. In particular, variations in reconnection flow-jet velocities are revealed in panels b, e, and h. Lastly, we note the decrease in dynamic pressure after 2205 UT results in an outward motion of the magnetopause causing THE to re-enter the boundary layer. Even though the sheath B_z has dropped to zero, reconnection flows are still evident at THE (2112–2117 UT). The degree of outward motion can also be partially quantified ($\sim$1100 km) since THC sees some hints of magnetospheric ions in the boundary layer. In summary, Fig. 1 demonstrates the capabilities of THEMIS to resolve structure and dynamics through multi-point measurements with 3 s resolution. Current plans are to have even closer separations for the inner three probes during the extended mission to allow investigations of coherence scales at the magnetopause.

Figure 2 shows an outbound magnetopause crossing on June 10, 2007 during a period of steady northward IMF. For this crossing, THD was leading THC by $\sim$1200 km and THE was trailing THC by $\sim$1800 km. The vertical bars show the outer edge of the boundary layer as identified from the electron distribution (not shown). Heated magnetosheath electrons are a signature of field lines that have reconnected in at least one hemisphere (Fuselier et al. 1995). As in the case of southward IMF, the boundary layer ions showed variations in structure even though the time between each crossing was less than 15 minutes. In particular the thickness of that portion of the boundary layer with fast flows ($\sim$100 km/s) in the M-direction seems to be increasing with time.

For this event, the majority of the fast-flow flux tubes inside the boundary layer had uni-direction electron heating (not shown) indicating only one end of the field line had reconnected in the lobes. These observations, located $\sim 4R_e$ from the sub-solar point, differ with other THEMIS observations near the sub-solar region (McFadden et al. 2008c) where the majority of flux in the boundary layer have bi-directional heated electrons during northward IMF indicating dual-lobe reconnection. A larger study of the Low Latitude Boundary Layer (LLBL) is needed to quantify the importance of dual-lobe and single-lobe reconnection. Section 2.4 provides a more detailed demonstration of THEMIS's ability to resolve structure in the LLBL, including electron heating. Additional insight into LLBL formation and evolution using THEMIS data can also be found in Oieroset et al. (2008).

2.1 Plasmaspheric Plumes and Cold Plasma

The plasmasphere consists of a torus of co-rotating cold plasma that, depending upon geomagnetic conditions, extends equatorially from two to five R_e along dipole field lines. It is a primary reservoir of magnetospheric plasma, filling with cold ions (primarily protons) escaping from the ionosphere and attaining densities of 10^2 to 10^3 cm^{-3}. Magnetospheric convection distorts and erodes the outer portions of the plasmasphere creating a bulge on the dusk side (Carpenter et al. 1993) and forming plasmaspheric plumes that stretch out to the magnetopause. During magnetic storms the loss of plasmaspheric plasma can be quite dramatic as demonstrated vividly by pictures from the IMAGE satellite (Goldstein et al. 2004). In situ measurements of plumes have been performed by Ogo 5 (Chappell 1974), by ISEE 1 and 2 (Gosling et al. 1990a, 1990b), and by the Dynamics Explorer spacecraft (Craven et al. 1997). Plumes are routinely monitored at geosynchronous orbit by the MPA instruments on the LANL satellites (Su et al. 2001). In addition, cold plume plasma's participation in reconnection at the magnetopause has been inferred by Gosling et al. (1990a, 1990b) from low latitude boundary layer measurements, and by Su et al. (2000) during times of high solar

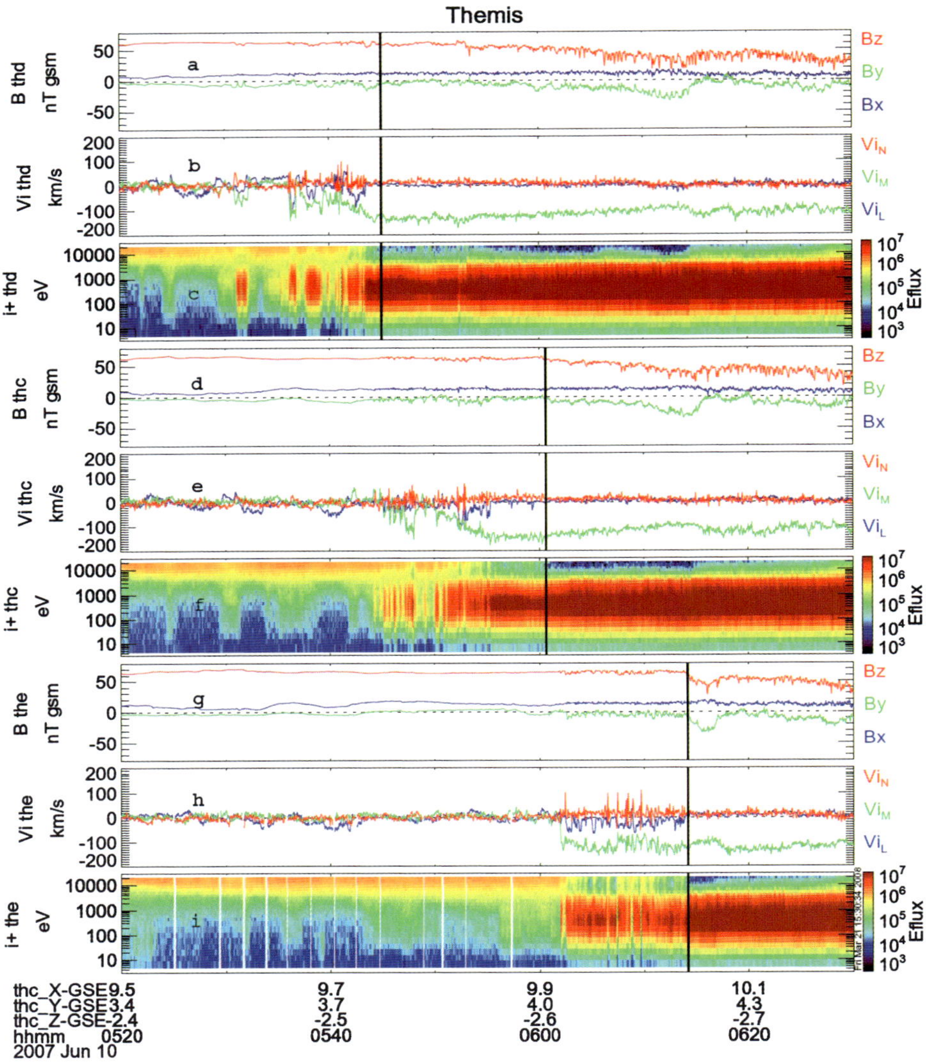

Fig. 2 Magnetic field in GSM, ion velocity in LMN, and ion spectrograms for the THD (panels **a–c**), THC (panels **d–f**) and THE (panels **g–i**) spacecraft. The plot shows three closely spaced crossing of the magnetopause during steady northward IMF

wind dynamic pressure at geosynchronous orbit. The Cluster spacecraft have also measured cold ions near the high latitude magnetopause (Sauvaud et al. 2001) and Chandler and Moore (2003) reported cold plasma densities as high as 70 cm^{-3} from Polar satellite observations near the equatorial magnetopause.

Single-spacecraft (Polar) statistical-studies of the distribution of plume plasma have been reported (Chen and Moore 2006), however multi-point measurements to quantify plume structure have not been undertaken. THEMIS, with its string or pearls configuration, provides a unique opportunity to observe the evolution of plasmaspheric plumes under a variety

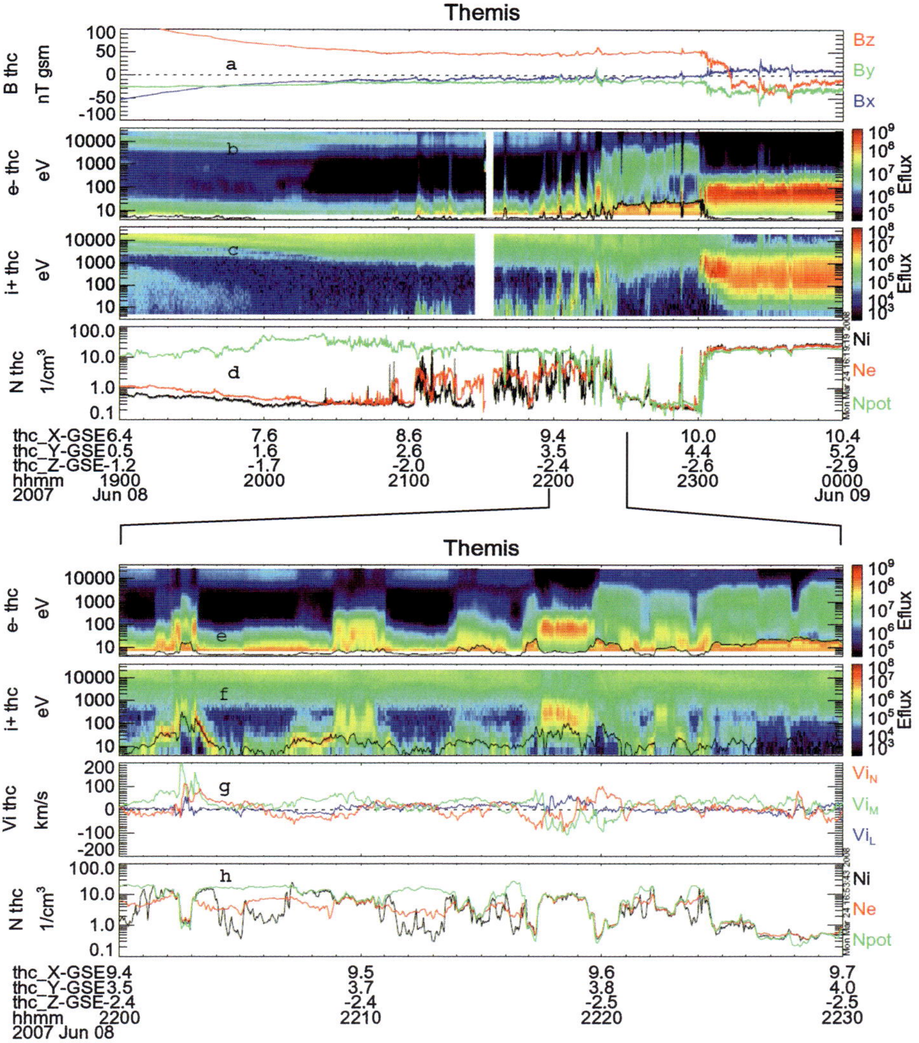

Fig. 3 **a** Magnetic field, **b** electrons **c** ions and **d** density. The three density curves include the measured ion (*black*) and electron (*red*) densities, and the inferred density (*green*) from spacecraft potential. A high density plume ($>10/\text{cm}^3$) extends from the plasmasphere to within $\sim 0.5 R_e$ of the magnetopause. The *lower panels* zoom in to the outer edge of the plume illustrating the **e** electrons, **f** ions, **g** ion velocity, and **h** density. The *black line* on panels **b** and **e** indicate the spacecraft potential and the black line on panel **f** indicates the energy of protons at the ion flow velocity minus $e\Phi_{sc}$. Figure adapted from McFadden et al. (2008b)

of solar wind conditions. In particular, the near equatorial orbits provide a large volume of data on these outflows, demonstrating the increasingly important role that cold plasma plays in magnetospheric dynamics.

Figure 3 illustrates a high-density, cold plasma plume observed by the THC spacecraft that extended from the plasmasphere to within $\sim 0.5 R_e$ of the magnetopause. The upper three panels show the magnetic field, and electron and ion spectrograms as the spacecraft traveled outbound from near geosynchronous to the magnetosheath. Panel d shows the measured ion (black) and electron (red) densities, and the density inferred from spacecraft potential (green). Due to measurement limitations discussed below, the inferred density (green) provides the best measure of actual plume density. The cold plasma plume dominates the density, varying between $10/\text{cm}^3$ and $50/\text{cm}^3$, before abruptly decreasing to less than the hot plasma density ($\sim 0.3/\text{cm}^3$) just before 2230 UT.

The lower panels of Fig. 3 zoom in on a region where cold ions can be measured. Panels e–h show electron and ion spectrograms, ion velocity, and density. Spacecraft potential is indicated by the black line on the electron spectrogram. The black line on the ion spectrogram is the proton energy at the drift velocity. Density and velocity calculations include corrections for spacecraft potential, and ion moments are calculated assuming only protons. Cold ions are seen in panel f when the plasma flow velocity is high enough (>50 km/s) so that protons can overcome the retarding barrier resulting from spacecraft charging. As seen between 2202 UT and 2204 UT, these ions often appear as a narrow spectral peak whose changing energy indicates acceleration of the bulk plasma. Flows this large are often observed near the magnetopause, where changes in solar wind dynamic pressure cause substantial motion of this boundary and the nearby plasma.

Although some agreement is observed between the electron (red), ion (black), and inferred (green) densities in panel h, there are many periods where calculated densities differ significantly from the inferred density. These disagreements illustrate two measurement limitations of the THEMIS ESA that result in missed cold plasma and demonstrate the importance of spacecraft potential inferred density. The first limitation is a consequence of spacecraft charging which prevents cold ions from reaching the sensor. Only when the convective flow is large enough so that cold protons can penetrate the spacecraft potential barrier will the ion density be correct. When the black line in panel b, which indicates the energy of protons at the ion flow velocity minus $e\Phi_{sc}$, drops near or below the lowest energy measured, as at 2204–2207 UT, cold ions will be missed by the ion sensor resulting in a measurement error.

The second measurement limitation occurs when the spacecraft potential, black line in panel a, drops below the lowest energy measured by the electron sensor, as at 2203–2209 UT and 2211–2214 UT. During these periods cold electrons, which neutralize the cold ions, are missed by the electron sensor. Missed cold electrons at 2207–2209 UT result in a measured ion density greater than the measured electron density. Of course both densities were actually equal, and the discrepancy results from our measurement limitations. This limitation to the electron measurement can be fixed by sweeping the sensor to lower energy (~ 2 eV as opposed to the current 7 eV), as is currently planned for the second year.

The multi-satellite THEMIS measurements provide global context for plasmaspheric plume studies. All 5 THEMIS spacecraft observed this plume, measuring high-densities ($>10/\text{cm}^3$) and simultaneous magnetopause motions. Similar high-density plumes have been observed on at least 15 orbits, with the highest density plume having $\sim 60/\text{cm}^3$ near the magnetopause. In addition, cold plasma with density $>2/\text{cm}^3$ (nearly an order of magnitude greater than the hot plasma) was observed at distances $>8R_e$ on all THC orbits in June, 2008, in the post-noon sector. A more in depth look at these cold plasma plumes can be found in McFadden et al. (2008b).

2.2 Auroral Ionospheric Outflows

Ionospheric outflows in the form of ion beams and conics are a significant source of plasma to the plasmasheet (Yau et al. 1985) and the lobes (Moore 1991). Most of these outflows originate in the low-altitude auroral oval where ions are energized and ejected into most regions of the magnetosphere. These ions provide a significant contribution to mass loading of the magnetosphere, and as such are an important part of the overall magnetospheric plasma circulation. Early contributions to identifying these outflows were made by the S3-3, ISEE and DE satellites which identified ion beams (Shelley et al. 1976) and ion conics (Sharp et al. 1977), quantified contributions from different plasma sources as a function of IMF (Lennartsson and Shelley 1986), determined solar cycle dependence (Yau et al. 1985), identified ion heating mechanisms including EMIC waves (Chang et al. 1986), and proposed innovative ideas including the pressure cooker heating mechanism (Gorney et al. 1985). Recent work has focused on results from the FAST and Freja satellites including the anti-correlation of conics and inverted-V electrons (Andre et al. 1998), the discovery of parallel electric fields in the downward current region and the associated electron solitary waves that form the pressure cooker mechanism (Carlson et al. 1998; Ergun et al. 1998), the identification of mass dependent ion acceleration (Moebius et al. 1998; Lund et al. 1998, 2001), the discovery of the role of Alfvén waves in ion acceleration (Stasiewicz et al. 2000; Chaston et al. 2004), the recognition of the impact of conic ions outflows on magnetic storms and substorms (McFadden et al. 2001; Tung et al. 2001), and the parameterization of ion outflows for global MHD models (Strangeway et al. 2005). A review of recent observations can be found in Auroral Plasma Physics (Paschmann et al. 2002). Although much work has been performed identifying and categorizing ionospheric outflows, the THEMIS mission offers the unique ability to capture and quantify these outflows using widely separated multi-point measurements. The following discussion provides a quick look at the general features of these outflows at THEMIS altitudes.

Unlike the cold ions (Sect. 2.1) which are primarily observed with velocities perpendicular to the magnetic field during enhanced convection, and which result from ambipolar outflow that maintains their low temperature and small parallel velocities, the warm, ionospheric ion outflows discussed in this section generally form a broader energy spectral component that consists of field aligned ions which are detected independent of convection. Since auroral ion conics and beams are produced at low altitudes, they both fold up in pitch angle into narrow field-aligned beams as they move to the equatorial regions at THEMIS making it virtually impossible to distinguish one from the other based on their angular signature. Conics and beams may be distinguished by their energy spectra, with conics generally much broader in energy, however time variations in beam energy coupled with dispersion may broaden the energy of ion beams. Since ions that make up ion beams generally started out as conics at lower altitudes, for the following discussion we do not attempt to differentiate between the two and refer to all these warm ions as "ionospheric outflows". A cursory examination of the dayside data indicates these outflows are less common than the cold ions, however we expect this to change as the THEMIS spacecraft move into the magnetotail where they can sample nightside outflows.

Figure 4 illustrates ionospheric outflows spanning 10:43–10:46 UT and 10:57–12:08 UT. The THD spacecraft is located at $\sim$1340 LT and traveling inward from the magnetopause. Ionospheric outflows appear as a broad, low-energy (5–200 eV) band in the ion spectrogram (panel b). They are observed nearly continuously for more than an hour, decreasing slowly in energy with time. Panel c shows that their appearance does not depend on high flow velocities. Panel a demonstrates that the ion outflows are accompanied by a low energy electron component. The bulk of these low energy electrons are relatively isotropic, however the >100 eV electrons include a field-aligned, counter-streaming component.

Panel d compares measured electron and ion densities, assuming only protons. The hot ion density is only $\sim 0.3/\text{cm}^3$, therefore the ionospheric outflows dominate the density. The

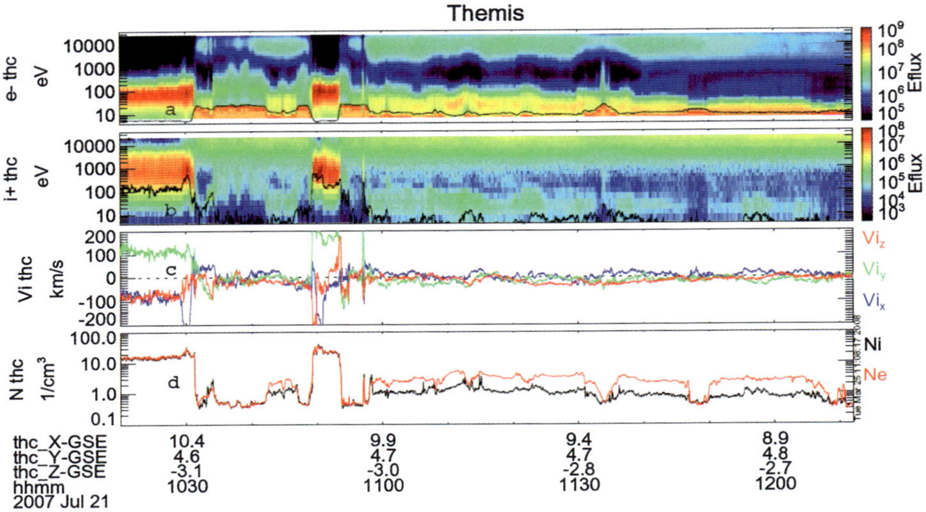

Fig. 4 Electron (panel **a**) and ion (panel **b**) spectrograms, ion velocity (panel **c**), and ion (*black*) and electron (*red*) densities (panel **d**) during an ionospheric outflow event with THD located at ~1340 LT as it traveled inward from the magnetopause. These outflows, which are most likely conics that have folded up in pitch angle, appear as a broad, low-energy (5–200 eV) band in the ion spectrogram (panel **b**). The mismatch in ion and electron densities in panel **d** is caused by an error in the ion density calculation resulting from a "protons only" assumption. These ion outflows are most likely O+

density mismatch illustrates a general problem interpreting THEMIS data when composition is not known. By assuming only protons when calculating the density, the computed fractional density of any non-proton component will be reduced by a factor of $(m_p/m_i)^{1/2}$, where m_i is the component's mass and m_p is the proton mass. However, the mismatch could also be due to missed cold ions below the lower energy cutoff of the sensor (~15 eV when spacecraft potential is taken into account).

To test for missed ions we plot the proton energy at the calculated flow velocity, less a correction for spacecraft charging, as a black line on the spectrogram in panel b. This curve allows one to determine whether cold protons below the sensor cutoff energy could explain the density mismatch, assuming "protons only" plasma. The fact that the mismatch occurs even when the black curve in panel b is above the sensor cutoff indicates the "protons only" assumption is incorrect and that either He+ or O+ are present. In addition, if the "protons only" assumption is invalid, then the calculated ion velocity will also be incorrect. (Note that this does not invalidate the above test.) The velocity is overestimated by an amount that depends on the fraction of higher mass components. For the example in Fig. 4, the ratio of ~3 for the calculated electron and ion densities indicates that the bulk of the ions are probably O+. Unfortunately it may be impossible to quantitatively untangle the fractional composition for this period. Nonetheless we hope this example illustrates the methodologies and complexities involved in analyzing data without composition information.

2.3 Pc 5 Field Line Resonances

Although magnetospheric oscillations and pulsations have been observed for more than a century, our understanding of these phenomena was limited until the pioneering work of Dungey (1954) that describes standing Alfven waves which result from a magnetosphere populated by plasma. He outlined the basic equations which define the "toroidal" (azimuthal flow and radial electric fields) and "poloidal" (radial flow and azimuthal electric fields) modes. Boundary conditions in the ionosphere quantize the wavelength, and therefore quantize the frequency. This model was eventually validated by Samson et al. (1971) who used a latitudinal chain of ground magnetometers to identify field line resonances (FLRs). Southwood (1974) and Chen and Hasegawa (1974) provided a theoretical basis for an energy input mechanism, where magnetopause motion produces fast mode waves that propagate inward and couple to shear Alfven waves producing the FLRs. Kivelson and Southwood (1985, 1986) added to this model by introducing cavity modes to explain the discrete spectra that results from broadband energy input. This description was then extended to a waveguide model by Samson et al. (1992), which removed the requirement for symmetry in the azimuthal direction. In addition to the ground magnetometer studies (Samson et al. 1971), radar studies (Walker et al. 1979; Ruohoniemi et al. 1991), and optical studies (Samson et al. 1996) of FLRs, satellite measurements by Ogo 5 (Kokubun et al. 1976), GOES (Lin and Barfield 1985), ISEE (Singer et al. 1982), DE1 (Cahill et al. 1986), AMPTE (Anderson et al. 1990), Geotail (Sakurai et al. 2001), Polar (Keiling et al. 2003), FAST (Lotko et al. 1998), and Cluster (Zheng et al. 2006) provide a wealth of in situ magnetospheric observations. For a review on FLRs see Glassmeier et al. (1999) or Takahashi (1998), or for a historical perspective see Hughes (1994).

Azimuthally polarized toroidal Pc 5 pulsations or FLRs are commonly observed by THEMIS in the dusk and dawn magnetosphere. These resonances have a magnetic node at the equator and therefore exhibit large amplitude plasma motions with only small magnetic perturbations at the THEMIS spacecraft. The multi-satellite measurements of THEMIS should provide observations that can identify the energy input mechanisms (Kelvin-Helmholtz instabilities, dynamic pressure variations, changes in reconnection, or flux transfer events) for FLRs. Glassmeier et al. (2008), in a case study, provides an analysis of this energy input and suggests most magnetospheric variations are "quasi-static responses to pressure induced magnetopause motions". These FLRs are often observed to grow in amplitude with one phase relationship between the velocity components, then decay away with a different phase relationship.

Figure 5 shows an unusually large toroidal Pc 5 event. Since the spacecraft is near the node at the equator, the magnetic field variations are small (panel a). The resonance motion is easily seen in the ion spectrogram (panel b) where cold plasma ($n_c \sim 1\ \mathrm{cm}^{-3}$) is accelerated to nearly keV energies by the wave. These resonances have been extremely helpful in identifying the cold plasma component near dawn, which often has a density between 0.1 cm^{-3} and 1.0 cm^{-3}. Panel c shows the ion velocity determined from the onboard moments (assuming only protons). The dominant amplitude during the wave growth period (prior to 1805 UT) is in GSM x-direction. The subsequent wave decay has nearly equal x and y components, as expected for a toroidal wave at a position of $(X, Y, Z)_{GSE} = (6R_e, -7R_e, 2R_e)$. Panel c shows that the hot electron component is strongly modulated by these waves, with nearly order of magnitude pressure variations being observed in the largest events. This pressure modulation may indicate a role that toroidal Pc 5 waves play in electron energization, or may just be the result of a pre-existing azimuthal pressure gradient. A modulation of the ion pressure can also be observed when ESA and SST data are combined.

For future studies of FLRs, there are some measurement problems that can limit the accuracy of the calculated plasma moments. First, the ion density may be overestimated and the ion velocity may be underestimated, due to a calculation error caused by the energetic electron flux (panel c). Energetic electrons scattering into the ion sensor, or their associated Bremsstrahlung X-rays, produce a small number of counts at low energy that mimic

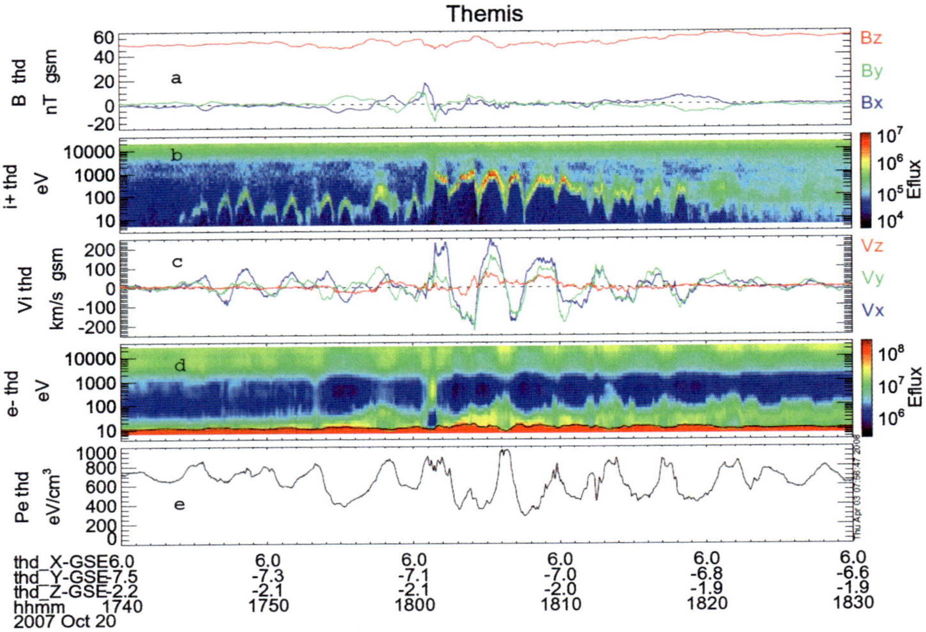

Fig. 5 **a** Magentic field, **b** ion spectrogram, **c** ion velocity, **d** electron spectrogram, and **e** electron total pressure measured by the ESA during a toroidal Pc 5 field line resonance (FLR). Cold ion stand out in the spectrogram, accelerated to keV energies by this wave. Ion velocity is slightly underestimated as explained in the text. Significant modulation of the electron pressure is observed in both the ESA and at higher energies by the SSTs

a tenuous, isotropic ion component. For example, at the peak in the flow velocity around 1806 UT, the background from energetic electrons adds ~25 counts/spin/energy. This background causes a ~0.4 cm^{-3} error in the calculated density, primarily from the lowest few energy bins. Since cold plasma (~1.4 cm^{-3}) and hot plasma (~0.14 cm^{-3}) have about four times this density, the velocity moment, given by flux/density, is underestimated by ~25% if this background is not removed. A second source of error is a consequence of the high energy cutoff of the ESA. When the ions are very hot, a substantial fraction of the ion flux may not be measured by the ESA resulting in an underestimation of density, velocity and pressure. For these events, the inclusion of SST measurements into the plasma distribution is generally required for accurate moment computations.

2.4 Low Latitude Boundary Layer during Northward IMF

The low latitude boundary layer (LLBL) is a dynamic region where magnetosheath and magnetospheric plasmas mix, and where mass, momentum and energy are exchanged between the solar wind and the magnetosphere. LLBL structure changes dramatically with local time and with orientation of the IMF (Mitchell et al. 1987). During northward IMF, the LLBL forms a thick layer with trapped magnetosheath plasma and small flows. During southward IMF, the LLBL is much thinner and contains reconnection flow jets and flux transfer events as described in Sect. 2.5. Gosling et al. (1990b) was the first to recognize the existence of a layered magnetopause, and Fuselier et al. (1995) and Le et al. (1996) have used observations of this layering to deduce its large scale structure and the dynamics of its formation. The monograph "Earth's Low-Latitude Boundary Layer" (Newell and Onsager 2003) provides a collection of articles describing the LLBL including historical and observational papers, models for its formation, and the impact of the LLBL on the plasmasheet, ionosphere and aurora.

In this section we provide an example of THEMIS observations in the sub-solar region during northward IMF when a plasma depletion layer forms and dual-lobe reconnection traps magnetosheath plasma onto closed field lines. The ability of THEMIS to provide simultaneous, closely spaced observations at the magnetopause offers a unique opportunity to resolve the LLBL layers. In particular, the ESA instrument is ideally suited to provide information on topology and history of the various layers while the multi-point measurement differentiates the spatial from temporal evolution of features within the LLBL.

Figure 6 illustrates the spatial and temporal structure of the sub-solar LLBL revealed by THEMIS during northward IMF. The figure shows the magnetic field, density, velocity, and ion and electron spectrograms as THC and THD traveled from the magnetosheath to magnetosphere. The spacecraft were closely spaced, separated by about $[-400, 700, 160]$ km in GSE coordinates. During this crossing the upstream magnetosheath plasma was highly striated, with large density and magnetic field variations associated with mirror mode waves (panels a, f). The initial crossings of the magnetopause at 0820 and 0826 UT showed a relatively thin boundary layer with no indications of a plasma depletion layer (PDL). THD observed a large magnetic hole (Turner et al. 1977) on the inbound crossing at 0820 UT, but not on the outbound crossing minutes later. THC may also have observed the edge of this magnetic hole on its inbound crossing, but measured a much smaller decrease in magnetic field strength.

After returning to the magnetosheath both spacecraft again approached the magnetopause, this time observing a PDL from ~0850 until ~0910 UT. 0910 UT marks the outer edge of the boundary layer as indicated by heated electrons (panels e and j). Multiple magnetic holes were observed by each spacecraft within the PDL, and the close proximity of these holes to the mirror mode waves suggests the holes evolved from the waves. The field decreases in the magnetic holes were correlated between the two spacecraft, with simultaneous large decreases for 3 events and combinations of large/small decreases for 6 other events, all between 0852 and 0908 UT. These correlations indicate the scale size of the holes is about the spacecraft separation perpendicular to the magnetic field, ~800 km. This is about the same scale size estimated from an average perpendicular flow velocity (~30 km/s) times the typical duration of the holes (~30 s). For the magnetic holes in Fig. 6, whose core fields are between 5 and 10 nT, the scale size of these holes is a few thermal ion (~100 eV) gyro-radii.

THC enters the boundary layer at ~0908 UT, as indicated by the tenuous energetic tail in the electrons (panels e), followed by THD at ~0912 UT. THC briefly crosses into the magnetosphere (0913–0916 UT) before re-entering and remaining in the boundary layer for ~10 minutes, followed by an final crossing into the magnetosphere (~0927 UT). For all but the outermost contact of the boundary layer, THC measured bi-directional heated electrons. Similarly, THD measured primarily bi-directional heated electrons in the boundary layer. These observations are consistent with other THEMIS observations of the sub-solar LLBL (McFadden et al. 2008c) which showed primarily bi-directional heated electrons during northward IMF indicating dual-lobe reconnection. These and other THEMIS observations at the magnetopause, which combine multi-satellite measurements with 3 s resolution of the plasma, should allow THEMIS to completely characterize the LLBL during its mission.

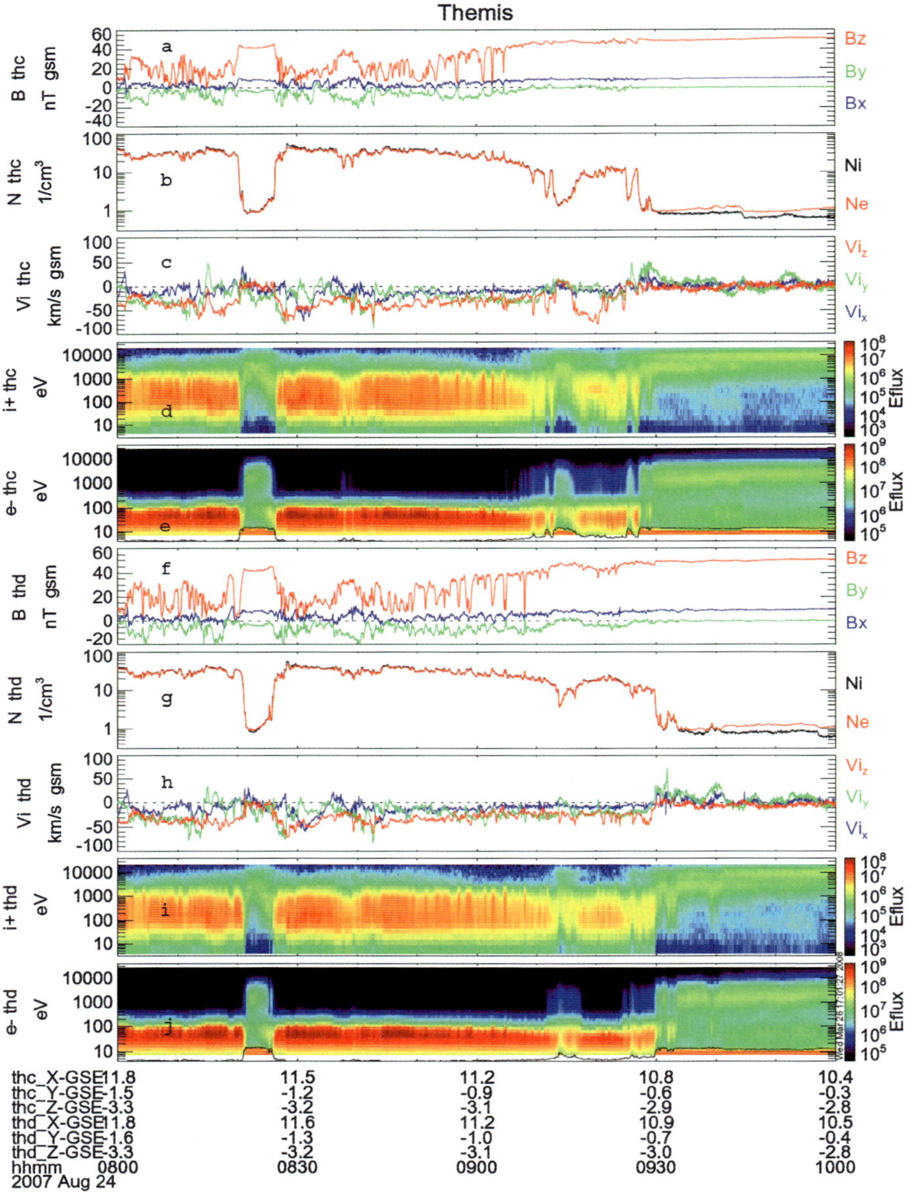

Fig. 6 Magnetic field (panel **a**), ion (*black*) and electron (*red*) densities (panel **b**), ion velocity (panel **c**), and ion (panel **d**) and electron (panel **e**) spectrograms during a LLBL crossing near the subsolar point by THC. Panels **f**–**j** are the same for THD

2.5 Magnetopause Reconnection and FTEs during Southward IMF

The early THEMIS mission, where all five probes crossed the equatorial magnetopause with spacecraft separations varying from a few hundred kilometers to ~10,000 km, provides an ideal data set for exploring ion-scale physics associated with dayside low-latitude reconnection. Spin period resolution of the plasma distribution functions, when combined with high resolution magnetic and electric field measurements, allows detailed studies of magnetopause structure. Early results published in a special issue of Geophysical Research Letters include: (1) multi-satellite observations of asymmetric reconnection demonstrating that the Hall electric fields are significant only on the magnetospheric side of the magnetopause (Mozer et al. 2008), (2) interactions of kinetic Alfven waves with particles at the magnetopause (Chaston et al. 2008), (3) multi-satellite observations of a crater flux transfer event (FTE) that is used to recreate the 2-D structure of the magnetopause (Sibeck et al. 2008), and (4) additional analyses of FTE structure (Lui et al. 2008; Liu et al. 2008).

FTEs are a common feature of the dayside magnetopause. They are believed to result from multiple reconnection lines forming plasmoids or flux ropes in a manner similar to plasmoid formation in the magnetotail. FTE signatures were initially observed in magnetometer data (Russell and Elphic 1979), which displayed a variety of signatures (Elphic 1995). The plasma signatures of FTEs were observed later (Paschmann et al. 1982a) and MHD simulations are now manifesting similar structures (Raeder 2006). Active FTEs, ones that are dynamically being generated, can be determined by the presence of large velocity variations within and near the FTE. Simultaneous measurements upstream and downstream should provide information on the boundary conditions necessary for their formation. In particular, THEMIS ESA measurements should provide the topology and history of the plasma populations which interacted during the formation of the FTE. Below we present THEMIS observations of a magnetopause crossing that demonstrates some of these capabilities.

Figure 7 shows a THEMIS magnetopause crossing during southward IMF. The current sheet crossing between 0834 and 0838 UT (panel a) was accompanied by reconnection flow jets (panel b) as expected for southward IMF. Prior to this crossing, evidence of earlier reconnection can be seen from at least nine FTE signatures. FTEs can be identified from both the magnetic field (panel a) and the plasma velocity (panel b), which are displayed in LMN coordinates. In addition, cold plasma adjacent to the magnetopause becomes visible in the ion spectrogram (panel c) due to bulk motion resulting from the passage of the FTEs. Cold plasma is also revealed by the mismatch of ion and electron densities (panel e) except during periods of bulk motion.

As shown in McFadden et al. (2008b), cold ions can be captured in the FTEs indicating that cold ions at the magnetopause do not suppress reconnection. The FTE at 0750 UT provides a particularly clear example of this capture. The reconnection of closed flux tubes containing cold ions is demonstrated from the electron spectrogram (panel d), whose energetic magnetospheric population is lost. Within the FTE, sheath electrons ($\sim$100 eV) replace the hot magnetospheric and cold ionospheric populations. The $\sim$100 eV electron component inside the FTE is anisotropic, with matching counterstreaming field aligned components, and with field aligned fluxes five times larger than the perpendicular flux. An interesting aspect of these FTEs is the factor of 2–3 higher density inside the FTE as compared to outside the FTEs (panel e). This increase appears to occur before sheath plasma arrives suggesting compression of the cold component. However, it is not simply compression of the flux tube since the magnetic field dominates the pressure and it remains relatively constant. Whether this is a snowplow effect of the FTE, or just a density gradient in the cold plasma between the location the FTE traps the cold plasma and is later observed, is not known and will require investigations of more events. It is anticipated that these and other high resolution observations during southward IMF will allow a comprehensive study of the structure of FTEs.

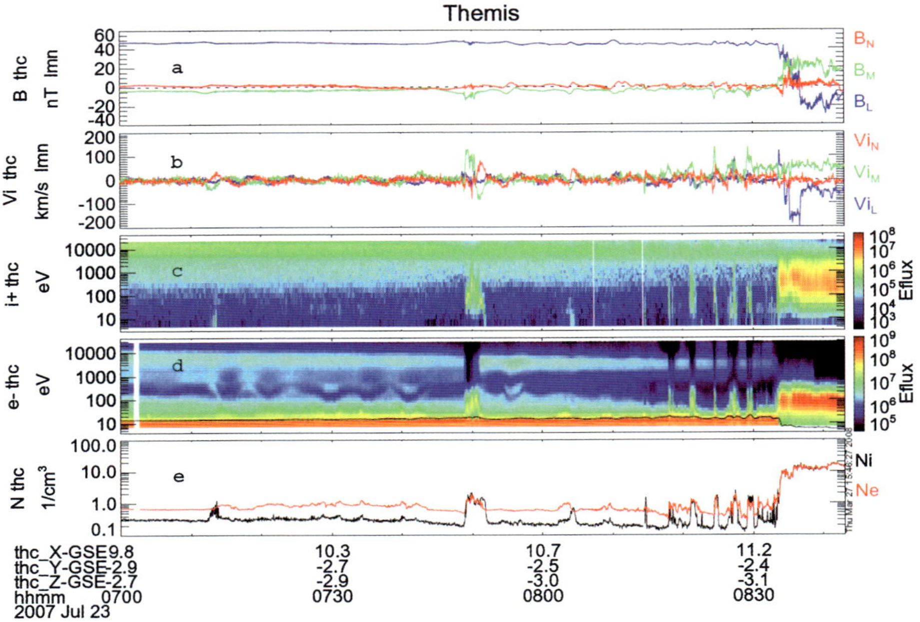

Fig. 7 **a** Magnetic field, **b** ion velocity, **c** ion spectrogram, **d** electron spectrogram, and **e** ion (*black*) and electron (*red*) densities. FTEs observed on the magnetospheric side of the magnetopause capture cold ions (panel **c**) and excite a weak FLR (panel **b**, 0715–0750 UT). The cold ions are only measured (panel **e**) when bulk motion allows them to penetrate the ~15 V spacecraft potential

An additional feature of this magnetopause crossing is the small but periodic modulation of the plasma velocity between 0710 and 0750. The magnetic field has a signature of an FTE at ~0712 UT, just at the start of the FLR ringing, suggesting the FTE is the driver. Additional evidence for an FTE source for this FLR can be found in the 3 minute periodic FTEs found adjacent to the magnetopause, twice the 6 minute period of the FLR. Finally, we note that the FLR interacts with the electron distribution (panel d) producing enhancements and dropouts in the few hundred eV electrons. Further investigation of these plasma signatures, and the interactions between FLRs and FTEs, should provide a much better understanding of the dynamics of a reconnecting magnetopause.

2.6 Bow Shock

The Earth's bow shock has been observed and cataloged by numerous satellites which revealed a complex structure whose dynamics is primarily controlled by the orientation and magnitude of the interplanetary magnetic field (IMF) relative to the shock normal. Comparisons of these measurements with theory have provided a foundation for understanding numerous other collisionless shock phenomena including planetary shocks, interplanetary shocks, the heliosphere's termination shock, interstellar shocks, and intergalactic shocks. For a review of shock physics see "Physics of Collisionless Shocks" (Russell 1995), or for recent observational and theoretical results see "The Physics of Collisionless Shocks" (Li et al. 2005). For applications of multi-point measurements to foreshock and bow shock observations, see Chaps. 2 and 4 of "Outer Magnetospheric Boundaries: Cluster Results" (Paschmann et al. 2005).

The variety of structure observed at the Earth's bow shock is impressive. It includes gyrating ions upstream of the quasi-perpendicular shock (Paschmann et al. 1982b), an electron foreshock extending far upstream of the bow shock (Anderson et al. 1979), an ion foreshock (Asbridge et al. 1968) consisting of reflected and diffuse ions (Gosling et al. 1978) that precondition the solar wind (Paschmann et al. 1979), Short Large Amplitude Magnetic Structures (SLAMS, Schwartz et al. 1992), Hot Flow Anomalies (HFAs, Schwartz 1995), ion density holes (Parks et al. 2006), 30 s period wave modulations of the upstream IMF (Greenstadt et al. 1968; Fairfield 1969), 3 s waves (Le et al. 1992), and a variety of higher frequency waves (whistler and Langmuir waves, see Gurnett et al. 1979) and wave structures (electron solitary waves, Bale et al. 1998). During the first 9 months of operations, all of these phenomena have been observed except for high frequency wave phenomena which THEMIS does not measure. THEMIS multi-point observations of bow shock phenomena, with separations varying from ~200 km to ~10,000 km should provide new and important information about shock formation and stability, and about the 3-D structure and propagation of various upstream phenomena. Cross-correlation of the particle and fields signatures on multiple satellites should resolve the propagation direction, wave growth rate, particle thermalization rate, and effect of energy transfer on the shock.

Figure 8 shows an example of THC observations at and upstream of the quasi-parallel bow shock. After crossing the shock at ~0053 UT, the spacecraft spent significant time in the upstream region sampling a plasma dominated by 30 s waves (panel a). Foreshock ions propagating upstream of the shock are seen at energies >2 keV (panel c). The shock moves outward and briefly touches the spacecraft at 0112 UT before retreating Earthward. The

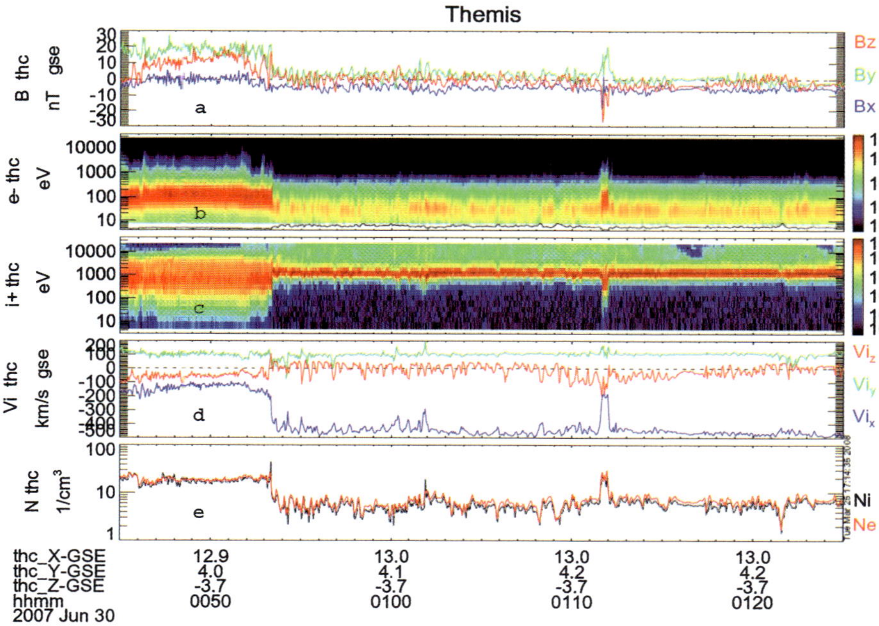

Fig. 8 Magnetic field (panel **a**), electron (panel **b**) and ion (panel **c**) spectrograms, ion velocity (panel **d**), and ion (*black*) and electron (*red*) densities (panel **e**) near a quasi-parallel bow shock. 30 s waves (panels **a** and **d**) dominate the turbulent upstream region, along with energetic gyrating ions (panel **c**)

solar wind density is a nominal $6/\text{cm}^3$ and the velocity nearly 500 km/s. These observations reveal the complicated structure upstream of a quasi-parallel shock.

The upstream observations contain significant density turbulence (panel e). This turbulence was virtually identical on the three inner probes whose separations were less than 300 km. Within this turbulence one can identify several density holes (Parks et al. 2006), the most prominent appearing at 0121:30 UT. This hole had a factor of $\sim$4 drop in density and is associated with a small shear in the magnetic field (panel a). Density holes may be related to hot flow anomalies (HFAs), another upstream structure that has been observed by THEMIS (Eastwood et al. 2008). HFAs are thought to result from a tangential discontinuity intersecting the bow shock (Schwartz 1995), which causes upstream reflection of ions and an associated pressure that creates a low density, low magnetic field channel. Multi-satellite observation by THEMIS, with 3 s resolution of the plasma, may be able to determine any connections between these disparate phenomena.

We end this discussion of shock observations with a caution about THEMIS observations within the solar wind. During the first 9 months, THEMIS plasma sensors were operated almost exclusively in magnetospheric mode, which has 32 sweeps per spin (11.25° resolution). To fully resolve the solar wind beam, which is generally $\sim$6° wide, the ion ESA sensor should be operated in solar wind mode which has 64 sweeps/spin (5.6° resolution). The observations in Fig. 8, which show excellent agreement between ion and electron densities, suggest that the nominal magnetospheric mode was sufficient for this event. The upstream turbulence has likely broadened the solar wind beam, allowing the instrument to resolve the beam with only 11.25° resolution. In addition, the solar wind density was low enough so that counter saturation was not a problem. However, this is not always the case during solar wind encounters. For those using THEMIS data, a good indication of instrumental problems with solar wind is a poor agreement between electron and ion density. An example can be found in the companion paper (Fig. 14, McFadden et al. 2008a), where ion density is underestimated, as compared to the electron density, due to a narrow beam that is not resolved.

3 Performance Issues

Before embarking on analysis of the THEMIS ESA data, it is important for the scientist to be aware of various performance issues with the ESA: sources of background or contamination, non-ideal response of the instrument, limitations due to missing information, and telemetry formatting problems. The data set is too large for a subset of problem data to be routinely corrected or purged from the files. Instead the scientist should be aware of these data limitations in order to avoid periods where non-geophysical interference or missing information can result in misinterpretation of the observations. In this section we outline known performance issues associated with the THEMIS ESA data that were uncovered during the first nine months of operations.

3.1 Sources of Sensor Background or Contamination Counts

When performing detailed analysis of data from the THEMIS ESAs, care should be taken to assure that non-geophysical sources of counts do not affect the result. There are several sources for this background or contamination: (1) Solar UV scattered directly into the detector, (2) photoelectrons produced by solar UV that reach the detector, (3) energetic electrons scattered through the sensor to the detector, and (4) penetrating radiation.

3.1.1 Scattered UV

Scattered solar UV to the detectors was estimated from a ~140 min period on January 29, 2008 when THB was in the lobes and fast survey data were collected. Ion counts were negligible during this period, and MCP detector background rates were ~0.8 Hz per 22.5 degree anode. Sunlight contamination rates were determined from sun-viewing solid angle sectors, with the spin dependence of sunlight mapped from the energy sweep. Peak rates were <20/s summed over all anodes, and sunlight contamination was confined to ~10 degrees of rotation. This compares well with previous instruments, such as Cluster-HIA which had peak sunlight count rates of ~50/s, or the FAST satellite where peak rates were ~650/s. For additional information on UV scattering in ESA detectors see Carlson and McFadden (1998).

3.1.2 Photoelectrons

Spacecraft photoelectrons are a source of non-geophysical counts in the electron ESA. This photoelectron contamination comes from three different sources: Langmuir sensors, spacecraft surfaces, and internal analyzer surfaces. This contamination can often be removed from the measurement when the spacecraft potential is known. Photoelectrons with energy greater than the spacecraft potential generally escape into the plasma, therefore the spacecraft potential provides a relatively clean separation, or cutoff energy, between reflecting photoelectrons and the in situ plasma. However, there are photoelectrons that appear at energies above the spacecraft potential that may require consideration if precise measurements are required.

The most prominent photoelectron contamination comes from the Langmuir sensors as illustrated in Fig. 9a. Spectral peaks at ~15 eV and ~28 eV are both due to this contamination which results from proper Electric Field Instrument (EFI) operation. The EFI electronics supplies a bias current to the Langmuir sensors roughly equal to ~30% of the sensor's photo-emission current (Bonnell et al. 2008). This bias current allows the Langmuir sensors to lose a significant fraction of their photoelectrons and to float slightly below the local plasma potential, Φ_{plasma}. Therefore Langmuir sensor photoelectrons are accelerated to nearly $e(\Phi_{sc} - \Phi_{plasma})$ when they reach the ESA. At large spacecraft potentials, Langmuir sensor photoelectrons dominate over spacecraft photoelectrons at energies of $\sim e\Phi_{sc}$, but can generally be eliminated from moment calculations since they are confined to only one or two energy bins. However, when a cold ionospheric population of electrons is present in the plasma, as often happens in the inner magnetosphere or in the lobes, it may be difficult to cleanly separate the spacecraft or Langmuir sensor photoelectrons from the plasma. Current observations suggest most cold ionospheric electrons at THEMIS altitudes are warmer (~5–10 eV) than the spacecraft photoelectron population (~2 eV), so modeling these populations may provide a reasonable separation in a moment computations.

In addition to the Langmuir sensor photoelectrons, the EFI has several other antenna surfaces (usher, guard and braid) that can be voltage biased relative to the spacecraft. On THC from May 26 to June 22, 2007, the usher and guard were biased 8 V negative relative to the adjacent Langmuir sensor producing photoelectrons 8 eV above Langmuir sensor potential, or ~5 eV above spacecraft potential (axial antenna). These are illustrated in Fig. 9b where the axial antenna introduces photoelectrons at ~10 eV into two of the polar angle bins of the ESA. Similar photoelectrons from the radial antenna are not apparent. Prior to May 26 the usher was 6 V positive, but the guard was 20 V negative relative to the adjacent Langmuir sensor, resulting in electrons at ~19 eV above spacecraft potential. Although most of these electrons escaped to space, some are observed in the electron ESA, especially in the polar

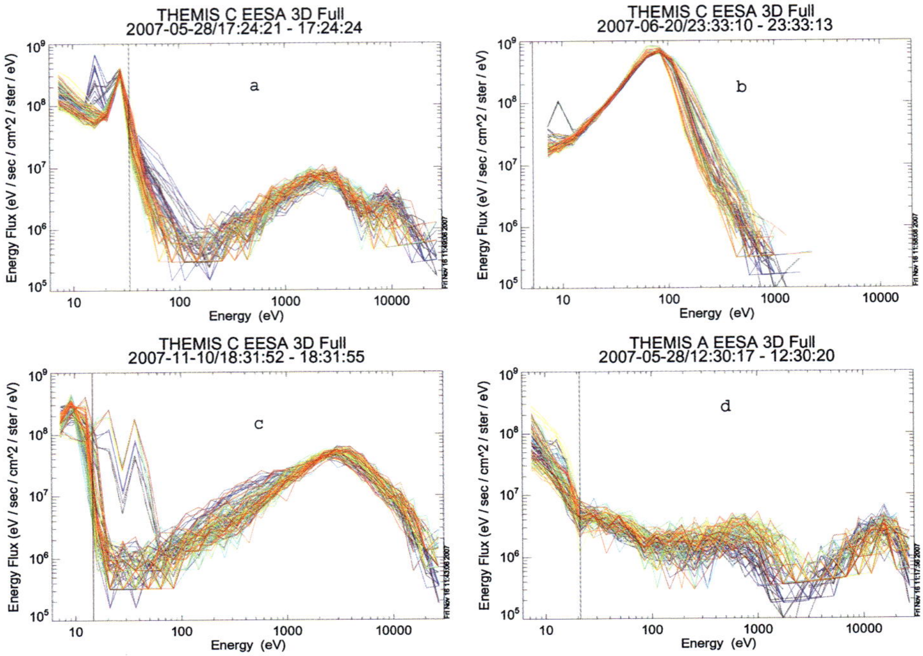

Fig. 9 Electron spectra from 88 different solid angle look directions. The spacecraft potential relative to the plasma is indicated by vertical lines. **a** Photoelectrons from the axial (15 eV) and radial (28 eV) Langmuir sensors. **b** A photoelectron peak (10 eV) above the spacecraft potential, and confined to just two of 88 angle bins, caused by the axial Langmuir sensor usher and guard being biased −8 V relative to the Langmuir sensor. **c** Photoelectrons (15–60 eV) produced in a 25 ms burst when EFI sensor shadowing resulted in the EFI braids charging to −85 V relative to the spacecraft. **d** Spacecraft photoelectrons at energies below the spacecraft potential measured prior to EFI antenna deployment

bins that look along the spacecraft surface and record photoelectrons from the axial antenna. After June 22, 2007, photoelectrons from the usher, guard, and braid generally appear at energies below the spacecraft potential. Similar photoelectrons were observed on THD and THE after their antenna deployments (June 2–6, 2007) until the bias changes on June 22, 2007.

Starting July 20, 2007 the braid on all radial antenna were driven at the same voltage as antenna 1's Langmuir sensor (on THC, THD, THE) to improve the EFI response. Starting about October 29, 2007 on THD and THE, and November 6 on THC, the spacecraft entered portions of the orbit where the spacecraft body shadowed the Langmuir sensors each spin for ∼25 ms. This resulted in the voltage on the shadowed sensor charging to −85 V relative to the spacecraft, the upper limit of the EFI power supplies. Since the braid on all four antenna were tied to Langmuir sensor 1, the sunlit braids on antennae 2, 3 and 4 would produce a ∼25 ms burst of photoelectrons that could be seen in the electron ESA as illustrated in Fig. 9c between 15 eV and 60 eV. The antenna shadowing ended by November 9, 2007 on THD and THE. By November 16, the braid on THC was switched back to ground to prevent further contamination. The current plans are to change EFI operating modes during shadowing periods to prevent this contamination in the future.

Spacecraft photoelectrons are the second most prominent contamination to the electron ESA. An electron spectrum taken prior to EFI antenna deployment is shown in Fig. 9d, with photoelectrons appearing below the spacecraft potential indicated by the vertical line at ~21 eV. Without EFI's measurement of Φ_{sc}, eliminating these photoelectrons from moment calculations is often difficult, with no clear spectral break to allow determination of Φ_{sc}. However, since photoelectrons are expected to have relatively constant spectra, Φ_{sc} could be estimated from the deviation between a measured low energy spectra and a characteristic photoelectron spectra.

Internally produced photoelectrons in the ESA aperture constitute a third form of photoelectron contamination. These electrons are produced over about 20° of spacecraft rotation centered on the sunward direction and generally enhance only the lowest energy (<10 eV) bins. These photoelectrons do not generally introduce errors to moment calculations unless the spacecraft potential is small. However, since small potentials occur during high densities, the impact of internally produced photoelectrons on moments is often minimized. Since these photoelectrons are confined to low energies and a few angular bins that look toward the sun, they can be easily removed from a distribution if needed for precise moment calculations.

3.1.3 *Scattered and Secondary Electrons, and Bremsstrahlung X-Rays*

A third source of contamination counts in ESA sensors results from internal scattering of primary plasma electrons, and the production of secondary electrons by these primaries. For shallow incidence angle, electrons have a ~50% chance of forward scattering with little energy loss. To reduce the fraction of these electrons that reach the detectors, the analyzer surfaces are scalloped and roughened with ebanol-C to reduce forward scattering. Although a small fraction of scattered primaries will reach the detectors, a larger problem is caused by incident primaries that strike surfaces near the analyzer entrance producing secondary electrons and degraded (in energy) primary electrons that can pass through the electron analyzer, mixing with lower energy incident primary electrons.

Figure 10 illustrates secondary electron production inside an electron ESA. The colored lines are electron spectra from the electron ESA (<30 keV) and electron SST (>30 keV). Different colors represent different look directions of the sensors, and the bunching of these spectra indicates the electrons were relatively isotropic. ESA data were formed by averaging six "snapshot" distributions, with each one spin snapshot taken every 128 spins. SST data were formed by averaging 12 snapshots over the same time interval. The black solid line that follows the electron spectra below 30 keV is the average omni-directional spectra over the interval. The majority of electrons below 200 eV are internally produced secondaries and energy-degraded primaries as shown below.

The data in Fig. 10 were taken while THB was in eclipse, so there are no photoelectrons contributing to the spectra. In addition the spacecraft had charged to about −2000 V, as determined from ion distribution functions, preventing low-energy in situ plasma electrons from reaching the sensor. Thus any primary electron registered at low energies, between the 0 to 1000 eV, must have initially been an electron between 2000 and 3000 eV in the plasma. Since the primary electrons are a good fit to a Maxwellian near the peak in the spectra, we use the dashed curve in the figure to show the expected differential energy flux of primaries after being retarded by the ~2000 V spacecraft potential. (Note: An energy shifted Maxwellian is still a Maxwellian). The deviation of the measured spectra from the dashed line at <1000 eV is due to secondary electrons and degraded primaries. Both of these populations are internally produced at the sensor aperture. Secondary electrons produced on spacecraft surfaces will not enter the ESA due to the large negative spacecraft potential.

The secondary and degraded primary spectra in Fig. 10 are similar to atmospheric secondary and degraded primary spectra calculated by Evans (1974) to explain the shape of the

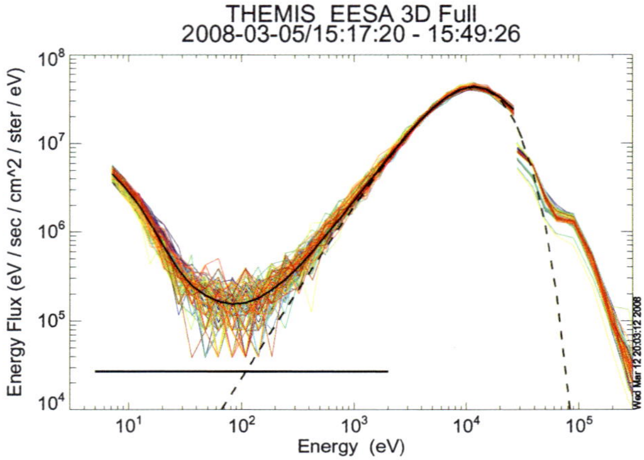

Fig. 10 Electron spectra measured by the ESA (<30 keV) and SST (>30 keV) during an eclipse when the spacecraft charged to about −2000 V. The *dashed curve* is the expected spectrum for an energy retarded Maxwellian. Deviation from the Maxwellian at low energies (<1000 eV) is due to internally produced secondary and degraded primary electrons. The *solid horizontal line* represents the expected contamination counts from scattered primaries and from Bremsstrahlung X-rays produced by energetic electrons striking the instrument and spacecraft

auroral electron spectra. The main difference is that atmospheric secondary electron fluxes are about an order of magnitude larger since they build up over multiple electron bounces between the auroral acceleration region and the atmosphere. Therefore internally produced secondary electrons add only a small error to auroral plasma measurements. However, for THEMIS high altitude measurements, where the fluxes of low-energy electrons can be very small, errors due to secondary electrons can be significant. For the distribution shown in Fig. 10, the error in the calculated density from including the internally produced secondaries is larger than the actual density.

Besides contamination due to secondary electrons, primary electrons can scatter through the analyzer or can produce Bremsstrahlung X-rays that penetrate through the analyzer, resulting in additional contamination counts in the detector. The black line at the bottom of Fig. 10 indicates the expected contamination from these sources as determined from contamination rates in the ion ESA. The ion sensor should record no ions at energies <2000 eV during the interval used in Fig. 10 since the spacecraft was charged to −2 kV. All ion sensor events recorded at <2 keV were therefore either background from scattered ∼10 keV electrons or from X-rays. (Ions have virtually no forward scattering.) As discussed below, correlation analysis indicates the majority of these contamination counts are from X-rays.

Energetic electrons and their Bremsstrahlung X-rays also create contamination in the ion ESA. One difference between electron and ion sensor sensitivity to these electrons results from the −2 kV potential at the front of the ion detector. This potential requires scattered electrons to be greater than 2 keV before they can be observed in the ion sensors, which virtually eliminates secondary electrons. Observations within the plasma sheet show that significant contamination in THEMIS ion sensors is primarily correlated with ∼100 keV electrons. This correlation with energetic electrons, rather than the much larger ∼10 keV electron fluxes, suggests that Bremsstrahlung x-rays are the primary source of this contamination.

Figure 11 shows an example of energetic electron contamination to an ion sensor. These observations are from THE within the plasma sheet. Panels a and b show the electrons and

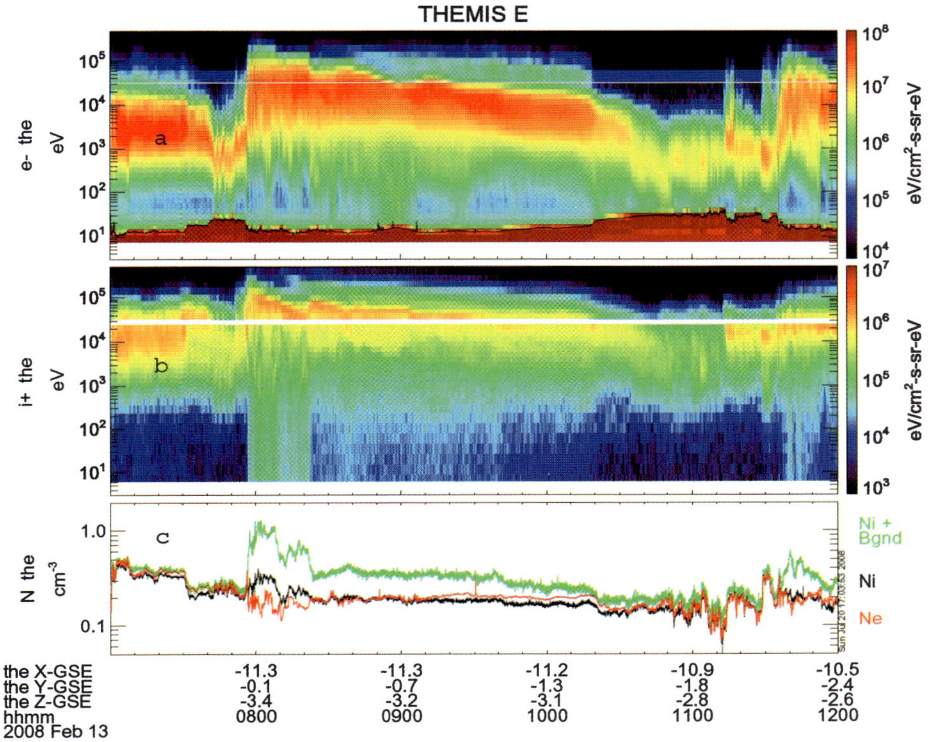

Fig. 11 Panel **a** shows electrons measured by the SST and ESA. Panel **b** is the same for ions. The figure illustrates problems caused by energetic >10 keV electrons scattering into the ion sensor and producing Bremsstrahlung X-rays that penetrate the ion sensor. This contamination is most easily seen at low energies (<300 eV) in panel **b**. Panel **c** illustrates the over-estimated ion density (*green*) when these background counts are included in the density integral. The *red curve* is the estimated electron density and the *black curve* is the estimated ion density integrated above 100 eV

ions measured by both the SST and ESA. During periods when the energetic electron flux is high, as at ∼8:00 UT, large increases in low-energy (<300 eV) ion counts are observed in the spectrogram (panel b). Most of the ion sensor's counts registered below 300 eV between 8:00 and 10:30 UT, and after 11:30 UT, are due to contamination. Panel c compares the uncorrected estimate of ion density (green), with the electron density (red) illustrating the large errors in estimated ion density when energetic electron contamination is present. The black curve in panel c is the estimated ion density integrated above 100 eV, which agrees much better with the electron density, showing that the density error is primarily due to low-energy counts. Algorithms are currently being developed to reduce or remove this contamination. Energetic electrons can also introduce errors in the ion flow velocity and ion temperature, therefore care must be taken to quantify any effects of this contamination before proceeding with moment computations.

3.1.4 Penetrating Radiation

A fourth source of background results from penetrating radiation as found in the inner magnetosphere. Generally these background counts become important at $\sim 6R_e$ geocentric. This background produces flat spectra, with increasing count rate that peaks at $\sim 4R_e$. There is often a dropout of these background counts at perigee, as the spacecraft dips below the radiation belts. Background subtractions that reduce this contamination have been developed for the FAST mission (K. Seki, private communication), and could be applied to THEMIS if required for science analysis.

3.2 Errors due to Measurement Limitations

In this section, we remind the reader of the primary measurement limitations that can prevent precise calculation of plasma distributions and moments of the distributions. We begin with the limitations to the ESA plasma measurement, describing the primary missing components and their impact on moment calculations. Measurement limitations are also associated with invalid information from other instruments, and we elaborate on problems with EFI potential measurements and spacecraft attitude.

3.2.1 Lack of Composition Information

Although the THEMIS ESA plasma instruments provide an accurate measurement of the bulk of the plasma in most regions of the magnetosphere, this instrument does not measure all of the plasma, nor does it measure all the properties of the plasma. Significant errors can occur due to composition changes. THEMIS software developed for ESA data analysis assumes the measured ions are protons. In a density calculation, a higher-mass ion's contribution to number density will be incorrectly underestimated by the factor $(m/q)^{1/2}$. For example, consider a magnetosheath measurement where an alpha to proton ratio, N_α/N_p, is 10%. N_e will equal $N_p + 2N_\alpha$. However, the calculated ion density assuming protons, N_{ci}, will be given by $N_p + N_\alpha/\sqrt{2}$. This will result in $N_e/N_{ci} \sim 1.12$, or a $\sim$12% difference in the calculated ion and electron densities. Mass density estimates can be skewed even more than number density, especially if significant oxygen is present.

Composition can also affect the velocity moment, with higher mass ions recorded as having higher velocity by the ratio of $(m/q)^{1/2}$. Therefore the fractional over-estimate of velocity will be similar to the fractional under-estimate in density for a plasma with non-proton components. Pressure is generally less affected by composition since pressure is proportional to particle energy and the sensor measures E/q. However temperature, given by the ratio of pressure to density, can be affected by the error in the density.

3.2.2 Spacecraft Charging

Spacecraft charging may prevent the measurement of portions of the plasma distribution. Since spacecraft generally charge positive to attract photoelectrons (when in sunlight and when the density is less than $\sim 300/\mathrm{cm}^3$), cold ions are often missed. As discussed in Sect. 2.1, the cold ions often dominate the density within the dayside magnetosphere. These missed ions will also affect other moment computations such as temperature or flux. However, computations of the pressure and velocity are generally not affected by this missing plasma since cold ions contribute little to the pressure and since the velocity can often be well determined by the asymmetry of hotter plasma components.

Although spacecraft charging does not prevent the measurement of various electron components in the plasma, it can still result in distortions of the electron distribution that are difficult to correct. In particular, when a cold (few eV) ionospheric electron component is present in the plasma, as often occurs on closed flux tubes, these electrons are difficult to resolve. For example, spacecraft charging of 20 V will cause the ESA sensor to measure cold electrons at ~20 eV, where the ESA's intrinsic energy resolution and energy sweep result in energy bins whose width is ~6 eV. The temperature of these cold electrons will not be resolved. In addition, if spacecraft photoelectrons are present, these wide energy bins will detect a mixture of cold plasma electrons and photoelectrons that is difficult, if not impossible, to separate. Since a cold electron component often accompanies a cold ion component, accurate density measurements in plasmaspheric plumes is doubly difficult.

3.2.3 Finite Energy Range

The upper limit to the ESA's energy range can be another source of error when calculating moments. Within the plasmasheet or inner magnetosphere, the ESA's upper energy limit of ~25 keV for ions, and ~30 keV for electrons, often results in a significant fraction of the plasma being missed. The limited energy range primarily impacts the higher order moments, such as pressure, but can also affect density and velocity calculations.

Figure 12 shows an example of plasma sheet measurements when both the ions and electrons extend above the ESA energy range (panels b and c). Panel d illustrates the fractional densities measured by the ESA and SST, while panel e illustrates the fractional pressures measured by these instruments. Clearly the ions measured by the SST are important for both the density and the pressure, and SST electrons occasionally have >10% of the density and pressure. To obtain better density estimates, ion integrals only counted >200 eV ions (to eliminate scattered electrons) and electron integrals only counted >30 eV electrons (to eliminate internally produced secondary electrons). We generally recommend that the SST data be combined with the ESA data when moment calculations are made in the plasma sheet or inner magnetosphere.

3.2.4 Limited Field of View

Two other measurement limitations in the ESA sensor are related to its field-of-view. Since the sensor only looks in a half-plane at any instant, time aliasing during a spin can skew the measurement. This is especially true for electron velocity moment calculations as illustrated in Fig. 13, panels f–g. A few percent density variation during a spin can result in large (~ 100 km/s) errors in the calculated flows, depending upon the electron temperature. For THEMIS these errors are primarily confined to the spin plane since the sensor continuously measures in both directions along the spin axis.

The second measurement limitation is the ~6° FWHM sensor field-of-view perpendicular to its 180° field-of-view. Since nominal energy sweeps occur during ~11.25° of rotation, narrow beams might be missed, especially if they are near the spin plane. Examples of such narrow beams would include counter-streaming field-aligned electron beams and the solar wind ion beam. The ESA instrument has a solar wind mode with 64 sweeps/spin (5.6° resolution), that can resolve the solar wind ions, however it has generally not been used in the solar wind.

3.2.5 Dead Time Corrections

For plasma sensors, lost counts due to instrument dead-time result from a combination of electronic and detector dead-time. The THEMIS electronic dead-time is 170 ± 10 ns for all Amptek A121 preamplifiers. Detector dead time, caused by a decrease in microchannel

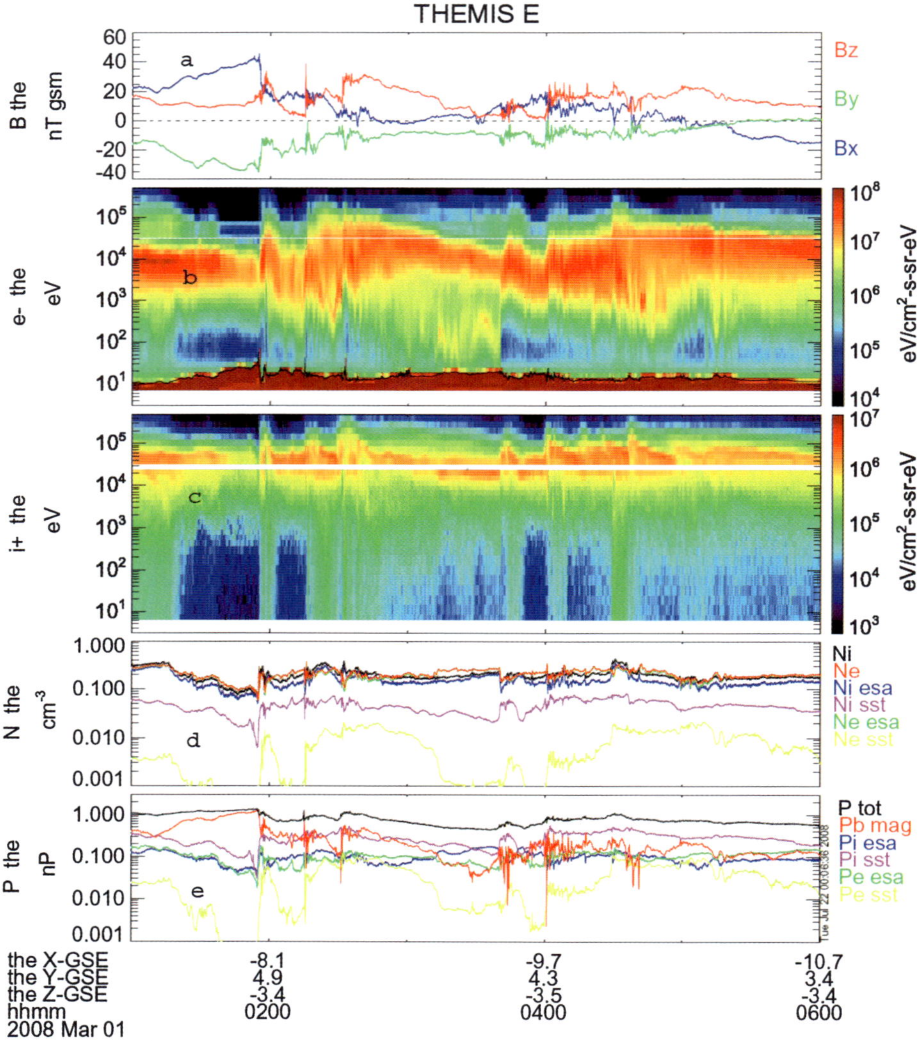

Fig. 12 **a** Magnetic field, **b** SST and ESA electrons, **c** SST and ESA ions, **d** density, and **e** pressure. Density and pressure are broken down by components to show the fractional contributions from each instrument. Ion moments are calculated for >200 eV ions to eliminate electron contamination at low energies

plate (MCP) gain at high count rates, which results in some events dropping below the preamplifier threshold, also adds to the overall instrument dead time. THEMIS ESAs were fitted with high current MCPs for fast recharging and therefore detector dead time is not expected to be important. An intense magnetosheath event shown in Fig. 12 of McFadden et al. (2008a) confirms that electronic dead time accounts for the majority of instrument dead time for fluxes distributed uniformly over a measurement sector. However, if a particle flux manifests as an intense narrow beam focused on a small portion of the detector, as may happen in the solar wind, then detector dead time may still be important.

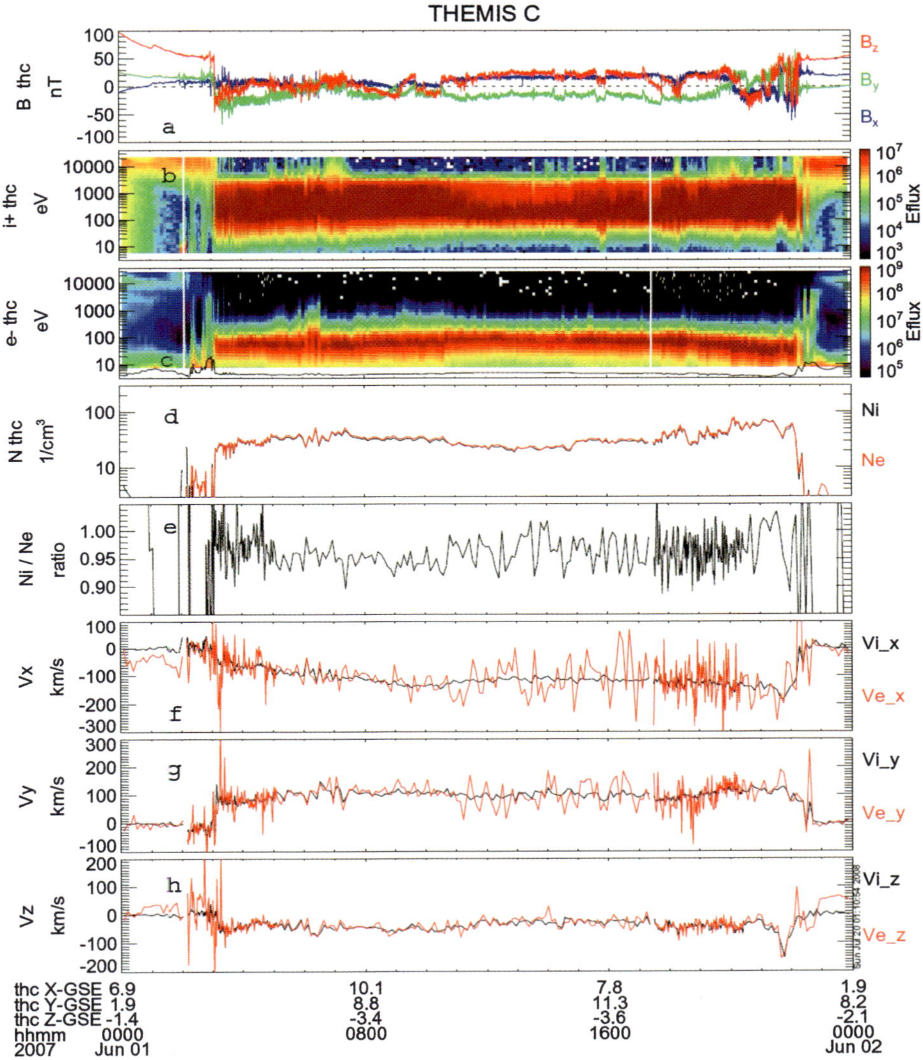

Fig. 13 **a** Magnetic field, **b** ion spectrogram, **c** electron spectrogram, **d** ion (*black*) and electron (*red*) density, **e** electron to ion density ratio, **f–g** components of velocity. Panels **f**, **g**, and **h** illustrates the errors in the calculated electron velocity caused by small density fluctuations during a spacecraft spin. The largest density fluctuations are due to mirror modes in the magnetosheath

For most magnetospheric and magnetosheath data, corrections to THEMIS ESA data for instrument dead-time are negligible. Significant dead time corrections (>10%) are only required for high particle fluxes as seen by the electron sensor in a high-density (>50 cm^{-3}) magentosheath, or as seen by the ion sensor in the solar wind. For solar wind ions, dead time corrections depend not only on the density, but also on the ion temperature and flow velocity. These corrections are generally important even at the lowest ($\sim$2 cm^{-3}) densities and can be a factor of 2 at densities of $\sim$10 cm^{-3}. If THEMIS project software is used, dead time corrections are handled as part of the conversion of counts to physical units.

3.2.6 EFI Availability

Since knowledge of spacecraft potential is essential for correct transformations of measured counts to plasma distribution function, EFI measurements are required. Start dates where valid EFI potential measurements are initially available are listed in Bonnell et al. (2008). Prior to EFI deployments, spacecraft potential will have to be estimated from the plasma measurements alone. Even after EFI sensor deployment, there are periods where the EFI data may be less than ideal. The EFI Langmuir sensors are designed for sunlit conditions, therefore when the spacecraft are in eclipse the sensor potential does not provide a good estimate of spacecraft-to-plasma potential. During sensor diagnostic sweeps, which are occasionally run and can take several hours, spacecraft potential measurements can be invalid. The bias currents to the Langmuir sensors, and the associated voltages on adjacent surfaces, have been changed several times during the first year resulting in different functional relationships between spacecraft-to-sensor potential and spacecraft-to-plasma potential. Lastly, during periods with restricted telemetry rates, EFI data may not be at the same cadence as the plasma data, resulting in time aliasing problems at steep density gradients. Since there are several sources of, or data products with, spacecraft-to-sensor potentials and since these sources have variable time resolution, care must be taken to select the appropriate data type as an input to the plasma data calculations.

3.2.7 Measurements in Eclipse

One additional source of error in the plasma measurements can occur during eclipse, when sun-sensor data is unavailable to organize spin-synchronous plasma data. As sun pulses disappear, the spacecraft electronics shifts into a "fly-wheel mode" that assumes the spin rate is constant. However, small changes in spin period due to thermal contraction of the antenna, magnetometer booms and fuel result in a drift in the orientation at the start of a spin. Although this drift is small, the accumulated error can be significant by the time the spacecraft exits the Earth's shadow. The primary error is therefore in the orientation of vector and tensor quantities such as the plasma velocity and pressure tensor. Errors to scalar quantities such as density and temperature are negligible. In principle these orientation errors could be corrected with modeling, but at the time of this publication there are no plans to develop this code since data collection during eclipse is not considered part of the baseline mission.

A second source of plasma measurement error during eclipse results from spacecraft charging. When a spacecraft encounters a hot ($\sim$10 keV) electron plasma while in eclipse, spacecraft potentials of several kilovolts negative can be observed. This charging will distort the measured distributions, eliminating low energy electrons and producing an ion spectral peak at $-e\Phi_{sc}$ as low energy ions are accelerated to the spacecraft. This charging is generally observed at distances of 6 to 10 R_e. Most eclipse passes do not result in this runaway charging, however even when spacecraft potentials are moderate, determination of the potential is difficult since EFI does not operate properly in shadow. In this case, features of the particle distributions will have to be used to identify Φ_{sc} for moment computations.

3.3 Errors due to Data Formatting Problems

At the start of the THEMIS mission, several data formatting problems in the ETC board were discovered. The ETC board processes the counts data from the ESA and SST, averaging and compressing these data before handing the data products to the instrument processor board for data packet formation. The ETC contains mapping tables that are loaded from a PROM at the start of a mode. For ESA 3-D data products (survey, burst and reduced data products), energy maps and an angle maps are used to direct and sum the counter readouts into data product arrays. Early in the mission it was discovered that an occasional bit error could occur in the table load if the processor was simultaneously performing other tasks. These bit errors generally caused some elements of the data product arrays to be zero and resulted in other data products receiving these counts. These bit errors were generally confined to only a few of the 30 3-D data products generated by the 5 satellites, with the subset of tainted products changing with each mode change. The misdirection of data was discovered early in commissioning phase, and software changes to eliminate processor conflicts during table loads were implemented on all spacecraft by April 27, 2007. Prior to this date, care should be taken in interpreting ESA data, especially any moment computations.

A second data formatting error occurs at the transition between instrument modes. Instrument mode transitions are associated with both configuration changes, such as transitions from magnetospheric mode to solar wind mode, and operational changes, such as transitions from Fast Survey to Slow Survey. In both cases, the transition between table maps and data packet formatting result in the loss of data. Depending upon the data product, and in particular the number of spin-snapshots in a data packet, these transitions can result in a data loss, or incorrectly formatted data, that lasts for a few seconds to a few minutes.

A third data formatting problem results from ETC counter saturation and is confined to reduced data products. During slow survey mode, reduced data packets are formed by averaging all counts over a spin into a single energy spectrum. High count rate data, such as electron data in the magnetosheath, often result in counter saturation at the peak in the spectrum. The ETC is designed not to overflow and saturation is easily recognized in the data as a flattened peak in the spectra. If the spectra are plotted as counts, the maximum count in any energy-angle bin is 65,535. During the most intense magnetosheath events, saturations can be observed in the ion, slow-survey, reduced data products, or in the electron, fast-survey and slow-survey, reduced data products.

Early in the mission, data formatting problems also plagued the onboard moment computations performed by the ETC. In particular the same table load bit errors seen in the 3-D products were present in the moment calculations. Since the moment tables are several orders of magnitude larger than the 3-D data product tables, detection and correction of errors is nearly impossible. In addition, an error in the PROM resulted in the loss of one of the components of the velocity moment (Vy), and incorrect ordering of higher moment components in the moment data packet. Although some onboard moment data can be extracted from these early data, we strongly recommend working with THEMIS team members before incorporating any early-mission moment data into science papers. Corrections to the flight software were not implemented until August 6, 2007, and additional problems with spacecraft potential corrections to on-board moments were present between November 18 and 22, 2007. Lastly, we point out that electron moments computed onboard are generally invalid until EFI antenna deployments allow corrections for spacecraft charging. For THB and THA these antenna deployments were completed on November 18, 2007 and January 13, 2008, respectively.

4 Summary

The THEMIS ESA plasma instruments measure the 3-D distribution function of electrons and ions at 3 second cadence. Instrument design and calibration can be found in the companion paper (McFadden et al. 2008a), which includes a description of the in-flight calibration which provides very accurate inter-calibration of these sensors. In this first results paper we demonstrate the capabilities of the ESA instruments including their ability to resolve ion scale phenomena, to separate spatial and temporal structure, and to reveal new details of the dayside magnetosphere. Observations are presented of plasmaspheric plumes, auroral ionospheric outflows, field line resonances, the low latitude boundary layer, flux transfer events, and structure at the quasi-parallel bow shock. Although the highest-quality ESA burst data is limited to ~30–60 minutes per orbit, coarser 3-D plasma distributions at spin resolution are available for ~12 hours each orbit with adequate resolution for moment computation or detection of field-aligned beams. Even in the slowest data collection mode, spin-resolution energy spectra and onboard moment computations provide adequate information for interpretation of plasma structure. Accurate calibrations allow the combined electron and ion data to be used to deduce additional features about the plasma, including mass composition or the presence of missed cold plasma. These first result observations illustrate the capabilities of the plasma sensors and the synergy of its measurements with the other THEMIS experiments, demonstrating the successful achievement of all measurement objectives.

Last of all, we point out that THEMIS has an open data policy that strives for an immediate release of data to the community. These data are made available before data quality can be determined, or before the data can be validated. Therefore, this paper includes discussions of various performance issues with the ESA instrument, such as sources of sensor background, measurement limitations, and data formatting problems. It is hoped that this discussion provides scientists with an adequate reference so that understanding and correcting for these performance issues will allow full use of the THEMIS measurement capabilities while avoiding any misinterpretation of the observations.

Acknowledgements The analysis of THEMIS data was supported by NASA NAS5-02099. Financial support for the work of the FGM Lead Investigator Team at the Technical University of Braunschweig by the German Ministerium für Wirtschaft und Technologie and the Deutsches Zentrum für Luft- und Raumfahrt under grant 50QP0402 is acknowledged.

References

B.J. Anderson, M.J. Engebretson, S.P. Rounds, L.J. Zanetti, T.A. Potemra, J. Geophys. Res. **95**, 10495–10524 (1990)

K.A. Anderson, R.P. Lin, F. Martel, C.S. Lin, G.K. Parks, H. Reme, Geophys. Res. Lett. **6**, 401 (1979)

M. Andre, P. Norqvist, L. Andersson, L. Eliasson, A.I. Eriksson, L. Blomberg, R.E. Erlandson, J. Waldemark, J. Geophys. Res. **103**, 4199 (1998)

V. Angelopoulos, Space Sci. Rev. (2008). doi:10.1007/s11214-008-9336-1

J.R. Asbridge, S.J. Bame, I.B. Strong, J. Geophys. Res. **73**, 5777 (1968)

H.U. Auster, K.H. Glassmeier, W. Magnes, O. Aydogar, D. Constantinescu, D. Fischer, K.H. Fornacon, E. Georgescu, P. Harvey, O. Hillenmaier, R. Kroth, M. Ludlam, Y. Narita, K. Okrafka, F. Plaschke, I. Richter, H. Schwarzi, B. Stoll, A. Valavanoglu, M. Wiedemann, Space Sci. Rev. (2008). doi:10.1007/s11214-008-9365-9

S.D. Bale, P.J. Kellogg, D.E. Larson, R.P. Lin, K. Goetz, R.P. Lepping, Geophys. Res. Lett. **25**, 2929 (1998)

J.W. Bonnell, F.S. Mozer, G.T. Delory, A.J. Hull, R.E. Ergun, C.M. Cully, Space Sci. Rev. (2008, this issue)

L.J. Cahill, N.G. Lin, M.J. Engebretson, D.R. Weimer, M. Sugiura, J. Geophys. Res. **91**, 8895–8908 (1986)

C.W. Carlson, J.P. McFadden, R.E. Ergun, M. Temerin, W. Peria, F.S. Mozer, D.M. Klumpar, E.G. Shelley, W.K. Peterson, E. Moebius, R. Ephic, R. Strangeway, C. Cattell, R. Pfaff, Geophys. Res. Lett. **25**, 2017 (1998)

C.W. Carlson, J.P. McFadden, Geophys. Monogr. **102**, 125 (1998)

D.L. Carpenter, B.L. Giles, C.R. Chappell, P.M.E. Decreau, R.R. Anderson, A.M. Persoon, A.J. Smith, Y. Corcuff, P. Canu, J. Geophys. Res. **98**, 19243 (1993)

M.O. Chandler, T.E. Moore, Geophys. Res. Lett. **30**(16), 1689 (2003). doi:10.1029/2003GL017611

T. Chang, C.B. Crew, N. Hershkowitz, J.R. Jasperse, J.M. Retterer, J.D. Winningham, Geophys. Res. Lett. **7**, 636 (1986)

C.C. Chaston, J.W. Bonnell, C.W. Carlson, J.P. McFadden, R.E. Ergun, R.J. Strangeway, E.J. Lund, J. Geophys. Res. **109**, A04205 (2004). doi:10.1029/2003JA010053

C. Chaston, J. Bonnell, J. McFadden, C. Carlson, C.M. Cully, O. LeContel, A. Roux, U. Auster, K.-H. Glassmeier, V. Angelopoulos, C.T. Russell, Geophys. Res. Lett. **35**, L17S08 (2008). doi:10.1029/2008GL033601

C.R. Chappell, J. Geophys. Res. **79**, 1861 (1974)

L. Chen, A. Hasegawa, J. Geophys. Res. **79**, 1024–1032 (1974)

S.-H. Chen, T.E. Moore, J. Geophys. Res. **111**, A03215 (2006). doi:10.1029/2005JA011084

P.D. Craven, D.L. Gallagher, R.H. Comfort, J. Geophys. Res. **102**, 2279 (1997)

D.W. Dungey, Sci. Rep. 69, Penn. State Univ., University Park, 1954

J.P. Eastwood, D.G. Sibeck, V. Angelopoulos, T.D. Phan, S.D. Bale, J.P. McFadden, C.M. Cully, S.B. Mende, D. Larson, S. Frey, C.W. Carlson, K.-H. Glassmeier, H.U. Auster, A. Roux, O. Le Contel, Geophys. Res. Lett. **35**, L17S03 (2008). doi:10.1029/2008GL033475

R.C. Elphic, Geophys. Monogr. **90**, 225–233 (1995)

R.E. Ergun, C.W. Carlson, J.P. McFadden, F.S. Mozer, G.T. Delory, W. Peria, C.C. Chaston, M. Temerin, M. Temerin, I. Roth, L. Muschietti, R. Elphic, R. Strangeway, R. Pfaff, C.A. Cattell, D.M. Klumpar, E. Shelley, W. Peterson, E. Moebius, L. Kistler, Geophys. Res. Lett. **25**, 2041 (1998)

D. Evans, J. Geophys. Res. **79**, 2853–2858 (1974)

D.H. Fairfield, J. Geophys. Res. **74**, 3541 (1969)

S.A. Fuselier, B.J. Anderson, T.G. Onsager, J. Geophys. Res. **100**, 11805 (1995)

K.-H. Glassmeier, C. Othmer, R. Cramm, M. Stellmacher, M. Engebretson, Surv. Geophys. **20**, 61–109 (1999)

K.-H. Glassmeier, H. Auster, D. Constantinescu, K. Fornacon, Y. Narita, F. Plaschke, V. Angelopoulos, E. Georgescu, W. Baumjohann, W. Magnes, R. Nakamura, C.W. Carlson, S. Frey, J.P. McFadden, T.D. Phan, I.R. Mann, I.J. Rae, J. Vogt, Geophys. Res. Lett. **35**, L17S01 (2008). doi:10.1029/2008GL033469

J. Goldstein, B.R. Sandel, M.F. Thomsen, M. Spasojevic, P.H. Reiff, J. Geophys. Res. **109**, A03202 (2004). doi:10.1029/2003JA010281

D.J. Gorney, Y.T. Chiu, D.R. Croley, J. Geophys. Res. **90**, 4205 (1985)

J.T. Gosling, J.R. Asbridge, S.J. Bame, G. Paschmann, N. Sckopke, Geophys. Res. Lett. **5**, 957–960 (1978)

J.T. Gosling, M.F. Thomsen, S.J. Bame, R.C. Elphic, Geophys. Res. Lett. **17**(12), 2245–2248 (1990a)

J.T. Gosling, M.F. Thomsen, S.J. Bame, R.C. Elphic, C.T. Russell, J. Geophys. Res. **95**, 8073 (1990b)

E.W. Greenstadt, I.M. Green, G.T. Inouye, A.J. Hundhausen, S.J. Bame, I.B. Strong, J. Geophys. Res. **73**, 51 (1968)

D.A. Gurnett, R.R. Anderson, F.L. Scarf, R.W. Fredricks, E.J. Smith, Space Sci. Rev. **23**, 103 (1979)

W.J. Hughes, Geophys. Monogr. **81**, 1 (1994)

A. Keiling, K.-H. Kim, J.R. Wygant, C. Cattell, C.T. Russell, C.A. Kletzing, J. Geophys. Res. **108**(A7), 1275 (2003). doi:10.1029/2002JA009340

M.G. Kivelson, D.J. Southwood, Geophys. Res. Lett. **12**, 49 (1985)

M.G. Kivelson, D.J. Southwood, J. Geophys. Res. **91**, 4345–4351 (1986)

S. Kokubun, R.L. McPherron, C.T. Russell, J. Geophys. Res. **81**(28), 5141–5149 (1976)

D. Larson, T. Moreau, R. Lee, R. Canario, R.P. Lin, Space Sci. Rev. (2008, this issue)

G. Le, C.T. Russell, M.F. Thomsen, J.T. Gosling, J. Geophys. Res. **97**, 2917 (1992)

G. Le, C.T. Russell, J.T. Gosling, M.F. Thomsen, J. Geophys. Res. **101**, 27239 (1996)

W. Lennartsson, E.G. Shelley, J. Geophys. Res. **91**, 3061–3076 (1986)

G. Li, G.P. Zank, C.T. Russell (eds.), in *"The Physics of Collisionless Shocks", 4th Annual IGPP International Astrophysics Conference*. AIP Conference Proceedings/Astronomy and Astrophysics, 2005

C.S. Lin, J.N. Barfield, J. Geophys. Res. **90**, 11075 (1985)

J. Liu, V. Angelopoulos, D. Sibeck, T. Phan, Z.Y. Pu, J. McFadden, K.H. Glassmeier, H.U. Auster, Geophys. Res. Lett. **35**, L17S07 (2008). doi:10.1029/2008GL033673

W. Lotko, A.V. Streltsov, C.W. Carlson, Geophys. Res. Lett. **25**, 4449–4452 (1998)

A.T.Y. Lui, D.G. Sibeck, T. Phan, V. Angelopoulos, J. McFadden, C. Carlson, D. Larson, J. Bonnell, K.-H. Glassmeier, S. Frey, Geophys. Res. Lett. **35**, L17S05 (2008). doi:10.1029/2007GL032933

E.J. Lund, E. Moebius, L. Tang, L.M. Kistler, M.A. Popecki, D.M. Klumpar, W.K. Peterson, E.G. Shelley, B. Klecker, D. Hovestadt, M. Temerin, R.E. Ergun, J.P. McFadden, C.W. Carlson, F.S. Mozer, R.C. Elphic, R.J. Strangeway, C.A. Cattell, R.F. Pfaff, Geophys. Res. Lett. **25**, 2049 (1998)

E.J. Lund, E. Moebius, K.A. Lynch, D.M. Klumpar, W.K. Peterson, R.E. Ergun, C.W. Carlson, Phys. Chem. Earth **26**, 161 (2001)

J.P. McFadden, Y.K. Tung, C.W. Carlson, R.J. Strangeway, E. Moebius, L.M. Kistler, Space Weather, Geophys. Monogr. **125**, 413 (2001)

J.P. McFadden, C.W. Carlson, D. Larson, V. Angelopoulos, Space Sci. Rev. (2008a, this issue)

J.P. McFadden, C.W. Carlson, D. Larson, J. Bonnell, F.S. Mozer, V. Angelopoulos, K.-H. Glassmeier, U. Auster, Geophys. Res. Lett. **35**, L17 (2008b). doi:10.1029/2008GL033677
J.P. McFadden, T.D. Phan, C.W. Carlson, V. Angelopoulos, K.-H. Glassmeier, U. Auster, Geophys. Res. Lett. **35**, L17 (2008c). doi:10.1029/2008GL033630
D.G. Mitchell, F. Kutchko, D.J. Williams, T.E. Eastman, L.A. Frank, C.T. Russell, J. Geophys. Res. **92**, 7394 (1987)
E. Moebius, L. Tang, L.M. Kistler, M. Popecki, E.J. Lund, D. Klumpar, W. Peterson, E.G. Shelley, B. Klecker, D. Hovestadt, C.W. Carlson, R. Ergun, J.P. McFadden, F. Mozer, M. Temerin, C. Cattell, R. Elphic, R. Strangeway, R. Pfaff, Geophys. Res. Lett. **25**, 2029 (1998)
T.E. Moore, Rev. Geophys. (Suppl.) 1039–1048 (1991)
F.S. Mozer, V. Angelopoulos, J. Bonnell, K.-H. Glassmeier, J.P. McFadden, Geophys. Res. Lett. **35**, L17S04 (2008). doi:10.1029/2007GL033033
P.T. Newell, T. Onsager (eds.), Geophys. Monogr. 133 (2003)
M. Oieroset, T.D. Phan, V. Angelopoulos, J.P. Eastwood, J. McFadden, D. Larson, C.W. Carlson, K.-H. Glassmeier, M. Fujimoto, J. Raeder, Geophys. Res. Lett. **35**, L17S11 (2008). doi:10.1029/2008GL033661
G.K. Parks, E. Lee, F. Mozer, M. Wilber, E. Lucek, I. Dandouras, H. Rème, C. Mazelle, J.B. Cao, K. Meziane, M.L. Goldstein, P. Escoubet, Phys. Plasmas **13**, 050701 (2006)
G. Paschmann, N. Sckopke, S.J. Bame, J.R. Asbridge, J.T. Gosling, C.T. Russell, E.W. Greenstadt, Geophys. Res. Lett. **6**, 209 (1979)
G. Paschmann, G. Haerendel, I. Papamastorakis, N. Sckopke, S. Bame, J. Gosling, C. Russell, J. Geophys. Res. **87**, 2159 (1982a)
G. Paschmann, N. Sckopke, S.J. Bame, J.T. Gosling, Geophys. Res. Lett. **9**, 88–91 (1982b)
G. Paschmann, S. Haaland, R. Treumann, (eds.), *Auroral Plasma Physics*. Reprinted from Space Sci. Rev. 103(1–4) (2002)
G. Paschmann, S.J. Schwartz, C.P. Escoubet, S. Haaland, (eds.), *Outer Magnetospheric Boundaries: Cluster Results*. Reprinted from Space Sci. Rev. 118(1–4) (2005)
J. Raeder, Ann. Geophys. **24**, 381–392 (2006)
J.M. Ruohoniemi, R.A. Greenwald, K.B. Baker, J.C. Samson, J. Geophys. Res. **96**, 15 (1991)
A. Roux, O. Le Contel, C. Coillot, A. Bouabdellah, B. de la Porte, D. Alison, S. Ruocco, M.C. Vassal, Space Sci. Rev. (2008, this issue)
C.T. Russell, R.C. Elphic, Geophys. Res. Lett. **6**, 33–36 (1979)
C.T. Russell (ed.), Adv. Space Res. 15 (1995)
T. Sakurai, Y. Tonegawa, Y. Shinkai, K. Yumoto, S. Kokubun, K. Tsuruda, T. Mukai, Earth Planets Space **53**, 843–849 (2001)
J.C. Samson, J.A. Jacobs, G. Rostoker, J. Geophys. Res. **76**, 3675–3683 (1971)
J.C. Samson, B.G. Harrold, J.M. Ruohoniemi, R.A. Greenwald, A.D.M. Walker, Geophys. Res. Lett. **19**, 441 (1992)
J.C. Samson, L.L. Cogger, Q. Pao, J. Geophys. Res. **101**, 17 (1996)
J.A. Sauvaud et al., Ann. Geophys. **19**, 1523 (2001)
S. Schwartz, Adv. Space Res. **15**, 107–116 (1995)
S.J. Schwartz, D. Burgess, W.P. Wilkinson, R.L. Kessel, M. Dunlop, H. Luhr, J. Geophys. Res. **97**, 4209 (1992)
R.D. Sharp, R.G. Johnson, E.G. Shelley, J. Geophys. Res. **82**, 3324 (1977)
E.G. Shelley, R.D. Sharp, R.G. Johnson, Geophys. Res. Lett. **11**, 654 (1976)
D.G. Sibeck, M. Kuznetsova, V. Angelopoulos, K.-H. Glassmeier, J.P. McFadden, Geophys. Res. Lett. **35**, L17S06 (2008). doi:10.1029/2008GL033568
H.J. Singer, W.J. Hughes, C.T. Russell, J. Geophys. Res. **87**, 3519 (1982)
D.J. Southwood, Planet Space Sci. **22**, 483–491 (1974)
K. Stasiewicz, P. Bellan, C. Chaston, C. Kletzing, R. Lysak, J. Maggs, O. Pokhotelov, C. Seyler, P. Shukla, L. Stenflo, A. Streltsov, J.E. Wahlund, Space Sci. Rev. **92**, 423 (2000)
R.J. Strangeway, R.E. Ergun, Y.-J. Su, C.W. Carlson, R.C. Elphic, J. Geophys. Res. **110**, A03221 (2005). doi:10.1029/2004JA010829
Y.-J. Su, J.E. Borovsky, M.F. Thomsen, R.C. Elphic, D.J. McComas, J. Geophys. Res. **105**, 7591 (2000)
Y.J. Su, M.F. Thomsen, J.E. Borovsky, R.C. Elphic, D.J. Lawrence, D.J. McComas, J. Atmos. Sol.-Terr. Phys. **63**, 1185 (2001)
K. Takahashi, Ann. Geophys. **16**, 787–803 (1998)
Y.-K. Tung, C.W. Carlson, J.P. McFadden, D.M. Klumpar, G.K. Parks, W.J. Peria, K. Liou, J. Geophys. Res. **106**, 3603 (2001)
J.M. Turner, L.F. Burlaga, N.F. Ness, J.F. Lemaire, J. Geophys. Res. **82**, 1921–1924 (1977)
A.D.M. Walker, R.A. Greenwald, W.F. Stuart, C.A. Green, J. Geophys. Res. **84**, 3373 (1979)
A.W. Yau, P.H. Beckwith, W.K. Peterson, E.G. Shelley, J. Geophys. Res. **90**, 6395–6407 (1985)
Y. Zheng, A.T.Y. Lui, I.R. Mann, K. Takahashi, J. Watermann, S.-H. Chen, I.J. Rae, T. Mukai, C.T. Russell, A. Balogh, R.F. Pfaff, H. Reme, Ann. Geophys. **24**, 707–723 (2006)

First Results of the THEMIS Search Coil Magnetometers

O. Le Contel · A. Roux · P. Robert · C. Coillot · A. Bouabdellah · B. de la Porte · D. Alison · S. Ruocco · V. Angelopoulos · K. Bromund · C.C. Chaston · C. Cully · H.U. Auster · K.H. Glassmeier · W. Baumjohann · C.W. Carlson · J.P. McFadden · D. Larson

Originally published in the journal Space Science Reviews, Volume 141, Nos 1–4, 509–534.
DOI: 10.1007/s11214-008-9371-y

Abstract We present the first data from the THEMIS Search Coil Magnetometers (SCM), taken between March and June 2007 while the THEMIS constellation apogee moved from the duskside toward the dawnside. Data reduction, especially the SCM calibration method and spurious noise reduction process, is described. The signatures of magnetic fluctuations in key magnetospheric regions such as the bow shock, the magnetopause and the magnetotail during a substorm, are described. We also discuss the role that magnetic fluctuations could play in plasma transport, acceleration and heating.

O. Le Contel (✉) · A. Roux · P. Robert · C. Coillot · A. Bouabdellah · B. de la Porte · D. Alison · S. Ruocco
Centre d'étude des Environnements Terrestre et Planétaires (CETP), 10-12 avenue de l Europe, 78140 Vélizy, France
e-mail: olivier.lecontel@cetp.ipsl.fr

V. Angelopoulos
IGPP/UCLA, Los Angeles, CA 90095, USA

K. Bromund
SP Systems, Inc. on contract to NASA/GSFC, Space Weather Laboratory, Code 674, Greenbelt, MD, USA

C.C. Chaston
Space Sciences Laboratory, University of California, Berkeley, CA, USA

C. Cully
Laboratory for Atmospheric and Space Physics, University of Colorado, Boulder, CO, USA

H.U. Auster · K.H. Glassmeier
Institut für Geophysik und extraterrestrische Physik der Technischen Universität Braunschweig, 38106 Braunschweig, Germany

W. Baumjohann
Space Research Institute, Austrian Academy of Sciences, Graz, Austria

C.W. Carlson · J.P. McFadden · D. Larson
Space Sciences Laboratory, University of California, Berkeley, CA, USA

Keywords THEMIS · Solar wind · Shock · Magnetosheath · Flux transfer event · Magnetopause · Substorm · Search-coil · ULF/ELF magnetic waves

1 Introduction

The identification of the instability leading to substorm breakup and expansion is a key issue for magnetospheric physics and beyond. Indeed similar explosive processes are known to occur in other astrophysical contexts, such as the solar corona, and in laboratory machines designed for controlled fusion. In all cases the accumulated magnetic energy is released explosively, thereby leading to fast changes in the magnetic configuration and particle acceleration. These plasmas being hot and dilute, binary collisions are rare and cannot supply the dissipation needed for the development of instabilities that can break the magnetic configuration, such as the collisionless tearing instability. Kinetic effects, in particular associated with the development of waves, are expected to take over the role that binary collisions cannot fulfill.

Together with the Electric field Instrument (EFI) (Bonnell et al. 2008), the THEMIS Search Coil Magnetometer (SCM) (Roux et al. 2008) will be used to identify the possible role of waves at substorm breakup and expansion phase. It will also allow the remote tracking of the motion of active regions, via ducted waves.

The association between substorm onset and intense emissions of ULF/ELF/VLF waves has been known for quite a long time (Gendrin 1970; Russel 1972; Gurnett et al. 1976). Later, the search coils magnetometers onboard the geostationary European satellite GEOS-2, launched in 1978, detected magnetic impulsive signals in the range 0.5–11.5 Hz (ULF) at substorm breakup (Robert et al. 1984), characterized by particle injection and changes from a tail-like to a more dipolar magnetic configuration. Robert et al. interpreted the Short Irregular Pulsations (SIPs) observed at breakup as the signatures of small scale field aligned current structures passing by the spacecraft (Robert et al. 1984). Due to their non-steadiness, these structures can also be interpreted as kinetic Alfvén waves in the proton gyrofrequency (0.1–1 Hz) range (Perraut et al. 1993). Kremser et al. showed that electron parallel acceleration takes place inside these structures (Kremser et al. 1988).

On the basis of AMPTE/CCE data ($X_{GSM} \simeq -8.8\ R_E$), Lui et al. suggested that the cross field current driven instability triggers substorm onset (Lui et al. 1991, 1996). Farther in the magnetotail ($X_{GSM} \simeq -15.2\ R_E$), observations from the Japanese Geotail satellite give evidence for the existence of electromagnetic waves in the lower hybrid frequency range during a small substorm (Shinohara et al. 1998). Yet the authors concluded that wave energy is too small to supply enough dissipation, via anomalous resistivity, to allow a resistive-mode instability to develop. On the theoretical side, Cheng and Lui proposed another solution to resolve the substorm onset enigma; they suggested that the coupling between high frequency (cross tail current instability) and low frequency (ballooning mode) electromagnetic fluctuations accounts for the fast occurrence of the breakup onset (Cheng and Lui 1998). As the Geotail apogee moved closer to the earth ($X_{GSM} \simeq -10$ to $-13\ R_E$), magnetic fluctuations were also observed during substorm in the proton gyrofrequency (0.1–1 Hz) range as well as in the lower hybrid frequency range (5–16 Hz). However, according to Sigsbee et al. these fluctuations reach their maximum amplitude after onset and could not therefore be considered as a substorm trigger (Sigsbee et al. 2001).

Further analysis of GEOS-2 data (at the geostationary orbit) suggested that substorm onset and impulsive plasma transport is controlled by a micro-instability with frequencies around the proton gyrofrequency such as a parallel current instability (Perraut et al. 2000a,

2000b; Le Contel et al. 2001a, 2001b). The regular occurrence at different radial distances (at the geostationary orbit with GEOS-2, at 10 R_E with Geotail and 20 R_E with Cluster) of wave activity around the proton gyrofrequency at substorm onset was also noted by Le Contel et al. (2002). Yet more recent analysis of Geotail observations in the near-earth plasma sheet ($X_{GSM} \simeq -8.3\ R_E$) gave evidence for large amplitude wave emissions (5–15 nT s^{-1}) in the lower hybrid frequency range (5–20 Hz), just prior to the dipolarization (Shiokawa et al. 2005). Thus the identification of the intense waves that develop at substorm onset, and the elucidation of their potential role at triggering breakup, is still a matter of debate.

In this context the determination of the radial distribution of wave characteristics is an important issue that should be resolved thanks to THEMIS wave instrumentation. The THEMIS orbit is well suited to achieve this goal (Angelopoulos et al. 2008; Sibeck and Angelopoulos 2008a), but the full characterization of the waves is made difficult by the rapid changes of the magnetic configuration during substorm and by the strong inhomogeneity of the medium. In order to avoid adding confusion in the debate it would be wise to use the same parameter, namely the amplitude B of the waveform, rather than its derivative (ωB). Indeed the use of the derivative can give the false impression that amplitudes are larger at larger frequencies.

At higher frequencies, whistler waves were identified in the magnetotail and proposed as a way to monitor energetic electrons and processes of reconnection (Zhang et al. 1999). More recently, Cluster observations from the STAFF instrument (Cornilleau-Wehrlin et al. 2003) gave evidence for very large amplitude ($\simeq$ 1 nT) waves in the whistler mode range ($f_{ci} < f < f_{ce}$), emitted in the magnetotail during substorms (Le Contel et al. 2006). These intense emissions last only a few seconds and are associated with very thin current sheets ($\leq \rho_i$, the proton Larmor radius). They occur in conjunction with accelerated electrons. Whistler mode waves are expected to be produced in some of the magnetic reconnection theories (Mandt et al. 1994). Yet, at present time, it is unclear whether they are a by-product of reconnection or whether they effectively play a crucial role as a trigger of the substorm process. Further studies are therefore needed in order to clarify the role of these different kinds of wave during various substorm phases.

Substorms are not the only space physics process for which waves could play a crucial role. Magnetic reconnection and plasma transport at the magnetopause can be modeled as a global process controlled by the level of ULF wave activity (see for instance Rezeau and Belmont 2001). Magnetosheath turbulence is fundamentally based on mirror mode wave activity (Sahraoui et al. 2003). Dissipative mechanisms for collisionless shocks are also thought to be strongly related to wave activity (Krasnosselskikh et al. 2003). Therefore the THEMIS SCM instruments should not only gather crucial data for substorm studies but also for all the main fundamental processes controlling the physics of collisionless plasmas, namely shocks, magnetic reconnection, turbulence, plasma acceleration, transport and heating. In Sect. 2, we briefly present the science objectives also detailed in a companion paper (Roux et al. 2008). Section 3 is devoted to the presentation of the data reduction especially the calibration and spurious noise reduction process needed for the scientific use of SCM data. Preliminary science results are discussed in Sect. 4 for different regions, from the solar wind to the magnetotail, via magnetosheath and magnetopause.

2 Science Objectives

The primary goal of the Time History and Macroscale Interaction during Substorms (THEMIS) mission is to establish when and where substorms start, and to use this information to resolve the controversy about what instability triggers them (Angelopoulos et al.

2008). The main difference between the two types of models is the sequence of events that leads to the breakup (see for instance Lui 2001; Baumjohann et al. 2007). For the first type of model, magnetic reconnection (MR), presumably associated with the development of the tearing instability, is the trigger. In the second type of model, labeled as "Current Disruption" or CD, the breakup is triggered by a reduction in the cross-tail current associated with the development of an instability. In both cases ULF and ELF waves are expected to play a critical role.

In recent versions of the MR models (Mandt et al. 1994), whistler mode waves are expected to accelerate electrons up to very large (super-Alfvénic) velocities, thereby enhancing the reconnection rate. It has also been suggested (Bulanov et al. 1992; Attico et al. 2002) that very thin current sheets can be destabilized directly by HF tearing in the whistler mode range. On the other hand, in CD models the cross tail current can be directly disrupted by HF cross field instabilities, or undergo a low frequency (ballooning mode) instability coupled via div $J = 0$ to parallel currents, which eventually drives ion cyclotron and/or whistler mode instabilities. The modeling of the latter instabilities is made difficult by the fact that electron bounce frequency in the magnetotail is around the proton gyrofrequency. Therefore the proper handling of HF frequency current driven instabilities should take into account the electron bounce motion and the corresponding electron bounce resonance (Karpman et al. 1977). Thus the identification of the waves, at substorm breakup (and during the expansion phase), is an important clue toward understanding what triggers substorms.

Together with the EFI instrument, which measures electric fields in the same frequency range, the Search Coil Magnetometer (SCM) will determine the nature of the waves that develop at breakup and during the expansion phase and, in the case of guided waves, remotely track the active region where breakup starts. The SCM is also needed to assess whether waves are electrostatic (such as lower hybrid waves) or electromagnetic (such as whistler mode waves). Thus wave observations provide a critical test to substorm scenarios and could provide a remote sensing of substorm dynamics during the expansion phase.

3 Data Reduction

3.1 Reminder on THEMIS Coordinate Systems

Raw data are recorded in the sensor magnetic coordinate system for which the axes are defined by the magnetic axes of the search coils (see also Roux et al. 2008). Figure 1 displays the SCM antennas at the tip of the boom and gives the angles between the antenna axes and the probe geometric axes. In order to express data in a geophysical frame such as the GSE we need to perform the following coordinate transformations by the application of appropriate rotation matrices (see for more details Quinn et al. 2006):

(1) Data are moved from the sensor magnetic coordinate system to the sensor mechanical coordinate system (SMC); actually these frames can be considered as identical.
(2) Then data are moved from this latter frame to the spinning probe geometrical coordinate system (SPG) for which X axis corresponds to the ProbeX axis (see Fig. 1) and the Z axis is along the geometric Z axis (U2G matrix, see Table 1), the origin being at the geometric probe center.
(3) Data are transformed from the SPG frame to the spinning sunsensor L-vectorZ (SSL) coordinate system, for which the X is directed toward the sun and the Z axis along the spin axis (G2S matrix, see Table 1).

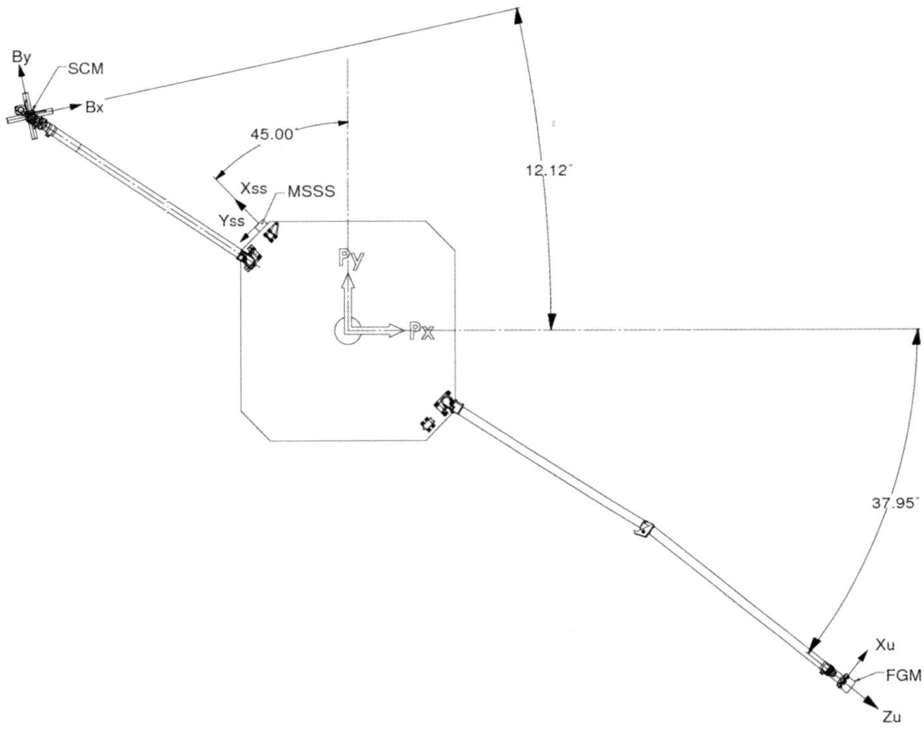

Fig. 1 Detail of sun sensor coordinates from SSL/UCB

Table 1 Matrices of rotations: Matrix **U2G** from Sensor mechanical coordinates (SMC) to probe geometric coordinates (SPG); Matrix **G2S** from Probe geometric coord. (SPG) to probe spin-Sun sensor coord. (SSL)

U2G			G2S		
0.9777	−0.2100	0.00000	−0.7071	0.7071	0.0000
0.2100	0.9777	0.00000	−0.7071	−0.7071	0.0044
0.0000	0.0000	1.00000	0.0031	0.0031	1.0000

(4) Finally, data are moved into a nonspinning system named Despun Sun L-vectorZ (DSL) obtained by a rotation about the spin axis by an angle equal to the opposite of the spin phase in the direction of the spin; the X axis is directed toward the sun and the Z axis corresponds to the spin axis.

3.2 SCM Calibration Method

The raw signal in Volts must be calibrated to physical units (nT). Each antenna response is characterized by its own transfer function, giving the ratio V/nT for a given frequency (for more details see Roux et al. 2008). These functions not being linear in frequency, a dedicated process must be applied to calibrate the raw waveforms. Basically two methods of calibration exist:

(1) Perform the Fourier Transformation (FT) of the signal, on a given time period, then divide by the complex transfer function to take into account amplitude correction and

phase shift; note that one has to fix a lower cut off frequency f_c as the transfer function goes to zero at null frequency. Finally perform an inverse FT to get the calibrated signal in the time domain.

(2) Deconvolve the instrument impulse response from the signal in the time domain.

The second method is implemented in the SCM calibration routine of the THEMIS software package; this method permits us to use optimized IDL convolution routines and gives better results in terms of computing time. The convolution is performed using a sliding window in order to provide a continuous calibration process.

Calibration of a SCM mounted onboard a spinning spacecraft with a spin frequency in the antenna bandwidth such as THEMIS probes requires additional steps compared with laboratory calibration. Indeed the components of the DC magnetic field perpendicular to the spin axis are measured as a sinusoid with a large amplitude at the spin frequency. Since this DC field ($\simeq$ 100 nT) is about 100 times larger than the wave amplitude ($\simeq$1 nT), it has to be removed from the raw signal before calibration to avoid undesirable effects. Furthermore the spinning motion at f_s introduces a Doppler shift: a circular wave at frequency f_L turning in the opposite direction of the satellite is measured by the sensor as a wave at frequency $f_L + f_s$ whereas a wave turning in the same direction at f_R is measured as a wave at frequency $f_R - f_s$. Therefore sensors mounted onboard a spinning spacecraft are not able to fully restitute spin-plane fluctuations with frequency around the spin frequency. Indeed any circular waves turning in the same direction as the spin rotation are not detected because their apparent frequency becomes null. This also means that the sensitivity of the experiment at low frequency depends on the polarization of the waves with respect to the spin axis. Practically it is recommended before analyzing waves in a fixed frame (DSL, GSE, GSM, ...) to low-pass filter the data at a minimum frequency higher than the spin frequency; typically Fmin could be fixed to the sum of the spin frequency f_s and the cut off frequency f_c.

We distinguish a few different steps in the SCM calibration process (also described in the header of the THEMIS IDL calibration routine for scm data called thm_cal_scm available in the THEMIS software package). Different kinds of output may be useful to different users, and can be obtained using the following values of the step keyword in the IDL function:

step 0: *waveform in counts unit*
Each data gap is time tagged and NaN symbols (Not a Number) are inserted for proper plotting. This step corresponds to raw data expressed in telemetry units.

step 1: *waveform in Volts, spinning sensor system, with DC field*
The conversion factor from telemetry units to volts is applied to the data but the spin modulation is still present.

step 2: *waveform in Volts, spinning sensor system, without DC field*
By using a sliding window of about two spin periods, we estimate the amplitude and phase of the spin signal on the three (x, y, z) components of the signal delivered by the sensors; the x-y amplitude and phase provides a measurement of the component perpendicular to the spin axis of DC magnetic field. This DC magnetic field can be expressed in the DSL system and can be compared to the FGM data, while the amplitude along Z allows the computation of the misalignment angle between the Z sensor axis and the spin axis. Note that the number of spin periods used for the sliding window can be fixed by the n_spinfit keyword. At this step, these DC field components are removed from the signal. Furthermore, a special detrend method can be applied to perform a more efficient rejection of the spin signal and its harmonics, but frequencies below the detrend frequency are strongly reduced. This treatment can be applied using the Fdet keyword that allows specification of the detrend frequency. Finally during this step, a special noise reduction method can be also

applied to reject spacecraft-related spurious tones characterized during the commissioning. The spurious noise reduction process is described in the next subsection.

step 3: *waveform in nT, spinning sensor system, without DC field*
Calibrated waveforms for which the DC magnetic field variation has been removed. Note that data are still in a spinning frame associated with the sensors. The calibration is performed using a convolution kernel with a number of points fixed by the nk keyword. The choice of the value of nk involves a trade-off between quality of the low-frequency calibration and data processing speed. By default, an optimal value is chosen for each of the data modes, as a multiple of the sample frequency. For fast survey mode (scf mode with $fe = 8$ S/s nominally, where fe is the sampling frequency), $nk = 8 \cdot fe$. For particle burst mode (scp mode with $fe = 128$ S/s nominally), $nk = 4 \cdot fe$ and for wave burst mode (scw mode with $fe = 8192$ S/s nominally), $nk = fe$. The mk keyword can be used to select a different multiple of the sampling frequency.

step 4: *waveform in nT, spinning SSL system, without DC field*
Same as step 3 but now data are in the SSL system using matrix of Table 1.

step 5: *waveform in nT, fixed DSL system, without DC field, filtered*
Data are now calibrated, without DC magnetic field variations, and are projected to a fixed frame common to all instruments (DSL). After this step, the SCM waveform can be transformed into any physical frame (GSE, GSM, etc.) using the cotrans routine available in the THEMIS software package. Note that after coordinate transformations the three original components will be mixed. Therefore it is important to ensure that before the change of frame the parallel and perpendicular components to the spin axis have undergone the same level of filtering. Thus at this step it is recommended to low-filter the three components with the same minimum frequency (fmin being fixed also as a keyword).

step 6: *waveform in nT, fixed DSL system, with xy DC field*
Same as step 5 but the calculated DC magnetic field components perpendicular to the spin axis are added to the waveform. X-Y components can be compared to the FGM data for cross-calibration. The Z component is unchanged.

3.3 Spurious noise reduction process

First data analysis during commissioning showed that two types of noise with unexpectedly high amplitudes were present in the SCM waveform data. First, the power system produces a tone at twice the spin frequency (1/3 Hz) and its harmonics. As expected, the level of these tones decreases during eclipse period. Secondly, tones at 8 and 32 Hz, and their harmonics, were found to dominate the spectrum at higher frequencies. The frequencies of these tones correspond to the frequencies of instrument clocks onboard the spacecraft. At the present time it has not been possible to identify the exact source of the spurious noise. However we know that the 8 and 32 Hz are due to radiated noise, as their levels strongly decreased after SCM boom deployment (−10 dB).

Fortunately both type of spurious noises are locked in phase and relatively constant in amplitude. Spin tones are locked to the spin phase whereas the 8/32 Hz tones are phase locked with the onboard 1 s instrument clock. Thanks to this property it was suggested that a superposed epoch analysis (SEA) could reduce the level of these two types of phase locked noise (C. Chaston; private communication). Therefore a noise reduction process was developed based on two successive SEAs. A SEA consists of cutting the waveform data of duration T into N windows of definite duration ($t_w \ll T$), named hereafter averaging windows. These N windows are superposed, or summed, which gives an average profile of the phase locked noise. Then a "noise waveform" of duration T is built by duplicating N

times the averaged noise. Finally the noise waveform at appropriate phase is subtracted from the raw waveform.

The spurious noise reduction process can be summarized as: (1) a first SEA with an averaging window equal to a multiple of the spin period, (2) a second SEA with an averaging window equal to a multiple of 1 s. The noise reduction process has been implemented in the current THEMIS software; two versions of the noise reduction routine are available and can be activated (or not) in the SCM calibration routine by fixing the keyword clnup_author (clnup_author = 'ole' by default or can be 'ccc'). Basically they give same results but algorithms are slightly different (see code sources for more details). Different levels of noise reduction can be selected for each version by using specific keywords (cleanup = 'spin' for only spin tone cleanup or 'full' for spin tones and $8/32$ Hz cleanup). Note that the duration in seconds of the first and second averaging windows can be fixed by the keywords wind_dur_spin and wind_dur_1s respectively.

Figure 2 displays the different stages of the noise reduction process on particle burst data (128 S/s). Note that some spikes remain after this process. They come from noise which is not phase locked therefore not eliminated by SEA.

Figure 3 displays spectra performed from SCM wave burst data (scw mode with 8192 S/s) on March 23rd 2007 between 135945 and 140216 UT. Data are despun and projected into the sensor coordinates system in order to be compared with data measured at the laboratory. As in the example of figure 2, 8/32 Hz tones are strongly reduced by noise reduction process but still present in the spectra as other spikes which are not phase locked. However, we see that these in flight measurements give the same Noise Equivalent Magnetic Inductions (NEMI) as reported in (Roux et al. 2008): 0.4 pT/$\sqrt{\text{Hz}}$ at 10 Hz, 0.08 pT/$\sqrt{\text{Hz}}$ at 100 Hz for Bz and 0.15 for Bx and By and 0.01 pT/$\sqrt{\text{Hz}}$ at 1 kHz. In-flight NEMIs can be smaller than NEMIs in the laboratory if the in-flight temperatures of the preamplifiers and sensors are lower than at laboratory (273 K). Note that the higher level of noise on Bx and By between 15 Hz and 1 kHz is found only on Bx in the SSL frame, and is therefore related to the sun sensor direction (Ludlam et al. 2008).

4 Science Results

In this section we present first examples of THEMIS data from the solar wind to the magnetotail.

4.1 Solar wind–magnetosheath–dayside magnetosphere regions: June 21st 2007 event

Figure 4 shows the position in the GSE frame of the THEMIS constellation on June 21st from 0800 to 1050 UT. Around 0800 UT, THEMIS-A (THA) is the farthest probe from the earth whereas THB is the closest one; probes c, d and e are slightly separated along Y but almost at the same X in between a and b. The selected time period displayed in Fig. 5 corresponds to a fast survey mode. In this mode, FGM data (Auster et al. 2008) have 0.25 s time resolution (panels a, b, c and d). Ion ESA data (McFadden et al. 2008) including moments (n_i, V_i, T_i) have 3 s time resolution (panels e, f, g, and h), and electron fluxes are available with the same time resolution (panel i). SCM waveform data have 0.125 s time resolution (scf mode) from which we construct spectra at 32 s resolution (panels j, k, l, and m).

For most of the selected time period THA is located in the solar wind as illustrated by Fig. 5. Indeed from 0800 to 1015 UT, the magnetic field modulus is smaller than 10 nT (panel d), the average ion energy is around 800 eV (panel e) with $n_i \simeq 10$ p·cm^{-3} (panel f), $V_x \simeq -350$ km·s^{-1} (panel g) and $T_i \leq 50$ eV (panel h).

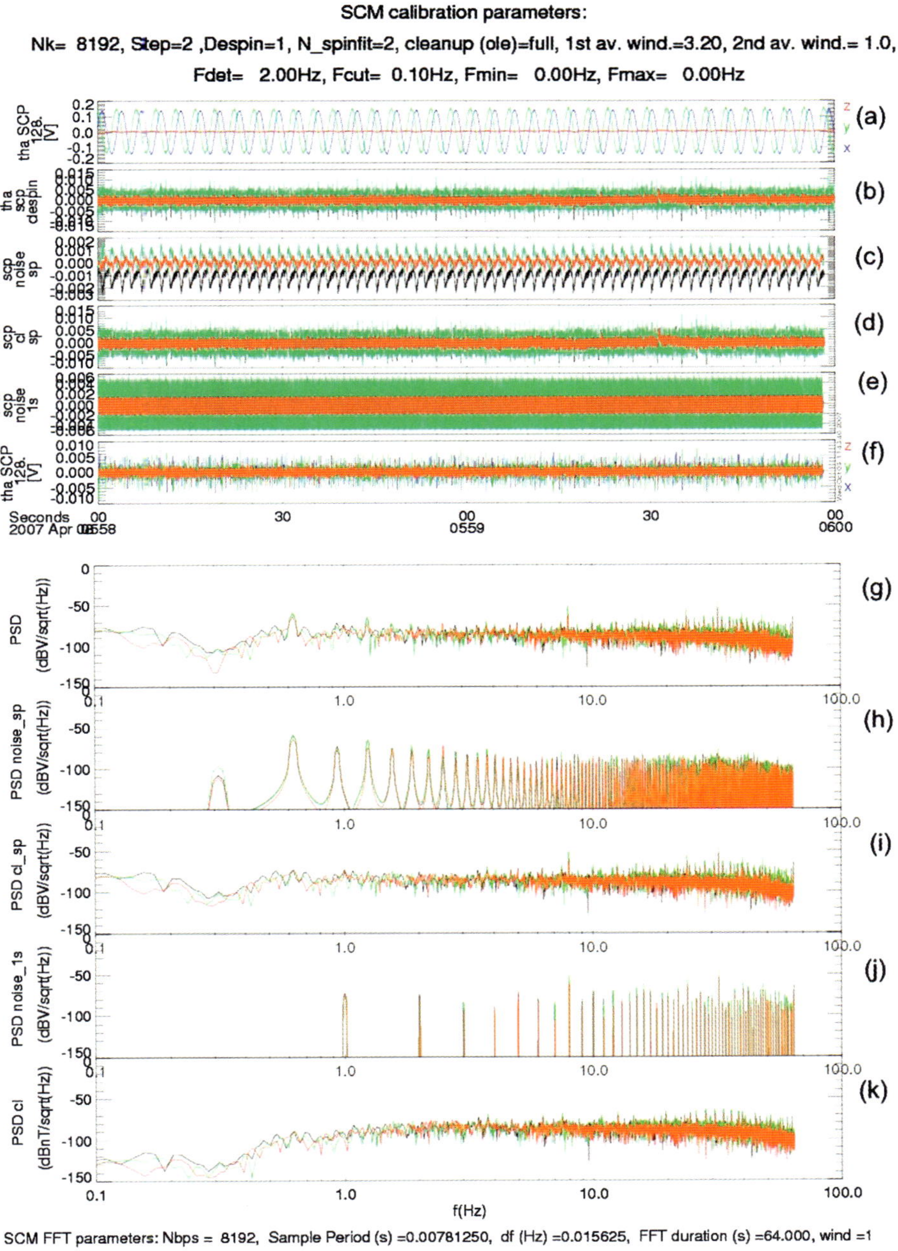

Fig. 2 Search coil particle burst data recorded by THEMIS-A (THA) on April 8th, 2007 between 0558 and 0600 UT. From top to bottom: (**a**) scm raw waveform in volts, (**b**) despinned waveform, (**c**) spin phase locked noise built by applying SEA, (**d**) cleaned (only power ripples) waveform, (**e**) 1 s phase locked noise waveform (SEA), (**f**) fully cleaned waveform, corresponding spectra in dBV/$\sqrt{\text{Hz}}$ follow in the same order except for despinned waveform data: (**g**) spectrum of (b), (**h**) spectrum of (c), (**i**) spectrum of (d), (**j**) spectrum of (e), and (**k**) spectrum of (f). Tones at $2f_S$, $4f_S$, 8 and 32 Hz are strongly reduced. However, note that at the end of noise reduction process (panel **k**) some spikes are still visible at high frequencies due to the fact that they come from noise which is not phase locked

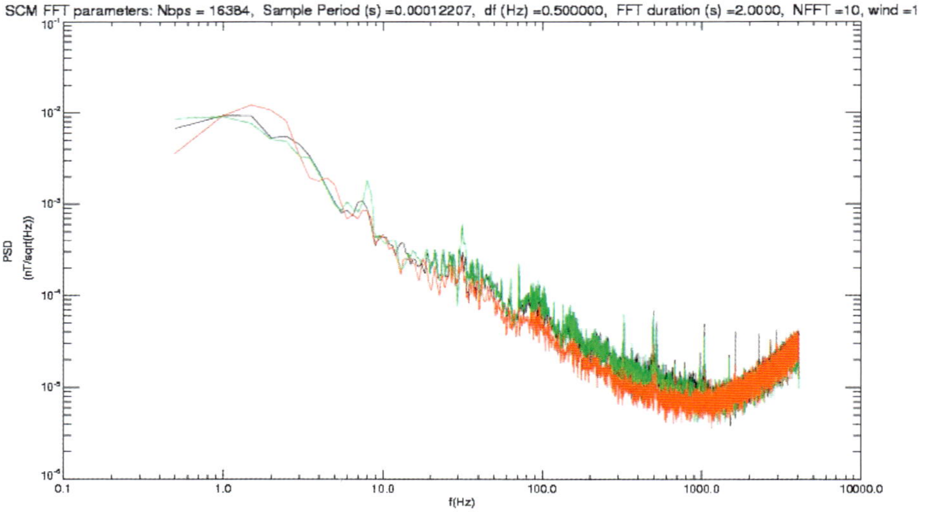

Fig. 3 Spectra of SCM data during a wave burst period (scw mode) on March 23rd 2007 from 1359:45 to 1400:16 UT. Data are in sensor coordinate system (step 3), with nk = 16384, Despin = 1, Fdet = 2, and a full noise reduction with wind_dur_1s =1

Average electron energy (panel i) is no more than 15 eV which is consistent with the characteristics of the slow solar wind around 1 astronomical unit. At 1015 UT THA crosses the shock and the modulus of B increases from 5 to 15 nT, the V_x component of the ion velocity decreases from 400 to 150 km$\cdot$s^{-1}, the ion density increases from 10 to more than 40 p$\cdot$cm^{-3} and the ion temperature increases from less than 70 eV to more than 200 eV. Therefore the solar wind slows down and is heated as it goes through the collisionless shock. The amplitude of magnetic fluctuations in the ultra low frequency (ULF) range is maximum at the shock crossing ($\simeq$few nT) with frequencies up to (at least) 4 Hz, while it is less intense ($\leq$ 1 nT) and mainly below 1 Hz in the magnetosheath after 1014 UT.

In addition to the shock crossing described above, wave and ESA data give evidence for smaller perturbations (0825, 0850, 0925 and 1000 UT) which look like very brief shock crossings or approaches. Notably the sharp signature around 0850 UT has the same characteristic as the shock crossing (velocity decrease, temperature and density increases, strong wave activity up to 4 Hz). It could correspond to a fast sunward and then earthward motion of the shock. Unlike the 1050 UT shock crossing, however, energetic ions are also detected together with the magnetic perturbations, which tells us that the probe penetrates into the foreshock before possibly crossing the shock.

In Fig. 6, the same set of data is displayed, but now gathered by THB (the closest to the earth). THB is almost always into the magnetosheath during the selected time period except between 0920 and 0930 UT when it briefly crosses the shock, and after 1020 UT when it enters into the magnetosphere. During the period where probe b is in the magnetosheath, the amplitude of ULF magnetic fluctuations varies strongly (panels j, k, l and m); probe b records three intensifications (0825, 0850, and 1000 UT) up to 15 nT, well above the average magnetosheath amplitude ($\leq$ 1–2 nT) in addition to the intensification associated with the shock crossing at 0925 UT. These intensifications seem to correspond in a one-to-one fashion with the small perturbations observed on THA data in the solar wind (discussed

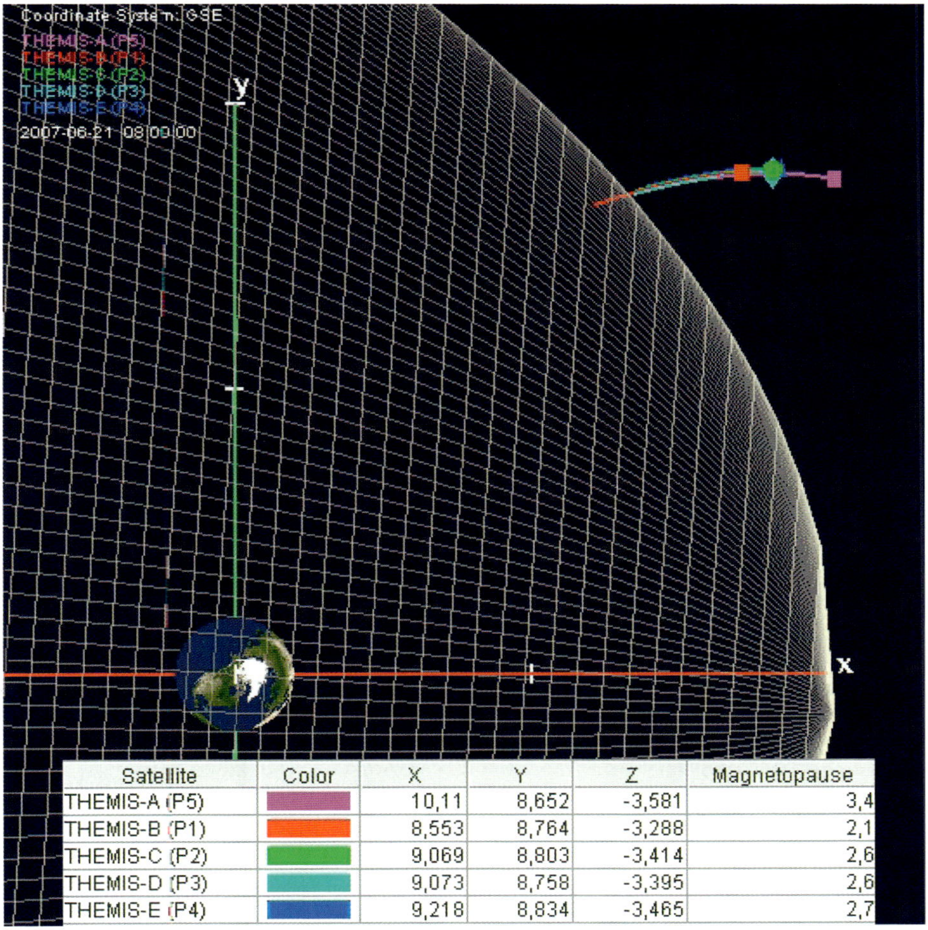

Fig. 4 THEMIS probe locations (GSE, R_E) on June 21st 2007 between 0800 and 1050 UT (This product is available at http://sscweb.gsfc.nasa.gov/tipsod). Satellite locations are also tabulated in GSE coordinates at 0800 UT. Magnetopause location for a solar wind dynamic pressure of 3.5 nPA (obtained from WIND survey) is also shown as well as distance to the magnetopause for each probe

above). Yet the amplitudes of the magnetic fluctuations are larger by a factor of 5–10 in the magnetosheath, and an amplification mechanism has to be invoked.

Also, the magnetopause crossing detected by THB (1018 UT) turns out to occur four minutes later than the shock crossing detected by probe a (1014 UT), which suggests a sunward global motion of both the magnetopause and the shock. In addition to this global outward motion, quasi-periodic perturbations convected by the solar wind lead to ion foreshock crossings and even skimming along the bow shock. Cluster and Geotail are located close to the magnetopause in the dawnside and dusk sector, respectively. They also detect magnetopause crossings during the same time period. Thus this event is particularly well suited to conduct a global analysis of the shock/magnetopause response to solar wind perturbations.

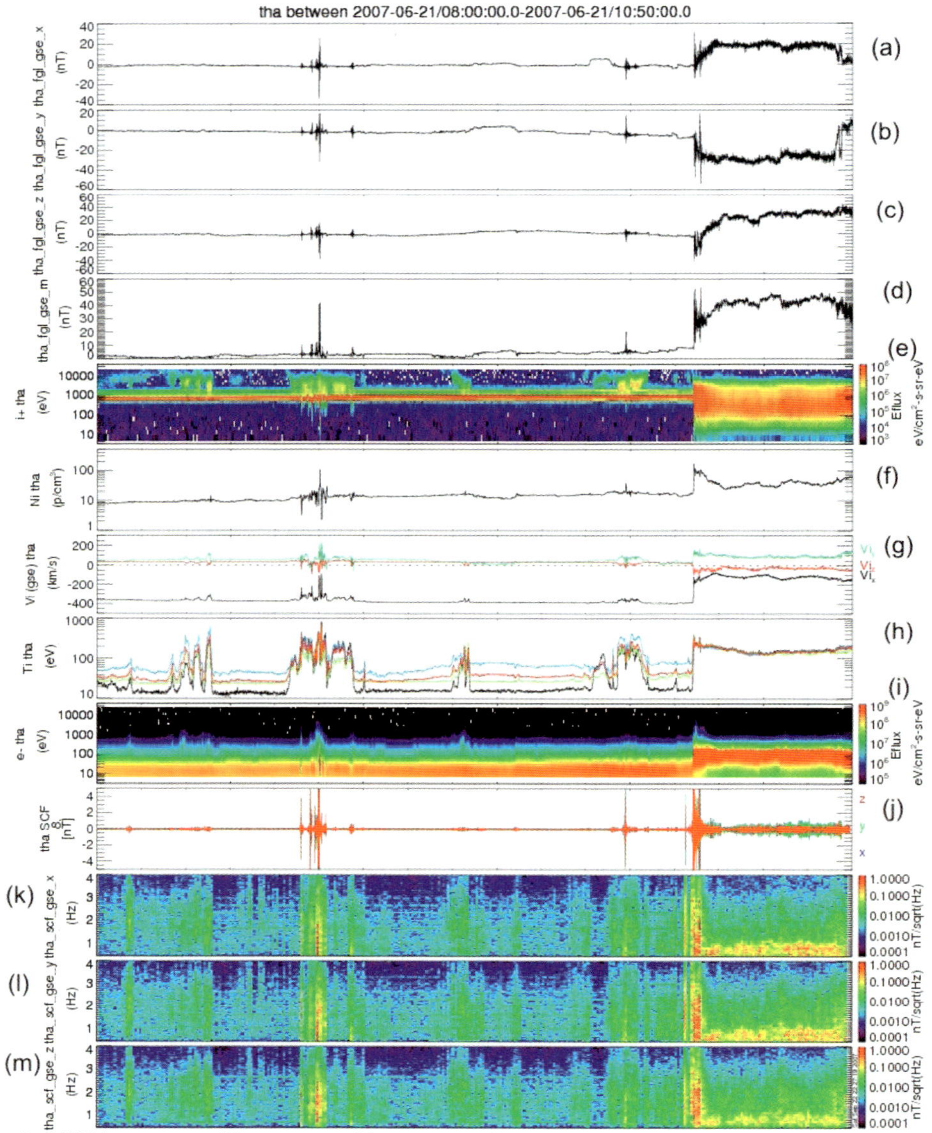

Fig. 5 From *top* to *bottom*: panels **a**, **b**, **c**, and **d** correspond to Bx, By, Bz and B (FGM data), ion energy spectrogram (panel **e**), density (**f**), velocity (**g**) and temperature (**h**), panel **i** corresponds to electron energy spectrogram (ESA data), panels **j**, **k**, **l**, and **m** display waveforms and spectrograms of Bx, By and Bz fluctuations from 0.45 to 4 Hz, (SCM data)

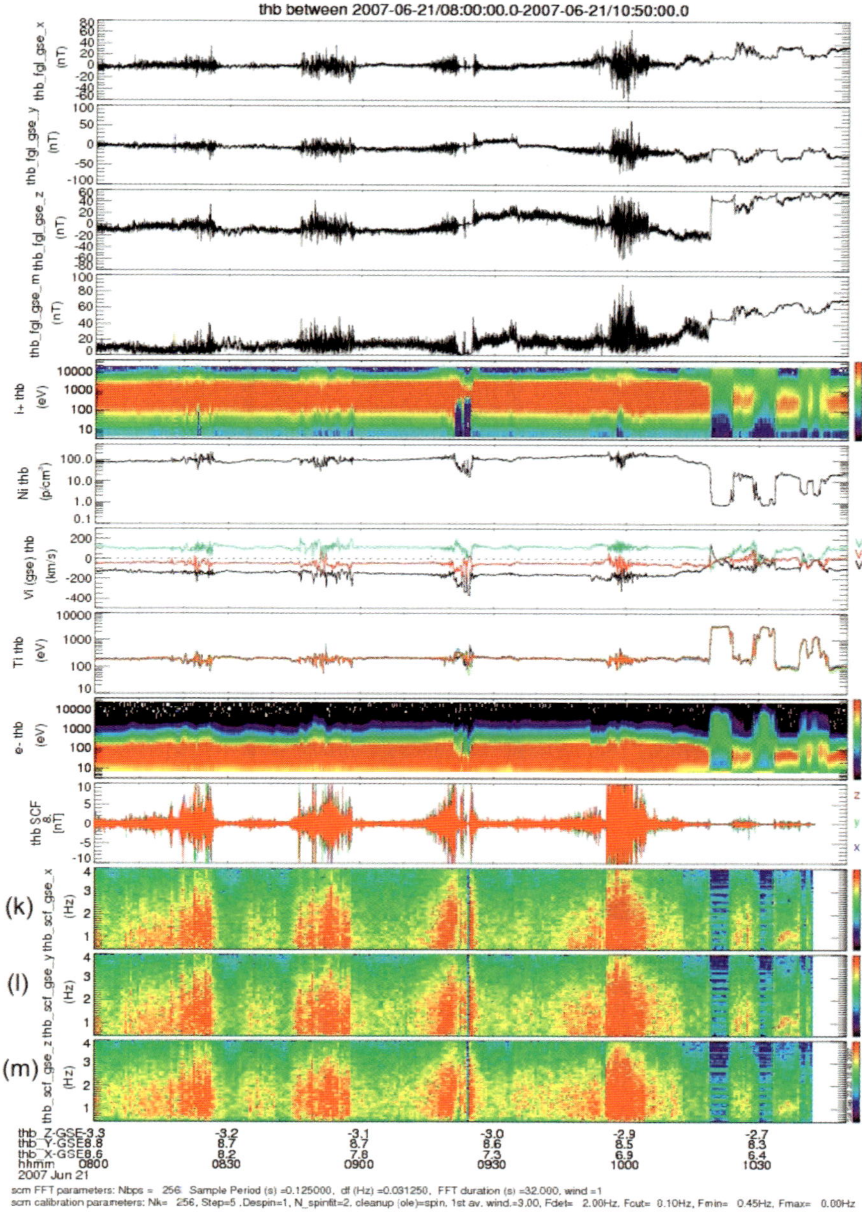

Fig. 6 Same legends as Fig. 5

4.2 Flux Transfer Event: May 20th 2007

On May 20th, 2007, the 5 THEMIS spacecraft encountered a Flux Transfer Event (FTE). Figure 7 shows the locations of the five spacecraft in GSE. Notice that Y is of the order of 12.5 R_E, while X is of the order of 5.5 R_E, so that the satellites were located in the afternoon

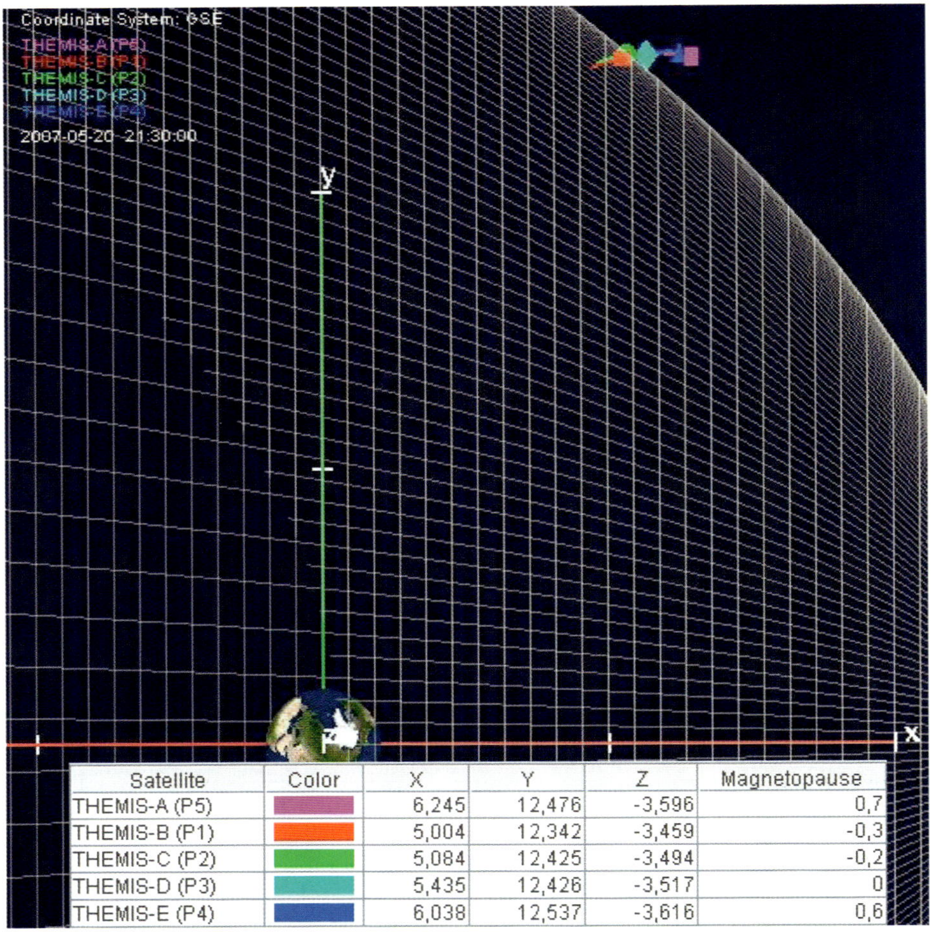

Fig. 7 THEMIS probe locations (GSE, R_E) on May 20th 2007 between 2130 and 2230 UT. Satellite locations are also tabulated in GSE coordinates at 2200 UT. Magnetopause location for a solar wind dynamic pressure of 1 nPA (obtained from WIND survey) is also shown as well as distance to the magnetopause for each probe

sector. This spacecraft configuration is particularly interesting since the five satellites bracket the FTE structure, with THB and THC on the magnetospheric side and THA and THE on the magnetosheath side, while THD is close to the magnetopause current layer as illustrated by the estimated distances to the magnetopause of each probe on Fig. 7.

The geometry of the FTE, and the general characteristics of this event will be discussed elsewhere (Sibeck and Angelopoulos 2008a; Sibeck et al. 2008b). Here we give a preliminary description of ULF wave observations inside and near the FTE, and suggest possible wave-particle interactions. For practical reasons we only show data from THC in Fig. 8; data from the other spacecraft are available and are discussed below. The magnetic structure of the FTE is clearer on THB, THC, and THD, located inside the magnetosphere and at the current layer, than on THA and THE, on the magnetosheath side. Panels a, b, and c of Fig. 8, show the 3 magnetic components in GSE, while panel d shows the modulus of B. THC, on

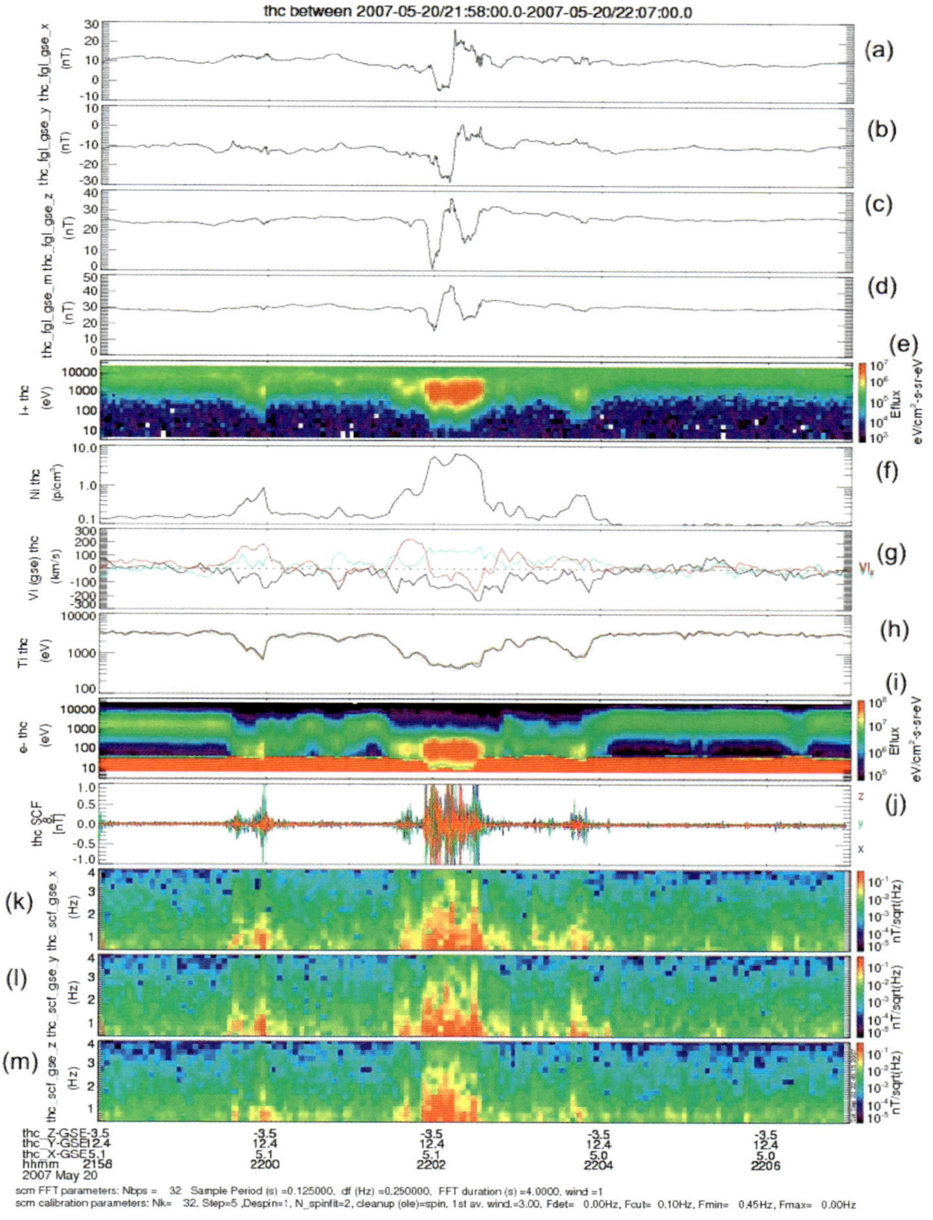

Fig. 8 Same legends as Fig. 5 but time resolution of spectra is 4 s

the magnetospheric side, observes a bipolar magnetic field signature of By, which is close to normal to the nominal magnetopause. A crater-like variation in the magnetic field strength is observed on THB (closest to the earth), and to a lesser extent on THC (see Fig. 8, panel d). On THD, at the current layer, a maximum in the modulus of B is found.

THC was first in the magnetosphere (before 2201:50), as evidenced by the large energies of electrons (panel i) and ions (panel e), and it returns to the magnetosphere after the FTE crossing (after 2202:30). As it penetrates the core of the FTE, THC observes increased ion flux, corresponding to: (i) heated magnetosheath plasma (panel e), (ii) enhanced densities, comparable to magnetosheath values (panel f), (iii) enhanced ion flow velocities (panel g), (iv) enhanced fluxes of accelerated or heated magnetosheath electrons (panel i) and (v) enhanced magnetic components of ULF waves (panels j to m). The simultaneity between the ULF waves and the particle acceleration/heating suggests that a wave-particle interaction process is at work. Can we use data from the other spacecraft to discriminate between possible wave interaction processes; in particular can we identify the particle species, electrons or ions, involved in the interaction with waves?

In the magnetosheath, THE and THA observe quasi-steady fluxes of magnetosheath ions. The flux and the energy bandwidth do not change as these spacecraft pass by the FTE, and ions are not correlated with variations in wave amplitude. On the other hand magnetosheath electrons (THE and THA), and current sheet electrons (THD) are also heated/accelerated in a region that is broader than the FTE, but this region coincides with wave observations.

In summary, ULF waves are observed together with accelerated/heated electrons. On the magnetospheric side these signatures coincide with that of the FTE, while on the magnetosheath side and at the current layer, ULF waves and heated/accelerated electrons are observed in a broader region (boundary layer). These preliminary observations suggest that electromagnetic waves interact with electrons inside the FTE on the magnetospheric side, and in a broader region comprising the current layer on the magnetosheath side. We will analyse wave characteristics, as well as the shape of the electron distribution function, and investigate possible signatures of wave particle interactions as a potential electron heating mechanism. More detailed studies based on the analysis of electron distributions are planned to confirm these preliminary results.

4.3 Magnetopause Crossing Event: June 19th 2007

Figure 9 displays the THEMIS constellation between 0800 and 1030 UT. THB is the leading probe, while THA is the trailing one; THC, THD, and THE follow almost the same trajectory. As the THEMIS constellation goes away from the earth it moves toward the duskside and out of the GSE equator.

The THEMIS instruments are in the fast survey mode during this time period (0800–1030 UT). THA is most of the time in the quiet magnetosphere, while the other probes leave the magnetosphere to enter the magnetosheath in the following order: THB, then THD and THC almost at the same time, and finally THE. Before staying definitely in the magnetosheath each probe undergoes multiple magnetopause crossings, hence the timing has to be considered in the averaged sense. Note that THD and THC undergo exactly the same number of magnetopause crossings although magnetic field or density profiles can be slightly different thereby indicating a temporal variation of the boundary during its motion.

Now we focus on the particle burst period recorded onboard THC between 0940:30 and 0944:00 UT shown on Fig. 10. In this mode FGM data are sampled at 128 S/s (panels a, b, c, and d), ion and electron moments from ESA data are available at 3 s resolution (panels e to l) and SCM data are sampled at the same rate as FGM (panels m to p).

All instrument signatures indicate that THC crosses the magnetopause and enters into the magnetosheath: the modulus of B decreases (from 45 nT to 30–35 nT), the particle den-

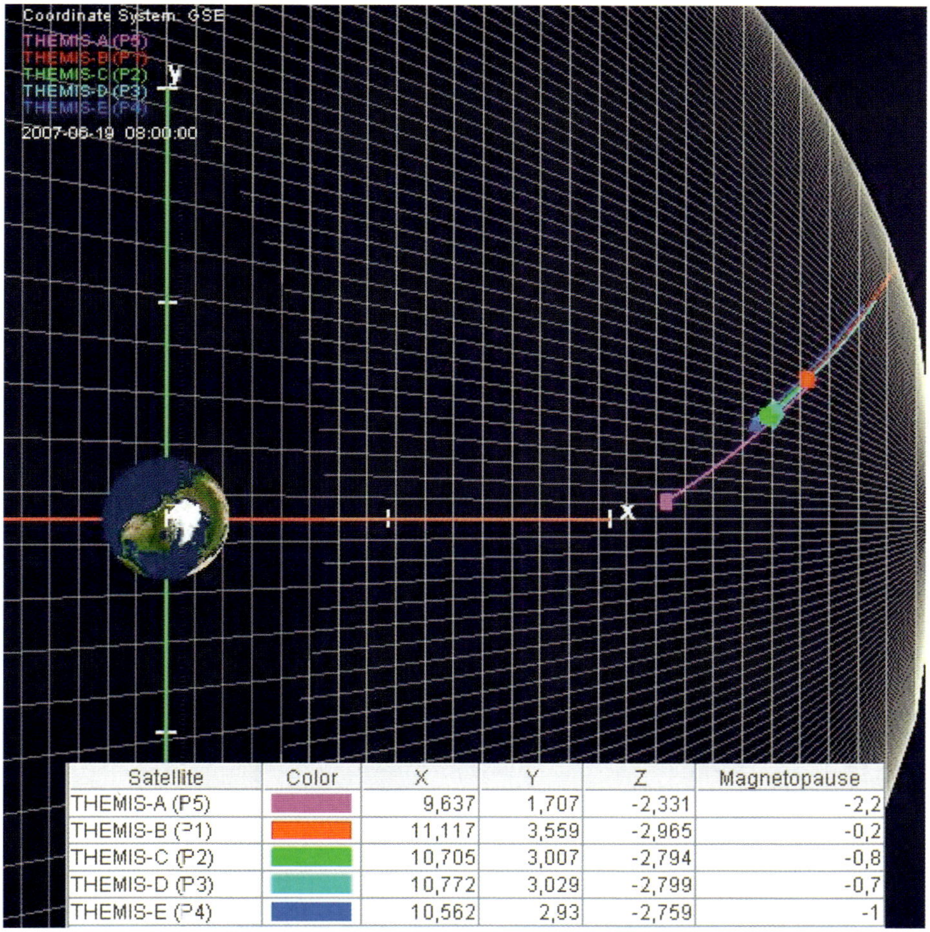

Fig. 9 THEMIS probe locations (GSE, R_E) on June 19th 2007 between 0800 and 1030 UT. Satellite locations are also tabulated in GSE coordinates at 0943 UT. Magnetopause location for a solar wind dynamic pressure of 1 nPA (obtained from WIND survey) is also shown as well as distance to the magnetopause for each probe

sity increases (from 0.2 to 10 p·cm^{-3}), while particle temperatures decrease (from 2 keV to 300 eV). The amplitude of the ULF magnetic fluctuations is maximum at the boundary; their frequencies increase up to 10 Hz as the probe crosses the magnetopause. Ion and electron fluxes as well as velocities are largest within the boundary and during the intense ULF magnetic wave activity. More investigations are needed to determine whether plasma transport occurs through the boundary and whether magnetic fluctuations play a crucial role in this transport. In particular, electron moments have to be carefully checked in such conditions where the density varies rapidly (see McFadden et al. 2008 for discussion on this instrumental issue). However we can already mention that an amplification of the magnetosheath turbulence at the magnetopause is expected by models and has been already observed by pre-

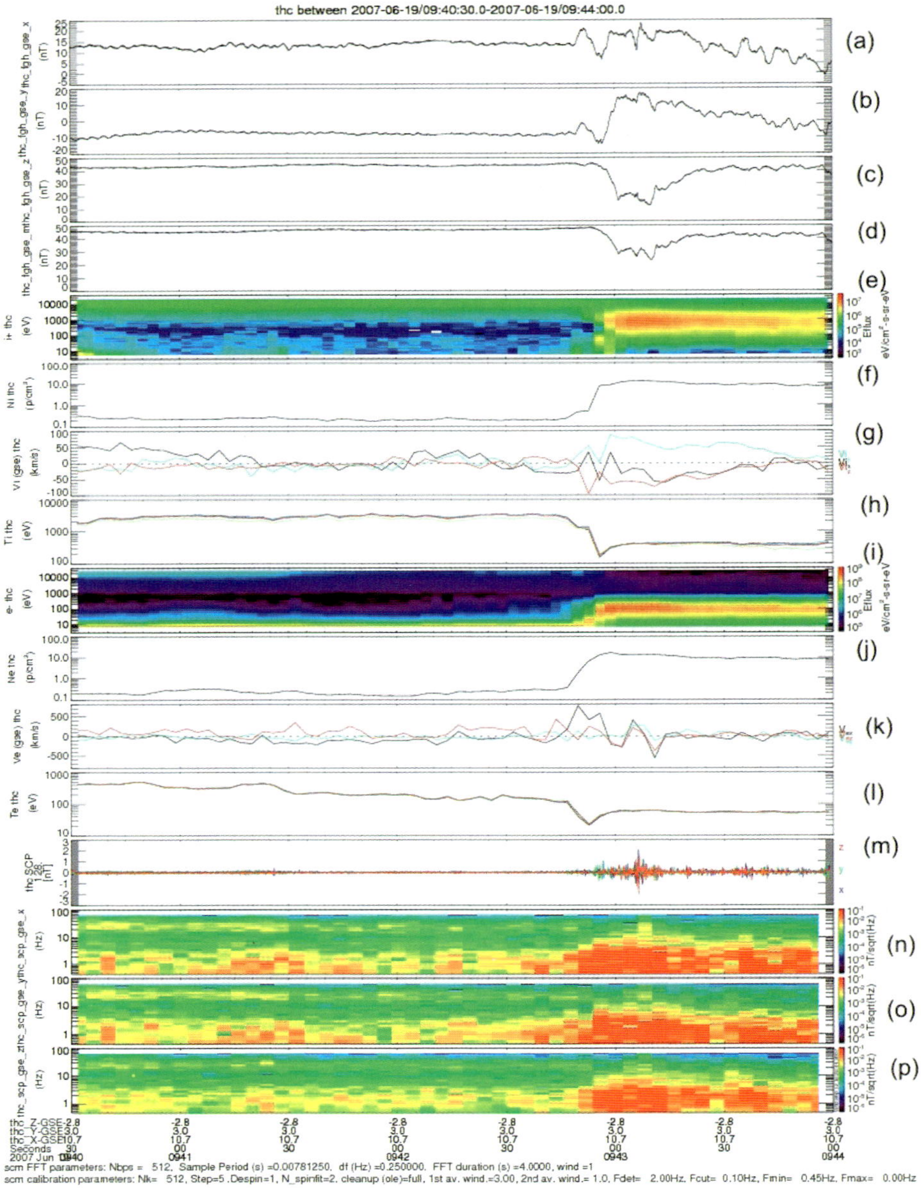

Fig. 10 From *top* to *bottom*: panels **a**, **b**, **c**, and **d** correspond to Bx, By, Bz and B (FGM data), ion energy spectrogram (panel **e**), density (**f**), velocity (**g**) and temperature (**h**), electron energy spectrogram (**i**), density (**j**), velocity (**k**) and temperature (**l**) (ESA data), panels **m**, **n**, **o**, and **p** display waveforms and spectrograms of Bx, By and Bz fluctuations from 0.45 to 64 Hz (SCM data)

vious magnetospheric missions. Models show that it can lead to a plasma transport across the boundary (see for instance Rezeau and Belmont 2001).

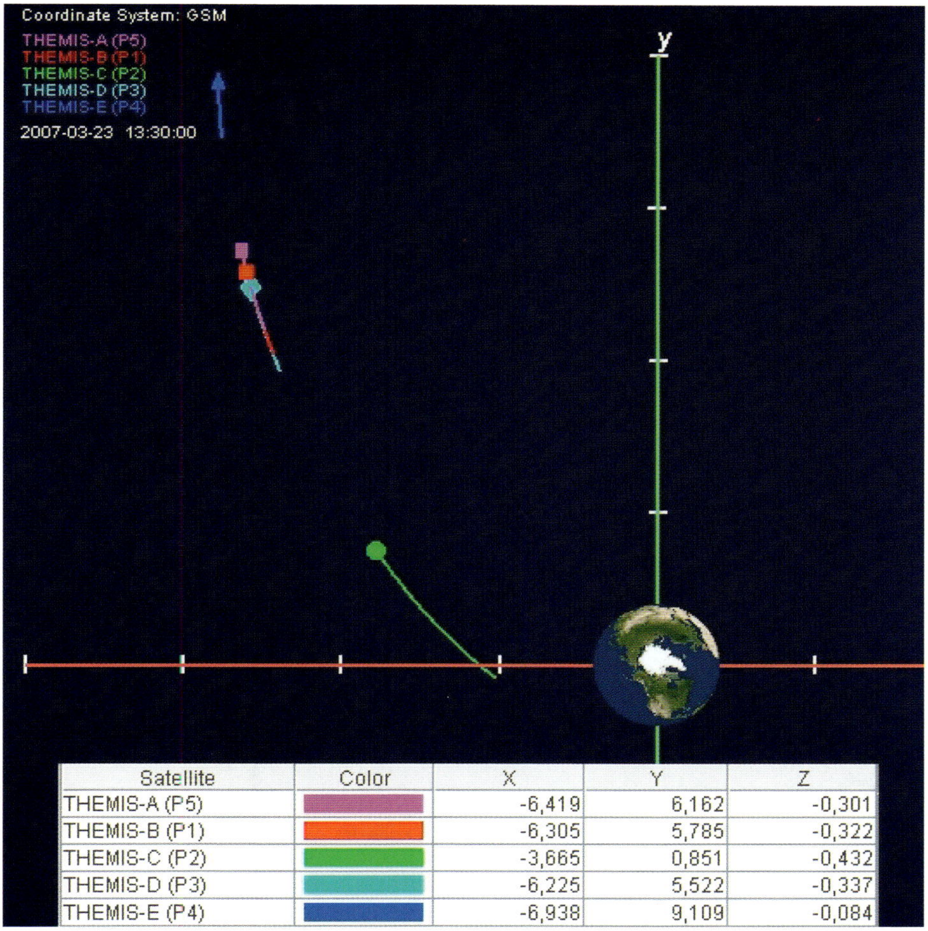

Satellite	Color	X	Y	Z
THEMIS-A (P5)		-6,419	6,162	-0,301
THEMIS-B (P1)		-6,305	5,785	-0,322
THEMIS-C (P2)		-3,665	0,851	-0,432
THEMIS-D (P3)		-6,225	5,522	-0,337
THEMIS-E (P4)		-6,938	9,109	-0,084

Fig. 11 THEMIS probe locations (GSM, R_E) on Mars 23rd 2007 between 1330 and 1430 UT. Satellite locations are also tabulated in GSM coordinates at 1400 UT

4.4 Substorm Event: March 23rd 2007

Two substorms were captured on March 23rd 2007. THEMIS positions are plotted in Fig. 11 in GSM coordinates between 1330 and 1430 UT around the second local dipolarization. THC is the leading probe moving from the duskside toward the earth with THE trailing; THD, THB and THA are in between and following THC in this order. During this period the THEMIS constellation is located in the south hemisphere.

The first local dipolarization is observed by THC at 1114:30 UT while ground based observations and Polar data indicate an auroral substorm onset at 1113 UT dawnward of the mapped point of THC. This first dipolarization is analyzed in detail in (Angelopoulos et al. 2008). The order of the probe positions is the same as for the second dipolarization displayed in Fig. 11 and discussed later; the THEMIS constellation was also located in the dusk side but farther from the earth than during the second event, slightly closer to the equator. Here

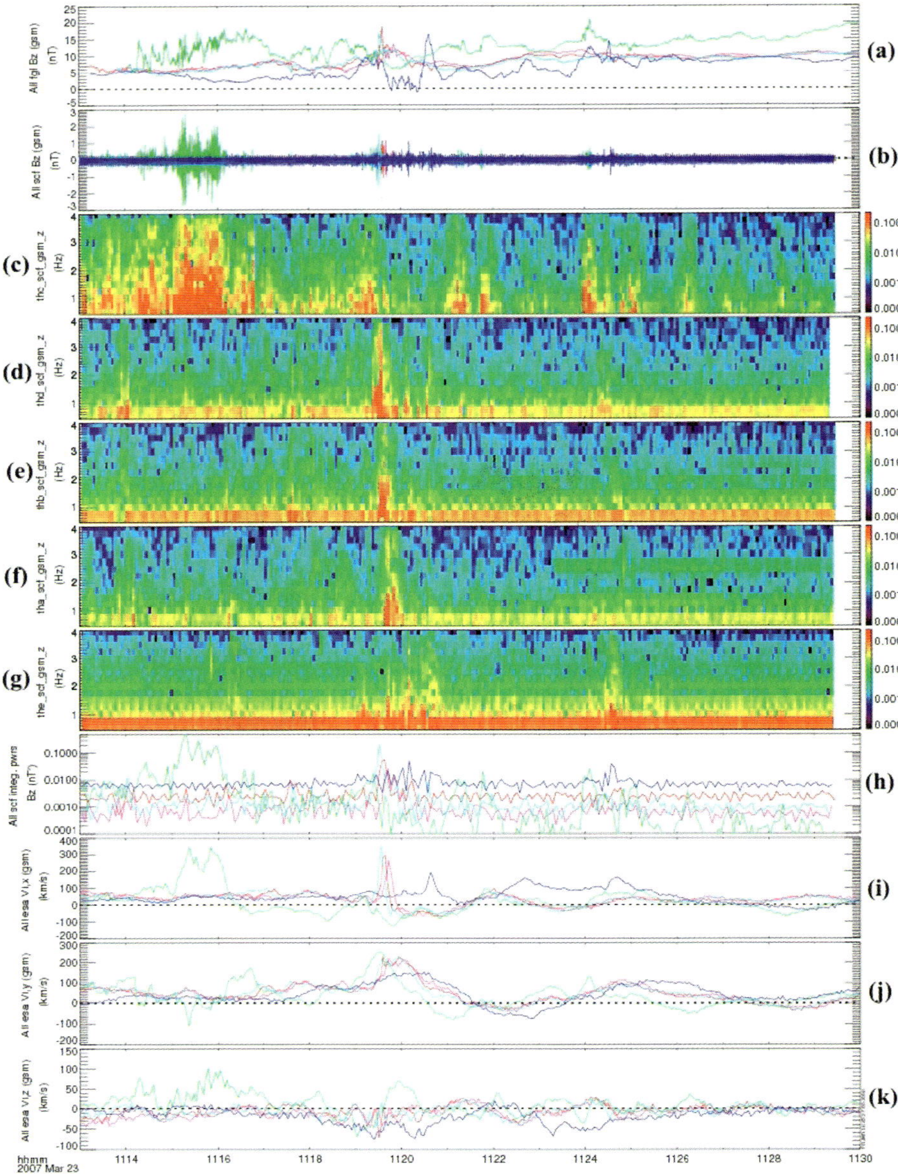

Fig. 12 Five THEMIS probes data in fast survey mode on March 23rd 2007 between 1113 and 1130 UT. From *top* to *bottom* Panel **a**: Bz from FGM data with 0.5 s time resolution, panel **b**: Bz from SCM waveform data between 0.45 to 4 Hz, panels **c** (THC), **d** (THD), **e** (THB), **f** (THA), **g** (THE): SCM Bz power spectral densities, panel **h**: integrated powers from 0.45 to 4 Hz with 4 s time resolution for the 5 probes, panels **i**, **j**, **k**: x, y, and z components of the ion velocity with 3 s time resolution

we just show a summary plot around the first local dipolarization onset (Fig. 12) to compare with the second event discussed later in more details.

In panel a, the Bz component of the magnetic field is shown in GSM (FGM data) for the five probes. THC (green) clearly detects the dipolarization first while THD (cyan), THB (red), THA (magenta) and THE (blue) detect it later. The delays ($\simeq 67$ s) between THD and THE ($\Delta Y_{d-e} \simeq 2.5\ R_E$) calculated from both Bz, from Vx or from Vy signatures (panel j) give an estimated azimuthal propagation velocity of about 240 km/s. This also agrees with the velocity inferred from ground-based and polar observations (for more details see Angelopoulos et al. 2008).

ULF Magnetic fluctuations amplitudes (panel b) as well as integrated powers (panel h) recorded by SCM in the range 0.45 to 4 Hz (scf data) increase at the dipolarization and are well correlated with the amplitude of the x component of the ion velocity (panel i). Most of the ULF magnetic activity is below 4 Hz (panels c, d, e, f, g) and around the proton cyclotron frequency ($f_{c,i} \simeq 0.6$ Hz). Note that a high level of spurious noise below 1.5 Hz is still present despite of the noise reduction process; it is due to particular conditions during the commissioning on March 23rd. The ion velocity increases suddenly at the dipolarization especially on the Vx component up to 350 km/s on THC and THD (panel i).

Now we deal with the second dipolarization which starts first at 1358:30 UT on THD. This dipolarization is fortunately caught during a particle burst period. Unfortunately the POLAR cameras were not in a position to detect the substorm. Figure 13 displays an overview of THA data on March 23rd between 1330 and 1430 UT during a fast survey period. Panels a, b, and c show that the magnetotail stretches from 1330 to 1358:30 UT, as expected for a growth phase; the Bx and By components are increasing in modulus (from 25 nT to 33 nT, and from 35 nT to 55 nT, respectively). The THEMIS constellation is located in the dusk sector, so both Bx and By correspond to the main field components of the magnetotail. The dipolarization occurs suddenly at 1359 UT as Bz increases from 3 nT to 10 nT and By decreases from 55 nT to 45 nT. Magnetic low-frequency fluctuations are recorded (5–15 mHz) until the end of the dipolarization around 1410 UT when the Bz component reaches 16 nT while By $\simeq 35$ nT. Panel e (SST data) and panels f, h, and i (ESA data) indicate that the ions are mostly accelerated and heated at the dipolarization onset while electrons seem to be accelerated in successive steps (panels j and k). Panel g shows a short lasting peak in the ion density profile at the beginning of the dipolarization onset and panel h gives evidence for low-frequency fluctuations on the three components of the velocity. Note that the oscillation on the Vy component of the ion velocity starts before the dipolarization. Finally panels l, m, n, and o show four intensifications of magnetic fluctuations during the dipolarization. Again most of the ULF magnetic fluctuations are below 4 Hz and around $f_{c,i} \simeq 0.64$ Hz. Figure 14 is a detail of the dipolarization period data gathered by THA (magenta), THB (red), THD (cyan) and THE (blue). THC is too close to the earth and misses the dipolarization. Panel a shows that the increase of the Bz component starts first on THD ($\simeq$1358:30 UT), then THB ($\simeq$1358:50 UT), THA ($\simeq$1359:00 UT) and finally on THE ($\simeq$1400:20 UT). The delay ($\simeq 110$ s) between THD and THE ($\Delta Y_{d-e} \simeq 3.6\ R_E$) calculated from both Bz signatures gives an estimated azimuthal propagation velocity of about 210 km/s; the same estimation using THA and THE gives 240 km/s. Panel b shows SCM Bz waveforms which give the same timing for the dipolarization onset. Note that the largest Bz amplitudes (SCM data) correspond to successive local dipolarizations (FGM data).

Panels c, d, e and f indicate for each probe (in the observed order of the dipolarization THD, THB, THA and THE) that the power spectral density increases suddenly at the dipolarization and is again maximum for fluctuations with frequencies below 4 Hz. Integrated power of the magnetic fluctuations between 0.45 and 4 Hz is plotted on panel g. These integrated powers are smaller ($\simeq 5 \times 10^{-3}$–3×10^{-2} nT^2) than those recorded closer to the magnetic equator at the geostationary orbit (see for instance Perraut et al. 2000a;

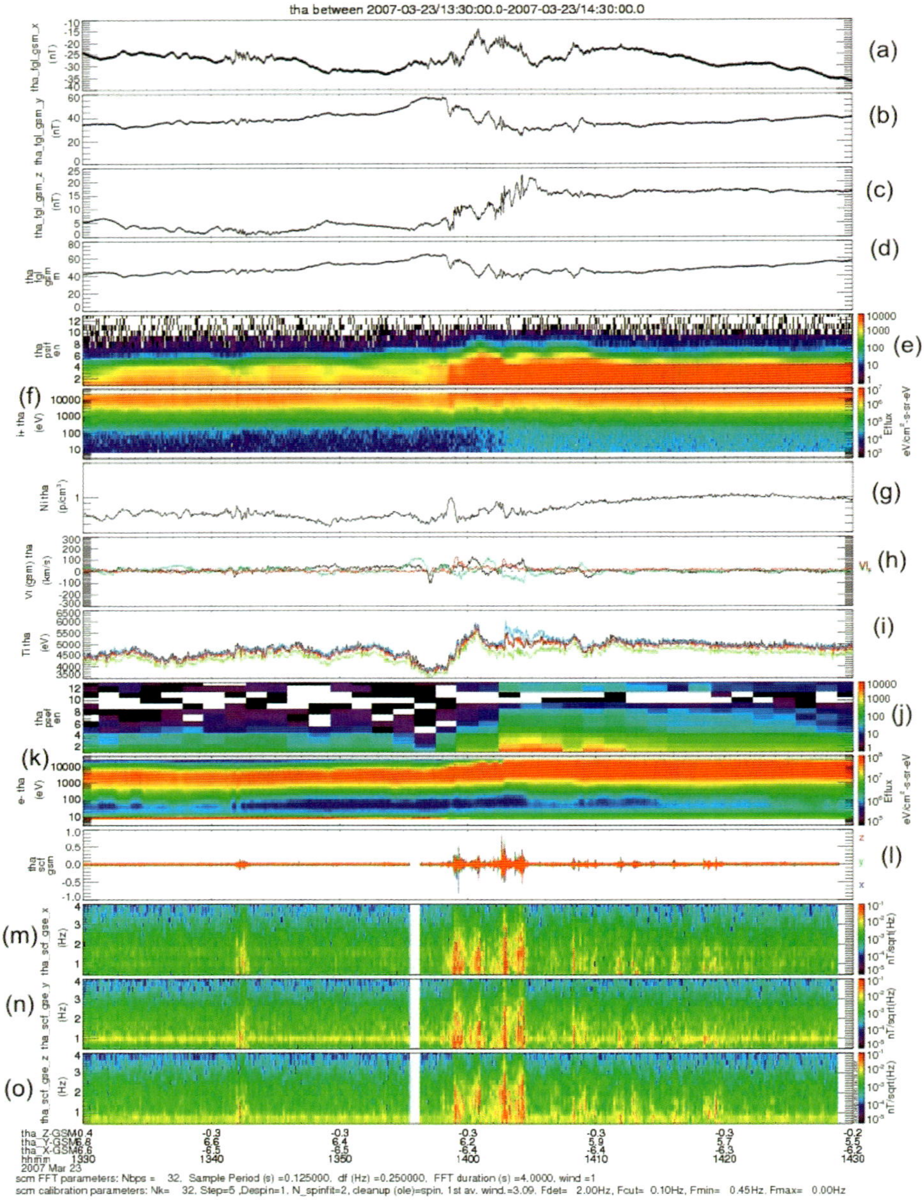

Fig. 13 Overview of THA (fast survey) data on march 23rd 2007 between 1330 and 1430 UT. From *top* to *bottom*: Bx, By, Bz and B (FGM data), ion energy spectrograms (SST and ESA data), density, velocity and temperature, electron energy spectra (SST and ESA data), waveform and spectrograms of magnetic fluctuations from 0.45 Hz to 4 Hz (SCM data)

Le Contel et al. 2001b). They are poorly correlated with the velocity profiles shown in the next panels as compared with the first dipolarization. It could be due to the fact that all probes are farther from the source of the waves. The last panels (h, i, j) show the three

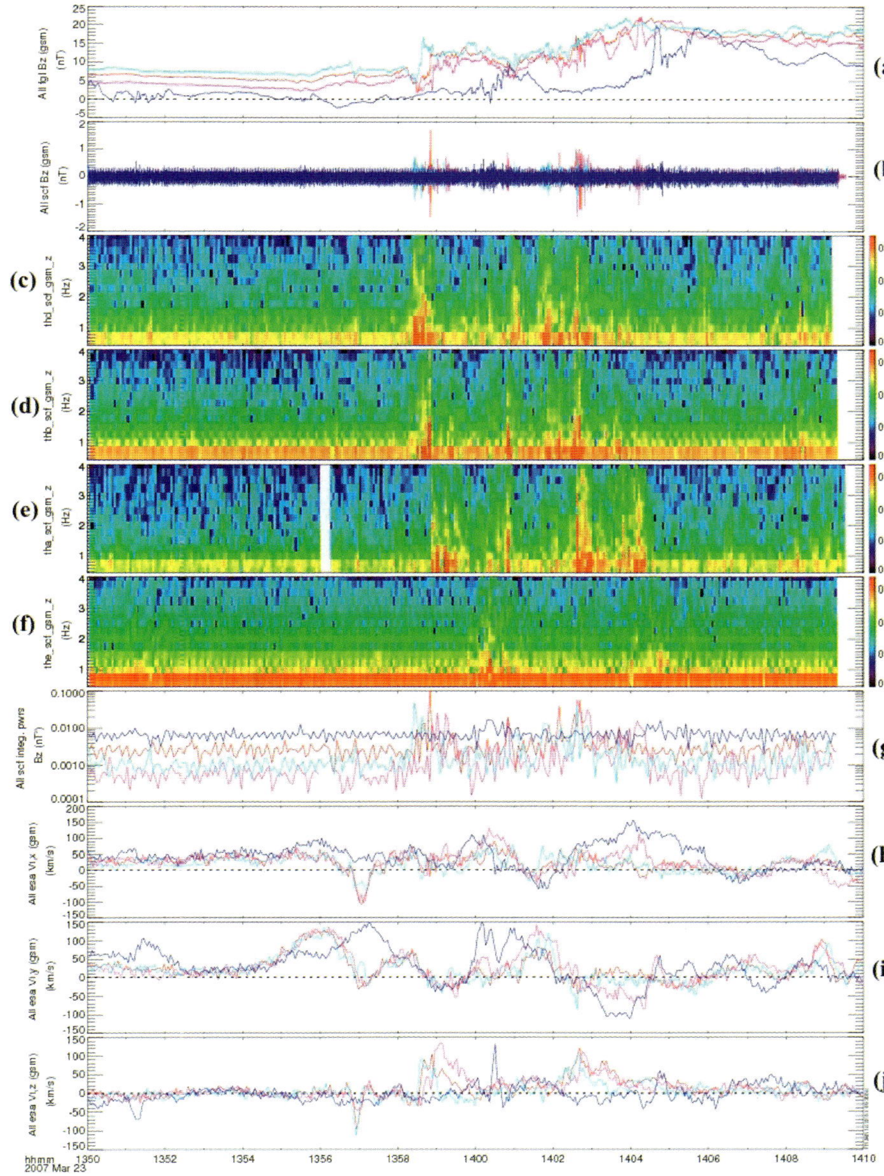

Fig. 14 Data from four THEMIS probes (**a**, **b**, **d**, and **e**) in fast survey mode on March 23rd 2007 between 1350 and 1410 UT. Panel **a**: Bz from FGM data with 0.5 s time resolution, panel **b**: Bz from SCM data from 0.45 to 4 Hz, panels **c** (THD), **d** (THB), **e** (THA), **f** (THE): SCM Bz power spectral densities and panel g: integrated powers from 0.45 to 4 Hz with 4 s time resolution, panels **h**, **i**, **j**: x, y, and z components of the ion velocity with 3 s time resolution

components of the ion velocity. Amplitudes are smaller than for the first event. These lower velocities may be due to an underestimate of the velocities by ESA (see McFadden et al. 2008 for more details). Indeed the ion energy exceeds 40 keV, the upper energy limit of

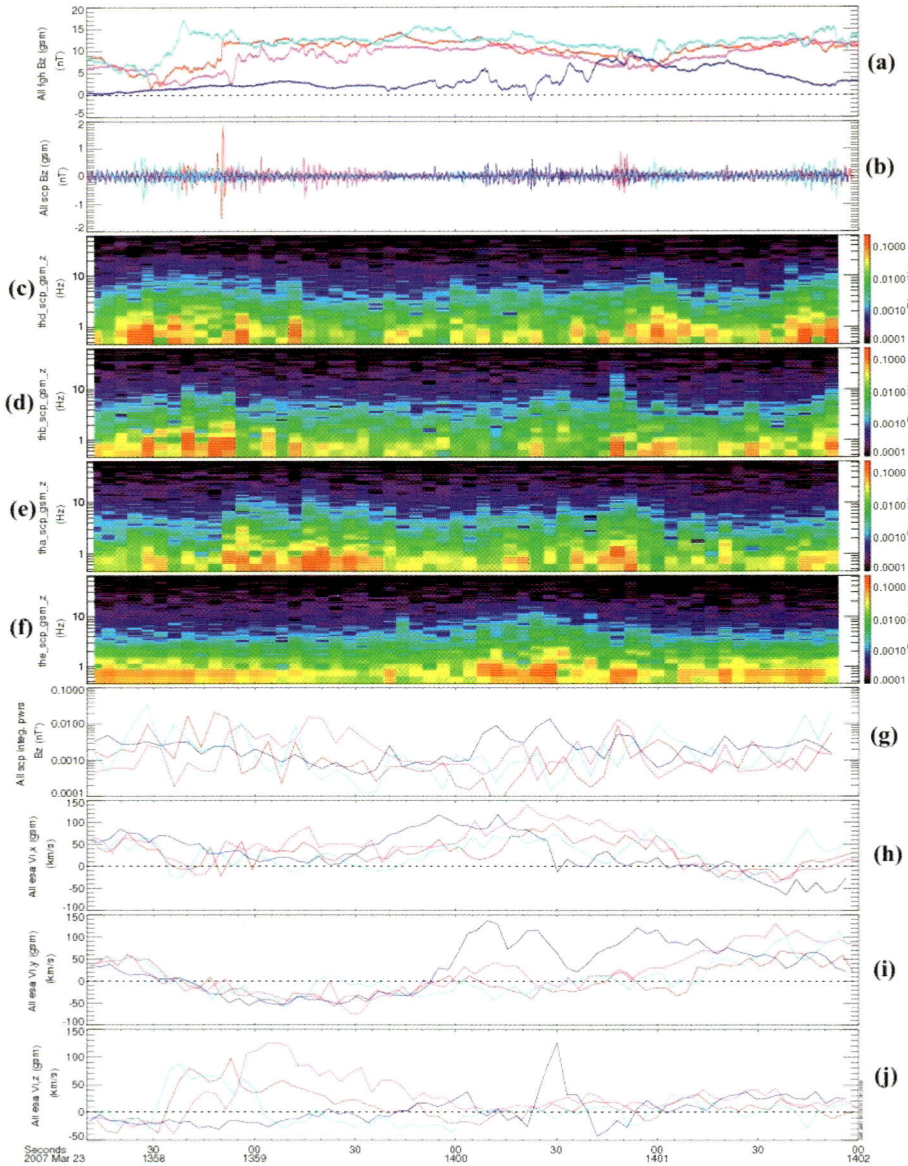

Fig. 15 Four THEMIS probes (**a**, **b**, **d**, and **e**) data from particle burst mode on march 23rd 2007 between 1358:10 and 1402:00 UT. Panel **a**: Bz from FGM data sampled at 128 S/s, panel **b**: Bz from SCM data at 128 S/s filtered between 0.45 and 64 Hz, panels **c** (THD), **d** (THB), **e** (THA), **f** (THE): SCM Bz power spectral densities, and panel **g**: integrated powers from 0.45 to 64 Hz with 4 s of time resolution, panels **h**, **i**, **j**: x, y and z components of the ion velocity with 3 s of time resolution

ESA and clearly enters in the energy range of the SST instrument (see panels e and f on Fig. 13).

While the first dipolarization onset is characterized mainly by a large Vx component up to 350 km/s, the second one has a different signature: large values of the Vz component at the dipolarization onset (panel j). These large Vz values toward the magnetic equator are recorded in the same order as the dipolarization (THD, THB, and THA) and could be interpreted as the motion of the magnetic field lines toward the earth during the dipolarization. Figure 15 is the same as Fig. 14 but for the particle burst period during 1358:10 and 1402:00 UT. The time resolution is therefore better (FGM as well as SCM at 128 S/s). We observe the same signatures: dipolarization on THD, THB, THA and THE (panel a), ULF/ELF magnetic fluctuations up to 64 Hz but with maximum intensities still below 4 Hz (panels b, c, d, e and f), integrated power up to 3×10^{-2} (panel g), and large ion Vz at the dipolarization onset (panel j). Again the integrated power of magnetic fluctuations is somewhat smaller than previously reported values from geostationary spacecraft.

Note that no intense wave emission is recorded around $f_{LH} \simeq 27$ Hz, the lower hybrid frequency which is well included in the range of the SCM. Finally we can remark that the SCM spectra performed from wave burst data shown in Fig. 3 are taken during this substorm period (135946–140016 UT). However the level of measured noise is similar to NEMI measured at ground. Therefore we can conclude that at least during this short lasting period no high-frequency waves are emitted notably in the range of whistler waves (between $f_{c,i} \simeq 0.7$ Hz to $f_{c,e} \simeq 1.4$ kHz).

5 Summary and Conclusions

The SCM and EFI are designed to characterize the electromagnetic fluctuations at substorm onset and to clarify their role in the different phases of the substorm expansion. We described the calibration method for the SCM instrument, as well as the spurious noise reduction process which recovers the sensitivity of the SCM instrument as it was measured on the ground.

First THEMIS SCM results, obtained in various regions, from the solar wind to the magnetotail, were presented and discussed. The usefulness of the magnetic fluctuations for the identification of the key regions of the magnetosphere was discussed, and their possible role in the different basic plasma processes was pointed out. Notably it was suggested that ULF fluctuations could heat electrons inside an FTE event (May 20, 2007). It has also been shown that the level of ULF magnetic fluctuations below 4 Hz (around $f_{c,i}$) is largely enhanced during the substorm-related dipolarizations (March 23rd, 2007, 1113 and 1358 UT).

These first THEMIS SCM results demonstrate that the 5 tri-axis instruments function nominally, and illustrate the capability of the THEMIS mission to provide a comprehensive set of data not only on substorms but also on the physics of key regions such as the magnetopause and bow shock.

Acknowledgements We are pleased to acknowledge the friendly collaboration and the help of other THEMIS team members, in particular, P. Harvey, R. Jackson, J. Lewis, M. Ludlam, D. Meilhan, H. Richard, and E. Taylor. The French involvement on THEMIS is supported by CNES and CNRS. Work in the US was supported by NASA contract NAS5-02099. The work of KHG and UA at the Technical University of Braunschweig was financially supported by the German Ministerium für Wirtschaft und Technologie and the German Zentrum für Luft- und Raumfahrt under grant 50QP0402.

References

V. Angelopoulos et al., Space Sci. Rev. (2008). doi:10.1007/s11214-008-9336-1

N. Attico, F. Califano, F. Pegoraro, Phys. Plasmas **9**, 458–464 (2002)
H.U. Auster, K.H. Glassmeier, W. Magnes, O. Aydogar, W. Baumjohann, D. Constantinescu, D. Fischer, K.H. Fornacon, E. Georgescu, P. Harvey, O. Hillenmaier, R. Kroth, M. Ludlam, Y. Narita, R. Nakamura, K. Okrafka, F. Plaschke, I. Richter, H. Schwarzl, B. Stoll, A. Valavanoglou, M. Wiedemann, Space Sci. Rev. (2008). doi:10.1007/s11214-008-9365-9
W. Baumjohann, A. Roux, O. Le Contel, R. Nakamura, J. Birm, M. Hoshino, A.T.Y. Lui, C.J. Owen, J.A. Sauvaud, A. Vaivads, D. Fontaine, A. Runov, Ann. Geophys. **25**, 1365–1389 (2007)
J. Bonnell et al., Space Sci. Rev. (2008 this issue)
S.V. Bulanov, F. Pegoraro, A.S. Sakharov, Phys. Fluids B **4**, 2499–2508 (1992)
C.Z. Cheng, A.T.Y. Lui, Geophys. Res. Lett. **25**, 4091 (1998)
N. Cornilleau-Wehrlin, G. Chanteur, S. Perraut, L. Rezeau, P. Robert, A. Roux, C. Villedary, P. Canu, M. Maksimovic, Y. Conchy, D. Hubert, C. Lacombe, F. Lefeuvre, M. Parrot, J.-L. Pinçon, P.M.E. Décréau, C.C. Harvey, P. Louarn, O. Santolik, H.S. Alleyne, M. Roth, T. Chust, O. Le Contel, STAFF team, Ann. Geophys. **21**, 437–456 (2003)
R. Gendrin, Space Sci. Rev. **11**, 54–130 (1970)
D.A. Gurnett, L.A. Frank, R.P. Lepping, J. Geophys. Res. **81**, 6059–6071 (1976)
V.I. Karpman, B.I. Meerson, A.B. Mikhailovsky, O.A. Pokholetov, Planet. Space Sci. **25**, 273 (1977)
V.V. Krasnosselskikh, B. Lembège, P. Savoini, V.V. Lobzin, Phys. Plasma **9**, 1192–1209 (2003)
G. Kremser, A. Korth, S.L. Ullaland, S. Perraut, A. Roux, A. Pedersen, R. Schmidt, P. Tanskanen, J. Geophys. Res. **93**, 14,453 (1988)
O. Le Contel, S. Perraut, A. Roux, R. Pellat, Space Sci. Rev. **95**, 415–426 (2001a)
O. Le Contel, A. Roux, S. Perraut, R. Pellat, Ø. Holter, A. Pedersen, A. Korth, J. Geophys. Res. **106**, 10,817–10,827 (2001b)
O. Le Contel, A. Roux, S. Perraut, R. Pellat, P. Robert, G. Chanteur, D. Fontaine, N. Cornilleau-Wehrlin, J.A. Sauvaud, C. Cully, G. Parks, D. Chua, M. André, A. Balogh, A. Fazakerley, H. Rème, T. Nagai, T. Mukai, H. Hayakawa, A. Matsuoka, R.R. Anderson, H. Matsumoto, in *Sixth International Conference on Substorms (ICS-6)*, ed. by R. Winglee. University of Washington, Seattle, 25–29 March 2002, pp. 326–333
O. Le Contel, F. Sahraoui, A. Roux, D. Fontaine, P. Robert, J.A. Sauvaud, C. Owen, A.N. Fazakerley, in *Eighth International Conference on Substorms (ICS-8)*, ed. by M. Syrjaeso, E. Donovan. University of Calgary, Canada, 27–31 March 2006, pp. 143–148
M. Ludlam et al., Space Sci. Rev. (2008 this issue)
A.T. Lui, R.E. Lopez, B.J. Anderson, K. Takahashi, L.J. Zanetti, R.W. McEntire, T.A. Potemra, D.M. Klumpar, E.M. Greene, R. Strangeway, J. Geophys. Res. **101**, 13,067–13,088 (1996)
A.T.Y. Lui, Rev. Geophys. **39**(4), 535–563 (2001)
A.T.Y. Lui, C.L. Chang, A. Mankofsky, H.K. Wong, D. Winske, J. Geophys. Res. **96**, 11,389–11,401 (1991)
M.E. Mandt, R.E. Denton, J.F. Drake, Geophys. Res. Lett. **21**(1), 73–77 (1994)
J.P. McFadden, C.W. Carlson, D. Larson, V. Angelopoulos, Space Sci. Rev. (2008 this issue)
S. Perraut, O. Le Contel, A. Roux, A. Pedersen, J. Geophys. Res. **105**, 21,097–21,107 (2000a)
S. Perraut, O. Le Contel, A. Roux, R. Pellat, A. Korth, Ø. Holter, A. Pedersen, Geophys. Res. Lett. 4041–4044 (2000b)
S. Perraut, A. Morane, A. Roux, A. Pedersen, R. Schmidt, A. Korth, G. Kremser, B. Aparicio, R. Pellinen, Adv. Space Res. **13**(4), 217 (1993)
T. Quinn, J.W. Bonnell, A. Roux, U. Auster, D. Larson, K. Khurana, M. Ludlam, P. Harvey, V. Angelopoulos, THEMIS Science Coordinate Systems Definition THM-SOC-110, 29 September 2006
L. Rezeau, G. Belmont, Space Sci. Rev. **95**, 427–441 (2001)
P. Robert, R. Gendrin, S. Perraut, A. Roux, A. Pedersen, J. Geophys. Res. **89**, 819 (1984)
A. Roux, O. Le Contel, P. Robert, C. Coillot, A. Bouabdellah, B. la Porte, D. Alison, S. Ruocco, M.C. Vassal, Space Sci. Rev. (2008 this issue)
C.T. Russel, Planet. Space Sci. **20**, 1541–1553 (1972)
F. Sahraoui, J.L. Pinçon, G. Belmont, L. Rezeau, N. Cornilleau-Wehrlin, P. Robert, L. Mellul, J.M. Bosqued, A. Balogh, P. Canu, G. Chanteur, J. Geophys. Res. **108**, 1335 (2003). doi:10.1029/2002JA009587
I. Shinohara, T. Nagai, M. Fujimoto, T. Terasawa, T. Mikai, K. Tsuruda, T. Yamamoto, J. Geophys. Res. **103**, 20,365 (1998)
K. Shiokawa, Y. Miyashita, I. Shinohara, A. Matsuoka, J. Geophys. Res. **110**, 219 (2005). doi:10.1029/2005JA011144
D.G. Sibeck, V. Angelopoulos, Space Sci. Rev. (2008a this issue)
D.G. Sibeck et al., Geophys. Res. Lett. (2008b). doi:10.1029/2008GL033568
K. Sigsbee, C.A. Cattell, F.S. Mozer, K. Tsuruda, S. Kokubun, J. Geophys. Res. **106**, 435–445 (2001)
Y. Zhang, H. Matsumoto, H. Kojima, J. Geophys. Res. **104**, 28,633–28,644 (1999)

OpenGGCM Simulations for the THEMIS Mission

Joachim Raeder · Douglas Larson · Wenhui Li ·
Emil L. Kepko · Timothy Fuller-Rowell

Originally published in the journal Space Science Reviews, Volume 141, Nos 1–4, 535–555.
DOI: 10.1007/s11214-008-9421-5

Abstract The THEMIS mission provides unprecedented multi-point observations of the magnetosphere in conjunction with an equally unprecedented dense network of ground measurements. However, coverage of the magnetosphere is still sparse. In order to tie together the THEMIS observations and to understand the data better, we will use the Open Geospace General Circulation Model (OpenGGCM), a global model of the magnetosphere-ionosphere system. OpenGGCM solves the magnetohydrodynamic (MHD) equations in the outer magnetosphere and couples via field aligned current (FAC), electric potential, and electron precipitation to a ionosphere potential solver and the Coupled Thermosphere Ionosphere Model (CTIM). The OpenGGCM thus provides a global comprehensive view of the magnetosphere-ionosphere system. An OpenGGCM simulation of one of the first substorms observed by THEMIS on 23 March 2007 shows that the OpenGGCM reproduces the observed substorm signatures very well, thus laying the groundwork for future use of the OpenGGCM to aid in understanding THEMIS data and ultimately contributing to a comprehensive model of the substorm process.

Keywords THEMIS · OpenGGCM · Magnetosphere · MHD · Simulation · Substorm

J. Raeder (✉) · D. Larson · W. Li · E.L. Kepko
Space Science Center, University of New Hampshire, Durham, NH, USA
e-mail: J.Raeder@unh.edu

D. Larson
e-mail: Douglas.Larson@unh.edu

W. Li
e-mail: wenhuil@cisunix.unh.edu

E.L. Kepko
e-mail: Larry.Kepko@unh.edu

J. Raeder
Physics Department, University of New Hampshire, Durham, NH, USA

T. Fuller-Rowell
CIRES, Colorado University, Boulder, CO, USA
e-mail: Tim.Fuller-Rowell@noaa.gov

1 Introduction

The substorm debate has been a central part of space physics for over four decades and centers on the question of what physical process(es) precipitate the sudden energy release in the magnetotail and the sudden auroral brightening and expansion (Akasofu 1977; Lui 1991; Fairfield 1992; Kennel 1992; McPherron 1991; Baker et al. 1999).

It is probably fair to say that it is widely accepted that substorms are ultimately powered by magnetic reconnection. Reconnection signatures are often observed in the tail during the course of substorms. However, the location of the associated x-lines is typically observed $\sim 20 R_E$ from Earth or further down the tail. Conversely, the initial brightening of the aurora maps much closer to Earth. Thus, the question is commonly posed as to whether reconnection causes the process that brightens the aurora or whether the process that brightens the aurora causes reconnection. The THEMIS mission (Sibeck and Angelopoulos 2008; Angelopoulos 2008) is designed to answer this question by providing simultaneous measurements at five locations in order to establish how events proceed in time and space.

However, in spite of the unprecedented coverage, ambiguities will likely remain because processes such as dipolarization of the field or earthward flows may not necessarily occur strictly radially but sweep azimuthally over the spacecraft, creating an apparent radial motion that does not correspond to the real one. Furthermore, substorms come in different sizes and shapes, and at this point it is only a hypothesis that they all follow the same scheme. It is well known that some substorms are triggered by various solar wind or IMF changes while others occur spontaneously. Furthermore, there are other forms of geomagnetic activity such as pseudo-breakups and Steady Magnetospheric Convection (SMC) events, that have some traits of substorms but differ in certain aspects. THEMIS will undoubtedly clarify the phenomena and the relationships between different forms of activity and substorm triggers. However, the physical processes will not be understood fully until we are able to model them.

We will thus complement the THEMIS mission with global simulations of the magnetosphere. While it is possible to use local models to study isolated processes such as reconnection in detail, it is not possible to apply local models to substorms. Substorms are inherently global and encompass physical processes ranging from the dayside magnetopause, the lobes, the plasma sheet, and the inner magnetosphere to the ionosphere and to the ground. There have been a few attempts in the past to model substorms with global models, such as the "GEM substorm challenge" (Slinker et al. 1995; Fedder et al. 1995; Wiltberger et al. 2000; Raeder and Maynard 2001; Raeder et al. 2001b). None of these simulations has been able to reproduce a substorm in its entirety. Some substorm-related phenomena such as particle injections are beyond the MHD description of the models. However, even the phenomena that global MHD based models should be able reproduce do often not come out well. For example, all models have a tendency to enter an SMC-like state, where nightside reconnection closely balances dayside reconnection and no loading-unloading cycle occurs. Models then often require tweaking of parameters for a substorm to occur (Raeder et al. 2001b). The necessity for such tweaking reflects the multi-scale nature of substorms, i.e., the effects of small-scale processes, such as anomalous resistivity, kinetic instabilities, or other processes that break the frozen flux condition. Such processes are not included self-consistently in the model, but they are represented, at least to some extent, by parameterizations. As long as self-consistent treatment of such small-scale processes in global models is not possible, one hopes that these parameterizations are good enough to capture the substorm physics correctly. In essence, these parameterizations constitute hypotheses concerning the underlying physical processes, and, by comparison of the model results with in situ data, we test them. We present one example of such a comparison later in this paper which shows a quite reasonable agreement with the data. However, many more such studies are needed to firmly establish the validity of the model.

We will thus use the OpenGGCM in at least three different ways to support THEMIS and to better understand substorms:

1. We will test and constrain the model by simulating a number of substorm events with THEMIS (and other) observations. These simulations will be driven with observed solar wind and IMF data which are usually available from solar wind monitors such as ACE or Wind. The output of these simulations will be critically compared, timetrace by timetrace, to the observations to find out what the model captures well and what it does not. Of course, the comparisons will never be perfect, and in many cases not good at all. We will run different simulations for one event with varying parameters, such as numerical resolution, M-I coupling parameters (see below), different anomalous resistivity, and different sub-models (ionosphere, ring current.) From these runs we expect learn what parameters are important and how to choose them to get the physics right. We will also learn where we can trust the model, and which outputs will most likely differ from reality. In Sect. 3 below we present an example, the 23 March 2007 substorm, which shows that the OpenGGCM correctly predicts several aspects of this substorm, but not all.
2. Based on verification of the results as outlined above we will use the OpenGGCM results to help interpret the THEMIS data. THEMIS observations are still spotty and leave large gaps in the spatial coverage, which can be filled with model results. The OpenGGCM can also provide relationships that are generally not observable, such as the mapping between the plasma sheet and the ionosphere. This mapping has also been done in the 23 March 2007 substorm example shown below. It turns out that magnetic mapping using the OpenGGCM explains the observations much better than mapping based on empirical models (Angelopoulos et al. 2008).
3. The magnetosphere cannot be controlled and manipulated like a laboratory experiment; it can only be observed passively. Simulations, on the other hand, can be controlled within certain limits. For example, it is possible to use different solar wind and IMF, but it is not possible to reduce diffusion or resistivity below the inherent numerical diffusivities. We will thus use the OpenGGCM for numerical experiments to test hypotheses; for example, how solar wind and IMF changes trigger substorms, and how the ionosphere controls convection and the substorm process.

In the end, we hope that a new and more coherent picture of the substorm process will emerge from the THEMIS data in conjunction with OpenGGCM simulations. Without the simulations, the THEMIS data will likely leave ambiguities, while without the data the simulations would be essentially speculation.

In the remaining sections we first describe the OpenGGCM in detail. Then we present first results of the 23 March 2007 substorm event, which is also discussed in a companion paper (Angelopoulos et al. 2008). That section also serves to illustrate some of the outputs that the OpenGGCM can produce. The last section summarizes our results and provides an outlook.

2 The OpenGGCM Model

The OpenGGCM is a global coupled model of Earth's magnetosphere, ionosphere, and thermosphere. The magnetosphere part solves the MHD equations as an initial-boundary-value problem. The MHD equations are only solved to within $\sim 3 R_E$ of Earth. The region within $3R_E$ is treated as a magnetosphere-ionosphere (MI) coupling region where physical processes that couple the magnetosphere to the ionosphere-thermosphere system are parameterized using simple models and relationships. The ionosphere – thermosphere system is modeled using the NOAA CTIM (Coupled Thermosphere Ionosphere Model, Fuller-Rowell et al. 1996; Raeder et al. 2001a). In the following we describe each part of the model in more detail.

2.1 Outer Magnetosphere

The physics of the outer magnetosphere is governed by the magnetohydrodynamic equations, which we use in their normalized, semi-conservative form:

$$\frac{\partial \rho}{\partial t} = -\nabla \cdot (\rho \mathbf{v}) \tag{1}$$

$$\frac{\partial \rho \mathbf{v}}{\partial t} = -\nabla \cdot (\rho \mathbf{v}\mathbf{v} + p\mathsf{I}) + \mathbf{j}\times\mathbf{B} \tag{2}$$

$$\frac{\partial e}{\partial t} = -\nabla \cdot (\{e + p\}\mathbf{v}) + \mathbf{j} \cdot \mathbf{E} \tag{3}$$

$$\frac{\partial \mathbf{B}}{\partial t} = -\nabla \times \mathbf{E} \tag{4}$$

$$\nabla \cdot \mathbf{B} = 0 \tag{5}$$

$$\mathbf{E} = -\mathbf{v}\times\mathbf{B} + \eta \mathbf{j} \tag{6}$$

$$\mathbf{j} = \nabla \times \mathbf{B} \tag{7}$$

$$e = \frac{\rho v^2}{2} + \frac{p}{\gamma - 1} \tag{8}$$

The symbols have their usual meaning, e.g., **B** and **E** are the magnetic and the electric field, respectively, **v** is the plasma velocity, ρ is the density, p is the pressure, **j** is the current density, η is a resistivity, I is the unit tensor, and γ is the ratio of specific heats.

The semi-conservative formulation is chosen because it allows for finite difference schemes that numerically conserve mass (ϱ), momentum ($\varrho \mathbf{v}$), and plasma energy (e), but with no strict conservation of total energy. Fully conservative schemes that conserve total energy ($U = \frac{p}{\gamma-1} + \frac{\rho v^2}{2} + \frac{B^2}{2}$) often suffer from instability in low β regions where the pressure must be computed as the difference of two large quantities (U and $B^2/2$). The semi-conservative form avoids this difficulty.

The solution of the MHD equations in the outer magnetosphere is accomplished using an explicit second-order predictor-corrector finite difference time stepping scheme. The spatial derivatives are also computed using finite differences. However, because the simulation involves super-magnetosonic flows and shocks, simple finite differences are not sufficient but flux-limited schemes must be used. In the case of the OpenGGCM, we use a hybrid scheme that was originally proposed by Harten (Harten and Zwas 1972), where we combine a fourth-order scheme with a minimal diffusion error (Zalesak 1979, 1981) with the diffusive first-order Rusanov scheme. The numerical switch ensures that we obtain a high-order solution in regions of smooth variation of the flow, i.e., where there are no discontinuities, which degrades to a low-order solution at discontinuities, such as shocks and contacts, where the high-order scheme would fail due to numerical dispersion. Such shock-capturing schemes

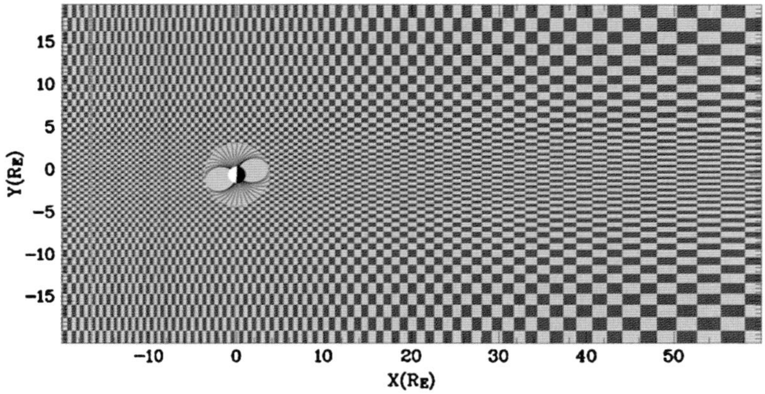

Fig. 1 Section of the OpenGGCM numerical grid. The grid is Cartesian but non-uniform in each of the coordinate directions. This figure shows only a fraction of the grid, which normally extends several hundred R_E anti-sunward, and $\sim 40R_E$ in each of the y- and z-directions. Also, the resolution is generally much finer (by a factor 2–3 in each direction) than shown here

are common in computational fluid dynamics of trans-sonic and supersonic flows (Hirsch 1990; Laney 1998).

Maxwell's equation states that $\nabla \cdot \mathbf{B} = 0$ at all times, since there are no magnetic monopoles. Strictly speaking, this is only an initial condition for $\mathbf{B}$ because Faraday's law demands that if $\nabla \cdot \mathbf{B} = 0$ at some time, it is to remain so as the magnetic field evolves, which can be seen from:

$$\nabla \cdot \frac{\partial \mathbf{B}}{\partial t} = \frac{\partial(\nabla \cdot \mathbf{B})}{\partial t} = -\nabla \cdot \nabla \times \mathbf{E} = 0 \tag{9}$$

Many numerical schemes do not a priori preserve $\nabla \cdot \mathbf{B}$. For such schemes the accumulation of $\nabla \cdot \mathbf{B}$ can lead to serious errors, in particular spurious parallel acceleration, wrong magnetic topology (field lines that are not closed), and significant errors in the shock jumps (Brackbill and Barnes 1980; Toth 2000). There are a few methods to clean the magnetic field of monopoles, for example the projection method, but none of these is perfect, and they also incur substantial additional cost (Toth 2000). The OpenGGCM uses the Constrained Transport (CT) method introduced by Evans and Hawley (Evans and Hawley 1988), which employs a staggered grid that allows near perfect (to roundoff error) preservation of $\nabla \cdot \mathbf{B}$. With CT, the magnetic field components are put on cell faces, and the electric field components for the right hand side of Faraday's law are put on the centers of the cells' edges. Such staggered grids require interpolation for the coupling terms $\mathbf{j} \times \mathbf{B}$ and $\mathbf{j} \cdot \mathbf{E}$; however, this is a small price to pay for magnetic flux conservation.

An important aspect of every MHD code is the spatial grid. Many choices are possible, ranging from equidistant Cartesian grids to structured adaptive mesh refinement (AMR) grids (see Raeder 2003, for an overview and discussion of grids). The OpenGGCM employs a stretched Cartesian grid. Figure 1 shows a cut through the grid in the x-y plane at $z = 0$. The figure shows only part of the grid; typically the grid extends to $\sim 20R_E$ in the sunward direction (to the left), several $100R_E$ in the anti-sunward direction (to the right), and $\sim 40R_E$ in the transverse (y and z) directions. Also, the grid resolution is substantially better than Fig. 1 indicates, typically 0.1–$0.2R_E$ at the sub-solar magnetopause and 0.2–$0.3R_E$ in the near-Earth tail, with a total of 10^7–10^8 grid cells.

The primary advantage of the OpenGGCM grid is that it allows for a well load balanced and efficient parallelized code, while it, for the most part, optimizes the resolution where it is needed. A uniform Cartesian grid would need 10^2–10^3 times the number of cells to achieve the same resolution in critical regions, such as the magnetopause or the plasma sheet. On the other hand, non-Cartesian or AMR grids may be able to optimize resolution better, but they also incur a higher computational cost and require much more complex codes.

2.2 Ionosphere and MI Coupling

As outlined above, the MHD calculation only extends to $\sim 3R_E$ from Earth. At that inner boundary the MHD part of the model is coupled with the ionosphere, mainly by the closure of field-aligned currents (FACs) in the ionosphere. The OpenGGCM uses a static dipole model to map the FACs into the ionosphere, which is possible for two reasons: (1) the current density obeys a continuity equation and (2) these currents typically do not close across field lines at this altitude. At the ionosphere end, a potential equation is solved on a sphere (or a section thereof) to yield the ionospheric convection potential (Fedder and Lyon 1987). The potential is then mapped back to the inner boundary of the MHD calculation where it is used as boundary condition for the flow and field integration ($\mathbf{v} = (-\nabla\Phi) \times \mathbf{B}/|B|^2$). Because the mapping originates at $3R_E$, it covers the latitudes from $\sim 58°$ to $90°$.

The dipole orientation is kept fixed in the OpenGGCM; i.e., the dipole does not rotate. Its orientation is set to the real geophysical dipole orientation at a given time, which is usually chosen to match a specific event. For studies of short-lived phenomena such as a substorm, there is no significant drawback in keeping the dipole orientation fixed, since the dipole does not rotate much during the period of interest. However, the fixed dipole orientation is less realistic for long duration events such as magnetic storms. During such events the largest error would occur at times that are an odd number of half-days different from the time that the dipole orientation corresponds to, and the error could be as large as twice the offset of the magnetic pole from the geographic pole, i.e., $\sim 22°$. This error affects mainly the dayside reconnection geometry because it alters the shear angle between the IMF and the dipole field.

The ionosphere-thermosphere model CTIM is described in detail elsewhere (see Fuller-Rowell et al. 1996, and references therein); thus we only provide a brief description here. CTIM is a global multi-fluid model of the thermosphere-ionosphere system with a long heritage. CTIM solves both neutral and ion fluid equations self-consistently from 80 to 500 km for the neutral atmosphere and from 80 to 10,000 km for the ionosphere on a spherical grid with 2° latitude resolution and 18° longitude resolution. The thermosphere part solves the continuity equation, horizontal momentum equation, energy equation, and composition equations for the major species O, O_2, and N_2 on 15 pressure levels. The ionosphere model part solves the continuity equations, ion temperature equation, vertical diffusion equations, and horizontal transport for H^+ and O^+, while chemical equilibrium is assumed for N_2^+, O_2^+, NO^+, and N^+. The horizontal ion motion is governed by the magnetospheric electric field. The coupled model includes about 30 different chemical and photo-chemical reactions between the species. Compared to the magnetosphere, the CTIM time scales are relatively long, allowing for numerical time steps of the order of one minute. Consequently, CTIM is computationally very efficient and runs considerably faster than real-time (>10 times) on a single CPU.

CTIM's primary input are the solar UV and EUV flux (parameterized by the solar 10.7 cm radio flux), the tidal modes (forcing from below), auroral electron precipitation parameters, and the magnetospheric electric field.

The electron precipitation parameters, energy flux F_E, and mean energy E_0 are computed separately for diffuse precipitation and for discrete precipitation, i.e., for electrons accelerated in regions of upward FAC.

Diffuse precipitation is parameterized by:

$$F_E = n_e(kT_e/2\pi m_e)^{\frac{1}{2}}, \qquad E_0 = kT_e \tag{10}$$

where T_e and n_e are the magnetospheric electron temperature and density, respectively, and k is the Boltzmann constant.

Discrete electron precipitation is modeled using the Knight relation (Knight 1972):

$$\Delta\Phi = K \max(0, -j_{\parallel}) \tag{11}$$

$$K = \frac{e^2 n_e}{\sqrt{2\pi m_e k T_e}} \tag{12}$$

$$F_E = \Delta\Phi_{\parallel} j_{\parallel}, \qquad E_0 = e\Delta\Phi_{\parallel} \tag{13}$$

where $\Delta\Phi$ is the parallel potential drop on an auroral field line. Because the MHD model cannot provide an electron temperature, we use the MHD single fluid temperature adjusted by a fudge factor.

The electric field in the ionosphere is assumed to be a potential field and is obtained from current conservation, which leads to the following potential equation (Vasyliunas 1970; Kelley 1989):

$$\nabla \cdot \Sigma \cdot \nabla\Phi = -j_{\parallel} \sin I \tag{14}$$

with the boundary condition $\Phi = 0$ at the magnetic equator. Because the ionosphere is a magnetized and partially ionized plasma, the ionospheric conductance is a tensor (Strangeway and Raeder 2001), given by:

$$\Sigma = \begin{pmatrix} \Sigma_{\theta\theta} & \Sigma_{\theta\lambda} \\ -\Sigma_{\theta\lambda} & \Sigma_{\lambda\lambda} \end{pmatrix} \tag{15}$$

$$\Sigma_{\theta\theta} = \frac{\Sigma_P}{\sin^2 I}, \qquad \Sigma_{\theta\lambda} = \frac{\Sigma_H}{\sin I}, \qquad \Sigma_{\lambda\lambda} = \Sigma_P \tag{16}$$

where Σ_H is the Hall conductance, Σ_P is the Pedersen conductance, θ is the magnetic latitude, λ is the magnetic longitude, and I is the magnetic field inclination.

The ionospheric Hall and Pedersen conductance is computed by CTIM from first principles, i.e., from the electron-neutral collision terms. In addition, the neutral wind dynamo is explicitly included in the solution of the electric potential. The neutral dynamo plays no significant role during substorms, but can produce a flywheel effect during storms, where the neutrals are accelerated by ion drag during the storm main phase, while the neutrals impart momentum on the ions during the recovery phase and thereby generate an electric field (Rishbeth et al. 1991). Using the CTIM conductances, as opposed to using conductances from empirical models, significantly affects the simulations. The effect of different conductance models in the OpenGGCM and its predecessors has been studied previously (Raeder et al. 1996, 2001a). The latter study showed that the model produced significantly more realistic ionosphere potentials in runs where the MHD model was coupled with CTIM. Although that study focused on storms, we have also conducted simulation runs with different

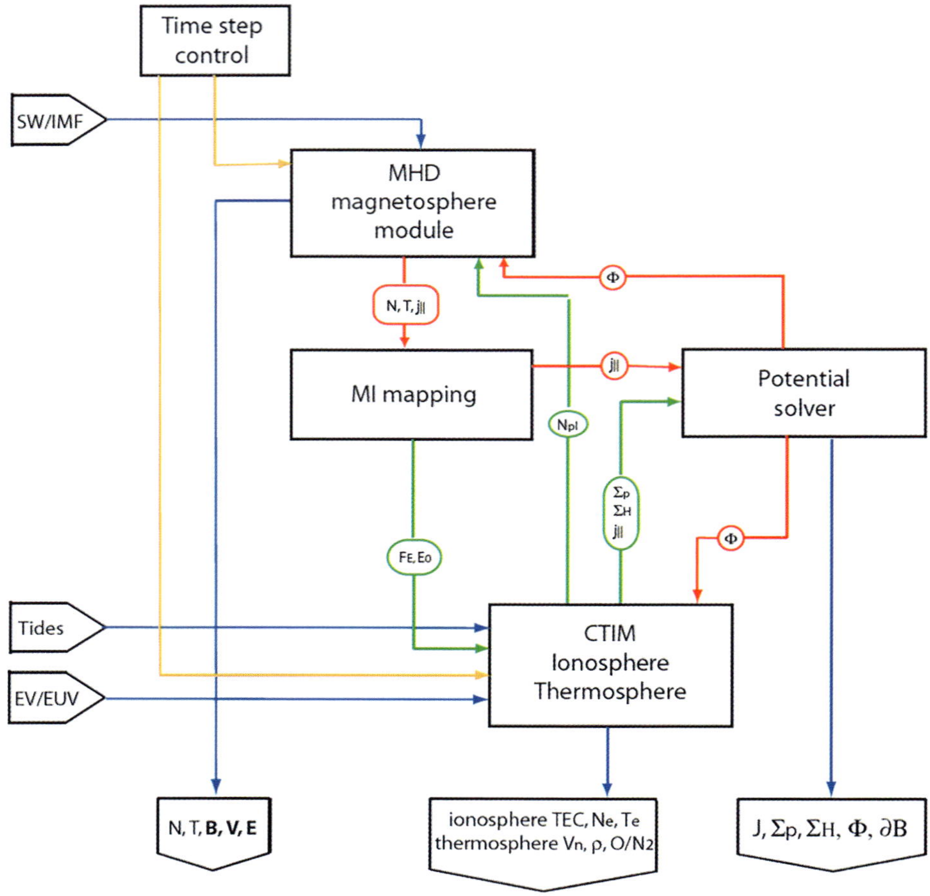

Fig. 2 Block diagram of the OpenGGCM with its models and with data/control flow. *Blue lines* denote model input and output. *Red lines* denote data flow with strong coupling. *Green lines* denote data flow with weak or slow coupling. *Orange lines* denote control flow. B, N, and T are the magnetospheric magnetic field, plasma density, and temperature, respectively. The field aligned current is $j_{\|}$, Φ is the ionosphere potential, F_E and E_0 the energy flux and mean energy of precipitating electrons, Σ_H and Σ_P the ionosphere Hall and Pederesen conductances, and ∂B is the ground magnetic perturbation

conductance models for the 23 March 2007 substorm presented later in this paper. Here we also find that the simulation results from the coupled model are in much better agreement when the CTIM conductances are used. Conversely, when uniform conductance is used, or when the nightside e^- precipitation is switched off in CTIM, only a weak substorm or no substorm at all may develop in the simulation.

Figure 2 shows a block diagram of the OpenGGCM elements and their relationships. The connection arrows indicate the flow of data. Note that the OpenGGCM only requires a minimal set of inputs. The solar wind and IMF are typically taken from a solar wind monitor such as ACE or Wind. Geotail and Cluster can also provide SW and IMF data when they are upstream of the bow shock. Such data taken closer to Earth are preferable because they better represent the solar plasma and fields that ultimately interact with the magnetosphere. However, even these data are not perfect as input for the OpenGGCM because they generally lack information about the three-dimensional solar wind structure, which must be known ideally, to specify the time-dependent MHD variables across the entire inflow boundary. We thus need to make an assumption as to the structure of the solar wind. One option is to assume that the solar wind parameters are independent of Y_{GSE} and Z_{GSE}. In that case, the IMF B_x component cannot change in time because that would violate $\nabla \cdot \mathbf{B} = 0$. If there are significant IMF B_x variations, the assumption of Y_{GSE} and Z_{GSE} independence cannot be true. In that case, we attempt to find a direction $\mathbf{N}$ in the solar wind such that the magnetic field component along that direction (B_N) does not change significantly. Since usually only one solar wind monitor is available, we employ the minimum variance method of Sonnerup and Cahill (1967, 1968) to find that direction. We call this the MINVAR method. If observations from multiple solar wind monitors are available, more precise methods are available (see, for example Russell et al. 2001). If B_N is fairly constant over the time interval of interest, we set B_N to be constant in time at the value of its average and then transform the field back into GSE coordinates and use it as input to the MHD model. In this case the solar wind and IMF convects into the model as sheets whose orientation is given by their normal vector $\mathbf{N}$.

In the case that B_N from the minimum variance transform is not nearly uniform (defined such that the variance of B_N is significantly smaller, say $<10\ \%$ of the total field) the solar wind does not have a simple sheet-like structure and there is also not enough information available to determine the structure. The options are to either ignore the IMF B_x component or to set it to some constant value that seems reasonable. This may in many cases not be a bad choice, because the IMF B_x component essentially does not contribute to the interplanetary electric field (IEF) and because the draping of the IMF around the magnetosphere normally reduces the B_x component before the field interacts with the magnetosphere. However, if the IMF B_x component dominates the IMF it may affect the reconnection geometry at the magnetopause and the simulation results must be carefully assessed for their validity.

Figure 3 shows the time series of the solar wind and IMF data observed by Wind for the 23 March 2007 substorm along with the data that have been processed using the MINVAR procedure. The top panel shows the B_L (maximum variance, red line), B_M (intermediate variance, green line), and B_N (minimum variance, blue line) IMF components. The eigenvector (direction) corresponding to the minimum variance $\mathbf{N}$ is given at the top of the figure. The direction of $\mathbf{N}$ is close to sunward but has a significant Z component. For this case the variance of the B_N component is small compared to the total field; thus the IMF must at least be ordered locally in sheets that are normal to $\mathbf{N}$. There are several discontinuities, i.e., rapid changes in the IMF direction, in this interval. Since B_N across these discontinuities is close to zero, they are most likely tangential discontinuities.

The following three panels show the three IMF components. The red lines show the Wind observations. The blue lines show the result from setting B_N to zero and transforming back to GSE. The green lines show the result from setting B_N to its average over the entire interval and transforming back to GSE. For each component the three lines nearly coincide. The near coincidence of the green and blue lines simply reflects the fact that the B_N average is nearly zero for this interval. The near coincidence of the red and green traces shows that the IMF model of inclined planar sheets whose normal is $\mathbf{N}$ is consistent with the data.

The bottom two panels of Fig. 3 show the solar wind plasma parameters. These time series are not affected by the MINVAR procedure. However, in order to be consistent with the treatment of the plasma parameters they are convected into the simulation box in the same manner as the field components; i.e., the MHD state vector $\mathbf{U} = (\mathbf{B}, \mathbf{V}, N, T)$ at the inflow boundary is not just a function of time $\mathbf{U}(t)$, but also a function of Y_{GSE} and Z_{GSE} ($\mathbf{U}(y, z, t)$) to take the inclined sheet structure of the solar wind and IMF into account.

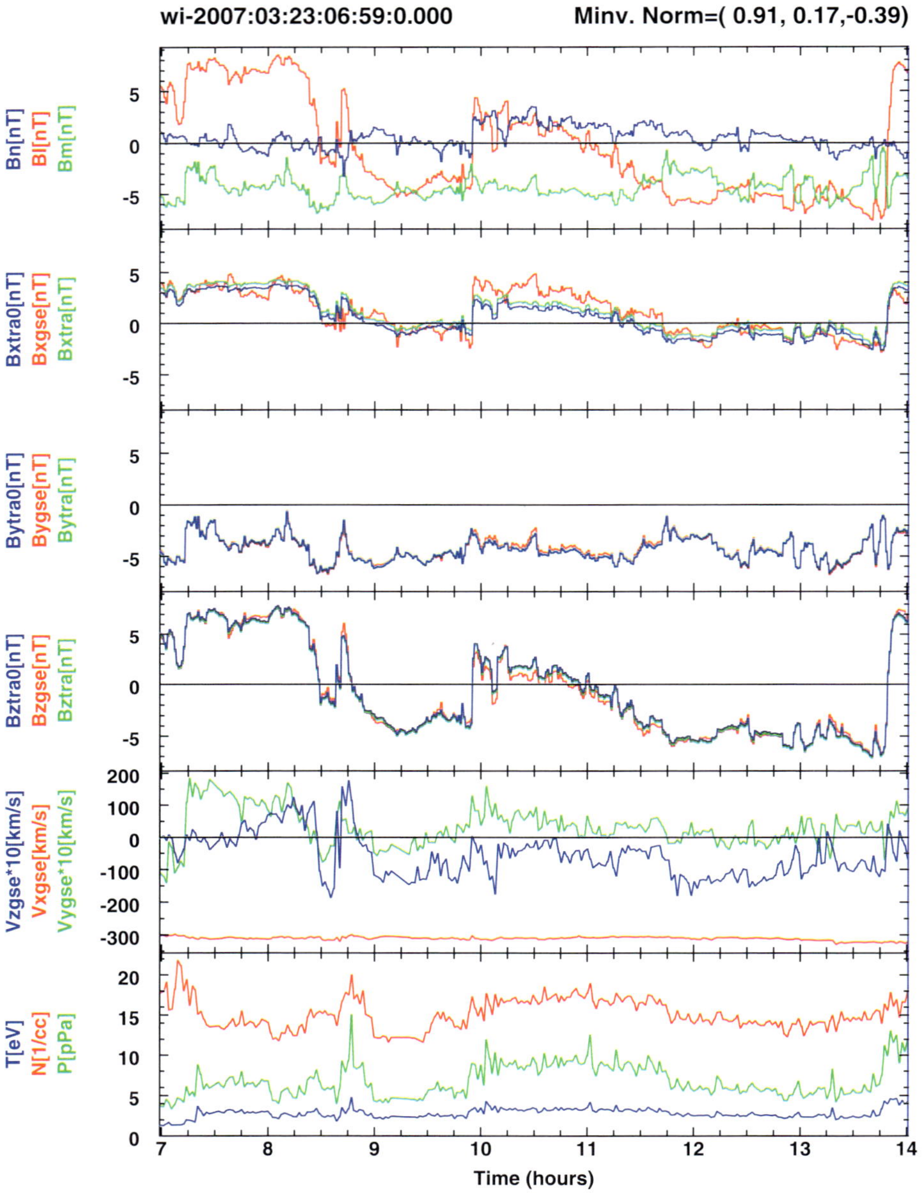

Fig. 3 Solar wind and IMF data processed by the MINVAR procedure and used as input for the simulation of the 23 March 2007 substorm event

3 OpenGGCM Products and the 23 March 2007 Substorm Example

The OpenGGCM produces the three-dimensional grids with the MHD state vector $(\rho, p, \mathbf{V}, \mathbf{B})$, along with a number of ionosphere and thermosphere quantities, such as the ionosphere potential, ionospheric currents, electron density, neutral composition, and neutral winds. These quantities describe the state of the system; however, they often cannot be directly compared to observations, and the information contained in the gridded fields is too overwhelming to see the physical processes that occur and to draw conclusions. The output of a simulation run can easily exceed one TB (terabyte, equal to 10^{12} bytes) of data. Thus, in order to extract useful information, a substantial amount of post-processing and visualization is necessary. There are many ways to extract information from the raw data, and new techniques are still being developed. In the following we demonstrate some of the most basic techniques and applications. We will use an OpenGGCM simulation of the 23 March 2007 substorm as an example. This substorm is also discussed elsewhere in this issue (Angelopoulos et al. 2008; Keiling et al. 2008), and thus we will not discuss the observations here but direct the reader to these papers.

3.1 Satellite Time Series

The most basic comparison is that of the moments and fields measured at the satellite with the time series taken in the simulation at the same location. We refer to these time series as "virtual satellites." In version 3.1 of the OpenGGCM, these time series are automatically generated for a number of satellites (made up or real) and their trajectories, which are input to the model. This approach has the advantage that the time series output can be generated at high cadence ($\sim$5 s) but it does not allow re-positioning the virtual satellite after the run is completed. The latter approach, i.e., generating time series from three-dimensional output at a location that is somewhat displaced from the true satellite location, is sometimes useful when a satellite is located close to a boundary. In that case a small error in the boundary location can lead to a complete mismatch between the observations and the virtual satellite. By placing a second virtual satellite at the other side of the boundary, often just by a fraction of $1R_E$, one can then show that the boundary location is primarily in error, not the MHD state variables themselves.

Figures 4, 5, and 6 show the comparison of the virtual satellites THEMIS C, THEMIS B, and THEMIS E with the in situ observations. THEMIS C is the closest to the tail center, THEMIS E is the closest to the dusk flank, and THEMIS D, B, and A are located very close together between C and E, and thus observe nearly the same, at least on the MHD time scale. The companion paper (Angelopoulos et al. 2008) discusses the locations in detail.

All three satellites observe a flow burst, both tailward and duskward, near the time of the substorm onsets, which was determined by Angelopoulos et al. (2008) to occur at 10:54 UT (minor activation) and 11:19 UT (major activation). At the same time the magnetic field becomes strongly deflected. That deflection is similar but not the same as a classical dipolarization. The B_z and B_y components increase as in a dipolarization; however, the magnitude of B_x also increases, which is opposite to dipolarization. Furthermore, the sunward flows sweep colder and denser plasma past the spacecraft.

The virtual spacecraft see essentially the same signatures, but with some significant differences. First, the substorm onset, as defined by the onset of fast flows here, occurs too early ($\sim$10:40 UT) as opposed to the 10:54 UT and 11:19 UT onsets and intensifications observed by THEMIS and the imagers (see Angelopoulos et al. 2008, this issue). This is also borne out in the aurora from the simulation discussed further below. However, the general pattern of the magnetic field and flow variations are quite well reproduced. Plasma density and temperature match least well, which can be understood as a memory effect of the magnetosphere. This simulation was started at 07:00 UT, i.e., four hours before the substorm onset, thus

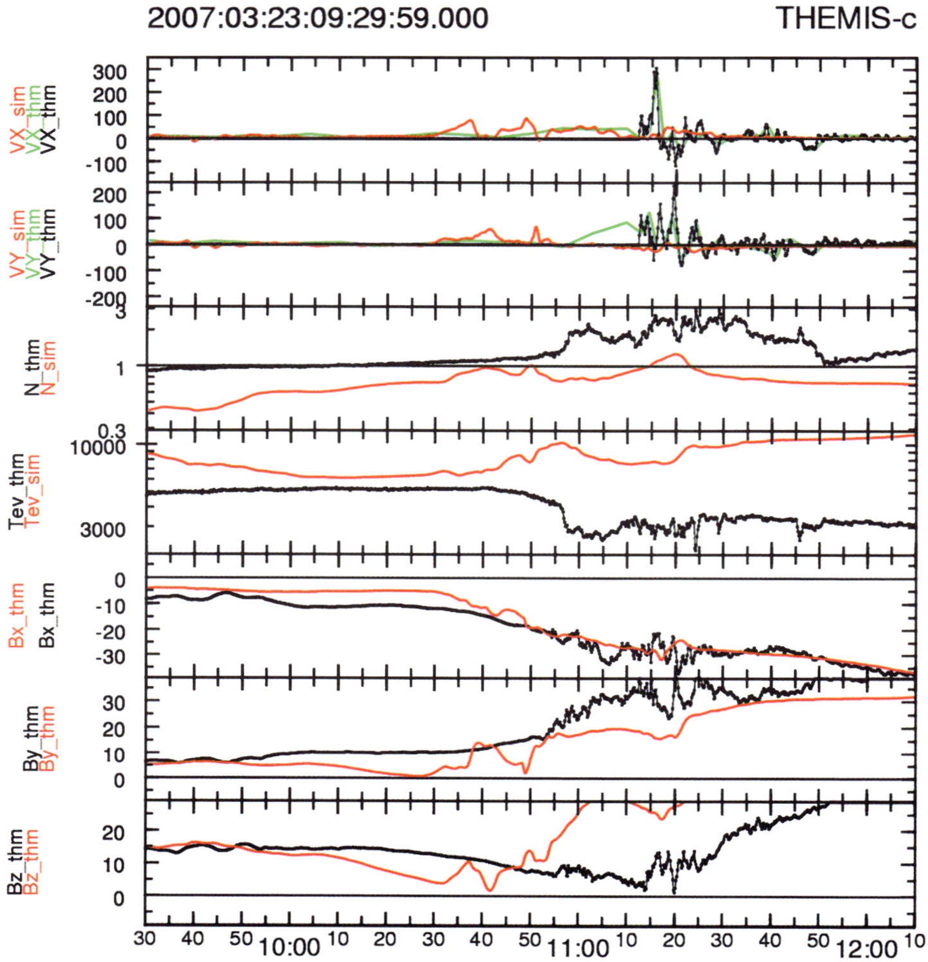

Fig. 4 Comparison of the MHD state variables measured by THEMIS C (*black* and *green lines*, the latter from on-board computation of moments) and from the OpenGGCM simulation (*red lines*). The panels show, from *top* to *bottom*: the three components of the velocity, the plasma number density, the ion temperature, and the three components of the magnetic field. All variables are in GSE coordinates. THEMIS C is the closest to the tail center

much of the plasma in the simulated magnetosphere may still be primordial, i.e., a remnant from the initial conditions. How long it takes for the magnetosphere to completely replenish all of its plasma from the solar wind and ionospheric sources is not well known and probably depends on the solar wind and IMF conditions.

Overall the simulation reproduces the key observational features well enough for there to be confidence in the results. In particular, a reasonable comparison like this one can be the starting point for a more detailed analysis of the simulation in order to elucidate the processes that lead to the observed phenomena.

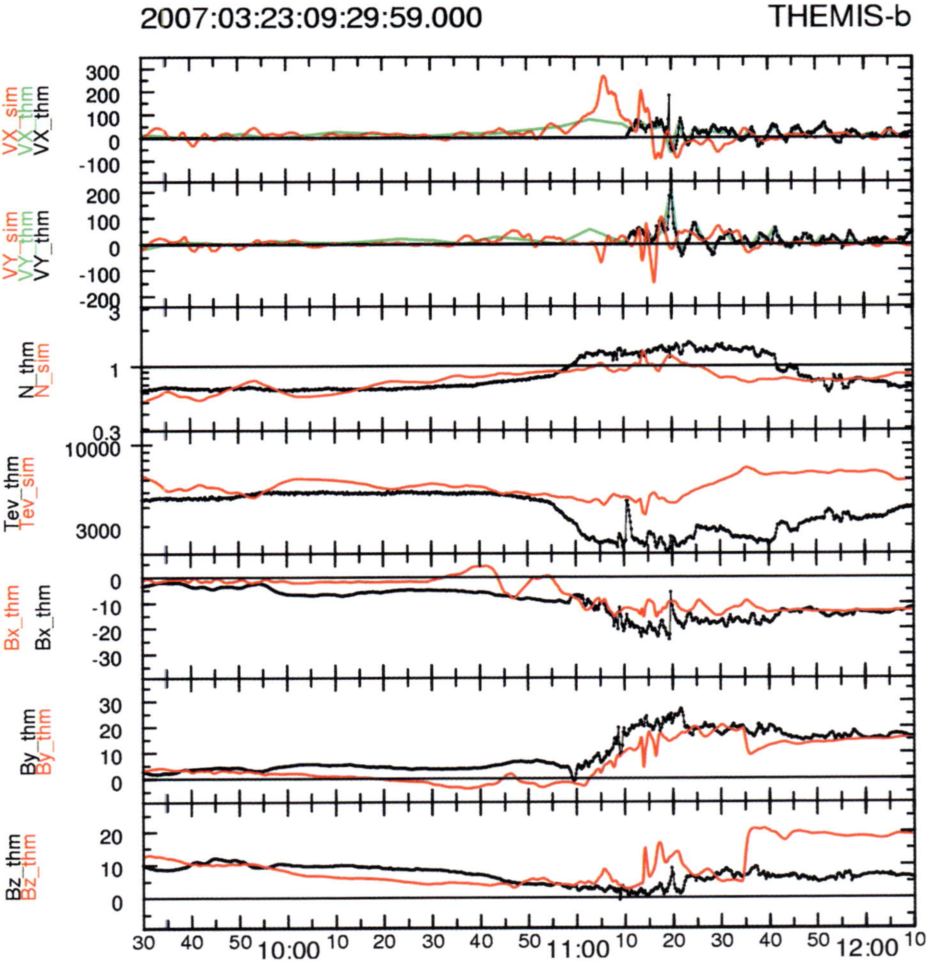

Fig. 5 Comparison of the MHD state variables measured by THEMIS B in the same format as Fig. 4. THEMIS B is the middle spacecraft between C and E

3.2 Ionosphere and Aurora

Of course, the defining characteristic of a substorm is the brightening of the aurora and the development of the westward traveling surge (WTS) (Akasofu 1964, 1977). The OpenGGCM does not produce auroral emissions; however, it does produce the energy flux and the mean energy of two populations of precipitating electrons. The first population is the thermal electron flux from the inner magnetosphere, which is unstructured and representative of the diffuse aurora. The second population is made up of electrons that have been accelerated in regions of upward flowing field-aligned current (FAC), as discussed in Sect. 2. This population is highly structured and it is considered generating the discrete aurora, although that distinction may not be made from an experimental view. In high resolution OpenGGCM runs such as the one presented here, features in the discrete precipitation as small as $\sim 0.5^\circ$ in latitude and $\sim 2^\circ$ in longitude can be resolved. In the plots discussed below

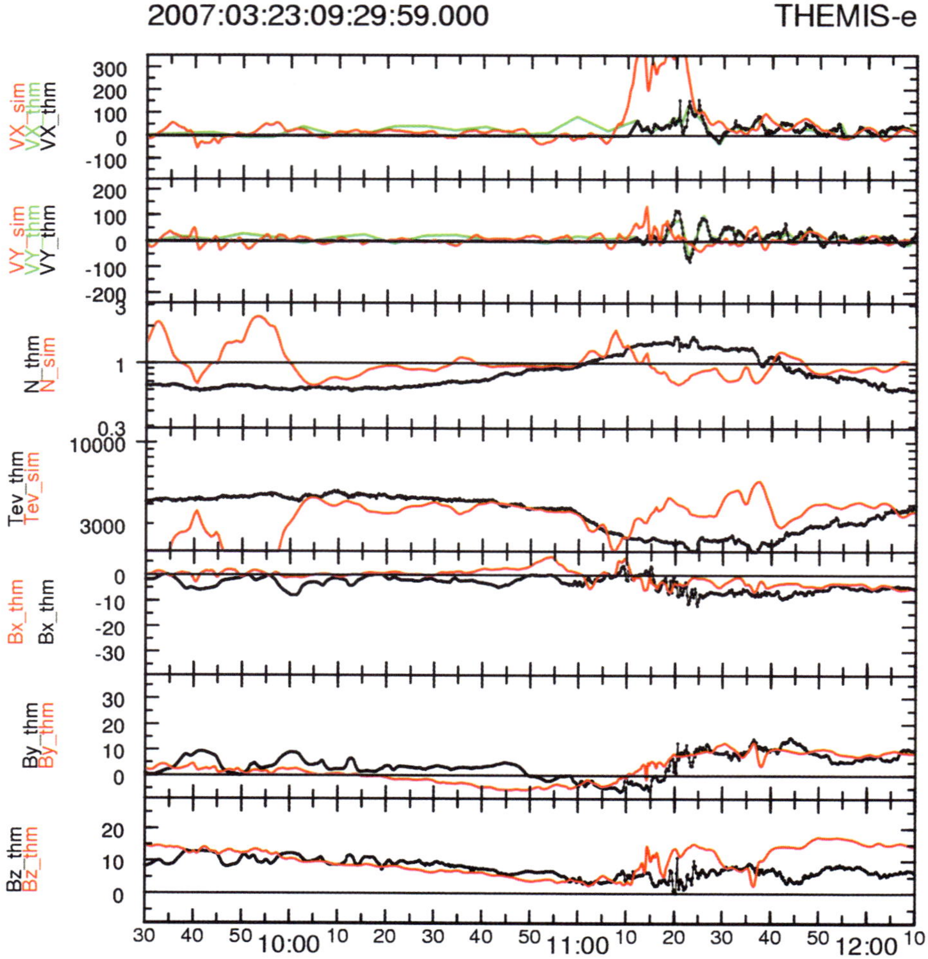

Fig. 6 Comparison of the MHD state variables measured by THEMIS E in the same format as Fig. 4. THEMIS E is the spacecraft the closest to the dusk flank

we show the energy flux of these accelerated electrons as a proxy for auroral emissions. In principal the emissions could be calculated (Emery et al. 1996; Germany et al. 1997; Lummerzheim et al. 1997). However, in order to be able to compare the emissions to data the specific instrument responses need to be modeled, which has not been done here.

Figure 7 shows a polar view of the northern hemisphere at six different times. Each of the six panels shows the energy flux of precipitating electrons color coded in units of mW/m^2. The thick black line shows the polar cap boundary, i.e., the boundary between open and closed magnetic flux. Each panel has the date and the UT time indicated in the upper left corner.

At 10:00 UT the IMF at the magnetopause is still northward, and thus the magnetosphere is in a geomagnetically quiet state. The polar cap (PC) is small, with the PC boundary (PCB) mostly located between 75° and 80° magnetic latitude. At 10:40 UT the IMF has turned southward at the dayside magnetopause. The PC has expanded considerably, with

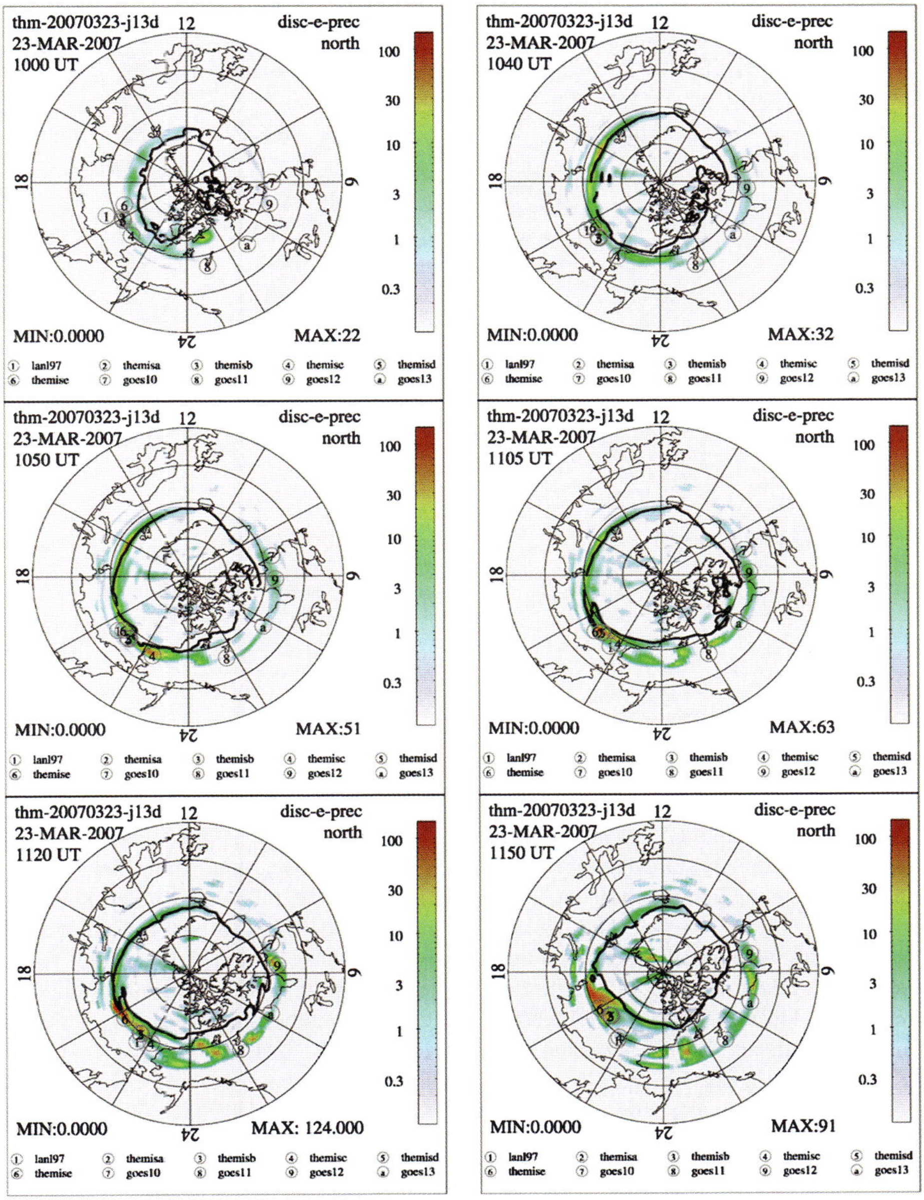

Fig. 7 Polar view of the northern hemisphere. The color coding shows the energy flux of accelerated electrons, which serves as a proxy for auroral emissions. The *thick black line* is the polar cap boundary

the PCB located just above 70° at most local times. This state represents the growth phase of the substorm, i.e., magnetic energy is being convected into and stored in the tail lobes. As the tail lobes expand, so does the PC, and at the same time the magnetic flux and magnetic energy in the lobes ($B^2/2\mu_0$) increases. During the quiet time and during the growth phase, the discrete aurora occurs primarily just equatorward of the PCB.

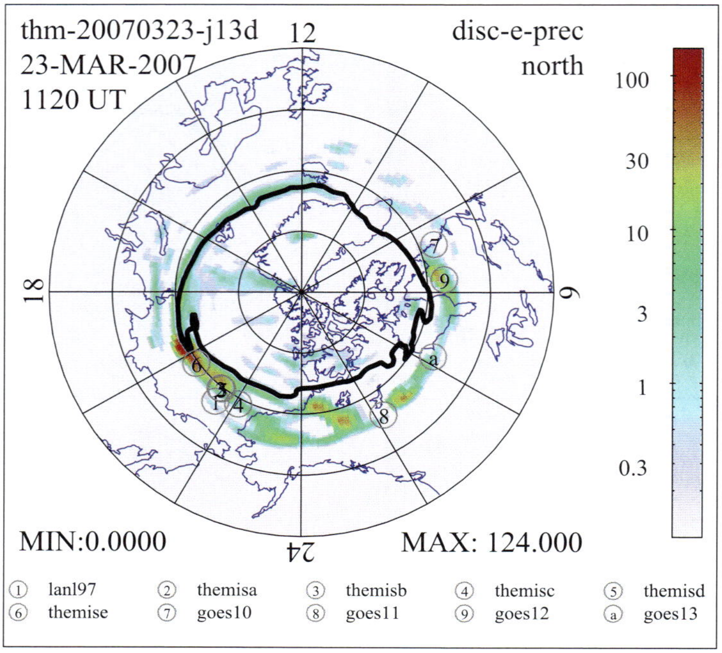

Fig. 8 Panel 5 of Fig. 7 at a larger size. The magnetic footpoints of several satellites show that the THEMIS probes map right into the westward traveling surge

At 10:50 UT the first indication of a substorm onset becomes visible near 22.5 MLT and 68° magnetic latitude. Closer inspection of the time series of polar plots shows that this intensification already started at 10:44 UT. This is considerably earlier than the first onset in the data, where the first intensification occurred at ~10:54 UT, followed by intensifications at ~11:10 UT and ~11:19 UT (Angelopoulos et al. 2008). The next panel shows that by 11:05 UT the aurora had expanded northward by ~3° and westward to ~21 MLT. Close inspection of the following 10 min (not shown here) shows that the expansion slows down and stops. However, between 11:15 UT and 11:20 UT, another intensification starts that causes a significant further expansion westward to ~19 MLT and ~77° magnetic latitude, as can be seen in the 5th and 6th panels. This latter expansion appears to correspond to the 11:19 UT expansion observed in the data. Comparison with the IMF data shown in Fig. 3 indicates that this intensification is likely caused by the sharp northward turn of the IMF at 09:55 UT. The Wind data need to be time shifted by ~75 min to account for the convection time from the Wind location ($X_{GSE} = 198R_E$) to the magnetopause, which means that the arrival of the northward turning at the magnetopause occurs at ~11:10 UT. There is considerable uncertainty in the actual arrival time because Wind is off the sun—Earth line by $\sim 30R_E$.

In Fig. 7 the magnetic mapping from the THEMIS (and several other) satellites to the ionosphere is marked. The 5th panel of Fig. 7 is shown as Fig. 8 for clarity. The THEMIS probes map magnetically into the path of the WTS. This result is expected because of the flow and field signatures observed by THEMIS and on the ground. However, magnetic mapping using an empirical magnetic field model places the THEMIS probes east and south of the WTS (Angelopoulos et al. 2008), and thus the OpenGGCM mapping is much more realistic.

3.3 Magnetosphere Flow and Field Evolution

In order to understand the physical processes that trigger the expansion phase onset and provide the power in the expansion phase, at least as they happen in the model, an analysis of the complete three-dimensional data sets from the simulation is needed. The huge data sets in 4 dimensions, i.e., 3 space dimensions and time, make this a difficult task. The simulation of the substorm shown here has $\sim 30 \times 10^6$ cells, and at least a few hundred snapshots in time are needed to obtain a complete picture. Furthermore, there are at least the eight MHD state variables per cell, and in practice a number of derived variables, such as current density or electric field require examination. In all, the simulation leads to 10^{11} to 10^{13} data values, which are impossible to comprehend altogether. Furthermore, contemporary data presentation techniques and tools are mostly restricted to a two-dimensional "flatland" (paper or computer screen) which requires a substantial reduction and projection of the data (Tufte 1990). Occasionally the time dimension can be added with movies or animations, but even in there only some 3-dimensional hyperspaces can be seen out of the 4-dimensional data that matter.

Figure 9 shows three three-dimensional renderings of the magnetosphere in the vicinity of the THEMIS probes. Each rendering is composed of three cut planes (at $X_{GSE} = 0R_E$, $Y_{GSE} = 0R_E$, $Z_{GSE} = -2R_E$), which are color coded according to a physical quantity, such as B_z in the $Z_{GSE} = -2R_E$ plane and $\log(T)$ in the other two planes. The $Z_{GSE} = -2R_E$ also shows arrows that depict the flow velocity. These arrows are scaled to $\sqrt{V}$ in order to cover a large dynamic range of speeds. The sphere centered on the origin has a radius of $3.5R_E$ and coincides with the inner boundary of the MHD simulation. Its surface is also color coded, in this case with the Hall conductance, which has been mapped along dipole field lines from the ionosphere. A number of field lines are drawn as pink tubes. One set of field lines originates from 65° magnetic latitude, at every 7.5° longitude. These field lines are all dipolar and mostly undisturbed. They provide a good idea of the dipole orientation. A second set of field lines is seeded along the X-axis in the tail. These field lines are obviously highly dynamic. Satellites of interest are pictured by spheres, and field lines are drawn through them. In Fig. 9, these satellites are, in order of decreasing distance, THEMIS E, A, B, D, C, and LANL97, which is a geosynchronous satellite and the closest to the Earth. Finally, there is a pink iso-surface that depicts sunward flows in excess of 180 km/s.

The top panel of Fig. 9 shows the magnetosphere at 10:30 UT, which is during the substorm growth phase. At this time the B_z in the plasma sheet is positive; i.e., the plasma sheet is on closed field lines. However, the field is not simply dipolar but is significantly stretched. Also, the negative IMF B_y exerts a twist on the tail which is clearly visible. The flows in the plasma sheet are mostly calm and of the order of a few 10s km/s.

The second panel shows the tail during the first substorm activation at 10:45 UT. There is now a significant patch of negative B_z in the plasma sheet at $X_{GSE} \sim -20R_E$ and somewhat duskward of the tail center. This negative B_z is also accompanied by strong tailward flows of several 100 km/s. Note that a negative B_z closer to Earth can also arise from the dipole tilt and the fact that the $Z_{GSE} = -2R_E$ plane does not coincide everywhere with the center of the plasma sheet. Such negative B_z comes from southern lobe field lines that bend toward the Earth into the southern polar cap. Coincident with the tailward flows are also earthward flows closer to Earth. These are not visible in the $Z_{GSE} = -2R_E$ plane, but in the iso-surface that extends earthward of the reconnection site. At this time not much has happened at the

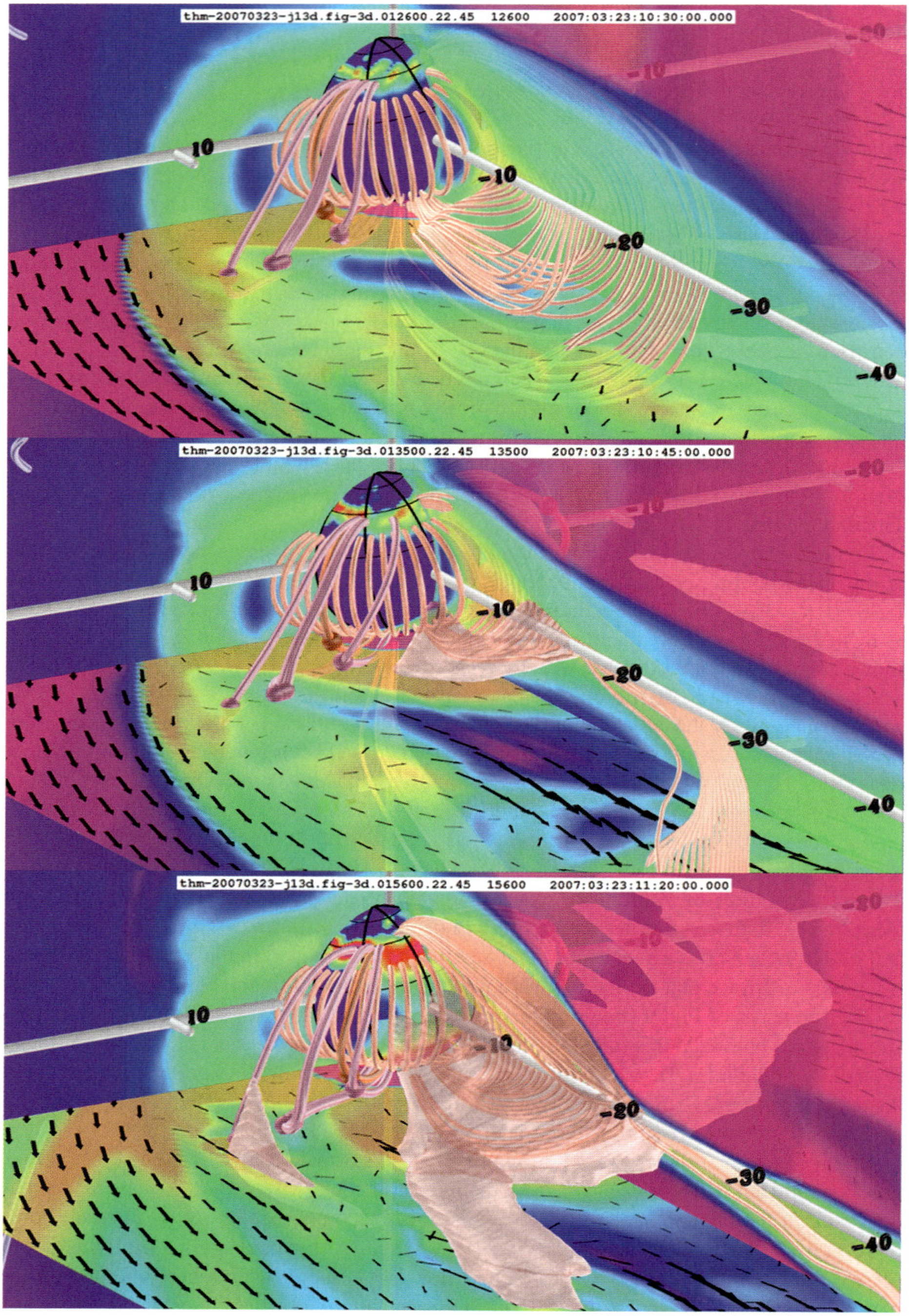

Fig. 9 Three-dimensional rendering of the magnetosphere configuration in the vicinity of the THEMIS probes during the 23 March 2007 substorm event. See text for a detailed explanation

THEMIS and LANL locations, but there are clear enhancements visible in the ionosphere Hall conductance. These occur because a substorm current wedge has formed that causes enhanced electron precipitation in the ionosphere at ~22 MLT.

The bottom panel shows the tail at 11:20 UT, just after the second intensification. At this time, the field near midnight has already dipolarized. There is now a new reconnection event further duskward. This event also creates a patch of strong southward IMF and both earthward/sunward and tailward flows. Most importantly, the sunward flows engulf the THEMIS spacecraft and cause the field at the spacecraft to dipolarize. Comparison with Figs. 4–6 shows that these flows and the dipolarization of the field are indeed observed.

Although Fig. 9 reveals a lot about the structure of the tail and its time evolution, it only shows some very limited aspects of the physical processes occurring in the simulation and will still require a substantial amount of work in the future. Replacing the sequence of the snapshots in time with a movie can give a much better impression of the dynamics of the plasma sheet. However, even then, the careful selection of the visualizing elements, such as cut planes or field lines is crucial to show the dynamical evolution.

4 Summary and Conclusions

In this paper we presented a description of the OpenGGCM and its role in supporting the THEMIS mission. For the simulations to be useful we first need to show that the model indeed produces the observed dynamical changes in the magnetosphere and ionosphere. The comparison of the model results with observations will never be perfect. However, as long as the model produces the essential features, the resulting confidence in the model allows conclusions to be drawn from the model results. We have presented here the first of such comparisons and shown that the model reproduces most of the salient features of the 23 March 2007 substorm. We have shown how the model can contribute to the mission and data analysis, in this case by providing the magnetic mapping between the ionosphere and the plasma sheet. We also provided a first glimpse at how the model results might be used to investigate the physical processes that ultimately lead to the observations.

In the future many more comparisons between the model and data need to be done to firmly establish what the model gets right and what it does not. With that background, the OpenGGCM can then be used to analyze the underlying physics, at least at the macro-scale, in more detail. Furthermore, the model can be used to conduct numerical experiments and to ask specific questions. For example, we may use the model to establish what mechanism leads to substorm triggering by northward turnings of the IMF. Comparing a run with a substorm triggered by a NBZ turn to one that is otherwise identical but without NBZ turn should provide the clues. Similarly, we may investigate the role that various parameters play in the onset mechanism, such as anomalous resistivity or ionosphere conductance. Such studies should eventually lead to a clearer and less controversial picture of the substorm process.

Acknowledgements This research was supported by NASA grant NAS5-02099. Development of the OpenGGCM has been supported by NASA grant NNG05GM57G and NSF grants ATM-0639658. Part of the simulations were performed at the San Diego Supercomputer Center and at the National Center for Supercomputer Applications.

References

S.-I. Akasofu, The development of the auroral substorm. Planet. Space Sci. **12**, 273 (1964)

S.-I. Akasofu, *Physics of Magnetospheric Substorms* (Reidel, Dordrecht, 1977)
V. Angelopoulos, The THEMIS mission. Space Sci. Rev. (2008, this issue). doi:10.1007/s11214-008-9336-1
V. Angelopoulos et al., First results from the THEMIS mission. Space Sci. Rev. (2008, this issue). doi:10.1007/s11214-008-9378-4
D.N. Baker, T.I. Pulkkinen, J. Büchner, A.J. Klimas, Substorms: A global instability of the magnetosphere-ionosphere system. J. Geophys. Res. **104**, 14,601 (1999)
J.U. Brackbill, D.C. Barnes, The effect of nonzero div B on the numerical solution of the magnetohydrodynamic equations. J. Comput. Phys. **35**, 426 (1980)
B.A. Emery et al., Comparing POLAR UVI imager data and other conductance sources in AMIE. Eos Trans. AGU **77**(46), 613 (1996)
C.R. Evans, J.F. Hawley, Simulation of magnetohydrodynamic flows: A constrained transport method. Astrophys. J. **332**, 659 (1988)
D.H. Fairfield, Advances in magnetospheric storm and substorm research: 1989–1991. J. Geophys. Res. **97**, 10,865 (1992)
J.A. Fedder, J.G. Lyon, The solar wind-magnetosphere-ionosphere current-voltage relationship. Geophys. Res. Lett. **14**, 880 (1987)
J.A. Fedder, S.P. Slinker, J.G. Lyon, R.D. Elphinstone, Global numerical simulation of the growth phase and the expansion onset for a substorm observed by Viking. J. Geophys. Res. **100**, 19,083 (1995)
T.J. Fuller-Rowell, D. Rees, S. Quegan, R.J. Moffett, M.V. Codrescu, G.H. Millward, A coupled thermosphere-ionosphere model (CTIM), in *STEP Report*, ed. by R.W. Schunk (Scientific Committee on Solar Terrestrial Physics (SCOSTEP), NOAA/NGDC, Boulder, 1996), p. 217
G.A. Germany, G.K. Parks, M. Brittnacher, J. Cumnock, D. Lummerzheim, J.F. Spann, L. Chen, P.G. Richards, F.J. Rich, Remote determination of auroral energy characteristics during substorm activity. Geophys. Res. Lett. **24**, 995 (1997)
A. Harten, G. Zwas, Self-adjusting hybrid schemes for shock computations. J. Comput. Phys. **9**, 568 (1972)
C. Hirsch, *Numerical Computation of Internal and External Flow*, vol. II (Wiley, New York, 1990)
A. Keiling et al., Correlation of substorm injections, auroral modulations, and ground Pi2. Geophys. Res. Lett. **35**, L17S22 (2008). doi:10.1029/2008GL033969
M.C. Kelley, *The Earth's Ionosphere* (Academic Press, New York, 1989)
C.F. Kennel, The Kiruna conjecture: The strong version, in *Substorms I*, vol. SP-335 (ESA, Noordwijk, 1992), p. 599
S. Knight, Parallel electric fields. Planet. Space Sci. **21**, 741 (1972)
C.B. Laney, *Computational Gasdynamics* (Cambridge University Press, Cambridge, 1998)
A.T.Y. Lui, Extended consideration of a synthesis model for magnetic substorms, in *Magnetospheric Substorms*, ed. by J.R. Kan, T.A. Potemra, S. Kokubun, T. Ijima. AGU Geophys. Monogr. Ser., vol. 64 (American Geophysical Union, Washington, 1991), p. 43
D. Lummerzheim, M. Brittnacher, D. Evans, G.A. Germany, G.K. Parks, M.H. Rees, J.F. Spann, High time resolution study of the hemispheric power carried by energetic electrons into the ionosphere during the May 19/20, 1996 auroral activity. Geophys. Res. Lett. **24**, 987 (1997)
R.L. McPherron, Physical processes producing magnetospheric substorms and magnetic storms, in *Geomagnetism*, vol. 4, ed. by J. Jacobs (Academic Press, New York, 1991), p. 593
J. Raeder, Global magnetohydrodynamics—A tutorial, in *Space Plasma Simulation*, ed. by J. Büchner, C.T. Dum, M. Scholer (Springer, Berlin, 2003)
J. Raeder, N. Maynard, Foreword. J. Geophys. Res. **106**, 345 (2001)
J. Raeder, J. Berchem, M. Ashour-Abdalla, The importance of small scale processes in global MHD simulations: Some numerical experiments, in *The Physics of Space Plasmas*, vol. 14, ed. by T. Chang, J.R. Jasperse (MIT Cent. for Theoret. Geo/Cosmo Plasma Phys., Cambridge, 1996), p. 403
J. Raeder, Y.L. Wang, T. Fuller-Rowell, Geomagnetic storm simulation with a coupled magnetosphere—ionosphere—thermosphere model, in *Space Weather*, ed. by P. Song, G. Siscoe, H.J. Singer. AGU Geophys. Monogr. Ser., vol. 125 (American Geophysical Union, Washington, 2001a), p. 377
J. Raeder et al., Global simulation of the geospace environment modeling substorm challenge event. J. Geophys. Res. **106**, 381 (2001b)
H. Rishbeth, W. Deng, R.G. Roble, T.L. Killeen, A.G. Burns, The flywheel effect: ionospheric currents after a geomagnetic storm. Geophys. Res. Lett. **18**, 1845 (1991)
C.T. Russell, Y.L. Wang, J. Raeder, The interplanetary shock of September 24, 1998: Arrival at Earth. J. Geophys. Res. **105**, 25,143 (2001)
D.G. Sibeck, V. Angelopoulos, THEMIS science objectives and mission phases. Space Sci. Rev. (2008, this issue). doi:10.1007/s11214-008-9393-5
S.P. Slinker, J.A. Fedder, J.G. Lyon, Plasmoid formation and evolution in a numerical simulation of a substorm. Geophys. Res. Lett. **22**, 859 (1995)

B.U.O. Sonnerup, L.J. Cahill, Magnetopause structure and attitude from Explorer 12 observations. J. Geophys. Res. **72**, 171 (1967)
B.U.O. Sonnerup, L.J. Cahill, Explorer 12 observations of the magnetopause current layer. J. Geophys. Res. **73**, 1757 (1968)
R.J. Strangeway, J. Raeder, On the transition from collisionless to collisional magnetohydrodynamics. J. Geophys. Res. **106**, 1955 (2001)
G. Toth, The $\nabla \cdot B$ constraint in shock-capturing magnetohydrodynamics codes. J. Comput. Phys. **161**, 605 (2000)
E.R. Tufte, *Envisioning Information* (Graphics Press, Cheshire, 1990)
V.M. Vasyliunas, Mathematical models of magnetospheric convection and its coupling to the ionosphere, in *Particles and Fields in the Magnetosphere* (Dordrecht, Reidel, 1970), p. 61
M. Wiltberger, T.I. Pulkkinen, J.G. Lyon, C.C. Goodrich, MHD simulation of the magnetotail during the December 10, 1996, substorm. J. Geophys. Res. **105**, 27,649 (2000)
S.T. Zalesak, Fully multidimensional flux-corrected transport. J. Comput. Phys. **31**, 355 (1979)
S.T. Zalesak, Very high order pseudospectral flux-corrected transport (FCT) algorithms for conservation laws, in *Proceedings of the Fourth IMACS International Symposium on Computer Methods for Partial Differential Equations*, ed. by R. Vichnevetsky, R.S. Stepleman (IMACS, Rutgers University, New Brunswick, 1981), p. 126

The Time History of Events and Macroscale Interactions during Substorms (THEMIS) Education and Outreach (E/PO) Program

L.M. Peticolas · N. Craig · S.F. Odenwald · A. Walker · C.T. Russell · V. Angelopoulos · C. Willard · M.B. Larson · W.A. Hiscock · J.M. Stoke · M.B. Moldwin

Originally published in the journal Space Science Reviews, Volume 141, Nos 1–4, 557–583.
DOI: 10.1007/s11214-008-9458-5

Abstract During the pre-launch phase of NASA's THEMIS mission, the Education and Public Outreach (E/PO) program successfully brought the excitement of THEMIS to the public, students and teachers through a variety of programs. The Geomagnetic Event Observation Network by Students (GEONS) was the main effort during this time, a project in which 13 magnetometers were placed in or near 13 rural schools across the country. High school teachers and a few middle school teachers at these and/or neighboring schools took part in a long-term professional development program based around space science and the

J.M. Stoke was previously at Space Telescope Science Institute, 3700 San Martin Drive, Baltimore, MD 21218, USA.

L.M. Peticolas (✉) N. Craig
Space Sciences Laboratory, University of California, Berkeley, CA 94720-7450, USA
e-mail: laura@ssl.berkeley.edu

S.F. Odenwald
Catholic University of America, Washington D.C., 20064, USA

A. Walker
Cornerstone Evaluation Associates LLC, 205 Peddler Place, Pittsburgh, PA 15212, USA

C.T. Russell · V. Angelopoulos · M.B. Moldwin
Institute of Geophysics and Planetary Physics, University of California, Los Angeles, CA 90095, USA

C. Willard
Lawrence Hall of Science, University of California, Berkeley, CA 94720-5200, USA

M.B. Larson
Utah State University, 1435 Old Main Hill, Logan, UT 84322, USA

W.A. Hiscock
Montana Space Grant Consortium, Montana State University, 416 Cobleigh Hall, Bozeman, MT 59717-3835, USA

J.M. Stoke
North American ALMA Science Center, National Radio Astronomy Observatory, 520 Edgemont Road, Charlottesville, VA 22903, USA

magnetometer data. The teachers created week-long to semester-long projects during which their students worked on THEMIS lessons that they, their colleagues, and the E/PO team created. In addition to this program, THEMIS E/PO also launched the only Lawrence Hall of Science (LHS) Great Explorations in Mathematics and Science (GEMS) site in Nevada. This site provides a sustainable place for teacher professional development using hands-on GEMS activities, and has been used by teachers around the state of Nevada. Short-term professional development for K-12 teachers (one-hour to two-day workshops), with a focus on the Tribal College and Society for the Advancement of Chicanos and Native Americans in Science (SACNAS) communities have reached hundreds of teachers across the country. A Space Telescope Science Institute (STScI) ViewSpace show on auroras and THEMIS was created and distributed, and shown in over a hundred science centers and museums nationwide. The THEMIS E/PO program developed and maintained a THEMIS E/PO Website for dissemination of (1) information and multimedia about the science and engineering of THEMIS, (2) updated news about the mission in language appropriate for the public, (3) the GEONS data, the GEONS teacher guides with classroom activities, and (4) information about the THEMIS E/PO program. Hundreds of thousands of visitors have viewed this website. In this paper, we describe these programs along with the evaluation results, and discuss what lessons we learned along the way.

Keywords Education · Outreach · Earth's Magnetosphere · Aurora · Magnetometer

PACS 01.00.00 · 01.30.la · 01.30.lb · 01.30.Os · 01.40.-d · 01.40.Di · 01.40.E- · 01.40.ek · 01.40.G- · 01.40.gb · 01.40.J- · 01.40.jh · 01.50.-i · 01.50.F- · 01.50.H- · 94.05.-a · 94.05.Sd

1 Introduction

While the five THEMIS spacecraft (probes) and instruments were being built and tested for launch and operation, the THEMIS Education and Public Outreach (E/PO) program was creating opportunities for students, teachers, and the public at large to learn about magnetic fields, electromagnetism, Earth's magnetism, auroras, substorms, solar storms, and the THEMIS mission itself. This article describes these *pre-launch E/PO activities* as part of the topical issue of Space Sciences Review describing the THEMIS Mission and Science, of which E/PO is a component. Reflecting on these activities, we hope to (1) provide science educators with ideas for ways to incorporate research and data into the classroom, (2) share some of the evaluation results of our programs from which space scientists and future E/PO professionals can learn, (3) provide scientists who have not participated in E/PO activities with an understanding of E/PO, and (4) let space scientists and E/PO professionals know what E/PO resources are available from the THEMIS mission. Of particular emphasis is our magnetometer-in-the-classroom program. This paper provides guidance for the success of similar efforts placing magnetometers at schools. Note that due to the large number of acronyms throughout the paper, we have created an acronym list in Appendix, which may be helpful when reading this paper.

1.1 Education and Public Outreach at NASA

The THEMIS E/PO program was funded as part of the NASA requirement that each satellite mission use a small percentage (around 1%) of the mission budget, excluding launch costs, for E/PO projects. For more on the history of this requirement, as well as a discussion on the general need in the U.S. of training more scientists and engineers, and the benefit of having education programs tied directly to science and engineering programs, see Peticolas et al. (2007) and Rosendhal et al. (2004).

During the early phases of THEMIS, the E/PO program was a part of a "White Paper" review of NASA education programs to "(1) Obtain a much more in-depth understanding of the quality of a subset of NASA Education Programs than was done during the 2003 Program Reviews; and (2) Test out a revised set of review procedures designed to examine programs both in more depth and in a more consistent way than has been done in the past" (Naik et al. 2004). THEMIS E/PO and the Mars Public Engagement were the two programs under NASA's then-called "Office of Space Science" (OSS) to be reviewed. The THEMIS E/PO was rated "Very Good" on all the review criteria: Customer Focus, Partnerships/Leverage/Sustainability, Evaluation, Content, Pipeline, and Diversity. Peticolas et al. (2007) discuss the meaning behind the 2006 NASA OSS E/PO proposal review criteria, which were the same as these 2004 criteria.

The review panel found that small programs that focused on doing a few things very well tended to suffer in the review process. There needs to be a way to review such programs that takes into account the size and cost of a program. The review panel also found that "with one or two exceptions, evaluation was a striking weakness in most programs" (Naik et al. 2004). Since THEMIS E/PO rated "Very Good" on these criteria, it is clear that it was one of these positive exceptions. Throughout the descriptions of the THEMIS programs in this paper, we include evaluation results of our programs.

Since this white-paper review in 2004, several new discussions about education have taken place at NASA. This resulted in attempts to gather data from all the education programs and use them to assess NASA's impact on educating students, teachers, and the public in science, technology, engineering, and mathematics (STEM) fields. As this NASA self-review of its educational programs continues, we hope that this paper provides some insight for scientists and E/PO professionals into (1) what is needed to create and sustain a successful NASA education program and (2) what influences these types of programs can have on teachers and their students.

1.2 Goals and Overview of the THEMIS E/PO Program

The THEMIS mission will determine the onset time and location of magnetic substorms in Earth's space environment. Its science is the science of auroras, Earth's magnetosphere, and sudden energy release in the form of plasma and electromagnetic fields. This energy release causes spectacular motions of auroras filling the night skies in the high latitudes of the northern and southern hemispheres. The THEMIS team, recognizing our country's need for improved STEM education (National Center for Education Statistics 2003), proposed a nation-wide partnership with science centers, K-14 educators, professional science organizations, and mission scientists to implement a comprehensive Education and Public Outreach (E/PO) program. The main goals for this program were to:

- Share the excitement of real-time measurements with science teachers and their students;
- Develop appropriate physical science and Earth and space science lesson plans that would be used in classrooms nationwide, adhere to appropriate grade-levels and National Science Education Standards (NSES), incorporate THEMIS data, and provide background content for teachers on the THEMIS magnetometers and mission;
- Share the awe of auroral substorms and the mystery of the trigger for the dynamical displays with the museum-going public around the country;

- Share THEMIS discoveries with teachers, students, and the general public through well-developed E/PO web pages;
- Share THEMIS science in the context of other NASA missions such as IMAGE, FAST, STEREO, and RHESSI;
- Motivate scientists involvement in E/PO;
- Use existing infrastructure in order to leverage THEMIS E/PO activities and to avoid duplication of effort;
- Partner with Tribal Colleges, schools on tribal lands, and the Society for Advancement of Chicanos and Native Americans in Science (SACNAS) to reach minority and underserved groups; and
- Provide teachers across the country with professional development opportunities to learn more about auroras and solar storms, and take appropriate lessons back to their classroom.

We have met all of these goals through five main projects:

1. The Geomagnetic Event Observation Network by Students (GEONS) in which 13 magnetometers are placed in or near 13 rural schools across the country. The primarily high-school teachers at these and/or neighboring schools take part in long-term professional development around space science and the magnetometer data. They can use their experience to inspire their students to learn about Earth's magnetic field and its changes related to substorm activity;
2. Launch of a new Lawrence Hall of Science (LHS) Great Explorations in Mathematics and Science (GEMS) site in Nevada to provide a sustainable teacher professional development site using the hands-on GEMS activities for elementary and middle school students;
3. Short-term professional development for K-12 teachers (1-hour to 2-day workshops), with a focus on the Tribal College and SACNAS communities;
4. Creation and dissemination of a Space Telescope Science Institute (STScI) ViewSpace show on auroras and THEMIS for the public and informal education venues;
5. Development and maintenance of a THEMIS E/PO Website for dissemination of (i) information and multimedia about the science and engineering of THEMIS, (ii) updated news about the mission in the language appropriate for the public, (iii) the GEONS data, the GEONS teacher guides with classroom activities, and (iv) information about the THEMIS E/PO program.

In addition to these five main programs, THEMIS has supported scientist and engineer visits to the classroom and the public relations efforts at NASA's Goddard Space Flight Center. In this paper we will describe all of these efforts, what was involved in starting the projects, what we learned along the way, and how we envision the future of these projects, in hopes of offering best practices and lessons learned to other scientists and E/PO professionals seeking effective programs.

1.3 Introduction to THEMIS E/PO Evaluation

The THEMIS E/PO partnerships, methods, activities, and visibility have been monitored and evaluated by Cornerstone Evaluation Associates (CEA), an established independent evaluation group with experience in evaluating the development of science learning resources and the use of technology in science education. CEA has assessed the effectiveness of the THEMIS E/PO effort as two major thrusts—(1) **Formative**—the documentation of partners' views of the strengths, weaknesses and necessary *improvements* of their programmatic contributions, and (2) **Summative**—assistance to partners in gathering outcomes data to measure program impact.

In addition to CEA's evaluation and assessments, all THEMIS E/PO products are submitted to NASA's education review. Summaries or highlights from many evaluations are included throughout this paper. For more details on the evaluation, please refer to the evaluation reports that CEA has written and NASA product reviews, which can be found from the THEMIS E/PO web site under "About Us" (see Sect. 4). Because at the time of writing this paper, the THEMIS E/PO program is still in progress, the overall summative report has not been completed. We anticipate this report being completed by August, 2009.

2 THEMIS in the Classroom: Formal Education

As mentioned in Sect. 1.2, there are several ways in which we have aimed to bring THEMIS successfully into the classroom, increasing teachers' understanding of magnetism, aurora substorms, and solar storms, as well as helping them to bring this content knowledge and excitement about the THEMIS mission to their students. In this section, we first discuss the THEMIS project that has spent the most time and resources on several dedicated teachers in rural and underserved regions with the goal of ultimately creating a long-term, in-depth and sustainable program for high school teachers (and others). Because this has been the "flagship" of the THEMIS E/PO program, most of this paper is dedicated to it. Next we discuss the THEMIS project that leverages an existing teacher professional development network of math and science teachers around an already-established curriculum mostly for elementary and middle school teachers. The remaining two aspects of the THEMIS E/PO follow: short-term teacher professional development workshops and scientists/engineers involvement in the classroom.

2.1 GEONS Program

The nature of the THEMIS science investigation, in particular the correlation of ground-based measurements of auroral activity with spacecraft-based measurements of changes in the magnetosphere, holds tremendous potential for inquiry-based instruction of pre-college students and teachers. In recognition of this, THEMIS E/PO has established thirteen ground-based magnetometer stations, each located in the proximity of a rural school in traditionally under-served, under-represented communities. These thirteen sites are located in the following ten states: Alaska, Oregon, Nevada, North Dakota, South Dakota, Montana, Wisconsin, Michigan, Pennsylvania, and Vermont, as shown in Fig. 1. Two schools are located in Michigan and three schools in Alaska. The two most northern Alaska sites also have all-sky cameras to observe the auroras in white light. A teacher at each of these schools is responsible for their magnetometer data and system as well as using the resulting data with their students through lesson plans that the THEMIS E/PO team and some of the teachers have developed. Table 1 provides the name of the schools, the location of the schools, and the local teachers who have been intimately involved in the project.

We provide yearly professional development opportunities for these teachers to help them understand the science of THEMIS and introduce new or modified classroom lessons around the science and the magnetometer data. The magnetometer data are located on the THEMIS E/PO website so teachers and students all across the country can take part in the program and schools can compare their data with those of other schools. The network of the 13 teachers, students, and magnetometers, together with other teachers and students who participate, is called the Geomagnetic Event Observation Network by Students (GEONS).

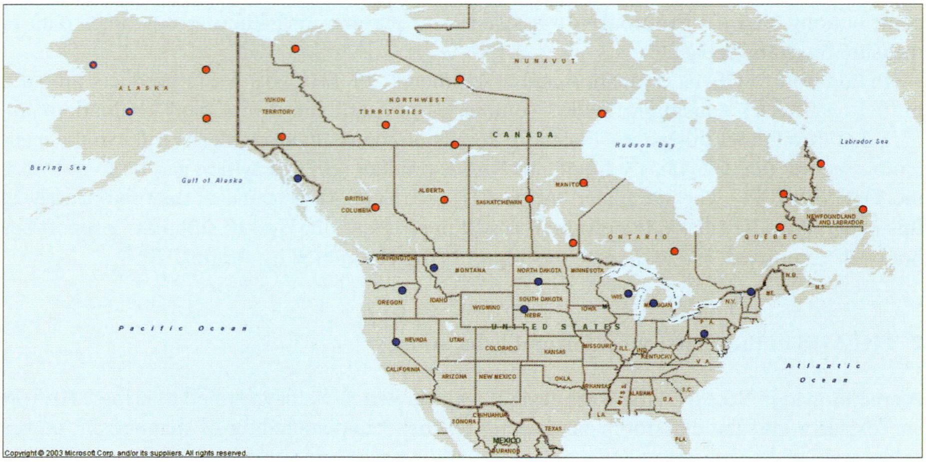

Fig. 1 Location of the GEONS magnetometers and schools are marked as *blue dots* and *dots* with *blue circles*. The *red dots* indicate the ground-based observatories (GBOs), which are part of the THEMIS science mission (Russell et al. 2008 and Mende et al. 2008). The *red dots* with *blue circles* are GBOs at schools and thus part of our E/PO program

Table 1 Name(s) of the school(s) involved in the GEONS project, location of the schools and magnetometers, and name(s) of the teacher(s) intimately involved in the GEONS project either presently or in the past

School(s)	City, State	Educator(s) *(past and present)*
Kiana School	Kiana, AK	Glenn Miller
McGrath School	McGrath, AK	Ray Benson
Petersburg City School	Petersburg, AK	Victor Trautman
Bay Mills Community College	Brimley, MI	Robert Dickinson, Michael Doyle
Standing Rock Public School	Fort Yates, ND	Harriet Howe, Frank Martin
Shawano Community High School	Shawano, WI	Wendy Esch
North Country Union Jr High School	Derby, VT	Holly Wyllie
Maine School of Science & Mathematics		Manju Prakash
Hot Springs High School	Hot Springs, MT	Sean Estil
Chippewa Hills High School	Remus, MI	Cris DeWolf
Red Cloud High School	Pine Ridge, SD	Wendell Gehman
Ukiah School	Ukiah, OR	Laura Orr
N. Bedford County High School	Loysburg, PA	Keith Little
Western Nevada Community College	Carson City, NV	Robert Collier
Carson City Middle School		Terry Parent
Carson City High School		Jim Bean

For scientists and E/PO professionals interested in finding schools to work with, we describe how the different magnetometer sites were chosen. We chose the Carson City, NV site to connect the magnetometer project with the THEMIS E/PO LHS GEMS site launch program. The coordinators of this program helped identify the community college as a good location and brought in the local middle and high school to participate in the program.

The NASA's Office of Space Science Support Network (Cooper et al. 2004) in Boston selected the magnetometer site in Vermont. This school was a NASA Explorers School (Ruberg et al. 2007).

A magnetometer site in each of eight states were identified by statewide competitions run by the Space Grant Consortia of Montana working with the Space Grant Consortia of seven other states—AK, OR, ND, SD, WI, MI, and PA. The selection criteria were (1) commitment of the school/teacher and availability of local infrastructure, (2) demonstrable advancements to the education process at the particular school with particular consideration towards reaching underserved students, (3) the potential for reaching a large community of students and teachers, and (4) the site's potential for science discoveries, based on its geographic location within the state, that may result in stronger interactions with the THEMIS research team. At all the sites, we required that the school's administration, either a superintendent or principal, support the project to ensure sustainability of the project. Eight of twenty-four schools/teachers were selected.

Two additional magnetometer sites in Alaska were chosen as part of the science ground-based observatory (GBO) network with magnetometers and all-sky cameras across Alaska and Canada (Russell et al. 2008; Mende et al. 2008). For these two sites, the science team selected the region of Alaska appropriate for these observatories and then approached area schools where there might be teachers and school administrators interested in taking part in the project. The Space Grant in AK helped to find these schools as well, though they were not part of the initial E/PO competition because the idea of placing these observatories in schools came about after the competition.

The thirteenth magnetometer site came about through a separate proposal process. The THEMIS E/PO team worked with the Bay Mills Community College, a tribal community college in Brimley, MI, on a proposal to do teacher professional development for 2005 and 2006 on the THEMIS and GLOBE projects. This proposal was funded through the Earth Explorers: SPHERE (Students as Professionals Helping Educators Research the Earth), with one of its goals to support undergraduate research participation especially Minority Serving Institutions (MSIs) in NASA Earth Science. This grant included the installation of another THEMIS magnetometer. Drs. Craig, Peticolas, and Odenwald, as well as GEONS teacher Cris DeWolf from Remus, MI, were presenters of the THEMIS program, science, and student activities. This collaboration increased the number of magnetometers in the GEONS program from 12 to 13. In addition to this site, four other THEMIS schools are on tribal lands: two in AK, and one in each of the states SD, ND, and MT.

From evaluation results from questionnaires and interviews discussed below in Sect. 2.1.2, we collected information on the demographics of many of these teachers. We learned about their school environment:

- Most GEONS teachers work at rural schools.
- The majority of schools in which GEONS teachers work are middle and high schools.
- The average number of students in GEONS schools is slightly over 600.
- The average number of faculty in GEONS schools is slightly under 40.
- On average, half of their students are female, with class composition ranging from 30% to 65% females.
- More than three-fifths (61%) of their classes are comprised of minority students—both male and female—ranging from 2% to 100% minorities.

We learned about the teachers' educational backgrounds, teaching experience and current teaching circumstances:

Table 2 Percentages of teachers offering various reasons for becoming involved with the THEMIS project

Motivation for Involvement with THEMIS Project	Teachers ($N = 9$)
Students—Opportunity to motivate and/or involve students in 'real science'; share materials with students—Saw this as a great opportunity to motivate students; the opportunity to share this material with my students was too good to pass up; the students get to be involved in real data/research... good stuff!	67%
Personal—Interest in astronomy/space science/THEMIS project—Always have been into space science; the subjects THEMIS covers are some of my favorites; astronomy interest; general interest in project itself	44
Personal—Opportunity to learn—The opportunity to learn; great opportunity to re-energized my enthusiasm for teaching	22
Personal—Opportunity to work with NASA/globally significant project—The opportunity to work with NASA in a research capacity; opportunity to involve students in actual research and fact gathering that has global implications was simply too awesome to pass up	22
Discussions/sharing with colleagues—Talking to fellow teachers; opportunity to share this material with other teachers	22

- Nearly all GEONS teachers have undergraduate degrees in the sciences, but only a few with physics degrees.
- The majority of GEONS teachers have science degrees beyond their bachelors degree.
- The average GEONS teacher has almost 17 years of teaching experience.
- All GEONS teachers are teaching at the middle and high school level.

We also learned *why* the teachers wanted to participate in this program. This is useful to know in developing future education programs. Nine teacher's responses are presented in Table 2. Nine teachers responded giving multiple responses, thus percentages sum to greater than 100%.

Placing magnetometers in or near schools and ensuring that the real-time data and archived data were available on a student-friendly website required additional support of several engineers and a system administrator. This needs to be factored in for any other educational programs bringing data to the classroom. Don Dearborn (UCLA—University of California, Los Angeles) was in charge of most of the magnetometer installations and took the opportunity to explain in person to local students and teachers about the THEMIS mission and how the magnetometer worked. This was an important way of connecting with the teachers and students at the schools at a personal level while also setting up the scientific instrumentation. At most schools, students helped Don Dearborn to dig the trench and hole for the magnetometer cable and sensor

We found quickly, that even though these schools are in rural communities, there were locations where the magnetometers could not be placed due to either high magnetic noise levels or Internet restrictions imposed by the school system. In these cases, the magnetometers were placed at nearby elementary schools or, in one case, at the superintendent's home. The teachers or administrators at the magnetometer sites have been instrumental in keeping the magnetometers running and sending data. More about the installations and workings of the magnetometers, as well as what has been required to keep the magnetometers running, can be found in Russell et al. (2008).

In order to engage students successfully with these data, one requirement was that students could be able to access the real-time data on-line in a student-friendly manner. Moreover, we soon realized that the archived data needed to be available as well, because times of

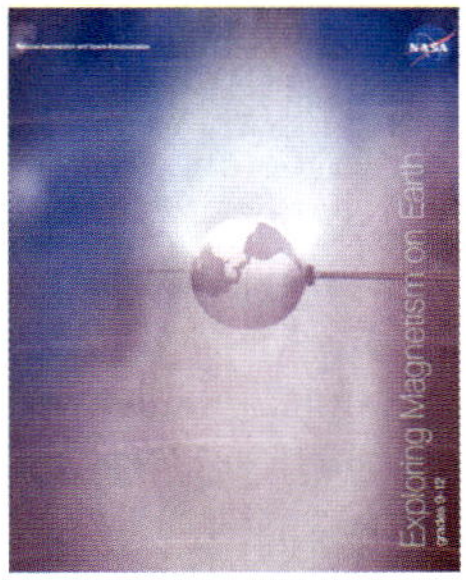

Fig. 2 Cover of "Exploring Magnetism on Earth"

active aurora occur when school is out. David Pierce (UCLA), Igor Ruderman (UCB—the University of California, Berkeley), and Tim Quinn (UCB) were instrumental in supporting this aspect of the GEONS project. The three types of data products produced are: the vector magnetic field—B_x, B_y, B_z (approximately geomagnetic coordinates); B_H, B_D, and B_{tot} (horizontal, declination, and total magnetic field); and power spectrograms of the magnetic field fluctuations. The teachers and students were most excited about the spectrograms because of the colors. But because spectrograms are not part of the local science education standards and because they are too complex for a high school level of science, these data have only been used as an indication of magnetic activity in a general sense. The B_x, B_y, B_z plots are used most frequently in the classrooms, in order to calculate local magnetic indices and to examine the total magnetic field variations over several months or even a year.

2.1.1 GEONS Teacher Guides

This GEONS project provides students and teachers with project-based activities that support access to real scientific data and at times inquiry, an important focus of the National Science Education Standards (NSES). We developed, in consultation with THEMIS Principal Investigator, THEMIS scientists, GEONS teachers, and existing technical magnetometer user manuals: (1) a ground magnetometer and background science "user manual" appropriate for high school teachers, (2) nationally-tested inquiry-based lesson plans in four theme-based teacher guides, and (3) learning materials on how to further utilize the magnetometer data to enhance classroom instruction in space science concepts. The teacher guides are titled: "Magnetism and Electromagnetism," "Exploring Magnetism on Earth," "Space Weather," and "Earth's Magnetic Personality." Figure 2 shows the cover of "Exploring Magnetism on Earth." The lesson plans within these guides include topics such as: ***Forces and Motion, Magnetic Induction, the Geomagnetic Field, Solar Storms and Space Weather and Data Analysis of the Magnetic Field Data.*** Teachers developed lessons for their students using the magnetometer data that went beyond the lessons in the fourth teacher guide, which contains all the lessons using data.

THEMIS utilizes this approach to introduce grade 8–12 students to themes of fundamental importance, such as space weather and its effects on the habitability of the near-Earth environment, on satellite communications, and on electrical power distribution on Earth. The THEMIS E/PC grade 8–14 module development leverages the resources from the Center for Science Education (CSE) at UCB and avoided duplication of products by coordinating with the existing IMAGE, FAST, STEREO/IMPACT, Science Education Gateway (SEGway), and selected theme-related Sun-Earth Connection Education Forum (SECEF) EP/O resources.

As any curriculum designer knows, creating teacher guides in such a way as to ensure they will be scientifically correct, useful, and well-employed is a challenge and can take years of testing and revisions and creating the THEMIS guides was no exception. Working closely with the teachers on these lessons and evaluating the lessons helped immensely with this process. As described below in Sect. 2.1.2, CEA interviewed the teachers in 2006. CEA dedicated a large portion of this interview to gathering feedback about the classroom lessons after the teachers had a chance to use actually them in the classroom. The feedback showed us that many lessons then needed to be altered to make them more effective to use with students. In two lessons, we found we needed to develop our own website with aurora and magnetospheric information rather than having the students go to non-THEMIS websites, because navigating through the other websites was too tedious for them and thus detracted from the goals of the lesson. Comments and feedback from these interviews about the use of the lessons in particular types of classes (astronomy, geology, physics) spurred the decision to break the one teacher's guide into four teacher's guide that still built on one another, but could also be used as stand-alone teacher guides. See the full report on the THEMIS E/PO website for more information about the teachers' feedback. Many curriculum designers use such evaluation techniques and we strongly recommend this for scientists or E/PO professionals new to creating lessons who cannot afford to work with curriculum designers directly.

After an initial successful summer (2006) with one teacher doing research with the magnetometer data for use in his classroom, in summer 2007, we increased the number of teachers to four whom we paid a stipend to do research with the data and produce lessons that could involve their students in similar type research. So far, this been the most successful way of involving the teachers and their students in the actual magnetometer data. All of these teachers regularly use the data in their classrooms after these summer research opportunities, whereas most of the other teachers do not use the data, but rather use the other THEMIS hands-on science and mathematics magnetism and space weather lessons.

We have had the entire set of teacher guides reviewed at least once. Individual guides were reviewed more than once by the NASA education review board. The review board provided valuable feedback from an outside perspective and helped to make the teacher guides appropriate for non-GEONS teachers to pick up and use. NASA strongly suggests that all materials developed for NASA education projects go through this review and we hope scientists or E/PO professionals who have not used this NASA service, do so.

All the teacher guides are located on the THEMIS E/PO website under "In the Classroom," and reviews from the NASA panels are located under "About Us/Evaluations" (see Sect. 4 below.) As we discuss below, these lessons have been used in all of the GEONS teacher's classrooms for at least two years and in many cases for longer, as we discuss in Sect. 2.1.2.

2.1.2 GEONS Impact on Teachers and Students

The THEMIS E/PO team maintains regular contact with the GEONS teachers through weekly emails that are part of an established THEMIS Yahoo Group, bi-monthly teleconference calls, and yearly professional development workshops. These means of communication act not only to provide support to the teachers, but also provide the E/PO team a way to determine how the program is succeeding, what impact it is having on teachers and students, and what changes need to be made as it progresses. Regular communication with the teachers has been shown to be very important to keeping the teachers engaged in this type of E/PO project.

In addition to these opportunities embedded in the program, Cornerstone Evaluation Associates (CEA) had two telephone interviews with a subset of the GEONS teachers.

Fig. 3 Photograph of the teachers and THEMIS E/PO team at the second annual GEONS teacher professional development workshop in Nevada at the magnetometer site (Western Community College)

Throughout the pre-launch E/PO phase, Cornerstone Evaluation Associates produced several reports on the GEONS program from surveys and interviews with the teachers. These are available on the THEMIS E/PO website under "About Us/Evaluations" (see the website Sect. 4) The reports are carefully written with attention to the evaluation tool and its responses. Without going into too much detail here, we try to give a sense of what these evaluation reports tell us in terms of the success of the program, the impact of the program on teachers and students, and what changes we have made as a result of these evaluation reports. We recommend working closely with such an evaluator to increase the effectiveness of an E/PO effort.

During the years of 2004, 2005, and 2007, the THEMIS E/PO team held professional development workshops with the GEONS teachers. The project covered the teachers' expenses to attend the workshops. These workshops provided science content, time to model the THEMIS activities from the teacher guides, and provided an opportunity for us to get further feedback from the teachers on the program. At the first two workshops, THEMIS principal investigator, Dr. Angelopoulos and THEMIS co-investigator, Dr. Bonnell, gave additional presentations to the teachers. The teachers provided feedback, both informally during the course of the workshop as well as formally by responding to questionnaires developed and analyzed by CEA. A photograph of the teachers and THEMIS E/PO team attending the second THEMIS GEONS workshop is shown in Fig. 3.

In 2004, the GEONS teachers did not yet have magnetometers and had just started with the project. Nine teachers attended. The agenda for the workshop is shown in Fig. 4. The teachers completed workshop questionnaires expressing that they learned about magnetism lessons usable in the classroom; that they had a better sense of the THEMIS project and their role in it; and that they enjoyed interacting with colleagues and experts on the THEMIS project. Their suggestions led to future workshop and program improvements. Improvements were to: (1) make the workshops one day longer, (2) make each day shorter, (3) increase opportunities to share and collaborate with other teachers, (4) provide updates on the progress of the mission, and (3) create an email-support-network for the teachers.

In the 2005 GEONS workshop in NV, nine teachers attended the workshop, together with the Principal Investigator (PI) of the grant, "Bay Mills Community College Charter School Science Teachers and Native American Youth Serving Organization THEMIS and GLOBE Training Project," that was funded to provide a magnetometer in Brimely, MI at a Tribal College. (GLOBE stands for Global Learning and Observations to Benefit the Environment.) The Western Nevada Community College astronomy professor, who manages the magnetometer near his observatory in Carson City, NV, hosted this workshop.

Fig. 4 GEONS 2004 teacher professional development workshop agenda

AGENDA

DAY 1
- Overview of THEMIS E/PO
- Introduction to THEMIS Mission science
- Activities—magnetism, building magnetic bottle magnetometers, soda bottle measurements
- Introduction to magnetic storms, magnetic fields in the universe
- Introduction to professional magnetic data

DAY 2
- Mission science talk
- Activities—soda bottle measurements, exploring magnetism
- Magnetometer, installation and exploring magnetometer data
- Solar storm statistics, NASAs Student Observation Network introduction

At this time, five of the initial ten magnetometers had been installed. At this workshop, the teachers shared how they were using the classroom lessons developed by the THEMIS E/PO team. These lessons were also modeled with all the teachers. Through discussions and feedback from the use of the lessons in science classrooms, we learned what modifications were needed. Four of the teachers indicated that there were some barriers to using all the lessons in the classroom, mostly trying to determine what to eliminate from their curriculum in order to "fit" it in and finding time to prepare and deal with the time between the workshop and class starting. We think EPO professionals would be well-served to have these kinds of practical discussions with teachers using their materials. All the teachers did implement the lessons in some of their classes over the next years, as indicated in the 2007 GEONS workshop.

Eight of the GEONS teachers attended the third GEONS workshop: a two-day St. Louis workshop in 2007 that was intentionally planned to coincide with the National Science Teachers Association (NSTA) conference, as is helpful to maximize a teacher's time and travel budget by providing extra opportunities for professional development. Five of the teachers had been participants since the 2004 inception of GEONS, two since 2005 and one joined the project in 2006. The focus of this workshop was on the presentation of activities included in the fourth guide being rolled out for classroom use. The fourth guide includes activities previously developed using magnetometer data, but revised based on feedback from GEONS teachers and research done by one teacher during the Summer 2006.

GEONS teachers indicated that presentations of the six activities included in the fourth guide were clear, offering a mean rating of 3.5 on a 4-point scale ranging from '1-not clear at all' to '4-very clear'. Nearly two-thirds of the GEONS teachers had not yet tried the activities before this workshop. Teachers who had not yet tried the activities signaled that they were very likely to do so in the future. When asked if they could foresee any barriers to implementation for these activities, most of the GEONS teachers cited their concerns about fitting them into the curriculum and time constraints.

Because this workshop took place a few months after the launch of the THEMIS satellites, the data are a nice summary of the state of the GEONS project at the time of launch. Many of the GEONS teachers at this point had as many as three years of project experience. They were asked to provide information about their successes in implementing all activities and student reactions to them, as well as their efforts in dissemination and professional development. We provide details from the CEA report based on these data in the following bulleted paragraphs

- **Implementation**—All of the teachers are using THEMIS materials, ideas, the Web site, etc. and do so primarily for courses in Earth Science/Geology (31%), Astronomy (23%) Physics/Physical Science (23%), General Science (15%) and Math (8%). Once the project ends, they reported that they will continue to use the materials and data from the magnetometer. Many (57%) feel they will not need additional help to do so, but would like to see teleconferences, updating activity guides and networking continue.
- Three-quarters of the teachers had used the THEMIS-GEONS Users Guide in the last school year. The majority (88%) reported using all or part of the four activity guides containing 20 THEMIS-related activities. On average, five of the eight teachers have tried each activity, with anywhere from one to eight teachers trying out any one activity.
- **Student impact**—The GEONS teachers who tried the activities report, on average, that the students have responded with interest as evidenced by a mean rating of 3.8 on a 5-point scale ranging from '1-extremely disinterested; bored' to '5-extremely interested; enthusiastic'. By activity, the mean ratings ranged from 3.0 to 4.0.
- Nearly three-fifths (57%) of the GEONS teachers reported seeing increased general interest in science among the elementary, middle and high school populations in their schools and school districts. Nearly three-fifths (57%) said that active participation in the project (real science) has sparked interest as students feel a vital connection to the mission. The materials and instructor's enthusiasm inspire students. Teachers have also reported science course enrollment increases.
- **Dissemination**—GEONS teachers engage in multiple means of disseminating THEMIS materials, both informally and formally. Most (86%) of the GEONS teachers said that they share THEMIS materials on an informal basis with their colleagues in department meetings, at lunch, in teachers' rooms, etc. More than two-fifths (43%) of the teachers shared THEMIS materials by making presentations at state teachers' conferences and within the community.
- Nearly three-fifths (57%) of the teachers have gained local or national media exposure—most notably with one teacher being featured on the Jim Lehrer News Hour on PBS. GEONS teachers also reported that they update the school's Web site with THEMIS news, make presentations at local community groups and are planning activities for future dissemination.
- **Professional development**—Inspired by involvement in THEMIS, the GEONS teachers have become involved in other NASA-related projects such as Cosmic Times and the NASA WISE Mission, in attending National Science Teachers Association (NSTA) conferences, in research/student activity projects and in the Teacher Leaders Research Based Science Education program (TLRBSE) at the National Optical Astronomy Observatory.

In each of the winters of 2005 and 2006, a subset of the GEONS teachers accepted invitations to participate in telephone interviews with CEA. These interviews provided information about

- teacher demographics
- how the teachers were informed about the competition for participation in the GEONS program and why they applied
- the communication with the THEMIS E/PO team and fellow teachers
- teacher outreach/dissemination efforts
- teacher involvement in other professional development activities
- feedback on the magnetometer installation process

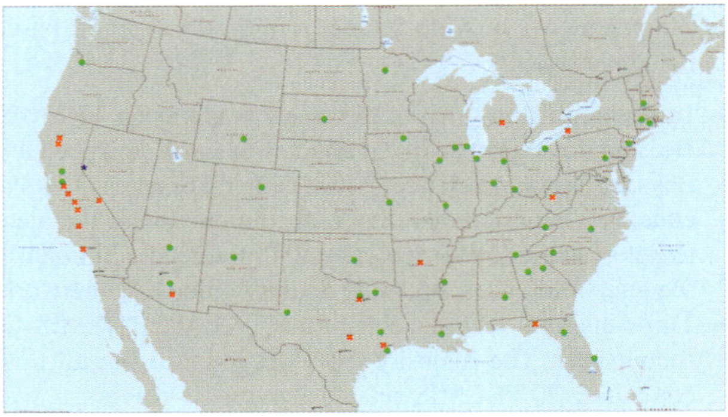

Fig. 5 GEMS sites (*green circles*) and centers (*red* ×'s) with the THEMIS-sponsored GEMS site in Nevada shown as a *blue star*

- plans for using or feedback on the use of the THEMIS-related classroom materials and activities

The full reports from these interviews can be found on the THEMIS E/PO web site. Together with the other reports mentioned above, they provide a sense of the perceptions of the program from the teacher perspectives, including how the teachers successfully anticipated their involvement in the program and in their classrooms. The ***PBS News Hour with Jim Lehrer*** report on the Petersburg, AK program provides an excellent window into the student's view of this program and its affect on their choices to become scientists. This is a goal of NASA E/PO in general. This PBS report can be seen on the THEMIS E/PO website from the "Gallery" page. A final evaluation report will be available on this program at the end of 2009.

2.2 Launch of a New LHS GEMS Site in Nevada

The THEMIS E/PO team, together with the Lawrence Hall of Science (LHS), launched a new GEMS Network site at the Carson City School District in Carson City, Nevada with a two-day Teacher Professional Development workshop in 2005. GEMS, the Great Explorations in Math and Science Teacher's Guide series, is a proven resource for excellence in inquiry-based mathematics and science. Developed at UCB's LHS, GEMS guides are used nationwide, from preschool through eighth grade. To support the growing number of teachers using GEMS materials, LHS GEMS maintains an international network of over 65 sites offering professional development and other services for teachers. The sites and centers as of 2008 are shown in Fig. 5 on the U.S.A. map as green circles and red ×'s respectively. The blue star marks the THEMIS-funded GEMS site. Centers offer more resources than do sites.

The Carson City GEMS Network Site serves teachers in northern Nevada. Many of these teachers are in very remote and under-served school districts, including ones on tribal lands. Carson City was selected as a GEMS site because it satisfies the conditions for being a prime candidate for E/PO magnetometer installation and the because of its strong ties with the GEMS effort.

Gail Bushey, a committed and active Associate of the GEMS Program is the lead at the new site with the strong support of District Assistant Superintendent Mike Watty. The Nevada State Science Coordinator and the Director of the new observatory at nearby Western Nevada Community College also lend their support. It is important in these types of programs to include this type of local support for sustainability of the program. Because of

Table 3 Percentages of teachers indicating various motivations for participating in GEMS site launch workshop

THEMIS GEMS SITE LAUNCH LEADERSHIP WORKSHOP 2005: Motivation to Participate	Teachers ($N = 38$)
Familiarity with GEMS—I have several GEMS guides and think they are very well written and explain the concepts very well	37%
Materials—Seeking new or free materials/resources/ideas—The six free handouts; I needed new ideas for teaching science; wanting more science information and experience	24
Colleague's encouragement—Encouragement from mentor/colleague/district—Judith Dragon's announcement at school; teaching with Gail Bushey kept me informed of GEMS activity; asked by the district to attend; encouragement from another teacher	18
Workshop convenience—Convenient time, date, location, no cost—I was already in Carson City for a workshop preceding this one; I was in the location; summer date; the fact it's free!	16
Credits—Recertification credit—I needed a credit for my teaching certificate; credit; I needed a recertification credit; in-service credit	16
Inquiry-based learning interest—I'm excited about inquiry-based science curriculum; my interest in hands-on science	16
Love of science—I enjoy science as a topic; interest in science; my love of science	13
Sharing knowledge—desire to share knowledge with other teachers or students—I wanted to teach a GEMS class to teachers	5

this strong local support, this GEMS site is continuing long after its initial funding provided by the THEMIS E/PO program.

Laura Tucker of LHS led the launching of the GEMS site in the Summer of 2005 with a 2-day leadership workshop, emphasizing space science, Earth science, and physical science. The THEMIS E/PO team gave two presentations at the workshop including Mapping the Magnetic Field and Living with a Star. The agenda for the workshop can be found in CEA's complete evaluation report of this site launch. The report is found on the THEMIS E/PO website. We provide some of the details from this report to give a sense of the impact of the workshop and the anticipated use by Nevada teachers of this GEMS site launch. These evaluation results suggest that partnering with the LHS for future GEMS sites is a worthwhile project for future NASA E/PO programs planned by scientists and E/PO professionals.

A total of 38 teachers, primarily from Nevada, attended the two-day workshop. Five of the teachers were also GEONS (Geomagnetic Event Observation Network by Students) teachers, who had attended a GEONS workshop in Carson City prior to this one. Most of the participants reported that they were teaching at pre-school/elementary level and/or at the middle school level, with minorities representing 30% of their students, on average, and that they were teaching on average 92 students. Of the 38 teachers at the workshop, 26 indicated that there are, on average, 39 teachers in their schools with whom they might share the GEMS materials in some way. The reason for teachers attending the workshop is shown in Table 3. Multiple responses were allowed, thus the percentages sum to over 100%.

Teachers indicated that all topics presented at the workshops were solidly in the 'somewhat likely to use' to 'very likely to use' range. The 'Mapping the Magnetic Field' and 'Living with a Star' presentations were some of the highest rated in anticipated use of materials but among some of the lowest mean ratings for understanding for GEMS presentations.

Table 4 Percentages of teachers indicating various ways they anticipate the Carson City GEMS site will assist them

THEMIS GEMS SITE LAUNCH LEADERSHIP WORKSHOP 2005: Anticipated Assistance from Carson City GEMS Site	Teachers ($N = 32$)
Training/resource person available—get questions answered—Contacts, more leadership trainings; training; Gail teaches two doors down, borrow and steal; It will be easy to get questions answered	50%
Materials/resources availability—GEMS site will be a great resource for finding information, borrowing kits, etc; borrowing materials for workshop; it will be easier to check out a guide I don't have; nice to know kits are available to present to staff and allow them to experience the activities	47
Facilitating connections with colleagues/support—Teacher support and dialogue; moral support; contact/network; love to collaborate with Gail Bushey; Be close to communicate with other teachers doing the same thing!; support with resources and awareness	28
Uncertain—I am not certain at this point; I will more likely use a GEMS center closer to me	6
Proximity—It's close!	3

These lessons were the lessons the most related to THEMIS science. This anti-correlation in expected use with understanding emphasizes the need for lessons and background science content around physics and space physics at the elementary level. This knowledge can help scientists and E/PO professionals create physical science workshops that elementary teachers need.

Just under half of the teachers responding to the question about what barriers would keep them from using these materials cited having limited time to prepare lessons and make kit components in advance as well as having difficulties in finding time to actually implement lessons. The THEMIS team anticipated this barrier in advance by providing 10 kits to the Carson City Site.

Table 4 summarizes the comments of the 32 teachers who responded to the question of how they might use the Carson City GEMS Site. Multiple responses result in the percentages totaling more than 100%.

Since the launch of this GEMS site, it continues to operate independently, maintaining contact with LHS and using district funds, fees and grants to support its work. In follow-up interviews, three teachers indicated that they really liked the GEMS lessons because they are user friendly with scripts and background science and because of the inquiry nature of the lessons. All three emphasized that aligning the GEMS lessons to the local state standards was very important and something they had to do themselves. One of these three teachers made an effort to determine which lessons met the state standards and she was able to use GEMS as the core of what she teaches. Due to the difficulty of obtaining responses from all 29 teachers, we do not know how many of them used the materials from the workshop. However, in 2006 and 2008, Gail Bushey informed us that several teachers had gone on to provide teacher professional development opportunities on GEMS guides in their region of Nevada and that many teachers have checked out GEMS kits throughout the years following the GEMS site launch.

2.3 Professional Development for Teachers with a Focus on the Tribal College and SACNAS Communities

THEMIS E/PO provides teacher professional development to teachers nationwide as well as those who are part of GEONS. These workshops have some of the same content elements as those mentioned in the GEONS section, but they are short-term workshops held at state and national conferences such as the NSTA, California Science Teachers Association (CSTA), and at the Center for Science Education (CSE) at UCB. As a part of these workshops, we aim to reach teachers other than the GEONS teachers who work with Native American students. In order to be more effective at such trainings, we took part in "One Earth, One Universe" workshops hosted by the Sun-Earth Connection Education Forum on understanding Native American science and culture in 2005. As one of many valuable lessons, these workshops helped the THEMIS E/PO team to understand NASA's involvement with tribal communities around the United States from the perspective of the tribal members. EPO professionals seeking to reach Native American populations may want to research the results of these workshops or other valuable materials, for example, see http://www.oneearthoneuniverse.com/.

In order to reach teachers whose students are Native American, we have presented THEMIS workshops at a SACNAS conference in 2004. The Society for Advancement of Chicanos and Native Americans in Science (SACNAS) has as its primary aim to promote careers in science and engineering to under-served minorities, particularly Chicanos and Native Americans. Its annual conference offers an educational component for K-12 teachers, providing useful professional development opportunities for participants teaching minority students.

At the October 2004 SACNAS conference, a total of twenty teachers attended a workshop offered in two similar sessions by the THEMIS E/PO team. This workshop focused on teaching magnetism and exploring the Earth's magnetic field. Across the twenty teachers, they reported that they would reach upwards of 1700 students with these materials and share this information with some 80 other teachers. The teachers rated all aspects of the session very highly—nearly three-quarters of the teachers said that this session was better than most. The complete report on this workshop is available on the THEMIS E/PO website.

The short-term teacher professional development workshops we gave nationally and locally (Berkeley, CA) on average reached teachers who indicated that 20% of their students were non-White and non-Asian male students. This indicates that these teachers had > 20% males who are underserved in science. Note that on our evaluation forms we did not ask about students who are White and Hispanic and so some of the 30% may come from this underserved population. We plan on revising this evaluation form to better gather demographic data on the students whose teachers we are reaching. On average, 50% of their students were female, also an underserved population in the physical sciences. The particular focus of the workshop determines what teachers will attend—elementary, middle, high school, or college. At high school and community college levels, we model the THEMIS teacher guides. For elementary teachers, the THEMIS E/PO team taps existing K-4 education resources at Center for Science Education at UCB, namely the "Eye on the Sky" program as well as other Sun-Earth Connection Education Forum (SECEF) resources. The majority of these workshops served as a partnership between THEMIS and three other NASA missions—FAST, STEREO-IMPACT and RHESSI—as well as SECEF at Berkeley.

We began teaching these short-term workshops at the beginning of the THEMIS E/PO program in 2004. We incorporated many lessons we learned through our experience over the course of the years from 2004 to 2007. Some such lessons that EPO professionals might make note of were to allow more time for discussion with the teachers, to provide and explicitly mention the state or national science education standards for the lessons we were presenting, to give a definition of magnetism, to help teachers focus on a couple key concepts in the content part of the workshop, to provide the presentation on a CD-ROM to the teachers as well as to provide notes during the session for teachers to write down extra information, and in the longer workshops, to provide time for teachers to contemplate how to incorporate the lessons into their curriculum.

CEA analyzed data from questionnaires from most of these workshops and for many of them created in-depth reports from the results, which can be found on the THEMIS website. In 2008, CEA analyzed most all the teacher workshops that THEMIS participated in (nine workshops) during FY2007 (Oct, 2006–Oct, 2007), which included the workshop in Florida at the time of the THEMIS launch. A total of 168 of the teachers attending these workshops completed questionnaires regarding their workshop experience. This report does not include the GEONS 2007 workshop that was discussed above, in Sect. 2.1. We quote the report created for these workshops in the following paragraphs.

General Workshops—The 168 teachers who attended the nine workshops offering topics related to the THEMIS mission told us a little about themselves and the environments in which they teach...

- **Experience**—$N = 156$. Teachers averaged 11.4 years experience ranging from 1 to 40 years.
- Grade Levels—$N = 168$. Over two-fifths taught at the elementary grade level and more than one-quarter at both the high school and middle school levels.
- **Setting**—$N = 154$. More than half taught in suburban schools, nearly one-third in urban schools and a little more than one-tenth in rural schools.
- **Student Population**—Two-fifths of the teachers responding said they were teaching in Title I schools—$N = 110$. On average, half of their students receive free or reduced lunches—$N = 106$. Teachers said that their classes included 47% females and 20% non-White, non-Asian males (a set of underserved populations in science)—$N = 117$.

Most of the teachers told us that they learned about the opportunity from e-mails that piqued their interest in the workshop topics. Thus, having a good email listserve interesting workshop topics are important to bringing in teachers to workshops. The nine workshops presented a total of 38 sessions related to the THEMIS Mission.

Teachers rated their *understanding of the topics* presented in these sessions as being 'clear.' This is in stark contrast to their prior knowledge of the topics. They told us that before attending the workshops, their knowledge of the topics was between 'just a little' and 'moderate.'

Teachers reported that they were 'very likely' to use the materials and ideas in their classroom. A full 61% of the teachers anticipated they would be using the information gleaned from these workshops primarily as integral parts of basic science courses and 46% envisioned using these materials as resources or supplements to basic science courses. This is dramatically different from their use of these topics prior to the workshops. Before THEMIS, an average of 29% said that they never taught the topics presented, 34% had used the topics as resources or supplements to basic science courses, and 31% had used the topics as integral parts of their courses.

Despite the high percentage of teachers eager to implement THEMIS materials and ideas, some expressed concern that their ability to use the materials would be constrained by a lack of financial support to purchase materials, scarce resources and a deficiency in classroom technology. They were also concerned about time constraints. EPO professionals should consider these kinds of concerns when developing curriculum and products.

These findings suggest that the workshops have presented complex materials to teachers in a clear manner that gives them the confidence to present the materials to their students. Additionally, they are now more likely to include the materials and ideas as integral parts of or resources to supplement their basic science courses.

3 THEMIS ViewSpace Show and Public Events: Informal Education

The E/PO plan also contributed (and at the time of this publication continues to contribute) THEMIS science discoveries through a visually captivating ViewSpace show produced in collaboration with the Space Telescope Science Institute (STScI), which contributed its services at no cost to the project. This show, for plasma screen displays and projection mini-theaters at science centers around the nation, consists of auroral images, all-sky movies, THEMIS animations, and interpretive text, woven together to tell the THEMIS story. STScI distributes it though a network of self-updating displays that STScI has developed over the past eight years. As of May, 2008, ViewSpace was showing in 188 venues around the world, with three to four new venues joining the network each month. Most of these venues are Planetariums and Science Centers. The THEMIS program, entitled "Exploring the Mysterious Aurora," receives in excess of 5,500 performances per month. The ViewSpace "Sun Report" news segment, which features the latest results from a variety of solar observatories and probes, including THEMIS, is performed some 14,000 times per month. The total number of visitors at the venues showing ViewSpace who see the THEMIS ViewSpace show depends on such factors as the quality of the installation, the quality and quantity of the "competition" within the venue, the degree to which a lengthy experience fits the desire of the visitor base, and the number of other ViewSpace shows being shown in a "loop" of shows. We do not have the impact evaluation of the show on venue visitors at this time. THEMIS programming will continue to be presented on ViewSpace through the duration of the mission. This indicates that working with STScI to create ViewSpace shows is an effective way of reaching a large audience through an E/PO program.

As per most NASA science E/PO programs, we have attended public events where we distribute THEMIS materials we've produced for the public—flyers about the THEMIS mission and education program, lithographs, stickers, and THEMIS pins. We have participated at Cal Day at the University of California, Berkeley Open House, Sun-Earth Day where we participate at local science museums, Space Weather Day in Maryland, and amateur astronomy club at Mt. Diablo, East of the San Francisco Bay Area in CA. At several of these events, THEMIS scientists have also given talks as part of the event. These events typically reached thousands of people, another effective form of outreach to many.

4 The THEMIS E/PO Website

The Internet is an incredibly rich resource of information, and almost all of the THEMIS E/PO efforts are documented on the web or have web components. The projects and activities described in the sections above and below can be found on the THEMIS E/PO website, URL: http://ds9.ssl.berkeley.edu/themis.

The education and public outreach and the mission websites of any NASA mission are their primary doors to the public. The THEMIS E/PO website has proved an excellent source for the general public to come and learn about THEMIS mission and its science in language appropriate for a general audience. At this website, one can learn how to bring THEMIS science and GEONS activities into the classroom, how to understand THEMIS data, and where the THEMIS public activities are taking place. All the GEONS data and information about the schools and teachers are located on these webpages. In addition, this site contains many images of aurora, auroral movies, and the THEMIS instrumentation in a gallery page. The public has great enthusiasm for this beautiful imagery, which they can download if desired. This gallery page includes videos of scientist and engineer interviews, several videos made by different public affairs offices (NASA and UCB) as well as the segment of the PBS News Hour with Jim Lehrer about student involvement in the GEONS program in Petersburg, AK. This webpage is accessible and friendly to the public, teachers, and to the students who visit it—very important for any outreach webpage—and it is designed to be easily updated. This webpage is also accessible to those with disabilities.

The THEMIS E/PO website has been up and running since December 2003. Our focus for the following paragraphs is on the website statistics for FY06 and FY07 during which time the actual mission launch was successfully completed on February 17, 2007.

Visitor Profile—In each of the fiscal years, the domain names for visitors to the Web site were catalogued. This offers an avenue for identifying visitors' countries of origin. We found that during the two years, about half of the site's visitors can be identified as residents of the United States. In FY06, 2% were from other countries—with Canada, Spain, and the United Kingdom leading the list. In FY07, 3% were from other countries—with Canada, Switzerland, and Germany leading the list. The remaining visitors to the E/PO website could not be identified.

We found that the visitors to the site had an 83% 'hit' rate in FY06 and FY07, that is, the percentage of times (requests) a visitor was successful in accessing the specific files of which a Web page is composed, and did so without receiving an error message. This is not a very good 'hit' rate and will absolutely improve upon this percentage in the remaining years of THEMIS.

A single Web page can be made up of any number of unique files (hundreds even). Since there may be multiple files making up a Web page—resulting in hundreds of 'successful requests' or 'hits'—counting those requests may not be the most accurate reflection of Web traffic. Consequently, for the remainder of this discussion we will refer to the Web site's activity levels in terms of requests for a page—a page that has been viewed by a visitor rather than all of the files that make up the Web page. Note that when comparing our web statistic numbers with other website statistics, the number of 'requests for pages' will be an order of magnitude smaller than the number of 'successful requests' or 'hits.'

Activity Levels—A general summary (Table 5) indicates activity levels as reflected in successful requests for pages. These numbers are not the millions that the most popular sites on the web get, but for a NASA E/PO website, the request for pages are relatively high. For the number counts and averages for FY06 and FY07, we will discuss specific patterns of activity by month, day-of-the-week and hour. We note that activity for FY07 was 46% higher than that for FY06, most likely due to the February 2007 launch.

- **Monthly**—In FY06, above average activity of 31,275 pages was noted in October 2005. The week of October 15 brought a partial eclipse of the Sun, which may have accounted for the above average activity.
- For FY07, it was the period from January to April 2007 that saw the greatest spike in activity as the build up to the launch generated higher than average requests, maximizing at in 41,130 requests in March. It is noteworthy that in addition to the February launch, the Sun-Earth Day Forum highlighted all missions during a March 22 Webcast. This exposure also may have contributed to the higher than average activity and is worth noting that launches are important times for E/PO programs to provide their materials and programs

Table 5 Website activity in FY06 (pre-launch) and FY07 (year of launch). See: http://ds9.ssl.berkeley.edu/themis/stats/themisstatfy06.html and http://ds9.ssl.berkeley.edu/themis/stats/themisstatfy07.html for more details on the FY06 and FY07 webpage statistics respectively

Requests for Pages	FY06	FY07
Total # Successful Requests for Pages	168,756	246,261
Average Successful Requests for Pages per Month	16,298	20,522
Average Successful Requests for Pages per Day	462	674
# of Page Requests in Peak Month for Entire Year—Oct/Mar	31,275	41,130
# of Page Requests on Peak Day for Entire Year—Thursday	33,772	42,474
# of Page Requests in Peak Hour for Entire Year—7amET	9417	14,657

on their website. It also helps to have the E/PO website as the "primary" website listed in press releases.

The daily- and hourly-use statistics are outlined in more details in the THEMIS evaluation report for FY06-07 found on the THEMIS website. These statistics indicate high traffic during the work/school week and during months in which school is in session—October and March. While these monthly and daily activity patterns immediately point to student traffic, the early morning pattern of peak requests is unlikely to be due to student activity.

We suspect that the early morning hours Eastern Time may be reflecting a strong European contingency that checked onto the THEMIS Web site between 9 am and 5 pm. With almost half of the THEMIS Web site users having unidentifiable addresses, we speculate that many of these may be Europeans, drawn to the site after seeing PR events—particularly in Germany, France, Austria and Great Britain. Furthermore, since the THEMIS science Web site was temporarily off-line at this time, all THEMIS traffic around launch time was coming to the THEMIS E/PO site. In the final report in 2009, we will provide a comparison of these statistics with those in 2004 and 2005—the early years of THEMIS, and 2008–2009—the dissemination years of THEMIS E/PO.

5 Discussion

Most of the THEMIS pre-launch activities have focused around formal education, which is bringing THEMIS and science, technology, engineering, and mathematics into the K-12 classroom. We described these efforts in Sect. 2. In this discussion section, we share the "White Paper comments" from NASA on the THEMIS program as a way to describe the strengths of the program and how we have addressed some of the initial concerns from this review, which includes our plans for dissemination and sustainability of the program. We then elaborate on the lessons we have learned from such an exciting and large E/PO project. And end with a short description of the future education and outreach plans for with the THEMIS satellites in orbit.

5.1 White Paper Review, Dissemination and Sustainability

As mentioned in Sect. 1.1 during the early phases of THEMIS, in 2004 the E/PO program was part of a "White Paper" review of ten NASA education programs. This review required that we provide a description of our program in light of several education criteria and then arrive in person to answer questions posed by the review panel. At the end of the review process, we were provided with comments from the committee. Their feedback was that the THEMIS E/PO program was 'Very Good' and that it was well defined, well managed, and successful. Only one education program was rated purely 'Excellent,' one 'Excellent/Very Good' and two were rated purely 'Good.' The other seven NASA education programs rated in this review obtained ratings 'Very Good/Good' or 'Very Good.' A few of the strengths mentioned in the final THEMIS E/PO review are included here.

- The program is certainly accessible to its intended audience, and it's based on a mutual need—educators' need for compelling teaching tools and NASA's need for well-educated future scientists.
- The creative partnerships that have been established for THEMIS add to the strength of this program.
- THEMIS uses lessons learned from previous missions and the program is not wasting time or resources to reinvent the wheel.
- The program clearly will promote improvement of STEM skills for the students who participate, and this is likely to inspire interest in STEM careers.
- The program makes a special effort to include Native American and Hispanic students

The full review can be found on the "Evaluations" page on the THEMIS E/PO website (see Sect. 4). One way we have measured our success in the accessibility to the educators in the GEONS program (first bullet above) is from feedback from the GEONS teachers. Several teachers have shared with us that they were planning on retiring a couple of years ago, but this program has kept them actively teaching.

The major concern of this NASA review committee was that we were spending a large number of resources on a small number of teachers. Our response to this concern at the time was that we felt these teachers would become "Magnetometer Ambassadors." By this, we meant that in addition to being experts in educating students about magnetism and THEMIS, especially using the magnetometer data, that they would then disseminate this information to other teachers in their states, reaching far more teachers than we could reach alone. This has indeed happened with several of our teachers. The GEONS teachers in OR, MI, ND, and WI have all given workshops on the THEMIS materials at state teacher conferences in their respective states. In addition to this statewide teacher training, most of the GEONS teachers have shared THEMIS materials with other teachers at their schools and neighboring schools. As the program continues, more teachers join the GEONS network and we anticipate this to continue throughout the coming years. We started with 10 teachers and have worked with a total of 18 teachers associated with the magnetometers throughout the span of the program.

Another concern in the white paper was that there was "no plan to promote the THEMIS (E/PO) website and attract large numbers of participants is apparent." As described in Sect. 4, we have actually attracted hundreds of thousands of visitors to this site annually by leveraging the Public Relations (PR) work based out of NASA Goddard Space Flight Center (GSFC) prior to and after launch and through our work with teachers across the nation. We also worked with the PR group to help develop a lithograph with one of our magnetism lessons on the back (THEMIS satellites on the front), which directed the public to the lessons on the THEMIS E/PO website. As the mission continues, the E/PO team will work with the NASA GSFC PR team to help bring the public to the THEMIS E/PO site, as well as distribute the URL at science teacher workshops nationally. In addition to these venues, we will distribute lithographs and postcards through the well-established "Night Sky Network" of amateur astronomers run by the Astronomical Society of the Pacific. These amateur astronomers distribute materials to hundreds of members of the public. We anticipate reaching over 10,000 members of the public through this venue.

In addition to these excellent dissemination venues, we have worked in other ways to ensure dissemination of the program and in particular the teacher guides on magnetism, space weather, and THEMIS magnetometer data. We have continued to improve the teacher guides, and to make sure they are accessible to people with disabilities and through these improvements, we anticipate that we will be able to provide these guides to other NASA education programs such as the Educator Resource Center programs and the Aerospace Education Services Project (AESP) programs. We have already placed the guides on NASA's Space Science Education Resource Directory, where most of NASA's space science curriculum is collected. And we have printed 16,000 CD-ROMs with the suite of magnetism guides developed not only by THEMIS but also by STEREO, RHESSI, and FAST Education programs. A full 10,000 of these CD-ROMs were distributed to educators nationally and internationally through the SECEF Sun-Earth Day packets for Sun-Earth Day, 2008.

We are also collaborating with the NASA-funded, Space Math Weekly Problems program of SpaceMath@NASA to develop simple math problems featuring THEMIS data (Odenwald 2008). SpaceMath@NASA has been in operation since 2004 and is the primary resource at NASA for creating and disseminating real-world math problems to the K12 community through a broad network of Listserves. Through this collaboration, we will extend our contact to the K12 community through our contribution of THEMIS-themed math problems to teacher workshops supported by SpaceMathg@NASA, in particular at national conventions for mathematics teachers (e.g. NCTM). Currently, all of the GEONS teachers participate in the SpaceMath@NASA program during the school year.

In addition to these types of dissemination and the teacher professional development workshops modeling lessons from the teacher guides, we have partnered with several other groups bringing magnetometer and radio wave data related to space weather into the classroom in the U.S. and in Canada. These programs are making use of the THEMIS magnetism guides as well. We are also working with a middle school teacher in southern CA to create a summer school course for middle-school students based around THEMIS and magnetism. We anticipate that these types of programs using the THEMIS materials will continue beyond THEMIS. It is important to think about how to sustain the projects in an E/PO program after the funding is no longer available.

Another concern mentioned in the review was that "Younger students—those who have no understanding of physics—are less likely to benefit from involvement in an effort that is over their heads." This is true with the GEONS program, which is why we partnered with the LHS with the GEMS program, reaching predominantly K-5 teachers as well as middle school teachers. In addition to the LHS GEMS program, we have also partnered in many teacher workshops with SECEF and in particular, R. Paglierani (UCB), who has several successful and widely used elementary lessons about the Sun and the solar system. We will continue to work with her to reach the younger students and will also work with the NASA-funded education group located at the University of Colorado, Boulder, which has developed elementary lessons around auroras. All of the above-mentioned elementary lessons incorporate reading, writing, and mathematics since this is the primary need for students at this young age.

5.2 Lessons Learned

What have we learned by running this program? What advice would we give for others expecting to place magnetometers in classrooms? What advice would we give other scientists and EPO professionals hoping to make significant increases in the breadth and depth of their program? The first lesson comes from listening to a teacher at the beginning of our GEONS program and following through on our initial promise to stay in close touch with the teachers at the schools where the magnetometers are located. A teacher at the very first GEONS workshop told us of a seismometer program where a seismometer was placed in his school. No teacher was trained on the seismometer, no lessons were provided to bring the data to the classroom, and no support was given to the teachers. He said that if our program turned out like that one—just placing a scientific instrument at the school and expecting the teacher to do all the rest of the work—he was not interested in it, and would quit. Although THEMIS always intended to more supportive than the seismic program, this teacher's comments did drive home with us the need to stay responsive to the teacher's needs—staying in regular contact, providing them with in-depth professional development and classroom lessons at yearly workshops, providing support when the magnetometers or computer servers had technical difficulties, and providing teachers an opportunity to do compensated research with the data that was appropriate for the classroom. As an addendum, we are happy to say that this aforementioned teacher took part in all of these activities and has since become our biggest advocates, and one of our strongest teachers! Several of his students have since gone on to college with the intention of becoming scientists.

For this type of success, it is imperative that programs involving complex data, such as magnetometer data, provide this kind of in-depth support to teachers. Only three of the 18 teachers shown in Table 1 had physics bachelor degrees. The others had degrees in other sciences and therefore did not have the content knowledge to support their students with the magnetometer data. We have needed to provide this support and training throughout the program. In part because of the request of teachers for more regular contact, we started bimonthly teleconference calls, weekly emails to the teachers, and a Yahoo Group. Because of the time needed for such regular contact, it has been necessary to have an E/PO team member designated to be in charge of this aspect of the program and we recommend having such a person for similar programs.

To our dismay, we learned that the impact of No Child Left Behind, which came into full-force soon after the THEMIS E/PO effort was begun, may have had a chilling effect upon teacher adoption of extensive new curriculum packages such as our six magnetometer guides. In the modern test-oriented classroom, where topics are covered rapidly and teachers may tend to "teach to the test," future curriculum guides may need to be simplified and greatly de-scoped to insure broader utilization.

Another lesson learned is to keep in contact not only with the teachers as previously stressed, but also the administrators (superintendents, principals, etc.) who were involved at the start in helping us find teachers for this program. We have lost teachers in the past due to personal issues and migrations to other school districts. But with the help of superintendents, we have replaced most of those teachers. We actively engage the new teachers to fill in when GEONS teachers leave the project and will continue to do so. Sometimes this process can take on the order of a year, however, depending on the responsiveness of the potential new teachers to the program. Continuing to bring in new teachers has taken much more time and effort than initially expected. For such programs in the future, it would be wise to have the partner who helps find teachers, also agree to help to fill in gaps when teachers or administrators leave.

We also like to add that when it comes to data that is complex, such as magnetometer data, it helps to fund the participating teachers in "research" with the raw magnetometer data. They can adapt lessons for use with their students either as labs, special projects, or science fair projects. From our own experience, there seems to be an assumption in the science community that teachers will jump at the chance to have real data associated with a NASA program in their classroom. This is consistent with the intention of some teachers, but in reality, most of the teachers in our program did not have their students actually work with the data until they were *funded* to be involved with the data themselves. We are still trying to find ways to help other teachers around the country to use the teachers' magnetometer data lessons without the financial incentive. This problem was only applicable to the magnetometer lessons, due to their complexity. Most all of the GEONS teachers used the basic magnetism and space weather lessons in their classrooms without the financial incentive. And non-GEONS teachers at our short-term teacher professional development workshops indicated on evaluations that they were very likely to use the basic science lessons in their classrooms—but only somewhat likely to use the magnetometer data lessons in their classrooms.

For the GEONS program, it has also been crucial to have the technical expertise and support of the THEMIS ground-based observatory hardware and software engineering team. Without their help with installations of the magnetometers, maintaining the magnetometer network, creating software for web-based displays and access of the real-time and archived data, maintaining the web presence of the magnetometer data, the use of this data in the classroom would simply not be possible. In several cases, teachers have had to ship hardware back to UCLA to have it fixed and returned. This exchange takes additional funding, as well as commitments from the teachers and the engineering team.

The final lesson we have learned, and one that is applicable for all those doing mission-related E/PO is that it is important to build in support for, and education of, the THEMIS scientists. In some cases, it really is necessary to train the scientists how to effectively and engagingly interact with the public or with teachers. The scientists need to understand what the public and teachers need and how to best provide it to them. The idea that if one can do science research, one can also teach it is not necessarily true. It has helped our program to involve scientists in as many projects as possible, giving them an opportunity to learn about education through project-based learning, just as we educate teachers and students through inquiry-based lessons. Luckily THEMIS scientists have willingly gone into classrooms, staffed tables at public events, given public talks at universities and amateur astronomy clubs, and provided input to our educational products, and been willing to understand the points-of-view of non-scientists. The THEMIS scientists provide the enthusiasm for the THEMIS science necessary to help bring the specific aurora and magnetospheric science to teachers, students and the public and have been an invaluable part of our program. We hope to bring in their generous time and enthusiasm to our future programs as well.

5.3 Future Programs

The future of the THEMIS E/PO program is to continue to disseminate the products developed and to bring new discoveries to teachers, students and the public. To disseminate resources nationally and prevent duplication of effort, our E/PO program is coordinated with the Sun-Earth Connection Education Forum, a UCB-GSFC collaboration, and with networks supported by NASA Education. Working within the Science Education Gateway, an organization at SSL@UCB that encompasses many NASA education and outreach programs, allows us to leverage these programs and existing partnerships. Other E/PO programs can make use of such strong support networks.

Particular programs coming in the following years include:

- mapping data to sounds (sonification)
- adding new THEMIS science updates to the ViewSpace THEMIS show
- supporting the update of SECEF's Space Weather Multimedia Viewer
- finalizing the THEMIS teacher guides and printing them

- broadening the impact of the GEONS program
- Collaborating with SpaceMath@NASA to develop more math problems featuring THEMIS
- creating and printing new materials to distribute through networks of amateur astronomers
- updating the THEMIS website with new science results and educational materials

We will continue evaluation of these programs and creating a summative evaluation report of the entire THEMIS E/PO program. This final report will be placed on the THEMIS E/PO website. And we will write articles about different aspects of the program in science education journals.

Acknowledgements This project would not have been possible without the support and dedication of the following GEONS teachers: V. Trautman, C. DeWolf, L. Orr, W. Gehman, T. Parent, W. Esch, J. Bean, S. Estill, H. Howe, R. Benson, F. Martin, M. Prakash and the support technicians at the magnetometer schools. We would like to thank them for their exemplary work and dedication to the education project. We also would like to thank the engineers who helped to make this project possible, D. Dearborn, D. Pierce, I. Ruderman, T. Quinn, and K. Rowe. D. Dearborn was instrumental in installing most of the magnetometers and giving talks to the community while he was at the installation site. D. Pierce developed the software to display the THEMIS data as the teachers and program required. I. Ruderman and T. Quinn made it possible to display the THEMIS data, both the real-time data and archived data, on-line and with an easy interface to access the data. K. Rowe interacted with the teachers and school support staff to ensure that the magnetometers were continuing to work and getting them back to UCLA when there were problems with the system that needed fixing. We would also like to thank Gail Bushey for her hard work creating a successful GEMS site in Carson City after we launched the site. And we would like to thank Laura Tucker for her work getting the GEMS site launch organized. We would also like to thank Karin Hauck for editing this paper. This paper was written with the support of NASA NAS5-02099.

Appendix: Acronym List

AESP	Aerospace Education Services Project
ASTC	Association of Science-Technology Centers
CEA	Cornerstone Evaluation Associates
CSE	Center for Science Education
E/PO	Education and Outreach
GEMS	Great Explorations in Mathematics and Science
GEONS	Geomagnetic Event Observation Network by Students
GLOBE	Global Learning and Observations to Benefit the Environment
GSFC	Goddard Space Flight Center
LHS	Lawrence Hall of Science
NSES	National Science Education Standards
NSTA	National Science Teachers Association
OSS	Office of Space Science
PBS	Public Broadcast System
PR	Public Relations
SACNAS	Society for the Advancement of Chicanos and Native Americans in Science
SECEF	Sun-Earth Connection Education Forum
SEGway	Science Education Gateway
STEM	Science, Technology, Engineering, and Mathematics
STScI	Space Telescope Science Institute
THEMIS	Time History of Events and Macroscale Interactions during Substorms
TLRBSE	Teacher Leaders Research Based Science Education program
UCB	University of California, Berkeley
UCLA	University of California, Los Angeles

References

L.P. Cooper, C.A. Morrow, R.A. Pertzborn, J. Rosendhal, P. Sakimoto, An explanatory guide to NASA Office of Space Science Education and public outreach evaluation criteria (2004). http://science.hq.nasa.gov/research/guide.pdf

S.B. Mende, S.E. Harris, H.U. Frey, V. Angelopoulos, C.T. Russell, E. Donovan, B. Jackel, M. Greffen, L.M. Peticolas, The THEMIS array of ground-based observatories for the study of auroral substorms. Space Sci. Rev. (2008). doi:10.1007/s11214-008-9380-x. ISSN0038-6308 (Print) 1572-9672 (Online)

N. Naik, B. Anderson, D. Conrod, B. Eisenhamer, P. Kassaie, B. Komisaruk, C. Person, D. Temple, S. Williams, NASA: August 2004 program review summary (2004)

National Center for Education Statistics (NCES) Trends in international mathematics and science study. Report (2003). http://nces.ed.gov/timss

S. Odenwald, SpaceMath@NASA (2008). http://spacemath.gsfc.nasa.gov

L.M. Peticolas, N. Craig, T. Kucera, D.J. Michels, J. Gerulskis, R.J. MacDowall, K. Beisser, C. Chrissotimos, J.G. Luhmann, A.B. Galvin, L. Ratta, E. Drobnes, B.J. Méndez, S. Hill, K. Marren, R. Howard, The STEREO education and public outreach program. Space Sci. Rev. (2007). doi:10.1007/s11214-007-9287-y. ISN 0038-6308 (Print) 1572-9672 (Online)

J. Rosendhal, P.L. Sakimoto, R. Pertzborn, L. Cooper, The NASA Office of Space Science Education and Public Outreach Program. Adv. Space Res. **34**(10), 2127 (2004) doi:10.1016/j.asr.2003.03.069

J. Ruberg, K. Chen, J. Huang Martin, NASA explorer schools project evaluation: Summer 2003 to Spring 2006. Final Report (2007). http://explorerschools.nasa.gov/pdf/202270main_2003-2006EvaluationSummary.pdf

C.T. Russell, P.J. Chi, D.J. Dearborn, Y.S. Ge, B. Kuo-Tiong, J.D. Means, D.R. Pierce, K.M. Rowe, R.C. Snare, THEMIS ground-based Magnetometers. Space Sci. Rev. (2008)